Thermodynamic Approaches in Engineering Systems

Thermodynamic Approaches in Engineering Systems

Stanisław Sieniutycz
Warsaw University of Technology
Faculty of Chemical and Process Engineering
Warsaw, Poland

AMSTERDAM • BOSTON • HEIDELBERG • LONDON
NEW YORK • OXFORD • PARIS • SAN DIEGO
SAN FRANCISCO • SINGAPORE • SYDNEY • TOKYO

Elsevier
Radarweg 29, PO Box 211, 1000 AE Amsterdam, Netherlands
The Boulevard, Langford Lane, Kidlington, Oxford OX5 1GB, UK
50 Hampshire Street, 5th Floor, Cambridge, MA 02139, USA

Copyright © 2016 Elsevier Inc. All rights reserved.

No part of this publication may be reproduced or transmitted in any form or by any means, electronic or mechanical, including photocopying, recording, or any information storage and retrieval system, without permission in writing from the publisher. Details on how to seek permission, further information about the Publisher's permissions policies and our arrangements with organizations such as the Copyright Clearance Center and the Copyright Licensing Agency, can be found at our website: www.elsevier.com/permissions.

This book and the individual contributions contained in it are protected under copyright by the Publisher (other than as may be noted herein).

Notices

Knowledge and best practice in this field are constantly changing. As new research and experience broaden our understanding, changes in research methods, professional practices, or medical treatment may become necessary.

Practitioners and researchers must always rely on their own experience and knowledge in evaluating and using any information, methods, compounds, or experiments described herein. In using such information or methods they should be mindful of their own safety and the safety of others, including parties for whom they have a professional responsibility.

To the fullest extent of the law, neither the Publisher nor the authors, contributors, or editors, assume any liability for any injury and/or damage to persons or property as a matter of products liability, negligence or otherwise, or from any use or operation of any methods, products, instructions, or ideas contained in the material herein.

British Library Cataloguing-in-Publication Data
A catalogue record for this book is available from the British Library

Library of Congress Cataloging-in-Publication Data
A catalog record for this book is available from the Library of Congress

ISBN: 978-0-12-805462-8

For information on all Elsevier publications
visit our website at https://www.elsevier.com/

Publisher: John Fedor
Acquisition Editor: Anita Koch
Editorial Project Manager: Sarah Watson
Production Project Manager: Anitha Sivaraj
Cover Designer: Maria Ines Cruz

Typeset by Thomson Digital

Contents

Preface .... *xi*
Acknowledgments .... *xvii*

Chapter 1: Contemporary Thermodynamics for Engineering Systems 1
    1.1 Introduction 1
        1.1.1 Energy Transformation Laws 2
        1.1.2 Property Relationship Laws 2
        1.1.3 Energy Conversion and Its Efficiency 2
    1.2 Basic Structure of Nonequilibrium Thermodynamic Theory 17
    1.3 Extremum Properties of Thermodynamic Systems 30
        1.3.1 Equilibrium Problems 31
        1.3.2 Jaynes's and Callen's Bridges From Equilibrium to Disequilibrium 31
        1.3.3 Steady-State Problems 38
        1.3.4 Nonstationary Nonequilibrium Problems 42
        1.3.5 Example of Dissipative Dynamics 43
            1.3.5.1 Introduction 43
            1.3.5.2 Kinetic Potentials, Optimization Criteria and Maximum Principle 46
            1.3.5.3 Reversible Motions 48
            1.3.5.4 Irreversible Motions 48
            1.3.5.5 Conserved Irreversible Hamiltonian 50
            1.3.5.6 Some Additional Remarks 51
    1.4 Classical Motion of Heat Quasiparticles in Thermal Field 53
    1.5 Thermodynamic Geometries 59
        1.5.1 Equilibrium Metrics 59
        1.5.2 Nonequilibrium Metrics 61
    1.6 A Finite Rate Exergy 65
    1.7 Thermodynamics of Transport and Rate Processes 75
        1.7.1 Introduction 75
        1.7.2 The Significance of Kinetic Equations 75
        1.7.3 Transport and Rate Phenomena 79
    1.8 Complex Thermodynamic Systems 91
        1.8.1 Introduction 91
        1.8.2 Classical and Quasi-Classical Complex Systems 91
        1.8.3 Progress in Extended Thermodynamics of Complex Systems 95

Chapter 2: Variational Approaches to Nonequilibrium Thermodynamics ... 115
- 2.1 Introduction: State of Art, Aims, and Scope ... 115
- 2.2 Systems with Heat and Mass Transfer Without Chemical Reaction ... 137
- 2.3 Gradient Representations of Thermal Fields ... 139
- 2.4 Inclusion of Chemical Processes ... 142
- 2.5 Change of Thermodynamic Potential ... 146
- 2.6 From Adjoined Constraints to Potential Representations of Thermal Fields ... 149
- 2.7 Action Functional for Hyperbolic Transport of Heat ... 151
- 2.8 Evolution Described by Single Poissonian Brackets ... 154
- 2.9 Extension of Theory to Multiphase Systems ... 158
- 2.10 Final Remarks ... 160

Chapter 3: Wave Equations of Heat and Mass Transfer ... 163
- 3.1 Introduction ... 163
- 3.2 Relaxation Theory of Heat Flux ... 166
- 3.3 Extended Thermodynamics of Coupled Heat and Mass Transfer ... 171
- 3.4 Various Forms of Wave Equations for Coupled Heat and Mass Transfer ... 176
- 3.5 Stability of Dissipative Wave Systems ... 180
- 3.6 Variational Principles ... 181
- 3.7 Other Applications ... 185
- 3.8 High-Frequency Behavior of Thermodynamic Systems ... 185
- 3.9 Further Work ... 186

Chapter 4: Classical and Anomalous Diffusion ... 193
- 4.1 Introduction ... 193
- 4.2 Classical Picture of Diffusion ... 194
- 4.3 Introducing Anomalous Diffusion ... 202
- 4.4 Compte's and Jou's (1996) Treatment ... 206
  - 4.4.1 Application of Tsallis Theory ... 206
  - 4.4.2 Fractal Entropy and Generalized Diffusion Equation ... 208
  - 4.4.3 Some Special Cases ... 214
  - 4.4.4 Conclusions of Compte and Jou (1996) ... 215
- 4.5 Zanette's (1999) Treatment ... 216
  - 4.5.1 Introduction ... 216
  - 4.5.2 Anomalous Diffusion and Lévy Flights ... 216
    - 4.5.2.1 Anomalous Diffusion in Nature ... 216
    - 4.5.2.2 Random-Walk Models of Anomalous Diffusion ... 219
  - 4.5.3 Maximum-Entropy Formalism for Anomalous Diffusion ... 222
    - 4.5.3.1 Traditional Formalism ... 223
    - 4.5.3.2 Generalized Formalism ... 224
  - 4.5.4 Conclusions of Zanette (1999) ... 227
- 4.6 Discussion of Selected Works ... 227
  - 4.6.1 General Statistical Mechanics and Stochastic Systems ... 227
  - 4.6.2 Diffusion ... 229

Chapter 5: Thermodynamic Lapunov Functions and Stability 233
- 5.1 Introduction 233
- 5.2 Qualitative Properties of Paths Around Equilibrium and Pseudoequilibrium Points 238
- 5.3 Stability of Steady States Close to Equilibrium 239
- 5.4 Chemically Reacting Systems, Fluctuations, and Turbulent Flows 244
- 5.5 Periodic States, Oscillatory Systems, and Chaotic Solutions 268
- 5.6 Stability of Thermal Fields in Resting and Flow Systems Far From Equilibrium 277
- 5.7 New Approach to Lapunov Functions and Functionals 282
- 5.8 Further Work 285
- 5.9 Concluding Remarks 288

Chapter 6: Analyzing Drying Operations in Thermodynamic Diagrams 291
- 6.1 Introduction 291
- 6.2 Modeling of Moisture Extraction in Drying Systems 294
- 6.3 Graphical Approach to Drying of Single Grain and Drying in Fluidized Beds 297
  - 6.3.1 Drying of Single Grain by Gas Whose State is a Function of Time 297
  - 6.3.2 Fluidized Drying of Solid by a Gas with Parameters Variable in Time 299
  - 6.3.3 Fluidizing Systems with Thermodynamic Equilibrium in the Outlet Stream 307
- 6.4 Grain Drying in Countercurrent, Crosscurrent, and Concurrent Gas Flows 308
- 6.5 Graphical Classification of Experimental Data 311
- 6.6 Concluding Remarks 314
- 6.7 Appendix: Some Associated Drying Problems 315

Chapter 7: Frictional Fluid Flow Through Inhomogeneous Porous Bed 321
- 7.1 Introduction 321
- 7.2 A Discrete Model Leading to a Bending Law 325
- 7.3 Bending of Fluid Paths in Inhomogeneous Porous System 328
- 7.4 Variational Approach to Nonlinear Darcy's Flow 331
- 7.5 Use of Hamilton–Jacobi–Bellman Theory 332
- 7.6 Extensions to Other Systems 335
- 7.7 Example 338
- 7.8 Final Remarks 340

Chapter 8: Thermodynamics and Optimization of Practical Processes 347
- 8.1 Introduction 347
- 8.2 Backgrounds for Optimizing in Thermodynamic Systems 356
- 8.3 Mathematical Methods of Optimization 364
  - 8.3.1 Static Methods 365
  - 8.3.2 Dynamic Methods 367
  - 8.3.3 Viscosity Solutions 376
- 8.4 Sorption Models for Minimal Catalyst Deactivation 376
- 8.5 An Excursion to Bejan's Constructal Theory 388
- 8.6 Discussion of Research Works 393

Chapter 9: Thermodynamic Controls in Chemical Reactors 421
- 9.1 Introduction 421
- 9.2 Stoichiometry of General Chemical Reactions 424
- 9.3 Driving Forces of Transport and Rate Processes 426
- 9.4 Nonlinear Macrokinetics of Chemical Processes 430
- 9.5 Heterogeneities, Affinities, and Chemical Ohm's Law 435
- 9.6 Other Important Results Advancing the Field 439
- 9.7 Stability and Fluctuations in Chemical Reactions 442
- 9.8 Instabilities and Limit Cycles 446
- 9.9 Chaos and Fractals in Chemical Word 449
- 9.10 Power Yield in Chemical Engines 452
  - 9.10.1 Introduction 452
  - 9.10.2 Principles of Modeling of Power Yield in Chemical Systems 455
  - 9.10.3 Thermodynamics of Power Production in Chemical Engines 457
  - 9.10.4 Entropy Generation in Steady Systems 460
  - 9.10.5 Characteristics of Steady Isothermal Engines 461
  - 9.10.6 Sequential Models for Computation of Dynamic Generators 466
  - 9.10.7 Further Problems 469

Chapter 10: Power Limits in Thermochemical Units, Fuel Cells, and Separation Systems 473
- 10.1 Introduction 473
- 10.2 Internal Dissipation in Steady Thermal Systems 476
- 10.3 Selected Results for Dynamical Thermal Systems 477
- 10.4 Radiation Engines by the Stefan–Boltzmann Equations 478
- 10.5 Hamiltonians and Canonical Equations 480
- 10.6 Simple Chemical and Electrochemical Systems 483
- 10.7 Power Yield and Power Limits in FCs 487
- 10.8 Exergy and Second Law Analyses of FC Systems 500
- 10.9 Limits on Power Consumption in Thermochemical Systems 504
- 10.10 Estimate of Minimum Power Supplied to Power Consumers 508
- 10.11 Final Remarks 511

Chapter 11: Thermodynamic Aspects of Engineering Biosystems 515
- 11.1 Introduction 515
- 11.2 Control of Biological Reactions and Decaying Enzymes 517
- 11.3 Biophysics and Bioenergetics 522
- 11.4 Power Yield and Exergy-Valued Biofuels 527
- 11.5 Ecology and Ecological Optimization 531
- 11.6 Fractals and Erythrocytes 536
- 11.7 Animal Locomotion and Pulsating Physiologies 537
- 11.8 Thermostatistics of Helix-Coil Transitions 538
- 11.9 Biochemical Cycles in Living Cells 539
- 11.10 Elucidation of Protein Sequence–Structure Relations 540
- 11.11 Complexity, Self-organization, Evolution, and Life 543

Chapter 12: Multiphase Flow Systems 561
12.1 Introduction 561
12.2 Aspects of Thermodynamics of Surfaces and Interfaces........................ 562
12.3 Heating or Cooling Policies Minimizing Entropy Production 563
12.4 Optimal Control in Imperfect Multiphase Systems 566
12.5 Turbulent Mixtures, Instabilities, and Phase Transitions 567
12.6 Multiphase Chemical Reactors and Regenerators 570
12.7 Final Remarks 579

Chapter 13: Radiation and Solar Systems 583
13.1 Introduction 583
13.2 Basic Problems in Radiation Thermodynamics 583
13.3 Conversion of Solar Flux Into a Heating Medium 588
13.4 Maximum of Exergy or Work Fluxes in Radiation Systems 597
13.5 Solar Buildings and Solar Systems 605
13.6 Closing Remarks 610

Chapter 14: Appendix: A Causal Approach to Hydrodynamics and Heat Transfer 613
14.1 Introduction 613
14.2 Action, Lagrangian and Thermohydrodynamic Potentials 617
14.3 Basic Information on Thermal Inertia 620
14.4 Matter Tensor of General Relativistic Theory 622
14.5 Thermal Mass and Modified Temperatures 623
14.6 Tensor of Matter Including Heat and Viscous Stress 625
14.7 Conclusions and Final Remarks 628

References 631
Glossary 695
Subject Index 703

Preface

The present volume responds to the need for a synthesizing book that throws light upon extensive fields of thermodynamics and entropy theory from the engineering perspective and applies basic ideas and results of these fields in engineering problems. The book is, in brief, a collection of reviews synthesizing important achievements of applied nonequilibrium thermodynamics in some important branches of mechanical, chemical, and biochemical engineering with a few excursions into environmental engineering.

Contemporary applications of thermodynamics, both reversible and irreversible, are broad and diverse. The range of thermodynamic applications is expanding and new areas finding a use for the theory are continually emerging. One of the main goals of this book is to outline the achievements in applied thermodynamics (both equilibrium and nonequilibrium) obtained within the recent 50 years. This goal is very broad since thermodynamics serves currently as a formal and substantive theory in applications involving analyses of physical, chemical, biological, and even economical systems. Dealing with such a broad goal within a single limited volume requires necessary elimination of many attractive mathematical aspects; consequently the book exploits the power of mathematical analysis to a lesser extent than many more specialized while narrower texts. Yet, this disadvantage is compensated by a vast literature analysis, which is a strong asset of the book to serve as a comprehensive reference text.

Let us shortly discuss several basic topics, most of which are analyzed in the present book. Thermodynamics provides theoretical tools to exclude improper kinetic equations or to improve empirical kinetic formulas. It may furnish thermodynamic equilibrium metrics and its disequilibrium extensions to measure a thermodynamic length. It may also help to describe reversible and irreversible dynamics in terms of energy-type Hamiltonians or related Lagrangians. The use of disequilibrium entropies serves to develop a wave theory of heat and mass transfer, applied to overcome the well-known paradox of infinite propagation speeds of thermal and material disturbances, implied by classical parabolic equations. Advanced thermodynamics can provide unifying theoretical framework for both classical and anomalous diffusion. The latter is usually treated by applying Tsallis's (1988) theory for a generalized definition of entropy, consistent with both thermodynamics and statistical mechanics. Applied

Preface

in a chemical context, thermodynamics provides substantiation for the kinetic law of mass action consistent with suitable chemical resistance, a quantity useful in the analysis of chemical networks.

In fact, both classical and nonequilibrium thermodynamics predict the global stability of the thermodynamic equilibrium. However, for disequilibrium singular points and distributed steady states, sufficient stability conditions cannot be formulated in terms of entropy or other thermodynamic potentials, although, occasionally, some excess thermodynamic or quasi-thermodynamic potential can prove their utility as stability criteria (Lapunov functions).

The problem of thermodynamic stability criteria in disequilibrium systems has not yet been completely solved, despite extensive research in this area. Without doubt, thermodynamics may establish exact stability criteria for system trajectories approaching an equilibrium singular point. In this case one may apply the negative excess entropy $W = S^{eq} - S$ as a Lapunov functional for which the entropy density ρ_s is the same function of the thermodynamic state as prescribed by classical thermodynamics (local equilibrium assumption). Invoking the local form of the second law $\sigma > 0$ one obtains in the gradient-less case the conditions $dW/dt < 0$, which proves the stability of all system trajectories approaching the equilibrium singular point. This means that thermodynamics ensures the global stability of the thermodynamic equilibrium (Tarbell, 1982).

A practical question can be asked: In what sense can it be said that the distribution of the driving forces in one industrial configuration is better than in another? To answer this question realistic engineering examples, involving heat exchangers, plate distillation columns, chromatographic separators, etc., may be analyzed. When "a duty" (useful effect) for a process is defined, the following equipartition principle for the entropy production is advanced: "for a given duty, the best configuration of the process is that in which the entropy production rate is most uniformly distributed" (Tondeur and Kvaalen, 1987). A generalization of the principle may also be proposed for systems with a distributed design variable: "the optimal distribution of an investment is such that the investment in each element is equal to the cost of the energy degradation in this element". Thus a uniform distribution of the ratio of these two quantities should be preserved. These results show the importance of irreversibility analysis in design.

Thermodynamics may help to predict selling prices and develop paraeconomic analyses and balances which involve unit exergies as measures of unit economic values. Interestingly, thermodynamics and especially exergy theory may be applied to define the so-called proecological tax, the quantity replacing actual personal taxes. A practical field related to thermodynamics is thermoeconomics. It defines the physical–mathematical background for energy–economic–ecological analyses which are made in different fields of knowledge. In this field the concept of exergetic cost is employed (Valero et al., 1994a,b); this is a concept close to that of embodied energy or cumulative exergy consumption (Szargut et al., 1988). The analysis of the exergetic cost focuses rigorously on the process of cost formation.

A relatively new subset of nonequilibrium thermodynamics, called finite-time thermodynamics (FTT), has proved its remarkable effectiveness in modeling the performance and power limits of thermal, chemical, and electrochemical engines, refrigerators and heat pumps. Limits for finite resources are associated with the notion of exergy. Classical thermodynamics is capable of providing energy limits in terms of exergy changes. However, they are often too distant from reality (real energy consumption can be much higher than the lower bound and/or real energy yield can be much lower than the upper bound). Yet, by introducing rate dependent factors, irreversible thermodynamics offers enhanced limits that are closer to reality. Limits evaluated for finite resources refer either to a sequential relaxation of a resource to the environment (engine mode) or to resource's upgrading in the process undergoing in the inverse direction (heat pump mode). The current research in the field of FTT includes fuel cells which are expected to play a significant role in the next generation of energy systems and road vehicles for transportation. However, substantial progress is required in reducing manufacturing costs and improving performance of FC systems.

Although the field of finite-time thermodynamics is both broad and vital, FTT is only briefly treated in the present book because there is already a number of sources in this area (Berry et al., 2000; Chen and Sun, 2004; Sieniutycz and Jeżowski, 2009, 2013; etc). This treatment refers to chemical reactors analyzed in Section 9.10, where a problem of power yield in thermochemical generators and fuel cells is presented with some detail in the spirit of FTT. A discussion in FTT style is developed for radiation systems (chapter: Radiation and Solar Systems) in view of recent results of Badescu (2014), who showed that that the upper bound for the reversible work extraction is not always Carnot efficiency and all upper-bound efficiencies depend on the geometric factor of the radiation reservoir.

Evaluation of the entropy lowering in reference to its equilibrium value yields a good indicator of disequilibrium (Ebeling and Engel-Herbert, 1986). Entropy lowering can be linked with the contraction of the occupied part of the phase space due to the formation of attractors. Oscillations in solids and turbulence in liquid flows may serve as examples. Ebeling et al. (1986) investigate these entropy properties in the turbulence context. Self-oscillations and flows in tubes (laminar and turbulent) are considered. The results show that under the condition of a fixed energy the entropy decreases with excitation. Engel-Herbert and Ebeling (1988a) study the Brownian motion of nonlinear oscillators in a heat bath. For the excitation of sustained oscillations, realized by van der Pol oscillators, the entropy, calculated in the limit of weak dissipation and weak noise, decreases monotonically with the feedback strength, at the fixed average energy.

Turbulent flows are common in nature. They are observed in many technological processes, hence they are of interest for thermodynamics. Turbulent flows are a manifestation of spatially extended, nonlinear dissipative systems in which diverse length scales are simultaneously excited and coupled. The significance of turbulence in our physical world cannot be overestimated. Without turbulence, mixing of air and fuel would not occur on suitable

Preface

time scales, the transfer and dispersion of chemical reagents, energy, momentum, and pollutants would be far less effective; that is, many observed positive effects would be impossible (Sreenivasan, 1999). Yet, turbulence has also negative consequences, for example, it increases energy consumption in airplanes, automobiles, ships, pipelines, etc., distorts the signal propagation in fluids, decreases safety of aircraft flight, etc. (Sreenivasan, 1999). For an engineer the main goal is prediction and control of turbulence effects, their suppression or enhancement depending on current needs.

In the present book we focus on thermodynamic aspects of turbulence, following mainly the results obtained by Klimontovich and Ebeling, and some coworkers of these researchers who compared with experiments the theory of hydrodynamic turbulence. Thermodynamic aspects of turbulent two-phase flows are also briefly outlined. The structure of a turbulent flow may be expressed by the collective mechanisms for the transfer of momentum, heat, and mass and by entropy lowering determined relative to an equilibrium system of the same energy. The transition from laminar to turbulent motion may be regarded as a nonequilibrium transition to a more ordered state.

One can evaluate the change of the entropy during the transition of a flow from laminar to turbulent motion. As the Reynolds number Re grows, a certain portion of the flow energy is transferred to the collective macroscopic motions of the molecules, which are reflected, for example, by hydrodynamic velocity, by turbulent pulsations, and randomness, which, "apparent from a causal observation, is not without some order" (Sreenivasan, 1999). Fixing the energy as the Reynolds number Re varies, this portion of the transferred energy is taken from the microscopic thermal motion. But the energy contained in the thermal degrees of freedom possesses a higher value of entropy. Thus, the transformation of energy from the microscopic scale of thermal motion to macroscopic scales of collective molecular motion at fixed energy lowers the entropy (Engel-Herbert and Ebeling, 1988b). Inspired by this result Engel-Herbert and Schumann (1987) studied the entropy behavior during a nonequilibrium phase transition. They observed that for the generation of sustained oscillations the entropy decreases monotonously if the average oscillator energy remains fixed. Soon afterward, Klimontovich (1991) showed the link between a turbulent motion and the structure of chaos.

Some problems are discussed here in a necessarily brief way, but the book will serve its purpose if the reader gets encouraged to further studies, to improve his knowledge and understanding of problems which could only be outlined here. In particular, the chapter on thermodynamic aspects of engineering biosystems will require updated studies soon, in view of rapid progress in this area.

This book can be used as a supplementary or reference text in the following courses:

- technical thermodynamics and industrial energetics (undergraduate)
- separation operations and separation systems (undergraduate)

- alternative and unconventional energy sources (graduate)
- stability of chemical reactions and transport processes (graduate)
- thermoeconomics of solar energy conversion (graduate)

The content organization of the book is as follows. Chapters: Contemporary Thermodynamics for Engineering Systems and Variational Approaches to Nonequilibrium Thermodynamics outline main approaches and results obtained in applied disequilibrium thermodynamics, focusing on the methods and examples considered in the whole book. Chapter: Wave Equations of Heat and Mass Transfer shows the role of thermodynamics in setting a simple theory of wave equations for heat and mass transfer. Chapter: Classical and Anomalous Diffusion describes the significance of Tsallis's (1988) theory for a generalized definition of entropy in the development of the generalized theory of diffusion, which includes diffusive phenomena, both classical and anomalous. Chapter: Thermodynamic Lapunov Functions and Stability displays some stability problems and related Lapunov's functions derived from thermodynamics. Chapter: Analyzing Drying Operations on Thermodynamic Diagrams presents drying operations with granular solids on thermodynamic diagrams. Chapter: Frictional Fluid Flow through Inhomogeneous Porous Bed describes basic aspects of frictional fluid flow through both inhomogeneous and anisotropic porous media in which fluid streamlines are curved by a location dependent hydraulic conductivity. Chapter: Thermodynamics and Optimization of Practical Processes exposes the role of thermodynamics in optimization of applied processes undergoing in nonreacting systems. Chapter: Thermodynamic Controls in Chemical Reactors extends the related control ideas and results to chemical reactors. Chapter: Power Limits in Thermochemical Units, Fuel Cells, and Separation Systems offers a generalized treatment of power limits in various thermochemical engines, fuel cells, and separation systems. Chapter: Thermodynamic Aspects of Engineering Biosystems considers some basic thermodynamic properties of engineering biosystems. Chapter: Multiphase Flow Systems presents valuable selected results for multiphase flow systems, whereas Chapter: Radiation and Solar Systems—some recent results for applied radiation and solar systems. Chapter entitled Appendix: A Causal Theory of Hydrodynamics and Heat Transfer constitutes a supplementary text which shows the role of entropy four-flux in the dissipative relativistic phenomena and variety of disequilibrium temperatures obtained within a causal approach to hydrodynamics and heat transfer.

Acknowledgments

The author expresses his gratitude to the Polish Committee of National Research (KBN) and the Ministry of National Education of Poland under whose auspices a considerable part of his own research discussed in the book was performed in the framework of two grants: grant 3 T09C 02426 (*Nonequilibrium thermodynamics and optimization of chemical reactions in physical and biological systems*) and grant N N208 019434 (*Thermodynamics and optimization of chemical and electrochemical energy generators and the related applications to fuel cells*). A critical part of writing any book is the process of reviewing, thus the authors are very much obliged to the researchers who patiently helped them read through subsequent chapters and who made valuable suggestions. In preparing this volume the authors received help and guidance from Viorel Badescu (Polytechnic University of Bucharest, Romania), Ferenc Markus (Budapest University of Technology and Economics, Hungary), Alina Jeżowska (Rzeszów University of Technology, Poland), Andrzej B. Jarzębski (Institute of Chemical Engineering of Polish Academy of Science, and Faculty of Chemistry at the Silesian University of Technology, Gliwice), Lingen Chen (Naval University of Engineering, Wuhan, P. R. China), Piotr Kuran, Artur Poświata, and Zbigniew Szwast (Faculty of Chemical and Process Engineering at the Warsaw University of Technology), Elżbieta Sieniutycz (University of Warsaw), Anatolij M. Tsirlin (System Analysis Research Center, Pereslavl-Zalessky, Russia), and Anita Koch (Elsevier). We also acknowledge the scientific cooperation of many colleagues of the Institute of Fluid Flow Machinery in Gdańsk, Poland, the former Publisher of *Archives of Thermodynamics*. Finally, appreciations also go to the whole book's production team in Elsevier for their cooperation, help, patience, and courtesy.

Stanisław Sieniutycz
Warszawa, August 2015

CHAPTER 1

Contemporary Thermodynamics for Engineering Systems

1.1 Introduction

We begin with some information about the nature of thermodynamics as a macroscopic theory and then describe basic concepts and definitions in thermodynamics. Thermodynamics is a science that includes the study of energy transformations and relationships among the physical properties of substances that are affected by these transformations. The properties which really sets thermodynamics apart from other sciences are energy transformations through heat and work. It is clear that the aforementioned definition is broad and vague, and that various users will apply in their work different aspects of the thermodynamic theory. Chemical engineers typically focus on phase equilibria, stoichiometry, chemical reactions, catalysis, and so forth. Mechanical engineers are more interested in power generators, fuel cells, refrigeration devices, and nuclear reactors. Both users can apply in their research some common methods, such as optimal control theory thus contributing to the field called thermodynamic optimization.

Thermodynamic properties can be determined by studying either macroscopic or microscopic behavior of matter. Classical thermodynamics treats matter as a continuum and studies the macroscopic behavior of matter. Statistical thermodynamics investigates the statistical behavior of large groups of individual particles. It postulates that observed physical property (eg, temperature T, pressure P, energy E, etc.) is equal to the appropriate statistical average of a large number of particles. For a statistical theory of disequilibrium systems, see, for example, Kreuzer (1981). Thermodynamics is based upon experimental observation of macroscopic systems. Its power and beauty follows from a basic property of macrosystems composed of sufficiently large number of particles. Namely, as a consequence of the law of large numbers, these macrosystems can be described well (modulo to fluctuations) in terms of only several variables, for example, temperature, pressure and mole numbers.

Conclusions stemming from observations have been cast as postulates or laws. The study of thermodynamics considers five laws or postulates. Two of them deal with energy transformation and three deal with the physical properties.

1.1.1 Energy Transformation Laws

Energy transformation laws comprise the first and the second law of thermodynamics. Both laws acquired many formulations which are available in numerous books on thermodynamics. Subsequently, we limit ourselves to the following formulations:

First Law of Thermodynamics: Energy is conserved, that is, system's energy can change only by transfer of heat, work or substances by the system's boundary, which means that there is no energy generation within the system. Note that this formulation may be invalid for *partial* energies of subsystems occupying the system's space.

Second Law of Thermodynamics: Entropy is not conserved and/or there is a non-negative entropy generation within the system. The important consequence of the second law is that all real processes spontaneously proceed in one direction of time. Another basic and useful conclusion stemming from the second law states that various forms of energy have different qualities and the notion of exergy (available energy) can be introduced as a sort of the value measure for the useful energy contained in the substance or attributed to the heat.

1.1.2 Property Relationship Laws

Property relationship laws include zeroth law of thermodynamics, third law of thermodynamics and the state postulate. They may be expressed as follows.

Zeroth Law of Thermodynamics: When each of two systems is in thermal equilibrium with a third system, they are also in thermal equilibrium with each other.

Third Law of Thermodynamics: The entropy of a perfect crystal equals zero at absolute zero temperature.

State Postulate: The state of a simple, single-phase thermodynamic system is completely specified by two independent variables which are the intensive properties.

1.1.3 Energy Conversion and Its Efficiency

Energy conversion and efficiency of this conversion are of primary concern in applied thermodynamics. For energy consuming or producing devices the thermal efficiency is:
η_{th} = energy obtained/(energy paid)

Example efficiencies are:

1. Solar cell: 12%
2. Gas turbine: 12–16%
3. Automobile: 12–25%
4. Steam power plant: 38–41%

5. Fuel cells: 40–60%
6. Electric motor: 90%

Now we shall pass to some definitions and say several words about the thermodynamic vocabulary.

1. *Thermodynamic system*

    Thermodynamic system is a three-dimensional region of the physical space bounded by arbitrary surfaces (which may be real or imaginary and may change size or shape). There are various types of hermodynamic systems, namely the following five.
    a. Closed system is a system which is impermeable with respect to the flow of matter, for example, a fixed, closed volume. A property of the closed system is its constant mass.
    b. Open system is a system open with respect to the flow of matter such as a flow reactor. The system is defined by a well-defined volume surrounding the region of interest. The surface of this volume is called the control surface. Mass, heat, work and momentum can flow across the control surface into the system and from the system.
    c. Isolated system is a system that is not influenced by the part of space which is external to the system boundaries. No heat, work, mass, or momentum can cross the boundary of an isolated system. Variables such as N, V and U are fixed and constant in isolated system.
    d. Simple system is a system that does not contain any *internal* adiabatic, rigid and impermeable boundaries and is not acted upon by external forces.
    e. Composite system is a system that is composed of two or more simple systems.

2. *Property*

    Property, by definition, is a characteristic of a system. There are several types of properties:
    a. Primitive property is a property that can in principle be specified by describing an operation or a test to which the system is subjected. Examples include mechanical measurements (eg, pressure, volume, and temperature T) and heat capacity.
    b. Derived property is a property mathematically defined in terms of primitive properties.
    c. Intensive property is a property that is independent of the extent of or mass of the system. Examples are T, P, density, ρ, ..., and so forth.
    d. Extensive property is a property whose value for the system is dependent upon the mass or extent of the system. Examples are the enthalpy, internal energy, volume, and so forth.
    e. Specific property is an extensive property per unit mass. Specific properties are intensive.
    f. State property is a property that only depends on the thermodynamic state of the system, not the path taken to get to that state.

3. *State of a system*
   State, by definition, is a characteristic of a system. There are several types of states:
   a. Thermodynamic state is the condition of the system as characterized by the values of its properties.
   b. Stable equilibrium state is a state in which the system is not capable of finite spontaneous change to another state without a change in the state of the surroundings. Several types of equilibrium must be fulfilled: thermal, mechanical, phase, and chemical.
   c. State postulate: The equilibrium state of a simple closed system can be completely characterized by two independent variables and the masses of the species within the system.
4. *Thermodynamic process*
   A thermodynamic process is a transformation from one equilibrium state to another.
   a. Quasistatic process is a process where every intermediate state is a stable equilibrium state.
   b. Reversible process is one in which a second process could be performed so that the system and surroundings can be restored to their initial states with no change in the system or surroundings.

   The following remarks can be made at this point. Reversible processes are quasistatic, but quasi-static processes are not necessarily reversible. A quasistatic process in a simple system is also reversible. Some factors which render processes irreversible are friction, unrestrained expansion of gasses, heat transfer through a finite temperature difference, mixing, chemical reaction, and so forth.
5. *Thermodynamic path*
   Thermodynamic path is the specification of a series of states through which a system passes in a process. Various paths can be specified, such as: isothermal, $\Delta T = 0$; isobaric, $\Delta P = 0$; adiabatic, $\Delta Q = 0$; polytropic, PV^k = constant; isochoric, $\Delta V = 0$; isentropic, $\Delta S = 0$ (adiabatic and reversible); isenthalpic, $\Delta H = 0$ (adiabatic and entirely irreversible, as in the throttling process). Path calculations usually involve the knowledge of the properties of gases and liquids (Reid et al., 1987).
6. *Heat engines, refrigerators and HPs*
   Work may be converted to heat directly and completely, but converting heat to work requires the use of a special device called a heat engine. Heat engines may vary considerably from one another, but share the following characteristics: They operate on a cycle, they receive the driving heat from a high temperature source, they convert part of this heat to work, they reject the remaining waste heat to a low-temperature sink.

   The work-generating device that fits the definition of a heat engine is the steam power plant, Fig. 1.1. The fraction of the heat input that is converted to network output is a measure of the performance of a heat engine, and is called the thermal efficiency. Most heat engines possess poor thermal efficiencies, with more than one half of the thermal energy supplied to the working fluid ending up in the environment (Naragchi, 2013).

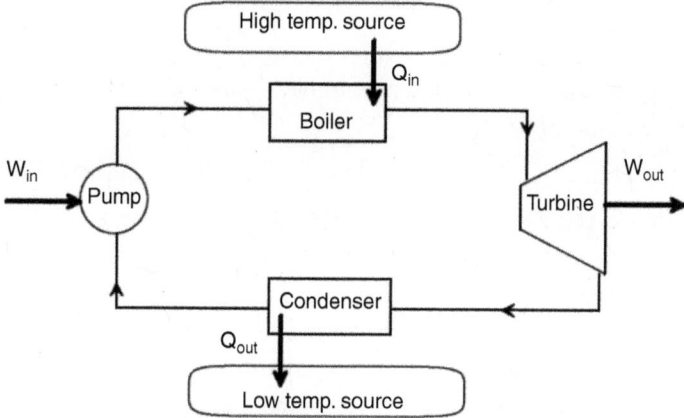

Figure 1.1: Steam power-plant as a heat engine.

In Nature, heat flows in the direction of decreasing temperature. The reverse process, however, cannot occur spontaneously and requires the use of special devices called refrigerators. Like heat engines, refrigerators are cyclic devices. The working fluid used in the refrigeration cycle is called a refrigerant. The most frequently used refrigeration cycle is called the vapor-compression refrigeration cycle, shown schematically in Fig. 1.2.

The refrigerant enters the compressor as a vapor and is compressed to the condenser pressure. It leaves the compressor at a relatively high temperature and cools down and condenses as it flows through the coils of the condenser by rejecting heat to the surrounding medium. Next, the refrigerant enters an expansion valve, where the Joule-Thompson effect takes place and the refrigerant's temperature and pressure decrease due to the throttling. The refrigerant then enters the evaporator, where it evaporates by absorbing heat from the refrigerated space (cooling effect). The cycle is completed as the refrigerant leaves the evaporator and returns to the compressor. The efficiency of a

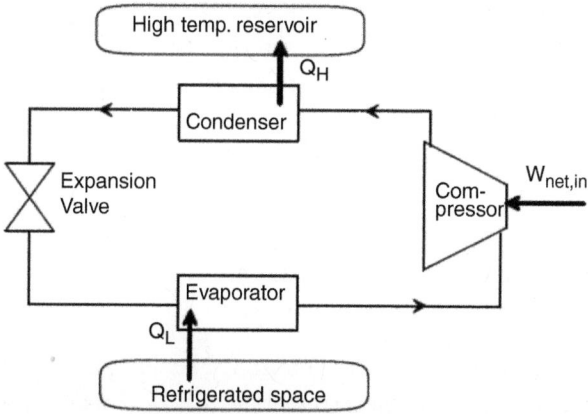

Figure 1.2: A vapor-compression refrigeration cycle.

refrigerator is expressed in terms of the coefficient of performance (COP), denoted by COP_R. The value of $COP_R = Q_L/W_{net,in}$ can be greater than unity, and usually is for most refrigerators. That is, the amount of heat removed from the refrigerated space can be greater than the amount of work input.

Another device that transfers heat from a low temperature medium to a high temperature one is the heat pump (HP). Refrigerators and heat pumps operate on the same thermodynamic cycle, but have different objectives. The objective of a refrigerator is to maintain the refrigerated space at a low temperature. The objective of an HP is to maintain a heated space at a high temperature. The coefficient of performance of an HP, COP_{HP}, is defined as: $COP_{HP} = Q_H/W_{net,in}$. Air-conditioning units are refrigerators whose refrigerated space is a room or building. The same air-conditioning unit can operate as an HP in the winter by installing it backwards. In this mode, which we call the HP mode, the air-conditioner absorbs heat from the refrigerated space (outdoors) and rejects it to the heated space (house). The performance indices of refrigerators and air-conditioning units are often expressed in the United States in terms of the Energy Efficiency Rating (EER), which is the amount of heat removed (in BTUs) for one watt-hour of electricity consumed. The relation between EER and COP is: $EER = 3.412 COP_R$

The work-energy theorem derived from Newton's second law is essential for the development of energy concepts in physics (Arons, 1999). The theorem applies to the displacement of a particle or the center of mass of an extended body treated as a particle. Because work, as a quantity of energy transferred in accordance with the first law of thermodynamics, cannot be calculated in general as an applied force times the displacement of center of mass, the work-energy theorem is not a valid statement about energy transformations when work is done against a frictional force or actions on or by deformable bodies. To use work in conservation of energy calculations, work must be calculated as the sum of the products of forces and their corresponding displacements at locations where the forces are applied at the periphery of the system under consideration. Failure to make this distinction results in errors and misleading statements prevalent in textbooks, thus reinforcing confusion about energy transformations associated with the action in everyday experience of zero-work forces such as those present in walking, running, jumping or accelerating a car. Without a thermodynamically valid work definition, it is also impossible to give a correct description of the connection between mechanical and thermal energy changes and of dissipative effects. The situation can be corrected and student understanding of the energy concepts greatly enhanced by introducing the internal energy concept, that is, articulating the first law of thermodynamics in a simple, phenomenological form without unnecessary mathematical encumbrances (Arons, 1999).

Before closing this section it is worth mentioning a beautiful book on energy history, development and current state of research (Badur, 2012). This treatise is along the borderline between classical and irreversible thermodynamics. Attention also should be paid to a concise and exact review of the theory and applications of classical thermodynamics and its relation to nonequilibrium theories (Mikielewicz, 1997, 2001).

7. *Irreversible phenomena*

The laws of irreversible evolution of macroscopic systems belong to the fundamental problems of contemporary science and engineering. They receive attention in a great number of papers in scientific journals with the topics ranging from the extensions and generalizations of the classical theory to diverse applications in fluid mechanics (Newtonian and non-Newtonian), heat and mass transfer, chemistry, electrochemistry, membrane technology, energy conversion, systems theory, biophysics, and so forth.

Entropy source or the amount of the entropy generated in the unit volume per unit time is the basic quantity characterizing the irreversibility in a continuum. In principle, the structure of the nonequilibrium thermodynamic theory can be analyzed in two ways.

The first (most popular) approach is based on the derivation of the entropy source by combining the conservation or balance laws and the Gibbs' equation. This combining can be applied in the framework of either classical or extended thermodynamics, provided that a generalized form of the Gibbs' equation is used in the extended case. The first approach is analyzed further on in this chapter. The first approach is well known, and it has been presented in detail in many books (Prigogine, 1947; Haase, 1969; Baranowski, 1974; Kestin, 1979; de Groot and Mazur, 1984; Keizer, 1987a, and many others). In particular, Kestin (1979, 1993) developed the theory of internal variables in the local equilibrium approximation, Kestin and Bataille (1980) set the thermodynamics of solids, and Kestin and Rice (1970) have described a number of paradoxes in the application of thermodynamics to strained solids. They gave an outline of a general method of the local state which allows us to extend our knowledge of uniform systems to include continuous systems. They paid particular attention to plastic deformation and to strain hardening, and concluded that the relation between the stress and the plastic strain rate does not seem to come within the scope of Onsager's relations restricted to the small departures from the equilibrium. Muschik and Brunk (1977) defined a thermodynamic analogue of the temperature by the vanishing heat exchange between a nonequilibrium system and the heat reservoir. Properties of this analogue called contact temperature were discussed and illustrated graphically. Muschik and Papenfuss (1993) developed an evolution criterion of nonequilibrium thermodynamics and show its application to liquid crystals. Muschik et al. (2001) gave a sketch of continuum thermodynamics. Muschik and Restuccia (2002) demonstrated that in general constitutive equations depend on the motion of the material with respect to a chosen frame of reference. Despite this dependence of motion, the constitutive equations are isotropic functions on the state space. Muschik (2008) surveyed a number of branches of thermodynamics and proposed a classification of thermodynamic schools.

The second (less popular) approach is developed from the point of view of an extremal behavior exhibited by thermodynamic systems. In this approach, equations of the nonequilibrium theory are derived subject to the assumption that the laws of equilibrium state, steady states, or even the whole time evolution can be derived from an extremum or variational principle. This variational theory usually applies given expressions for dissipation functions (Rayleigh, 1945) to formulate a thermodynamic integral which is next extremized subject to the given conservation laws. The result is equations of change and phenomenological equations (ie, the same equations which are attempted to be derived in the first approach). We list here only some propagators of the variational theory of thermodynamics (Lebon, 1976; Gyarmati, 1977; Anthony, 1988; Scholle, 1994; Gambar and Markus, 1994; Shiner, 1984; Sieniutycz, 1981e, 1994; Fernández-Pineda and Velasco, 2001; Wagner, 2005; Lin and Wang, 2013, and others). This theory is analyzed in detail in chapter: Variational Approaches to Nonequilibrium Thermodynamics, where a more complete list of references is given. Zhuravlev (1979) published a nice book on problems and solutions in thermodynamics of irreversible processes, where the two approaches to NT discussed previously can be introduced through examples.

Lebon and Mathieu (1983) provided a compilation of diverse theories of nonequilibrium thermodynamics. They emphasize the basic hypotheses and relationships between the theories. Criticisms of these theories are also discussed. Fluxes of mechanical or electric work are quite frequently neglected in thermodynamic balances; see, for example de Groot and Mazur (1984) for a basic description. A very popular branch of applied thermodynamics, purposely designed for systems with explicit work flux (power), is the finite-time thermodynamics (Berry, 1979, 1989a; Berry et al., 1978, 2000; Andresen, 1983, 2008; Andresen et al., 1983; 1984b; Mozurkewich and Berry, 1981; Orlov and Berry, 1990; 1993; Salamon and Nitzan, 1981; Salamon et al., 1980, 1984, 1985; Carrera-Patino and Berry, 1991, Hoffmann et al., 1985; Feidt, 1999; Tsirlin and Kazakov, 2002a, b; Tsirlin and Leskov, 2007; Tsirlin et al., 1998, 2005; Chen, 2005; Chen and Sun, 2004; Chen et al., 2008, 2010; Wu et al., 2000; Xia et al., 2009, 2010a, b, and many others). It deals with thermal and chemical generators of mechanical or electrical energy (engines) and the apparata or devices in which the energy is consumed (refrigerators, HPs, and separators). In principle, any complete irreversible theory is suitable to evaluate power limits for processes occurring in finite time and in systems of a finite size. However, classical irreversible thermodynamics seldom attack systems with explicit work flux (power). A typical thermodynamics analysis of an energy system refers to a topological structure that belongs in the irreversible thermal networks (diverse units connected by appropriate links). For the purpose of energy systems, the thermodynamic theory is most often applied in a discrete rather than a continuum form. The discrete structures could, in principle, be treated by the thermodynamic network theory, a general field that transfers meanings and tools of electrical circuit theory to macrosystems described by discrete models. Peusner (1986, 1990) presented a review of network thermodynamics in his topological graph approach. Thermodynamic networks (TNs) are seen as specialized forms of directed graphs in which Kirchhoff's laws hold. The approach has several advantages: it can incorporate both the equilibrium Gibbsian theory and the theory of engines in a single representation. It exposes well the theory of thermodynamic transformations and the thermostatic metric, it integrates statistical and kinetic views, and it leads readily to nonequilibrium extensions. Examples show that a variety of reversible and irreversible problems can be presented in a unified way.

Yet in view of the strongly theoretical nature and complexity of NT in the systems with power yield, its two "practical derivatives" are currently applied. The first is the (aforementioned) method of finite time thermodynamics (FTT) considered in the research groups of R.S. Berry in the United States and A. Tsirlin in Russia (Berry et al., 2000; Band et al., 1980, 1981, 1982a, b; Rubin, 1980; Tsirlin et al., 2005; Tsirlin and Kazakov, 2002a, b; Tsirlin and Leskov, 2007; Tsirlin et al., 1998), whereas the second is A. Bejan's method of entropy generation minimization (EGM), (Bejan, 1982, 1987). A significant contribution in France has been made by M. Feidt (1996, 1997, 1999) and his coworkers Costea et al. (1999). In China the main contribution to the field has been made by Lingen Chen research group (Chen, 2005; Chen and Sun, 2004; Chen et al., 2008, 2010; Wu et al., 2000; Xia et al., 2009, 2010a, b). At this point the reader is referred to many other authors and some original works in FTT, for example: Andresen (1983, 2008), Andresen and Gordon (1991); Andresen et al. (1983, 1984a, b), De Vos (1991, 1992, 1993a, b); Salamon and Nitzan (1981); Salamon et al. (1980); Sieniutycz (1973a, b, 1998a, b), and in EGM: Bejan (1982, 1987, 1996c). In his early book Bejan (1982) considers technical and physical consequences of entropy generation through heat and fluid flow. Bejan (1996c) discusses the history of the methods of entropy generation minimization and FTT, stressing especially the role of engineering researchers in the field. Boyle (2004) gives a nice review of applications of renewable energy systems. Demirel's (2014) *Nonequilibrium Thermodynamics* emphasizes the unifying role of thermodynamics in analyzing the natural phenomena that consist of simultaneously occurring transport processes and chemical reactions. They both may interact with each other and may lead to self-organized structures, fluctuations, instabilities, and evolutionary systems.

In a famous paper by Curzon and Ahlborn (1975), the efficiency of a Carnot engine is treated for the case where the power output is limited by the rates of heat transfer to and from the working substance. It is shown that the efficiency, η, at maximum power output is given by the expression $\eta = 1 - (T_2/T_1)^{1/2}$ where T_1 and T_2 are the respective temperatures of the heat source and heat sink. It is also shown that the efficiency of some existing engines is well described by the previous result. Chen et al. (2001a) discuss a basic problem of FTT, that is, the link between the Curzon–Ahlborn efficiency and the efficiencies of real heat engines. Chen and Sun (2004) review advances in FTT in the context of system's analysis and optimization. Chen (2005) reviews achievements of FTT of irreversible processes and cycles. Bracco and Siri (2010) give an example of exergetic optimization of gas–steam power plants considering different objective functions.

Considering adiabatic processes in monoatomic gases, Carrera-Patino (1988) applies a kinetic model to predict the temperature evolution of a monatomic ideal gas undergoing an adiabatic expansion or compression at a constant finite rate, and it is then generalized to treat real gases. The effects of interatomic forces are considered, using as examples the gas with the square–well potential and the van der Waals gas. The model is integrated into a Carnot cycle operating at a finite rate to compare the efficiency's rate–dependent behavior with the reversible result. Limitations of the model, rate penalties, and their importance are evaluated. Efficiency of the finite-rate process is expressed in the form $\eta = \eta_c - \Delta$, where Δ is the correction retated to the finite rate. For a Carnot cycle operating with the piston velocity 20m/s, the efficiency of the cycle follows as 56% of the Carnot efficiency. This clearly shows the decreasing effect of the finite rate on the thermodynamic efficiency. This is the significant paper that introduces the effects of FTT by statistical approaches.

In FTT (Berry et al., 2000; Feidt, 1999), which served the present author as the topic for the complementary books (Sieniutycz and Jeżowski, 2009, 2013), extremal solutions for control thermodynamic processes constitute usually the main task of a mathematical analysis. In that case, extremum conditions lead to important in-principle limits (bounds) for control processes occurring in finite time and in systems of a finite size. The related analyses can lead to improved design for thermodynamic processes in engineering systems ranging from heat engines (Andresen et al., 1984; Andresen, 1983) to chemical plants (Månsson and Andresen, 1985; Månsson, 1985). Many other optimization problems set and solved within the FTT are discussed in chapter: Thermodynamics and Optimization of Practical Processes.

Modern approaches involving extremum problems such as the maximum entropy formalism (Jaynes, 1957, 1963; Ingarden and Urbanik, 1962; Levine and Tribus, 1979; Jaworski, 1981; Grandy, 1983, 1988; Karkheck, 1986, 1989; Nettleton and Freidkin, 1989), thermodynamic geometry and related metric ideas (Ingarden, 1982; Weinhold, 1975, 1976; Ruppeiner, 1979; Nulton and Salamon, 1985; Sieniutycz and Berry, 1991), minimum entropy production theorems (Prigogine, 1947, 1949; Glansdorff and Prigogine, 1971; Essex, 1984; Salamon et al., 1980; Sieniutycz, 1980, 1994, 1997), and various macroscopic variational principles for heat conduction, hydrodynamics and thermohydrodynamics (Serrin, 1959; Biot, 1970; Bhattacharya, 1982; Caviglia, 1988; Muschik and Trostel, 1983; Markus and Gambar, 1991; Gambar and Markus, 1994; 2003; Gambar et al., 1991; Nyiri, 1991; Van and Nyiri, 1999; Vázquez et al., 2009; Finlayson, 1972; Farkas et al., 2005; Kupershmidt, 1990, 1992; Lebon, 1976, 1986; Mornev and Aliev, 1995; Schechter, 1967; Schmid, 1970a, b; Seliger and Whitham, 1968;

Salmon, 1987a, b; Shiner and Sieniutycz, 1994; Shiner et al., 1996; Sieniutycz, 1977, 1979, 1980, 1982, 1983, 1984a, b, 1985, 1986, 1987a, b, 1988a, b, 1990, 1992a, b, 1994, 1995, 1997, 2000f, 2001d, f, 2002b) can lead to general results. In particular, significant progress has been achieved in extremum approaches to dynamic analysis of various thermal, hydrodynamic, radiative and porous systems (Markus and Gambar, 1991; Gambar and Markus, 1994; Salmon, 1987a; Essex, 1984; Sieniutycz, 2007a, 2008c). The essential role of action-type variables in the field theories of thermohydrodynamics have been recognized, making it possible to incorporate the entropy source and chemical changes into an extended Gibbs equation (Sieniutycz, 1994). The progress in variational and extremum formulations has been synthesized in a book (Sieniutycz and Farkas, 2005).

The topological network approach in network thermodynamics has proved to be a powerful tool improving many classical results and asking new questions (Peusner, 1986, 1990). A review of network thermodynamics in the context of the topological graph approach gives good information about the effectiveness of the method (Peusner, 1990). TNs are seen as specialized forms of directed graphs in which Kirchhoff's laws hold. The approach has several advantages: it can incorporate both the equilibrium Gibbsian theory and the theory of engines in a single representation. It incorporates nicely the theory of thermodynamic transformations and the thermostatic metric, it integrates statistical and kinetic views, and it leads readily to nonequilibrium extensions. Examples are given which show that a variety of reversible and irreversible problems can be imbedded in a unified way using this approach.

Important generalizations in the context of maximum entropy approaches have been obtained for the statistical foundations of nonequilibrium thermodynamics (Jaynes, 1957; Lewis, 1967; Bedeaux, 1986, 1992; Callen, 1988; Garcia-Collin et al., 1984; Grandy, 1983, 1988; Jou, 1989; Jou and Llebot, 1980; Jou et al., 1988, 2001; Kreuzer, 1981; Keizer, J. 1987; Lavenda, 1985a, b; Nulton and Salamon, 1985, 1988; Nulton et al., 1985; Salamon et al., 1985; Wang, 2003) and in the stochastic theory of diffusion (Schlögl, 1974, 1980; Lavenda, 1985a, b). Benefits of nonequilibrium approaches are particularly visible in the diffusion processes (Compte and Jou, 1996). A generalized Onsager principle was found and its applications were presented for complex diffusion problems governed by memory kernels (Shter, 1973). Mechanism of stress-induced diffusion of macromolecules in solutions was worked out (Tirrell and Malone, 1977). Diffusion engines were designed and analyzed (Tsirlin et al., 2005). Balance equations of extensive quantities for a multicomponent fluid taking into account diffusion stresses were formulated and analyzed (Wiśniewski, 1984). A critical appraisal of the earlier theories of irreversible thermodynamics is available (Lavenda, 1978).

Phenomenological theory for polymer diffusion has been developed for diffusion phenomena in nonhomogeneous velocity gradient flows and applied to continuum modeling of polyelectrolytes in solutions (Drouot and Maugin, 1983, 1985, 1987, 1988;

Drouot et al., 1987; Maugin and Drouot, 1992). Equations of diffusion and electric conduction in multicomponent electrolyte systems have been formulated (Ekman et al., 1978). Transport models for multicomponent diffusion through capillaries in the transition region between gas diffusion and Knudsen diffusion have been derived (Hesse, 1974).

A dissipative Lagrangian formulation of the network thermodynamics of reaction—diffusion systems has been constructed (Shiner, 1984), and an algebraic symmetry was shown to hold in chemical reaction systems at stationary states far from thermodynamic equilibrium (Shiner, 1987). The Lagrangian formulation has been extended for chemical dynamics far from equilibrium (Shiner, 1992). The chemical results were generalized to describe chemical dynamics of biological systems in terms of variational and extremum formulations (Shiner and Sieniutycz, 1994). It was shown that certain phenomenological macroscopic symmetries in dissipative nonlinear systems are effective (Shiner and Sieniutycz, 1996). A unification of the dynamics of chemical reactions with other sorts of processes (eg, mechanical and electrical) has been found which can treat electrochemical systems in a natural fashion. In a synthesizing approach, Shiner, Fassari and Sieniutycz generalized Lagrangian and network formulations to nonlinear electrochemical systems (Shiner et al., 1996). On this basis several biological applications were found, in particular the mechanical and chemical equations of motion of the muscle contraction (Shiner and Sieniutycz, 1997).

Inertial terms in mass diffusion equations were shown to be a consequence of finite molecular mass of transported particles (Sandler and Dahler, 1964; Altenberger and Dahler, 1992). Nonequilibrium thermodynamics has been defined as the basis for the theoretical explanation of anomalous diffusion (Compte and Jou, 1996). The concept of thermodynamic length has been linked with the dissipated availability (Salamon and Berry, 1983); and link has been established between entropy and energy versions of thermodynamic length (Salamon et al., 1984). Notion of the thermodynamic length was transferred to the realm of statistical thermodynamics (Salamon et al., 1985).

Progress has been achieved in deriving evolution equations for a system in a heat bath as regards the dependence of the fluctuation-dissipation relations on the nature of the bath. The investigation of the dynamic behavior of a system interacting with the bath, based on general statistical mechanical and stochastic properties, provides a powerful tool toward understanding limitations of the usual assumptions of statistical mechanics. Particularly interesting are forms of the fluctuation-dissipation relations that ensure eventual equilibration of the system with the bath; see, for example, Lavenda (1985a, b, 1990) and Lindenberg (1990) and a discussion (Sieniutycz and Salamon, 1990c).

Gyarmati's research had a considerable influence on contemporary irreversible thermodynamics (Gyarmati, 1961a, b, 1970). His book (Gyarmati, 1970) appeared in Hungarian, English and Russian languages. He generalized the variational principles in irreversible thermodynamics, proposing a general one whose integral form is called Governing Principle of Dissipative Processes (GPDP; Gyarmati, 1969). Gyarmati's principle applies in the realm of the linear Onsager theory and for some nonlinear phenomena; a number of well-known variational principles, for example, those of plasticity, are special cases. The nonlinear generalization of Onsager's reciprocal relations—the Gyarmati-Li reciprocal relations—proves a necessary and sufficient condition for the validity of the variational principle. GPDP provides a background for numerical methods concerning dissipative processes, as well. It is quite useful in the thermodynamic theory of liquid crystals. Where Onsager's equations are clumsy to formulate both in tensorial and scalar form, the variational principle provides graceful expressions. He also contributed to the development of extended irreversible thermodynamics (Gyarmati, 1977). In his work with Lengyel he made valuable efforts to integrate chemical reaction kinetics into irreversible thermodynamics (Lengyel and Gyarmati, 1986). Although the usual theory of chemical reactions fits, in linear order, to Onsager's thermodynamics, it is too restrictive, so a nonlinear theory is desirable. Of special value are corresponding Lengyel papers (Lengyel, 1988, 1989, 1992) on deduction of the Guldberg–Waage mass action law from Gyarmati's principle. They describe progress in the field of consistency of chemical kinetics with thermodynamics, show the restrictiveness of chemical affinity and call for the modification of its definition.

The macroscopic principle known as the "governing principle of thermodynamics" (Gyarmati, 1969, 1970) has been applied for a variety of thermodynamic processes, including even involved examples of chemical kinetics (Lengyel, 1988; Lengyel and Gyarmati, 1986), provided that chemical dissipation functions are defined in a new, proper way. A general two-brackets formalism to describe dynamics and thermodynamics of complex fluids has been developed, based on the reversible (Poissonian) bracket and a dissipative (irreversible) bracket, the latter constituting a functional counterpart of the classical dissipation function (Grmela and Öttinger, 1997). However, in spite of many valuable and involved applications (Öttinger and Grmela, 1997), the two-bracket formalism cannot be generally attributed to an extremum of a definite physical quantity; for this purpose a single Poissonian bracket or symmetry of the Frechet derivative for the extremized functional are necessary. Various relationships between thermodynamic formalisms for complex fluids have been compared (Edwards et al., 1997).

The understanding of the flow behavior of polymeric liquids is of great interest from a practical as well as a theoretical point of view. An important part of the research in this field consists of the development of suitable models, describing the rheological properties of the materials (Jongschaap, 1990). Depending upon its purpose, such a model may be based upon empirical knowledge of the macroscopic flow behavior or on information about the microstructure of the materials. Moreover, for a given system, different types of modelling may be possible. In order to provide an overview of the various approaches in this area,

Jongschaap (1990) discusses the basic principles of some important models: continuum, bead-rod-spring, transient network, reptation and configuration tensor models. Emphasis has been put on a consistent treatment of the fundamentals of the various models and their interrelationship, rather than considering any of them in much detail. These synthesizing analyses performed in groups of applied mathematicians contributed significantly to the setting and development during the past 15 years of a two-bracket formalism of dissipative continuum physics called GENERIC (Grmela and Öttinger, 1997; Öttinger and Grmela, 1997; Öttinger, 2005). The formalism has been applied to different problems of continuum thermodynamics, often in this way, that a well-known problem was reformulated in GENERIC formalism. To learn some more about the GENERIC procedure, Muschik et al. (2000) attempted to compare GENERIC with rational nonequilibrium thermodynamics and other general methods. They consider a gas which is contained in a cylinder closed by a piston moving with friction. They treat this simple discrete system with rational nonequilibrium thermodynamics by using Liu's method of Lagrange multipliers (Liu and I-Shih, 1972) and, for comparison, also with the GENERIC formalism. Both different procedures yield the same results, especially in the same entropy production. The researchers compare and discuss differences, similarities, and fundamental presuppositions of both formalisms. In particular, they show that GENERIC is more redundant than the rational theory, although the results for both are the same for the example under consideration.

An improved understanding of relaxation transients and oscillators has been obtained for both hydrodynamical and chemical systems (Velarde and Chu, 1990, 1992; Sieniutycz and Salamon, 1990c). Chaos theory emerged and received its universal form since Feigenbaum's (1979) development showing the role of functional equations to describe the exact local structure of bifurcated attractors of $x_{n+1} = \lambda f(x_n)$ independent of specific f. His discovery of universal metric properties of nonlinear transformations has prompted the understanding of chaos problems along with appropriate corrections to the turbulence theory in the newer editions of fluid mechanics (Landau and Lifshitz, 1987). Moreover, Zalewski (2005) and Zalewski and Szwast (2006) worked out a method to determine regimes of chaos for heterogeneous catalytic reactions in the presence of deactivating catalysts. Using Bateman's Lagrangian (Bateman, 1929, 1931), containing its usual exponential term, Steeb and Kunick (1982) showed that a class of dissipative dynamical systems with limit-cycle and chaotic behavior can be derived from a Lagrangian function. Breymann et al. (1998) reviewed entropy balance, time reversibility and mass transport in dynamical systems modelling transport of mass or charge. The key ingredient of understanding entropy balance is the coarse graining of the local phase-space density. Their irreversible entropy production is proportional to the average growth rate of the local phase-space density. The average phase-space contraction rate measures the irreversible entropy production. They also show consistency of their results with thermodynamics. Gorban et al. (2001) developed a general method of constructing dissipative equations, following Ehrenfest's idea of coarse graining. The approach resolves the major issue of discrete time coarse graining versus continuous time macroscopic equations. Proof of the H theorem for macroscopic equations is given, several examples supporting the construction are presented, and generalizations are suggested. Coveney (1988) reviews the second law of thermodynamics from the standpoint of entropy, irreversibility, and dynamics. Dettmann et al. (1997) show how the irreversibility arises from reversible microdynamics. (See also "Classical and Anomalous Diffusion)

Nonequilibrium thermodynamics of surfaces has been developed with applications in catalysis, adsorption theory and interface phenomena (Bedeaux, 1986, 1992). These improvements resulted in an amended description of thermal, hydrodynamic, and chemical behavior for many physical and industrial systems.

Progress in applications accompanies progress in the theory. With the help of theories of irreversible processes, many nonequilibrium systems have successfully been modelled and implemented, including, for example, complex heat and mass exchangers, thermal, acoustic and solar-driven engines, convection cells, radiation-emitting fluids, viscoelastic continua (especially polymeric fluids), ion-selective membranes, diffusional separators, thermocouples, semiconductor devices, and so forth. These applications have promoted a deeper understanding of the irreversible theory, and have in turn led to valuable improvements in the design of related industrial devices. Extended thermodynamics has been formulated (Garcia-Collin et al., 1984; Jou, 1988, 1989) as a theory based on extended Gibbs equations including fluxes as independent variables. It has evolved to the status of a mature theory, summarized in a book (Jou et al., 2001) and capable of dealing with a vast variety of nonequilibrium continua. An extended thermodynamic description is important especially for systems involving high fluxes, steep gradients, and highly unsteady transients.

Progress in the theoretical and applied irreversible thermodynamics has been summarized in the corresponding project of Advances in Thermodynamics Series (series editor A. Mansoori), in the form of several edited volumes (Sieniutycz and Salamon, 1990a, c, 1992a, c). Their contents are outlined subsequently.

Volume 3 of the series entitled *General Theory and Extremum Principles* (Sieniutycz and Salamon, 1990a) is the first of four multiauthored volumes on the theory and applications of nonequilibrium thermodynamics. This volume offers a collection of papers dealing with nonequilibrium theory and extremum principles. Various theories of nonequilibrium thermodynamics are discussed in a quasi-historical context. The significance of the extremum, variational and geometric formulations for nonequilibrium theory is analyzed. A variety of nonequilibrium theories are considered including Onsagerian, rational, Meixnerian, network, information-theoretic, Mullerian, finite-time and extended. A perspective is given on the phenomenon of spacial and temporal structure formation in self-organizing systems, and self-organizing criteria are presented which show that information is an important concept. Some recent results in network thermodynamics are also reviewed. Geometric theories originating from Riemannian thermostatic curvature are extended to nonequilibrium situations. Selected statistical and stochastic aspects of the theory of nonequilibrium systems are analyzed, and their macroscopic consequences are shown in the form of generalizations of the so-called governing variational principle of dissipative processes and the associated extremum principles. An emphasis is put on variational principles of considerable generality, involving nondissipative and dissipative nonequilibrium continua containing matter or radiation. Radiation processes are treated on both statistical and phenomenological levels and the nonequilibrium thermodynamic theory of emitting and absorbing media is developed.

Volume 4 entitled *Finite-Time Thermodynamics and Thermoeconomics* (Sieniutycz and Salamon, 1990c) deals with applications of nonequilibrium thermodynamics and availability theory to mostly lumped systems where a certain external control can be applied in order to achieve improved performance. Approaches from nonequilibrium thermodynamics are capable of providing quite realistic performance criteria and bounds for real processes occurring in a finite time. Model systems have been developed which incorporate friction, heat loss, inertial effects, and finite heat conductance for real energy conversion processes. Finite-time thermodynamics, exergy analysis, and thermoeconomics seek the best adjustable parameters of various engines (thermal, solar, combustion, acoustic, convection cells, etc.), unit operations and unit processes (distillation, evaporation, chemical reactions, etc.) and systems of these operations or processes in chemical plants working under definite operational constraints. Optimal paths (or a set of optimal steady-state parameters) and optimal controls (for instance temperatures maximizing chemical efficiency) have been found. The role of optimization approaches and in particular optimal control theory is essential when solving these problems. The optimization results for a few important ecological and industrial systems are given.

Volume 5 of the series entitled *Analytic Thermodynamics with Applications in Energy Conversion Systems* supplements the previous results and gives a number of extra examples.

Volume 6 entitled *Flow, Diffusion, and Rate Processes* (Sieniutycz and Salamon, 1992a) reviews valuable results obtained for the nonequilibrium thermodynamics of transport and rate processes. Kinetic equations, conservation laws and transport coefficients are obtained for multicomponent mixtures. Thermodynamic principles are used in the design of experiments predicting heat and mass transport coefficients. Highly nonstationary conditions are analyzed in the context of transient heat transfer, nonlocal diffusion in stress fields and thermohydrodynamic oscillatory instabilities. Unification of the dynamics of chemical systems with other sorts of processes (eg, mechanical) is given. Thermodynamics of reacting surfaces is developed. Admissible reaction paths are studied and a consistency of chemical kinetics with thermodynamics is shown. Oscillatory reactions are analyzed in a unifying approach showing explosive, conservative or damped behavior. A comprehensive review of transport processes in electrolytes and membranes is given. Applications to thermoelectric systems and ionized gas (plasma) systems are reviewed.

Volume 7 entitled *Extended Thermodynamic Systems* (Sieniutycz and Salamon, 1992c) sketches the applications of nonequilibrium thermodynamics to complex systems. These are characterized by an involved form of the Gibbs equation and include systems such as solutions of macromolecules, magnetic hysteresis bodies, viscoelastic fluids, polarizable media, fluids under stresses and in the presence of essential nonstationarities and high temperature gradients. As a rule, the so-called internal variables (ie, variables not directly controllable through boundary conditions) and/or dissipative fluxes are essential in the thermodynamic description of such systems. Extended irreversible thermodynamics (EIT), the recent theory which incorporates dissipative fluxes (heat, diffusion, viscous tensor and viscous pressure) into the formalism of the Gibbs equation, is reviewed in the context of compatibility with kinetic theory, ideal and real gases, equations of state, and so forth. Applications to rheology, fluids subject to shear, shear-induced phase changes and so forth, are discussed. Generalized reciprocity relations are shown to be useful in calculating transport coefficients in liquids. Magnetohydrodynamic systems and relativistic thermal inertia systems are considered. The latter leads to EIT-like effects even in the limiting case of the low velocities. A novel theory of thermodynamic transformations is developed in which the extended Legendre transformations play the double role characteristic of mechanics and thermodynamics.

In the present chapter we shall occasionally cite and consider some detailed results published in chapters of the Advances in Thermodynamics Series.

1.2 Basic Structure of Nonequilibrium Thermodynamic Theory

It is well known that the classical macroscopic theory of equilibrium systems is just thermostatics while the word "thermodynamics" would be the proper choice for a word that expresses jointly the dynamic and thermal aspects of nonequilibrium behavior. However, this word is commonly used for what is, in fact, the thermostatics. Accordingly, nonequilibrium theories have received the expanded name "thermodynamics of nonequilibrium processes" or the simpler "nonequilibrium thermodynamics."

In the absence of the electric and magnetic fields, thermostatics is imbedded in the well-defined state space of the "classical" variables S, V, N_i. Thermostatics can rigorously be derived from statistical mechanics (Warner, 1966; Huang, 1967; Lewis, 1967; Landau and Lifshitz, 1975; Lavenda, 1985a, b; Keizer, 1987) by using the classical ensemble concept (Landau and Lifshitz, 1975) or by the method of information theory (Jaynes, 1957, 1963; Schlögl, 1974, 1980; Jaworski, 1981; Grandy, 1988; Månsson and Lindgren, 1990). Transfer of routine methods of statistical mechanics to disequilibrium to describe diverse nonequilibrium phenomena (Baranowski, 1974; Kestin, 1979; De Groot and Mazur, 1984; Censor, 1990; Demirel, 2002; Rowley, 1992; Bilicki, 1996; Sieniutycz, 2001c) is possible while frequently it is not easy (Bedeaux, 1986, 1992; Kreuzer, 1981; Lavenda, 1985a, b, 1990; Keizer, 1987; Grandy, 1988; Altenberger and Dahler, 1992). The two fundamental laws of thermostatics are the law of conservation of energy, usually involving the internal energy (the only kind of the energy in resting frames) and the law expressing the positiveness of the entropy production. The latter constitutes, in fact, a bridge between thermostatics and thermodynamics. It is perceivable that whenever the entropy production rate σ is finite, an irreversible behavior is involved, sometimes associated with a certain dynamics. By postulating that the entropy production σ can never be negative, the second law comprises both irreversible transients and dissipative steady states, for positive σ, and equilibrium steady states (equilibrium statics), for vanishing σ. But since we require to treat any irreversible situation quantitatively, the problem of the state space appropriate for treating of irreversible dynamics arises immediately.

Even for the simplest problems (eg, for pure heat exchange between two homogeneous systems separated by a membrane) the state space (in which the second law inequality should govern) is far from obvious in an exact sense. Doubts still persist about the appropriateness of parameter spaces for describing irreversible processes. Also, the problem of fluctuations close and far from the thermodynamic equilibrium becomes important (Onsager and Machlup, 1953; Machlup and Onsager, 1953; Schofield, 1966; Glansdorff and Prigogine, 1971; Jou and Llebot, 1980; Grabert et al., 1983; Keizer, 1987; Pavon et al., 1983; Jou, 1989; Jou et al., 1988, 2001; Lindenberg, 1990; Janyszek and Mrugała, 1990; Casas-Vazquez and Jou, 1992; Ross, 2008).

The classical Onsagerian theory of irreversible processes (Onsager, 1931a, b; Onsager and Machlup, 1953; Machlup and Onsager, 1953) is known from many sources (Haase, 1969; Fitts, 1962; Kluitenberg, 1966; de Groot and Mazur, 1984; Glansdorff and Prigogine, 1971; Landau and Lifshitz, 1974; Forland et al., 1988; Wisniewski et al., 1973). It extrapolates concepts and methods of statistical mechanics to nonequlibrium situations. Its basic hypothesis is that of stable local equilibrium (Prigogine, 1949; Vilar and Rubi, 2001). All variables defined in Onsagerian thermodynamics are the same as those in thermostatics, hence there is no problem with the definition of temperature and entropy outside of equilibrium; they merely become field variables with local values. The Gibbs equation for the perfect differential of the internal energy (entropy) holds locally in its unchanged form. Thus, there is no distinction between the nonequilibrium and equilibrium internal energies (entropies) which may differ due to the presence of the finite irreversible fluxes inside the system. Vilar and Rubi (2001) describe the difficulties and ambiguities associated with the absence of the local thermal equilibrium. They analyze explicitly the inertial effects in diffusion and show how the main ideas of an improved description can be applied to other situations.

Kinetic theory (Enskog, 1917; Burnett, 1935; Grad, 1958; Lifshitz and Pitajevsky, 1984) makes it possible to determine the limitations on the domain of validity of the local equilibrium hypothesis (Prigogine, 1949). Experiments show that the hypothesis applies frequently quite far from equilibrium in the case of such transport phenomena as heat and mass transfer. However, when chemical reactions, viscoelastic media, rarefied gases, shock waves and high-frequency phenomena are in question, the hypothesis is inapplicable. To describe such systems it is necessary to include so-called hidden or internal variables (Kestin, 1979; Bampi and Morro, 1984). Simple local nonequilibrium effects appear in the macroscopic description of diffusion where the diffusion velocities contribute to the kinetic energy (de Groot and Mazur, 1984). The kinetic energy of diffusion does not vanish in the barycentric frame. Its effect can modify the flux-force relationship (classical Fick's law) by an additional force of inertial type. This force involves the time derivative of diffusion velocity and accompanies the classical gradient of chemical potential in the phenomenological equation of diffusion (de Groot and Mazur, 1984; Sieniutycz, 1983).

Let's discuss some well and some less well-known shortcomings of the classical theory. Using the Eulerian (field) description of a continuum, we begin by formulating the conservation laws. For the case of a one-component fluid they can be written in the form:

$$\frac{\partial \rho}{\partial t} + \nabla(\rho \mathbf{u}) = 0 \tag{1.1}$$

$$\frac{\partial \rho \mathbf{v}}{\partial t} + \nabla(\rho \mathbf{u}\mathbf{u} + 1P + \Pi) = \rho \mathbf{F} \tag{1.2}$$

$$\frac{\partial \rho e}{\partial t} + \nabla(\rho \mathbf{u} e) = -\nabla \mathbf{q} - P\nabla \mathbf{u} - \Pi : \nabla \mathbf{u} \tag{1.3}$$

describing respectively the conservation of mass and the balances of the momentum and the internal energy. Here ρ is the mass density, $\mathbf{v}$ the barycentric velocity, P the pressure, Π the viscous shear tensor, $\mathbf{q}$ the heat flux vector and e the specific internal energy. $\mathbf{F}$ is an external force acting per unit mass of the fluid. Extracting from these equations the substantial time derivatives of ρ, $\mathbf{u}$ and e and substituting them into the Gibbs equation written in the form

$$\rho \frac{ds}{dt} = \rho T^{-1} \frac{de}{dt} - \rho^{-1} P T^{-1} \frac{d\rho}{dt} \tag{1.4}$$

leads to the well-known expression for the entropy balance:

$$\rho \frac{ds}{dt} = -\nabla \left(\frac{\mathbf{q}}{T} \right) - \mathbf{q} \nabla T^{-1} - T^{-1} \Pi : \nabla \mathbf{u} \tag{1.5}$$

The right-hand side of this expression contains a divergence term, related to the conductive entropy flow, and two source terms. These source terms postulate the forms of the phenomenological relationships, linking fluxes and forces and obeying the second law (positiveness of the entropy source σ).

Criticisms appear immediately (Muschik, 1981, 1986; Muschik et al., 2001; Lebon and Mathieu, 1983; Sieniutycz and Salamon, 1990a, b). First, the separation between the flux and source term is not unique and the theory cannot ensure unique separation. Formally, an arbitrary vector can be inserted into div($T^{-1}\mathbf{q}$) and its divergence can be subtracted from σ. Second, the phenomenological relations are postulated, not derived, and they may deviate from the true structures on account of the nonuniqueness described in the first flaw. They do not lead to an expression for the entropy source. It is only the expression for σ which furnishes a hint about the variety of the possible structures for the phenomenological laws. Third, no information is given by the classical theory about the phenomenological coefficients L_{ik} linking various fluxes and forces. These coefficients must be found experimentally or calculated from the microscopic theory of nonequilibrium statistical mechanics. Fourth, in spite of some experimental confirmation of the Onsager relations, $L_{ik} = L_{ki}$ (Rowley, 1992; Rowley and Horne, 1978; Rowley and Gruber, 1978), their validity is not assured in the nonlinear domain, and no indication is given for how a nonlinear theory should be constructed. Rowley and Gruber (1978), Rowley and Horne (1978), and Rowley (1992) use thermodynamic principles to predict values of heat-mass transport coefficients in mixtures and to design experiments measuring such values. The approach is a combined experimental and theoretical examination of diffusion, thermal conductivity, and heat of transport in nonelectrolyte liquid mixtures. Conclusions of a methological nature are drawn predicting a lasting role for molecular dynamic simulations.

Fifth, the local equilibrium postulate is certainly not allowable for continua with complex internal structure and even for simple materials in the case of high-frequency phenomena and fast transients. Sixth, the theory based on this postulate yields some physical paradoxes. For example, it leads to infinite propagation speed for heat and mass diffusion and viscous shear motion in a linear case.

Bedeaux et al. (2006) explored the possibility that the velocities that follow from the mass flux are not equal to those following from the translational momentum. According to Landau and Lifshitz (1974a, 1987) the inequality of these two velocities could lead to the violation of the conservation laws in the fluid. However, Bedeaux et al. (2006) conducted their analysis subject the assumption that the translational momentum velocity and the barycentic (mass flux) velocity in a fluid can differ without violation of the fundamentals of nonequilibrium thermodynamics. Their analysis shows that the internal energy, the total heat flux and the hydrostatic pressure are in general different in both descriptions. In an equilibrium fluid the difference disappears.

Moreover, Meixner (1961, 1966, 1968, 1974) has criticized the concepts of temperature and entropy out equilibrium. He argued that the thermometer can never be arbitrarily small and will always measure an average of T over a certain time interval and a certain volume. He also pointed out the ambiguity of the definition of entropy flow in the classical theory. He has given macroscopic arguments for various inequalities obeyed in a system that changes its state in response to an environment. He called these inequalities passivity conditions by analogy with those occurring in the theory of neural networks. The purpose of his approach was to avoid the concept of the entropy out of equilibrium. The derivations of these equalities were enhanced by Schlögl (1974) by statistical methods exploiting the properties of Kullback (1959) relative information.

As an example, one of these conditions is presented subsequently

$$\int_{-\infty}^{t} dt[(Y_i^0 - Y_i)X_i - Y_i(X^{0i} - X^i)] \geq 0, \qquad (1.6)$$

where the time derivative of X_i (energy, mole number, etc.) is the thermodynamic flux in the entropy representation. Meixner has given several inequalities of this type describing fluid systems with memory on the macroscopic time scale. By eliminating the heat flux from a nonequilibrium entropy production formula he obtained the fundamental inequality in a form not containing the nonequilibrium entropy (Meixner, 1972, 1974)

$$\int_{-\infty}^{t} dt[\rho^{-1}\mathbf{J}_q \nabla T^{-1} + T^{-1}(\rho^{-1}\sigma_e - e) + s_0] \geq 0. \qquad (1.7)$$

Here σ_e is the source of the internal energy and s_0 is the equilibrium value of the specific entropy. The derivation of this formula as well as a comparison of various thermodynamic approaches in continuum mechanics is contained in the Verhas' (1990) paper. His presentation is in the context of the Gyarmati's "Governing Principle of Dissipative Processes" (GPDP) and its extension to nonlinear constitutive equations (Gyarmati, 1969, 1970; Verhas, 1983, 1990). By applying Onsager's ideas to the passivity inequalities, Meixner was able to deduce the structure of the phenomenological equations and symmetries of the transport coefficients.

Criticisms of the classical theory were also raised by researchers in rational thermodynamics (Truesdell, 1962, 1969) who also constructed a different theory. It rests on several principles which should be met when formulating the constitutive equations. The principle of equipresence assures that, if a variable appears in one constitutive equation it has full right to be present in all remaining equations of a model (Jou et al., 1988, 2011). The principle of memory asserts that the behavior of the system at time t is influenced by all the past values of the state variables. The principle of objectivity (principle of material frame-indifference) requires that the constitutive equations are invariant with respect to all movements of the reference system. The second law inequality is expressed in the form

$$-(\dot{f} + s\dot{T}) + \sigma_{ik}\nabla v_{ik}^0 - T^{-1}\mathbf{q}\nabla T \geq 0 \qquad (1.8)$$

(note the presence of the volumetric entropy, s, and free energy, f). In this approach, temperature and entropy are primary quantities which are not introduced on physical grounds (Truesdell, 1962, 1969). The distinction between the state variables and phenomenological equations is lost. The memory effects involved result in a mathematical structure in terms of functionals rather than functions. Gibbs equations of the theory can be derived rather than postulated (Waldmann, 1958; Truesdell, 1969; Lebon and Mathieu, 1983). Relaxation phenomena may be investigated via hidden variable thermodynamics (Morro, 1985). The formalism of rational thermodynamics may be used to analyze extremum problems of viscoelasticity (Morro, 1985, 1992; Fabrizio et al., 1989; Bampi and Morro, 1984).

Despite its rigorous mathematical framework, rational thermodynamics has also been the subject of a number of objections (Lebon and Mathieu, 1983). First, it does not define either temperature or entropy and the latter quantity can be nonunique. Second, Eq. (1.6), pertains to nonequilibrium changes whereas in the original version of the Gibbs–Duhem inequality, a transition between two equilibrium states is considered. One convenient feature is that the entropy flux expression is the same as in thermostatics. Third, some constitutive equations lead to physical paradoxes and peculiarities as, for example, infinite propagation speed or negativeness of an effective heat capacity.

The equipresence principle of the rational theory has some value for extensions of the classical theory. For example, on the basis of the principle, the mass (heat) diffusion flux present in the first Fick's (Fourier's) law can also appear in the Gibbs equation. That this is indeed so was shown by de Groot and Mazur (1984) for diffusive mass flows, Grad (1958) for heat flow, Sieniutycz and Berry (1989) for diffusive entropy flow and by many other researchers in the extended irreversible thermodynamics for diffusive flows of heat, mass and momentum. See a review of this work (Jou et al., 1988, 2011).

Müller (1966, 1971a, b) proposed a version of the rational theory which gets rid of the concept of temperature independent of an instrument which measures it. He introduces an empirical temperature θ (measured for equilibrium situation or not) and a universal function $\Gamma(\theta, d\theta/dt)$ which he calls the "coldness." At equilibrium, Γ is equal to the inverse of the absolute temperature T. He assumes that the second law equality (1.5) should be replaced by the following inequality

$$\rho \frac{ds}{dt} + \nabla \mathbf{j}_s \geq 0 \qquad (1.9)$$

containing the entropy flux $\mathbf{j}_s$ which has to be described by a constitutive equation. $\mathbf{j}_s$ is shown to obey the formula

$$\mathbf{j}_s = \Gamma(\theta, \dot{\theta}) \mathbf{q} \qquad (1.10)$$

where Γ is the reciprocal of the temperature only at equilibrium. This theory leads to the finite propagation speed of thermal signals provided that one assumes that $\delta e / \delta(d\theta/dt)$ vanishes. This is criticized from the viewpoint of the equipresence principle (Jou et al., 2001). The role of $d\theta/dt$ as an extra independent variable of the theory seems to be physically unjustified and doubts were expressed about existence of a coldness function having universal character (Lebon and Mathieu, 1983; Meixner, 1974). However, the idea of the heat flux or entropy flux described by a constitutive equation survived and has found its accomplishment in other Müller approaches (Müller, 1985, 1992; Liu and Müller, 1983) as well as in the works of others culminating in the theory called extended irreversible thermodynamics (EIT) (Luikov, 1966a; Nettleton, 1969; Luikov et al., 1976; Bubnov, 1976a, b, 1982; Sieniutycz, 1979, 1980a, 1981a, c, d, e, 1982, 1983, 1984a, 1985, 1987b, 1988b, 1989a, 1992a, b, 1994; Jou, 1989; Jou and Llebot, 1980; Llebot, 1992; Jou et al. 1988, 2011; Lebon, 1986, 1992; Lebon and Boukary, 1982, 1988; Lebon and Casas-Vazquez, 1976; Lebon and Mathieu, 1983; Lebon et al., 1980, 1986, 1988, 1990; Garcia-Collin et al., 1984; Müller, 1992; Kremer, 1985, 1986, 1987, 1989, 1992; Perez-Garcia and Jou, 1986, 1992; Nettleton, 1986, 1987a, b, 1992; Nettleton and Freidkin, 1989).

In that theory, the awkward independent kinetic variable dθ/dt has been replaced by the more natural heat flux **q** (Jou et al., 1988, 2001; Garcia-Collin et al., 1984). Critical analyses of the diverse theories for nonequilibrium thermodynamics have been necessary to formulate the new theory. Hidden variable approaches contributed to improve the extended thermodynamic theory (Kestin, 1979; Maugin, 1979, 1987; Muschik, 1981, 1990a, b, 2007; Bampi and Morro, 1984; Morro, 1985; Maugin and Drouot, 1983a, 1992). The theory was also set within the maximum entropy formalism (Nettleton and Freidkin, 1989; Ruggeri, 1989). It has also been recommended to transfer to the realm of EIT some results in classical relativistic theories of fluids, plasmas, and quantum systems (Schmid, 1967a, b, 1970a, b; Jackson, 1975; Hiscock and Lindblom, 1983; Landau and Lifshitz, 1974b, 1984).

Jou and Casas-Vazquez (1992) have suggested an experiment to check the physical reality of a nonequilibrium absolute temperature previously proposed from theoretical grounds in the framework of extended irreversible thermodynamics. The idea involves the two thermodynamic systems connected through a highly conducting metallic plate. One system is at equilibrium at temperature T. Another system is in a nonequilibrium steady state under a heat flux q. EIT predicts a heat flow q from the equilibrium system to the nonequilibrium one because the nonequilibrium thermodynamic temperature Θ is not equal to the local equilibrium absolute temperature T. See also Jou et al. (2001).

Casas-Vazquez and Jou (1994) discuss the conceptual differences between a nonequilibrium absolute temperature (defined as the partial derivative of the steady-state nonequilibrium entropy) and the local-equilibrium absolute temperature. They explore two situations in which this difference could be observed in molecular-dynamical situations. By using a simple model for the nonequilibrium entropy, they compute the difference between both temperatures for gases, metals, and electromagnetic radiation. They analyze the compatibility of both temperatures in two simple examples in the kinetic theory of gases and in an information-theoretic analysis of harmonic chains. Finally, they compare their own results with some other works which have proposed nonequilibrium temperatures on several different grounds.

In the meantime, the general theory of irreversible thermodynamics developed in many directions. Lindenberg (1990) has accomplished a mesoscopic analysis of the dynamic behavior of a system interacting with a heat bath in both classical and quantum formulations. The analysis shows that the thermostatic equilibrium can be attained by many dynamics and that the specific route depends on the detailed nature of the system-bath interactions. Treating system and bath as a "universe," the bath variables can be systematically eliminated from the "universe Hamiltonian" to end up with the system equations of motions which contain appropriate fluctuating and dissipative forces. This type of analysis has been essential for understanding the limitations of the usual assumptions of statistical mechanics. In particular, departures from Langevin features have been considered. Particular attention has been directed to the forms of the fluctuation-dissipation relations which ensure eventual equilibration of the system with bath. It has been shown that the universality of these relations for classical systems may be violated in quantum systems. Romero-Salazar and Velasco (1997) proposed a microscopic model to construct a quantum Langevin equation for a system in a heat bath with temperature varying in time. The results were applied to the electron-phonon system.

Lavenda (1990) has developed a general approach to statistical thermodynamics based on Gauss's law of errors and on concavity properties of the entropy. Gauss's law has been shown to replace Boltzmann's principle relating the entropy to thermodynamic probability for isolated systems. Legendre transforms of the entropy lead to dual distributions for intensities. Their invariance determines the physical state equations. As an illustration, dual intensities can be used to derive the thermodynamic uncertainty relations. A kinetic derivation of Gauss's law of errors holds, which involves the optimal path for the growth of fluctuations. Lavenda has found that, without correlations between nonequilibrium states, deterministic rates maximize the path probability. In order to take these correlations into account, a diffusional limit shoud be analyzed. Limitations on Gauss's distribution can be determined by relating the entropy and its derivative to a quasipotential governing the evolution of diffusion. A number of inaccuracies in the literature have been explained by this unifying approach.

Verhas (1990) reviewed recent progress concerning the so-called governing principle of dissipative processes. This macroscopic principle, especially in its Gaussian form, can be associated with the classical Onsager–Machlup analysis and its extensions (Lavenda, 1990, discussed previously). The principle is based on the two dissipation potentials $\phi(\mathbf{J})$ and $\psi(\mathbf{X})$, known to Onsager, whose respective gradients are equal to the force $\mathbf{X}$ and the flux, $\mathbf{J}$. Introducing the Lagrangian $L = \sigma - \psi - \phi$ containing the (bilinear) entropy source σ and varying fluxes and forces, yields phenomenological equations. Verhas discusses quasilinear and nonlinear extensions of the principle in local and integral forms. He shows that the state dependence of dissipation potentials can be included in the quasilinear case making the principle of unrestricted type. He also shows that the principle can be reformulated in the nonlinear case even if the reciprocity relations fail. Applications presented refer to nematic liquid crystals and plastic flow theory. Other applications, involving nonlinear chemical kinetics in reaction systems, can also be found (Lengyel, 1988, 1992; Lengyel and Gyarmati, 1986; Cukrowski and Kolbus, 2005).

Hamiltonian and Lagrangian extended thermodynamics have been formulated (Grmela and Teichman, 1983; Grmela and Lebon, 1990; Sieniutycz, 1989b, 1990, 1992b, 1994). Some of these papers were limited to reversible processes. For example, Jezierski and Kijowski (1990) have presented a new exposition of formal field theory for the hydro-thermodynamics of reversibly flowing fluids. The authors' formalism gets rid of continuity equations as an explicit constraint of the classical kinetic potential. Their approach is technically simpler and more elegant than earlier approaches to this problem. Although only nondissipative systems are analyzed, the reader can gain valuable information about the connections between the Lagrangian and the Eulerian representations of fluid motion and the structures of the energy-momentum tensor in the two representations. The authors discuss also the Poisson bracket structure of thermohydrodynamics. Their formulation of thermohydrodynamics in terms of the kinetic potential is based on the free energy rather than the internal energy. It involves an important quantity, τ, defined such that its total time derivative is equal to the absolute temperature (up to a scale constant). They call τ the "material time" although it was known earlier as thermasy (Schmid, 1970). Grmela and Teichman (1983) have shown the effectiveness of a thermal coordinate for Lagrangian formulation of the Maxwell–Cattaneo hydrodynamics. An alternative interpretation of thermasy as a thermal action or a thermal phase has been given in Sieniutycz's (1990, 1992b) papers, where its crucial role for the description of irreversibility has been shown. Kupershmidt's (1990a) publication treats Hamiltonian formalism for reversible nonequilibrium fluids with heat flow, whereas his other paper develops a variational theory of magneto-thermohydrodynamics in terms of the canonical Hamiltonian approach rather than the Lagrangian approach. The author advocates the Hamiltonian approach for continuous mechanics systems. He specifies several arguments proving the superiority of the Hamiltonian approach over the Lagrangian for describing the time evolution of continua. He also presents recipes for determining the relation between the two formalisms. These recipes are first used to develop the Hamiltonian theory of magneto-hydrodynamics, and then to pass to the Lagrangian description. The bonus is the new Lagrangian for both relativistic and nonrelativistic cases. One conclusion drawn by the author is that no Lagrangians are possible (in the physical space-time) without the presence of extra variables of the Clebsch type. Kupershmidt's original research is summarized in his book on variational dynamics of perfect fluids in Eulerian (field) representation of fluid's motion (Kupershmidt, 1992).

Papers by Sieniutycz (1990, 1992b) present an enhanced version of Kuppershmidt's conclusion: for substantially irreversible processes, no physical Lagrangians are possible without the *explicit* presence of action-type (Clebsch) variables. The basic variables of this type are shown to be the material or thermal actions for the transferred species and entropies.

They can also be related to macroscopic quantum phases. They obey (vector) partial differential equations of the Hamilton–Jacobi type, and their characteristic equations define nonequilibrium (flux dependent) temperature and chemical potentials. These intensities turn out to be in a direct relation with Einstein's relativistic transformation of the temperature. The paper by Sieniutycz (1990) proposes a theoretical setting for the classical and extended thermodynamics. In this setting, many issues bearing on the thermodynamics of irreversibility can be seen in a new light. It shows that several aspects of irreversibility can be obtained from the conservation laws of the extended theory without the explicit introduction of the entropy source into the Lagrangian. It gives the Legendre transforms of classical mechanics and classical thermodynamics a common role within the thermohydrodynamic formalism. It shows that the *standard* conservation laws pertain to the particular quasireversible situation where the explicit effect of the material and thermal actions can be ignored. Finally, it shows that, for highly irreversible processes, these laws may be modified to take account of these action variables and of the entropy production.

Methods and ideas of classical theory of thermodynamic fluctuations (Onsager and Machlup, 1953; Machlup and Onsager, 1953; Schofield, 1966; Landau and Lifshitz, 1975; Lavenda, 1985a, b; Keizer, 1987; Grabert et al., 1983; Janyszek and Mrugała, 1990; Lindenberg, 1990) has been transferred into the framework of EIT including the electric current fluctuations (Jou and Llebot, 1980; Casas-Vazquez and Jou, 1992; Jou et al., 2001; Guy, 1992). However, the application of EIT to describe fluctuations in chemical and other systems was not always successful, and other approaches to disequilibrium fluctuations were proposed (Ross, 2008). Extended irreversible thermodynamics removed some doubts persisting in the contemporary theory of rheological materials (Lebon, 1992). A geometric model for the thermodynamics of continuous media was constructed, providing a clearer meaning to the commonly used concept of "processes" and "transformations" in the spirit of rational thermodynamics (Dolphin et al., 1999). The aim was to elucidate a clear ground suited to analyze thermodynamic transformations outside equilibrium. Relativistic aspects of extended thermodynamics have been investigated (Israel, 1976; Pavon et al., 1983; Sieniutycz, 1994, 2002b). These efforts have also been stimulated by many research papers in engineering and science, aimed to remove paradoxes of infinite propagation speed in classical equations of heat and mass diffusion. Number of results have led to wave equations of heat conduction and mass diffusion (Cattaneo, 1958, Nettleton, 1960; Kaliski, 1965, Grigoriew, 1979; Sieniutycz, 1979, 1981a, b) and many other papers (Bubnov, 1976a, b, 1982; Gyarmati, 1977; Bhattacharya, 1982). These effects are pronounced especially in the media with large memory effects (Builtjes, 1977; Sieniutycz, 1992a). Soon a number of basic problems for wave equations of heat and mass diffusion have been attacked, namely: thermodynamic stability problems for coupled heat and mass transfer (Sieniutycz, 1981c, d, 1985); variational principles (Sieniutycz, 1987a, 1981e, 1982, 2001d, f); quantitative role of inertial terms in Brownian and molecular diffusion (Sieniutycz, 1983, 1984a, b); chemical aspects (Sieniutycz, 1987b, 1995, 1997a); relativistic aspects (Kranys, 1966; Israel, 1976; Kupershmidt, 1990b, 1992; Scholle, 1994; Sieniutycz, 1986, 2002b, d); drying relaxation problems (Sieniutycz, 1989a). Energy-momentum tensor and conservation laws have been formulated for extended multicomponent fluids with the transport of heat, mass and electricity (Sieniutycz, 1989b, 1990, 1992b, 1994; Wiśniewski, 1984, 1992; Wiśniewski et al., 1973). Progress in EIT research has improved the current status of nonequilibrium thermodynamics of emitting and absorbing media and thermodynamics of ionized systems (Wiśniewski, 1990, 1992). Wiśniewski (1995) gave a concise, exact treatment of classical thermodynamics of elastic solids. This topic is also analyzed elsewhere, Syczew (1973), Callen (1988). Wiśniewski (1990) and Wiśniewski et al. (1973) presented an Onsagerian-type and Glansdorff–Prigogine type analyses of optically active continua, capable of emitting and absorbing radiation. Radiative transfer in such continua (semitransparent media) is important in atmospheric physics, astrophysics, radiation gas dynamics, and combustion. Effects such as radiation viscosity, radiative heat, and radiation pressure must be taken into account at high temperatures. Photochemical reactions may be included. All these effects require simultaneous treatment of the balance equations for matter and radiation. The authors show how the exchange of the radiative energy modifies the known classical expressions for the entropy source and the phenomenological equations. For example, the radiant energy source strength depends not only on the temperature difference but also on the velocity divergence (bulk viscosity effect) and the chemical affinity. Expressions for the excess entropy production and the local potential are obtained and the stability conditions are determined. These results may be complemented by balance equations of extensive quantities for multicomponent fluids with diffusion stresses (Wiśniewski, 1984), and extended to the realm of thermodynamics of ionized systems (Wiśniewski, 1992).

In a group of papers, associated with minimum entropy production in radiative transfer and failure of the bilinear formalism, Essex and his coworkes point out the exceptional nature of radiation systems as thermodynamic systems (Essex, 1984, 1990; Holden and Essex, 1997; Essex and Kennedy, 1999; Essex et al., 2003; Li et al., 2003). They argue that radiation is typically out of equilibrium, even if the matter which may interact with radiation is locally in equilibrium. Also, they identify ambiguities and paradoxes that arise in the case of radiation from the interpretation of bilinear forms expressed in terms of thermodynamic fluxes and forces. Roughly stated, majority of these effects are caused by the anisotropy and nonlocality of nonequilibrium radiation. Their analyses are directed toward showing that the entropy production is not adequately described by any of the bilinear forms thus far advanced.

The nonclassical role of **q** as an extra (vector) state coordinate, crucial or EIT, has already been observed in the kinetic theory (Enskog, 1917; Burnett, 1935; Grad, 1958; Eu, 1987, 1988, 1989, 1992), where the concept of the heat flux **q** as an independent state variable has been in complete agreement with the Grad's moment analysis (Grad, 1958). In the general case all diffusive fluxes (heat, mass, momentum) constitute extra independent variables in the EIT. The origins and present status of the EIT theory has been reviewed (Garcia-Collin et al., 1984; Jou et al., 1988, 2001; Sieniutycz and Salamon, 1992; Sieniutycz, 1994).

Bhalekar (1990) proposes a general phenomenological theory of irreversible thermodynamics which involves the identification of a new quantity designated as thermodynamic heat and the adoption of a Universal Inaccessibility Principle which yields a simple extended Gibbs relation. Bhalekar (1996) shows that the Clausius' inequality analogue cannot exist for open systems. Also the Clausius–Duhem inequality is not assured by the Clausius' inequality and may not be applicable in all nonequilibrium situations. Bhalekar (1998) discusses at length implications of universal inaccessibility principle. Bhalekar (1999) claims that, for both reversible and irreversible processes, no thermodynamic description can be constructed without a Gibbs–Duhem equation. His analysis involving this equation shows that the temperature and pressure do not change in going from reversible to irreversible processes.

The second law of thermodynamics directly provides an entropy function for equilibrium states (Callen, 1988). For nonequilibrium situations, however, one has only the Clausius inequality, and hence an entropy function does not directly follow from the second law of thermodynamics. This difficulty is sidelined in CIT by postulating a local equilibrium assumption. Therefore it is natural to ask whether an IT theory can be developed for situations beyond the local equilibrium assumption. Bhalekar (2000a) presents an argument to show that the differential form of Clausius' inequality contains the entropy function for nonequilibrium states. Bhalekar (2000b) describes a methodology to develop an irreversible thermodynamics framework by applying it to a spacially uniform chemically reacting closed systems with no irreversibility in the energy exchange. Bhalekar (2000c) proposes a generalized zeroth law of thermodynamics which takes in its fold all nonequilibrium and equilibrium states of a thermodynamic system.

Considering heat conduction in a rigid solid Bhalekar and Garcia-Colin (1998) claim that the practice in EIT of raising the physical fluxes (such as heat flux density, $\mathbf{q}$) to the status of independent thermodynamic variables does not satisfy the basic thermodynamic requirements. Garcia-Colin and Bhalekar (1997) focus on two irreversible thermodynamics descriptions of extended type, showing, in particular, different implications of local thermodynamic equilibrium in various extended theories. Garcia-Colin and Green (1966) discuss the problem of the appropriate definition of temperature in the kinetic theory of dense gases. On the road they resolve an apparent contradiction in the value of the bulk viscosity between two methods of making the transition from the kinetic to the hydrodynamic stage. They show that the definitions of temperature through the kinetic energy and through the total energy are equivalent. Also, they discuss the general question of the appropriate choice of macroscopic variables in nonequilibrium statistical mechanics. The form of the equations of hydrodynamics is covariant with the definition of the macroscopic variables, but the molecular-distribution functions are invariant to this choice. In spite of the equivalence of temperature definition between a quantity which is a local integral of motion and one which is not, the principle that macroscopic observables are properly chosen to be "approximate single-valued integrals of motion" can still be maintained. The question of which temperature or bulk viscosity is measured is discussed in the context of specific experimental situations. Garcia-Colin et al. (1984) analyze the theory of extended irreversible thermodynamics at three levels of description: macroscopic (thermodynamic), microscopic (projection operators) and mesoscopic (fluctuation theory). Extended irreversible thermodynamics is understood as the theory of macroscopic systems which takes as independent variables the slow (conserved) ones plus the fast dissipative fluxes. Such description is compared with the memory function approach based only on the conserved variables. They find that EIT is reacher and wider than the memory function approach.

The previous results assumed that solely the Gibbs equation is modified by the presence of diffusive fluxes, whereas the form of the conservation laws remains unchanged. However, it has been shown (Sieniutycz, 1988a, b, 1989a, 1990; Sieniutycz and Berry, 1989) that extended Gibbs equations can be obtained from a Lagrangian approach (Hamilton's stationary action) where the extended internal energy can be extracted from the total energy appearing as time component G^{tt} of energy-momentum tensor G^{ik}. The matrix representing this tensor includes a term for generalized kinetic energy of diffusion. This term is responsible for finite propagation speed effects as well as other effects predicted by EIT. The form of this tensor corresponds with the classical conservation laws only when the sources of mass and entropy become negligible (Sieniutycz, 1990, 1994). However for substantially irreversible processes these sources may be large, and then the structure of G^{ik} deviates from the classical structure. Presumably, some new variables may play a role in G^{ik} and in the energy E. These new variables might describe the so-called thermal phase and matter phase, the latter being averaged microscopic quantum phase (Sieniutycz, 1992b, 1994). A major theorem of phase dependent thermodynamics is that the entropy source is equal to the derivative of the energy with respect to the thermal phase. Perhaps, in strongly irreversible processes, the analytical form of the conservation laws could be changed due to contribution of the thermal phase, or rather certain invariant quantities constructed on its basis. These matters are discussed more thoroughly in a paper (Sieniutycz, 1992b), yet the theory is still immature since it is linked with an evergreen problem of hypotheses for thermal mass and thermal inertia (Sieniutycz and Berry, 1997). Quantum theories, such as de Broglie micro-thermodynamics (de Broglie, 1964, 1970; Sieniutycz, 1992b, 1994), may accomodate EIT effects. An approach toward constructing the Hamiltonian formulation as a basis of quantized thermal processes (Markus, 2005) has turned out to be quite successful. The approach is based on a generalized Hamilton–Jacobi equation for dissipative heat processes (Markus and Gambar, 2004) as the fundamental component of the variational theory developed by these authors (see also chapter: Variational Approaches to Nonequilibrium Thermodynamics). Chen and Su (2003) consider discontinuous solutions for Hamilton–Jacobi equations. The basic results of the variational theory are quasi-particles in thermal processes (Markus and Gambar, 1996, 2005; Vazquez et al., 2009), siblings of phonons or quasiparticles of quantized acoustic fields. Previously Markus and Gambar (1996) have pointed out the possibility of existence of quasiparticles in the heat conduction. In their next paper Markus and Gambar (2005) introduce an abstract scalar field and a covariant field equation, by which they make an attempt to connect the Fourier heat conduction and wave-like heat propagation. The link with Bohm's (1952) interpretation of the quantum theory in terms of "hidden" variables seems inevitable.

In Bohm's (1952) paper quantum-mechanical probabilities are regarded as their counterparts in classical statistical mechanics, that is, as only a practical necessity and not as a manifestation of an inherent lack of complete determination in the properties of matter at the quantum level. As long as the present form of the Schrödinger's equation is retained, the physical results obtained with the alternative (Bohm's) interpretation are the same as those obtained with the usual interpretation. Bohm and Vigier (1954) presented a causal interpretation of the quantum theory in which a set of fields act on a particle-like inhomogeneity which moves with the local stream velocity of the equivalent fluid. They introduce the hypothesis of a very irregular and effectively random fluctuation in the fluid motion and prove that arbitrary probability density ultimately decays into a square of the Schrödinger's wave function ψ. As Goldstein (2009) states, many recent results suggest that quantum theory is about information, and the quantum theory is best understood as arising from principles concerning information and information processing. At the same time, by far the simplest version of quantum mechanics, Bohmian mechanics, is concerned, not with information but with the behavior of an objective microscopic reality given by particles and their positions. According to Goldstein's own statement, he would like to examine whether, and to what extent, the importance of information, observation, and the like in quantum theory can be understood from a Bohmian perspective. Also, he would like to explore the hypothesis that the idea that information plays a special role in physics naturally emerges in a Bohmian universe.

1.3 Extremum Properties of Thermodynamic Systems

Mathematically, two classes of extremum problems can be distinguished: those involving extrema of functions and those dealing with extrema of functionals (Sieniutycz and Farkas, 2005). Usually we refer to the second class as variational principles (see chapter: Variational Approaches to Nonequilibrium Thermodynamics). Physically however, it is more natural to divide the approaches according to whether they pertain to equilibrium, to steady state, or to nonstationary disequilibrium. While many problems whose solutions terminate at equilibria or steady states involve extrema of functions, some problems in distributed systems can involve variational principles. Thus the mathematical and physical classifications differ. Subsequently, we use the physical classification.

1.3.1 Equilibrium Problems

Boundary conditions play essential role in classification of thermodynamic systems; they define constraints imposed on a thermodynamic system in the physical space–time.

Problems involving optimization of thermodynamic potentials in the presence of spatially homogeneous, time independent constraints at the boundaries constitute the extremum problems of classical thermostatics. Since these problems are treated in many textbooks (Werle, 1967, Landau and Lifshitz, 1974, 1975, 1984, 1987), we limit the discussion to listing a few examples subsequently.

Equilibrium conditions at interfaces in multiphase systems and local equilibrium curvature of interfaces can be obtained as minimum conditions of the grand thermodynamic potential (Landau and Lifshitz, 1974a). Equilibrium concentrations of species in chemical systems can be found by minimizing free energy subject to the linear constraints resulting from stoichiometry (White et al., 1958). In some textbooks the popular Callen postulates which characterize final equilibria achieved in thermodynamic systems are also included to the class of equilibrium problems (Callen, 1988). Since however these equlibria are attained in systems with unconstrained internal variables (the notion from nonequilibrium thermodynamics), Callen postulates constitute in fact the bridge from disequilibrium to equilibrium (see subsequently). Clearly, in purely equilibrium problems, any optimization procedure should deal with merely equilibrium potential functions, which is, in fact, a seldom situation limited to quasistatic variations.

Equilibrium problems can also involve variational principles. Equilibrium shapes of membranes, layers, and so forth encountered in the theory of elasticity can be found from the minimum of the energy or the complementary energy (Castigliano principle). An equilibrium distribution of the liquid covering the grains of a packed bed or wetting a solid skeleton corresponds to the minimum of the free energy functional (Amar et al., 1989). This constitutes a generalization of the Plateau problem of the calculus of variations (Salamon et al., 1990; Osserman, 1986; Gelfand and Fomin, 1970).

1.3.2 Jaynes's and Callen's Bridges From Equilibrium to Disequilibrium

Jaynes' MEP approach. The important formalism involving the search for maximum of the information-theoretic entropy (Jaynes, 1957, 1963; Ingarden and Urbanik, 1962) was originally developed for processes in lumped systems whose final states are at the thermodynamic equilibrium. Study of classical thermodynamic ensembles provide a way for deriving basic relationships of equilibrium thermodynamics (Levine and Tribus, 1979).

Grandy (1983, 1988) extended the information-theoretic approach to nonequilibrium phenomena.

The formalism pertains to physical systems whose states are described by a probability distribution and asserts that the physical states maximize the informational entropy under the constraints which describe expectation values of some microscopic quantities (eg, energy and mole numbers). It follows that the extremum distribution at equilibrium is a Gibbs distribution. A natural generalization is to incorporate constraints for statistical moments of any order (Jaworski, 1981). The basic features of the extremum solution can be derived by using both absolute and relative (Kullback, 1959) information (Moszynski, 1983; Månsson and Lindgren, 1990).

Moreover, many information-theoretic approaches have been developed to characterize systems with nonequilibrium transitions (Levine and Tribus, 1979; Schlögl, 1974, 1980; Ingarden, 1982; Haken, 1984; Grandy, 1983, 1988). Schlögl's systematic approach leads to all basic results of classical thermodynamics, including the Onsagerian nonequilibrium formalism, fluctuation theory, Meixner's passivity conditions and the Glansdorff–Prigogine criterion (Schlögl, 1974). Thermodynamic geometries based on the information theoretic approach have been introduced in R.S. Berry and R. Ingarden's research groups (Salamon et al., 1985; Ingarden, 1982; Mrugala et al., 1990; Sieniutycz and Salamon, 1990b; Sieniutycz and Berry, 1991). More recent developments encompass the response of the nonequilibrium dynamical systems to both thermal and dynamical forces. They use a correlation function description to derive the equations of hydrodynamics and rheology (Grandy, 1983, 1988).

Ingarden (1976) states that the information theory is not only the theory of the entropy concept itself (in this aspect the information theory is most interesting for physicists), but it is also a theory of transmission and coding of information, that is, a theory of information sources and channels. He points out that the classical information theory (Shannon and Weaver, 1969) cannot be directly generalized to the usual quantum case, because there is no general concept of joint and conditional probabilities in the usual quantum mechanics. Using, however, the generalized quantum mechanics of open systems and a generalized concept of observable it is possible to construct a quantum information theory which is a straightforward generalization of Shannon's theory (Ingarden and Urbanik, 1962; Ingarden, 1982; Ingarden et al., 1997).

Many introductory physics texts introduce the statistical basis for the definition of entropy in addition to the Clausius definition, $\Delta S = q/T$. Schoep (2002) uses a model based on equally spaced energy levels to present a way that the statistical definition of entropy can be developed at the introductory level. In addition to motivating the statistical definition of entropy, he also develops statistical arguments to answer the following questions: (1) Why does a system approach a state of maximum number of microstates? (2) What is the equilibrium distribution of particles? (3) What is the statistical basis of temperature? (4) What is the statistical basis for the direction of spontaneous energy transfer? Finally, he shows the correspondence between the statistical and the classical Clausius definitions of entropy.

One of the basic problems of statistical physics is how to find a probability distribution fitting in a best way our incomplete knowledge of a physical system. The method is based on the Maximum Entropy Principle (MEP) which was promoted by Jaynes (1957, 1963) and Ingarden and Urbanik (1962). Appling MEP systematic derivations of all findings of classical thermodynamics are available (Tribus, 1970). Using the concepts of IT and MEP Mrugala (1996) and Janyszek and Mrugała (1990) developed Riemannian metric on contact thermodynamic spaces. They also compared Riemannian and Finslerian geometries and developed the fluctuation theory in thermodynamic systems. They showed that the Riemannian metric tensor can be defined using the second fluctuation moments of the stochastic variables. The curvature tensor and the scalar curvature were calculated, and it was verified that the Riemannian scalar be adopted as a measure of thermodynamic stability. Jaworski (1981) applied the information thermodynamics to define the second order temperatures for the simplest classical systems. Ingarden (1992) has presented a higher-order generalization of the usual thermodynamic formalism for systems with the number of particles N large in comparison with 1, but sufficiently small that one or many higher-order fluctuations of energy (or other observables) are measurable, directly or indirectly. This yields a mesoscopic thermodynamics for small systems in higher-order states.

Nulton and Salamon (1985) work out the geometry of ideal gas and statistical mechanics of combinatorial optimization. Mijatovich and Veselinovic (1987) synthesize the results in differential geometry of equilibrium thermodynamics. Nulton et al. (1985) treat quasistatic processes as step equilibrations. Ruppeiner (1979) introduces a Riemannian geometric model and considers the thermodynamic curvature in the thermodynamic space. Ruppeiner (1990) reviews the thermodynamic metric mainly in the context of fluctuation theory. (See chapter: Thermodynamic Lapunov Functions and Stability) Salamon et al. (1984) and Mrugala (1996) determine the relations between entropy and energy versions of thermodynamic length. Salamon et al. (1985) and Mrugala et al. (1990) transfer the length concept to the realm of statistical thermodynamics. Schindler et al. (1998) propose a characterization of systems represented by macromolecules (enzymes, ion channels, etc.) in terms of their distance from equilibrium (DFE). Standard chemical and physical approaches are not applicable because they need too detailed kinetic and thermodynamic information. They suggest a general measure for DFE in systems described by fluxes. Each connection between subsystems corresponds to two unidirectional fluxes, f and b. These fluxes form forward and backward vectors, **F** and **B**, respectively. The forward direction is defined by a net flux, so that $0 < b < f$. The twice dimensional flux vectors $\mathbf{R} = \{\mathbf{F},\mathbf{B}\}$ and $\mathbf{U} = \{\mathbf{F},-\mathbf{B}\}$ are then introduced. DFE is defined as $P = (\mathbf{RU}/\mathbf{UU})^2 = \cos^2\alpha$, α being the angle between the **R**- and **U**-vectors, with $0 < \alpha < \pi/2$. The distance from strict irreversibility, E, is defined symmetrically $E = \cos^2(\pi/2 - \alpha) = 1 - P$.

For a single-stage reaction, parameter P is strictly equivalent to the chemical affinity. For multistage reactions P reflects DFE better than chemical affinity. Near equilibrium P is equivalent to excess Gibbs energy. Away from equilibrium, P scales DFE better.

The MEP dictates that one should look for a distribution, consistent with available information, which maximizes the entropy. However, as stated by Ingarden (1976), Ingarden and Urbanik (1962), Ingarden et al. (1982, 1997), and Harremoes and Topsoe (2001), this principle focuses only on distributions and it appears advantageous to bring information theoretical thinking into play by also focusing on the observer and on coding. This leads to the consideration of a certain game, the Code Length Game and, via standard game theoretical thinking, to a principle of Game Theoretical Equilibrium (Topsoe, 1993). He believes that that sort of principle is more basic than the MEP in the sense that the search for one type of optimal strategies in the Code Length Game translates directly into the search for distributions with maximum entropy. Harremoes and Topsoe (2001) propose a self-contained, treatment of both principles mentioned, based on a study of the Code Length Game. The promising results tempt the authors to speculate over the development and stability of languages. Grendar and Grendar (2001) describe several mystifications and mysteries pertinent to MaxEnt and propose a new formulation of the problem. Lucia's (2010) develops an analytical and rigorous formulation of the maximum entropy generation principle. The result is suggested as the Fourth Law of Thermodynamics.

Styer (2000) asks what is the qualitative character of entropy. He presents several examples from statistical mechanics including liquid crystal reentrant phases, two different lattice gas models, and the game of poker. They demonstrate facets of this difficult question and point toward an answer. The common answer of "entropy as disorder" is regarded here as inadequate. An alternative but equally problematic analogy is "entropy as freedom." Neither simile is perfect, but if both are used cautiously and not too literally, then the combination provides considerable insight. The author does not accept the entropy as a measure of disorder. The most important failing, is that the analogy between entropy and disorder invites us to think about a single configuration rather than a class of configurations.

Struchtrup and Weiss (1998) propose a plausible principle for stationary thermodynamic processes: The maximum of entropy production becomes minimal in stationary processes. The authors show the usefulness of the principle on the example of 1D stationary heat transfer in monoatomic gases for which they solve a set of extended moment equations following from the Boltzmann equation. These equations require boundary conditions for all moments considered, and these are constructed by means of the principle. Importantly, the authors show that the minimum principle for the global entropy production does not lead to acceptable results in the case considered.

The tendency of the entropy to a maximum as an isolated system is relaxed to the equilibrium (the second law of thermodynamics) has been known since the mid-19th century. However, as stated by Martyushev and Seleznev (2006), independent theoretical and applied studies, which suggested the maximization of the entropy production during nonequilibrium processes (the so-called maximum entropy production principle, MEPP), appeared in the 20th century. The recognition and the use of MEPP were considerably delayed. The first part of this review is concerned with the thermodynamic and statistical basis of the principle (including the relationship of MEPP with the second law of thermodynamics and Prigogine's principle). Various applications of the principle to analysis of nonequilibrium systems are discussed in the second part. The main conclusions state that MEPP is a logical generalization of the second law of thermodynamics to nonequilibrium processes, and it is confirmed in studies of various systems of the physical, chemical or biological origin on different observation scales (both microscopic and macroscopic). MEPP and the minimum entropy production principle do not contradict one the other. The latter is a consequence of the former. MEPP shows greatest promise in biology (the theory of evolution, ecology, etc.) and astrophysics (problems of the origin of the Universe, the evolution of galaxies, etc.). Some researchers point to a relation between MEPP and such frequent manifestations of a nonequilibrium development as self-organized critical phenomena, dendrite/fractal growth (Mandelbrot, 1982), sigmoidal growth curves and relaxation laws. The development of these ideas is promising from the viewpoint of the substantiation of the principle and better understanding of laws underlying the evolution of systems far from the equilibrium (Martyushev and Seleznev, 2006).

Nettleton (1987b, 1992) and Nettleton and Freidkin (1989) postulate nonlinear reciprocity relations on the basis of the maximum entropy formalism. Nettleton (1992) treats EIT problems from the viewpoint of the general reciprocity (and antireciprocity) relations in nonequilibrium systems and predicted by extended theories. These relations result from the introduction of diffusional fluxes of mass, energy, and momentum, as (extra) state variables of the nonequilibrium Gibbs equation of EIT. They are shown to be useful in calculating transport coefficients in liquids. Reciprocity reduces the number of coefficients which must be estimated from molecular approaches. By using the reciprocity, generalizations of the Cattaneo (1958) equation of heat and the Maxwell equation of viscoelasticity can be obtained. These equations govern coupled flows of heat and matter and viscoelasticity under several creep mechanisms. Nonlinear reciprocity is derived from the Fokker–Planck equation for a distribution g of values of certain microscopic operators whose averages are variables such as heat and mass fluxes. The equation for g is obtained by applying Zwanzig or Grabert projection operators to the Liouville equation. The maximum entropy formalism provides an ansatz for the solution of the Liouville equation which yields a finite number of moments. From this ansatz one can derive kinetic equations exhibiting nonlinear reciprocity by taking moments of the equation for g. Nettleton (1992) also shows that a Lagrangian formulation can lead to the reciprocity and antireciprocity properties.

In the famous and very efficient approach called the method of Lagrange multipliers, Liu and I-Shih (1972) forms an inequality by adding the standard entropy inequality and a linear combination of the field equations. The factors multiplying the field equations in this linear combination are called Lagrange multipliers. Restrictions on constitutive equations then follow by the observation that this inequality is linear in certain derivatives of the fields, which are now considered to be entirely arbitrary. Many examples of application are given in this important paper and in the corresponding literature (Liu and Müller, 1983).

Callen's postulates. Callen's (1988) postulates provide another bridge from disequilibrium to equilibrium. A basic notion in his postulates is the Gibbs equilibrium submanifold (GES). Classical thermostatics deals principally with changes through equilibrium states, located on the GES which is a submanifold in a space of n extensive variables and a potential function. The potential function is usually energy or entropy depending on the "representation" used (Callen, 1988; Gibbs, 1948;). A postulate which characterizes Gibbs's equilibrium submanifold in the space of n extensive variables and a potential (energy or entropy), provides the statement about the extremum properties of equilibrium states (Callen, 1988). However, the limitation of consideration to the $n + 1$ dimensional space (of n extensities and a potential) leads to narrowed interpretation of the problem. A suitable notion to avoid this restriction is the Gibbs equilibrium manifold (GEM) or an extremum surface which emerges in a space of higher dimension than the classical space. Callen's postulates formulated on this ground (Callen, 1988; Sieniutycz and Salamon, 1990b) provide the formal prescription for that emergence. For concreteness, we restrict here to the energy representation.

In Callen's postulates, the Gibbs equilibrium manifold GEM (as distinguished from the submanifold GES in the narrowed $n + 1$ dimensional space of classical variables), arises as the result of minimizing of the energy E in an extended space involving not only the n classical variables and a potential but also a certain number, say m, of free internal or nonequilibrium variables, **i**. The mathematical structure of the minimization problem in the extended space of $n + m + 1$ coordinates should be understood. The internal variables **i** are, in fact, unrestricted controls or free decisions. The minimization problem involves the variables **i** (dim **i** = m), parameters **X** (dim **X** = n), and the minimization criterion $E(\mathbf{X}, \mathbf{i})$. In that context, classical thermostatic coordinates **X** are the nonvaried variables (in the optimization theory such variables are usually called "parameters"). While suitable, it is not necessary that the optimal internal variables, **i**, must vanish at equilibrium. An example is the degree of advancement of a chemical reaction, ξ, whose reference value is usually set to zero at the beginning of the reaction, not at the thermodynamic equilibrium.

For a nonequilibrium system, the classical coordinates **X** frequently represent certain global quantities (eg, total entropy, volume and numbers of moles) which are kept constant when minimizing the energy E. Certain other variables such as the energy of a selected subsystem are of **i** type; their values which minimize the potential E are determined by the optimization. Examples of variables in the **i** space are flow velocities or irreversible fluxes; degrees of advancements of chemical reactions ξ, electric or magnetic susceptibilities, viscous pressure, and so forth. Sometimes, some of coordinates of vector **i** vanish at equilibrium. However, in general, only the relation $\mathbf{i}_{eq} = f(\mathbf{X})$ follows after the optimization is performed.

It may be shown that the information-theoretic model, with **X** as the expected values (macroscopic averages) and **i** as independent probabilities or functions (functionals) of these probabilities exhibits a formal similarity with the scheme described previously. This similarity is masked, however, by the presence of Lagrange multipliers, absent in the Callen's postulates. Yet, in the entropy representation, equilibrium distribution of probabilities is a function of the average energy and other expectations. In fact, Jaynes' and Callen's postulates describe the origin of equilibrium thermodynamic structures from nonequilibrium ones (physical or artificial), and, therefore, they bridge equilibrium thermodynamics and nonequilibrium thermodynamics.

Since $\mathbf{i}^{eq} = f(\mathbf{X})$, the equilibrium (extremum) potential E^{eq} is a function of constraints **X**. The previously analysis shows that the pair of the following relationships holds

$$\mathbf{i}^{eq} = f(\mathbf{X}) \tag{1.11}$$

and

$$\min_\mathbf{i} E(\mathbf{i}, \mathbf{X}) = E^{eq}(\mathbf{X}). \tag{1.12}$$

These simple relations stress that that the equilibrium values of the nonclassical (internal) variables $\mathbf{i}^{eq}$ are determined by the classical thermodynamic coordinates and the extremum (equilibrium) energy E^{eq} is a function of the classical variables exclusively.

Minimizing the total energy E with respect to **i**, keeping **X** frozen at $\mathbf{X} = \mathbf{X}_0$, gives a locus of equilibrium states in the extended space (**i**, **X**, E), rather than in the classical space (**X**, E). This leads to the Gibbs equilibrium manifold, GEM. It is the projection of the GEM on the classical space (**X**, E) which yields the Gibbs equilibrium submanifold (GES) of classical thermostatics. In conclusion, the equilibrium Gibbs manifold is generated in the space of dimension $n + m + 1$ (involving X, **i** and E).

In fact, the two figures from Callen's book (Callen, 1988; his Figs. 5.1 and 5.2) illustrate graphically such "extended" situation where an equilibrium state, say A, is a point of minimum E with respect to **i**, for each preassigned classical state $\mathbf{X} = \mathbf{X}_0$. Locus of points A for various $\mathbf{X} = \mathbf{X}_0$ is GEM. We can appreciate significance of Callen's postulates for nonequilibrium thermodynamics and, in particular, for extended thermodynamic theories. These theories use nonextremal energies (or entropies) which explicitly incorporate the nonequilibrium variables **i**.

Having one thermodynamic potential (TP), we can introduce others. These nonequilibrium potentials should be obtained from $E(\mathbf{X}, \mathbf{i})$ [or $S(\mathbf{X}', \mathbf{i}')$] via various Legendre transformations with respect to **X** and, possibly, **i**. Those with respect to coordinates **X** lead to various (**i** dependent) nonequilibrium generalizations of classical TPs, whereas those with respect to variables **i** provide a modification group for these generalized potentials. A suitable example is a viscoelastic fluid, where nonequilibrium variables **i** are represented by the components of the viscous pressure tensor, $\mathbf{P}^v$. For this tensor, Jou, Casas-Vazquez, and Lebon present in their book an example of a Legendre transform involving $\mathbf{i} = \mathbf{P}^v$; see problem 9.8 on page 221 in Jou et al. (2001).

Only equilibrium TP's are unaffected by (changes in) the choice of variables **i**. Yet, the Legendre transforms of $E(\mathbf{X}, \mathbf{i})$ depend on the choice of **i**. The use of the so-called "natural" or physical variables, selected amongst possible **X** and **i**, should be preferred, to abandon the redundant variables **i** and secure the completeness of the physical information (Callen, 1988). The knowledge of the relation between the Lagrangian and canonical formalism is helpful in this selection; it leads to field TP's (Sieniutycz and Berry, 1989, 1991). For example, it follows that in order to have a self-consistent description of heat flow in the least action context, the natural variable of the nonequilibrium energy should be neither the heat flux nor the entropy flux but a "thermal momentum," defined as the partial derivative of the heat Lagrangian, L, with respect to the velocity of the transferred entropy, $\mathbf{u}_s$ ($\mathbf{u}_s = \mathbf{v}_s = \mathbf{j}_s/\rho_s$ in the frame of resting continuum). The thermal momentum may also be applied in the stability conditions for a fluid with heat flow, predicted by a nonequilibrium metric (Section 1.5.2). As opposed to the classical TP's discussed previously Graham (1985) derived entropy-like potentials in nonequilibrium systems with coexisting attractors. Graham and Schenzle (1983) and Graham and Tel (1984) derived nonequilibrium potentials for dissipative dynamical systems and stationary probability distributions for some models without manifest detailed balance.

1.3.3 Steady-State Problems

In stationary systems all state variables are independent of time and, in particular, the system entropy S is constant. Using this constancy in the second law equality,

$$\frac{dS}{dt} = \frac{d_e S}{dt} + \sigma_s \tag{1.13}$$

we conclude that, in view of positiveness of σ_s, there must be a negative entropy flow from the system to its environment such that the total entropy change vanishes. Consequently, the entropy of a stable steady state is smaller than that of the initial state. The original statement of this property (Prigogine, 1947) didn't stress the system's stability as the necessary condition for the validity of the statement. Later works, however, imply the proper correction (Glansdorff and Prigogine, 1971; Nicolis and Prigogine, 1977).

Stationary systems close to equilibrium are frequently associated with various formulations involving the theorem of minimum entropy production. Again, the reference state stability and linearity of dynamics are necessary for the validity of the theorem. Both continuous and lumped formulations are common (de Groot and Mazur, 1984; Prigogine, 1947, Glansdorff and Prigogine, 1971; Bejan, 1982, 1996c; Mornev and Aliev, 1995; Baclic, and Seculic, D. 1978; D'Isep and Sertorio, 1982, 1986; Sieniutycz, 2002b, e; Sieniutycz and Shiner, 1993). In spite of doubts concerning the theorem validity (Ross, 2008), some nontrivial affirmative examples are provided by radiation problems where the conservation laws may be obtained by minimization of the entropy production (Essex, 1984, 1990). Examples involving engines, flames, combustion systems and radiation are also known (Salamon et al., 1980; Arpaci, 1987, 1990; Arpaci and Selamet, 1988). This leads to applications minimizing entropy production in the class of finite-time heating and cooling systems. They show that, often, the optimal spatial distribution is close to homogeneity, the property called "the equipartition of entropy production" (Tondeur, 1990; Tondeur and Kvaalen, 1987; Andresen and Gordon, 1991, 1992a, b, 1994; Kjelstrup et al., 1999; Chen, 2005; Chen et al., 1999b, 2009).

Arpaci (1987, 1990) has introduced the lost heat as opposed to lost work. The approach is reminiscent of de Donder's "uncompensated heat" (de Donder, 1920). Arpaci presents analyses of practical engineering problems leading to quantitative evaluations of the entropy production. His study is based on an assumption of local equilibrium. His balance equations include radiative processes. He shows that the irreversible part of all nonthermal forms of energy is dissipated into heat and then to a part of entropy production while the irreversible part of the heat energy is balanced by the other part of the entropy production. The theory is illustrated by two illustrative examples (Arpaci, 1990). The first involves the entropy production on the wall of a thermal boundary layer. This production is found to be proportional to the square of the Nusselt number. The second involves the entropy production σ in the luminous zone of a quenched flame. This entropy production is found to be inversely proportional to the Peclet number. The tangency condition, which determines the quench distance, is related to an extremum in entropy production. The relation between turbulent energy dissipation and σ is developed. In a complementary work, entropy production in flames is considered (Arpaci and Selamet, 1988).

There are serious doubts concerning the general validity of the theorem of minimum entropy production (Månsson and Lindgren, 1990; Ross, 2008). The theorem is generally invalid for nonstationary states and even for stationary states far from equilibrium. Månsson and Lindgren (1990) gave some examples involving limit cycles. To this class belong also various extremum formulations of nonequilibrium network thermodynamics (Meixner, 1963; Peusner, 1990; Shiner et al., 1996). As a rule, steady network systems are not associated with the minimum entropy generated within the system. In lumped systems having inertial terms in series with resistances, the entropy production can even attain a maximum (Peusner, 1986).

Electrical networks are an excellent example for the application of irreversible thermodynamics. In particular, they provide one of the simplest cases where, besides Onsager coefficients, one has Casimir coefficients in the phenomenological equations. Meixner (1963) gives two thermodynamic formulations of the network equations, which closely resemble the Lagrangian and the Hamiltonian formalism of classical mechanics. In the first formulation, only Onsager coefficients occur, but the thermodynamic forces are of a peculiar type in that they are Lagrangian derivatives. Meixner also shows that Casimir reciprocal relations can generally be replaced by Onsager reciprocal relations if the independent variables in the linear phenomenological relations are chosen in a proper way. As a generalization of the network equations, Maxwell's equations in continuous matter with dielectric, magnetic, and Joulean heat losses are considered. Matter is isothermal, but not necessarily uniform nor isotropic. Under the influence of imposed electric fields, current distributions arise. The connection of these fields is expressed by a generalized admittance function. A well-known reciprocity theorem for electromagnetic fields is seen to hold even if all types of losses are present. This is due to the Onsager–Casimir reciprocal relations for the dielectric tensor, the permeability tensor, and the resistivity tensor. From the reciprocity theorem, a symmetry relation can be derived for the generalized admittance function. A generalized version is given in the presence of a static magnetic field.

The progress in application of irreversible thermodynamic approaches to chemical reactions is relatively slow. Nummedal et al. (2003) describe the practical consequences of minimizing the entropy production rate of an exothermic reactor with the ammonia reaction. Moroz (2009) formulates a variational framework for nonlinear chemical thermodynamics employing the maximum energy dissipation principle. Månsson and Lingren (1990) outline the role of the transformed information for creation processes by showing how the destructive processes of entropy production are related to the creative processes of formation. The role of information for the theory of thermodynamic equilibrium and relative (Kullback, 1959) information for the definition of the exergy are reviewed. Superiority of the information theoretic approach over traditional thermodynamics is advocated for analyzing systems with emergent spatial structures, for example, spatially self-organizing chemical systems. Using examples of oscillatory reactions, it is shown that the average entropy production in an oscillating system is not necessarily lower than the entropy production of an (unstable) steady state. They conclude that there cannot exist a general principle of minimum entropy production for oscillatory systems.

Attempts to extend the minimum entropy production theorem beyond the linear regime resulted in nonclasical variational formulations involving the method of local potential (Glansdorff and Prigogine, 1971). That method is of Rosen type as it involves "restricted variations" (Rosen, 1953). Some of the variables are frozen and subsidiary conditions force consistency between the frozen and the unfrozen variables. The applications of the local potential method and similar approaches are described elsewhere (Schechter, 1967). They can lead to general equations of change and hence serve as computational tools for obtaining solutions by direct variational methods (Kantorovich and Krylow, 1958). A nontrivial local potential formulation involves radiation in optically active media (Wiśniewski, 1990).

It is unknown whether the methods using extremum thermodynamic potentials can be suitably modified to include nonequilibrium phenomena. Some corresponding problems are discussed in chapter: Variational Approaches to Nonequilibrium Thermodynamics, where the consequence of the passage from the entropy picture to the free energy picture is analyzed. In nonequilibrium systems the diversity of constraints increases. Working either in the spirit of Callen's principles (Callen, 1988), or in the spirit of information theory, one could expect that perhaps the entropy of the system rather than the entropy production should attain a restricted maximum. To have a chance, such entropies would, of course, need to differ from the usual equilibrium entropies.

To illustrate the difficulties that appear in the problems of entropy maxima, we briefly consider a simple problem of stationary heat transfer in a continuum. We introduce the nonequilibrium entropy density

$$S_v(E,V,\mathbf{q}) = S_v^{eq}(E,V) - (1/2)(L_q)^{-1}\tau\mathbf{q}^2 \tag{1.14}$$

where $L_q = \lambda T^2$ is Onsager coefficient, τ is the thermal relaxation time, V-volume, and λ is the heat conductivity. The unconstrained maximum of S_v with respect to $\mathbf{q}$ is, of course, S_v^{eq}, achieved for $\mathbf{q} = 0$. This is the true result, which, however, is not easy to extend to constrained noneqilibrim systems.

When a finite stationary heat flux flows across the system, we search for the constrained maximum of $S(E, V, \mathbf{q})$ with respect to $\mathbf{q}$, under the energy conservation constraint, div$\mathbf{q} = 0$. Using a Lagrange multiplier $-\beta$ to handle this constraint, that is, maximizing

$$S_v(E,V,\mathbf{q}) = S_v^{eq}(E,V) - (1/2)(L_q)^{-1}\tau\mathbf{q}^2 - \beta\text{div}\mathbf{q} \tag{1.15}$$

we obtain as the extremum condition the Fourier law in the form

$$\mathbf{q} = L_q \tau^{-1} \text{grad}\,\beta \tag{1.16}$$

with $\beta = \tau T^{-1}$ describing the preassigned field of $T(\mathbf{x})$. The severe necessary assumption in this derivation include constancy of the coefficients, the limitation of the conservation law to its stationary form, and—to the worse—the necessity to know $T(\mathbf{x})$. The restrictivity of these conditions is similar to those encountered when the entropy production is minimized (de Groot and Mazur, 1984). Furthermore, generalizations of these results to unsteady states no longer involve entropy and the physical interpretations of the integrands are no longer clear (Sieniutycz, 1983). The results are only slightly better than those which one can obtain from thermodynamics based on equilibrium entropies. However, in chapter: Variational Approaches to Nonequilibrium Thermodynamics we shall see that many improved reformulations of the present problem can be obtained by inclusion of the information about the phenomenological equations and after using the nontruncated (nonstationary) conservation law. Moreover, by applying the kinetic theory Karkheck (1986, 1989) has shown that one must go beyond hydrodynamic quantities in order to obtain expressions for the transport coefficients. These kinetic results indicate that when the constraints involve the one-particle distribution function as well as the potential energy density then the transport coefficients appear naturally as the outcome of the formalism. Consequently, both conservation laws and transport coefficient expressions can be obtained via the generalized maximum entropy approach. See Karkheck (1986, 1989) for more details.

1.3.4 Nonstationary Nonequilibrium Problems

Physical and biological systems are often quasi-stationary or nonstationary. Respiratory calorimeter measurements of energy and mass fluxes lead to calculation of entropy production of a living human. This quantity shows the nonmonotonic change in human's life-time, with a maximum for age of young child (Aoki, 1995, 2001). Biological organisms and ecosystems show many attractive collective properties, of interest for thermodynamics and ecological modelling (Jorgensen, 2001; Shiner and Sieniutycz, 1994, 1996, 1997). Entropy and entropy production are essential in investigations of system's self-organization and evolution (Wicken, 1979, 1980, Ebeling and Klimontovich, 1984; Ebeling and Feistel, 1992; Ebeling et al., 1986).

Variational formulations of dynamical problems may involve perfect and real fluids in both Lagrangian and Eulerian (field) representations of fluid motion (chapter: Variational Approaches to Nonequilibrium Thermodynamics). The description of the thermohydrodynamic systems usually corresponds to the field representation (Rund, 1970). The classical Hamiltonian formulation of thermohydrodynamics (least action principle) allows the most rigorous description. However, general Lagrangians are known for perfects fluids only (Herivel, 1955; Stephens, 1967; Schmid, 1967a, b, 1970a, b; Berdicevskii, 1983; Salmon, 1987; Nonnenmacher, 1986). Heat flow effects can be included in the reversible case of very high conductivity (Sieniutycz, 1988a, b). However, nobody has thus far succeeded in deriving the complete structure of irreversible thermo and hydrodynamics from a general Lagrangian. Reversible thermohydrodynamics and relativistic magneto-hydrodynamics are represented, between others, by the research of Jezierski and Kijowski (1990) and Kupershmidt (1990a, b, 1992). The significance of variational formulations for the extended nonequilibrium thermodynamics is being recognized in the context of conservation laws for fluids with heat conduction (Sieniutycz and Berry, 1989) and for stability problems (Anthony, 1988). See chapters: Variational Approaches to Nonequilibrium Thermodynamics and Multiphase Flow Systems, for more information.

The paper of Sieniutycz (1990) attempts to lay a bridge between reversible and irreversible variational formulations. It proposes that taking irreversibility into account can be achieved by an explicit presence of action type (or phase type) variables (Clebsch potentials) in an action integral. A genuine nonequilibrium energy density E is obtained from the kinetic potential L through the generalized Legendre transform involving not only the hydrodynamic velocity **u** but also the velocities of diffusion. Exact matter tensor and nontruncated conservation laws are also found from this variational approach. The entropy source(s) and chemical production terms appear as extra thermodynamic variables of the extended Gibbs equation (Sieniutycz and Berry, 1993). A brief application of phase-dependent thermodynamics is given in Section 1.4, where a Fermat's type formulation for heat pulses in a rigid solid is developed. A different theory of Fermat's principle in lossy media has also been proposed (Censor, 1990). His paper treats an extension to absorptive media based on the concept of a complex wave vector. Extremum principles in viscoelasticity are reviewed by Morro (1992). A Lagrangian theory of chemical networks is developed by Shiner (1992). In view of the great difficulties involved in formulating irreversible transport processes in terms of exact Hamiltonian principles, principles constructed on different bases have been advanced. Occasionally, they incorporate Rayleigh's (1945) dissipation function into a functional or a variational expression (Biot, 1970; Vujanovic, 1971, 1992; Vujanovic and Djukic, 1972; Vujanovic and Jones, 1988, 1989; Sieniutycz, 1983, 1987), see also chapter: Variational Approaches to Nonequilibrium Thermodynamics.

Stochastic diffusion theory offers nonequlibrium Lagrangians which take the irreversible fluxes of energy and matter into account. To this class belongs the well-known Onsager–Machlup (1953) approach to the fluctuation theory and its nonlinear generalizations. These "kinetic analogues of Boltzmann principle" (Lavenda, 1985a, b) involve Lagrangians based on the entropy production and on dissipation potentials. The results can be presented in terms of path integrals (Lavenda, 1985a, b). The so-called governing principles of irreversible thermodynamics (Gyarmati, 1969) are their macroscopic counterparts. These approaches are discussed by Verhas (1990). Dissipative variational principles of this sort complement reversible theories leading to the hydrodynamics of perfect fluids (Berdicevskii, 1983).

1.3.5 Example of Dissipative Dynamics
1.3.5.1 Introduction

There are the two broad categories of approaches where methods of the contemporary mathematical theory of optimization can be used for analyzing thermodynamic systems. In the first category, one wants to *optimize* these systems by changing some external parameters (decisions or controls). In this case both extremal and nonextremal solutions pertain to real physical situations, and the resulting equations describe the best dynamical changes of the internal state and the best controls imposed through the system boundaries. This first category includes studies which seek in-principle limits to the operation of thermodynamic processes and studies which seek to improve existing engineering systems. In the second approach, one wants to predict the system behavior (reversible or irreversible) under prescribed external conditions and therefore seeks to derive its governing equations from certain variational or extremum principles. It may be said that in this case the extremal control is set by nature rather than by man.

In this section we discuss the second approach in the context of motion of a particle or a quasiparticle in a dissipative medium, for example in a viscous fluid, porous system, heat conducting solid, and so forth. Consider, for example, an aerosol particle treated as a macroscopic body moving in the atmosphere, where the presence of the dissipative phenomena results in the appearance of certain frictional forces (Davies, 1966). Can we describe the system in terms of a variational principle taking into account dissipation? Despite of many negative opinions regarding this question expressed earlier (eg, Milikan, 1929; Ray, 1979; Yourgrau and Mandelstam, 1968), various kinetic potentials generalizing the classical, reversible Lagrangian in a potential field $V(\mathbf{x})$

$$L = m\mathbf{u}^2/2 - V(x), \tag{1.17}$$

have been advanced (Bateman, 1929; Denman and Buch, 1962; Kiehn, 1975; Sieniutycz, 1976, 2008c; Vujanovic and Jones, 1988; Vujanovic, 1992). These older approaches made use the assumption that the kinetic potential of a irreversible motion is a time-dependent quantity, so that the reversible L of Eq. (1.17) should be multiplied by the term $\exp(t/\tau)$ to get irreversible equations of motion from the least action principle involving $L_\tau = L \exp(t/\tau)$ as the integrand of an action integral. This concept, however, lead to the troublesome conclusion about the violation of the resulting energy conservation in dissipative systems, the conclusion that is not accepted by the contemporary thermodynamics.

Let us characterize the energy terms appearing in Eq (1.17) restricting ourselves to simple mechanical problems. For a motion of a particle suspended in the fluid, mass m should be understood as an effective mass defined as the sum of the actual mass of the particle, m_p, and the virtual mass of the accreted fluid, m_f. The hydrodynamic considerations yield $m_f = (1/2)\rho_f v_p$ that is, the virtual mass is one half of the fluid mass contained in the volume v_p occupied by the particle. The particle volume v_p is also the parameter of potential energy V because the potential energy of the particle-fluid system equals $(\rho_p - \rho_f)v_p\mathbf{gx}$, that is, it contains the Archimedes term related to the buoyancy force. Thus the explicit form of the particle Lagrangian is

$$L = \frac{(\rho_p + \rho_f/2)v_p \mathbf{u}^2}{2} - (\rho_p - \rho_f)V_p\mathbf{gx} \tag{1.18}$$

where the product of the (constant) gravitational field strength and radius vector $\mathbf{gx}$ represents the potential energy per unit effective mass. However Eqs. (1.17) and (1.18) refer to nondissipative motions only, that is, they do not predict frictional (Stokes) forces in the system.

To date, kinetic potentials generalizing (1.1) or (1.2) to irreversible motions have had very limited success even for simple systems. One of such methods, originated by Bateman (1929), uses a modified, time dependent Lagrangian $L_\tau = L\exp(t/\tau)$, where τ is a time constant describing an interaction of the body with the fluid and L is the classical (Hamilton's) kinetic potential, Eq. (1.17). The reader can verify that for $L = m\mathbf{u}^2/2 - V(\mathbf{x})$, the nonautonomous Lagrangian

$$L_\tau = L\exp(t/\tau) \tag{1.19}$$

leads to the proper "frictional" force $\tau^{-1}\partial L/\partial \mathbf{u} = \tau^{-1}m\mathbf{u}$ in the equation of motion corresponding to the laminar friction.

However the physical interpretation of the related Hamiltonian in energy terms is awkward since the nonautonomous Lagrangian (1.19) leads to the following Hamiltonian H_τ

$$H_\tau = \frac{\partial L_\tau}{\partial \mathbf{u}}\mathbf{u} - L_\tau = \left(\frac{\partial L}{\partial \mathbf{u}}\mathbf{u} - L\right)\exp(t/\tau) = E\exp(t/\tau) \tag{1.20}$$

which is not a constant of the irreversible motion. On the other hand, the standard energy E understood as the Legendre transform of a reversible autonomous L is the constant of motion. The violation of the constancy property for the irreversible Hamiltonian H_τ poses a serious physical problem since the autonomous nature of the system involved dictates that its Hamiltonian should be a constant of motion.

Therefore, we abandon here the older approaches quoted previously. Recalling the explicit appearance of partial actions of entropy and components in equations of variational thermohydrodynamics (Sieniutycz, 1994), we shall make an assumption that the kinetic potential of an irreversible motion L is action dependent or phase dependent, so that the system's zeroth coordinate x_0 that represents these quantities appears explicitly in a mathematical model. Specifically we assume that the irreversible L contains explicitly either $x_0 \equiv -h\psi$, where ψ is the thermal phase, or $x_0 \equiv -\eta k_B$, where η is the unit entropy action. The consideration of variable x_0 with the identification $x_0 \equiv -\eta k_B$ (η is unit entropy action, k_B Boltzmann constant and h Planck constant) is substantiated by the importance of thermal motions (motions of heat quasiparticles) in heat conduction problems. Since in the present form of L does not contain explicitly time t, the process irreversibility is no longer associated with the time dependent properties of the kinetic potential L, so that the resulting energy can be conserved.

Technically, the present approach is accomplished by adding extra additive term $\tau^{-1}x_0$ to the classical Lagrangian L. That structure of L allows one to secure the constancy of autonomous Hamiltonians, H, and has some other virtues that are discussed in the further text. Applying the approach one shall be able to secure the energy conservation and achieve a satisfactory set of equations of motion. The present model is helpful to describe the Brownian diffusion in concentrated aerosols (Mori, 1965), and to formulate a generalized variational setting for the case of concentrated suspensions governed by behavior of the chemical potential (Lavenda, 1978, 1985a, b, 1990, Sieniutycz, 1984a, b). In the context of heat conduction processes the application of x_0 dependent (phase dependent or action dependent) kinetic potential L leads to irreversible variational equations of heat flow in thermal fields, thus generalizing earlier variational results in which these equations are restricted to thermal systems of very high thermal conductivity (Kupershmidt, 1990a, b, 1992; Sieniutycz, 1988b, 1994; Sieniutycz and Berry, 1989; Gavrilyuk and Gouin, 1999).

It should also be stressed that the present variational formulation, and also other ones considered in this section, is appropriate for a particle or systems of particles which move in a resting dissipative medium (environment, solid, porous bed, etc.). For heat conduction in solids the identification $\mathbf{u} = \mathbf{v}_s$, where $\mathbf{v}_s$ is the velocity of entropy diffusion (related to conductive heat flux, not to total energy flux), is therefore necessary. All approaches here will require elementary modification based on use of Gallilean transformation to apply them in moving media.

1.3.5.2 Kinetic potentials, optimization criteria and maximum principle

Euler equations of a variational problem can be obtained by investigating an extremum (maximum) of the objective, x_0 at a final time t^f. By definition, the variable x_0 is the negative action variable, such that $x_0(t) = -h\psi(t)$ at each time t. For the process of heat conduction, maximum of the objective coordinate $x_0(t^f) = -h\psi(t^f) = -k_B\eta(t^f)$ at a final time t^f is required. $\psi(t)$ is the thermal phase variable and $\eta(t)$ is a thermal action corresponding with the entropy flow. The variational problem corresponds with a minimum of the entropy action variable, $\eta(t^f)$. Writing the objective in the form of the integral

$$x_0(t^f) = x_0(t^i) + \int_{t^i}^{t^f} f_0(x_0, \mathbf{x}, \mathbf{u}, t)\, dt \equiv -h\psi(t^f) = -k_B\eta(t^f) \tag{1.21}$$

(Eq. (1.34) shows the differential counterpart of this formula), we search for the necessary optimality conditions for maximum of $x_0(t^f)$ or, equivalently, for minimum of $\psi(t^f)$ or $\eta(t^f)$, under the simple differential constraints

$$\frac{d\mathbf{x}}{dt} = \mathbf{u} \tag{1.22}$$

The n-component vector $\mathbf{x} = (x_1, x_2, ..., x_n)$, is the subset of the action including $n + 1$ dimensional vector $\mathbf{X} = (x_0, x_1, x_2, ..., x_n)$. In this problem dim $\mathbf{u}$ = dim $\mathbf{x} = n$.

For reversible problems of analytical mechanics, the negative of f_0, designated as $L(\mathbf{x}, \mathbf{u}, t)$, is the Hamilton's kinetic potential L equal (nonrelativistically) to the difference between the kinetic and potential energy. Kinetic potential L of the reversible case is exemplified by Eqs. (1.17) and (1.18).

In the irreversible case we shall use the following action-dependent kinetic potential L,

$$L(\mathbf{u},\mathbf{x},x_0,t) = \frac{m\mathbf{u}^2}{2} - V(\mathbf{x}) + \tau^{-1}x_0 = \frac{m\mathbf{v}_s^2}{2} - V(\mathbf{x}) - h\tau^{-1}\psi \quad (1.23)$$

(t independent kinetic potential L is applied). The second expression refers to the heat conduction in a resting cmedium, where $L(\mathbf{u},\mathbf{x},x_0,t)$ equals $L(\mathbf{v}_s,\mathbf{x},\psi,t)$. (See the remark at the end of the previous section.) This case is considered in detail in Section 1.4. In the contrast, classical action $x_0 = -h\psi$, and its kinetic potential L depend only on the coordinates of the enlarged state vector $\tilde{\mathbf{x}} = (x_1,...., x_n, t)$ which does not contain the variable x_0. This results in the constancy of the adjoint $z_0 = 1$ along the extremal path in the classical (reversible) models, autonomous or time-dependent.

The maximum principle (Pontryagin et al., 1962; Fan, 1966; Halkin, 1966; Cadenillas and Karatzas, 1995) reduces the global problem of the maximizing a functional to the problem of maximizing the Hamiltonian function H at each instant of time. Whenever finding of an analytical solution is impossible or difficult, this method also allows us to organize an approximating numerical scheme using a large but finite number of time instants which give a discrete approximation for the optimal path and the corresponding optimal control.

The Hamiltonian of the generalized process (characterized by an action dependent L) has the structure

$$H = z_0 f_0 + z_1 u_1 + \cdots z_n u_n = z_1 u_1 + \cdots z_n u_n - z_0 L \quad (1.24)$$

Briefly, $H = \mathbf{zu} - z_0 L$, in contrast to the classical $H = \mathbf{zu} - L$ (Landau and Lifshitz, 1976). Extremum conditions of the objective (1.21) are contained in the canonical set, which i

$$\frac{dx_v}{dt} = \frac{\partial H}{\partial z_v}, \quad v = 0,1,2...n \quad (1.25)$$

$$\frac{dz_v}{dt} = -\frac{\partial H}{\partial x_v}, \quad v = 0,1,2...n \quad (1.26)$$

$$\frac{dH}{dt} = \frac{\partial H}{\partial t} \quad (1.27)$$

and

$$z_t + \max_{\mathbf{u}} H(x_0,\mathbf{x},z_0,\mathbf{z},t,\mathbf{u}) = 0 \quad (1.28)$$

(n is the dimensionality of the state space without the action coordinate). Equation (1.25) reproduces the state equations (1.22) and (1.23), Eq. (1.26) describes equations of motion, and Eq. (1.27) describes the nonconservative property of the extremum Hamiltonian in nonautonomous fields. Equation (1.28) is of Hamilton–Jacobi-Bellman type, and states that an optimal control must maximize H at each instant of time in the interval $t^i \leq t \leq t^f$ (Pontryagin's Maximum Principle). In what follows we shall keep in mind the following property of optimal criteria: $\max x_0(t^f) = -\min h\psi(t^f)$, whenever $x_0(t) = -h\psi(t)$.

1.3.5.3 Reversible motions

In accordance with the Maximum Principle (Pontryagin et al., 1962), the maximum of the negative action $x_0(t^f) = -h\psi(t^f)$ or the minimum of final phase variable

$$\psi(t^f) = -h^{-1}x_0(t^f) \tag{1.29}$$

requires the stationarity condition $\partial H/\partial u_k = 0$ for the Hamiltonian with respect to free controls u_k. For L independent of x_0 this extremum requirement leads to the well-known classical relations

$$z_k = -\frac{\partial f_0}{\partial u_k} = \frac{\partial L}{\partial u_k}, \qquad k = 1,\ldots,n, \tag{1.30}$$

which define generalized momenta z_k or state adjoints. Yet, the result is limited to reversible motions.

Equation (1.30) is usually applied for the reversible microscopic motion of elementary particles or the frictionless motions of macroscopic bodies in inviscid media or in a vacuum.

Only for "reversible" momenta (1.30) the canonical equations (1.27) and (1.30) yield

$$\frac{d}{dt}\left(\frac{\partial L}{\partial \dot{x}}\right) = \frac{\partial L}{\partial x} \qquad v = 1,2\ldots n \tag{1.31}$$

which are the well-known Euler equations of classical variational calculus, written in a vector form.

Equation (1.29) and the condition $z_0 = 1$ are limited to a reversible process; they prove that, for a nondissipative motion, extremum Hamiltonian H is the standard energy satisfying the usual definition $E = zu - L$ (Landau and Lifshitz, 1976). For autonomous systems, where $L = -f_0$ does not contain the time t explicitly, H is constant along an optimal arc. We associate that motion with "conservative" or "reversible" behavior. Yet, neither standard energy E nor time-dependent Hamiltonian H_t, Eq. (1.20), are constants in the irreversible model described by Eqs. (1.25) – (1.28).

1.3.5.4 Irreversible motions

On the other hand, when the responsibility to handle the process irreversibility is taken over by the objective coordinate of Eq. (1.21), definitions of generalized momenta z_k will be affected by the nonconstant variable z_0. Indeed, applying in this (irreversible) case the stationarity condition for the Hamiltonian (1.24) with respect to controls u_k, we obtain the relations $z_k = -z_0 \partial f_0/\partial u_k = z_0 \partial L/\partial u_k$, or in the vector form

$$\mathbf{z} = -\left(\frac{z_0 \partial f_0}{\partial \mathbf{u}}\right) = \left(\frac{z_0 \partial L}{\partial \mathbf{u}}\right) \qquad (1.32)$$

and thus from (1.24) and (1.26)

$$\frac{d}{dt}(z_0 \partial L/\partial \dot{\mathbf{x}}) = z_0 \frac{\partial L}{\partial \mathbf{x}} \qquad (1.33)$$

For irreversible processes Eqs (1.32) and (1.33) replace the well-known "reversible" Eqs. (1.30) and (1.31). Note that the classical method of variational calculus cannot handle action integrals in which action variable appears explicitly in L, so the use of the Pontryagin's method was necessary. Allowing dissipative motions, Eq. (1.32) can be regarded as a generalization of the traditional Euler-Lagrange equation. The results obtained show that the generalized momentum of an irreversible motion is $z_0 \partial L/\partial \mathbf{u}$ rather than the 'classical' $\partial L/\partial \mathbf{u}$ of a reversible motion (Landau and Lifshitz, 1974b). The consequence of this property are forces of frictional type, discussed subsequently.

Equations (1.32) and (1.33) should be used together with the differential equation for the optimization criterion

$$\frac{dx_0}{dt} = f_0(x_0, \mathbf{x}, \mathbf{u}, t) \qquad (1.34)$$

Let us analyse the generalized adjoints which allow irreversible motions. The canonical Eq. (1.25) recovers state equations (1.22) and differential counterpart of Eq. (1.23), whereas the canonical Eq. (1.26) yields the dynamical equations

$$\frac{d\mathbf{z}}{dt} = -\frac{\partial H}{\partial \mathbf{x}} = z_0 \frac{\partial L}{\partial \mathbf{x}}, \qquad (1.35)$$

$$\frac{dz}{dt} = -\frac{\partial H}{\partial x_0} = z_0 \tau^{-1}, \qquad (1.36)$$

The condition of the stationary extremum of the Hamiltonian with respect to the control $\mathbf{u}$, Eq. (1.28) has in the irreversible case form of Eq. (1.32) which shows that the generalized momenta $\mathbf{z}$ of the irreversible process are $z_0 \partial L/\partial \mathbf{u}$ rather than $\partial L/\partial \mathbf{u}$, where z_0 depends on time t. Using the transversality condition $z_0(t^f) = 1$ we can evaluate the time behavior of z_0. From Eq. (1.36), z_0 is

$$z_0(t) = \exp\left(\frac{t-t_f}{\tau}\right) \qquad (1.37)$$

Since z_0 is everywhere a multiplier, the term $-\exp(t^f/\tau)$ is inessential and a simpler expression $z_0(t) = \exp(t/\tau)$ may be used equally well. This leads to the generalized momenta in the form

$$\mathbf{z} = z_0 \,\partial L/\partial \mathbf{u} = \exp(t/\tau)\partial L/\partial \mathbf{u}. \qquad (1.38)$$

This result explains the limited success of the modified kinetic potential $L\exp(t/\tau)$, mentioned earlier. Incidentally, both modifications of L lead to the same momenta $\mathbf{z}$. Substituting Eqs. (1.37) and (1.38) into Eq. (1.33) yields a simple equation of motion with a frictional term

$$\frac{d}{dt}\left(\frac{\partial L}{\partial \dot{\mathbf{x}}}\right) + \tau^{-1}\frac{\partial L}{\partial \dot{\mathbf{x}}} - \frac{\partial L}{\partial \mathbf{x}} = 0 \qquad (1.39)$$

In terms of the classical momenta $\partial L/\partial \dot{\mathbf{x}} = md\mathbf{x}/dt = m\mathbf{u}$ and the potential energy V

$$\frac{d}{dt}(m\dot{\mathbf{x}}) + \tau^{-1}m\dot{\mathbf{x}} = -\frac{\partial V}{\partial \mathbf{x}} \qquad (1.40)$$

When a spherical solid particle of diameter δ and (effective) mass m moves in the fluid of viscosity ν, the coefficient $\tau^{-1} = 3\pi\nu\delta/m$, which corresponds with the Stokes frictional force exerted by mass m.

1.3.5.5 Conserved irreversible Hamiltonian

Consider the Hamiltonian H along an extremum trajectory. It is defined on the space $(\mathbf{X}, \mathbf{Z})$ or $(\mathbf{x}, x_0, \mathbf{z}, z_0)$. The extremal Hamiltonian $H^e(\mathbf{x}, x_0, \mathbf{z}, z_0)$ is obtained after substituting $\mathbf{u}$ expressed in terms of the $(\mathbf{x}, x_0, \mathbf{z}, z_0)$ into $H = \mathbf{z}\mathbf{u} + z_0 f_0$. Since for the model $\mathbf{z} = -z_0 \partial f_0/\partial \mathbf{u} = z_0 \partial L/\partial \mathbf{u} = z_0 m\mathbf{u}$ and, hence, $\mathbf{u} = \mathbf{z}/(mz_0)$, we obtain for L of Eq. (1.23)

$$H^e(\mathbf{x}, x_0, \mathbf{z}, z_0) = \frac{\mathbf{z}^2}{2mz_0} + z_0 V(\mathbf{x}) - \frac{z_0 x_0}{\tau} = \exp\left(\frac{t}{\tau}\right)\left[\frac{m\mathbf{u}^2}{2} + V(\mathbf{x}) - \frac{x_0}{\tau}\right] \qquad (1.41)$$

This optimal Hamiltonian is the constant of an "irreversible" (damped) motion governed by the canonical equations written for H^e in the space of the natural variables of H, that is, $\mathbf{x}$, x_0, $\mathbf{z}$, and z_0. These canonical equations are equivalent with the dynamical equations (1.35), (1.26) and (1.39) describing for $H = H^e$ the time evolution of the generalized momenta and the velocities expressed in terms of the canonical variables

$$\frac{d\mathbf{x}}{dt} = -\frac{\mathbf{z}}{mz_0} \qquad (1.42)$$

and

$$\frac{dx_0}{dt} = -\frac{\mathbf{z}^2}{2mz_0^2} + V(\mathbf{x}) - \frac{x_0}{\tau} \qquad (1.43)$$

Thus we have found a variational formulation for an irreversible system. The price for taking the irreversibility into account is not low. First, the usual space of the mechanical variables (**x**, **z**) had to be replaced by the enlarged space (**x**, x_0, **z**, z_0). Second, the extremum Hamiltonian (1.41) is not the classical energy.

However, H^e is a constant of motion of irreversible canonical equations (1.35), (1.36), (1.42) and (1.43). Also, any generalized momentum z_i is the constant of motion when the external field V(**x**) vanishes or is independent of the coordinate x_i. This proves that the (autonomous) irreversible process is in fact conservative if the macroscopic energy and momenta are properly defined. The process only behalves like nonconservative when we are using the standard definitions of the energy and formulated for reversible processes. Furthermore, in the limiting case of infinite τ, the Hamiltonian becomes the classical energy of the body. This limit corresponds to the vanishing frequency τ^{-1} of collisions between the body and the molecules of the reservoir or a solid skeleton. In this limit, the adjoint variables **z** become the usual, classical momenta. Hence the correspondence of the formalism with the frictionless case is assured. In conclusion, variational treatments may be formulated for at least some irreversible systems. These treatments require extended definitions of the energy and momenta and give explicit role to the individual actions of entities involved. Some further details can be obtained for complex systems which require a vector of actions (Rund, 1970, Sieniutycz, 1990).

An extension of this model, based on an earlier work (Sieniutycz, 1984a, b), may deal with concentrated suspensions where the potential energy term should be properly generalized so as to include the specific chemical potential as a statistical measure of the internal energy of the particle which moves along a Lagrangian trajectory. This may enable one to describe the Brownian diffusion in the concentrated aerosols, to formulate a variational setting for the case of concentrated suspensions governed by behavior of the chemical potential.

1.3.5.6 Some additional remarks

The variational model proposed in this work has some essential virtues in comparison with time dependent models originated from nonconservative kinetic potentials (Bateman, 1929), which involve explicitly time and forces the researcher to attribute the irreversibility to the nonconservative properties of the Hamiltonian H_τ. In the action model proposed previously the momentum variable z_0, associated with action $h\psi(t^f)$ or action coordinate $x_0(t) = -k_B \eta(t)$, is essential. While the momentum-type variable z_0 is constant in both reversible models and models based on nonconservative (eg, Bateman's) Lagrangians, in the action-involving models z_0 growths monotonically in time, thus exhibiting the property of an evolution coordinate. As contrasted with the nonconservative model, the time adjoint z_t of the present model remains constant in time, hence the generalized (action involving) energy of the system H^e

$$H^e = \exp\left(\frac{t}{\tau}\right)\left[\frac{m\mathbf{u}^2}{2} + V(x) - \frac{x_0}{\tau}\right] \quad (1.44)$$

is also the constant of the irreversible motion of the system. In the case when the time between collisions tends to infinity or the parameter describing the frequency of collisions τ^{-1} tends to zero the Hamiltonian obtained here approaches the classical reversible Hamiltonian, corresponding with the classical energy

$$H^e = \frac{m\mathbf{u}^2}{2} + V(x) \tag{1.45}$$

The Hamiltonian function (1.44) obtained via optimization can thus be seen as extension of the standard energy for imperfect systems with friction. Setting to zero of the total time derivative of the conserved Hamiltonian (1.44) leads to the conclusion that the classical reversible energy (1.45) decreases in time in the irreversible process in accordance with the formula

$$\frac{d}{dt}\left(\frac{m\mathbf{u}^2}{2} + V(x)\right) = -\tau^{-1}\left(\frac{m\mathbf{u}^2}{2} + V(x)\right) \tag{1.46}$$

The right-hand side of this equation with an opposite sign defines also the rate of growth of the classical internal energy of the system caused by the friction phenomena. Remember that m is the *effective* mass.

When heat quasiparticles transfer through a nonequilibrium solid, the potential $V(x)$ follows from Eq. (76) on p. 178 in Sieniutycz (1994). In the absence of an external gravitational field and for reversible dynamics, potential V is the thermal energy $k_B T(\mathbf{x},t)$, see (next) Section 1.4. This led us to develop a theory describing the classical dynamics of thermal quasiparticles in thermal fields.

Equations (1.32) – (1.41) can be applied to describe motion of heat quasiparticles in the thermal field of a resting dissipative medium. Motion of heat in a continuum, that is, in a resting conductive solid, is an irreversible phenomenon manifested by frictional-like forces in process equations. Looking forward for a variational derivation of these equations we realize that the physical interpretation of the Hamiltonian (1.20) in energy terms is awkward and not admissible from the physical viewpoint since the considered system is autonomous whereas the Hamiltonian H_τ is not constant. In Section 1.4 we confirm the effectiveness of the approach which introduces the extra additive term $\tau^{-1}x_0$ into f_0.

Let us make some necessary preparations. Assuming that the objective variable x_0 is explicit in f_0 we shall apply the kinetic potential $L = -f_0$ in the form (1.23). In agreement with Eq. (1.21) one has in terms of the entropy coordinate $\mathbf{x} = \mathbf{x}_s$ and x_0

$$\frac{dx_0}{dt} = f_0(x_0,\mathbf{x},\mathbf{u},t) = -h\frac{d\psi}{dt} = -k_B\frac{d\eta}{dt} = -L(x_0,\mathbf{x}_s,\mathbf{v}_s,t) \tag{1.47}$$

For the heat flow in a resting continuum, $\mathbf{x} = \mathbf{x}_s$ and $\mathbf{u} = \mathbf{v}_s$, where $\mathbf{v}_s$ is the velocity of the entropy diffusion. We shall verify that the potential $V(\mathbf{x},t)$ is the thermal energy $k_B T(\mathbf{x},t)$. Consequently the kinetic potential of the thermal motion is

$$L = \frac{m\mathbf{v}_s^2}{2} - V(\mathbf{x},t) + \tau^{-1}x_0 = \frac{m\mathbf{v}_s^2}{2} - k_B T(\mathbf{x},t) + \tau^{-1}h\psi \quad (1.48)$$

and the Hamiltonian

$$H^e = \exp\left(\frac{t}{\tau}\right)\left(\frac{m\mathbf{v}_s^2}{2} + k_B T(\mathbf{x},t) + \tau^{-1}h\psi\right) \quad (1.49)$$

This property is achieved by Hamiltonian $H = \mathbf{z}\mathbf{u} + z_0 f_0$, Eq. (1.24), which contains the variable adjoint z_0. We shall exploit Eq. (1.49) and the second expression of Eq. (1.48) in the next section.

1.4 Classical Motion of Heat Quasiparticles in Thermal Field

The search for analogies between the motion of the light rays, heat quasiparticles and acoustic phonons (quasiparticles of quantized acoustic fields) constitutes another point of contact between nonequilibrium thermodynamics and variational methods (Luikov et al., 1976; Predvolitelev, 1981; Bubnov, 1976a,b; Sieniutycz, 1979, 1990; Tan and Holland, 1990; Markus and Gambar, 2004, 2005; Vázquez et al., 2009). The approach originated partly from the theory of characteristics of partial differential equations (Wiggert, 1977) which correspond physically to the motion of pulses or disturbances (acoustical, optical, thermal, etc.).

The classical theory of disturbance and pulse transfer in terms of Hamilton's equations involves formulations of the Fermat's principle type (Gelfand and Fomin, 1970; Tan and Holland, 1990). The difficulties of finding Fermat type formulations are particularly acute in the case of heat flow and other irreversible phenomena (Tan and Holland, 1990; Censor, 1990; Bejan, 2003; Sieniutycz, 2000b; Farkas et al., 2005).

Censor (1990) reviews the present status of the theory of pulse propagation. His treatment includes reversible propagation in lossless media and dissipative propagation in absorptive media. The applications of pulse propagation appear in various branches of physics involving electromagnetism, acoustics, elastodynamics and heat transmission (eg, radar, sonar, geophysical exploration, and thermal waves, respectively). The transmission of mass and information also involves pulses. In lossless systems a consistent and physically important mathematical theory has been developed. It is based on the canonical Hamiltonian equations describing the rays (pulse trajectories) in physical space-time rather than in space only. Standardly, the proper time is the quantity that attains an extremum on natural paths in physical space-time. This constitutes a generalization of the classical Fermat problem which was originally formulated, as a minimum time problem for stationary rays in physical space. The canonical equations result from the dispersion (or eiconal) equation describing the pulse phase in terms of a (quasilinear, homogeneous) partial differential equation of Hamilton–Jacobi type. In absorptive media, the pulse is deteriorated; it loses its identity and its information. The extension of the theory to absorptive media depends on the definition of the group velocity in the canonical equations. However, this case is quite important for the theory of dissipation. The author proposes a formalism based on the complex wave vector and the complex phase. Subsequently, we propose another extension based on the Hamiltonian which contains explicitly the phase and its canonical adjoint (without complex quantities) and leads to a non-Fourier phenomenological equation of heat.

In the absence of matter flow, heat transfer is associated with the flow of entropy. Eq. (1.50) describes an ideal "lossless" case of a nonrelativistic theory. Ignoring external fields, the partial differential equation of the reversible thermal eikonal in the Gallilean space-time $x' = (\mathbf{x}, t)$ has a Hamilton–Jacobi form

$$F(\partial_x \eta, \partial_t \eta) \equiv \frac{\partial \eta}{\partial t} + \frac{(\partial_x \eta)^2}{2\theta} + T(\mathbf{x},t) = 0 \tag{1.50}$$

In this equation θ is an inertial coefficient which has units of mass per unit of entropy and the order of magnitude m/k_B. η is the Lagrange multiplier of the entropy balance in the thermohydrodynamic least action principle (Sieniutycz, 1994). Since η has units of action per unit of entropy it is called thermal action. Approximate data of θ can be evaluated from the results of Grad's analysis of the Boltzmann equation (Grad, 1958). It fact, $\theta = gs^2$ where g is the inertial coefficient used earlier (Sieniutycz and Berry, 1989). Relativistic consideration (chapter: Multiphase Flow Systems) forces us to state that it is θ rather than (approximate) g which is constant. Limiting the analysis to a medium at rest, for example, a rigid immobile solid, no global momentum balance is necessary when modelling the thermal motion.

In terms of the phase variable ψ such that $\psi(t) = -h^{-1}x_0(t)$ and $\psi(t) = k_B h^{-1}\eta(t)$ the reversible equation (1.50) takes the form

$$F(\partial_x \psi, \partial_t \psi) \equiv \frac{\partial \psi}{\partial t} + \frac{a(\partial_x \psi)^2}{2} + k_B h^{-1} T(\mathbf{x},t) = 0 \tag{1.51}$$

where a new coefficient $a = h(\theta k_B)^{-1}$, which has meaning and units of the specific action (action per unit mass), is introduced.

To generalize Eq. (1.51) for irreversible heat transfer we use the first expression for optimal H of Eq. (1.41) in the form

$$H^e(\mathbf{x}, x_0, \mathbf{z}, z_0) = \frac{\mathbf{z}^2}{2mz_0} + z_0 k_B T(\mathbf{x},t) - vz_0 x_0 \tag{1.52}$$

($\mathbf{u} = \mathbf{v}_s$), where $v = \tau^{-1}$. In terms of wave four-vector, $\mathbf{K} = (\mathbf{k}, k_0)$, and after inclusion of time adjoint k_t (to secure vanishing of the enlarged Hamiltonian)

$$\tilde{H}(\mathbf{K},\mathbf{X}) = a\frac{\mathbf{k}}{2k_0^2} + k_0 k_B T(\mathbf{x},t) - vk_0 x_0 + k_t \equiv 0. \quad (1.53)$$

In an eikonal equation, z and z_0 should be expressed in terms of spatial and time derivatives of the phase variable ψ. In terms of thermal phase, $x_0 = -h\psi$ (or $x_0 = -k_B\eta$) and its time and spatial derivatives, extremum Hamiltonian of Eq. (1.53) yields the following equation:

$$F(\partial_x\psi, \partial_t\psi, \psi) \equiv \frac{\partial \psi}{\partial t} + v\psi + \frac{a(\partial_x\psi)^2}{2} + k_B h^{-1} T(\mathbf{x},t) = 0. \quad (1.54)$$

Here $T(\mathbf{x}') = T(\mathbf{x}, t)$ is a nonstationary temperature field.

Since we work with nondimensional phase variables, thermal wave four-vector $\mathbf{k}' = (\mathbf{k}, k_t) = (\partial\psi/\partial\mathbf{x}, \partial\psi/\partial t)$ appears in Eq. (1.54), and coefficient of specific action, a, appears. The accompanying picture for phase of matter might be obtained from the substance action ϕ; its phase is $\Phi = mh^{-1}\phi$. Phase based approaches work with the dimensionless quantities $k_B h^{-1}\eta$ and $mh^{-1}\phi$. The matter-related quantity $\Phi = mh^{-1}\phi$ is an averaged quantum phase in the hydrodynamic representation of the Schrödinger equation (Nonnenmacher, 1986). The entropy-related phase, $\psi = k_B h^{-1}\eta$, considered here, is a new notion associated with the role of entropy in thermohydrodynamics (Sieniutycz, 1994).

In the phase picture, temperature T is given in frequency units $k_B T/h$ (mean frequency of thermal agitations). The dissipative generalization of Eq. (1.50) in the form (1.54) takes into account "lossy" or "dissipative" nature of the medium, and involves thermal phase ψ explicitly in the form of the product $v\psi$, where v is a frequency constant or the reciprocal of a relaxation time τ. The coefficient v characterizes the intensity of the energy loss by the heat quasiparticle. Models of dissipative phenomena containig explicitly phases Φ and ψ, or actions ϕ and η, replace the less efficient, earlier models which contained explicitly the time t. (Bateman, 1931; Sieniutycz, 1976, 1977, 1980b).

Using method of characteristics it can be shown that no canonical formulation exists for the partial differential equation (1.53) in the original phase space-time involving the variables $\mathbf{k}' = \partial\psi/\partial\mathbf{x}'$ and $\mathbf{x}'$. The explicit presence of variable ψ requires a shift of the Hamiltonian formulation and the related Fermat principle from the original (eight-dimensional) phase space-time $(\mathbf{k}', \mathbf{x}')$ to the extended (10-dimensional) phase space-time $(\mathbf{K}, \mathbf{X}) = (k_0, x_0, \mathbf{k}', \mathbf{x}')$. Here $\mathbf{x}' = (x_1, ..., x_n, t)$ is the vector of the particle coordinates in the space-time. In Eq. (1.54), x_0 equals $-h\psi$ or $-k_B\eta$, and k_0 is its adjoint in an enlarged Hamiltonian, which involves time t

$$\tilde{H}(\mathbf{K},\mathbf{X}) = \frac{k_0 dx_0}{d\tau} + \frac{\mathbf{k}' d\mathbf{x}'}{d\tau} \quad (1.55)$$

The rates in $\tilde{H}(\mathbf{K},\mathbf{X})$ may be based on an arbitrary parameter τ, a semblance of a "proper time." τ may be called "thermodynamic time" in an arbitrary situation where it does not coincide with the coordinate time t. Here we have used $\tau = t$ by definition. We refer the reader to the previous section for considerations leading to the extremum Hamiltonian H, where a final action variable $x_0(t^f) = -h\psi(t^f)$ or $-k_B\eta(t^f)$, is maximized. Since the optimal control theory is the suitable mathematical tool in this case, considerations may be performed in the context of Pontryagin's maximum principle (Pontryagin et al., 1962). Here we will exploit the enlarged extremum Hamiltonian (1.53) corresponding to the definition, Eq. (1.55), to obtain the equations of motion for the heat pulse.

The Fermat principle describing the thermal motion conforms to the general structure (Censor, 1990)

$$\int_{x^i}^{x^f} \left[\mathbf{KX} - \tilde{H}(\mathbf{K},\mathbf{X})\right] d\tau = 0 \tag{1.56}$$

At this stage the knowledge of the Euler equation from the calculus of variations is sufficient. It has to be applied to the aforementioned equation treating $\mathbf{X}$ and $\mathbf{K}$ as independent variables. This yields, of course, the canonical set of equations. For the Hamiltonian (1.53) the canonical set has the form

$$\frac{dx_0}{d\tau} = \frac{\partial \tilde{H}}{\partial k_0} = -a\frac{\mathbf{k}^2}{2k_0^2} + k_B T(\mathbf{x},t) - \nu x_0 \tag{1.57}$$

$$\frac{d\mathbf{x}}{d\tau} = \frac{\partial \tilde{H}}{\partial \mathbf{k}} = a\frac{\mathbf{k}}{k_0} \tag{1.58}$$

$$\frac{dt}{d\tau} = \frac{\partial \tilde{H}}{\partial k_t} = 1 \tag{1.59}$$

$$\frac{dk_0}{d\tau} = -\frac{\partial \tilde{H}}{\partial x_0} = \nu k_0 \tag{1.60}$$

$$\frac{d\mathbf{k}}{d\tau} = -\frac{\partial \tilde{H}}{\partial \mathbf{x}} = -k_0 k_B \frac{\partial T(\mathbf{x},t)}{\partial \mathbf{x}} \tag{1.61}$$

$$\frac{dk_t}{d\tau} = -\frac{\partial \tilde{H}}{\partial t} = -k_0 k_B \frac{\partial T(\mathbf{x},t)}{\partial t} \tag{1.62}$$

The previous canonical set may be regarded as the set of classical equations of motion for the quasiparticles of quantized thermal fields (Veinik, 1966, 1973; Anthony, 1989, 1990;

Scholle, 1994; Markus, 2005; Markus and Gambar, 2004, 2005; Sieniutycz and Berry, 1997; Vazquez et al., 2009). The first of the previous equations defines the negative intensity of the thermal phase change, $dx_0/dt = -hd\psi/dt$, along the extremal path. The second defines the group velocity. The third shows that the proper time is the coordinate time; this is a consequence of the nonrelativistic (Galilean) description. The fourth equation describes the canonical adjoint of the thermal phase. The last two equations of the set describe respectively the motion of the pulse and its energy change. When the temperature field is time independent, the energy (time adjoint) is constant along the stationary arc. To have a complete solution, equations of the particle or quasi-particle motion should be supplemented by the field equations (Landau and Lifshitz, 1974b). For the present problem purpose, partial differential equations of change describing the temperature field $T(\mathbf{x},t)$ may serve (Baumeister and Hamil, 1968). An example is Eq. (1.69).

The previous set of equations describes the extremum conditions for any system governed by the Hamiltonian (1.54). Therefore, the equations are the same for any action function $\zeta[\mathbf{x}(t^f), x_0(t^f)]$ extremized at the final time t^f. The transversality conditions may however change.

The physical action A of a distributed multicomponent system (species $k = 1, 2, \ldots n$) obeys in the single-temperature case the general differential formula (Sieniutycz, 1990).

$$dA = \sum_k \mathbf{P}_k d\mathbf{x}_k + \mathbf{I} d\mathbf{x} - \sum_k M_k d\phi_k - S d\eta - H dt \tag{1.63}$$

In Eq. (1.63) $\mathbf{x}_k$ and $\mathbf{x}$ are the vectors of material and thermal Lagrangian coordinates and $\mathbf{P}_k$ and $\mathbf{I}$ the corresponding momenta. Note that $\mathbf{k}$ of the canonical set obeys $\mathbf{k} = \mathbf{I}/h$. ϕ_k and η are specific material and thermal actions. M_k and S are masses of the species k and the system entropy, respectively. Recall that x_0 equals $-h\psi$ or $-k_B\eta$. From Eq. (1.63), $\partial A/\partial \eta = -S$, or, since $x_0 = -k_B\eta$, we find $\partial A/\partial x_0 = S/k_B$.

Let us return to the maximizing of coordinate x_0. The transversality condition for $k_0 = \partial x_0^f$ $(x_0,\mathbf{x}',t)/\partial x_0$ is $k_0(t^f) = 1$. From Eq. (1.60), $k_0(t) = k_0(t^f) \exp[v(t-t^f)]$. But since $k_0(t^f) = 1$, $k_0(t) = \exp[v(t-t^f)]$ and $k_0(t=0) = \exp[-vt^f]$. This means that $k_0(t)$ grows exponentially in time along the extremal path, changing between $\exp[-vt^f]$ and 1. When the frequency of interaction, v, between the pulse and its environment vanishes, $k_0(t)$ becomes a constant of motion. For constant k_0 the wave vector $\mathbf{k}$ is proportional to the group velocity, Eq. (1.58). This case refers to the classical relation between the momentum and velocity.

The effect of exponential grow of k_0 is accompanied by the damped behavior of the velocity. Equations (1.58), (1.60) and (1.61) describe the effect. Equations (1.58) and (1.60) yield

$$\mathbf{k} = k_0(t) a^{-1} \frac{d\mathbf{x}}{dt} = k_0(t^f) \exp[v(t-t^f)] a^{-1} \frac{d\mathbf{x}}{dt} \tag{1.64}$$

Substituting this result into Eq. (1.61) leads to the equation of motion

$$\frac{d^2\mathbf{x}}{dt^2} + v\frac{d\mathbf{x}}{dt} = -\frac{ak_B}{h}\frac{\partial T}{\partial \mathbf{x}} \tag{1.65}$$

where coefficient $a = h(\theta k_B)^{-1}$ appears. Multiplying this equation by the entropy density ρ_s and introducing the heat flux $\mathbf{q} = T\mathbf{j}_s = T\rho_s d\mathbf{x}/dt$, yields a Cattaneo type equation of heat conduction

$$T\rho_s v^{-1}\frac{d}{dt}\left(\frac{\mathbf{q}}{T\rho_s}\right) + \mathbf{q} = -\lambda\frac{\partial T}{\partial \mathbf{x}} \tag{1.66}$$

with the thermal conductivity

$$\lambda = \frac{k_B T}{hv}a\rho_s = \frac{T\rho_s}{\theta v} \tag{1.67}$$

since $a = h/k_B\theta$. For the ideal hard sphere gas, the coefficient θ can be evaluated from Grad's solution of the Boltzmann equation (Sieniutycz and Berry, 1989). This yields $\theta = gs = (2/5) m^2 k_B^{-2}\rho s\rho^{-1}$ where ρ is the mass density. The resulting expression for the thermal conductivity, $\lambda = (5/2)v^{-1}Pk_B m^{-1}$, is in agreement with kinetic theory, where P is the pressure.

A hypothesis has been advanced to the effect that $\theta = T\rho_s c_p^{-1}G^{-1}$ in the general case. The hypothesis can be confirmed for ideal gases and is based on a thermal wave analysis. The generalized formula contains the shear modulus G instead the pressure P (Sieniutycz, 1981a, b, d). Thus, Eq. (1.67) may serve to evaluate the coefficient v so that it will match experimental data for λ. By linearizing Eq. (1.65) the usual Cattaneo form is obtained with the term $v^{-1}\partial\mathbf{q}/\partial t$. Clearly, v^{-1} equals the thermal relaxation time τ.

When the resulting linearized form of Eq. (1.65) is combined with the energy balance

$$\rho c_p \partial T/\partial t = -\text{div}\mathbf{q}, \tag{1.68}$$

the well-known hyperbolic equation is obtained for the change in the temperature field

$$\rho c_p \frac{\partial T}{\partial t} = \lambda\left(\nabla^2 T - \frac{\partial^2 T}{c_0^2 \partial t^2}\right) \tag{1.69}$$

where

$$c_0 = \left(\frac{a_h}{\tau_h}\right)^{1/2} = \left(\frac{\lambda}{\rho c_p \tau_h}\right)^{1/2} \tag{1.70}$$

is called the propagation speed of the thermal wave. This equation describes damped thermal waves propagating with a finite speed c_0 such that $(c_0)^2 = Ts(\theta c_p)^{-1} = G/\rho$.

The previous analysis is not a sole example which shows that irreversible models involve explicitly thermal and material phases. The world of chemical reactions provides many other examples. It may be shown that for a kinetic potential L containing explicitly phases of entropy and components and expressed in terms of their natural variables in the sense of Callen (1988), the stationarity conditions of the L-based action functional are the following (Sieniutycz, 1990, 1994):

$$\frac{\partial \rho_k}{\partial t} + \nabla(\rho_k \mathbf{u}_k) = -\frac{\partial L}{\partial \phi_k} \quad (1.71)$$

and

$$\frac{\partial \rho_s}{\partial t} + \nabla(\rho_s \mathbf{u}_s) = -\frac{\partial L}{\partial \eta} \quad (1.72)$$

Thus the partial derivatives of kinetic potential L with respect to the component phases ϕ_k represent the mass sources of the species in all chemical reactions, whereas those with respect to phase η represent the source of the entropy.

1.5 Thermodynamic Geometries

1.5.1 Equilibrium Metrics

Since the entropy of an isolated macroscopic system achieves the maximum at the thermodynamic equilibrium, Hessian matrix composed of the second derivatives of entropy d^2S with respect to the thermodynamic state variables is negative (Glansdorff and Prigogine, 1971). When the components of the hydrodynamic velocity are included into the set of thermodynamic coordinates, the (enlarged) second differential of the entropy takes over the role of d^2Z, the quantity which was originally designed by Glansdorff and Prigogine to replace d^2S (Oono, 1976). The definite signs of entropy and energy Hessians (and hence d^2S) in spaces of their natural variables prompted the efforts toward the construction of metric structures on the thermodynamic ground. Weinhold (1975, 1976) showed that the first and second laws of thermodynamics endow the second-derivative matrix of the (equilibrium) internal energy with the positivity required to make it a metric, or first fundamental form, on the Gibbs equilibrium surface GES, wherever the matrix is nonsingular.

A physical line of reasoning leading to the concept of a metric structure on the set of equilibrium states and to its applications to fluctuation theory are presented in Ruppeiner's reviews who revealed the significance of the related metric defined in the entropy representation by formulating a theory of large fluctuations around equilibrium

(Ruppeiner, 1979, 1990). He reviews the thermodynamic metric in the context of fluctuation theory. A covariant fluctuation theory is proposed extending the classical theory of thermodynamic fluctuations beyond the Gaussian approximation. A path integral formulation is presented. His research shows that the scalar Riemannian curvature plays a special role as the volume where the classical theory of fluctuations breaks down. Physically, this is expected to be the correlation volume at the critical point. Examples are presented discussing the thermodynamic curvature for fluid and magnetic systems. The value of the curvature is related to interactions in nonideal systems.

For the theory of nonequilibrium thermodynamics the metric structure is important by its intimate connection to fluctuation and dissipation. Salamon and Berry (1983) showed that the dissipated availability of a finite-time endoreversible process is bounded from below by the square of the length of the corresponding path, L^2, divided by the number of relaxations. While the number of relaxations is clearly defined for a process which consists of a finite sequence of relaxations to equilibrium states (Nulton et al., 1985) it is defined for processes continuous in time as the total time divided by a mean relaxation time. If the process proceeds by states of local thermodynamic equilibrium, then each local system traverses a different path with different length L. The fact that L^2 is an extensive quantity then leads to a corresponding bound which involves d^2, the square of the thermodynamic distance that is, the length of the shortest path.

Levine (1986) gave a rigorous treatment of Euclidean geometry for classical thermodynamics. The central physical idea is that it is useful to characterize the system in terms of a number of mean values of "relevant" observables. These mean values are written, as usual, as an expectation $\Sigma A_i p_i$ over a classical probability distribution. The expectation value is then interpreted as a scalar product between vectors belonging to dual spaces. A metric is introduced via the transformation from one space to another. In terms of the metric, the scalar product of two vectors belonging to the same space (eg, two probability distributions or two observables) can be defined. In the space of all states the metric does not depend on the state of the system and the curvature tensor vanishes, that is, the space is Euclidean. It was shown (Salamon et al., 1984; Mrugala et al., 1990) that the two metrics based on the energy and the entropy representations are conformally equivalent. Geometries based on the information-theoretic entropy were introduced (Salamon et al., 1985; Ingarden et al., 1997) and shown equivalent under suitable conditions (Salamon et al., 1985). Possible changes in the thermodynamic curvature caused by heat and other transport phenomena were pointed out (Casas-Vazquez and Jou, 1985; Sieniutycz and Berry, 1991).

Janyszek and Mrugala (1990) have introduced Riemannian and Finslerian geometries for thermodynamic spaces by using statistical mechanics augmented by the concepts of relative information and the dispersion of relative information. They show that the Riemannian metric tensor can be defined using the second fluctuation moments of the stochastic variables.

The curvature tensor and the scalar curvature are calculated and it is proposed that the Riemannian scalar be adopted as a measure of thermodynamic stability. For nonequilibrium thermodynamics, it is shown how an arbitrary nonhomogeneous Lagrangian can be transformed into a homogeneous one and how this can be used to introduce a Finslerian (direction dependent) metric. Finslerian geometry is essentially the geometry of a Lagrangian and these developments have significance for both variational and geometric theories of dissipation.

Sieniutycz and Salamon (1990b) presented a phenomenological approach to the problem of the nonequilibrium metric. The approach is presented in the context of the field (Eulerian) description of the fluid motion. Initially, heat flux (or entropy flux) is the only nonequilibrium variable in the extended energy formula. Later mass diffusion and fluid flow effects are introduced and the metric is defined using the total nonequilibrium energy of the flowing fluid. This approach allows one to compute the length of a nonequilibrium thermodynamic path, bound the related dissipation and determine stability conditions for the given differential equations describing the transient modes of the system. These conditions can be found by testing the total energy and its time derivative by the second method of Lapunov. However, this energy is not necessarily convex for systems out of equilibrium. Subsequently, we outline a nonequilibrium generalization of the fundamental equation and the metric geometry. Related applications to the stability of heat and mass diffusion and nonequilibrium fluctuations are considered in" Thermodynamic Lapunov Functions and Stability. See also Gillmore (1984) on length and curvature in the geometry of thermodynamics.

1.5.2 Nonequilibrium Metrics

Analyzing the role of Callen's postulates, we find that an extended thermodynamics is necessary for describing changes through nonequilibrium states in the enlarged space $(E, \mathbf{X}, \mathbf{i})$ rather than in the classical space $(E, \mathbf{X})$. This observation means that both $\mathbf{X}$ and $\mathbf{i}$ variables should appear in the fluid energy E; it will help us to construct an example of the possible extensions of metric analyses to nonequilibrium.

We shall limit ourselves to the metric for heat conduction in a resting fluid. Solution of the Boltzmann kinetic equation by Grad's or Enskog's methods (Grad, 1958) can be exploited to obtain the nonequilibrium energy density of the heat conducting fluid in the form $E(\mathbf{X}, \mathbf{i}) = E^{eq}(X, 0) + \Delta E(\mathbf{X}, \mathbf{i})$. Here $\mathbf{X} = (v, s)$ or (ρ, s) where v, ρ, s are the specific volume, the mass density and the specific entropy respectively, and $\mathbf{i} = \mathbf{j}_s$ is the vector of the conductive flux of entropy. This entropy flux is related to the conductive flux of the energy or heat $\mathbf{q}$ by the well-known expression, $\mathbf{j}_s = \mathbf{q}/T$. The specific nonequilibrium energy, $e = E/\rho$, of a fluid with heat flux is (Sieniutycz and Berry, 1989):

$$e = \frac{1}{2} g \rho^{-2} \mathbf{j}_s^2 + e^{eq}(v,s) = \frac{1}{2} g s^2 \mathbf{v}_s^2 + e^{eq}(v,s) \tag{1.73}$$

where $g = (2/5)m^2/k_B^2$ for hard sphere gas. m is the mass of particle, k_B is Boltzmann constant, e^{eq} is the specific equilibrium internal energy, and $\mathbf{v}_s = \mathbf{j}_s/\rho_s = \mathbf{j}_s/\rho_s$ is the velocity of the entropy transfer.

When the set of variables is extended to include velocity ($\mathbf{v}_s$, v, s), the governing thermodynamic potential is the Lagrange kinetic potential L rather than the energy (Sieniutycz, 1990, 1992b, 1994). In the case of heat flow the kinetic potential per unit mass is

$$l = \frac{1}{2}gs^2\mathbf{v}_s^2 - e^{eq}(v,s) \tag{1.74}$$

The partial derivative of the kinetic potential with respect to a velocity is a momentum (Landau and Lifshitz, 1976). Equation (1.74) yields a partial momentum (thermal momentum)

$$\mathbf{i} = \frac{\partial l}{\partial \mathbf{v}_s} = gs^2\mathbf{v}_s \tag{1.75}$$

(Since the fluid is at rest, its convective momentum $\mathbf{J} = \rho\mathbf{u} = 0$.) The previous result allows one to express the energy e in the form of the Hamiltonian H containing only the natural (or canonical) variables v, s, and $\mathbf{i}$. Hence the fundamental equation for the resting fluid conducting heat is:

$$H(v,s,\mathbf{i}) = e = \frac{1}{2}a(v,s)\mathbf{i}^2 + e^{eq}(v,s) \tag{1.76}$$

where $a(s,v) = [g(s,v)s^2]^{-1}$. This energy formula is associated with the extended or nonequilibrium Gibbs equation for the fluid with heat:

$$de = Tds - Pdv + \mathbf{v}_s d\mathbf{i} \tag{1.77}$$

where the nonequilibrium temperature and pressure are:

$$T(v,s,\mathbf{i}) = \left(\frac{\partial e}{\partial s}\right)_{v,\mathbf{i}} = T^{eq}(v,s) + a_s\mathbf{i}^2 \tag{1.78}$$

and

$$P(v,s,\mathbf{i}) = -\left(\frac{\partial e}{\partial v}\right)_{v,\mathbf{i}} = P^{eq}(v,s) - a_v\mathbf{i}^2 \tag{1.79}$$

where $a_s = \partial a/\partial s$, $a_v = \partial a/\partial v$, $T^{eq} = \partial e^{eq}/\partial s$ and $P^{eq} = -\partial e^{eq}/\partial v$. Also, $\mathbf{v}_s = \partial e/\partial \mathbf{i} = a\mathbf{i} = \mathbf{i}/gs^2$, in agreement with (1.75).

In this approach, the heat flux density can be computed as a thermodynamic quantity. For an arbitrary function $e(s,v,\mathbf{i})$ the familiar formula $\mathbf{q} = T\mathbf{j}_s$ holds in the form

$$\mathbf{q} = T\rho s \mathbf{v}_s = T\rho_s(\partial e / \partial \mathbf{i}) \tag{1.80}$$

where for our model $\mathbf{q} = Tv^{-1}sa\mathbf{i}$ or $\mathbf{i} = gsT^{-1}\rho^{-1}\mathbf{q}$.

It is now easy to calculate extended thermodynamic metric tensors in the energy representation. The covariant tensor based on the Hessian of $e(s, v, \mathbf{i})$ is the Jacobian matrix $g_{ij} = \partial(T, -P, \mathbf{v}_s)/\partial(s, v, \mathbf{i})$. For one-dimensional system in a Galilean space

$$g_{ij} = \begin{bmatrix} \dfrac{\partial T}{\partial s} & \dfrac{\partial T}{\partial v} & \dfrac{\partial T}{\partial n} & \cdots & \dfrac{\partial T}{\partial i} \\ -\dfrac{\partial P}{\partial s} & -\dfrac{\partial P}{\partial v} & -\dfrac{\partial P}{\partial n} & \cdots & -\dfrac{\partial P}{\partial i} \\ \dfrac{\partial \mu}{\partial s} & \dfrac{\partial \mu}{\partial v} & \dfrac{\partial \mu}{\partial n} & \cdots & \dfrac{\partial \mu}{\partial i} \\ \cdots & \cdots & \cdots & \cdots & \cdots \\ \dfrac{\partial v_s}{\partial s} & \dfrac{\partial v_s}{\partial v} & \dfrac{\partial v_s}{\partial n} & \cdots & \dfrac{\partial v_s}{\partial i} \end{bmatrix} \tag{1.81}$$

For the model with arbitrary $a(s, v)$ this becomes

$$g_{ij} = \begin{bmatrix} \left(\dfrac{\partial T}{\partial s}\right)^{eq} + \dfrac{a_{ss}i^2}{2} & \dfrac{\partial T}{\partial v} + \dfrac{a_{sv}i^2}{2} & \dfrac{\partial T}{\partial n} + \dfrac{a_{sn}i^2}{2} & \cdots & a_s i \\ -\left(\dfrac{\partial P}{\partial s}\right)^{eq} + \dfrac{a_{vs}i^2}{2} & -\dfrac{\partial P}{\partial v} + \dfrac{a_{vv}i^2}{2} & -\dfrac{\partial P}{\partial n} + \dfrac{a_{vn}i^2}{2} & \cdots & a_v i \\ \left(\dfrac{\partial \mu}{\partial s}\right)^{eq} + \dfrac{a_{ns}i^2}{2} & \dfrac{\partial \mu}{\partial v} + \dfrac{a_{nv}i^2}{2} & \dfrac{\partial \mu}{\partial n} + \dfrac{a_{nn}i^2}{2} & \cdots & a_n i \\ \cdots & \cdots & \cdots & \cdots & \cdots \\ a_s i & a_v i & a_n i & \cdots & a[1] \end{bmatrix} \tag{1.82}$$

[1] means unit matrix. For an ideal gas $g = 2m^2/5k_B^2$ and since $a = (gs^2)^{-1}$ one obtains

$$\begin{bmatrix} \dfrac{2mT}{3k_B} + \dfrac{a_{ss}i^2}{2} & -\dfrac{2T}{v} + \dfrac{a_{sv}i^2}{2} & \cdots & a_s i \\ -\dfrac{2T}{3v} + \dfrac{a_{vs}i^2}{2} & \dfrac{5kT}{3mv^2} + \dfrac{a_{vv}i^2}{2} & \cdots & a_v i \\ \cdots & \cdots & & \\ a_s i & a_v i & & \dfrac{5k_B^2}{2m^2 s^2}[1] \end{bmatrix} \tag{1.83}$$

Note that (local) equilibrium corresponds to $\mathbf{i} = 0$. At equilibrium $(s, v, 0)$ the process must be stable and the positiveness of g_{ij} around this equilibrium implies

$$c_v(v,s,0) > 0, \qquad \kappa = -\left(\frac{\partial P}{\partial \ln v}\right)_{T,0} > 0, \qquad g > 0. \tag{1.84}$$

The additional inequality, $g > 0$, means that the presence of heat flux can only increase the organization of the system around equilibrium (increase the energy e).

Inclusion of mass diffusion and convection is based on the extended Gibbs equation

$$de = Tds - Pdv + \sum_k \mu_k dy_k + v_s d\mathbf{i} + \sum_k v_k d\mathbf{w}_k + \mathbf{u}d\mathbf{u} \tag{1.85}$$

where $k = 1, 2, \ldots n$ indexes the species, $\mathbf{v}_k$ is the mass diffusion velocity, y_k is the mass fraction of kth species and $\mathbf{w}_k = \Sigma(y_k \mathbf{v}_k)$ is the momentum of kth species per unit mass of the mixture.

Covariant and contravariant thermodynamic tensors:

$$g_{ij} = \frac{\partial(T, -P, \mu_k, \mathbf{v}_s, \mathbf{v}_k, \mathbf{u})}{\partial(s, v, y_k, \mathbf{i}, \mathbf{w}_k, \mathbf{u})} \tag{1.86}$$

$$g^{ij} = \frac{\partial(s, v, y_k, \mathbf{i}, \mathbf{w}_k, \mathbf{u})}{\partial(T, -P, \mu_k, \mathbf{v}_s, \mathbf{v}_k, \mathbf{u})} \tag{1.87}$$

can be computed as Hessian matrices of the thermodynamic potential $\Gamma = \Sigma \mu_k y_k$. Γ is the complete Legendre transform of e; it is related to the Gibbs free energy $G = h - T_s$ by

$$\Gamma = G - \mathbf{v}_s \mathbf{i} - \mathbf{v}_k \mathbf{w}_k - \mathbf{u}\mathbf{u} \tag{1.88}$$

These results enable us to conclude that metric tensors may be extended off (but not far off) the Gibbs equilibrium surface to incorporate nonequilibrium variables such as fluxes, rates, intensity deviations, and so forth. Convexity of the energy and concavity of the entropy is assured provided that the principal minors of the metric matrix in Eq. (1.83) are positive. In the entropy representation, one may find generalized criterion d^2Z of Glansdorff and Prigogine (1971) to incorporate macroscopic motion and diffusion of heat and mass. This is achieved by adopting a generalized entropy whose second differential is just d^2S expressed appropriately in terms of the total energy and momentum variables, allowing mechanical degrees of freedom. The quadratic forms based on the Hessians of the nonequilibrium energy and entropy can serve as Liapounov functions to generate sufficient conditions of stability. Positiveness of the (nonequilibrium) energy Hessian or negativeness of the corresponding entropy Hessian around equilibrium provides a set of both necessary and sufficient conditions for stability.

1.6 A Finite Rate Exergy

One of recent achievements of applied irreversible thermodynamics, more commonly FTT, is the notion of the so-called finite time exergy or finite rate exergy (FTE) (see Figs. 1.3 and 1.4). The notion refers to the process between a body (a medium, or a resource) and the environment, and it is defined as an extremum work released (supplied) with a minimum dissipation possible under given operational conditions, prescribed or free duration, and for some fixed end coordinates of the medium. As contrasted with the classical exergy, an extremum process for which a maximum or a minimum work is sought is not reversible but rather satisfies a weak condition of minimal irreversibility. Therefore a FTE yields more realistic restriction for a practical processes than the classical exergy function.

The generalized exergy is quantified in terms of a finite amount or flow of the resource and other parameters which are: end states of the resource, process direction, reference state of the environment, mean rate index (Hamiltonian h), and an imperfection factor of the thermal machine, Φ. To find extremum work at flow and the associated exergies, two optimization problems are considered, one for a maximum of work delivery [max W] and one for a minimum of the work supply [min $(-W)$]. Absolute values of these two works coincide only for reversible transitions. Generalized exergies constitute two extremum works related to these two minimally irreversible, finite-rate (finite duration) processes, of which the first extremum work refers to the engine mode (work production) and the second one to the HP mode (work consumption). To date, most of the postclassical terms in engine-mode

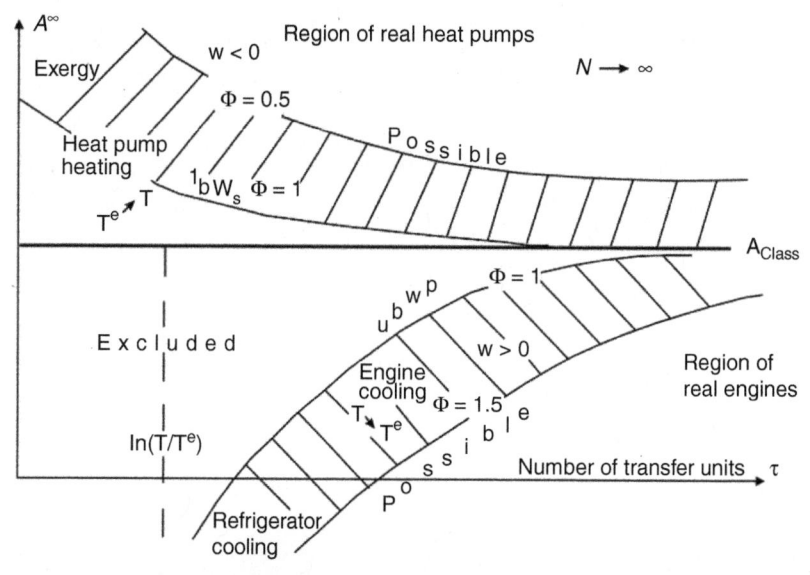

Figure 1.3: Finite rate exergy and influence of internal irreversibilities Φ on limiting finite-rate work generated in engines and consumed in HPs (the reference temperature $T_0 = T^e$).

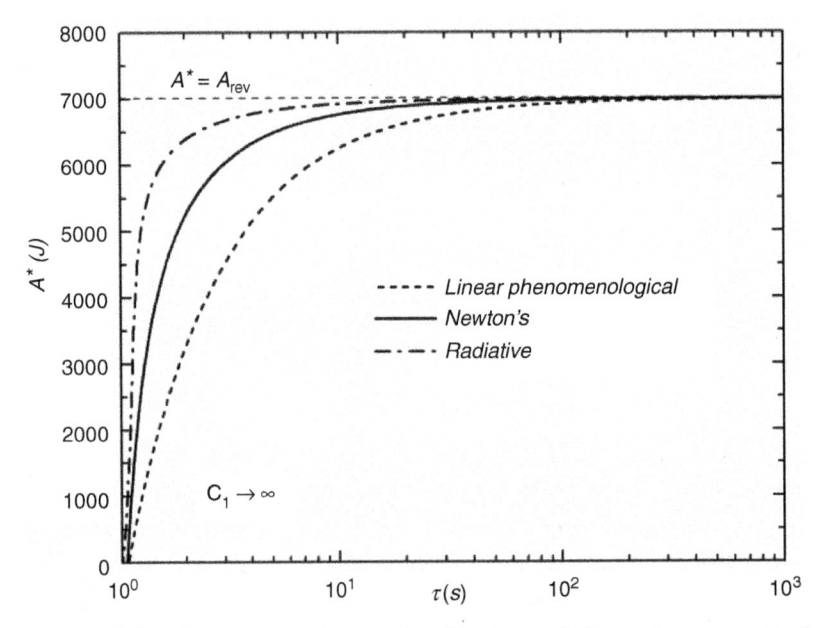

Figure 1.4: Finite-time exergy of an engine for three different heat transfer laws (Xia et al., 2010c).

exergies were evaluated under the assumption of the exponential relaxation, consistent with linear dynamics (Sieniutycz, 1997, 2004). However, for radiation fluids, which are nonlinear thermodynamic systems, fluid's properties vary along the path, and the optimal relaxation curve is nonexponential. Still the shape of the optimal curve can be determined from the optimum conditions. Various differential models for relaxation dynamics may be studied depending on the equation of energy exchange with the environment.

Here is a brief outline of the theory leading to the finite-rate exergy of radiation. Because of analytical difficulties associated with the use of Stefan–Boltzmann equation, approaches using the so-called pseudo-Newtonian model are applied (Sieniutycz and Kuran, 2005, 2006). In these approaches thermal conductances are used in the temperature dependent form $\alpha_1 = \alpha_{10} T_1^3$ and $\alpha_2 = \alpha_{20} T_2^3$, and, by assumption, each temperature dependent conductance is attributed to the driving force defined as the simple temperature difference. The virtue of the pseudo-Newtonian approach is its potential of an analytical solution. This is briefly characterized subsequently (Sieniutycz and Kuran, 2006; Kuran, 2006).

For an imperfect engine the first-law efficiency of each elementary unit is given by the pseudo-Carnot formula,

$$\eta = 1 - \Phi \frac{T_{2'}}{T_{1'}}, \tag{1.89}$$

where $\Phi = \Delta S_{2'}/\Delta S_{1'} = J_{s2'}/J_{s1'}$ is the parameter of internal irreversibility. The factor of internal irreversibilities, Φ, satisfies inequality $\Phi > 1$ for engine mode and $\Phi < 1$ for HP mode of the system. Temperatures of circulating fluid, $T_{1'}$ and $T_{2'}$, are not independent but connected by the entropy balance

$$\frac{g_2(T_{2'})(T_{2'} - T_2)}{T_{2'}} - \Phi \frac{g_1(T_1)(T_1 - T_{1'})}{T_{1'}} = 0. \tag{1.90}$$

We invert pseudo-Carnot formula (1.89) to substitute $T_{2'} = 1 - \eta \Phi^{-1} T_{1'}$ into the previous entropy balance. We then obtain an equation for $T_{1'}$

$$T_{1'} = (g_1 + \Phi^{-1} g_2)^{-1}(g_1 T_1 + (1-\eta)^{-1} g_2 T_2) \tag{1.91}$$

With this result the flux of the driving radiation energy,

$$q_1 = d\dot{Q}_1 \cong g_1(T_1^3)(T_1 - T_{1'}), \tag{1.92}$$

follows in the pseudo-Newtonian formalism in the form

$$q_1 = g'[T_1 - \Phi T_2/(1-\eta)]. \tag{1.93}$$

In Eq. (1.93) an overall conductance appears; it is defined in terms of all g_i as a function of respective bulk temperatures of reservoirs

$$g' \equiv g_1 g_2 (\Phi g_1 + g_2)^{-1} = (g_1^{-1} + \Phi g_2^{-1})^{-1}. \tag{1.94}$$

The previous equation defines the operational conductance for heat transfer which is suitably modified by the coefficient of internal dissipation.

The work flux (power) follows in the form

$$p = \eta q_1 = \eta q_1 g'(\Phi T_1, T_2)\left(T_1 - \frac{\Phi T_2}{1-\eta}\right). \tag{1.95}$$

The expression

$$T' = \Phi T_2 (1-\eta)^{-1} \tag{1.96}$$

appearing in Eqs. (1.93) and (1.95) describes the so-called Carnot temperature in terms of the efficiency (Sieniutycz and Jeżowski, 2009). The reference adduces also the thermodynamic definition of Carnot temperature, $T' \equiv T_2 T_{1'}/T_{2'}$. We note that despite temperature dependent conductances g_1 and g_2, Eq. (1.95) yields the same maximum power efficiency

$$\eta_{mp} = 1 - \sqrt{\frac{\Phi T_2}{T_1}} \tag{1.97}$$

as for engines driven by usual fluids where the heat flux is governed by the (linear) Newtonian law of cooling.

The differential of total entropy produced has a kinetics independent form

$$dS_\sigma = \frac{dQ_1}{T_2}\left(\Phi\frac{T_{2'}}{T_{1'}} - \frac{T_2}{T_1}\right) = dQ_1\left(\frac{\Phi-1}{T'} + \frac{1}{T'} - \frac{1}{T_1}\right) = dQ_1\left(\frac{\Phi}{T'} - \frac{1}{T_1}\right). \quad (1.98)$$

In a pseudo-Newtonian process conductances g_i are functions of temperatures of respective reservoirs.

For the exergy determining, consideration of dynamical changes of temperature of a finite reservoir (representing a body, medium, fluid, etc.) is necessary. This temperature changes (decreases in engine mode) because of the reservoir's finite capacity. This is the case with a finite resource, appropriate to define an exergy. By integration of power expressions over time, functionals of work generation (consumption) and related exergies are obtained.

The dynamical process is the passage of the vector $\mathbf{T} = (T, \tau)$ from its initial state $\mathbf{T}^i$ to its final state $\mathbf{T}^f$. In the absence of frictional effects the power functional corresponding to efficiency (1.89) is

$$\begin{aligned}\dot{W} &= -\dot{V}\int_T^{T_0}\left[c_v(T)\left(1 - \Phi\frac{T_0}{T + \chi dT/dt}\right) + P_T\right]dT \\ &= -\dot{V}\int_T^{T_0}\left[c_h(T) - c_v(T)\Phi\frac{T_0}{T + \chi dT/dt}\right]dT\end{aligned} \quad (1.99)$$

where $P_T \equiv dP/dT = (4/3)\,aT^3$ and $c_h(T) \equiv c_v(T) + P_T = (16/3)aT^3$. By definition $c_h(T) \equiv c_v(T) + dP/dT = (4 + 4/3)\,aT^3 = (16/3)aT^3$. A more transparent form of the previous power integral is obtained after transforming it so as to extract from it the effect of the reversible power

$$\dot{B}_f = -\int_{T,P}^{T_0,P_0}\dot{V}\left[c_v(T)\left(1 - \frac{T_0}{T}\right) + \frac{dP}{dT}\right]dT \quad (1.100)$$

and the associated efficiency term. For the pseudo-Newtonian model we obtain

$$\begin{aligned}\dot{W}_f &= -\dot{V}\int_T^{T_0}\left[c_h(T) - c_v(T)\frac{T_0}{T}\right]dT \\ &\quad - T_0\dot{V}\int_T^{T_0}\left[c_v(T)\left(\frac{\chi(dT/dt)^2}{T(T + \chi dT/dt)} + (1-\Phi)\frac{dT/dt}{T + \chi dT/dt}\right)\right]dt\end{aligned} \quad (1.101)$$

Associated entropy production per unit flowing volume can be evaluated as the difference between the outlet and inlet entropy fluxes. In terms of Carnot temperature $T' = T_1 + \chi dT_1/dt$ and after using $q_2 = q_1 \Phi T_0/T'$ we find for the pseudo-Newtonian model

$$\sigma_s = \frac{q_2}{T_2} - \frac{q_1}{T_1} = \frac{g'(T_1 - T')^2}{T_1 T'} + q_1 \frac{\Phi - 1}{T'}. \tag{1.102}$$

This is consistent with general Eq. (1.98) and the model-related equation

$$d\dot{Q}_1 \equiv d\gamma'[T_1 - \Phi T_2/(1-\eta)], \tag{1.103}$$

which describes the driving heat. Comparison of Eqs. (1.101) and (1.102) shows that the term multiplied by T_0 in the power expression (1.101) is the entropy production of the pseudo-Newtonian model. It is here split into the sum of two nonnegative terms. The first term, related to the approximate description of the effect of emission and adsorption of radiation, is obviously positive. The second term, or the product $q_1(\Phi - 1)/T'$, is always non-negative as the signs of q_1 and $\Phi - 1$ are the same (positive for engine and negative for HP). A concise form of power functional (1.101)

$$\dot{W}_f = -\dot{V} \int_T^{T_0} \left(c_h(T) - c_v(T) \frac{T_0}{T} \right) dT - T_0 \dot{V} \int_T^{T_0} \sigma_v \, dt, \tag{1.104}$$

(for photons $c_v(T) = 4aT^3$) is in agreement with the Gouy–Stodola law. This form is quite general and not restricted to the pseudo-Newtonian model. However, the (first) reversible term of this equation apparently shows a disagreement between the reversible efficiency and the Carnot efficiency. Therefore we stress that the reversible thermal efficiency of the radiation conversion is Carnot. (See, however, chapter: Radiation and Solar Systems for some complex radiation systems.) Indeed, comparison of classical equations describing the reversible exergy of the flowing radiation

$$\dot{B}_f \equiv \dot{B} + (P - P_0)\dot{V} = [h_v - h_{v0} - T_0(s_v - s_{v0})]\dot{V} \tag{1.105}$$

and

$$\dot{B}_f = -\int_{T,P}^{T_0,P_0} \dot{V} \left[c_v(T)\left(1 - \frac{T_0}{T}\right) + \frac{dP}{dT} \right] dT \tag{1.106}$$

shows that the apparent disagreement is caused by the additive, work-related term $P_T \equiv dP/dT$ in the exergy formulae, Eqs. (1.106), (1.99), and the like. For radiation fluids, the pressure contribution to the exergy in the form $P_T \equiv dP/dT$ is masked by the dependence of P on T.

Integration of the first (reversible) part of integral (1.104) and calculation of $\dot{W}_f^{rev}/\dot{V}$ yields the classical exergy of flowing radiation fluid per unit volume, Eqs. (1.105)–(1.107),

$$\dot{B}_f = \left[\frac{4}{3}a(T^4 - T_0^4) - T_0 \frac{4}{3}a(T^3 - T_0^3)\right]\dot{V} \qquad (1.107)$$
$$= [h_v - h_{v0} - T_0(s_v - s_{v0})]\dot{V} = \dot{B} + (P - P_0)\dot{V}$$

When the environment temperature is not necessarily an integration limit, the specific work of flowing radiation between two arbitrary states is obtained as the exergy difference. We obtain

$$\dot{W}_f^{rev}/\dot{V} = \int_{T^i}^{T^f} [(16/3)aT^3 - 4aT^2 T_0]dT = h_v^i - h_v^f - T_0(s_v^i - s_v^f) = \Delta b_v \qquad (1.108)$$

These results are in agreement with general thermodynamics. They confirm that the first term of power functional (1.101) is path independent. Thus, whenever a power extremum is sought, only the second, irreversible term contributes to the optimization solution.

In terms of nondimensional time $\tau = t/\chi$ and per unit system volume, entropy production functional of the pseudo-Newtonian model, Eq. (1.102), is

$$S_{V_{gen}} \equiv \dot{S}_{gen}/\dot{V} = \int_T^{T_0}\left\{c_v(T)\left[\frac{\dot{T}^2}{T(T+\dot{T})} + (1-\Phi)\frac{\dot{T}}{T+\dot{T}}\right]\right\}d\tau \qquad (1.109)$$

The additive structure of two parts in Eq. (1.101) is an important property that causes that the two problems of extremum work (1.101) and the associated problem of minimum entropy generation (1.109) have the same solutions whenever end states are fixed.

The optimization problem can thus be stated as a variational problem for either functional of work or of entropy production. When work W is an optimization criterion, the problem is that of maximum W for engine mode and that of minimum of $(-W)$ for heat-pump mode. When optimization of the entropy production is considered, a minimum is sought for each process mode. A generalized exergy is an extremum of W with appropriate integration limits ($T^i = T$ and $T^f = T_0$ for engine mode and $T^i = T_0$ and $T^f = T$ for heat-pump mode). In the quasistatic limit (zero rates, $\dot{T} = 0$), Eq. (1.101), always leads to *classical exergy*. Moreover, it leads to the same classical exergy for each mode when proper integration limits, stated previously, are used. Yet, in a general case the absolute value of work (1.101) describes the change of a generalized exergy of radiation in operations with imperfect thermal machines and when dissipative phenomena due to the radiation emission and adsorption are essential.

We focus here on the minimum entropy production for functional (1.109) when $\Phi = 1$. An equation for the optimal temperature follows from the condition $\varepsilon = h$, where $\varepsilon = (\partial L/\partial \dot{T})\dot{T} - L$ is the first integral for Lagrangian L contained in equations of entropy production, and h is a constant value of ε determined from the boundary conditions for T and τ. The present h has units of entropy density or specific heat per unit volume, and should be distinguished from Hamiltonians used occasionally with respect to the energy. Our $h = H/VT_0$, where H is a Hamiltonian expressed in the energy units and V is the volume. For any rate independent Φ the first integral for L of Eq. (1.109) is

$$\varepsilon(T,\dot{T}) = \frac{\partial L}{\partial \dot{T}}\dot{T} - L = \Phi c_v(T)\frac{\dot{T}^2}{(T+\dot{T})^2} = h \tag{1.110}$$

From this equation one can obtain optimal trajectories in terms of h. After introducing the function

$$\xi\left(\frac{h}{\Phi c_v(T)}\right) \equiv \pm\sqrt{\frac{h}{\Phi c_v(T)}}\left(1 \pm \sqrt{\frac{h}{\Phi c_v(T)}}\right)^{-1} = \left(\pm\sqrt{\frac{\Phi c_v(T)}{h}} - 1\right)^{-1} \tag{1.111}$$

(upper sign refers to the heat-pump mode, lower one to the engine mode) the family of pseudo-exponential extremals satisfies an equation

$$\dot{T} = \xi(h, \Phi, T)T. \tag{1.112}$$

In this case the slope of the logarithmic rate $\xi = dlnT/d\tau$ is a state dependent quantity. The slope ξ is the rate indicator, positive for the fluid's heating and negative for fluid's cooling.

Application of extremal (1.112) in Eq. (1.109) leads to the minimum entropy production in the form

$$S_{v_{gen}} = \int_{T^i}^{T^f} \frac{c_v(T)}{T}\left[\pm\sqrt{\frac{h}{\Phi c_v(T)}} + (1-\Phi)\left(1 \pm \sqrt{\frac{h}{\Phi c_v(T)}}\right)\right]dT \tag{1.113}$$

With this result and the Gouy–Stodola law we obtain the density of generalized exergy for the fluid at flow

$$\begin{aligned}a_v(T,T_0,h) &= b_v(T,T_0,0) \pm T_0 s_{v_{gen}} = \frac{4}{3}aT^4\left(1 - \frac{T_0}{T}\right) \\ &\pm T_0\int_T^{T_0}\frac{c_v(T)}{T}\left[\pm\sqrt{\frac{h}{\Phi c_v(T)}} + (1-\Phi)\left(1 \pm \sqrt{\frac{h}{\Phi c_v(T)}}\right)\right]dT,\end{aligned} \tag{1.114}$$

where the second-line term is nonclassical. For radiation $c_v = 4aT^3$, thus, since $e_v = 3P$, the expression $h_v = (4/3)aT^4$ describes the enthalpy of black radiation. Therefore the classical term in the previous exergy equation is the reversible flow exergy of black radiation per unit volume. This flow exergy is recovered from Eq. (1.114) in the reversible limit when the

Hamiltonian $h = 0$. It is the classical exergy of fluid at flow which satisfies the alternative formula

$$b_v \equiv \dot{B}_f / \dot{V} = h_v - h_{v0} - T_0(s_v - s_{v0}) \tag{1.115}$$

consistent with the general equation for exergy of the matter with vanishing chemical potential.

For the black-radiation fluid the optimal trajectory which solves Eqs. (1.110)–(1.112) satisfies an equation

$$\pm (4/3) a^{1/2} \Phi^{1/2} h^{-1/2} (T^{3/2} - T^{i^{3/2}}) - \ln\left(\frac{T}{T^i}\right) = \tau - \tau^i. \tag{1.116}$$

The integration limits refer to initial (i) and current state (no index) of the radiation fluid, that is, to temperatures T^i and T, corresponding with τ^i and τ. Fig. 5.2 in Sieniutycz and Jeżowski (2009) shows an example of optimal paths for radiation in both process modes. Relaxation of radiation to the equilibrium occurs in the engine mode, whereas the radiation utilization with associated escape from the equilibrium—in heat-pump mode. Clearly, radiation does not relax exponentially. Qualitative difference of relaxation curve from those describing exponential relaxation in linear processes is observed.

Equations (1.114) and (1.116) are associated with the entropy production (1.109) and the generalized availability of radiation

$$a_v(T, T_0, h) = b_v(T, T_0, 0)$$
$$\pm \frac{4}{3} a^{1/2} h^{1/2} \Phi^{1/2} T_0 (T^{3/2} - T_0^{3/2}) + \frac{4}{3} a T_0 (1 - \Phi)(T^3 - T_0^3) \tag{1.117}$$

The classical availability of *radiation at flow* resides in the previous equation in Jeter's (1981) form

$$b_v(T, T_0, 0) = h_v - h_{v0} - T_0(s_v - s_{v0}) = h_v\left(1 - \frac{T}{T_0}\right) = \frac{4}{3} a T^4 \left(1 - \frac{T}{T_0}\right) \tag{1.118}$$

As two modes are included, the common symbol T in generalized availability (1.117) refers to the initial temperature of engine mode or the final temperature of heat-pump mode. Qualitative properties of this availability function with respect to nondimensional duration or the overall number of heat transfer units $\Delta \tau$ are illustrated in Fig. 5.2 of Sieniutycz and Kuran (2005).

The slope ξ is constant in traditional Newtonian fluids with a constant c_v. ξ is positive for fluid's heating (HP mode) and negative for a fluid's cooling (engine mode). The numerical value of Newtonian ξ characterizes a constant logarithmic intensity satisfying the dynamical equation for the temperature logarithm

$$d \ln T = \xi d\tau. \tag{1.119}$$

Its integration for the fixed-end boundary conditions leads to exponential dynamics

$$T(\tau) = T^i \exp[\xi(\tau - \tau^i)]. \tag{1.120}$$

The associated Carnot temperature control ensuring extremum work is

$$T'(\tau) = T(\tau)(1+\xi) = T^i \left(\frac{T^f}{T^i}\right)^{(\tau-\tau^i)/(\tau^f-\tau^i)} \left(1 + \frac{\ln(T^f/T^i)}{\tau^f - \tau^i}\right). \tag{1.121}$$

Exponential decay of T implied by Eq. (1.120) for the engine-mode relaxation (negative ξ) in a strictly Newtonian process, can be compared with the temperature decrease in the radiation relaxation process described by Eq. (1.116). Newtonian ξ can be determined from the boundary conditions of the fixed-end problem; the result is

$$\xi = \frac{\ln(T^f/T^i)}{\tau^f - \tau^i}. \tag{1.122}$$

In the Newtonian process with a constant c_v the density of entropy generated is described by the formula

$$s_{v_{\text{gen}}} = c_v \left[\pm\sqrt{\frac{h}{\Phi c_v}} + (1-\Phi)\left(1 - \pm\sqrt{\frac{h}{\Phi c_v}}\right)\right] \ln\left(\frac{T^f}{T^i}\right). \tag{1.123}$$

For an endoreversible processes ($\Phi = 1$) we recover from the previous equation the special formula

$$s_{v_{\text{gen}}} = \pm c_v \int_{T^i}^{T^f} \sqrt{\frac{h}{c_v}} d\ln T = \pm c_v \sqrt{\frac{h}{c_v}} \ln\left(\frac{T^f}{T^i}\right) \tag{1.124}$$

This equation can be exploited to obtain the generalized exergy of a compressible Newtonian fluid in which viscous friction is ignored. The result is

$$a_v(T, T_0, h) = b_v(T, T_0, 0) + c_p T_0 \left(\frac{\xi}{1+\xi}\right) \ln\left(\frac{T}{T_0}\right) = c_p T_0 \left(\frac{T}{T_0} - 1\right) - \ln\frac{T}{T_0} + \ln\left(\frac{P}{P_0}\right)^{k-1/k}$$
$$+ c_p T_0 \left[\pm\sqrt{\frac{h}{\Phi c_p}} + (1-\Phi)\left(1 - \pm\sqrt{\frac{h}{\Phi c_p}}\right)\right] \ln\left(\frac{T}{T_0}\right) \tag{1.125}$$

The last line describes the rate-related term. In the classical case pressure P contributes to the exergy with a separate term. This is not a surprise since, for usual fluids, T and P are two independent variables, whereas, in the case of radiation the specification of T already defines the pressure P.

Similarly as the classical exergy, the finite rate exergy is a quasi-thermodynamic function because it depends not only on thermodynamic state parameters but also on the reference temperature (temperature of the environment) and the process rate or its duration. The finite rate exergy may also be presented in terms of functions or measures of these quantities, such as: holdup time, number of transfer units, extensive resistance, and even a cumulative entropy production (entropy generation intensity integrated over process time).

The finite rate exergy, Fig. 1.3, prohibits processes from operating below the heat-pump mode line for $\Phi = 0.5$ (lower bound for work supplied) and above the engine mode line for $\Phi = 1.5$ (upper bound for work produced). Weaker limits of classical exergy are represented by the straight line $A = A_{class}$. Dashed lines mark regions of possible improvements when imperfect thermal machines are replaced by those with better performance coefficients. Note a lowest limit for the number of transfer units below of which no power production is possible in the system.

Fig. 1.4 depicts the influence of heat transfer law on the shape of the curve of finite time exergy versus process duration, according to data obtained by Xia, Chen and Sun in L. Chen's research group (Xia et al., 2010c). As shown in Fig. 1.4 three different heat transfer laws yield different data of the finite-rate exergy (Xia et al., 2010c).

Interestingly, scalar fields of finite-rate exergy may be considered at each point of a continuum (D'Isep and Sertorio, 1983).

Landsberg's (1990b) paper is a convincing exposition of the thermodynamic quantum effects that can be quantitatively described by the quantum statistics and applied to radiation. He reviews statistical and thermodynamical problems of energy conversion from radiation treated as a quantum system. He points out the need for nonequilibrium entropy. This entropy is obtained as a statistical quantity for the ideal quantum gases. Maximization of this entropy leads to the equilibrium Bose distribution for photons. By analysing electron transition rates the condition for population inversion is obtained, and the nonequilibrium chemical potential of photons is determined. Fluxes of energy and particle number are defined using continuous spectrum and some applications are given. Various efficiency definitions are discussed for the diluted black body radiation with application to photothermal conversion in solar cells.

Our review for nonequilibrium thermodynamics will be further focused on thermodynamics of transport and rate processes.

1.7 Thermodynamics of Transport and Rate Processes
1.7.1 Introduction

In this section the role of irreversible thermodynamics and kinetic theory is considered in the context of various transport phenomena and rate problems (Sieniutycz and Salamon, 1992a). In these considerations the main role is played by the statistical mechanical approaches to the phenomenological quantification of nonequilibria. The role of the theory for treating highly nonstationary transients is quite essential. The significance of the consistency of chemical kinetics and thermodynamics and the unification of the chemical dynamics with other processes is important. Relative merits of the approaches which involve kinetic aspects in thermodynamics are discussed.

Some topics of nonequilibrium behavior, which require the knowledge of statistical mechanical and stochastic properties, are already described in Sections 1.1 and 1.2 (refer especially to the papers by Lavenda and Lindenberg discussed in Section 1.2). Here we shall outline the standard fundamental approach that applies the kinetic theory. This approach is exceptionally suitable for dealing with transport and rate phenomena. The kinetic substantiation of the nonequilibrium principles has been mostly limited to dilute gases and metals. However, substantial progress has been made by extending Enskog's method (Chapman and Cowling, 1973), Grad's moment method (Grad, 1958; Jou et al., 1988), and other methods of kinetic theory to dense and real gases (Schmidt et al., 1981).

Short discussion of the basic properties of kinetic equations (Section 1.7.2) the theory of transport and rate phenomena (Section 1.7.3). Papers included in this review pertain to the rate processes in both homogeneous and heterogeneous systems, with possible electric effects. They are quite important for applications in chromatography, membrane technology, and catalysis.

1.7.2 The Significance of Kinetic Equations

The typical ingredients of kinetic theories are the distribution functions (functions $f_i(\mathbf{r}, \mathbf{c}_i, t)$ for each component of a multicomponent mixture, $i = 1, 2, \ldots n$) and the governing kinetic equation (equations) describing its time evolution in the phase space. Here $\mathbf{r}$ is the radius vector and $\mathbf{c}_i$ is the molecular velocity of i-th species at time t.

The best-known example of such kinetic equations is the basic integro-differential equation of Boltzmann. Their use is usually preceded by a summary of the macroscopic properties that should result from any physically acceptable kinetic equation. It has to be shown that an H-theorem holds, the macroscopic expression for the entropy source follows, and that the kinetic equations lead to the Onsager reciprocity relations in the linear case. In fact, these ingredients reflect the fact that the irreversibility is already built into the fundamental kinetic equation whether this involves the Boltzmann integro-differential equation or its extensions.

For the standard derivation of the Boltzmann equation based on the assumption of the molecular chaos, the reader is referred to many texbooks on kinetic theory and nonequilibrium statistical mechanics (eg, Chapman and Cowling, 1973; Huang, 1967; Keizer, 1987). For a one-component system, the Boltzmann equation can be written in the form

$$\frac{\partial f}{\partial t} + \mathbf{c}\nabla_r f + \mathbf{F}\nabla_c f = \int \sigma_T g[f'f_1' - ff_1] d\mathbf{c}_1^3 \tag{1.126}$$

In the previous formula, the simplified notation is introduced: $f(\mathbf{r}, \mathbf{c}', t) = f'$, $f(\mathbf{r}, \mathbf{c}_1', t) = f'_1$, σ_T for a linear operator corresponding to the differential scattering cross section $\sigma(\Omega, g)$, where Ω is the solid angle and g the constant magnitude of a relative velocity. Gradient operators involve derivatives with respect to position and velocity (Keizer, 1987). Since the collision term of Eq. (1.126) tends toward the thermal equilibrium, the relaxation approximation is occasionally applied.

$$\frac{\partial f}{\partial t} + \mathbf{c}\nabla_r f + \mathbf{F}\nabla_c f = -\frac{f - f^{eq}}{\tau} \tag{1.127}$$

This approximation, which contains the relaxation time τ, leads frequently to quite good estimates of the transport coefficients (Warner, 1966).

When investigating the nonequilibrium phenomena, one must predict the time evolution of the distribution function by solving the kinetic equation with given initial conditions. This is a mathematically involved problem (Enskog, 1917; Chapman and Cowling, 1973; Cercignani, 1975). Yet, some rigorous properties of any solution to the Boltzmann equation follow from the fact that in each molecular collision there are some conserved dynamical quantities. Hence the macroscopic conservation laws follow for mass, energy, and momentum (Huang, 1967; Kreuzer, 1981). Summing up, the conservation laws appear in the kinetic theory as the result of collisional invariants for mass, momentum and kinetic energy of molecules. When they are combined with the equations describing the irreversible fluxes of mass, heat and momentum, the so-called equations of change are obtained which describe the hydrodynamic fields of velocity, temperature, and concentrations. The formalism is very useful when describing a variety of reacting and nonreacting systems on the hydrodynamic level (Hirschfelder et al., 1954; Bird et al., 1960).

The kinetic theory leads to definitions of the temperature, pressure (scalar and tensor), internal energy, heat flux density, diffusion fluxes, entropy flux, and entropy source in terms of definite integrals of the distribution function with respect to the molecular velocities. The inequality $\sigma > 0$, which expresses the fact that the entropy source strength must be positive in any irreversible process, constitutes, within the framework of the kinetic theory, a derivation of the second law. This is known as Boltzmann's H-theorem. One of the most important benefits of kinetic theories is that they provide us with expressions for the transport coefficients.

In the limit of an infinite time and provided the boundary conditions are compatible with equilibrium, the entropy source vanishes identically, detailed balance follows, and logarithms of the distribution functions become summational invariants. They must therefore be equal to a linear combination of the microscopic quantities m_i, $m_i c_i$ and $(1/2)m_i c_i^2$. This observation allows one to recover the Maxwell equilibrium distribution, the definition of hydrostatic pressure, and the ideal gas law. It also shows that the equilibrium value of the kinetic entropy density coincides with the thermodynamic one. It follows that the diffusion fluxes and the nondiagonal elements of the pressure tensor be nonzero only if the entropy source σ does not vanish. In particular, we get that the diffusional part of the entropy flux vanishes at equilibrium.

Importantly, in disequilibrium, the classical phenomenological form of expressions for the irreversible entropy flux and entropy source (involving the products of fluxes and forces) can be obtained from the approximate solution of the Boltzmann kinetic equation. This corresponds to the linear phenomenological laws, and can be shown by the Enskog iterational method (Enskog, 1917; de Groot and Mazur, 1984). More adequately, if one evaluates fluxes in the hydrodynamic equations by retaining in the distribution functions f_i only the first order perturbations $\Phi^{(1)}_i$, then it follows that the diffusion fluxes $\mathbf{j}_i$, the heat flux $\mathbf{q}$ and the off-diagonal elements of the pressure tensor $\mathbf{P}$ are linear functions of the gradients of the macroscopic functions μ_i, T and $\mathbf{u}$.

Onsager's symmetry results from the kinetic theory in a straightforward manner under the essential condition of microscopic reversibility (de Groot and Mazur, 1984). Indeed, the first approximation of Enskog corresponds to the linear laws of the phenomenological theory. On the other hand, if terms quadratic in $\Phi^{(1)}_i$ or linear in $\Phi^{(2)}_i$ are retained in the entropy density expansion $\rho_s = \rho_s^0 + \rho_s^1 + \rho_s^2$ .... this entropy will be a function of the macroscopic gradients in the Enskog formalism (Enskog, 1917). Consequently, the nonlinear terms in ρ_s correspond to nonequilibrium entropies.

The Enskog method only enables one to calculate the deviations from the local Maxwellian distribution which are related to the spatial nonuniformities in the system. This is not the most general solution of the Boltzmann equation. Higher-order hydrodynamic theories (Burnett, 1935) and/or analyzes of short-time effects require more detailed approaches. It can be shown that the homogeneous (spatially uniform) perturbations have a relaxation time, τ, of the order of a few collision times. Only for times larger than τ can the true solution of the Boltzmann equation be approximated by the Enskog solution.

In Grad's formalism the quadratic correction terms contain the diffusive fluxes rather than gradients; this is exploited now in extended thermodynamic theories involving nonequilibrium entropies which contain such fluxes. The relaxation times can be evaluated in Grad's moment approach (Grad, 1958). He introduces velocity tensors of order n, v_{ijk}^n ...., and the corresponding moments given by integrals of the products of v_{ijk}^n ... and fd**v**. The Boltzmann equation is multiplied successively by each of the v_{ijk}^n ... and integrated over the velocities. Successive equations for the moments are found. The first two moment equations and the contraction of the third are the usual equations of change describing the density mass-average velocity and the temperature. For these equations, the collision terms vanish due to properties of the summational invariants. In general, the equation of each moment involves the moment of higher order explicitly as well as the entire distribution function. Such a set of moment equations is equivalent to the original Boltzmann equation. The method of solution is based on the trial distribution functions; the Maxwellian distribution multiplied by a finite series of multidimensional Hermite polynomials with a number of arbitrary parameters which may then be written in terms of the first scalar moments (Grad, 1958). This method justifies the first approximation of Enskog and permits any level of description ranging from the hydrodynamic description to a complete description in terms of f_i (**r**, **c**$_i$, t). Grad's treatment is efficient for the study of gases at densities that are sufficiently low that the gas ceases to behave as a continuous medium.

The discussion of kinetic theory has thus far been limited to nonreacting mixtures of simple molecules. It has been assumed that the molecules are spherical and have no internal degrees of freedom. A nice analysis of systems with internal degrees of freedom (polyatomic gases) is available (Resibois and de Leener, 1977). It may be shown that the integral term of the Boltzmann equation represents the net number of molecules of a particular kind gained by a "group" of molecules in the phase space because of collision processes (Hirschfelder et al., 1954).

Consequently, if the collisions result in a chemical reaction, the right-hand side of the Boltzmann equation has to be changed. In general, a modification of the Boltzmann equation affects the integral properties used in obtaining the equations of change. Modifications of the kinetic approach for chemically reacting systems and extensions to monoatomic gases are not discussed here. See, for example, Hirschfelder et al. (1954) where these problems are analyzed extensively in the context of quantum theory, reaction paths, saddle points and chemical kinetics. However, mass, momentum and energy are conserved even in a collision which produces a chemical reaction.

Karlin et al. (2003) present details of the recently introduced systematic approach to derivations of the macroscopic dynamics from the underlying microscopic equations of motions in the short-memory approximation. The essence of this method is a consistent implementation of Ehrenfest's idea of coarse-graining, realized via a matched expansion of both the microscopic and the macroscopic motions. Applications of this method to a derivation of the nonlinear Vlasov–Fokker–Planck equation, diffusion equation and hydrodynamic equations of the fluid with a long-range mean field interaction are presented in full detail. The advantage of the method is illustrated by the computation of the post–Navier–Stokes approximation of the hydrodynamics which is shown to be stable, unlike the Burnett hydrodynamics.

Finally, it is worth pointing out Altenberger and Dahler's (1992) approach in which they have proposed the statistical mechanical theory of diffusion and heat conduction based on an entropy source-free method. The approach allows one to answer unequivocally questions concerning the symmetry of the kinetic coefficients. It rests on nonequilibrium statistical mechanics and it involves a linear functional generalization of Onsagerian thermodynamics. It extends some earlier attempts (Schofield, 1966; Shter, 1973) to generalize fluctuation dynamics and Onsager's relations using convolution integrals to link thermodynamic fluxes and forces. By using the Mori operator identity (Mori, 1965), applied to the initial values of the currents of mass and heat, constitutive equations linking the thermodynamic fluxes and forces are obtained. Correspondence of the results of the functional theory with the classical one is verified by the use of Fourier representations for the fields and for the functional derivatives. This approach provides a consistent description of transport processes with temporal and spatial dimensions and yet enables one to recover the classical results in the long-wave limit, where the gradients in the system are small. The possibility of extending this linear theory to situations far from equilibrium is postulated. The extension is based on the probability densities of the phase-space variables determined for stationary nonequilibrium reference states. The roles of external fields and various reference frames are investigated, with the conclusion that the laboratory frame description is as a rule the most suitable for the theory. Transformations of the thermodynamic forces are thoroughly analyzed, focusing on violations of the Onsagerian symmetries.

1.7.3 Transport and Rate Phenomena

Important understanding of irreversibility can be achieved by embedding the kinetic theory relationships into the maximum entropy formalism. Maximization of entropy has been employed for obtaining kinetic equations for both dilute and dense fluids. These approaches constitute significant progress; they should be distinguished from the previous ones where the entropy was maximized under constraints imposed on quantities at the hydrodynamic level, for example, the macroscopic energy (Lewis, 1967). It was shown relatively recently that one must go beyond hydrodynamic quantities in order to obtain expressions for the transport coefficients (Benin, 1970; Karkheck, 1986, 1989).

Karkheck (1986, 1989) develops the Boltzmann equation (Cercignani, 1975) and other kinetic equations in the context of the maximum entropy principle and the laws of hydrodynamics. He uses a general approach based on the maximum entropy principle, and provides a specific example with a constraint imposed on the one particle distribution function. Hydrodynamics emerges from the observation that a vast amount of detailed microscopic information is macroscopically irrelevant. His interpretation of minimum information (maximum entropy) is not the conventional interpretation of information which is missing but rather information which is irrelevant to a chosen level of observing the behavior of a system. His approach yields generalizations of Boltzmann and Enskog theories. The generalizations are applicable to real fluids and dense gases. An interparticle potential is used which can be adapted to real fluids by using perturbation theory of equilibrium statistical mechanics (Mansoori and Canfield, 1969). A sequence of kinetic equations is generated in the maximum entropy approach by imposing different constraints to yield different forms of closure. Examples of closed kinetic theories for simple liquids are given, and a correspondence with Boltzmann and Vlasov ensembles is shown when the interparticle potential is neglected. Then, from the moment approach, the conservation laws of mass, momentum and energy follow, with very general flux expressions. By expansion of the fluxes in powers of the gradients of density, velocity and temperature, the transport coefficients are obtained. Existence of two distinct temperatures out of equilibrium is shown. To include extra degrees of freedom modified Chapman-Enskog procedure (Enskog, 1917; Chapman and Cowling, 1973) is proposed for solving the kinetic equations. A comparison with molecular dynamics is presented; the agreement is good. Karkheck's (1986, 1989) papers should be studied by everyone who is interested in the origin and nature of irreversibility.

Memory functions for transport coefficients are important physical quantities (Perez-Garcia and Jou, 1986, 1992). Transport phenomena in multicomponent mixtures can be investigated via a statistical mechanical method for the constitutive relations that does not require the introduction of an entropy source and that allows one to establish with certainty the molecular definitions and symmetry properties of the Onsager coefficients (Altenberger and Dahler, 1992). These results indicate that when the constraints involve the one-particle distribution function as well as the potential energy density then the transport coefficients appear naturally as the outcome of the formalism. Consequently, both conservation laws and transport coefficient expressions can be obtained via the generalized maximum entropy approach. Such an approach is analyzed in detail in Karkheck's research (Karkheck, 1992). A knowledge of the numerical values of the transport coefficients (Waldmann, 1958) as well as their qualitative behavior as the thermodynamic state changes is crucial for understanding many macroscopic effects in continua. These include nonlinear effects and the dynamics of fluctuations (Ross, 2008). Standard approaches based on kinetic equations are the most popular. However, entropy-free and/or entropy source-free approaches have also been proposed (Schofield, 1966).

A thermomechanical theory, based on the isoenergy density (Sih, 1985, 1992), has been proposed to describe irreversible continua. The author promotes a new approach to nonequilibrium thermomechanics based on the so-called isoenergy density theory. He discusses a variety of examples in terms of the predictions of the theory which emphasizes the exchange of volume and surface energy. Nonequilibrium behavior is characterized by the rate at which a process can no longer be described in terms of physical parameters representing the system as a whole. Thermal changes are synchronized with the motion of mass elements. The isoenergy density theory is employed for characterizing irreversible behavior in gases, liquids, solids, and phase transformations. It can take into account both laminar and turbulent flows as well as the transition between them. The conditions for a fluid to separate from an immersed object are predicted along with the position of the separation. Nonuniform condensation of a gas in a confined region, fluid-solid separation, and heating of an uniaxial specimen are other examples analyzed. A critical analysis is presented concerning the application of concepts and notions established from equilibrium theories to nonequilibrium phenomena.

Thermodynamic and kinetic principles can be used to predict numerical values of heat-mass transport coefficients in mixtures and to design experiments measuring such values (Rowley and Gruber, 1978). This approach is never trivial but it is important for the design of many engineering devices, for example, heat and mass exchangers. It has the advantage of providing both micro and macroscopic information, and allows comparison with results from molecular dynamics. A review of the approach is available (Rowley, 1992; Rowley and Gruber, 1978). In this context Onsager's reciprocity has been experimentally verified (Miller, 1974; Rowley and Horne, 1978). Rowley's (1992) paper reviews an approach using thermodynamic principles to predict values of heat–mass transport coefficients in mixtures and to design experiments measuring such values. The approach is a combined experimental and theoretical examination of diffusion, thermal conductivity and heat of transport in nonelectrolyte liquid mixtures. Mutual diffusion coefficients are measured in a variety of liquids. In an effort to develop a predictive technique for such diffusion coefficients, the mixtures are modeled as Lenard-Jones fluids and the values of the mutual diffusion coefficients are obtained from molecular dynamics simulations. Onsager coefficients, heats of transport, and diffusion coefficients are found from experiments near liquid-liquid critical point, and Onsager reciprocity is verified. The thermal effect of diffusion is measured and developed into an accurate experimental technique for measurements of heat of transport. A new predictive method free of adjustable parameters is developed for the thermal conductivity. Conclusions of a methological nature predict the lasting role for molecular dynamic simulations.

With the help of the moment theories of Grad's type, the relaxation times have been evaluated, so that the highly nonstationary conditions may be analyzed. It has been shown that the kinetic theory approach leads to relaxation terms in the phenomenological equations implying damped-wave diffusion rather than the standard parabolic behavior (Jou et al., 1988). Furthermore, it has turned out that, in an exact sense, the natural variable of the extended Gibb's equation is neither heat flux nor entropy flux but a momentum variable called thermal momentum (Sieniutycz and Berry, 1989). The formulation which involves the thermal momentum allows one to define the nonequilibrium temperatures and pressures unambiguously and to develop the theory of nonequilibrium thermodynamic potentials in the field representation of thermodynamics (Sieniutycz and Berry, 1991). In addition, in a macroscopic approach, certain variational principles of Hamilton's type for transient heat transfer and mass diffusion have been formulated and their approximate solutions were found by direct variational methods (Vujanovic and Jones, 1989; Sieniutycz, 1979, 1983, and chapter: Variational Approaches to Nonequilibrium Thermodynamics). Morris and Rondoni (1999) combined the definition of temperature for a Hamiltonian dynamical system with the Hamiltonian representation of a nonequilibrium isokinetic steady state to obtain an expression for the temperature away from equilibrium. These authors report results of numerical simulations, performed to assess the validity of this approach for color field systems. They observe a strong correlation between the kinetic temperature orthogonal to the color current and the ratio of the averages of two given phase variables.

It has been shown that the competition of dissipation and damping with restoring forces leads to oscillations and limit cycles (Nicolis and Prigogine, 1977; Gray and Scott, 1990). Oscillatory chemical reactions have been analyzed in a collection of treatments with sequentially enhanced degrees of unification (Gray and Scott, 1990; Kawczyński, 1990; Kawczyński et al., 1992).

Interfacial instabilities and dissipative thermohydrodynamic oscillators have systematically been treated (Velarde and Chu, 1990, 1992). Their papers present a unifying picture of various thermohydrodynamic oscillatory instabilities in both Newtonian and viscoelastic fluids. They show that the competition of dissipation and damping with restoring forces can lead to oscillations and limit cycles. They explain how nonlinearity helps sustain these oscillations in the form of limit cycles. In particular, Velarde and Chu (1992) review the oscillatory Rayleigh-Benard convection driven by buoyancy forces. Then, interfacial oscillators are investigated. Starting with the perturbation equations, various effects are analyzed, such as Marangoni instabilities, the onset of oscillatory motion, transverse gravity-capillary oscillations, neutral stability conditions, longitudinal waves, interfacial electrohydrodynamics, and so forth. Solitons excited by Marangoni effect are introduced. It is shown how the Korteweg-de Vries soliton description can be used as a building block for some nonlinear traveling wave phenomena. The role played by dissipation in triggering and eventually sustaining nonlinear motions is shown. This acquaints the reader with the beauty of the methodological unification of various nonlinear phenomena in hydrodynamics. Marangoni instabilities are also analyzed in an applicative contex during absorption accompanied by chemical reaction (Warmuzinski et al., 1995).

In the past four decades many articles have appeared treating non-Fickian diffusion and, in particular, the transport of mass or energy described in terms of (damped) wave equations rather than classical parabolic equations. The topics pursued in the literature range from theoretical developments (based usually on the phenomenological non-Onsagerian thermodynamics or nonequilibrium statistical mechanics), to a growing number of applications in fluid mechanics, polymer science, environmental engineering and chemical engineering.

It has been recognized that non-Fickian theories of diffusive mass transport often provide a more realistic description of transfer phenomena than classical theories. In particular, such descriptions avoid the paradox of infinite propagation speed by incorporating inertial effects resulting from the finite mass of the diffusing particles. In fact, for concentration disturbances, the paradox can be removed in a classical way (Baranowski, 1974; Fitts, 1962; de Groot and Mazur, 1984). For thermal inertia, however, the situation is more subtle, and calls for the Boltzmann equation or certain extended quantum approaches (Kaliski, 1965; Sieniutycz, 1990, 1992b, 1994; Kaniadakis, 2002). The nonequilibrium energy obtained from the relaxation time approximation of the Boltzmann equation has lead to an extended Gibbs equation involving irreversible fluxes. The corresponding thermodynamic field theory is based on nonequilibrium field potentials which include the diffusion fluxes and hydrodynamic velocity as state variables (Sieniutycz and Berry, 1989, 1991).

The nonclassical effects are particularly strong in diffusion of aerosols containing heavy particles and in polymers characterized by large molecular weight. Dispersed systems, common in chemical engineering, exhibit enhanced nonclassical effects of inertial type which have been experimentally confirmed in drying processes (Luikov, 1966a, b). Other experiments show that the non-Fickian models of diffusion of mass and energy are more appropriate in the case of highly unsteady-state transients, occurring, for example, when laser pulses of high frequency are acting on the surfaces of solids (Domański, 1978a, b) or in the case of high-frequency acoustic dispersion (Carrasi and Morro, 1972, 1973). Diffusion in polymers shows, as a rule, severe nonclassical effects related to the complex structure of polymeric media (Bird et al., 1977; Sluckin, 1981; Carreau et al., 1986; Drouot and Maugin, 1983; Drouot et al., 1987; Lebon et al., 1990; Baranowski, 1992). The phenomenological thermodynamics of diffusion links these nonclassical effects with the absence of local equilibrium in the system. An extension of the chemical potential to stress fields and nonlocal diffusion effects were experimentally verified (Baranowski, 1992). Fractional diffusion has been linked with the irreversibility and entropy production (Compte and Jou, 1996; Hoffmann et al., 1998; Davison and Essex, 2001; Li et al., 2003). Anomalous diffusion on fractal objects was analytically described (Saied, 2000). Tsallis's (1988) general approach to nonextensive statistical mechanics and thermodynamics has been worked out and applied, in particular, to anomalous diffusion problems (Tsallis, 1988; Salinas and Tsallis, 1999; Casas et al., 2002). The fractional Boltzmann transport equation has been proposed as a tool to describe fractional diffusion and other fractional transport phenomena (El-Nabulsi, 2011). Tsallis (1999a, b) discusses the domain of validity of standard thermodynamics and Boltzmann-Gibbs statistical mechanics, and then formally enlarges this domain in order to cover a variety of anomalous systems. The generalization involves nonextensive thermodynamic systems. Casas et al. (2002) include thermodynamics into the framework of the Tsallis variational problem.

The growing field of research concerning nonextensive thermodynamics has a hard task to do. In order to have nonextensivity it could be necessary to release the well-known Boltzmann–Gibbs–Shannon (BGS) entropy and, perhaps, to substitute it by another entropic form. Therefore as stated by Curado (1999), we are obliged to relax some of Khinchin's (1957) axioms, because they lead only to the BGS entropy. Relaxing the fourth axiom, related to additivity, the possible entropy of a nonextensive system must obey only the first three Khinchin axioms, but these axioms are so weak that an enormous amount of entropic forms are allowed. With the proposition by Tsallis (1999) of a different entropic form that has been shown to satisfy the whole thermodynamic formalism and to have very interesting properties, the study of alternative forms of entropy in the study of nonextensive physical problems has begun. These alternative forms have been used principally in problems where the standard statistical mechanics fails, as for example in some problems with long-range interactions, long-term memory, or some kind of intrinsic fractality. This relatively new field of research on nonextensive generalizations of standard Boltzmann-Gibbs statistical mechanics is very promising but has been centered on applications of Tsallis statistics (Salinas and Tsallis, 1999) and not many papers have been dedicated to the general aspects of the nonextensive statistical mechanics. Curado (1999) paper discusses these general aspects starting with the Khinchin (1957) axioms, that provide some of the conceptual tools of information theory. He shows some new results concerning general aspects of the thermodynamical formalism and exhibits some general properties of entropic forms based, essentially, on the symmetry of permutation of probabilities that an entropic form must satisfy. This symmetry indicates the simplest way to write general entropies and the requirements of simplicity, and explicitly allow us to derive some basic entropic forms that satisfy the whole thermodynamical formalism.

A Lagrangian formulation of chemical dynamics preserving the mass action kinetics has been developed for reaction-diffusion systems (Shiner, 1984, 1987, 1992). It represents a unification of the dynamics of chemical systems with other processes (eg, mechanical) and simplifies the treatment of complex couplings. The unification is based on a concept of a nonlinear chemical resistance (Shiner, 1992; Sieniutycz, 1987). New symmetries in chemical systems have been discovered corresponding to the nonlinear properties of these systems (Shiner, 1987; Lengyel and Gyarmati, 1986). Shiner's (1992) formulation of chemical dynamics is based on the idea of a nonlinear chemical resistance and a corresponding Rayleigh dissipation function (Rayleigh, 1945). These quantities can be applied to chemical processes in a fashion which is formally similar to the use of the well-known resistances and dissipation functions of physical processes. The author's approach yields a unification of the dynamics of chemical systems with other sorts of processes (eg, mechanical) in the framework of Lagrangian dynamics and yet simplifies the treatment of complex couplings.

A major virtue of the work is a physical Lagrangian based on the Gibbs energy. The author's formulation stresses a natural algebraic symmetry in the chemical reaction system at stationary states far from equilibrium. This symmetry can be viewed as a more general case of Onsager's reciprocity, and can explain the complex nature of the symmetry properties in chemical systems. The formalism leads naturally to the network representation of chemical systems; this allows a treatment using standard methods of network thermodynamics. When the inertial effects are negligible, the Lagrangian dynamics can be set in terms of a variational formulation which simplifies near equilibrium to the theory of minimum power dissipation.

The thermodynamics of the reacting surfaces has been developed and basic relations on surfaces have been formulated (Bedeaux, 1986, 1992). Thermodynamically permitted reaction paths in open catalytic systems have also been studied. Gorban (1984) and Bykov et al. (1990) have worked out a qualitative method for the thermodynamic analysis of the equations of chemical kinetics. An estimate of the thermodynamic function called the convexity margin has been introduced for the ideal chemical system. An algorithm has been proposed for defining regions of accessible composition based on a given reaction mechanism, location of equilibrium and the original composition. Allowable reaction paths have been studied and open catalytic systems have been investigated. The method suggests a new procedure for planning unsteady-state experiments. Also, it can solve the problem of boundary equilibria, and evaluate the region of admissibility for linear estimation. The relation between the times of relaxation to equilibrium and equilibrium flows can be found.

New procedures for planning steady and unsteady-state kinetic experiments have been proposed. An analysis of the reacting systems displaced from equilibrium has been made showing that, in general, total dissipation is not minimized at steady states. Namely, Månsson and Lindgren (1990) have described the role of the transformed information for creation processes by showing how the destructive processes of entropy production are related to the creative processes of formation. The role of information for the theory of thermodynamic equilibrium and relative (Kullback, 1959) information for the definition of the exergy has been reviewed. Superiority of information theoretic approach over traditional thermodynamics has been indicated when analyzing systems with emergent spatial structures, for example, spatially self-organizing chemical systems. Using examples of oscillatory reactions, the authors have shown that the average entropy production in an oscillating system is not necessarily lower than the entropy production of an (unstable) steady state. They conclude that a general principle of minimum entropy production cannot exist for oscillatory systems.

By integrating the kinetic mass action law into nonequilibrium thermodynamics, a consistency of chemical kinetics with thermodynamics has been demonstrated in the framework of the governing principle of dissipative processes (Lengyel, 1988, 1992; Lengyel and Gyarmati, 1986). However, as shown in these papers, this takes place at the expense of rejecting the standard affinity. We focus here on two Lengyel's analyses which contain the derivation of the Guldberg–Waage mass action law from Gyarmati's governing principle of dissipative processes and investigate the consistency between chemical kinetics and thermodynamics (Lengyel, 1988, 1992). Lengyel's approach involves expressions for the entropy production as a bilinear form in the stoichiometrically dependent and independent elementary chemical reactions. For independent reaction systems, generalized reciprocity relations are satisfied (the so-called Gyarmati–Li relations; Li, 1958). Dissipation potentials are introduced and the kinetic law of mass action is obtained from Gyarmati's principle. This result is of certain importance since chemical reaction kinetics is frequently considered as a phenomenon which only reluctantly obeys a standard thermodynamic formalism. However, the traditional concept of chemical affinity has to be rejected; rather the partial Gibbs free energies of the reactants should be used. The corresponding flux potential is a function of both the partial Gibbs free energy of the reactants and of the products. It is shown that the theory can solve some difficult problems of chemical kinetics which were unsolved by other methods. This approach offers several surprises, but its consistency is firmly established by the examples.

A synthetic review of chemical stability properties has been presented by Farkas and Noszticzius (1992) for the three basic groups of the systems showing explosive, conservative or damped behavior (depending on parameters of a kinetic model). Their unified treatment of chemical oscillators considers these oscillators as thermodynamic systems far from equilibrium. The researchers start with the theory of the Belousov–Zhaboutinski (BZ) reaction pointing out the role of positive and negative feedback (Zhabotinsky, 1974). Then, they show that consecutive autocatalytic reaction systems of the Lotka–Volterra type, Fig. 1.5, play an essential role in the BZ reaction and other oscillatory reactions. Consequently, they introduce and then analyze certain (two-dimensional) generalized Lotka–Volterra models. They show that these models can be conservative, dissipative, or explosive, depending of the value of a parameter. The transition from dissipative to explosive behavior occurs via a critical Hopf bifurcation. The system is conservative at certain critical values of the parameter. The stability properties can be examined by testing the sign of a Liapounov function selected as an integral of the conservative system. By adding a limiting reaction to an explosive network, a limit cycle oscillator surrounding an unstable singular point is obtained. Some general thermodynamic aspects are discussed pointing out the distinct role of oscillatory dynamics as an inherently far-from-equilibrium phenomenon. The existence of the attractors is recognized as a form of the second law. The paper gives a pedagogical perspective on the broad class of problems involving the time order in homogeneous chemical systems.

A rigorous nonequilibrium thermodynamics of surfaces has been formulated (Bedeaux, 1986, 1992). He considers a boundary layer between two phases, which may change its position and curvature. The analysis requires time dependent curvilinear coordinates. At the surface the field properties, densities, fluxes, and sources exhibit discontinuities which are quantitatively described by appropriate excess functions. The

$$\frac{dX_2}{dX_1} = \frac{X_2(a_2 X_1 - b_2)}{X_1(a_1 - b_1 X_2)}$$

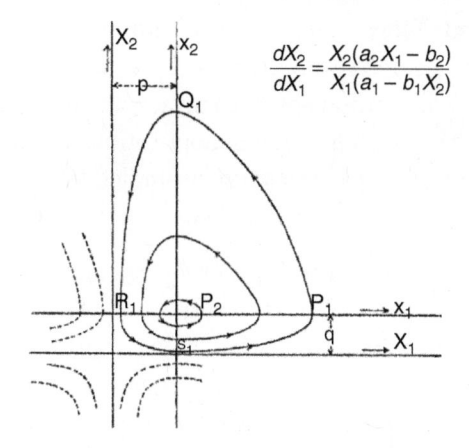

Figure 1.5: Lotka–Volterra model is a classical scheme describing interactions between populations of predators and preys.

The time variability of the populations of predators (foxes) and preys (rabbits) provides the ecological interpretation of the model. The Lotka diagram shows the general character of integral curves of his equation which describes ratio dX_2/dX_1 in terms of X_1 and X_2. In the positive quadrant of $X_1 X_2$ the curves are closed, contained entirely within the quadrant, and intersecting the axes of $x_1 x_2$ orthogonally. Near origin curves are very nearly elliptical (Lotka, 1920).

excesses of the densities and currents are described as singular interfacial contributions to the distributions of the densities and currents. After formulating detailed balances of mass, momentum and energy, the interface contributions are singled out from global equations. Finally, the entropy production for interfaces, σ_s, is found and the phenomenological equations postulated. Onsager relations are formulated on surfaces. The general expression for σ_s involves a variety of fluxes and forces and the author presents reasonable simplifications. Applications to membrane transport, evaporation, surfactant physics and chemistry are discussed. The virtue of the relatively simple analysis achieved is a unified approach to bulk and interface, that is, to continuous and discontinuous properties.

A matrix method for calculating surface reactions has been developed allowing effective calculations on lattice gases (Myshlyavtsev and Yablonski, 1992). The authors offer the reader extensive studies in the thermodynamics and kinetics of surface reactions. Their method is called "method of transfer matrix." It allows one to evaluate the thermostatistical and kinetic properties of surfaces. These properties are essential in the theories of catalysis, grain boundaries, and adsorption. They are also of value to a general understanding of the thermodynamics of surfaces. The transfer matrix method turns out to be the most effective for calculating with models of lattice gases. It allows one to determine the statistical thermodynamic properties from the grand partition function fitting a one-site, which is treated as a stage along a chain. Every site along the chain is found in one of m possible states, where m is any non-negative integer. It follows that the largest eigenvalue of the transfer matrix is equal to the grand partition function of a one-site along the chain. The probability that the site is in a definite state is found. Kinetic rate constants and coefficients of surface diffusion are determined. The authors give sample applications of the method in variety of involved cases: monomolecular adsorption, desorption spectra, surface diffusion coefficients, and desorption from a reconstructed surface. They also predict applications to more complicated lattices, with several kinds of centers. This paper should be studied by everyone who wants to apply thermodynamics to practical problems of surfaces.

A simple, comprehensive treatment of transport processes in electrolytes and membranes have been given by Forland et al. (1988) and Ratkje et al. (1992). The authors present a new, simple, approach to the thermodynamics of irreversible transport processes in electrolytes and membranes. Their description is lumped (ie, discontinuous) and preserves many features of the classical Onsagerian formalism as, for example, local equilibrium, bilinearity of the entropy production, reciprocity relations, and so forth. On the other hand, the classical chemical affinity is not applied in the formalism (see also Lengyel's papers giving up the affinity in a similar context; Lengyel, 1988, 1992; Lengyel and Gyarmati, 1986). In such systems scalar-vector couplings can take place. Only vectorial fluxes and forces (gradients of the chemical potentials) are used. The treatment yields valuable insight concerning some important processes, as, for example, active transport, or muscle contraction. Many examples are discussed including glass electrodes and cells with liquid junctions, cation transport in water in a cation exchange membrane, and transport-reaction interaction in a battery and in muscle contraction. Extensions for nonisothermal systems are reviewed. This minimum-mathematics approach provides an introduction to the irreversible problems of physiology, biophysics and physical chemistry of electrolytes.

Analysis of thermoelectric coefficients in metal couples have been presented with a new insight into the contact potential (Guy, 1992). He analyzes thermoelectric effects in metal couples. His analysis points out the necessity to distinguish the electrochemical potential of the electrons from the electrochemical potential itself. On this basis, he discusses in sequence: thermoelectric effects in a single metal, equilibrium in an isothermal couple, and thermoelectric effects in a metal–metal couple. His analysis links the Seebeck coefficient with the electron enthalpy in an unconventional way, describes quantitatively equilibrium conditions in metal couples, and specifies some thermoelectric characteristics. His conclusion is that thermoelectric phenomena in thermocouples are determined by the thermoelectric behavior of single isolated metals. The contact potential has no influence on the thermoelectric behavior of metal–metal couples and hence the Seebeck and Peltier coefficients for a couple are postulated to depend in a simple way on the corresponding coefficients of single metals.

Baranowski's (1992) paper explores the influence of stress fields on diffusion in isotropic elastic media. The thermodynamics of solids with stress fields is developed. An extension of the chemical potential to stress fields is given under (experimentally proved) assumption neglecting the role of off-diagonal elements in the stress tensor. Attention is directed to stresses developed by a diffusing component, causing displacements, and resulting stresses, in the solid. Plastic deformations are neglected, that is, an ideal elastic medium is assumed. Diffusion equations are derived in a one-dimensional case. It is shown that the resulting equation of change for concentration is a nonlinear integro-differential equation. Nonlocal diffusion effects are verified by an original experiment. Experimental results of hydrogen diffusion in a PdPt alloy are compared with the theory. The general importance of the stress field to diffusion in solids is outlined. Effect of stresses on the entropy production is discussed with the conclusion that a stress-free steady-state obeys the minimum entropy production principle as a stationarity criterion. The reader may consult Wiśniewski's (1984) analysis showing the role of diffusion stresses rather than mechanical stresses in a general multicomponent system.

Applications of thermodynamics to the ionized gas (plasma) systems have been summarized in the context of plasma formation, thermionic diodes and MHD generators (Wiśniewski, 1992). This is a nice, tutorial-style, presentation of nonequilibrium problems in plasma systems in which the author introduces the reader to the thermodynamics of solutions of ionized gases. When a large fraction of the species present is ionized, such solutions are called plasmas. Plasma formation is analyzed from the viewpoint of terms like: ionization potential, conditions of ionized gas equilibrium, Saha equation, nonideal plasmas, equilibrium composition, and degree of ionization. Thermionic generators, that is, devices which convert internal energy into electric energy, are analyzed with the thermionic diode serving as the example. Energy balance is given, entropy source is obtained, and the phenomenological equations are postulated for the flows of energy and electric current. The thermal efficiency of thermionic energy conversion treated as a heat engine is determined. Equations of plasma in an electrodynamic field are given along with Onsager–Casimir relations. The Hall effect in a plasma is quantitatively described. Also, the main types of magnetohydrodynamic generators are described by taking the Hall effect into account.

A synthesis of the statistical thermodynamics of Onsager and the kinetic molecular theory of Boltzmann has been achieved (Keizer, 1987). This theory provides a mechanistic foundation for thermodynamics. It is based on the idea of elementary molecular processes (indexed by i)

$$(n_{i1}^+, n_{i2}^+, \ldots\ldots) \leftrightarrow (n_{i1}^-, n_{i2}^-, \ldots\ldots) \tag{1.128}$$

with the extensive thermodynamic variables represented by the vector $\mathbf{n} = (n_1, n_2, \ldots, n_k)$. The superscripts + and – refer to the forward and backward steps in generalized reactions. It can be shown that these processes provide a natural description of bimolecular collision dynamics and chemical kinetics. The transition rate of an elementary process, that is the number of times per second that it occurs in the forward or reverse direction, is given by the canonical form

$$V_i^+ = \Omega_i^+ \exp\left(-\sum_l Y_l^+ n_{il} k_B^{-1}\right) \tag{1.129}$$

and likewise

$$V_i^- = \Omega_i^- \exp\left(-\sum_l Y_l^- n_{il} k_B^{-1}\right) \tag{1.130}$$

The constants Ω_i^{+-} are the intrinsic rates of the forward and reverse steps of the elementary process (i) and are the basic transport coefficients in the canonical theory. Y_l are intensive variables which are functions of the extensive variables; they are the derivatives of a local equilibrium entropy $S(\mathbf{n})$. Using the canonical representation for the rates of the elementary processes, one can develop the statistical thermodynamics of molecular processes in a form applicable to systems close or far from equilibrium. Oláh's thermodynamic theory of transport and rate processes, called thermokinetics, is also in this spirit and has many similar features (Oláh, 1997a, b, 1998). Both transport and rate processes can be treated in a unified way; an extension of the idea already pursued in Eyring's works (Eyring, 1962, Eyring and Jhon, 1969) imbed in rather different formalisms. The same basic approach may be used for molecular collisions, hydrodynamics and chemical or electrochemical processes. Nonlinear molecular mechanism gives rise to critical points, instabilities, bifurcations, chaotic behavior, Fig. 1.6, and oscillations, Fig. 1.5. It leads, in general, to various forms of organization (Ebeling, 1985; Ebeling and Klimontovich, 1984; Ebeling and Feistel, 1992; Ebeling et al., 1986; Rymarz, 1999; Nicolis and Prigogine, 1977; Turner, 1979; Zainetdinov, 1999). More information about these phenomena can be found in chapters: Thermodynamic Lapunov Functions and Stability and Thermodynamic Controls in Chemical Reactors.

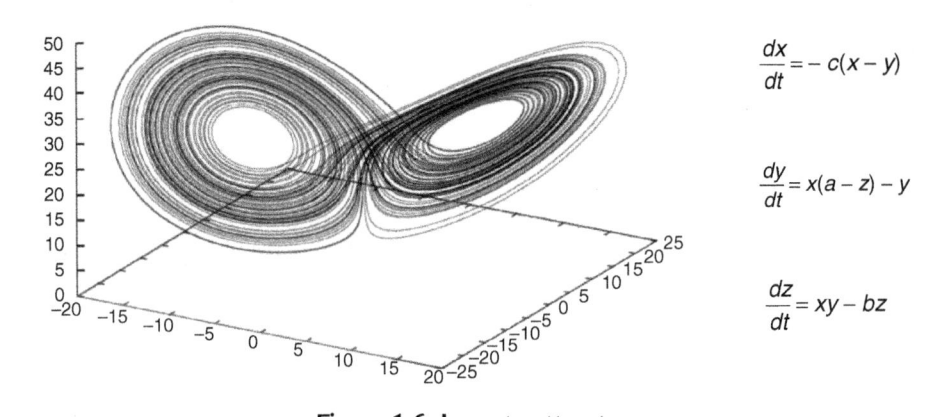

$$\frac{dx}{dt} = -c(x-y)$$

$$\frac{dy}{dt} = x(a-z) - y$$

$$\frac{dz}{dt} = xy - bz$$

Figure 1.6: Lorentz attractor.

1.8 Complex Thermodynamic Systems

1.8.1 Introduction

This section sketches the applications of nonequilibrium thermodynamics to complex systems such as solutions of macromolecules, magnetic hysteresis bodies, viscoelastic fluids, polarizable media, and so forth. These systems require some extra variables (so-called internal variables) to be introduced into the fundamental equation of Gibbs. Irreversible thermodynamic systems understood as EIT systems are also analyzed. In this approach, the dissipative fluxes are the independent variables treated on an equal footing with the classical variables of thermostatics. The significance of systems with Gibbs equations which incorporate additional internal variables is studied. Such systems include viscoelastic and viscoplastic materials, and fluids with high-diffusive fluxes of heat, mass, and momentum. Rational and extended thermodynamic theories are apropriate tools for the analysis of these systems. The relative merits of these approaches is discussed and corresponding research papers are reviewed.

1.8.2 Classical and Quasi-Classical Complex Systems

We use the adjective complex to refer to systems such as solutions of macromolecules, magnetic hysteresis bodies, viscoelastic fluids, polarizable media, and so forth. which exhibit involved thermodynamic behavior and for which the role of internal variables is essential in a thermodynamic description.

The concept of an internal state variable, that is, a variable not controllable through external conditions, originated from the analysis of some models in rheology (Lemaitre and Chaboche, 1988; Maugin, 1987, 1990a) and electromagnetic bodies (Meixner, 1961, Kluitenberg, 1966, 1977, 1981). The goal was a theoretical framework for studying relaxation processes such as a shock wave passing through an electrically polarizable substance. However, relaxation phenomena are not the only ones for which the concept of internal variables is useful. Theories of plasticity and fracture and description of dissipative effects in electrodeformable solids can also make good use of this concept (Maugin, 1990a, b). Furthermore, some nontrivial analogies exist between plastic behavior and magnetic hysteresis; hence the developments in the plasticity theory have implications for magnetic hysteresis systems.

It follows that hysteresis processes can be described by using the concept of the internal variables of state provided one takes into account some residual fields at vanishing load. However, one should be careful to distinguish between the relaxational recovery of equilibrium and hysteresis processes. In relaxation, time scale is an essential issue; in hysteresis the relatively slow response exhibited by a ferromagnetic sample is practically rate independent. Similarly, the low-temperature plastic effects are practically independent of the rate of strain. A detailed comparison of these effects is available (Maugin, 1992). For the theory of internal variables, see also Kestin (1979), Kestin and Rice (1970), Kestin and Bataille (1980), Muschik (1981, 1986, 1990a, b, 2007, 2008).

Maugin (1992) treats magnetic hysteresis as a thermodynamic process. As contrasted with many previous formal approaches, hysteresis is cast in the framework of nonequilibrium thermodynamics. The process is described at the phenomenological level, in terms of the constitutive equations, internal variables of state and the dissipation function which is homogeneous of degree one. The model allows one to account for the cumulative effect of the residual magnetization and it describes magnetic hardening in a natural way. Both local and global stability criteria for hysteresis loops are obtained. The model, being incremental, provides an operational way to construct hysteresis loops from a virgin state by alternate loads with increasing maximum amplitude. The interaction between the magnetic properties and stress or temperature variations are taken into account. An interesting analogy between solid mechanics (viscoplastic Bingham fluid) and magnetism is shown while a clear distinction is made between relaxation processes and hysteresis-like effects.

The continuum modelling of polarizable macromolecules in solutions is the next example showing the role of internal variables (Maugin and Drouot, 1983a, 1992; Maugin, 1987). Their use allows one to construct unknown constitutive equations and evolution equations describing mechanical, electrical, and chemical behavior. Also, the effects that do not contribute to dissipation can be singled out from the general formalism. Flow induced and electrically induced phase transitions can be studied as well as diffusion, migration and mechanochemical effects. In addition Maugin and Drouot (1983b) treat thermomagnetic behavior of magnetically nonsaturated fluids whereas Maugin and Sabir (1990) determine nondestructive testing in mechanical and magnetic hardening of ferromagnetic bodies.

Rheology of viscoelastic media (Leonov, 1976; Lebon et al., 1988), isotropic-nematic transitions in polymer solutions (Sluckin, 1981; Giesekus, 1986) and stress-induced diffusion (Tirrell and Malone, 1977) are other applications of internal variables. A number of reviews describe the requisite nonequilibrium thermodynamic framework (Bampi and Morro, 1984; Morro, 1985; Maugin, 1987). Bampi and Morro (1984) developed a hidden or internal variable approach to nonequilibrium thermodynamics. An application of the internal variable concept using rational thermodynamics to describe viscoelasticity is also available. Morro (1992) imbeds the thermodynamics of linear viscoelasticity into the framework of rational thermodynamics and extremum principles. First, a general scheme for thermodynamics of simple materials is outlined. This scheme is applicable to materials with internal variables and those with fading memory. Rigorous definitions are given for states, processes, cycles and thermodynamic laws. His thermodynamic approach is then applied to viscoelastic solids and fluids. The approach is essentially entropy-free in the sense that the dissipative nature of the theory is assured by the negative definiteness of the half-range Fourier sine transform of a (Boltzmann) relaxation kernel. The fact that such negative definiteness yields all conditions derived so far by various procedures is a new result. In addition, such a condition has been proved to be sufficient for the validity of the second law. Necessary and sufficient conditions for the validity of the second law are derived for fluids. Evolution equations are analyzed. The Navier-Stokes model of a viscous fluid follows as a limiting case when some relaxation functions are equal to delta-functions. The extremum principles are shown to be closely related to thermodynamic restrictions on the relaxation functions. The treatment is quite involved from the mathematical viewpoint, but the reader gets to reexamine many basic features of the rational theory in a brief and precise way. In this aspect the reader is referred to the paper by Altenberger and Dahler (1992) which discusses applications of convolution integrals to the entropyless statistical mechanical theory of multicomponent fluids.

Historically, the starting point for extended thermodynamic theories was the paradox of infinite propagation rate, encountered in the classical phenomenological equations describing the transport of mass, energy, and momentum. Attempts to overcome the paradox resulted in equations of change in terms of (damped) wave equations rather than classical parabolic equations. This has been demonstrated in a number of articles. The topics ranged from theoretical developments (based on phenomenological thermodynamics or nonequilibrium statistical mechanics), to applications in fluid mechanics, chemical engineering and aerosol science. It has been recognized that taking into account local nonequilibrium and/or inertial effects (resulting from the finite mass of the diffusing particles) the constitutive equations become non-Fourier, non-Fick, and non-Newtonian.

A tutorial review of the historical development of the elimination of the paradox is available (Sieniutycz, 1992a, and chapter: Wave Equations of Heat and Mass Transfer). The concept of relaxation time has been substantiated and its simplicity of evaluation for heat, mass and momentum has been shown for ideal gases with a single propagation speed. For nonideal systems, a simplest extended theory of coupled heat and mass transfer has been constructed based on a nonequilibrium entropy. In the approach discussed, the entropy source analysis yields a formula for the relaxation matrix, τ. This matrix describes the coupled transfer and corresponds to the well-known values of relaxation times for pure heat transfer and for isothermal diffusion. Various simple forms of the wave equations for coupled heat and mass transfer appear. Simple extensions of the second differential of entropy and of excess entropy production can be constructed. They allow one to prove the stability of the coupled heat and mass transfer equations by the second method of Liapounov. Dissipative variational formulations can be found, leading to approximate solutions by a direct variational method. Applications of the hyperbolic equations to the description of short-time effects and high-frequency behavior can be studied.

Some related ideas dealing with a variational theory of hyperbolic equations of heat can be found in Vujanovic's papers where a class of variational principles of Hamilton's type for transient heat transfer has been formulated and analyzed, and valuable approximate solutions have been found by direct variational methods (Vujanovic, 1971, 1992; Vujanovic and Djukic, 1972; Vujanovic and Jones, 1988).

Flux relaxation effects are particularly strong in the diffusion of aerosols containing particles of large mass and in polymers characterized by large molecular weight. Experiments with laser pulses of high frequency acting on solid surfaces show that nonclassical models of heat transfer and diffusion of mass and energy are more appropriate in the case of highly nonstationary transients (Domański, 1978a, b). Acoustic absorption at high frequency is another example (Bubnov, 1976a; Carrasi and Morro, 1972, 1973; Luikov et al., 1976). The phenomenological thermodynamics of diffusion links some relaxation effects with the absence of local thermodynamic equilibrium in systems with finite kinetic energy of diffusion (Truesdel, 1962, 1969; Sandler and Dahler, 1964; Sieniutycz, 1983, 1984a, b, 1992a). Banach and Piekarski (1992) attempt to give the meaning of temperature and pressure beyond the local equilibrium and obtain nonequilibrium thermodynamic potentials. To achieve these goals they apply the following maximum principle: among all states having the same values of conserved variables, the equilibrium state gives the specific entropy its greatest value. See chapter: Wave Equations of Heat and Mass Transfer.

The classical theory of nonequilibrium thermodynamics (de Groot and Mazur, 1984) assumes the local equilibrium. Consequently, the Gibbs equation for the entropy differential is used in the well-known classical form which contains the derivatives of the internal energy, volume (or density) and concentrations of chemical constituents. Application of the conservation laws for energy, mass and momentum yields the well-known expression for entropy balance:

$$\rho \frac{ds}{dt} = -\nabla \left(\frac{\mathbf{q}}{T} \right) + \mathbf{q} \nabla T^{-1} - T^{-1} \mathbf{\Pi} : \nabla \mathbf{u} \qquad (1.131)$$

The right-hand side of this equation includes a source term and a divergence term which is related to the conductive entropy flow. The considered equation serves to postulate phenomenological forms for these terms. These postulated forms are in agreement with the second law. We have already given a brief account of the problems which arise in the classical theory in Section 1.2. We do not repeat the criticisms here; for more details the reader is referred to the literature (Meixner, 1949, 1974; Müller, 1966, 1967, 1971a, b, 1985; Muschik, 1981, 1986, 1990a, b, 2004, 2007, 2008; Muschik et al., 2001; Lebon and Mathieu, 1983). Yet, we recall two arguments which have been essential for the development of extended theories.

First; the local equilibrium postulate is not valid for continua with complex internal structure and for simple materials in the presence of high-frequency phenomena and fast transients. Second, the principle of equipresence of rational thermodynamics (Jou et al., 1988, 2001) assures that if a variable appears in one constitutive equation it has full right to appear in all the remaining equations of the theory. For example, since the mass (heat) diffusion flux is present in the first Fick's (Fourier's) law, this flux should also appear in the Gibbs equation. Indeed, the need for such terms was approximately outlined by de Groot and Mazur (1984) for diffusive mass flows in the context of the kinetic energy of diffusion, by Grad (1958) for heat flow, by Sieniutycz and Berry (1989) for diffusive entropy flow and by a number of other researchers for other flows. The reader is referred to recent reviews of these approaches by Jou, Casas-Vazquez, and Lebon (Jou et al., 1988, 2001).

1.8.3 Progress in Extended Thermodynamics of Complex Systems

Extended thermodynamics uses nonequilibrium (flux dependent) entropies. Kinetic theory supports the existence of such entropies which were first introduced by Grad (1958). The general conservation laws retain their classical form in EIT. When they are substituted into the expression for the differential of the nonequilibrium entropy, they lead to modified expressions for the entropy flux and entropy source. In the simplest version of the extended thermodynamics of a one-component system, the second law inequality can be expressed in the form

$$\rho \frac{ds}{dt} = -\nabla \cdot \left(\frac{\mathbf{q}}{T} + \mathbf{K}(\overset{0}{\Pi} \mathbf{q}) \right) + \mathbf{q} \cdot \left(\frac{\mathbf{C}^{-1}}{G} \frac{d\mathbf{q}}{dt} + \nabla T^{-1} \right) - T^{-1} \overset{0}{\Pi} : \left(\frac{d \overset{0}{\Pi}}{2G dt} + \left(\overset{0}{\nabla} \mathbf{u} \right)^s \right) \qquad (1.132)$$

Phenomena associated with bulk viscosity are neglected here, and the tensor $\overset{0}{\Pi}$ is the deviator of the viscous pressure tensor. From Eq. (1.132), the stress $\overset{0}{\Pi}$ is related to the symmetrized velocity gradient, and the time derivative of the stress itself, the structure which leads to the viscoelastic behavior. G is the shear modulus which is the product of the mass density and the square of the propagation speed, c_0^2. $\mathbf{C}$ is the entropy capacity matrix which reduces to the scalar $C = -c_p T^2$ for the pure heat transfer. The extra term in $\mathbf{K}$ appears in the entropy flux expression in accordance with the kinetic theory. The problem of frame invariance is ignored in Eq. (1.131). Also chemical reactions are ignored. In fact, Eq. (1.132) is still approximate.

A more exact extended theory has been constructed on the basis of the rational formalism (Müller, 1985; Lebon and Mathieu, 1983; Lebon and Boukary, 1988; Kremer, 1985, 1986, 1987, 1989, 1992; Liu and Müller, 1983) or based on some additional postulates added to the kinetic theory (Eu, 1982, 1988).

The historical aspects of these ideas, from early simplified descriptions to the contemporary theory is presented in Müller's (1992) paper as the development of a theory with hyperbolic field equations. Here only a few introductory remarks are given. Essential for the development of EIT has been a simple and basic form for the entropy inequality introduced by Müller (1966, 1967, 1985). He assumed the second law in the form

$$\rho \frac{ds}{dt} + \nabla \mathbf{j}_s \geq 0 \qquad (1.133)$$

containing the entropy flux $\mathbf{j}_s$ which has to be described by a constitutive equation. The idea was improved in Müller's later approaches (Müller, 1985) as well as in works by others. An advanced theory (Liu and Müller, 1983) uses the method of Lagrange multipliers to take into account the entropy inequality (Liu and I-Shih, 1972).

As a simplest example we consider a one-component rigid heat conductor. The set of the constitutive equations is assumed in the form

$$\begin{aligned} f &= f(T, T_{,i}, q_i) \\ s &= s(T, T_{,i}, q_i) \\ Q_i &= Q_i(T, T_{,i}, q_i) \\ j_{si} &= T^{-1} q_i + K(T, T_{,i}, q_i) \end{aligned} \qquad (1.134)$$

where f is the free energy. These equations are substituted into the entropy inequality (1.133). The resulting inequality yields the classical thermodynamic conditions, for example, $\partial f/\partial T = -s$, as well as information about the structure of the equations describing the fluxes of entropy and heat. Nonclassical terms appear in the expression for $\mathbf{j}_s$, and the Cattaneo (1958) equation is recovered in the linear case. Free energy formula contains the heat flux as an independent variable.

Lebon (1992) develop the thermodynamics of rheological materials in the context of the extended irreversible theory. They show that EIT has capability of a general unifying formalism which is able to produce a broad spectrum of viscoelastic constitutive equations. In particular, linear viscoelasticity is easily interpreted within EIT. These works show how the classical rheological models of Maxwell, Kelvin–Voight, Poynting–Thompson, and Jeffreys are obtained as special cases of the formalism developed. Lebon (1992) offers a general Gibbs equation for viscoelastic bodies, and discusses its special cases, pertaining to the particular models specified previously. Then the author poses the question: what are the consequences of introducing a whole spectrum of relaxational modes for the pressure tensor, instead of working with one single mode. For that purpose the Rouse–Zimm relaxation models (Rouse, 1953; Zimm, 1956) are analyzed. These models are based on a whole relaxational spectrum; such spectra are frequently predicted by kinetic models. These models have proved useful for describing dilute polymeric solutions. It is shown that EIT is capable of coping with the relaxational spectrum of the Rouse–Zimm model. Finally, the analysis is extended to fluids described by nonlinear constitutive equations (non-Newtonian). The considered,

EIT-based approach to viscoelasticity can be compared with other thermodynamic approaches, based on other theories: classical (Meixner, 1949, 1961, 1966, 1968, 1972, 1974; Kluitenberg, 1966, 1977, 1981; Kestin, 1979; Kestin and Rice, 1970; Kestin and Bataille, 1980) or rational (Rivlin and Ericksen, 1955; Koh and Eringen, 1963; Huilgol and Phan-Thien, 1986).

A series of reviews is available devoted to extended irreversible thermodynamics (EIT). Müller's (1992) review starts with the analysis of imperfections in Cattaneo's (1958) treatment and then outlines a simplified extended theory based on a nonequilibrium (flux dependent) entropy (Müller, 1966, 1967, 1985). Its compability with the entropy resulting from Grad's 13-moment theory (Grad, 1958) is shown and the problem of material frame indifference is discussed. The role of rational thermodynamics as a starting point for constructing an improved extended theory is also shown. This is based essentially on Liu and Müller's approach (Liu and Müller, 1983) which uses Liu's method of Lagrange multipliers to take into account the entropy inequality (Liu and I-Shih, 1972). The nonrelativistic principle of relativity is used to make the fields explicit in velocity (Ruggeri, 1989). Constitutive equations, inertial frame balance laws and equilibrium conditions are obtained. A detailed comparison of the results of EIT with those of rational thermodynamics is made. The relativistic generalization of the theory is given and relativistic analogues of the phenomenological equations of Fourier and Navier-Stokes are derived. The structure of EIT may be analyzed from the viewpoint of mathematics (Bass, 1968).

In the general case, the approaches outlined previous results in a nonequilibrium theory of extended irreversible thermodynamics, with independent variables that include the heat flux $\mathbf{q}$, diffusion fluxes, and viscous stresses (Jou et al., 1988, 2001). The classical role of $\mathbf{q}$ as a (dependent) state variable has already been abandoned in the kinetic theory by Grad (1958); the use of the heat flux $\mathbf{q}$ as an independent variable is in complete agreement with his moment analysis. In the general case, all diffusive fluxes (of heat, mass, momentum) constitute extra independent variables in EIT.

The previous results were obtained by assumming that the Gibbs equation is modified by the presence of diffusive fluxes, whereas the form of the conservation laws remains unchanged. However, it was recently shown (Sieniutycz, 1988a, b, 1989, 1990; Sieniutycz and Berry, 1989), that extended Gibbs equations can be derived from a Lagrangian approach (Hamilton's stationary action) where an extended internal energy is extracted from the total energy. The extended internal energy is obtained within the time component G^{tt} of an energy-momentum tensor G^{ik}. The structure of this tensor corresponds to the classical conservation laws only when the sources of mass and entropy play a negligible role. For substantially irreversible processes, where these sources are large, the structure of G^{ik} deviates from the classical structure, and new important variables play a role in G^{ik} and in energy E. These new variables are called thermal phase and matter phase. For substantially irreversible processes, the energy formula and the conservation laws can be changed to take the sources of mass and entropy into account. These matters are discussed by two papers (Sieniutycz, 1990, 1992b) of which the first presents an extended hydrothermodynamics and the second an extended theory.

The concept of internal variables has been succesfully used to analyze relaxation in strained solids by Kestin and Rice (1970); dielectric and magnetic relaxation (Kluitenberg, 1977, 1981; Meixner, 1961); spin relaxation in ferromagnets (Maugin, 1975), magnetoelasticity (Maugin, 1979), thermoplastic fracture effects (Maugin, 1990), electromechanical and other couplings (Maugin, 1990a) and magnetically nonsaturated fluids (Maugin and Drouot, 1983a). As contrasted with previous approaches, magnetic hysteresis (Maugin, 1987, Maugin and Sabir, 1990) has been cast in the framework of nonequilibrium thermodynamics describing this phenomenon exactly in terms of the constitutive equations, internal variables and dissipation function. An interesting analogy between solid mechanics, viscoplastic Bingham fluids and magnetism has been shown (Maugin, 1990a). A clear distinction has been made between relaxation processes and hysteresis effects. A thermodynamic framework incorporating internal variables is summarized in the reviews by Coleman and Gurtin (1964) and Bampi and Morro (1984).

Thermodynamics is a core part of most science and engineering curricula. However, most texts available to students still treat thermodynamics very much as it was presented in the 19th century, generally for historical rather than pedagogical reasons. Modern Thermodynamics (Kondepudi and Prigogine, 1998) takes a different approach, and deals with the relationship between irreversible processes and entropy. The relationship between irreversible processes and entropy is introduced early on, enabling the reader to benefit from seeing the relationship in such processes as heat conduction and chemical reactions. This text presents thermodynamics in a contemporary and exciting manner, with a wide range of applications, and many exercises and examples.

Another model based on nonequilibrium thermodynamics is extended so as to account for the new phenomena related to the presence of electrically polarizable macromolecules in solution (Aubert and Tirrel, 1980; Bird et al., 1977; Bird and Öttinger, 1992; Carreau et al., 1986; Drouot and Maugin, 1983, 1985, 1987, 1988; Drouot et al., 1987). The role of dissipative effects is essential although certain less known phenomena are shown to be gyroscopic in a sense that they do not contribute to dissipation (Maugin, 1990a, b). This is discussed in the paper by Maugin and Drouot (1992), where a relatively simple account of electromechanical couplings and mechanochemical effects is presented. See also Maugin (1990a). Maugin and Drouot (1992) develop an internal variable based thermodynamics of solutions of macromolecules. The relevant internal variables of state are the components of the conformation tensor. This quantity models some properties of macromolecules by treating them as "material" particles undergoing deformations. The dissipation inequality is obtained in the form of the Clausius–Duhem inequality involving the free energy. Two different types of dissipative processes are distinguished. The first type involves only the observable variables of the thermo-electro-mechanics. The second are the relaxation processes (of the type encountered in chemical kinetics) involving electric and conformational relaxation. Processes of a third type (plasticity), while essential in the previous work discussed previously, do not appear. Evolution equations for conformation, electric polarization and chemical processes are given. An important effect for gyroscopic type reversible processes is developed. Applications to equilibrium conformations, flow-induced and electrically induced conformational phase transitions, electric polarization and mechanochemical effects are discussed. Possible generalizations accounting for diffusion and interaction between vortex fields and conformations are given. This paper should be read by everyone who has to deal with complex fluids, such as liquid crystals and emulsions.

Viscoelastic bodies (solids and fluids) have been analyzed in the framework of the rational thermodynamics using internal variables (Morro, 1985). Restrictions on the relaxation functions have been established (Day, 1971; Boven and Chen, 1973; Fabrizio and Morro, 1988; Fabrizio et al., 1989). Convexity of the governing functionals has been proved and minimum principles have been established. The relaxation functions satisfy the dissipation inequality. This makes the corresponding functional convex (Fabrizio et al., 1989) and hence the thermodynamic limitations imply the convexity of the governing functional. An analogous result has recently been found for the nonequilibrium energy function of a flowing fluid in the field representation of thermohydrodynamics (Sieniutycz and Berry, 1991). Extended thermodynamics of viscoelastic and non-Newtonian bodies has been formulated by Lebon and coworkers (Lebon et al., 1986, 1988, 1990, 1993). Evren-Selamet (1995) explored both diffusional and the hysteretic dissipation of the mechanical energy into entropy, and, consequently postulated both diffusive and hysteretic foundations of entropy production. Examples involve Kelvin solid and Maxwell viscoelastic fluid. Lavenda (1995) formulated basic principles in the thermodynamics of extreme and rare events.

Statistical thermodynamics is usually concerned with most probable behavior which becomes almost certainty if large enough samples are taken. But sometimes surprises are in store where extreme behavior becomes the prevalent one. Turning his attention to such rare events Lavenda's (1995) interest lies in the formulation of a thermodynamics of earthquakes which is gaining increasing attention. By properly defining entropy and energy, a temperature can be associated to an aftershock sequence giving it an additional means of characterization. A new magnitude-frequency relation is predicted which applies to clustered after-shocks in contrast to the Gutenberg–Richter law which treats them as independent and identically distributed random events.

Extended irreversible thermodynamics (EIT), has grown to the status of a mature theory. Typical problems investigated include: compatibility with Grad's 13-moment method (Grad, 1958), frame invariance, relation to rational thermodynamics, the role of Lagrange multipliers and speeds of propagation (Müller, 1985; Boukary and Lebon, 1986; Lebon and Boukary, 1988). Three levels of description have been analyzed: macroscopic (thermodynamic), mesoscopic (fluctuation theory) and microscopic (projection operators). The EIT description has been compared with the memory function approach based only on the conserved variables (Garcia-Colin et al., 1984; Jou et al., 1988). The superiority of the extended theory over the classical one has been shown. The correspondence of the theory with classical results for ideal and real gases as regards thermal and caloric equations of state has been established. Explicit results have been obtained for the classical ideal gas, for degenerate Bose and Fermi gases, and for molecular gases and mixtures (Müller, 1985; Kremer, 1985, 1986, 1987, 1989).

The theory was given kinetic theory foundations by means of the Boltzmann equation for a mixture of dilute monatomic gases and the generalized Boltzmann equation for a mixture of dense simple fluids (Eu, 1980, 1982, 1988). The theory has been applied to rheology, to fluids subject to shear and to shear induced phase changes (Eu, 1989). Additional reciprocity conditions resulting from introduction of heat, matter, and momentum fluxes as state variables have been shown to be useful in calculating transport coefficients in liquids and in the estimation of the nonlinear effects in heat flow and diffusion (Nettleton, 1987). These conditions have also been obtained in the framework of the maximum entropy formalism (Nettleton and Freidkin, 1989). The relation between EIT and fluctuation theory has been discussed (Jou et al., 1988). It has also been shown that EIT provides a natural framework for the interpretation of memory functions and generalized transport coefficients (Perez Garcia and Jou, 1986). Extended chemical systems have been treated (Garcia Colin et al., 1986; Garcia Colin, 1988; Sieniutycz, 1987). Astrophysical and cosmological applications of EIT have been pursued in the framework of a relativistic theory (Pavon, 1992; Pavon et al., 1980, 1983). Eu (1992) analyzes the relation between kinetic theory and irreversible thermodynamics of multicomponent fluids. His approach is based on an axiomatic formulation (Eu, 1980, 1982, 1987) and on nonequilibrium entropy (Eu, 1987, 1988, 1989), that is, on exactly those features supported by kinetic theories of gases and fluids. His emphasis is on the nonlinear constitutive equations compatible with the statistical mechanical theories of matter in nonequilibrium. The Boltzmann equation of dilute gases and a modified moment method are used to construct an irreversible theory. Approximation methods are employed to obtain the distribution functions. The requirement that the second law be rigorously satisfied by the approximate solution of the Boltzmann equation is the tenet under which the modified moment methods are developed. The equations for nonconserved moments are the constitutive equations. A consistency condition (in the form of a partial differential equation) is coupled to the evolution equations for the Gibb's variables. The Onsager–Rayleigh quadratic dissipation and classical phenomenological laws follow in the limit of the linear theory close to equilibrium. The theory is extended to dense multicomponent fluids, based on a set of the kinetic equations for a dense mixture of charged particles in a solvent. Tests of the constitutive equations are outlined in the framework of rheology (polymer viscoelasticity). The formal theory of nonequilibrium thermodynamic potentials is given with an extended Gibbs–Duhem relation. Applications for fluids under shear and shear-induced depression of the melting point are relevant.

Kremer's (1992) review extends EIT theory to a broader class of processes. He investigates EIT of ideal and real monatomic gases, molecular gases, and mixtures of ideal monatomic gases. For the ideal and real gases the formalism of 14 fields is developed where the 13 fields of mass density, pressure tensor, and heat flux are supplemented by an additional scalar field associated with nonvanishing volume viscosities. Coefficients of shear and volume viscosities, thermal conductivity, three relaxation times, and two thermal-viscous coeeficients are identified in the theory. The ideal gas theory results as a limiting case from the real gas approach. Explicit results are given for classical ideal gases and for degenerate Bose and Fermi gases. Then, an extended theory of molecular gases is formulated which contains 17 fields of density, velocity, pressure tensor, heat flux, intrinsic energy, and intrinsic heat flux. The intrinsic energy pertains to the rotational or vibrational energy of the molecules. Coefficients of thermal and caloric equations of state as well as transport or rate coefficients are identified (shear viscosity, thermal conductivity, self-diffusion, absorption and dispersion of sound waves). Finally, the EIT theory for mixtures of monatomic ideal gases is analyzed. Generalized phenomenological equations follow corresponding to the classical laws of Fick, Fourier, and Navier-Stokes. Onsager reciprocity follows without statistical arguments. The unusual breadth of EIT theory and its capability to synthesize static, transport and rate properties of nonequilibrium systems in one unifying approach is shown in this paper.

Nettleton (1992) treats EIT problems from the viewpoint of the general reciprocity (and antireciprocity) relations encountered in nonequilibrium systems and predicted by extended theories. These reciprocity relations result from the introduction of diffusional fluxes of mass, energy and momentum, as (extra) state variables of the nonequilibrium Gibbs equation. They are shown to be useful in calculating transport coefficients in liquids. Reciprocity reduces the number of coefficients which must be estimated from molecular approaches. By using reciprocity, certain generalizations of the Cattaneo equation of heat and the Maxwell equation of viscoelasticity can be obtained. These equations govern coupled flows of heat and matter and viscoelasticity under several creep mechanisms. The use of reciprocity in the estimation of nonlinear effects associated with the flow of heat and diffusion is shown. Nonlinear reciprocity is derived from the Fokker–Planck equation for a distribution g of values of certain microscopic operators whose averages are variables such as heat and mass fluxes. The equation for g is obtained by applying Zwanzig or Grabert projection operators to the Liouville equation. The maximum entropy formalism provides an ansatz for the solution of the Liouville equation which yields a finite number of moments. From this ansatz one can derive kinetic equations exhibiting nonlinear reciprocity by taking moments of the equation for g. It is shown that a Lagrangian formulation can lead to the reciprocity and antireciprocity properties.

Casas-Vazquez and Jou (1992) review the theory of fluctuations in the framework of EIT. Mutual contributions of the two theories are analyzed. Fluctuation theory enables one to evaluate the values of the coefficients of the nonclassical terms in the entropy and in the entropy flux, from a microscopic basis. The fluctuation-dissipation relations focus on the time correlation function of fluctuations in the fluxes. Equilibrium fluctuations of dissipative fluxes are considered and the second moments of the fluctuations are obtained. Connections with the Green-Kubo transport formulae are shown. Microscopic evaluations of the coefficients of the nonequilibrium entropy are presented for ideal gases, dilute real gases, and dilute polymer solutions. An expression is given for the entropy flux in terms of fluctuations. Next, the nonequilibrium fluctuations of dissipative fluxes are analyzed. The contribution of the irreversible fluxes to the entropy leads to explicit expressions for the second moments of nonequilibrium fluctuations.

EIT predicts nonequilibrium contributions to the fluctuations of variables as a consequence of the nonvanishing relaxation times of the fluxes. An example is given discussing fluctuations of the heat flux in a rigid heat conductor. The second differential of the nonequilibrium entropy (with nonvanishing mixed terms) is introduced into the Einstein formula to compute the second moments of fluctuations in a nonequilibrium steady state. It is shown that the correlations of energy and heat flux and those of volume and heat flux, which vanish in equilibrium on account of the different parity of u and q and of v and q, are different from zero in nonequilibrium. This expresses the breaking of the time-reversal symmetry in nonequilibrium states. The results of the second moments from the nonequilibrium entropy and from microscopic (kinetic) models are of the same order but not coincide. This open problem points out the status of the present theory of fluctuations in EIT.

Jou and Salhoumi (2001) propose a Legendre transform between a nonequilibrium thermodynamic potential for electrical systems using the electric flux as a nonequilibrium variable and another one using the electric field. The transforms clarify the definition of the chemical potential in nonequilibrium steady states and may be also useful in other contexts where the relaxation times of the fluxes play a relevant role. The relation between their proposal and some Legendre transforms used in the context of dissipative potentials is examined. See Sieniutycz's (1994) book for other examples of using the Legendre transform. Frieden et al. (1990) show that the Legendre-transform structure of thermodynamics can be replicated without any change if one replaces the entropy S by Fisher's information measure I. Also, the important thermodynamic property of concavity is shown to be obeyed by I. By this use of the Fisher information measure the authors develop a thermodynamics that seems to be able to treat equilibrium and nonequilibrium situations in a manner entirely similar to the conventional one. Much effort has been devoted to Fisher's information measure (FIM), shedding much light upon the manifold physical applications (Frieden and Soffer, 1995). It can be shown that the whole field of thermodynamics (both equilibrium and nonequilibrium) can be derived from the MFI approach. In the paper by Frieden and Soffer (1995) FIM is specialized to the particular but important case of translation families, that is, distribution functions whose form does not change under translational transformations. In this case, Fisher measure becomes shift-invariant. Such minimizing of Fisher's measure leads to a Schrödinger-like equation for the probability amplitude, where the ground state describes equilibrium physics and the excited states account for nonequilibrium situations.

Both nonequilibrium and equilibrium thermodynamics can be obtained from a constrained Fisher extremization process whose output is a Schrödinger-like wave equation (SWE). Frieden et al. (2002a). Within this paradigm, equilibrium thermodynamics corresponds to the ground state (GS) solution, while nonequilibrium thermodynamics corresponds to the excited state solutions. SWE appears as an output of the constrained variational process that extremizes Fisher information. Both equilibrium and nonequilibrium situations can thereby be tackled by one formalism that clearly exhibits the fact that thermodynamics and quantum mechanics can both be expressed in terms of a formal SWE, out of a common informational basis. The method is a new and powerful approach to off-equilibrium thermodynamics. Frieden et al. (2002a) discuss an application to viscosity in dilute gases and electrical conductivity processes, and thereby, by construction, show that the following three approaches yield identical results: (1) the conventional Boltzmann transport equation in the relaxation approximation, (2) the Rumer and Ryvkin Gaussian–Hermite–algorithm, and (3) the authors Fisher–Schrödinger technique.

Information measures (IM) are the most important tools of information theory. They measure either the amount of positive information or of "missing" information an observer possesses with regards to any system of interest. The most famous IM is the so-called Shannon entropy (Shannon and Weaver, 1969), which determines how much additional information the observer still requires in order to have all the available knowledge regarding a given system S, when all he/she has is a probability density function (PD) defined on appropriate elements of such system. This is then a "missing" information measure (Plastino et al. (1996, 2005). The IM is a functional of the PD only. If the observer does not have such a PD, but only a finite set of empirically determined mean values of the system, then a fundamental scientific principle called the Maximum Entropy one (MaxEnt) asserts that the "best" PD is the one that, reproducing the known expectation values, maximizes otherwise Shannon's IM.

Fisher's information measure (FIM), named after Ronald Fisher, is another kind of measure, in two respects, namely, (1) it reflects the amount of (positive) information of the observer, and (2) it depends not only on the PD but also on its first derivatives, a property that makes it a local quantity (Shannon's is instead a global one). The corresponding counterpart of MaxEnt is now the FIM-minimization, since Fisher's measure grows when Shannon's diminishes, and vice-versa. The minimization here referred to (MFI) is an important theoretical tool in a manifold of disciplines, beginning with physics. In a sense it is clearly superior to MaxEnt because the later procedure yields always as the solution an exponential

PD, while the MFI solution is a differential equation for the PD, which allows for greater flexibility and versatility (Plastino et al. (2005).

Plastino et al. (2005) analyze Fisher's variational principle in the context of thermodynamics. Their text, which shows applications of variational approaches in statistical physics and thermodynamics, constitutes the Chapter II.1 of book by Sieniutycz and Farkas (2005). Plastino et al. (2005) prove that standard thermostatistics, usually derived microscopically from Shannon's information measure via Jaynes' Maximum Entropy procedure, can equally be obtained from a constrained extremization of Fisher's information measure that results in a Schrödinger-like wave equation. The new procedure has the advantage of dealing on an equal footing with both equilibrium and off-equilibrium processes. Equilibrium corresponds to the ground-state solution, nonequilibrium to superpositions of the gas with excited states. As an example, the authors illustrate these properties with reference to material currents in diluted gases, provided that the ground state corresponds to the usual case of a Maxwell–Boltzmann distribution. The great success of thermodynamics and statistical physics depends crucially on certain necessary mathematical relationships involving energy and entropy (the Legendre transform structure of thermodynamics). It has been demonstrated that these relationships are also valid if one replaces S by Fisher's information measure (FIM). Much effort has been focused recently upon FIM. Frieden and Soffer (1995) have shown that FIM provides one with a powerful variational principle that yields most of the canonical Lagrangians of theoretical physics. Markus and Gambar (2003) discuss a possible form and meaning of Fisher bound and physical information in some special cases. They suppose that a usual choice of bound information may describe the behavior of dissipative processes.

In view of the recognition that equilibrium thermodynamics can be deduced from a constrained Fisher information extemizing process Frieden et al. (2002b) show that, more generally, both nonequilibrium and equilibrium thermodynamics can be obtained from such a Fisher treatment. Equilibrium thermodynamics corresponds to the ground-state solution whereas nonequilibrium thermodynamics corresponds to the excited-state solutions of a Schrödinger wave equation (SWE). That equation appears as an output of the constrained variational process that extremizes the Fisher information. Both equilibrium and nonequilibrium situations can therefore be tackled by one formalism that exhibits the fact that thermodynamics and quantum mechanics can both be expressed in terms of a formal SWE, out of a common informational basis. As an application, the authors discuss viscosity in dilute gases. In conclusion, it is becoming increasingly evident that Fisher information I is vital to the fundamental nature of physics. The authors are clearly aware of the fact that the I concept lays the foundation for both equilibrium thermodynamics and nonequilibrium thermodynamics, the notion which subsumes the horizon envisioned in contemporary theoretical works. The main result of all these works is the establishment, by means of Fisher information, of the previous connection. Frieden et al. (2002b) state that the emphasis lies in the word "connection" and ask why would such a link be of interest? They answer: "Because it clearly shows that thermodynamics and quantum mechanics can both be expressed by a formal SWE, out of a common informational basis." Applications of these ideas are also visible in some works on variational thermodynamics attempting to set a Hamiltonian formulation as a basis for the quantization of thermal processes (Markus, 2005; Vázquez et al., 2009). The unification of thermodynamics and quantum mechanics is also postulated by Gyftopoulos (1998a), Beretta (1986), and in other publications of these authors.

Markus (2005) applies a Hamiltonian formulation to accomplish the quantization of dissipative thermal fields on the example of heat conduction. He introduces the energy and number operators, the fluctuations, the description of Bose systems and the q-boson approximation. The generalized Hamilton–Jacobi equation and special potentials, classical and quantum–thermodynamical, are also obtained. The author points out an interesting connection of dissipation with the Fisher and the extreme physical information. To accomplish the thermal quantization the variational description of the Fourier heat conduction is examined. The Lagrangian is constructed by a potential function that is expressed by the coefficients of Fourier series. These coefficients are the generalized coordinates by which the canonical momenta, the Hamiltonian, and the Poisson brackets are calculated. This mathematical construction opens the way toward the canonical quantization that means that the operators can be introduced for physical quantities, for example, energy and quasiparticle number operators for thermal processes. The elaborated quantization procedure is applied for weakly interacting boson systems where the nonextensive thermodynamical behavior and the q-algebra can be taken into account and can be successfully built into the theory. The Hamilton–Jacobi equation, the action and the kernel can be calculated in the space of generalized coordinates. This method, called Feynman quantization, shows that repulsive potentials are acting in the heat process, similar to a classical and a quantum–thermodynamical potential.

Perez-Garcia and Jou (1992) present EIT theory as a suitable macroscopic framework for the interpretation of generalized transport coefficients. The classical thermodnamic theory holds only in the hydrodynamic limit of small frequency and small-wave vector (vanishing ω and k) for any disturbances. A generalized hydrodynamics (GH) is postulated which replaces the dissipative transport coefficients by general memory functions which are justified by some microscopic models. Memory terms constitute important ingredients considered in many advanced thermodynamic approaches (Nunziato, 1971; Shter, 1973; Builtjes, 1977; Altenberger and Dahler, 1992; Morro, 1985, 1992, and many others). A phenomenological theory involving higher-order hydrodynamic fluxes is given for the description of GH processes. The additional variables playing the role of the currents of the traceless pressure tensor and viscous pressure are introduced into the extended entropy and balance equations. The entropy flux contains the most general vector that can be constructed with all the dissipative variables. As specific applications, velocity autocorrelation functions are calculated around equilibrium, and ultrafast thermometry is analyzed. A parallelism of the macroscopic theory with the projection-operator techniques is outlined. Memory functions and continued fraction expansions for transport coefficients are interpreted in terms of the generalized thermohydrodynamics.

Llebot (1992) presents extended irreversible thermodynamics of electrical phenomena. He studies a rigid isotropic solid conductor in the presence of flows of heat and electricity. By combining the balance equations for the internal energy and electron concentration with the nonequilibrium entropy an expression for the entropy production is obtained containing the generalized thermodynamic forces with the time derivatives of the energy and the electron flux. He next considers nonstationary electrical conduction. He shows that only EIT can preserve the positiveness of the entropy production σ in general. For sufficiently large frequencies, the classical theory can yield an erroneous expression allowing negative σ during part of the cycle. However, EIT always gives the correct result implying $\sigma > 0$ at each time instant. Nonequilibrium fluctuations of the electric current are analyzed. Instability of the electric current in plasma is studied in the framework of EIT. The critical value of the electric field obtained within the EIT formalism is in good agreement with that obtained on the basis of kinetic considerations (Fokker–Planck equation). Jou and Llebot (1980) offer the treatment of fluctuations of electric current in the extended irreversible thermodynamics.

Garcia-Colin (1988, 1992) analyzes chemical kinetics as described by the classical kinetic mass action law from the viewpoint of the compatibility of this law with EIT. Consistency of the kinetic mass action law with thermodynamics is also discussed by Garcia-Colin et al. (1986). The difficulty lies, of course, in the nonlinearity of the kinetics. A single step homogeneous chemical reaction is considered with the well-known exponential relation between the chemical flux and the ratio of the affinity and the temperature. Nonequilibrium entropy is postulated to depend on the conventional variables, as well as on diffusion fluxes, heat flux and chemical flux. Using the formalism of extended thermodynamics, an expression for the entropy source is obtained and extended phenomenological relations are postulated. It is shown that a correspondence between classical kinetics and the implications of EIT can be made provided that an ansatz is adopted concerning the appropriate form of a phenomenological function appearing in the generalized relaxation formula. Fluctuations of the chemically reacting fluid around an equilibrium reference state are discussed in the framework of EIT. It is shown that their time decay obeys an exponential rather than a linear equation due to nonvanishing relaxation times in the fluctuation formulae. The open question concerning the existence of general thermodynamic criteria suitable for predicting how a chemical reaction evolves in time is discussed.

Ebeling (1983) compared with experiments the theory of hydrodynamic turbulence formulated by Klimontovich (1982, 1986), which yields the effective turbulent viscosity as a linear function of the Reynolds number. Klimontovich's theory seems to be able to describe the observed effective viscosity of hydrodynamic vortices and, after introducing some modifications, the fluid flow in tubes up to very high Reynolds numbers. The good agreement with experimental data from tube flow measurements dementrates the usefulness of the theory and soundness of its assumptions. Yet, in conclusion, the author states that the theory is still semi-phenomenological since there appear phenomenological constants. The development of a statistical theory would be highly desirable. The entropy of dissipative and turbulent structures, defined by Klimontovich is discussed by Ebeling et al. (1986). According to Beck (2002a, b) statistical properties of fully developed hydrodynamic turbulence can be well described using methods from nonextensive statistical mechanics (see chapter: Thermodynamic Lapunov Functions and Stability).

Relativistic thermal inertia systems (leading to EIT effects even in the limiting case of the low velocities) have been investigated in terms of Lagrangian and Hamiltonian structures. A theory of thermodynamic transformations has been constructed in which the extended Legendre transformations play the double role characteristic of mechanics and thermodynamics (Sieniutycz, 1986, 1988b, 1989b, 1990, 1992b). Similar qualitative effects have been found for the theory constructed on the basic of Grad's (1958) solution. Invariant thermodynamic temperatures have been found and a theory of thermodynamic transformations developed in the context of field thermohydrodynamics (Sieniutycz and Berry, 1989, 1991, 1992, 1997). Valuable variational principles for irreversible wave systems were found and successfully applied to obtain approximate solutions by direct variational methods (Vujanovic and Jones, 1989; Sieniutycz and Farkas, 2005).

Sieniutycz's (1992b) development of extended thermodynamics continues his earlier analysis (Sieniutycz, 1990) which views the EIT theory as a macroscopic consequence of the "micro-thermodynamic" theory of de Broglie (1964, 1970). The treatment is in the context of Hamilton's action and the conservation laws for multicomponent fluids with source terms in the continuity equations (Sieniutycz, 1994). The de Broglie theory is microscopic, quantum, and relativistic; however, its nonrelativistic and macroscopic counterpart is pursued. This is sufficient to preserve the effect of thermal inertia, inherent in that theory. In this extended thermodynamics, sources of entropy and matter result as partial derivatives of a generalized energy, E, with respect to the action-type variables (phases) under the condition of constancy of momenta and other natural variables of E. These sources can also be obtained by differentiation of generalized thermodynamic potentials (Legendre transforms of E) with respect to the phases. Internal symmetries of the theory are shown to yield the global mass conservation law and ensure chemical stoichiometry. Vector external fields and charged systems can be considered in a generalized variational scheme. Quite general equations of motion are obtained. They correspond to the phenomenological equations of irreversible thermodynamics under a suitable kinetic potential or generalized energy function. Since the usual assumption allowing the energy-momentum tensor be independent of the sources of the entropy and matter cannot generally be met, the conservation laws of irreversible processes seem to be modified by the presence of dissipation (explicit phase variables). These issues are summarized (Sieniutycz, 1994).

Bennett (1982) considers the thermodynamics of computation and defines algorithmic entropy as a microscopic analog of ordinary statistical entropy in the following sense: if a macrostate **p** is concisely describable, for example, if it is determined by equations of motion and boundary conditions describable in a small number of bits, then its statistical entropy is nearly equal to the ensemble average of the microstates' algorithmic entropy. Alicki and Fannes (2005) present line of ideas centered around entropy production and quantum dynamics, emerging from von Neumann's work on foundations of quantum mechanics and leading to current research. The concepts of measurement, dynamical evolution and entropy were central in von Neumann's work. Further developments led to the introduction of generalized measurements in terms of positive operator-valued measures, closely connected to the theory of open systems. Fundamental properties of quantum entropy were derived and Kolmogorov and Sinai related the chaotic properties of classical dynamical systems with asymptotic entropy production. Finally, entropy production in quantum dynamical systems was linked with repeated measurement processes and a whole research area on nonequilibrium phenomena in quantum dynamical systems seems to emerge. A stochastic parameter which appears to be related to the Kolmogorov entropy has been computed by Casartelli et al. (1976) for a system of N particles in a line with the nearest-neighbor Lennard-Jones interaction. It has been found that the parameter depends on the initial conditions, and is equal to zero or to a positive value which depends on the specific energy u. A limit seems to exist for the parameter at fixed u when N→∞, as shown by computations from N = 10 to 200. See also Landauer (1961) and Bag and Ray (2000) for further investigations.

Altenberger and Dahler (1992) propose an entropy source-free method which allows one to answer unequivocally questions concerning the symmetry of the kinetic coefficients. It is based on nonequilibrium statistical mechanics and it involves a linear functional generalization of Onsagerian thermodynamics. It extends some earlier attempts (Schofield, 1966; Shter, 1973) to generalize fluctuation dynamics and Onsager's relations using convolution integrals to link thermodynamic fluxes and forces. By using the Mori operator identity (Mori, 1965), applied to the initial values of the currents of mass and heat, constitutive equations linking the thermodynamic fluxes and forces are obtained. Correspondence of the results of the functional theory with the classical one is verified by the use of Fourier representations for the fields and for the functional derivatives. This approach provides a consistent description of transport processes with temporal and spatial dimensions and yet enables one to recover the classical results in the long-wave limit, where the gradients in the system are small. The possibility of extending this linear theory to situations far from equilibrium is postulated. The extension is based on the probability densities of the phase-space variables determined for stationary nonequilibrium reference states. The roles of external fields and various reference frames are investigated, with the conclusion that the laboratory frame description is as a rule the most suitable for the theory. Transformations of the thermodynamic forces are thoroughly analyzed, focusing on violations of the Onsagerian symmetries.

While endoreversible heat-to-power conversion systems operating between two heat reservoirs have been intensely studied, systems with several reservoirs have attracted little attention. However Sieniutycz and Szwast (1999) and Amelkin et al. (2005) have analyzed the maximum power processes of the latter systems with stationary temperature reservoirs. Amelkin et al. (2005) have found that, regardless the number of reservoirs, the working fluid uses only two isotherms and two infinitely fast isentropes/adiabats. One surprising result is that there may be reservoirs that are never used. This feature has been explained for a simple system with three heat reservoirs. However, Gyftopoulos (1999b) criticizes the finite time thermodynamics adducing several reasonable and a number of invalid arguments.

In PhD thesis of Andresen (1983) most of his earlier papers on FTT can be found. Andresen et al. (1984b) formulate basic principles of thermodynamics in finite time. Andresen et al. (1984c) criticize the claim that, among the several possible spacial structures, the system tends to prefer the structure producing the entropy at the largest rate. In other words, the most stable stationary state is unnecessarily associated with the largest rate of entropy production. Andresen (1990) regards finite-time thermodynamics (FTT) as a specific branch of irreversible thermodynamics which uses aggregated macroscopic characteristics (eg, friction coefficients, heat conductances, reaction rates, etc.) rather than their microscopic counterparts. Treating processes which have explicit time or rate dependences, he discusses finite-time thermodynamic potentials and the generalization of the traditional availability to the case of finite-time, involving entropy production and constraints. Paths for endoreversible engines are analyzed from the viewpoint of various performance criteria and comparison is made of their properties with those of traditional engines (eg, Carnot engines). Andresen (2008) postulates the need for disequilibrium entropy in finite-time thermodynamics and elsewhere. In his review of the later advances in the FTT field, Andresen (2011) describes the concept of finite-time thermodynamics, which, according to his opinion, can be applied not only to optimize chemical and industrial processes but also, when appropriate variables are replaced, to solve economic and possibly even ecological problems.

One of the classic problems of thermodynamics has been the determination of the maximum work that might be extracted when a prepared system is allowed to undergo a transformation from its initial state to a designated final state (Andresen et al., 1983). When that final state is defined by the condition of equilibrium between the system and an environment, the maximum extractable work is generally known now as the availability A (other name is exergy). This is a convenient shortening of "available work." In the paper by Andresen et al. (1983) the concept of availability as an upper bound to the work that can be extracted from a given system in connection with specified surroundings is extended to processes constrained to operate at nonzero rates or to terminate in finite times. The authors use some generic models which describe a whole range of systems in such a way that the optimal performance of the generic model is an upper bound to the performance of the real systems. The effects of the time constraint are explored in general and in more detail for a generic model in which extraction of work competes with internal relaxation. Extensions to nonmechanical systems are indicated. The authors wish to establish the finite-time availability as a standard of performance more useful than the traditional availability of reversible processes. For further work along this line, especially in the context of finite-time exergy via HJB equations of dynamic programming, see Sieniutycz (1997b, 1998a, b, 1999, 2000a, d, i, 2001a, b, e, 2013a, b), Xia et al. (2010b, c, 2011a, b), and some others. Andresen et al. (1984a) have pointed out the inadequacy of reversible thermodynamics to describe processes which proceed at nonvanishing rates and have pointed out a number of new methods to treat this situation. The primary goal has been to obtain bounds of performance which are more realistic than the reversible ones. Some of the finite-time procedures are generalizations of traditional quantities, like potentials and availability; others are entirely new, like the thermodynamic length. Since the central ideas of reversible thermodynamics are retained in finite-time thermodynamics, the authors are continuing their attempts to generalize traditional concepts to include time. Especially important are connections to statistical mechanics and irreversible thermodynamics, for example, investigating the finite-time content of Keizer's Σ-function and processes far from equilibrium (Keizer et al., 1988). The most exciting results emerging at the moment are applications of the thermodynamic length which seems to be able to simplify calculations on such diverse systems as lasers, separation by diffusion, and signal encoding. Badescu (2013) treats radiation systems and states that links between lost available work and entropy generation cannot be obtained in the general case, that is, for systems in contact with arbitrary heat and/or radiation reservoirs. Reservoirs in local (full or partial) thermodynamic equilibrium are needed (see chapter: Radiation and Solar Systems).

A main tenet of empirical sciences is the possibility of inferring the basic features of studied systems by means of the outputs generated by the system itself in the form of a measurable quantity. This implies the possibility of relating some features of the output with some other relevant feature of the generating system. Working with this premise Trulla et al. (2003) present the derivation of a general "distance-from-randomness" index for time series scaling with the relative complexity of an unknown generating system. This index ($SVSE_n$) corresponds to the Shannon entropy of the distribution of the eigenvalues of the correlation matrix derived by the embedding matrix of the series. The efficacy of the proposed index is tested by a number of simulations, demonstrating the possibility to derive a sensible complexity index without any strong theoretical constraint. Moreover $SVSE_n$ was demonstrated to have a direct relation with the relative complexity of the network system producing the series so allowing for a system identification approach on a strictly data-driven manner without any theoretical assumption.

A group of papers deals with entropy and relativistic extended thermodynamics (RET). The reader interested in causal relativistic theories is referred to the papers by Israel (1976) and Hiscock and Linblom (1983), and to chapter: Multiphase Flow Systems. Pavon et al. (1980) consider the heat conduction in ERT. Pavon et al. (1983) treat equilibrium and nonequilibrium fluctuations in relativistic fluids. Pavon (1992) deals with the applications to astrophysics and to cosmology, rather than dealing with the general theory. The second moments of fluctuations are determined from the Einstein probability formula involving the second differential of the nonequilibrium entropy. These moments are next used to determine the transport coefficients of heat conductivity, bulk viscosity and shear viscosity of a radiative fluid. By exploiting correlation formulae, second-order coefficients of transport are determined. Nonequilibrium fluctuation theory, based on the assumption of the validity of the Einstein formula close to equilibrium, is used to determine the nonequilibrium corrections of the bulk viscous pressure. An analysis of survival of the protogalaxies in an expanding universe completes the presentation of astrophysical applications of ERT. Cosmological applications are discussed next. It is shown that nonequilibrium effects and transport properties (such as, eg, bulk viscosities) play a certain role in the cosmic evolution of the universe. In that context, the following issues are outlined: cosmological evolution (described by a Bianchi-type model with the bulk viscosity), production of entropy in the leptonic period, the inflationary universe and FRW cosmology. Some cosmological aspects of the second law are also discussed.

In a highly theoretical and general paper Williams (2001) shows that the classical laws of thermodynamics require that mechanical systems must exhibit energy that becomes unavailable to do useful work. In thermodynamics this type of energy is called entropy. It is further shown that these laws require two metrical manifolds, equations of motion, field equations, and Weyl's quantum principles. Weyl's quantum principle requires quantization of the electrostatic potential of a particle and that this potential be nonsingular. The interactions of particles through these nonsingular electrostatic potentials are analyzed in the low velocity limit and in the relativistic limit. An extension of the gauge principle into a five-dimensional manifold, then restricting the generality of the five-dimensional manifold by using the conservation principle, shows that the four-dimensional hypersurface that is embedded within the 5-D manifold is required to obey Einstein's field equations. The 5-D gravitational quantum equations of the solar system are presented.

Gyftopoulos (1998a) discusses the statistical interpretation of thermodynamics and finds that is inadequate and incomplete to explain thermodynamic phenomena. He summarizes and compares two alternative approaches to the statistical interpretation, one purely thermodynamic and the other quantum-theoretic and thermodynamic, and the meaning of the nonstatistical entropy. References are given to the past research directed toward nonstatistical interpretations of thermodynamics. According to Gyftopoulos, recent understanding of thermodynamics casts serious doubts about the conviction that, in principle, all physical phenomena obey only the laws of mechanics.

The volume edited by Amoroso et al. (2002) has a unique perspective in that the chapters, the majority by world-class physicists and astrophysicists, contrast both mainstream conservative approaches and leading edge extended models of fundamental issues in physical theory and observation. For example in the first of the five parts: Astrophysics and Cosmology, papers review Bigbang Cosmology along with articles calling for exploration of alternatives to a Bigbang universe in lieu of recent theoretical and observational developments. This unique perspective continues through the remaining sections on extended EM theory, gravitation, quantum theory, and vacuum dynamics and space-time.

The question, why there are so many schools of thermodynamics, was answered concisely as follows (Muschik, 2007): There is no unique natural extension from thermostatics to thermodynamics. This extension from thermostatics to dynamics seems to be easy: one has to replace the reversible processes of thermostatics by real ones and to extent the balances of continuum mechanics by the appropriate thermal quantities such as heat flux density, internal energy and entropy. But it is not so easy as supposed, because usually thermostatics is formulated for discrete systems, whereas thermodynamics can be presented in two forms, as a nonequilibrium theory of discrete systems or in field formulation extending the balances of continuum mechanics. Both descriptions are used in practice which is widespread for thermodynamics. Its methods are successfully applied in various different disciplines such as Physics and Physical Chemistry, Mechanical and Chemical Engineering, Heat and Steam Engine Engineering, Material Science, Bio-Sciences, Energy Conversion Techniques, Air Conditioning and Refrigeration. Therefore, it is impossible to mention the different terminologies, methods and schools completely in a brief survey. Presupposing that classical thermostatics is well-known Muschik (2007) proceeds along a way sketched in Fig. 1.7.

We shall terminate with an original summary by Jou et al. (1988) in which they introduce a new formulation of nonequilibrium thermodynamics, known as extended irreversible thermodynamics (EIT). The basic features of this formalism and several applications are reviewed focusing on points which has fuelled increasing attention. Extended irreversible thermodynamics includes dissipative fluxes (heat flux, viscous pressure tensor, electric current) in the set of basic independent variables of the entropy. Starting from this hypothesis, and by using methods similar to classical irreversible thermodynamics, evolution equations for these fluxes can be obtained. These equations reduce to the classical constitutive laws in the limit of slow phenomena, but may also be applied to fast phenomena, such as second sound in solids, ultrasound propagation or generalized hydrodynamics (Jou et al., 1988). In contrast with the classical theory, extended thermodynamics leads to hyperbolic equations with finite speeds of propagation for thermal, diffusional and viscous signals. Supplementary information about the macroscopic parameters is provided by fluctuation theory. The results of the macroscopic theory are confirmed by the kinetic theory of gases and nonequilibrium statistical mechanics (Lebon, 1978). The theory is particularly useful for studying the thermodynamics of nonequilibrium steady states and systems with long relaxation times, such as viscoelastic media or systems at low temperatures. There is no difficulty in formulating the theory in the relativistic context (Israel, 1976). Applications to rigid electrical conductors as

Original irreversible thermodynamics

> Thermodynamics of irreversible processes

Nonequilibrium thermodynamics

> Rational thermodynamics
>> Large or small state spaces
>> Constitutive theory
>> Exploitation of dissipation inequality
>
> Extended thermodynamics
>> Extended irreversible thermodynamics
>> Extended rational thermodynamics
>
> Endoreversible thermodynamics
> Mesoscopic theory
> GENERIC
> Evolution criteria
> Quantum thermodynamics

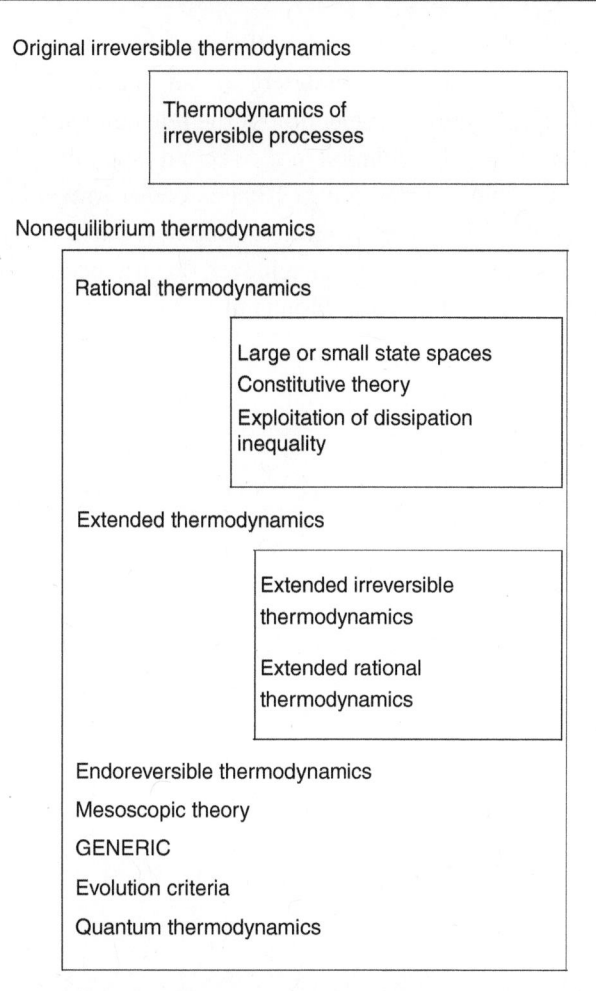

Figure 1.7: Muschik's classification of various theories of thermodynamics.

well as several generalizations including higher-order fluxes are also presented. A section is devoted to the formulation of extended irreversible thermodynamics within the framework of the rational thermodynamics (Perez-Garcia and Jou, 1986, 1992). Further, Jou et al. (1999) review the progress made in extended irreversible thermodynamics during the ten years that have elapsed since the publication of their first review on the same subject (Jou et al., 1988). During this decade much effort has been devoted to achieving a better understanding of the fundamentals and a broadening of the domain of applications. The macroscopic formulation of EIT is reviewed and compared with other nonequilibrium thermodynamic theories. The foundations of EIT are discussed on the bases of information theory, kinetic theory, stochastic phenomena, and computer simulations. Several significant applications are presented, some of them of considerable practical interest (nonclassical heat transport, polymer solutions, non-Fickian diffusion, microelectronic devices, dielectric relaxation), and some others of special theoretical appeal (superfluids, nuclear collisions, cosmology). They also outline some basic problems which are not yet completely solved, such as the definitions of entropy and temperature out of equilibrium, the selection of the relevant variables, and the status to be reserved to the H-theorem and its relation to the second law. In writing their review, the authors had four objectives in mind: to show (1) that extended irreversible thermodynamics stands at the frontiers of modern thermodynamics; (2) that it opens the way to new and useful applications; (3) that much progress has been achieved during the decade 1988–1998, and (4) that the subject is far from being exhausted (Jou et al., 1999).

CHAPTER 2
Variational Approaches to Nonequilibrium Thermodynamics

2.1 Introduction: State of Art, Aims, and Scope

To derive basic equations of nonequilibrium thermodynamics a standard method is usually applied that combines balance equations for mass, energy and momentum with the Gibbs equation. The method leads to the entropy source expression and, in the next stage, to admissible forms of linear kinetic equations. Yet, for the purpose of derivation of such equations another method can be used that applies a variational approach, where a functional is minimized. In the case when the entropy production functional is applied for assumed balance equations and suitable dissipation functions, the variational method produces classical mass action kinetics and familiar kinetics of accompanying transport processes in the form of a nonlinear set of equations describing diffusion and chemical reaction. A search proves that the variational method can be superior over the traditional one in the case of nonlinear systems and systems in which the local thermodynamic equilibrium is violated. The variational method does not postulate linearity, as it rests on given dissipation functions that depend on instantaneous state of the system. That method allows one to predict nonlinear kinetic laws for assumed balance equations, and also to treat surface and interphase reactions. Entropy functionals, which can be extremized by the so-called direct variational methods, can be useful to treat nonlinear effects, local discontinuities and local disequilibria responsible for instabilities inside phases and on interfaces. These functionals can also be useful to simplify transforming models of investigated processes to other reference systems. In this text we are focused on irreversible dynamics, so we neglect variational evaluations of equilibrium thermodynamic properties of fluids (Mansoori and Canfield, 1969).

The search for a variational principle is equivalent to the inverse problem of the calculus of variations. It consists in finding the functional whose stationary point is the required set of differential equations. Gelfand and Fomin (1970) give the basic mathematics of variational calculus. Leech (1958) offers excellent, lucid treatment of continuous mechanical systems allowing easy understanding of analytical and variational descriptions of distributed parameter systems with several independent variables. Logan (1977) presents a self-consistent treatment of invariant variational principles and conservation laws following from Nöther's fundamental theorems (Nöther, 1918, 1971). The significance of the

Hamilton–Jacobi theory in the calculus of variations is discussed in a readable book of Rund (1966, 1970).

Brownstein (1999) discusses two types of partial derivatives which occur in the Euler–Lagrange equations. To avoid confusion, a different notation and a new name (whole–partial derivative) are suggested for what is usually written as $\partial F/\partial x^{\mu}$ but acts on both the explicit and implicit dependence of F upon the coordinate x^{μ}. The reader should note first of all beautiful variational formulations in various fields of theoretical physics (Landau and Lifshitz, 1967, 1974a, b, 1976) and Osserman's (1986) survey of minimal surfaces.

The inverse problem was solved by Vainberg (1964) within the context of the modern functional analysis. He gives the necessary and sufficient conditions for a variational formulation based on the symmetry of Frechet derivative. Answering the question of existence of a variational principle is equivalent to determining whether or not an operator is the potential operator or it is self-adjoint (see further). If the operator is a potential operator then the investigated set is self-adjoint and the related variational principle is found by a straightforward calculation. Investigation of variational problems invariant with respect to space–time translations (Nöther, 1918, 1971) leads to the conservation laws for energy and momentum, and, consequently, to the matter tensor (energy–momentum tensor). Perlick (1992) discusses conditions that must be satisfied in order to put any finite dimensional dynamical system into Hamiltonian form locally. He concludes that the introduction of dummy variables is not necessary (at least at the final stage), yet the system must be even dimensional and one should allow an explicitly time-dependent symplectic structure.

Let us begin with a few variational results for lumped dissipative mechanical systems in which thermodynamics play a role (Sieniutycz, 1976a; Lin and Wang, 2013). A least action principle for damping motion has been proposed with a Hamiltonian and a Lagrangian containing the energy dissipated by friction (Lin and Wang, 2013). Due to the space–time nonlocality of the Lagrangian, mathematical uncertainties persist about the appropriate variational calculus and the nature (maxima, minima, and inflection) of the stationary action. The model is a small particle subject to conservative and friction forces. Two conservative forces and three friction forces are considered. The comparison of the actions of perturbed paths with that of the Newtonian path reveals the existence of extrema of action which are minima for zero or very weak friction and shift to maxima when the motion is overdamped. In the intermediate case, the action of the Newtonian path is neither least nor most, meaning that the extreme feature of the Newtonian path is lost. In this situation, however, no reliable evidence of stationary action can be found from the simulation result. Lin's (2013) thesis is devoted to the study of path probability of random motion on the basis of an extension of Hamiltonian/Lagrangian mechanics to stochastic dynamics. The path probability is first investigated by numerical simulation for Gaussian stochastic motion of nondissipative systems. This ideal dynamical model implies that, apart from the Gaussian random forces, the system is only subject to conservative forces. This model can be applied to underdamped real random motion in the presence of friction force when the dissipated energy is negligible with respect to the variation of the potential energy.

Herivel (1955) first formulated a variational principle for the irrotational motion of the perfect fluid in the field (Eulerian) representation. Yet, fluid motions limited to the potential flows are too restrictive. The dilemma of vanishing rotation was solved by Lin (1963) who introduced an extra constraint resulting from the fact that fluid particles are distinguishable due to their different initial positions $\mathbf{X} = \mathbf{x}(t_0)$. Lin's constraint ensures fluid flows with a nonvanishing vorticity, allowing the treatment of perfect fluids in the general case. It follows that Lin's constraint preserves the Lagrangian definition of the velocity (Salmon, 1988).

Bateman (1929, 1931) was probably the first who was able to describe some dissipative systems by variational principles of classical type, that is, principles attributed to a well-defined functional. Dissipative terms appeared in equations of motion because of an exponential term in the action functional. Denman and Buch (1962) described solutions of the Hamilton–Jacobi equation for some dissipative mechanical systems. Using Bateman's Lagrangian, Sieniutycz (1976a) proposed a stationary action principle for a laminar motion in a viscous fluid. Steeb and Kunick (1982) showed that a class of dissipative dynamical systems with limit-cycle and chaotic behavior can be derived from the Lagrangian function. Tan and Holland (1990) gave a variational setting for the tangent law of refraction when a heat flux crosses an interface.

Sieniutycz (1980b, 1981e, 1982, 1987a, 1988a) used an exponential matrix of Bateman's type to formulate action-type functionals for nonstationary wave dissipative systems with coupled heat and mass transfer. Entropy of the diffusive fluxes of mass and energy has been evaluated and then used in these variational principles (Sieniutycz, 1982). This entropy it is associated with the tendency of every element of a continuum to recover the thermodynamic equilibrium during the relaxational evolution of the fluxes. After incorporating the exponential relaxation matrix into the Lagrangian stationarity conditions of the thermodynamic action were found as the phenomenological and conservation equations (constituting a hyperbolic set). The practical significance of these variational principles consists in finding approximate fields of transfer potentials and fluxes by using the direct variational methods. For wave dissipative systems with coupled heat and mass transfer extremized action functionals lead to the fields of transfer potentials, T, μ_i, in the presence of inertial (non-Fourier) effects. A significant point and physical insight obtained is that the irreversibility can be related to time reversal in (nonautonomous) action functionals as well as that a generalization of the theorem of minimum entropy production can be found for high-rate nonstationary processes. A least action principle has also been set for simplified Maxwell–Stefan equations in the Lagrangian representation of fluid's motion (Sieniutycz, 1983) and applied to Brownian diffusion governed by hyperbolic equations (Sieniutycz, 1984a). Finally, passing to chemical systems Sieniutycz (1987b, 1995) derived basic laws that govern nonlinear chemical kinetics and diffusion from an action functional of Hamilton's type in a far-from-equilibrium nonstationary regime. Use of Nöther's theorem and reversible Lagrangian of extended thermodynamics has lead to the Hamiltonian energy–momentum tensor of one-component and multicomponent fluids (Sieniutycz, 1988b, 1989b). Further work lead to a macroscopic extension of the de Broglie microthermodynamics (de Broglie, 1964). The first part (Sieniutycz, 1990) deals with the multicomponent fluids with sourceless continuity constraints. The second part, (Sieniutycz, 1992b) analyzes multicomponent fluids with sources. Microrelativistic de Broglie thermodynamics (de Broglie, 1964) is extended to macroscopic systems so as to preserve the usual thermostatistical effects ignored in the de Broglie theory. Many of these issues are summarized in a book (Sieniutycz, 1994).

Wagner's (2005) contribution also uses Bateman's structures to describe hydrodynamics of fluids that are under the influence of frictional force densities proportional to the flow velocity. Such models approximately describe the seepage of fluids in porous media (Darcy's law). Bateman's structures help to incorporate such linearly damped hydrodynamic systems into the framework of Lagrangian field theory, that is, the construction of variational principles for these systems. The search for variational principles highly benefits from the fact that there exists a far-reaching analogy between the field equations of classical hydrodynamics and of the so-called hydrodynamic (Madelung) picture of quantum theory. Introduction of linear frictional force densities into the Madelung picture leads to certain well-known phenomenological extensions of the ordinary Schrödinger equation such as the Caldirola–Kanai equation and Kostin's nonlinear Schrödinger–Langevin equation. Despite of several failed attempts in the literature, the construction of variational principles for these equations turns out to be relatively straightforward. The main tool used by Wagner (2005) consists of the method of Frechet differentials. Utilizing the analogy between hydrodynamics and the Madelung picture, the results for the dissipative quantum systems can immediately be carried over to the analogous frictional systems in classical hydrodynamics. However, some additional considerations are still needed due to the possibility of rotational (vorticity) degrees of freedom which can be present in ordinary hydrodynamics but are usually foreign to the Madelung picture.

The importance of abstract Vainberg's work has been appreciated (Atherton and Homsy, 1975; Bampi and Morro, 1982). They derived operational formulas to check the self-adjoiness of differential operators, defined by an arbitrary number of nonlinear equations in an arbitrary number of independent variables and of arbitrary order. By application of the Vainberg's (1964) theorem Atherton and Homsy (1975) treated the inverse problem, that is, the existence and formulation of variational principles for systems of nonlinear partial differential equations. Three types of principles are discussed: potential, alternate potential, and composite principles. The results of the development are used to treat many well-known equations in nonlinear wave mechanics. These include the Burgers, Korteweg–deVries, and sine-Gordon equations and evolution and model equations for interfacial waves. The Navier–Stokes equations and the Boussinesq equations for an infinite Prandtl number fluid are briefly treated; these latter equations do not possess a potential principle. See also discussion of the inverse problem by Sieniutycz (1994), Sieniutycz and Farkas (2005), and Liu and He (2004).

Bampi and Morro (1982) considered the inverse problem of the calculus of variations applied to continuum physics. They set necessary and sufficient conditions for a differential system of equations to admit a variational formulation by having recourse to Vainberg's (1964) theorem. The use of the theorem also provided a method for producing the sought functional. Application of the method to the Lagrangian description of fluid dynamics lead to a new variational principle which, while fully general, revealed a hierarchy between variational approaches to fluid dynamics. Next, the method was applied to obtain new variational formulations in various areas of continuum physics: waterwave models, elasticity, heat conduction in solids, dynamics of anharmonic crystals, and electromagnetism. Owing to the power of the method, permissible variational formulations can be found. The paper places particular emphasis on equations which have, or are supposed to have, soliton solutions.

Searching for Lagrangian formulations of second order systems Santilli (1977) rediscovered the self-adjoiness condition characterizing potential operators, that is, determined necessary and sufficient conditions for the existence of a Lagrangian in a field theory which assure the self-adjointness for tensorial field equations. Examples of applications cover the vast region of perfect fluid dynamics, continuum mechanics, heat conduction, anharmonic crystals, electromagnetism (Bampi and Morro, 1982), thus complementing earlier applications of Schechter (1967). Besides involving an extensive application of Vainberg's theorem, their work derives a number of new Lagrangians in a systematic way, thus proving the power of the theorem. Salmon (1987a, b) reviewed the state-of-art of Hamiltonian fluid mechanics. Books by Yourgrau and Mandelstam (1968), Berdicevskii (1983), and Kupershmidt (1992) treat the variational principles of reversible dynamics of fluids. Especially Kupershmidt (1992) provides an original and systematic treatment of Eulerian (field) variational principles phenomena in terms of the Hamiltonian approach. His analyses substantiate the superiority of the Hamiltonian approaches over the Lagrangian ones. The theory is developed for perfect fluid hydrodynamics, magneto-hydrodynamics and some other reversible systems. Variables of Clebsch type are necessary for formulation of general Lagrangians: Herivel (1955); Serrin (1959); Lin (1963); Stephens (1967); Seliger and Whitham (1968); Sieniutycz (1994).

Jezierski and Kijowski (1990) worked out an interesting variational approach to the field theory for the hydrothermodynamics of reversibly flowing fluids. The authors' formalism gets rid of continuity equations as an explicit constraint of the classical kinetic potential. Their approach is technically simpler and more elegant than earlier approaches to this problem. Although only nondissipative systems are analyzed, the reader can gain valuable information about the connections between the Lagrangian and the Eulerian representations of fluid motion and the structures of the energy–momentum tensor in the two representations. The authors discuss also the Poisson bracket structure of thermohydrodynamics. Their formulation of thermohydrodynamics in terms of the kinetic potential is based on the free energy rather than the internal energy. It involves an important quantity, τ, defined such that its total time derivative is equal to the absolute temperature (up to a scale constant). They call τ the "material time" although it was known earlier as thermasy (Van Dantzig, 1940; Schmid, 1970). An alternative interpretation of this quantity as a thermal action or a thermal phase is given by Sieniutycz (1990), where its role for the description of irreversibility is shown. Grmela and Teichman (1983) were the first who introduced a thermal coordinate in the Lagrangian formulation of the Maxwell–Cattaneo hydrodynamics. Ingarden (1996) pointed out the relevance of Finsler geometry as the geometry of an arbitrary Lagrangian.

Caviglia (1988) describes the determination of conservation laws in the context of composite variational principles. Sieniutycz (1994) presents a systematic treatment of conservation laws in variational thermohydrodynamics. Badur (2012) considers the history of energy, its development, and current aspects. Badur (2005) also demonstrates that a postulate for balances of mass, momentum and energy in continuum thermomechanics can be derived from primary variational principles. These principles are partially known in the literature. Nevertheless, more consistent and legible conclusions and reinterpretations of primary variational principles are possible only after their deeper reevaluation. In the course of his derivation he reconstructs the principles set by Rankine, Gibbs, Tait, Natanson, Jaumann, and Cosserats. They help to provide a clearer picture for the structure of the set of basic balances in thermomechanics, especially in view of continual efforts to improve mathematically an old variational proposal of Natanson (1896). Further, Badur and Badur (2005) present the reconstruction and revalorization of a few basic, pioneering variational principles in thermohydrodynamics; most of their attention is aimed at Gibbs', Natanson's and Eckart's principles. They also present a method to obtain momentum equations from variational principles. This study is completed with analysis of variational principles for a "Generalized Lagrangean Mean (GLM) approach" recently applied in atmospheric and ocean turbulence description.

Anthony (1981, 1986, 1988), Scholle (1994), and Scholle and Anthony (1997) apply complex fields and state the variational problem as follows. Let $l = l(\psi, \partial\psi)$ be the Lagrange density function of an irreversible physical system depending on the set $\psi = (\psi_1(x, t),\ldots, \psi_n(x, t))$ of fundamental field variables and their first order derivatives. The time evolution of the state ψ of the system is called a process. Let V be the spatial region and $[t_1, t_2]$ the time interval the process takes place. Among all admissible processses the real processes are distinguished as solutions of Hamilton's variational principle: An action integral

$$I := \int_{t_1}^{t_2} \int_V l(\psi, \partial\psi) d^3x\, dt$$

assumes an extremum by free and independent variation of all fields ψ in V with fixed values at the beginning and the end of the process. Depending on the physical situation variations on the system's boundary ∂V are included. Hamilton's principle yields the set of Euler–Lagrange equations:

$$\partial_t \frac{\partial l}{\partial(\partial_t \psi_i)} + \text{div} \frac{\partial l}{\partial(\nabla \psi_i)} - \frac{\partial l}{\partial \psi_i} = 0$$

Applying the idea of basic complex fields (yet passing then to the associated real variables) Anthony (1981, 1986) is concerned with the setting of the Lagrange-formalism applied to phenomenological thermodynamics of irreversible processes. The general method is exemplified by the special process of heat conduction. The total information concerning the irreversible processes is included into a Lagrange density function. From this all fundamental field and balance equations and all constitutive equations are derived by means of straightforward methods of the field theory. In this respect universal invariance requirements and Noether's theorem play a most important rule. The entropy concept is included into the formalism in quite a natural way. Especially this theory allows for a methodical unification of dissipative and nondissipative physics within Lagrange formalism. Onsager's linear theory can be reproduced, which is demonstrated for the special case of heat conduction. The dynamical stability in the entropy context as well as the unification of continuum mechanics and thermodynamics by means a Lagrange formalism can be considered Anthony (1988, 1989). His aim is to take into account explicitly internal variables which are physically identified on the microlevel. The possible applications of the theory to defect dynamics in solids are discussed in some detail. As stated by Anthony (1990), Lagrange formalism allows for setting complete theory of phenomenological thermodynamics of irreversible processes. This statement is exemplified for heat transport, diffusion and chemical reactions. Traditional formulations of thermodynamics are included in the Lagrange formalism. However, the formalism goes beyond the scope of traditional thermodynamics. By means of this most general field theoretical approach a unified phenomenological theory of material behavior including thermodynamics, mechanics and electromagnetism might be obtained. A principle of least entropy production can be discussed in the framework of Lagrange formalism (Anthony, 2000). According to the author, this principle is verified for stationary processes of very special systems only. A general formulation is still missing. Yet, in the Lagrange formalism of TIP, a general concept of entropy is involved in quite a natural and straightforward way. It is implicitly defined by an inhomogeneous balance equation—the second law of thermodynamics—which essentially results from an universal structure of Lagrange formalism and from the invariance of the Lagrangian with respect to a group of common gauge transformations applied to the complex-valued fundamental field variables. Within the Lagrange formalism a general principle of least entropy production is associated with the entropy concept. According to many works of

Anthony (1981, 1988, 1989, 1990, 2000, 2001) Hamilton's action principle provides a unifying procedure for both reversible and irreversible processes. As a remarkable fact, the first and the second law of thermodynamics are "derived in" Lagrange formalism by means of straightforward procedures. The whole information on the dynamics of a system is included in one function only, namely in its Lagrangian. Scholle and Anthony (1997) attempt to find a combined Lagrangian for material flow with heat conduction. For this purpose Seliger's and Whitham's (1968) Lagrangian is adapted to Anthony's concept at first by substituting the six real–valued fundamental fields by three complex–valued ones, namely the matter field, the field of thermal excitation, and the vortex potential. Based on these substitutions, an ansatz for a combined Lagrangian is possible. In Anthony's (2001) paper the theory is presented in two courses. The first one offers an almost qualitative insight into the structure of the theory; the second one is concerned with associated mathematics. The theory is illustrated by examples to show the state-of-art of nonequilibrium thermodynamics of complex fluids. Anthony's and Scholle's results are particularly important for nonstationary processes for which finding of variational formulations was always difficult because of presence of both first and second derivatives in equations of motion. Variational principles for stationary fields are easier to formulate (Kirschner and Törös, 1984).

Some modification of Anthony variational approach is presented by Kotowski (1989). The evolution criterion, as formulated by Glansdorff and Prigogine, allows one to find a kinetic potential describing a dissipative process. A paper by Kotowski and Radzikowska (2002) deals with the construction of the kinetic potential for micropolar media embedded in the dielectric fluid and subjected to the coupled field interactions. They extend the results derived by Muschik and Papenfuss (1993), for the liquid crystals immersed in the perfect fluid, playing there the role of the thermostat. Generalization deals with the dissipative coupled mechanical, thermal and electromagnetic processes occurring in the micropolar medium. The formulated criterion has been applied to obtain the equilibrium conditions for micropolar elastic solids, micropolar electromagnetic fluids, and liquid crystals. Bhattacharya (1982) proposed a restricted variational principle for thermodynamic waves.

Sieniutycz and Berry (1989) treated reversible extended thermodynamics of heat-conducting fluids to obtain explicit formulas for nonequilibrium energy density of an ideal gas expressed as functions of classical variables and the diffusive entropy flux (a nonequilibrium variable). The Lagrangian density, obtained as the Legendre transform of the energy density with respect of all velocities, was used to obtain components of energy-momentum tensor and corresponding conservation laws on the basis of Hamilton's principle and Noether's theorem. The heat flux appeared naturally as a consequence of a free entropy transfer (independent of mass transfer) and a momentum transport was associated with tangential stresses resulting from this entropy transfer. The compatibility of the description with the kinetic theory was shown. This work extends Hamilton's principle so that the flux of entropy as well as the fluxes of mass are varied independently. It points out the role of Clebsch variables (as some potentials and Lagrange multipliers of balance constraints) and introduces the concept of thermal momentum as the derivative of the kinetic potential with respect to the entropy flux. The latter quality plays a basic role in the extension of Gibb's equation to describe a nonequilibrium fluid with heat flux.

Also, for reversible nonequilibrium fluids with heat flow Kupershmidt (1990a) developed a Hamiltonian formalism as opposed to the Lagrangian formalism of Sieniutycz and Berry (1989). In addition Kupershmidt (1990b) presented a variational theory of magneto-thermo-hydrodynamics in terms of the canonical Hamiltonian approach rather than the Lagrangian approach. This author advocates the use of Hamiltonian approach for continuous mechanics systems. He specifies several arguments proving the superiority of the Hamiltonian approach over the Lagrangian one to describe the time evolution of continua. He also presents recipes for determining the relation between the two formalisms. These recipes are first used to develop the Hamiltonian theory of magneto-hydrodynamics, and then to pass to the Lagrangian description. The bonus is a new Lagrangian for relativistic and nonrelativistic cases. One basic conclusion drawn by this author is that no Lagrangians are possible (in the physical space-time) without the presence of extra variables of the Clebsch type.

In a synthesizing work by Sieniutycz (1997a) a method was discovered, based on equivalent variational problems, which makes it possible to show the equivalence of thermodynamic potentials at nonequilibrium, and, in particular, the practical usefulness of free energy functionals. Also, an extended action approach was worked out for thermomechanical chemical kinetics in distributed systems. This should be important for biophysical systems such as, for example, contracting muscles, whose dynamics are described by dissipative Lagrange equations containing the mechanical equation of motion and equations of chemical kinetics (Guldberg and Waage, 1867), that is, mass action law and its nonlocal generalizations. Also, for the field or Eulerian description of heat conduction, Sieniutycz (2000f, 2001d, f) described thermal fields by a variational principle involving suitably constructed potentials rather than original physical variables. The considered processes were: simple hyperbolic heat transfer and coupled parabolic transfer of heat, mass and electric charge. By using various (gradient or nongradient) representations of original physical fields in terms of potentials, similar to those which Clebsch used in his representation of hydrodynamic velocity, suitable action-type criteria were found, and corresponding Lagrangian and Hamiltonian formalisms were developed (Sieniutycz, 2005c, d). Symmetry principles were also considered and components of energy-momentum tensor evaluated for a gauged Lagrangian. The limiting reversible case appeared as a suitable reference frame. These results suggest that thermodynamic irreversibility does not change neither kinetic potential nor action functional; it only complicates potential representations of physical fields in comparison with those describing the reversible evolution. Associated work (Sieniutycz, 2002a) discusses optimality of nonequilibrium systems and problems of statistical thermodynamics. Still another variational approach (Sieniutycz and Berry, 1993), which lead to a canonical formalism, fundamental equation, and thermomechanics for irreversible fluids with heat flux, was not really successful because the coincidence of irreversible equation of thermal motion with the Cattaneo equation occurred only in a very narrow regime in which effects of spatial derivatives of Clebsch potentials are negligible. These issues are summarized in books (Sieniutycz, 1994; Sieniutycz and Farkas, 2005). Variational theory for thermodynamics of thermal waves was constructed (Sieniutycz and Berry, 2002).

Other variational principles (Sieniutycz and Berry, 1992; Sieniutycz and Ratkje, 1996a; Sieniutycz and Farkas, 2005), also embedded in the physical space-time, substantiate the role of the entropy and entropy production rather than action. For the prescribed state and/or fluxes at the boundaries, these principles imply a least possible growth of entropy under the constraints imposed by conservation laws, which suggests that the entropy plays in thermodynamics a role similar to that of action in mechanics. These principles serve to derive nontruncated sets of the phenomenological equations, equations of change and bulk overvoltage properties in complex systems. The nonequilibrium temperatures and chemical potentials are interpreted in terms of its Lagrangian multipliers. These multipliers converge to the classical thermodynamic intensities when the local equilibrium is attained in the system. Variational Lagrangian and Hamiltonian-like formulations were compared with minimum dissipation approaches in the context of lumped chemical kinetics (Sieniutycz and Shiner, 1992, 1993).

Gavrilyuk and Gouin (1999) presented a new form of governing equations of fluids arising from Hamilton's principle. Gouin and Ruggeri (2003) compared the theory of mixtures based on rational thermomechanics with the one obtained by Hamilton principle. They proved that the two theories coincide in the adiabatic case when the action is constructed with the intrinsic Lagrangian. In the complete thermodynamical case they showed that coincidence occurs in the case of low temperature when the second sound phenomena arises for superfluid helium and crystals. Qiao and Shan (2005) introduced application of variational principles into the stability analysis of composite structures, in particular, local and global buckling of fiber-reinforced polymer (FRP) structural shapes, from which explicit solutions and equations for design and analysis are developed.

Gambar and Markus (1994) have developed the Hamilton–Lagrange formalism within nonequilibrium thermodynamics. Their theory follows from their earlier Lagrangian (Markus and Gambar, 1991) approach from which equations of thermal field were derived as Euler–Lagrange equations of the variational problem. They identified the canonically conjugated quantities and the related Hamiltonian. Then they deduced the canonical field equations and explained the Poisson-bracket structure. See Nonnenmacher (1986) for fluid mechanics applications of functional Poisson brackets. From the Poisson-bracket expression of the entropy density and the Hamiltonian, Gambar and Markus (1994) found that the entropy-produced density is a bilinear expression of the current densities and the thermodynamic forces. They also considered the invariance properties of derived irreversible thermodynamics. They showed that geometrical transformations do not lead to new conserved quantities. Finally they gave a dynamical transformation leading to the Lagrangian and showed that the reciprocity relations are the consequences of this inner symmetry. See also Gambar and Markus (2003) for their treatment of Onsager's regression and the field theory of parabolic processes and Gambar et al. (1991) for a variational principle describing the balance and constitutive equations in convective systems. Nice physical analyses incorporated in this research may be contrasted with rather formal variational approaches to the theory of nonlinear heat and mass transport (Mikhailov and Glazunov, 1985) and to the minimum dissipation in dynamical diffusion-reaction systems (Mornev and Aliev, 1995). Yet in this group, Ziegler (1963), presented a number of interesting extremum principles in irreversible thermodynamics with application to continuum mechanics. See also Muschik and Trostel (1983) for a review and classification of variational principles in thermodynamics.

Moroz (2008) attempts to revise and revaluate the applicability of the optimal control and variational approach to the maximum energy dissipation (MED) principle in nonequilibrium thermodynamics. The optimal control analogies for the kinetic and potential parts of thermodynamic Lagrangian (the sum of the positively defined thermodynamic potential and positively defined dissipative function) are considered. Interpretation of thermodynamic momenta is discussed with respect to standard optimal control applications, which employ dynamic constraints. Also included is an interpretation in terms of the least action principle. In effect, the maximum energy dissipation related to maximum entropy production hypothesis has been revised by Moroz (2008) in terms of the pure variational approach and the Pontryagin maximum principle in optimal control. The use of minimization functional as the sum of the positively defined thermodynamic potential and positively defined dissipation function corresponds to a least action principle. In this respect, Ziegler's principle of achieving the maximum energy dissipation rate (MED principle) can be interpreted as coinciding with the least action principle. The least action principle in such a sense is a methodological principle according to which, physical and chemical processes in the system are directed to the extremely fast elimination of the physical nonequilibrium, as far as the system's structural variety and kinetical mechanisms dependent on it, allow. Free energy, which is an energetic measure of how far a system is removed from its equilibrium, achieves its minimum in the way that minimizes the area under the dissipation curve—the physical action of the dissipative process, Moroz (2008). The costate variables, or thermodynamic momenta, could be interpreted as the marginal energetical dissipative losses for the partial alteration of the dissipative mechanisms (expressed by the dynamical system for the state variables) from optimal (maximal) processes. Moroz (2009) constructed also a variational framework for chemical thermodynamics employing the principle of maximum energy dissipation.

Much effort has been devoted to Fisher's information measure (FIM), shedding much light upon the manifold physical applications (Frieden and Soffer, 1995). It can be shown that the whole field of thermodynamics (both equilibrium and nonequilibrium) can be derived from the MFI approach. In the paper by Frieden and Soffer (1995) FIM is specialized to the particular but important case of translation families, that is, distribution functions whose form does not change under translational transformations. In this case, Fisher measure becomes shift-invariant. Such minimizing of Fisher's measure leads to a Schrödinger-like equation for the probability amplitude, where the ground state describes equilibrium physics and the excited states account for nonequilibrium situations. This agrees with Goldstein (2009) who states that many recent results suggest that quantum theory is about information and it is best understood as arising from principles concerning information and information processing.

Lin's (2013) model can be applied to underdamped real random motion in the presence of friction force when the dissipated energy is negligible with respect to the variation of the potential energy. The author finds that the path probability decreases exponentially with increasing action, that is, $P(A) \sim e^{-\gamma A}$, where γ is a constant characterizing the sensitivity of the action dependence of the path probability, the action is the time integral of the Lagrangian $L = K - V$ over a fixed time period T, K is the kinetic energy, and V is the potential energy. This result is a confirmation of the existence of a classical analog of the Feynman factor $e^{iA/\hbar}$ for the path integral formalism of quantum mechanics of Hamiltonian systems. The result is next extended to the real random motion with dissipation. For this purpose, the least action principle has to be generalized to damped motion of mechanical systems with a unique well-defined Lagrangian function which must have the usual simple connection to Hamiltonian. This goal is achieved with the help of the Lagrangian $L = K - V - E_d$, where E_d is the dissipated energy. By variational calculus and numerical simulation, the author proves that the action A is stationary for the optimal Newtonian paths. More precisely, the stationarity is a minimum for underdamped motion, a maximum for overdamped motion and an inflexion for the intermediate case. On this basis, the author studies the path probability of Gaussian stochastic motion of dissipative systems. He finds that the path probability still depends exponentially on Lagrangian action for the underdamped motion, but depends exponentially on kinetic action $A = \int_0^T K \, dt$ for the overdamped motion.

In spite of numerous works in recent years and a progress in the field, no exact variational formulation was achieved to date for the Navier–Stokes equation. The results obtained recently (Sciubba, 2005; Sieniutycz and Farkas, 2005) confirm earlier negative statements (Milikan, 1929; Atherton and Homsy, 1975). Yet, it seems that the hope lies in stochastic action principles. Eyink (2010) formulates a stochastic least-action principle for solutions of the incompressible Navier–Stokes equation. In the case of vanishing viscosity this stochastic principle formally reduces to the Hamilton's principle for incompressible Euler solutions. At the present time a number of papers on stochastic Lagrangian representation of the incompressible Navier–Stokes equations are available (Constantin and Iyer, 2008; Antoniouk et al., 2013; Gomes, 2005; Eyink, 2010).

Eyink (2010) uses his stochastic least-action principle to give a new derivation of a stochastic Kelvin theorem for the Navier–Stokes equation, recently established by Constantin and Iyer (2008), which shows that this stochastic conservation law arises from particle-relabelling symmetry of the action. Eyink (2010) discusses irreversibility, energy dissipation and the inviscid limit of Navier–Stokes solutions in the framework of the stochastic variational principle. In particular, he discusses the connection of the stochastic Kelvin theorem with his previous "martingale hypothesis" for fluid circulations in turbulent solutions of the incompressible Euler equations. He also analyzes another variational principle for fluid equations, namely, Onsager's principle of least dissipation (Onsager, 1931b; Onsager and Machlup, 1953). It determines the probability of molecular fluctuations away from hydrodynamic behavior in terms of the dissipation required to produce them. A modern formulation, presented in Eyink (1990), gives a rigorous derivation of Onsager's principle for incompressible Navier–Stokes equation in a microscopic lattice-gas model. It is still an open question if any relation exists between the least-action and least-dissipation principles.

Variational principles are also formulated at the statistical mechanical level (Lewis, 1967). Working in the context of variational principles for Boltzmann equation, Karkheck (1992) developed the Boltzmann equation and other kinetic equations in the context of the maximum entropy principle and the laws of hydrodynamics. Benin (1970) obtained variational principles for transport coefficients. A progress was achieved in the variational principles describing the nonequilibrium thermodynamics of complex fluids (Edwards and Beris, 2004). Gaveau et al. (2005) investigated master equations and path-integral formulations of variational principles for reaction-diffusion systems. The mesoscopic nonequilibrium thermodynamics of a reaction-diffusion system was described by the master equation. The information potential was defined as the logarithm of the stationary distribution. The authors show that the Fokker–Planck approximation and the Wentzel–Kramers–Brillouin method give very different results. The information potential obeys a Hamilton–Jacobi equation, and from this fact general properties of this potential are derived. The Hamilton–Jacobi equation has a unique regular solution. Using the path integral formulation of the Hamilton–Jacobi approximation to the master equation it is possible to calculate rate constants for the transition from one well to another one of the information potential and give estimates of mean exit times. In progress variables, the Hamilton–Jacobi equation has always a simple solution which is a state function if and only if there exists a thermodynamic equilibrium for the system. An inequality between energy and information dissipation is studied, and the notion of relative entropy is investigated. A specific two-variable system and systems with a single chemical species are investigated in detail, where all the relevant quantities are calculated.

The approximation scheme named "the method of weighted residuals" has been extended to systems of differential equations and vector differential equations (Finlayson and Scriven, 1965). The restricted (nonexact) variational principles proposed by Rosen (1953) and Biot (1970) were all shown to be applications of the method of weighted residuals. The von-Kármán–Pohlhausen method and the method of moments are also shown to be special cases. The method was illustrated by application to the problem of unsteady heat transfer to a fluid in ideal stagnation flow (Finlayson and Scriven, 1965). Several attempts to formulate variational principles for nonself-adjoint and nonlinear systems were examined by Finlayson and Scriven (1967) in their search for variational principles. The variational formulations were found to lack the advantages of genuine variational principles, chiefly because the variational integral is not stationary or because no variational integral exists. The corresponding variational methods of approximation were shown to be equivalent to the more straightforward Galerkin method or another closely related version of method of weighted residuals. The methods due to Rosen (restricted variations), Glansdorff and Prigogine (local potential), and Biot (Lagrangian thermodynamics) were considered and it was concluded that there is no practical need for pseudovariational formulations of this sort (Finlayson and Scriven, 1967).

Constraints in variational problems are usually treated by the method of Lagrange multipliers. Yet, the so-called semiinverse method appeared (He, 1997) successful in omitting Lagrange multipliers in composite principles. On this basis a variational approach to chemical reaction was constructed (Liu and He, 2004). Kennedy and Bludman (1997) reformulated the four basic equations of stellar structure as two alternate pairs of variational principles. Different thermodynamic representations lead to the same hydrodynamic equations. They also discussed the difficulties in describing stellar chemical evolution by variational principles. Yet the variational description of dissipative systems and chemical reactions is still invariably difficult. The difficulties persist when exact variational formulation is sought in the context of Gyarmati's (1969) "Governing principle of dissipative processes."

Quantization of classical variational principles received some attention. In his first paper on wave mechanics, Schrödinger presented a heuristic argument which led from the Hamilton–Jacobi equation through the quantum variational principle to his famous wave equation. In his second paper, Schrödinger withdrew this heuristic argument as 'incomprehensible'. Gray et al. (1998) showed by using a recently generalized form of Maupertuis's principle that Schrödinger's original heuristic argument can be made more logical. Aside from pedagogical interest, this path is useful as a method of quantization of general mechanical systems. Progress in classical and quantum variational principles was reviewed by Gray et al. (2004). Partial results on the Lagrangians for dissipative systems and the role of conformal invariance and Hamilton–Jacobi theory for dissipative systems are available (Kobe et al., 1985; Kiehn, 1975).

Approximation techniques and variational principles represent vital tools for solving partial differential equations. The popular book of Finlayson (1972) introduces the reader to solution methods at the freshman's level, before progressing through increasingly challenging problems. The book describes variational principles, including how to find them, and how to use them to construct error bounds and create stationary principles. Following an earlier treatise (Kantorovich and Krylow, 1958) devoted to approximate methods of higher analysis, Finlayson's (1972) book illustrates how to use simple methods to find approximate solutions, how to use the finite element method for complex problems, and how to ascertain error bounds. Applications to fluid mechanics and heat and mass transfer problems are emphasized. With problem sets included, this book is a self-study guide on the numerical analysis.

Wang and Wang (2012) extended the least action principle of mechanics to damped system by considering the Lagrangian function of a total system which contains a damped moving body and its environment, both coupled by friction. Although the damped body is a nonconservative system, the energy of the total system is conserved since energy is transferred only by friction from the body to the environment. Hence the total system can be considered as a Hamiltonian system whose macroscopic mechanical motion should a priori obey the least action principle. The essential of the formulation is to find a Lagrangian of the total system for the calculation of action. This is solved thanks to a potential like expression of the dissipated energy whose derivative with respect to the instantaneous position of the damped body yields the friction force. This Lagrangian was also derived from the virtual work principle which is not limited to Hamiltonian system. The authors think that this extension opens a possible way to study the relationship between two independent classes of variational principles, that is, those of Hamiltonian/Lagrangian mechanics and those associated with energy dissipation. An application to quantum dissipation theory can be expected via the path integral formulation of quantum mechanics.

The aim of this work is to formulate for dissipative system a least action principle that can keep all the main features of the conventional Hamiltonian/Lagrangian mechanics such as the Hamiltonian, Lagrangian, and Hamilton–Jacobi equations, three formulations of the classical mechanics, with a unique single Lagrangian function which has the usual close energy connection and Legendre transformation. We consider a whole isolated conservative system containing a damped body and its environment, coupled to each other by friction. The Lagrangian is $L = K - V - E_d$ with an effective conservative Hamiltonian $H = K + V + E_d$ where K is kinetic energy of the damped body, V its potential energy and E_d the negative work of the friction force. This Least action principle can also be derived from the virtual work principle. It is shown that, within this formulation, the least action principle can be equivalent to a least dissipation principle for the case of Stokes damping or, more generally, for overdamped motion with more kinds of damping. This extension opens a possible way to study the general relation between two independent classes of variational principles, that is, those of Hamiltonian/Lagrangian mechanics and those associated with energy dissipation. The effective action of the whole system on a given path is given in the form

$$A := \int_{t_1}^{t_2} \left(K_1 - V_1 - E_d \right) dt = \int_{t_1}^{t_2} \frac{1}{2} \left(m\dot{x}^2 - V_1 - \lambda \int_0^{x(t)} \dot{y} \, dy \right) dt$$

Despite the Lagrangian nonblocality, the usual variational calculus of least action principle through a tiny variation of $x(t)$ leads to the Lagrangian equation

$$m\ddot{x} = -\frac{\partial (V_1 + E_d)}{\partial x} = -\frac{\partial V_1}{\partial x} - \lambda \dot{x}$$

In summary, Wang and Wang (2012) formulated a possible answer to a longstanding question of classical mechanics about the least action principle for damped motion, in keeping all four conventional formulations of mechanics, that is, Newtonian, Lagrangian, Hamiltonian, and Hamilton–Jacobi equations. This work is based on the model of a conservative system composed of the moving body and its environment coupled by friction. It was shown that this system with "internal dissipation" satisfies both Lagrangian and Hamiltonian mechanics, leading to correct equation of damped motion in a general way. It was also shown that, within this formulation, the Maupertuis principle is equivalent to a least dissipation principle in the case of Stokes damping. A more general least dissipation principle is also discussed for the overdamped motion. We hope that these results are helpful for further study of the relations between the variational principles of energy dissipation and the fundamental principles of Lagrangian and Hamiltonian mechanics (see the efforts for stochastic dissipative systems). It is also hoped that the present result is useful for the study of quantum dissipation in view of the role of action in the quantum wave propagator $\psi = e^{iA}/h$ and the close relationship between the Schrödinger equation and the Hamilton–Jacobi equation.

Variational approach consists in the transformation of information contained in the Lagrangian Λ_σ which includes balance equations of energy, momentum and mass. By minimizing the space–time integral containing this quantity, equations of system's evolution in time and space can be determined. In nonequilibrium thermodynamics the Lagrangian incorporates the entropy production S_σ expressed in a suitable way, or another physical quantity. The variational principle is based on minimizing the functional Λ_σ in which the entropy production is expressed as the sum of two dissipation functions; the first one depends on diffusive fluxes and the second on thermodynamic forces. The method of entropy production minimization should be distinguished from approaches that modify variational principles of perfect systems (eg, Hamilton's principle) to include irreversible processes. Frequently the variational approach synthesizes the dynamics of chemical processes and phase changes regarded until now as "particular" in comparison with dynamics of other processes (mechanical, thermal, etc.).

It is commonly known that oversimplified chemical dynamics following linear nonequilibrium thermodynamics is practically useless to treat chemical reactions with strong nonlinearities, feedback, instabilities, chaos, and so forth. Introduction of nonlinear chemical resistance (Shiner, 1987; Sieniutycz, 1987) has proven, however, the invalidity of an earlier opinion that the classical chemical kinetics is inconsistent with thermodynamics. This concept has resulted in the so-called chemical Ohm law and showed that the classical kinetics of Guldberg and Waage (Sieniutycz, 1987) can be the subject of nearly routine nonequilibrium thermodynamics. Our goal here is to present various variational methods ensuring nonlinear chemical kinetics along with basic transport phenomena. We shall also consider applications of variational approaches to multiphase systems.

The purpose of this chapter is to expose various general variational methods of derivation of nonlinear chemical kinetics and accompanying thermal and diffusional kinetics at mechanical equilibrium, terminating the treatment with an example of inclusion of mechanical motion. The chemical kinetics corresponds to the Guldberg–Waage law as the "chemical Ohm's law" and the mechanical example to an extended Euler equation with a frictional term linear with respect to the velocity.

The concept of the chemical resistance is simple, general, and crucial for our purposes. To illustrate this concept let us consider an open, multicomponent system, of n components undergoing N chemical reactions

$$v_{ij}^f[i] \leftrightarrow v_{ij}^b[i] \quad i = 1,\ldots,n;\ j = 1,\ldots,N \tag{2.1}$$

Quantities v_{ij}^f and v_{ij}^b are respectively the forward and backward stoichiometric coefficients, respectively, for species i in reaction j. Resulting stoichiometric coefficient, commonly used in the physical chemistry is $v_{ij} = v_{ij}^b - v_{ij}^f$. The advancement of the jth reaction is denoted by ξ_j; and its rate $d\xi_j/dt$ or r_j. Since the chemical equilibrium constant for the jth reaction, K_j, is given by the ratio of the forward and backward reaction rate constants, k_j^f/k_j^b, the classical chemical affinity of the jth reaction in the energy representation can be expressed in the form

$$A_j = \sum_{i=1}^{N}(v_{ij}^f - v_{ij}^b)\mu_i = -\sum_{i=1}^{N} v_{ij}\mu_i = RT\ln\left\{(k_j^f \sum_{i=1}^{N} a_i^{v_{ij}^f})/(k_j^b \sum_{i=1}^{N} a_i^{v_{ij}^b})\right\} \tag{2.2}$$

The system contains species i with chemical potential μ_i; T is the local temperature, and R the gas constant. An associated meaning is the action of a chemical reaction, or "chemaction," which contains the molar actions of the components, ϕ_i, which are the Lagrange multipliers of mass balance constraints in the action approach. The chemical potentials μ_i are related to the ϕ_i and their time derivatives. When the frequency of the elastic collisions vanishes ($\omega = \tau^{-1} = 0$) on the chemical time scale, $d\phi_i/dt = -\mu_i$ on this scale; a result known from theories of perfect fluids (Seliger and Whitham, 1968). The operator $d_i/dt = \partial_i/\partial t + (\mathbf{u}_i\nabla)$

is based on the absolute velocity of ith component in a laboratory frame. When diffusion is ignored all velocities $\mathbf{u}_i$ are equal to common barycentric velocity $\mathbf{u}$, and the operator becomes barycentric. The net stoichiometric coefficient v_{ij} follows the standard convention (De Groot and Mazur, 1984). However, the definition of affinity and that of the associated chemaction used here, follow the network convention (MacFarlane, 1970). The dissipation inequality holds in the network convention in the form $A_j r_j > 0$. This convention insures that the rate of an isolated reaction and its affinity always have the same sign. This in turn allows one to write a "chemical" Ohm's relation with the positive resistance R_j of the jth reaction. This resistance is the ratio of the chemical force A_j to the chemical flux r_j

$$R_j = \frac{A_j}{Tr_j} = \sum_{i=1}^{N} \frac{(v_{ij}^f - v_{ij}^b)\mu_i}{Tr_j} = -\sum_{i=1}^{N} \frac{v_{ij}\mu_i}{Tr_j} \qquad j = 1, 2, \ldots, N \qquad (2.3)$$

We use here the entropy representation for the chemical force and resistance, as the most suitable in thermodynamics. The expression of R_j in terms of concentrations is crucial for any treatment of nonlinear kinetics. For the Guldberg–Waage kinetics of mass action the rate formula and the affinity formula (2.2) lead to the logarithmic resistance (Shiner, 1987; Sieniutycz, 1987). In the entropy representation

$$R_j(\mathbf{C}) = R \frac{\ln\left[(k_j^f \sum_{i=1}^{N} a_i^{v_{ij}^f}) / (k_j^b \sum_{i=1}^{N} a_i^{v_{ij}^b})\right]}{k_j^f \sum_{i=1}^{N} a_i^{v_{ij}^f} - k_j^b \sum_{i=1}^{N} a_i^{v_{ij}^b}} \qquad (2.4)$$

where R is the universal gas constant, $\mathbf{C}$ is the vector of molar concentrations with components c_i. In thermodynamic and variational approaches to chemical systems it is the resistance formula (2.4) rather than the mass action kinetics which is the starting point in majority of basic considerations. One should appreciate the versatility of the chemical resistances and the related inertial coefficients $I_j = \tau R_j$ when treating chemical steady states, transients, and nonlinearities. In particular, Eq. (2.4) leads immediately to the chemical dissipation function $\phi_j = (1/2)R_j r_j^2$ with the state dependent R_j, and this dissipation function satisfies Onsager's local principle of minimum dissipation (Sieniutycz, 1987). The steady states can be treated by a routine procedure leading to unknown rates of resulting reactions as ratios of the overall affinities and the overall resistances (Shiner, 1992). Guldberg–Waage kinetics follow from these formalisms *a posteriori* ($r_j = A_j/R_j$); however their generalizations are also possible as long as the standard phenomenological laws of Ohm, Fourier, Fick, and Guldberg and Waage are only asymptotic formulae (Jou et al., 1988). Ignored are, in this case, changes of potentials caused by the absence of the local thermal equilibrium (Sieniutycz and Berry, 1991, 1992; Keizer, 1987a; Sieniutycz, 1994; Shiner and Sieniutycz, 1997). For very fast transients of frequency comparable with that of elastic collisions, τ^{-1}, extended kinetics correspond to a generalization of the Guldberg and Waage kinetics, which involves a term with the time derivative of the reaction rate. More importantly, in general formulations, chemical kinetics are obtained in an unified treatment along with kinetics of various nonchemical processes, that is, transport and exchange processes. Our example in Section 2.4 treats isobaric reaction-diffusion fields along this methodological line.

Since Onsager's extremum principles (Onsager, 1931a, b; Onsager and Machlup, 1953; Machlup and Onsager, 1953; Hesse, 1974) and Glansdorff's and Prigogine's extremum and variational principles for steady linear irreversible processes (Prigogine, 1947; Glansdorff and Prigogine, 1971) there have been numerous approaches to generalized principles which could describe nonstationary nonlinear evolutions. Gyarmati's quasilinear generalizations (Gyarmati, 1969, 1970; Verhas, 1983) while of a considerable generality, belong to the class of the restricted principles of Rosen's type (Rosen, 1953) or local potential type (Glansdorff and Prigogine, 1971). There some variables and/or derivatives are subjectively "frozen" to preserve a proper result. Essex (1984a, b) has showed a potential of minimum entropy formulations to yield nonlinear balance equations for radiative transfer. Mornev and Aliev (1995) have formulated a functional extension of the local Onsager's principle. With a caloric coordinate Grmela and Teichman (1983) stated a negative-entropy-based H theorem as a proper setting for the maximum entropy in Lagrangian coordinates. Grmela (1985, 1993a, b), Grmela and Lebon (1990), and Grmela and Öttinger (1997) have worked out an important two-bracket formalism, with Poissonian brackets and dissipative brackets (Grmela and Öttinger, 1997), the latter being the functional extension of the Rayleigh dissipation function (Rayleigh, 1945). With applications to rheology, bracket approaches have systematically been exposed in a book (Beris and Edwards, 1994). Yet, two-bracket theorems are not associated with an extremum of a definite physical quantity; for that purpose the single Poissonian bracket and an exact Hamiltonian system are necessary.

Vujanovic (1992) has developed a variational theory related to the transient heat transfer theory and transport phenomena with a finite propagation rate. The theory is usually applied in the context of wave equations of heat conduction but it can be generalized to diffusion problems involving both isothermal and coupled diffusion. The variational approach is intended to obtain approximate solutions for problems of classical (Fourier's) transient heat conduction. The approximation is obtained by a limiting procedure in which the relaxation time tends to zero (variational principle with a vanishing parameter). The essential value of that approach lies in its ability to handle a variety of nonlinear problems with temperature-dependent thermal conductivities and heat capacities. A nice representation of the theory in terms of Biot's heat vector (Biot, 1970) can also treat some nonlinearities. Related aspects of the theory and its modifications are discussed in Vujanovic' book and papers (Vujanovic, 1971, 1992; Vujanovic and Djukic, 1972; Vujanovic and Jones, 1988, 1989).

Many illustrative examples using the approximate "method of partial integration" have been presented. This method is based essentially on the Ritz-Kantorovich direct method in variational calculus (Gelfand and Fomin, 1963) and works effectively for complex boundary value problems of radiation, melting, heating of slabs, and so forth.

However another approach (Gambar and Markus, 1991; Nyiri, 1991) has introduced certain potentials, similar to those known in the theory of electromagnetic field, and related integrals. Their application to nonreacting systems has proved considerable utility (Gambar et al., 1991; Gambar and Markus, 1993, 1994). Vázquez et al. (1996) substantiated the existence of Hamiltonian principles for nonself-adjoint operators. Moreover, Vázquez et al. (2009) described quantized heat transport in small systems.

Yet, the inclusion of chemical reactions has required treating of chemical sources as given/known functions of time and position. Recent approaches to coupled transport processes, based on a functional expression for the second law, with Lagrange multipliers absorbing balance constraints, have been, so far, restricted to nonreacting systems, both parabolic (Sieniutycz and Ratkje, 1996a, b) and hyperbolic (Sieniutycz and Berry, 1992). These approaches may be regarded as equality counterparts of Liu's multiplier method (Liu, 1972) which deals with the second law in an inequality form, and, as such, leads to qualitative rather than quantitative results. Anthony's variational method (Anthony, 1981, 1989, 1990; Kotowski, 1989; Schoelle, 1994) which uses a "field of thermal excitation," a thermal analogue of quantum wave function, and certain reaction potentials, remains, an exclusive, general treatment of distributed reaction-diffusion systems via a variational principle, at the expense, however, of undefined physical origin of these potentials and their relation to the mass action law (Guldberg and Waage, 1867).

Chemical reactions with nonlinear kinetics have long reminded excluded from exact variational formulations. An important step toward their inclusion was introduction of nonlinear chemical resistances (Grabert et al., 1983; Shiner, 1987, 1992; Sieniutycz, 1987, 1995; Shiner and Sieniutycz, 1996; Shiner et al., 1996). With these resistances, which correspond with the mass action law incorporated, in the works quoted, nonlinear variational and extremum formulations for *lumped* systems have become possible (Sieniutycz and Shiner, 1992, 1993; Shiner and Sieniutycz, 1934; Shiner et al., 1994). In the present chapter these formulations are generalized to *distributed* reaction-diffusion systems in the field (Eulerian) representation of the accompanying transport phenomena.

The admissibility of a variational formulation calls, however, for substantiation in case when the irreversibility enters into the issue. A system of differential equations admits a variational formulation if and only if it is self-adjoint, that is, when stringent conditions for partial derivatives of the related differential operator are satisfied (Atherton and Homsy, 1975; Bampi and Morro, 1982; Santilli, 1977; Perlick, 1992; Finlayson, 1972; Caviglia, 1988). It is also known that typical equations of irreversible processes are, as a rule, not self-adjoint (Finlayson, 1972; Morse and Feshbach, 1953). A pertinent argument is valid though: while equations of irreversible processes do not admit variational formulation in the state space spanned on their own dependent variables (this is the situation where the nonself-adjointness applies), the so-called composite variational principles in the extended space spanned on their state variables and certain new variables, called state adjoints, is always possible (Atherton and Homsy, 1975; Caviglia, 1988).

In fact, all successful recent variational formulations for irreversible continua always involve the space expansion. This claim refers also to formulations which use higher-order functionals (Nyiri, 1991; Gambar et al., 1991; Gambar and Markus 1993, 1994) which can be broken down to those based on the first-order functionals in an enlarged space, as well as some action-based approaches to irreversible continua (Anthony, 1990; Sieniutycz and Berry, 1993) and reversible fluids (Seliger and Whitham, 1968; Serrin, 1959; Jezierski and Kijowski, 1990; Nonnenmacher, 1986; Sieniutycz, 1988; Sieniutycz and Berry, 1989). Reviews and books are available (Muschik and Trostel, 1983; Jou et al., 1988; Lebon, 1986; Salmon, 1987; Sieniutycz and Salamon, 1994; Kupershmidt, 1992). With these results in mind, the conservation laws and the idea of minimum dissipation for thermal fields (De Groot and Mazur, 1984; Schechter, 1967), we search here for extrema of thermodynamic potentials (Callen, 1985) of inhomogeneous diffusion–reaction systems.

In recent works for lumped systems (Sieniutycz and Shiner, 1995) and continuous systems (Sieniutycz and Berry; Sieniutycz and Ratkje), the necessity of distinction between the standard (Gibbsian) thermodynamic intensities $\mathbf{u}$ (the temperature reciprocal and Planck chemical potentials), which are the partial derivatives of the entropy with respect to energy and mole numbers, and their transport counterparts or the Lagrangian multipliers of balance equations, $\mathbf{w}$, has been discovered. While both $\mathbf{u}$ and $\mathbf{w}$ may play a role in these functionals, the distinction is important only of local thermal equilibrium. We stress that the distinction is necessary only when a dynamic behavior is included into consideration, as only then instantaneous transfer potentials at a definite point of a continuum are not necessarily equilibrium quantities. The distinction is not relevant for situations and models which ignore dynamic development, such as early models of the maximum entropy formalism (Levine and Tribus, 1979), in which the Lagrange multipliers refer to the final state of equilibrium rather than to variety of intermediate states. Due to the commonness of the local equilibrium in real physical systems, two quantities $\mathbf{u}$ and $\mathbf{w}$ coincide for large majority of physical situations. Therefore, the condition $\mathbf{u} = \mathbf{w}$ should *a posteriori* be imposed for most models of heat and mass transport.

As commonly known (De Groot and Mazur, 1984), in the limiting local equilibrium situations, one can combine the conservation laws and phenomenological equations to yield the equations of change. The latter are therefore equations dependent with respect to the former ones. On the other hand, off local equilibrium, equations of change follow independently of conservation laws and phenomenological equations. But, it is just the condition **u** = **w** that drives one back to the local equilibrium situation (more restricted from the physical viewpoint than any local disequilibrium). All these effects link equilibrium and disequilibrium descriptions of continua in a natural way. They both hold with respect to the temperatures and chemical potentials in one component and multicomponent cases.

In the present analysis we discuss these effects through a few novel variational principles of nonequilibrium thermodynamics, involving both temperature and chemical potentials, in the contexts such as:

1. Evolution criteria based on the produced entropy and integrals of action
2. Gradient and nongradient representations of physical fields using Biot's vector of thermal displacement (Biot, 1970) and the potentials of thermal field (Gambar et al., 1991; Gambar and Markus, 1993, 1994; Nyiri, 1991)
3. Inclusion into a variational extremum principle of chemical reactions and chemical nonlinearities governed by the mass action kinetics (Guldberg and Waage, 1867).
4. Significance of various thermodynamic potentials at nonequilibrium, and, in particular, the practical role of free energy functionals in thermally inhomogeneous systems.
5. Relation between Lagrangian and Hamiltonian descriptions (the latter applying a bracket formalism).
6. Physical limitations imposed on results of variational analyses.

From the practical viewpoint item 2 is of special importance as it makes it possible to solve complex partial differential equations of reaction-diffusion processes by direct variational methods (Kantorovich and Krylow, 1958). It appears that we also succeeded in discovering of a method, based on equivalent variational problems, which makes it possible to show equivalence of various thermodynamic potentials at nonequilibrium. The problem is known to be difficult one, and unsolved to date in an exact way. The equivalence condition is restricted to small deviations of thermodynamic potentials from a global equilibrium, in agreement with classical statistical mechanics (Landau and Lifshitz, 1975).

The system considered is composed of components with various transport phenomena and chemical reactions in the bulk. The components are reacting but neutral (Sundheim, 1964; Ekman et al., 1978; Forland et al., 1988), obeying the phase rule (Van Zeggeren and Storey, 1970). As shown by Sundheim (1964) this setting leads to the independent fluxes of mass, energy, and electric current. For the ionic description see Newman (1973).

In short, one of main reasons of using variational approaches (leaving aside, of course, their computational virtues) is the general research direction towards extending Callen's postulational thermodynamics to inhomogeneous thermodynamic systems.

2.2 Systems With Heat and Mass Transfer Without Chemical Reaction

In this section we consider a nonreacting multicomponent fluid at rest. The macroscopic motion of the system is neglected by choice of the vanishing barycentric frame and assumption about the constancy of the system density, ρ, consistent with the mechanical equilibrium assumption. This is an assumption which makes transparent the effects considered. With this assumption, the total mass density, ρ, is a constant parameter rather than the state variable, and a reference frame in which the whole system rests easily follows. In the entropy representation the conservation laws refer to the conservation of energy, mass and electricity

$$\frac{\partial \mathbf{C}(\mathbf{u})}{\partial t} + \nabla \mathbf{J} = 0 \qquad (2.5)$$

The matrix notation (De Groot and Mazur, 1984) comprises all conservation laws consistent when $\mathbf{J}$ is the matrix of independent fluxes

$$\mathbf{J} = (\mathbf{J}_e, \mathbf{J}_1, \mathbf{J}_2, \ldots, \mathbf{J}_{n-1}, \mathbf{i})^T \qquad (2.6a)$$

(the superscript T means the transpose of the matrix). The corresponding column vector of densities $\mathbf{C}$ describes molar densities

$$\mathbf{C} = (e_e, c_1, c_2, \ldots, c_{n-1}, 0)^T \qquad (2.6b)$$

The nth mass flux $\mathbf{J}_n$ has been eliminated by using the condition $\Sigma \mathbf{J}_i M_i = 0$ for $i = 1, 2, \ldots, n$. The last component of $\mathbf{C}$ vanishes because of the electroneutrality. The independent transfer potentials are described by the vector

$$\mathbf{u} = (T^{-1}, \tilde{\mu}_1 T^{-1}, \tilde{\mu}_2 T^{-1}, \ldots, \tilde{\mu}_{n-1} T^{-1}, -\phi T^{-1})^T \qquad (2.7)$$

Reduced chemical potentials $\tilde{\mu}_k \equiv \mu_n M_k M_n^{-1} - \mu_k$ appear because the nth mass flux $\mathbf{J}_n$ has been eliminated. Their gradients: $X = \nabla \mathbf{u} = (\nabla T^{-1}, \nabla \mu_1 T^{-1}, \ldots, \nabla \phi_1 T^{-1})$ are independent forces. The densities (2.6b) and the transfer potentials (2.7) are the two sets of variables in the Gibbs equation for the entropy density $s_v = \rho s$ of an incompressible system which has the mass density $\rho = \Sigma M_i c_i$. The differential of the entropy density in terms of the vector $\mathbf{C}$ is $ds_v = \mathbf{u} \, d\mathbf{C}$. Conservation laws (2.5) are built into the entropy functional, Eq. (2.8), with the help of the vector of Lagrangian multipliers $\mathbf{w} = (w_0, w_1, w_2, \ldots, w_{n-1}, w_n)$. The *extremum* value of the multiplier $\mathbf{w}$ in the entropy functional (2.8) is the vector of the *kinetic* conjugates for the extensities $\mathbf{C}$, Eq. (2.6b). On the extremal surfaces of the entropy functional (2.8) the vector $\mathbf{w}$ coincides with the transport potential vector $\mathbf{u}$, Eq. (2.7), in the limiting situation of the local equilibrium.

The coincidence **u** = **w** does not occur off any extremal (nonequilibrium) solution and, therefore, **w** and **u** are generally two distinctive field variables in the entropy functional (2.8). Whenever no constraint **w** = **u** is imposed, they constitute two fields independent of each other. They may be interpreted, respectively, as the kinetic (Onsagerian) and thermodynamic (Gibbsian) intensities which coincide in a stable extremal process with local equilibrium. Any kinetic intensity is the Lagrangian multiplier of the related conservation law, whereas any Gibbsian intensity is the appropriate partial derivative of the entropy with respect to the adjoint extensity. For the extremum solution **w** = **u**, meaning that the extremal Lagrangian multipliers coincide with the components of the entropy gradient in the state space of coordinates C_i. In the limiting local equilibrium situation the multipliers w_i converge to the static (equilibrium) intensities (2.7), otherwise they converge to certain nonequilibrium intensities that are still the partial derivatives of an extended entropy although they then depend on both C_i and J_i (extended thermodynamics).

To simplify the notation we use the single-integral symbols for multiplied integrals in the physical space-time. The governing functional describes the second law between the two fixed times t_1 and a subsequent t_2

$$S(t_2) = \min \Big\langle S(t_1) + \int_{t_1,A}^{t_2} -\mathbf{J}_s,(\mathbf{J},\mathbf{u})d\mathbf{A}\, dt$$
$$+ \int_{t_1,V}^{t_2} \Big[\frac{1}{2}\mathbf{L}^{-1}(\mathbf{u}):\mathbf{JJ} + \frac{1}{2}\mathbf{L}(\mathbf{u}):\nabla\mathbf{u}\nabla\mathbf{u} + \mathbf{w}\Big(\frac{\partial \mathbf{C}(\mathbf{u})}{\partial t} + \nabla\mathbf{J}\Big) \Big] dV\, dt \Big\rangle, \quad (2.8)$$

where $\mathbf{J}_s(\mathbf{J}, \mathbf{u})$ is simply the product $\mathbf{J}\cdot\mathbf{u}$. A simple derivation of such functional structures from an error criterion has been given in an earlier work (Sieniutycz and Shiner, 1992; Sieniutycz and Berry, 1992). By the simple application of divergence theorem to the **w** term of Eq. (2.8), it was also shown for the steady system with **w** = **u** that the functional (2.8) can be broken down to the Onsagerian functional

$$\min \int_V \Big(\frac{1}{2}\mathbf{L}^{-1}:\mathbf{JJ} - \mathbf{J}\nabla\mathbf{u} \Big) dV \quad (2.9)$$

At the steady state, when only **J** is varied and **w** converges to **u** on the surfaces of the extremal solution, the Onsager's functional (2.9) and the kinetic equation

$$\mathbf{J} = \mathbf{L}\nabla\mathbf{u} \quad (2.10)$$

follow from Eq. (2.8) as the only steady-state formulae.

In the unsteady situation, Eq. (2.8) yields a more general result, as the Euler–Lagrange equations with respect to the variables **u**, **J**, and **w**. It is a quasilinear set, representing (at **w** = **u**) the standard model of the unsteady transfer of heat, mass and electric charge. The Euler–Lagrange equations of the entropy functional with respect to the variables **w**, **J**, and **u** are respectively

$$\frac{\partial C(u)}{\partial t} + \nabla J = 0 \tag{2.5}$$

$$L(u)^{-1} J = \nabla w \tag{2.11}$$

$$a(u) \frac{\partial w}{\partial t} = \nabla (L \nabla u) \tag{2.12}$$

where $\mathbf{a}(\mathbf{u}) = -\partial \mathbf{C}(\mathbf{u})/\partial \mathbf{u}$ is the thermodynamic capacitance matrix or the negative entropy hessian. From the viewpoint of the completeness of the physical equations regarded as extremum conditions of S, the second law Lagrangian (ie, the integrand of space–time integral in Eq. (2.8) does a good job since it leads to *all* pertinent equations, the property which is essentially not obeyed for older variational models. The conservation laws (2.5) are recovered as given constraints, whereas the remaining results are the quasilinear phenomenological equations and the equations of change. In fact, the vector Eq. (2.12) is the Fourier–Kirchhoff type matrix equation of change which links the fields of temperature, chemical potentials, and electrical potential. At the local equilibrium, when all states of the system are located on Gibbs manifold, the equality **w** = **u** holds and the three Eqs. (2.5), (2.11) and (2.12) become dependent. This is the well-known classical situation, in which only a subset of possible solutions is realized in the practice.

2.3 Gradient Representations of Thermal Fields

Let us now consider a different variational formulation for the same system which uses the so called Biot's (1970) thermal displacement vector, $\mathbf{H}_e$, in the integrand of the entropy functional. Since the system is multicomponent, one has to use a set of such vectors.

$$\mathbf{H} = (\mathbf{H}_e, \mathbf{H}_1, \mathbf{H}_2, ..., \mathbf{H}_{n-1}, \mathbf{H}_{el})^T \tag{2.13}$$

They are associated with the energy flux, component fluxes and electric current, and refer to each vector component of the matrix (2.6a). Since each flux and each density satisfy the gradient representations

$$\mathbf{J} = \frac{\partial \mathbf{H}}{\partial t} \tag{2.14}$$

and

$$\mathbf{C} = -\nabla \mathbf{H} \tag{2.15}$$

the conservation laws are identically satisfied. Indeed, the addition of the gradient of Eq. (2.14) to the partial time derivative of Eq. (2.15) yields Eq. (2.5). This formalism will serve to argue that the Nature may use only a part of solutions offered by the mathematical theory.

To derive the Euler–Lagrange equations from a functional based on the representations (2.14) and (2.15) is suffices to substitute them into the production part of the entropy functional (2.8). Restricting for brevity to linear system (the case of constant coefficients) yields

$$S_\sigma = \int_{t_1,V}^{t_2} \left(\frac{1}{2} \mathbf{L}^{-1} : \frac{\partial \mathbf{H}}{\partial t} \frac{\partial \mathbf{H}}{\partial t} + \frac{1}{2} \mathbf{W} : \nabla(\nabla \mathbf{H}) \nabla(\nabla \mathbf{H}) \right) dVdt, \qquad (2.16)$$

where the positive symmetric matrix $\mathbf{W}$ has been defined as

$$\mathbf{W} \equiv \mathbf{a}^{-1T} \mathbf{L} \mathbf{a}^{-1} \qquad (2.17)$$

The matrix Euler–Lagrange equation for the aforementioned functional is

$$\mathbf{L}^{-1} : \frac{\partial^2 \mathbf{H}}{\partial t^2} - \mathbf{W} \nabla^2 (\nabla^2 \mathbf{H}) = 0 \qquad (2.18)$$

After using the representations (2.14) and (2.15), Eq. (2.18) can be given the form

$$\mathbf{L}^{-1} \frac{\partial \mathbf{J}}{\partial t} + \mathbf{W} \nabla^2 (\nabla \mathbf{C}) = 0 \qquad (2.19)$$

Eq. (2.19) holds with the conservation laws (2.1) contained in the representations (2.14) and (2.15). Thus the system is characterized by Eqs. (2.14), (2.15), and (2.19). [Note that Eq. (2.19) followed from Eq. (2.16) without any prior recursion to the condition $\mathbf{w} = \mathbf{u}$, and that (2.19) can also be obtained by eliminating the multipliers $\mathbf{w}$ from Eqs. (2.11) and (2.12)]. However, when one wants to pass to the set (2.5), (2.11), and (2.12) from this model, a more general set is admitted by Eqs. (2.14), (2.15), and (2.19)

$$\frac{\partial \mathbf{C}(\mathbf{u})}{\partial t} + \nabla \mathbf{J} = 0 \qquad (2.5)$$

$$\mathbf{L}(\mathbf{u})^{-1} \mathbf{J} = \nabla \mathbf{w} + \mathbf{c}(\mathbf{x}) \qquad (2.20)$$

$$\mathbf{a}(\mathbf{u}) \frac{\partial \mathbf{w}}{\partial t} = \nabla(\mathbf{L} \nabla \mathbf{u}) + h(t) \qquad (2.21)$$

where $\mathbf{c}(\mathbf{x})$ and $h(t)$ are arbitrary functions. Based on the present result, the claim that the physical behavior should be described by Eqs. (2.14), (2.15), and (2.18) rather than by any other set, would be not entirely appropriate from the stand point of anyone who (preferring the traditional form of heat equations) wants to operate with all three Eqs. (2.5), (2.11), and (2.12). Again, the only reasonable way to resolve the dilemma is to restrict the structure of the independent Eqs. (2.5), (2.20), and (2.21) to the range implied by an experiment or microscopic transport theories. This, in turn, means their dependence, associated with taking **w** = **u** along with **c**(**x**) = 0 and $h(t)$ = 0. Thus, it is the physics of the problem which requires one to restrict to the case when **c**(**x**) = 0, $h(t)$ = 0 and **w** = **u**.

Another example illustrating this sort of problems is also in order. The origin of integrand of Eq. (2.8) lies in phenomenological equations, which have been squared in the form $\mathbf{L}^{-1}(u)\mathbf{J} - \mathbf{u} = 0$, to generate an error expression involving the resistance matrix (Sieniutycz and Shiner, 1992; Sieniutycz and Berry, 1992). For the same parabolic problem, with the concept of potentials in the linear case, Gambar and Markus (1994) have applied a dissipative Lagrangian obtained by squaring the variational adjoint of the equation of change [Eq. (2.12) for **u** = **w**]. This variational adjoint, which is next identified with **u**, is, in our notation

$$\mathbf{u} = -\mathbf{a}\frac{\partial \varphi}{\partial t} - \mathbf{L}\nabla^2 \varphi \tag{2.22}$$

(their $\rho S_{ik} = -\mathbf{a}^{-1}{}_{ik}$). Their Lagrangian is of the second order

$$L(\varphi, \varphi_{,t}, \varphi_{,xx}) = -\frac{1}{2}\left(\mathbf{a}\frac{\partial \varphi}{\partial t} + \mathbf{L}\nabla^2 \varphi\right)^2 \equiv \frac{1}{2}\mathbf{u}^2 \tag{2.23}$$

This Lagrangian leads to the Euler–Lagrange equation

$$\mathbf{a}\frac{\partial}{\partial t}\left(\mathbf{a}\frac{\partial \varphi}{\partial t} + \mathbf{L}\nabla^2 \varphi\right) - \mathbf{L}\nabla^2\left(\mathbf{a}\frac{\partial \varphi}{\partial t} + \mathbf{L}\nabla^2 \varphi\right) = 0 \tag{2.24}$$

In terms of the representation (2.22) the Euler–Lagrange Eq. (2.24) has the form

$$-\mathbf{a}\frac{\partial \mathbf{u}}{\partial t} + \mathbf{L}\nabla^2 \mathbf{u} = 0 \tag{2.25}$$

This is the linear form of the Fourier–Kirchhoff equation for heat or mass diffusion. That approach has resulted in a number of original interpretations of nonequilibrium thermodynamic systems (Gambar et al., 1991; Gambar and Markus, 1993, 1994). See also other papers in this direction (Nyiri, 1991; Van and Nyiri, 1999; Vázquez et al., 1996, 2009). These ingenious analyses and a field theory constructed on the basis of the Lagrangian (2.23) provide equations of change, and extra physical conditions (obtained from certain invariance requirements) should be satisfied to get the phenomenological equations or conservation laws from Eq. (2.23). This conforms with our earlier conclusion that physical analyses should accompany the variational results. Indeed, it is easy to see that Eq. (2.24) is satisfied in the form (2.25) not only by Eq. (2.22) but, also, by the more general representations containing a function $\mathbf{f}(\mathbf{x}, t)$

$$\mathbf{u} = -\mathbf{a}\frac{\partial \varphi}{\partial t} - \mathbf{L}\nabla^2 \varphi - \mathbf{f}(\mathbf{x}, t) \tag{2.26}$$

provided that the function $\mathbf{f}(\mathbf{x}, t)$ obeys the equation

$$\frac{\partial}{\partial t}\mathbf{f}(\mathbf{x}, t) + \nabla^2 \mathbf{f}(\mathbf{x}, t) = 0 \tag{2.27}$$

An example of such function for a one-potential process is $f = (1/6)x^2 - t$. In this example, the representations (2.26) have been rejected on account of the simpler representations (2.22) regarded as, perhaps, "more physical" since they obey the gradient invariance. One may argue that the adjoint quantities are "nonphysical" anyhow, so there is no matter which representation (2.22) or (2.26) is used. However, this argument is not always true since some adjoint quantities have, with no doubts, a well-defined physical meaning. One example is momenta of classical mechanics, which are adjoints of coordinates. Another example is rates of dynamic equations with inverted signs of resistances, which describe fluctuations around equilibrium. Therefore, restrictions imposed on definitions of potentials may have also physical reasons (consider, for example, restrictions on the structure of electromagnetic potentials (Jackson, 1975).

2.4 Inclusion of Chemical Processes

For systems in mechanical equilibrium consider now a generalization of the entropy production functional with two chemical dissipation functions

$$S_\sigma = \int_{t_1, V}^{t_2} \langle [\, \frac{1}{2}\mathbf{L}^{-1}(\mathbf{u}) : \mathbf{JJ} + \frac{1}{2}\mathbf{L}(\mathbf{u}) : \nabla\mathbf{u}\nabla\mathbf{u} \right. \\ \left. + \frac{1}{2}\mathbf{R} : \mathbf{rr} + \frac{1}{2}\mathbf{R}^{-1} : (\nu'^T\mathbf{u})(\nu'^T\mathbf{u}) + \mathbf{w}(\frac{\partial \mathbf{C}(\mathbf{u})}{\partial t} + \nabla\mathbf{J} - \nu'\mathbf{r})\,]\rangle dV\, dt\, \rangle, \tag{2.28}$$

where the term $-\nu'^T\mathbf{u}$ represents an extended vector of the chemical affinities in the entropy picture, $\mathbf{A}^s$, defined further, and both $\mathbf{R}$ terms refer to chemical dissipation described by the nonlinear chemical resistances $\mathbf{R}$. The chemical resistances (of the entropy representation) obey the logarithmic formula (2.4), where $a_i = a_i(\mathbf{C})$ are activities as state functions of the process.

When the system is not in mechanical equilibrium, which means that effects of macroscopic motions are essential, a general thermodynamic functional can be obtained (Sieniutycz, 1994)

$$S_\sigma = \int_{t_0,V}^{t} \langle \{\frac{1}{2}\sum_{i,k}^{n} R_{ik}(\mathbf{y},e,\rho): j_i j_k + \sum_{i}^{n} R_{ie}(\mathbf{y},e,\rho): j_i j_e + \frac{1}{2}\sum_{i}^{n} R_e(\mathbf{y},e,\rho): j_e^2$$

$$+ \frac{1}{2}\sum_{i,k}^{n} g_{ik}(\mathbf{y},e,\rho): \nabla y_i \nabla y_k + \sum_{i}^{n} g_{ie}(\mathbf{y},e,\rho): \nabla y_i \nabla e + \frac{1}{2}\sum_{i}^{n} g_e(\mathbf{y},e,\rho): \nabla e^2$$

$$+ T^{-1}\nu(\nabla_0^s \mathbf{v})^2 + \frac{1}{2}T^{-1}\nu_r(\nabla \times \mathbf{v} - 2\omega)^2 + \frac{1}{2}T^{-1}\nu_v(\nabla \mathbf{v})^2 + \frac{1}{4}T^{-1}\nu^{-1}(\Pi^s)^2 \quad (2.29)$$

$$+ \frac{1}{2}T^{-1}\nu_r^{-1}(\Pi^a)^2 + \frac{1}{2}T^{-1}\nu_v^{-1}(p_v)^2 + \frac{1}{2}\mathbf{R}:\mathbf{rr} + \frac{1}{2}\mathbf{R}:^{-1}(\mathbf{v}^T\mathbf{u})(\mathbf{v}^T\mathbf{u})$$

$$+ \Theta\{\frac{\partial(\rho v^2/2 + \rho e)}{\partial t} + \nabla[(\rho v^2/2 + \rho e)\mathbf{v} + P(\rho,e) + p_v\mathbf{v} + \Pi\mathbf{v} + \mathbf{j}_e)]\}$$

$$\upsilon\{\frac{\partial \rho \mathbf{v}}{\partial t} + \nabla[\rho \mathbf{vv} + \mathbf{I}P(\rho,e) + \mathbf{I}p_v\mathbf{v} + \Pi]\} + w_0\{\frac{\partial \rho}{\partial t} + \nabla(\rho \mathbf{v})\}$$

$$+ \mathbf{w}[\frac{\partial \rho y}{\partial t} + \nabla(\rho \mathbf{v}y + \mathbf{j}) - \mathbf{v}r\mathbf{r}]\}dVdt\rangle.$$

The three last lines represent balance equations multiplied by corresponding Lagrange multipliers, for energy (Θ), momentum (υ) and matter (w_0 and $\mathbf{w}$). Matrices R_{ik} and g_{ik} are the resistances and conductivities of transport processes, and $\mathbf{R}$ is the matrix of chemical resistances. Average angular velocity of fluid particles is accepted in the form $\mathbf{w} = \mathbf{w}_0(\mathbf{x}, t)$; in this sense Eq. (2.29) contains a simplification. Matrix $\mathbf{y} = (y1,..., ys)$ represents the column vector composed of concentrations whereas g_{ik} i g_{ie} are suitable submatrices of the Onsager's static matrix $\mathbf{g} = \mathbf{C}^{-1}\mathbf{T}\mathbf{R}^{-1}\mathbf{C}^{-1}$ based on the entropy hessian,

$$\mathbf{C} = \frac{\partial \mathbf{y}}{\partial \mathbf{u}} = \left(\frac{\partial^2 s}{\partial y_i \partial y_j}\right)^{-1} \leq 0.$$

Matrix $\mathbf{u}$ describes the vector of transfer potentials. Explicit inclusion of the continuity equation into (2.29) ensures independence of diffusive fluxes described by the matrix $\mathbf{j} = (\mathbf{j}_1, \mathbf{j}_2,..., \mathbf{j}_{s-1})$. In Eq. (2.29) effects of first and second viscosity were taken into account: total disequilibrium flux of momentum, $\mathbf{P}$, is decomposed into the sum of spherical part $\mathbf{I}p_v$ (where $p_v = 1/3\,\text{tr}\mathbf{P} = 1/3\mathbf{P}{:}\mathbf{I}$) and traceless part $\mathbf{P}^0$, where the latter part is next decomposed into the symmetric part $\mathbf{P}^s$ and the antisymmetric part $\mathbf{P}^a$. In a similar way the gradient of velocity $\mathbf{v}$ is decomposed. Functiononal (2.29) describes transport processes described by the matrix of total fluxes $\mathbf{J} = (\mathbf{J}_e, \mathbf{J}_1, \mathbf{J}_2,..., \mathbf{J}_{n-1}, \mathbf{J}_{el})$ and transfer potentials, $\mathbf{u}$. $\mathbf{R}$ is a nonlinear diagonal matrix of chemical resistances, and the members containing $\mathbf{R}$ refer to chemical dissipation. Vector $\mathbf{r}$ is composed of reaction rates, and, since $\mathbf{v}^T\mathbf{u} = -\mathbf{v}^T\boldsymbol{\mu}/T$ (in view of conservative stoichiometry), the product $\mathbf{v}^T\mathbf{u}$ describes the vector of chemical affinities in the entropy representation, $\mathbf{A}^s$. Balance equations of mass and electricity are adjoined to total dissipation expression L_σ with the help of Lagrange multipliers $\mathbf{w} = (w_1, w_2,..., w_{s-1}, w_{el})$, where w_{el} serves to adjoint balances of electric charge.

Eq. (2.28) and its generalization (2.29) may be applied to a homogeneous mixture or to each phase of a multiphase system that has well-defined and known interfaces. In this latter case the considered equations need to be supplemented with indices characterizing separate phases. Yet, the aforementioned equations may also serve to describe a multiphase mixture as a whole; they then constitute the proposal to describe complex multiphase mixtures by a pseudohomogeneous model that is based on operational transport coefficients (elements of resistance matrix, or conductivity matrix). It is essential that in the case of pseudocontinuous description of multiphase systems each phase is treated as a pseudocontinuum with its own sources of mass and energy. In such description phase change processes behave formally as chemical reactions; they are accompanied by interphase exchange of mass and energy. Eq. (2.29) is very complicated, thus in a concrete applications its simplified forms are usually applied, such as, for example, Eq. (2.28). Taking this into account, further considerations will be restricted to Eq. (2.28).

Now by using the standard mass conservation law for each chemical reaction,

$$\sum_{i=1}^{n} v_{ij} M_i = 0, \tag{2.30}$$

an appropriately extended stoichiometric matrix

$$\mathbf{v}'_{ij} \equiv \begin{bmatrix} v_{01} & \cdots & v_{0j} & \cdots & v_{0N} \\ \cdots & \cdots & \cdots & \cdots & \cdots \\ v_{i1} & \cdots & v_{ij} & \cdots & v_{iN} \\ \cdots & \cdots & \cdots & \cdots & \cdots \\ v_{n1} & \cdots & v_{nj} & \cdots & v_{nN} \end{bmatrix} \tag{2.31}$$

and the independent transfer potentials

$$\mathbf{u} = (T^{-1}, \tilde{\mu}_1 T^{-1}, \tilde{\mu}_2 T^{-1} \ldots \tilde{\mu}_{n-1} T^{-1}, -\phi T^{-1}) \tag{2.32}$$

where $\tilde{\mu}_k \equiv \mu_n M_k M_n^{-1} - \mu_k$, $k = 1, \ldots, n$ (the tilde potential of the nth component vanishes and plays no role in the definition of u), the classical affinity definition of jth reaction

$$A_j = -\sum_{i=1}^{N} v_{ij} \mu_i = -(\mathbf{v}^T \boldsymbol{\mu})_j \tag{2.33}$$

can be transformed into its entropy-representation counterpart (the first line of Eq. (2.34), and then extended [transfer mark $\Rightarrow$ in Eq. (2.34)] to an expression which can deal with all components of the vector (2.32)

$$A_j^s = -\sum_{i=1}^{n} T^{-1} v_{ij} \mu_i = -\sum_{i=1}^{n} T^{-1} v_{ij} (\mu_n M_i M_n^{-1} - \tilde{\mu}_i) = \sum_{i=1}^{n} T^{-1} v_{ij} \tilde{\mu}_i$$
$$\Rightarrow (\mathbf{v}'^T \mathbf{u})_j \tag{2.34}$$

where

$$\mathbf{v}'^{T}_{ij} \equiv \begin{bmatrix} v_{01} & \cdots & v_{i1} & \cdots & v_{n1} \\ \cdots & \cdots & \cdots & \cdots & \cdots \\ v_{0j} & \cdots & v_{ij} & \cdots & v_{nj} \\ \cdots & \cdots & \cdots & \cdots & \cdots \\ v_{0N} & \cdots & v_{iN} & \cdots & v_{nN} \end{bmatrix} \qquad (2.35)$$

In this extended affinity the component index changes from 0 (energy), through 1, 2,..., $n-1$ (independent components), to n (electric current). To illustrate how does the extended vector of affinity work, let us multiply the transformed stoichiometric matrix (2.35) by the transport potential vector (2.32). As long as all extra stoichiometric coefficients v^{0j} and v^{nj} are assumed to vanish, the considered product is the vector

$$\begin{bmatrix} v_{01} & \cdots & v_{i1} & \cdots & v_{n1} \\ \cdots & \cdots & \cdots & \cdots & \cdots \\ v_{0j} & \cdots & v_{ij} & \cdots & v_{nj} \\ \cdots & \cdots & \cdots & \cdots & \cdots \\ v_{0N} & \cdots & v_{iN} & \cdots & v_{nN} \end{bmatrix} \begin{bmatrix} T^{-1} \\ \cdots \\ \tilde{\mu}_i T^{-1} \\ \cdots \\ -\phi T^{-1} \\ \cdots \end{bmatrix}$$

$$= \begin{bmatrix} v_{01}T^{-1} + .. v_{i1}\tilde{\mu}_i T^{-1} - v_{n1}\phi T^{-1} \\ \cdots \\ v_{0j}T^{-1} + .. v_{ij}\tilde{\mu}_i T^{-1} - v_{nj}\phi T^{-1} \\ \cdots \\ v_{0N}T^{-1} + .. v_{iN}\tilde{\mu}_i T^{-1} - v_{nN}\phi T^{-1} \\ \cdots \end{bmatrix} \Rightarrow -T^{-1}\mathbf{v}^T\boldsymbol{\mu} = T^{-1}\mathbf{A} \qquad (2.36)$$

whose components are ratios of the classical affinities A_j and the temperature T.

Consequently, the extension (2.34) can be used in the variational principle associated with minimum of the integral (2.28). A gauged form, obtained by applying the divergence theorem to the **w** term of Eq. (2.28) can also be used

$$S^T_\sigma = \int_{t_1,V}^{t_2} \left\langle \begin{bmatrix} \frac{1}{2}\mathbf{L}^{-1}(\mathbf{u}):\mathbf{JJ} + \frac{1}{2}\mathbf{L}(\mathbf{u}):\nabla\mathbf{u}\nabla\mathbf{u} + \frac{1}{2}\mathbf{R}^{-1}:(\mathbf{v}'^T\mathbf{u})(\mathbf{v}'^T\mathbf{u}) \\ +\frac{1}{2}\mathbf{R}:\mathbf{rr} - \mathbf{C}(\mathbf{u})\frac{\partial \mathbf{w}}{\partial t} - \mathbf{J}\nabla\mathbf{w} - \mathbf{w}\mathbf{v}'\mathbf{r}) \end{bmatrix} \right\rangle dV\, dt, \qquad (2.37)$$

As shown in Section 2.8 this form is the most suitable for setting the Hamiltonian formalism.

The Euler–Lagrange equations of the functionals (2.28) or (2.37) with respect to the variables **w**, **J**, **r**, and **u** are respectively

$$\frac{\partial C(u)}{\partial t} + \nabla J = v'r \qquad (2.38)$$

$$L(u)^{-1}J = \nabla w \qquad (2.39)$$

$$R(u)r = v'^T w \qquad (2.40)$$

$$a(u)\frac{\partial w}{\partial t} = \nabla(L\nabla u) - v'R^{-1}A^s(u) \underset{u=w}{\Rightarrow} \nabla(L\nabla u) - v'r \qquad (2.41)$$

where $a(u) = \partial C(u)/\partial u$ and $A^s = v'^T u$. The conservation laws (2.38) contain the production terms which are nonvanishing for $i = 1,..., n$. For the energy and the electric charge ($i = 0$ and $i = n$) the production terms do not appear because these quantities are conserved. [The production terms do not influence the form of the Onsagerian transport equations, Eq. (2.39).] The chemical Ohm's law is described by Eq. (2.40). The matrix equation of change (2.41), which links the fields of the temperature, chemical potentials and electrical potential, does contain sources associated with chemical reactions. One should also keep in mind that the numerator of Eq. (2.4) is the chemical affinity $A^s = v'^T u$ (Shiner, 1992). Thus, provided that the local thermal equilibrium limit takes place, that is, $u = w$, and the chemical resistances satisfy Eq. (2.4), Eq. (2.40) is equivalent with the mass action kinetics of Guldberg and Waage. Modulo the sign, the limiting sources in Eqs. (2.38) and (2.41) are described by the vector

$$\sigma = \begin{bmatrix} v_{01} & \cdots & v_{0j} & \cdots & v_{0N} \\ \cdots & \cdots & \cdots & \cdots & \cdots \\ v_{i1} & \cdots & v_{ij} & \cdots & v_{iN} \\ \cdots & \cdots & \cdots & \cdots & \cdots \\ v_{n1} & \cdots & v_{nj} & \cdots & v_{nN} \end{bmatrix} \begin{bmatrix} r_1 \\ .. \\ r_j \\ .. \\ r_N \end{bmatrix} = \begin{bmatrix} 0 \\ .. \\ \Sigma v_{ij} r_j \\ .. \\ 0 \end{bmatrix} \qquad (2.42)$$

In the aforementioned treatment, the energy and electricity play the role of massless components. This assumption can be removed (Sieniutycz, 1997a).

2.5 Change of Thermodynamic Potential

It is interesting to consider the role of various thermodynamic potentials in the description of dissipative processes. The entropy function is the potential in the governing functional, Eq. (2.8). Generalizing this functional to chemical reactions and then using its equivalent expression in the form of the vanishing integral

$$\min \int_{t_1, V}^{t_2} \left\langle \left[\frac{1}{2} L^{-1}(u) : JJ + \frac{1}{2} L(u) : \nabla u \nabla u + \frac{1}{2} R : rr + \frac{1}{2} R^{-1} : (v'^T u)(v'^T u) \right. \right.$$
$$\left. \left. + w\left(\frac{\partial C(u)}{\partial t} + \nabla J - v'r\right) - \left(\frac{\partial s_v(u)}{\partial t} + \nabla J_s(J, u)\right) \right] dV\, dt \right\rangle = 0 \qquad (2.43)$$

one can pass to various thermodynamic potentials associated with appropriate constraints. The key observation, stemming from the structure of Eq. (2.43), is that, for any new thermodynamic potential, there must be no Lagrange multiplier before the four-divergence of the new thermodynamic potential in its governing functional. Consequently, one can yield the multiplier-free four-divergence of a new thermodynamic potential by taking the product of the integrand of Eq. (2.43) and the reciprocal of a pertinent Lagrange multiplier. There are some practical limitations when using this rule though. Concentrations of individual components are neither typical nor really suitable to be new potentials because of the presence of mass source terms in mass balances. However, the balances of energy-type quantities, such as energy, free energy, and so forth are convenient for this purpose.

Let us illustrate how this approach works by constructing the governing functionals for the thermodynamic potentials of energy or free energy. Applying the conserved property of the energy we are free to introduce its vanishing four-divergence into Eq. (2.43), and thus write the integrand of Eq. (2.43) as follows

$$\min \int_{t_1,V}^{t_2} T^{-1} \left\langle \begin{bmatrix} \frac{1}{2}T\mathbf{L}^{-1}(\mathbf{u}):\mathbf{JJ} + \frac{1}{2}T\mathbf{L}(\mathbf{u}):\nabla\mathbf{u}\nabla\mathbf{u} + \frac{1}{2}T\mathbf{R}:\mathbf{rr} + \frac{1}{2}T\mathbf{R}^{-1}:(v'^T\mathbf{u})(v'^T\mathbf{u}) \\ + \frac{\partial e_v(\mathbf{u})}{\partial t} + \nabla \mathbf{J}_e + \sum_{k=1}^{n} T\mathbf{w}_k \left(\frac{\partial c_k(\mathbf{u})}{\partial t} + \nabla \mathbf{J}_k - (v'\mathbf{r})_k \right) - T\left(\frac{\partial s_v(\mathbf{u})}{\partial t} + \nabla \mathbf{J}_s \right) \end{bmatrix} dV\,dt \right\rangle = 0 \qquad (2.44)$$

One can now pass to the energy representation of thermodynamics. In such representation the entropy flux $\mathbf{J}_s$ replaces the energy flux $\mathbf{J}_e$ of the entropy representation as an independent variable

$$\mathbf{J}_e = T\mathbf{J}_s + \sum_{k=1}^{n} \mu_k \mathbf{J}_k + \phi \mathbf{i} = T\mathbf{J}_s + \sum_{k=1}^{n-1} \tilde{\mu}_k \mathbf{J}_k + \phi \mathbf{i} \qquad (2.45)$$

($\tilde{\mu}_k \equiv \mu_n M_k M_n^{-1} - \mu_k$.), whereas the other fluxes remain the same. Apparently, in Eq. (2.44), we work with a source term, σ_e, that represents the dissipation of energy rather than the production of entropy. The Gouya–Stodola law links this term with the local entropy production

$$\sigma_s = \mathbf{J}\nabla \mathbf{u} + \mathbf{A}^s \mathbf{r} = T^{-1}(\mathbf{J}'\nabla \mathbf{u}' + \mathbf{A}\mathbf{r}) = -T^{-1}\sigma_e \qquad (2.46)$$

With the components of the entropy four-flux (s_v, $\mathbf{J}_s$) as the independent variables, the matrix of independent fluxes is now

$$\mathbf{J}' = (\mathbf{J}_s, \mathbf{J}_1, \mathbf{J}_2 \ldots \mathbf{J}_{n-1}, \mathbf{i})^T \qquad (2.47)$$

and the corresponding column vector of new densities $\mathbf{C}'$ is

$$\mathbf{C}' = (s_v, c_1, c_2 \ldots c_{n-1}, 0)^{\mathrm{T}}. \tag{2.48}$$

The new independent transfer potentials are

$$\mathbf{u}' = (T, \tilde{\mu}_1, \tilde{\mu}_2 \ldots \tilde{\mu}_{n-1}, -\phi) \tag{2.49}$$

($\tilde{\mu}_k \equiv \mu_n \mathbf{M}_k \mathbf{M}_n^{-1} - \mu_k$.) Their gradients are independent forces of the transport processes. They are associated with the Gibbs differential for the energy density $e_v = \rho_e$, written in the form

$$de_v = -\mathbf{u}' d\mathbf{C}'. \tag{2.50}$$

This expression can be compared with that describing the entropy differential, $ds_v = \mathbf{u} \cdot d\mathbf{C}$. In order not change signs of Lagrange multipliers, a new vector $\mathbf{u}'$ has been defined in Eq. (2.49), so that it appears in Eq. (2.50) with the minus sign.

We will now show an essential result which shows that because of transferring the Lagrange multipliers $\mathbf{w}$ to the new frame (primed variables $\mathbf{w}'$), corresponding with new constraints, the energy density e_v becomes a new potential of the system subject to the constraint of the apparently sourceless entropy. In the new representation, with the state variables $\mathbf{C}'$, $\mathbf{J}'$, the governing functional (2.43) becomes

$$\min \int_{t_1, V}^{t_2} T^{-1} \left\langle \left[\frac{1}{2} \mathbf{L}'^{-1}(\mathbf{u}') : \mathbf{J}'\mathbf{J}' + \frac{1}{2} \mathbf{L}'(\mathbf{u}') : \nabla \mathbf{u}' \nabla \mathbf{u}' \right.\right.$$
$$+ \frac{1}{2} \mathbf{R}' : \mathbf{rr} + \frac{1}{2} \mathbf{R}'^{-1} : (v'^T \mathbf{u}')(v'^T \mathbf{u}') + \frac{\partial e_v(\mathbf{u}')}{\partial t} + \nabla \mathbf{J}_e(\mathbf{J}', \mathbf{u}') \tag{2.51}$$
$$\left.\left. + \sum_{k=1}^{n} w'_k \left(\frac{\partial c_k(\mathbf{u}')}{\partial t} + \nabla \mathbf{J}'_k - (v'\mathbf{r})_k \right) + w'_0 \left(\frac{\partial s_v}{\partial t} + \nabla \mathbf{J}'_s \right) \right] dV \, dt \right\rangle = 0.$$

Here $\mathbf{J}_e = -\mathbf{J}'\mathbf{u}'$, $w_0' = -T$, in agreement with Eq. (2.45). The transformation of Onsagerian conductivity matrices $\mathbf{L}$ follows the established rule $\mathbf{L}' = \mathbf{P}^{-1\mathrm{T}}\mathbf{L}\mathbf{P}^{-1}$ when the transformation of fluxes is of the form $\mathbf{J}' = \mathbf{P}\mathbf{J}$; see, for example, De Groot and Mazur (1985). Note that the Lagrange multiplier of the entropy constraint equals the temperature.

In view of the positivity of the temperature coefficient T^{-1} before the large bracket of Eq. (2.51) it is clear that an analogous integral obtained by taking T^{-1} off the integral vanishes on its extremal surfaces as well; this procedure shows how one is lead to the energy functional. Such functional contains, as its integral, the energy dissipated plus the product of the new Lagrange multiplier vector and the appropriate balance constraints (including the conserved entropy)

$$-E(t_2) = \min \langle -E(t_1) + \int_{t_1,A}^{t_2} \mathbf{J}_e(\mathbf{J}', \mathbf{u}') d\mathbf{A} \, dt$$

$$+ \int_{t_1,V}^{t_2} \left[\frac{1}{2} \mathbf{L}'^{-1}(\mathbf{u}') : \mathbf{J}'\mathbf{J}' + \frac{1}{2} \mathbf{L}'(\mathbf{u}') : \nabla\mathbf{u}'\nabla\mathbf{u}' + \frac{1}{2} \mathbf{R}' : \mathbf{rr} + \frac{1}{2} \mathbf{R}'^{-1} : (v'^T \mathbf{u}')(v'^T \mathbf{u}') \right.$$

$$\left. + \mathbf{w}' \left(\frac{\partial \mathbf{C}'(\mathbf{u}')}{\partial t} + \nabla \mathbf{J}' - v'\mathbf{r} \right) \right] dV \, dt \rangle. \quad (2.52)$$

Consequently, it is the energy, not the entropy, which is the potential function for the primed set of constraints and associated variables. This result represents the extension of Callen's (1988) postulational thermodynamics to nonequilibrium, spatially inhomogeneous systems. The constraints, which now comprise the balances of mass, electric charge and the sourceless entropy, are multiplied by $\mathbf{w}'$ rather than by $-\mathbf{w}'$. This assures the identification $\mathbf{w}' = \mathbf{u}'$ rather than $-\mathbf{w}' = \mathbf{u}'$ at the local equilibrium. New balance laws are built into the energy functional (2.52) with the help of the vector of the Lagrangian multipliers $\mathbf{w}' = (w'_0, w'_1, w'_2, ..., w'_{n-1}, w'_n)$.

2.6 From Adjoined Constraints to Potential Representations of Thermal Fields

We shall now outline a simple and general procedure that generates various (gradient or nongradient) representations for a given set of physical fields satisfying definite relations or constraints. The procedure provides an effective way to construct evolution functionals. Its general principle is derived from the observation that extremizing arbitrary criterion versus given constraints generates for Lagrange multipliers of these constraints a set of equations that is canonically coupled with the set represented by the constraints. The procedure is illustrated by the examples given further. Note that second-order functionals should be treated.

Assuming that the phenomenological equations are satisfied for the source term of Eq. (2.8), one could instead minimizing this term minimize the functional

$$S_\sigma = \int_{t_1,V}^{t_2} \left(\mathbf{L} : \nabla\mathbf{u}\nabla\mathbf{u} + \lambda(-\mathbf{a}\frac{\partial \mathbf{u}}{\partial t} + \mathbf{L}\nabla^2\mathbf{u}) \right) dV dt, \quad (2.53)$$

where $\mathbf{a}(\mathbf{u})$ is thermodynamic capacity matrix represented by the negative hessian of entropy. The first term under integral (2.53) describes "an entropy production of a real process." Constraint (2.25) or the linear Fourier–Kirchhoff equation that is adjoined to such an entropy production by the multiplier λ takes into account transfer Eq. (2.11) and the equality $\mathbf{u} = \mathbf{w}$, that is, the phenomenological equation $\mathbf{J} = \mathbf{L}\nabla \mathbf{u}$. For the pure heat transfer, the condition

(2.25) represents exactly what we call the equation of change for the temperature reciprocal. Variation of Eq. (2.53) with respect to **u** yields the Euler–Lagrange equation

$$\mathbf{L}\nabla^2 \mathbf{u} = \mathbf{a}\frac{\partial \lambda}{\partial t} + \mathbf{L}\nabla^2 \lambda \qquad (2.54)$$

This relationship proves that, in the stationary state, the equality **u** = λ is a simple and appropriate representation for **u** in terms of λ. In this case the substitution λ instead of **u** into Eq. (2.53) defines a well-known functional with the integrand **L**: $\nabla \lambda \nabla \lambda$, for which the stationarity condition is a familiar form of the equation of change, $\mathbf{L}\nabla^2 \lambda = 0$ or $\mathbf{L}\nabla^2 \mathbf{u} = 0$, a correct while the quite restrictive result. However, except the steady state it is really difficult work with the vector **u** as a function of λ that satisfies Eq. (2.54). This difficulty suggests trials towards the change of the optimization criterion that we shall describe further.

We shall now show an essential methodological change in order to produce a variational principle for Eq. (2.25). It consists in minimizing, instead of S_σ, an arbitrary functional containing a positive integrand to which the constraint (2.25) is adjoined by a Lagrange multiplier. Consider, for example, the functional (2.55), with a symmetric positive matrix **B**

$$S' = \int\limits_{t_1, V}^{t_2} \left(\frac{1}{2}\mathbf{B} : \mathbf{uu} + \boldsymbol{\varphi} \cdot \left(-\mathbf{a}\frac{\partial \mathbf{u}}{\partial t} + \mathbf{L}\nabla^2 \mathbf{u} \right) \right) dVdt, \qquad (2.55)$$

The Euler–Lagrange equation of this second order variational problem with respect to vector **u** provides the following representation for the field vector **u**

$$\mathbf{u} = -\mathbf{B}^{-1}\left(\mathbf{a}\frac{\partial \boldsymbol{\varphi}}{\partial t} + \mathbf{L}\nabla^2 \boldsymbol{\varphi} \right) \qquad (2.56)$$

For the unit matrix **B** = **I**, the representation of **u** in terms of potentials found earlier follows, (Nyiri, 1991; Markus and Gambar, 1991; Gambar et al., 1991; Gambar and Markus, 1994); their $\rho S_{ik} = -a_{ik}^{-1}$. In terms of $\boldsymbol{\varphi}$ action (2.55) becomes

$$\begin{aligned} S' = &\int\limits_{t_1, V}^{t_2} \left(\frac{1}{2}\mathbf{B}^{-1} : \left(\mathbf{a}\frac{\partial \boldsymbol{\varphi}}{\partial t} + \mathbf{L}\nabla^2 \boldsymbol{\varphi} \right)\left(\mathbf{a}\frac{\partial \boldsymbol{\varphi}}{\partial t} + \mathbf{L}\nabla^2 \boldsymbol{\varphi} \right) \right) dVdt \\ &+ \int\limits_{t_1, V}^{t_2} \left(\boldsymbol{\varphi}\left[\mathbf{a}\mathbf{B}^{-1}\frac{\partial}{\partial t}\left(\mathbf{a}\frac{\partial \boldsymbol{\varphi}}{\partial t} + \mathbf{L}\nabla^2 \boldsymbol{\varphi} \right) - \mathbf{L}\mathbf{B}^{-1}\nabla^2\left(\mathbf{a}\frac{\partial \boldsymbol{\varphi}}{\partial t} + \mathbf{L}\nabla^2 \boldsymbol{\varphi} \right) \right] \right) dVdt \end{aligned} \qquad (2.57)$$

Since, however, the constraint expression in Eq. (2.55) must vanish as the result of the stationarity of S' with respect to $\boldsymbol{\varphi}$, we are left with

$$S' = \int\limits_{t_1, V}^{t_2} \left(\frac{1}{2}\mathbf{B}^{-1} : \left(\mathbf{a}\frac{\partial \boldsymbol{\varphi}}{\partial t} + \mathbf{L}\nabla^2 \boldsymbol{\varphi} \right)\left(\mathbf{a}\frac{\partial \boldsymbol{\varphi}}{\partial t} + \mathbf{L}\nabla^2 \boldsymbol{\varphi} \right) \right) dVdt. \qquad (2.58)$$

Consequently, for any nonsingular **B**, vanishing vector constraint (2.22) is produced as an Euler–Lagrange equations of functional (2.58). Indeed, the varying of Eq. (2.58) yields

$$\mathbf{B}^{-1}\left[\mathbf{a}\frac{\partial \varphi}{\partial t}\left(\mathbf{a}\frac{\partial \varphi}{\partial t}+\mathbf{L}\nabla^2\varphi\right)-\mathbf{L}\nabla^2\left(\mathbf{a}\frac{\partial \varphi}{\partial t}+\mathbf{L}\nabla^2\varphi\right)\right]=0. \quad (2.59)$$

which means that, for any nonsingular **B**, vector Eq. (2.24) is satisfied in its original form

$$\mathbf{a}\frac{\partial}{\partial t}\left(\mathbf{a}\frac{\partial \varphi}{\partial t}+\mathbf{L}\nabla^2\varphi\right)-\mathbf{L}\nabla^2\left(\mathbf{a}\frac{\partial \varphi}{\partial t}+\mathbf{L}\nabla^2\varphi\right)=0. \quad (2.24)$$

In other words, whenever **u** is represented by Eq. (2.56), Eq. (2.24) ensures the satisfaction of the original equation of change

$$-\mathbf{a}\frac{\partial \mathbf{u}}{\partial t}+\mathbf{L}\nabla^2\mathbf{u}=0. \quad (2.25)$$

Thus we have shown that the variational principle for the Fourier–Kirchhoff Eq. (2.25) is represented by the functional

$$S'=\int_{t_1,V}^{t_2}\left(\frac{1}{2}\mathbf{B}:\mathbf{u}\mathbf{u}\right)dVdt \quad (2.60)$$

which is minimized for **u** represented by Eq. (2.56). The result stating that the "representation of the state **u** in terms of the potential **φ** is needed as defined by Eq. (2.56)," sets an analogy with the well-known variational principle of electromagnetic field, in which one uses the electromagnetic potentials (**A** and *ϕ*) to state a variational principle for electric and magnetics fields (**E** and **B**) by using their representations defined by the first pair of the Maxwell equations.

The crucial role of Lagrange multipliers in constructing variational adjoints is well known in Pontryagin's maximum principle, but it seems to be overlooked in the literature of field variational principles. The results obtained here prove that an arbitrarily large number of diverse functionals and related variational principles can be found by the described approach.

2.7 Action Functional for Hyperbolic Transport of Heat

In this section we shall continue the technique of potential representations by constructing a variational formulation for the linear process of pure heat conduction (heat flux **q**) in a rigid solid at rest. We assume that the process is described by the Cattaneo equation and energy balance. The set of constraints has the form

$$\frac{\partial \mathbf{q}}{c_0^2 \partial t}+\frac{\mathbf{q}}{c_0^2 \tau}+\nabla \rho_e = 0 \quad (2.61)$$

$$\frac{\partial \rho_e}{\partial t} + \nabla \mathbf{q} = 0, \tag{2.62}$$

where ρ_e is the density of the thermal energy, c_0 propagation speed for the thermal wave, τ thermal relaxation time, and the product $\mathcal{D} \equiv c_0^2 \tau$ the thermal diffusivity. The thermal energy density ρ_e satisfies the familiar relation, $d\rho_e = \rho c dT$.

An action functional which adjoints constraints (2.61) and (2.62) by the Lagrange multipliers, the vector $\boldsymbol{\psi}$ and the scalar ϕ, is tested in the form

$$A = \int_{t_1, V}^{t_2} \varepsilon^{-1} \left[\frac{1}{2} \mathbf{q}^2 c_0^{-2} - \frac{1}{2} \rho_e^2 + \boldsymbol{\psi} \left(\frac{\partial \mathbf{q}}{c_0^2 \partial t} + \frac{\mathbf{q}}{c_0^2 \tau} + \nabla \rho_e \right) + \phi \left(\frac{\partial \rho_e}{\partial t} + \nabla \mathbf{q} \right) \right] dV dt, \tag{2.63}$$

where ε is the density of equilibrium energy at a reference state, the constant which assures the action dimension for A, but is otherwise unimportant. We shall call the multiplier-free part of Eq. (2.63) the kinetic potential, L. The kinetic potential of an indefinite sign is chosen as the structure suitable from the viewpoint of the energy conservation in a limiting reversible process (see further considerations).

Vanishing first variations of A with respect to multipliers $\boldsymbol{\psi}$ and ϕ recover constraints, whereas those with respect to state variables $\mathbf{q}$ and ρ_e yield the following representations of state variables in terms of $\boldsymbol{\psi}$ and ϕ

$$\mathbf{q} = \frac{\partial \boldsymbol{\psi}}{\partial t} - \frac{\boldsymbol{\psi}}{\tau} + c_0^2 \nabla \phi \tag{2.64}$$

$$\rho_e = \nabla \boldsymbol{\psi} - \frac{\partial \phi}{\partial t} \tag{2.65}$$

In a limiting reversible process (undamped or wave heat conduction for $\tau \to \infty$) the process is described by pure gradient representations; the representation for $\mathbf{q}$ has then the structure of the electric field $\mathbf{E}$ expressed in terms of electromagnetic potentials.

In terms of the potentials $\boldsymbol{\psi}$ and ϕ the action A, Eq. (2.63), is

$$A = \int_{t_1, V}^{t_2} \varepsilon^{-1} \left[\frac{1}{2 c_0^2} \left(\frac{\partial \boldsymbol{\psi}}{\partial t} - \frac{\boldsymbol{\psi}}{\tau} + c_0^2 \nabla \phi \right)^2 - \frac{1}{2} \left(\nabla \boldsymbol{\psi} + \frac{\partial \phi}{\partial t} \right)^2 \right] dV dt. \tag{2.66}$$

Its Euler–Lagrange equations with respect to $\boldsymbol{\psi}$ and ϕ are respectively

$$\frac{\partial}{\partial t} \left[\frac{1}{c_0^2} \left(\frac{\partial \boldsymbol{\psi}}{\partial t} - \frac{\boldsymbol{\psi}}{\tau} + c_0^2 \nabla \phi \right) \right] + \frac{1}{\tau c_0^2} \left(\frac{\partial \boldsymbol{\psi}}{\partial t} - \frac{\boldsymbol{\psi}}{\tau} + c_0^2 \nabla \phi \right) - \nabla \left(\nabla \boldsymbol{\psi} + \frac{\partial \phi}{\partial t} \right) = 0 \tag{2.67}$$

$$-\frac{\partial}{\partial t} \left(\nabla \boldsymbol{\psi} + \frac{\partial \phi}{\partial t} \right) + \nabla \left(\frac{\partial \boldsymbol{\psi}}{\partial t} - \frac{\boldsymbol{\psi}}{\tau} + c_0^2 \nabla \phi \right) = 0 \tag{2.68}$$

Applying the representations (2.64) and (2.65) in Eqs. (2.67) and (2.68) is easy to see that Eqs. (2.67) and (2.68) are the basic Eqs. (2.61) and (2.62) in terms of the potentials ψ and ϕ.

The equivalent form of Eqs. (2.67) and (2.68), obtained after their simplification, may be analyzed. For the Cattaneo Eqs. (2.61) or (2.67), in terms of the potentials ψ and ϕ

$$\frac{\partial^2 \psi}{c_0^2 \partial t^2} - \frac{\psi}{\tau^2 c_0^2} + \frac{\nabla \phi}{\tau} - \nabla(\nabla \psi) = 0 \tag{2.67'}$$

and, for the energy conservation Eqs. (2.62) or (2.68),

$$-\frac{\partial^2 \phi}{\partial t^2} - \frac{\nabla \psi}{\tau} + c_0^2 \nabla^2 \phi = 0. \tag{2.68'}$$

Interpreting τ as an average time between the collisions, we can regard the reversible process (with an infinite τ) as the collisionless one. This is a particular case when only vector potential ψ is necessary. To prove that statement let us as start with the action (2.63) in the special case when $\phi = 0$ that is, when potential ψ is the only variable. This case is still irreversible because of the finiteness of τ

$$A = \int_{t_1, V}^{t_2} \varepsilon^{-1} \left[\frac{1}{2} \mathbf{q}^2 c_0^{-2} - \frac{1}{2} \rho_e^2 + \psi \left(\frac{\partial \mathbf{q}}{c_0^2 \partial t} + \frac{\mathbf{q}}{c_0^2 \tau} + \nabla \rho_e \right) \right] dV dt. \tag{2.63'}$$

The stationarity condition of this action with respect to $\mathbf{q}$ are the representations

$$\mathbf{q} = \frac{\partial \psi}{\partial t} - \frac{\psi}{\tau} \tag{2.64'}$$

$$\rho_e = -\nabla \psi \tag{2.65'}$$

which constitute a truncated form of general representations (2.64) and (2.65). Yet, the truncated representations are invalid in the general irreversible case. Indeed, while the ψ representation of functional (2.63'), or

$$A = \int_{t_1, V}^{t_2} \varepsilon^{-1} \left[\frac{1}{2c_0^2} \left(\frac{\partial \psi}{\partial t} - \frac{\psi}{\tau} \right)^2 - \frac{1}{2} (\nabla \psi)^2 \right] dV dt, \tag{2.69}$$

assures the correct Cattaneo Eq. (2.61) for Eqs. (2.64') and (2.65'), the energy conservation (2.62) is violated by the oversimplified representations (2.64') and (2.65'). (In fact, they imply a source term $\nabla \psi / \tau = -\rho_e / \tau$ for the four-divergence of conserved energy.)

If, however, the process is reversible ($\tau \to \infty$) the simplest gradient representations hold, $\mathbf{q} = \partial \psi / \partial t$ and $\rho_e = -\nabla \psi$, which apply the multiplier ψ as the so-called Biot's heat vector (Biot, 1970). They then yield the truncated Cattaneo Eq. (2.61) without the irreversible $\mathbf{q}$ term and describe undamped thermal waves governed by the collisionless limit of action (2.69), which is

$$A = \int_{t_1, V}^{t_2} \varepsilon^{-1} \left[\frac{1}{2c_0^2} \left(\frac{\partial \psi}{\partial t} \right)^2 - \frac{1}{2} (\nabla \psi)^2 \right] dVdt, \qquad (2.70)$$

With the simplest (Biot, 1970) representations, the energy conservation law is satisfied identically. Functional (2.70) then refers only to undamped thermal waves, which propagate with the speed c_0 and satisfy d'Alembert's equation for the energy density ρ_e or temperature T.

In the developed analysis, Biot's potentials have either been generalized or they have been replaced by other potentials which lead to more general, correct results. This shows that the result stronger than Biot's is obtained in the context of the present variational analysis.

Finally, let us note that the choice of an inappropriate kinetic potential (eg, with the plus sign at ρ_e^2 in Eqs. (2.63) or (2.63′) would result in representations violating energy conservation even in the reversible case, should the Cattaneo equation be taken as the sole adjoined constraint.

2.8 Evolution Described by Single Poissonian Brackets

We return here to the entropy functional (2.8) or, speaking more precisely, to its production term. Thermodynamic functionals gauged by subtraction of the four-divergence ($\partial/\partial t$, ∇.) of a vector, here $\partial/\partial t(w_k \mathbf{c}_k) + \nabla \cdot (w_k \mathbf{J}_k)$, or linear combinations of such four-divergences, have found some useful applications (Sieniutycz, 1994; Sieniutycz and Salamon, 1990). For fields they lead directly to dissipative Hamiltonians identical with those of Onsager's discrete theory, thus making possible Hamiltonian formalisms and bracket formulations with only one type of brackets (Poissonian brackets). This seems interesting and important since the bracket approaches which stem from dissipative generalizations of concepts of ideal continua use two sorts of brackets (Grmela, 1985, 1993a, b; Beris and Edwards, 1994; Grmela and Öttinger, 1997; Öttinger and Grmela, 1997) and, as such, are not directly associated with extrema of definite physical quantities.

In the entropy representation, the thermodynamic functional, gauged as described earlier, has a general form (the superscript T refers to a gauged structure)

$$S_\sigma^T = \int_{t_1, V}^{t_2} \Lambda_\sigma^T [\mathbf{J}, \mathbf{u}, \mathbf{r}, \mathbf{w}] dVdt. \qquad (2.71)$$

It takes into account all constraints (2.5). For these constraints the "thermodynamic Lagrangian," or the integrant Λ_σ^T has the form

$$\Lambda_\sigma^T [\mathbf{J}, \mathbf{u}, \mathbf{r}, \mathbf{w}] \equiv \frac{1}{2} \mathbf{L}^{-1}(\mathbf{u}) : \mathbf{JJ} + \frac{1}{2} \mathbf{L}(\mathbf{u}) : \nabla \mathbf{u} \nabla \mathbf{u}$$
$$+ \frac{1}{2} \mathbf{R} : \mathbf{rr} + \frac{1}{2} \mathbf{R}^{-1} : (v'^T \mathbf{u})(v'^T \mathbf{u}) - \mathbf{w}^T v' \mathbf{r} - \mathbf{C}(\mathbf{u}) \frac{\partial \mathbf{w}}{\partial t} - \mathbf{J} \nabla \mathbf{w}. \qquad (2.72)$$

We also use the (**w** independent) quantity containing the sum of two dissipation functions, for transports and chemical reactions

$$L_\sigma[\mathbf{J},\mathbf{u},\mathbf{r},] \equiv \frac{1}{2}\mathbf{L}^{-1}(\mathbf{u}):\mathbf{JJ}+\frac{1}{2}\mathbf{L}(\mathbf{u}):\nabla\mathbf{u}\nabla\mathbf{u}$$
$$+\frac{1}{2}\mathbf{R}:\mathbf{rr}+\frac{1}{2}\mathbf{R}^{-1}:(v'^T\mathbf{u})(v'^T\mathbf{u}) \quad (2.73)$$

which, *per analogiam* with its classical mechanical counterpart, plays the role of a kinetic potential of the process.

Minimizing of the functional (2.71) with respect to fluxes **J** and rates **r** yield Euler–Lagrange equations which contain representations (in terms of **w**) of certain flux adjoints for a concrete L_σ. Quantities of this sort were first investigated by Vojta (1967), although for L_σ different than (2.73). The adjoints stemming from stationarity of Eq. (2.72) and Eq. (2.73) may be called the "dissipative" or "thermodynamic" momenta

$$\mathbf{p}_\sigma \equiv \frac{\partial L_\sigma}{\partial \mathbf{J}} = \mathbf{L}^{-1}\mathbf{J} = \nabla\mathbf{w} \quad (2.74)$$

$$\pi_\sigma \equiv \frac{\partial L_\sigma}{\partial \mathbf{r}} = \mathbf{R}\mathbf{r} = v'^T\mathbf{w} \quad (2.75)$$

At the local thermal equilibrium both these quantities become the thermodynamic forces, the transports driving gradients of **u** and the chemical affinities (2.34).

One can also define the "thermodynamic Hamiltonian," a quantity related to the entropy production, as the time component G^{tt} of the energy–momentum tensor associated with the gauged Lagrangian Λ_σ^T

$$H_\sigma \equiv \frac{\partial \Lambda_\sigma^T}{\partial \mathbf{w}}\mathbf{w} - \Lambda_\sigma^T. \quad (2.76)$$

For L_σ of Eq. (2.73), via extremum conditions (2.74) and (2.75), an extremal H_σ follows from Eq. (2.76). Such an extremal quantity is represented by the second line of Eq. (2.77). It is the field representation of the Onsager's thermodynamic Hamiltonian (Lavenda, 1985a, b), which describes a nonlinear chemical kinetics. Eqs. (2.72) and (2.76) yield

$$H_\sigma = \frac{1}{2}\mathbf{L}^{-1}(\mathbf{u}):\mathbf{JJ}-\frac{1}{2}\mathbf{L}(\mathbf{u}):\nabla\mathbf{u}\nabla\mathbf{u}$$
$$+\frac{1}{2}\mathbf{R}:\mathbf{rr}-\frac{1}{2}\mathbf{R}^{-1}:(v'^T\mathbf{u})(v'^T\mathbf{u}). \quad (2.77)$$

Clearly, this dissipative Hamiltonian vanishes at the local equilibrium, as its Onsagerian counterpart. In terms of the canonical variables

$$H_\sigma = \frac{1}{2}\mathbf{L}(\mathbf{u}):\mathbf{p}_\sigma\mathbf{p}_\sigma - \frac{1}{2}\mathbf{L}(\mathbf{u}):\nabla\mathbf{u}\nabla\mathbf{u}$$
$$+\frac{1}{2}\mathbf{R}^{-1}:\boldsymbol{\pi}_\sigma\boldsymbol{\pi}_\sigma - \frac{1}{2}\mathbf{R}^{-1}:(v'^T\mathbf{u})(v'^T\mathbf{u})$$
(2.78)

The direct imbedding of Onsager's Hamiltonian H_σ in the context of thermodynamic fields with nonlinear chemical kinetics and quasilinear transports is meaningful. To achieve this goal it was efficient to use the gauged structure (2.72) rather than the original functional (2.28). Let us also note that the extremum Lagrangian, Eq. (2.72), is connected with the extremum H_σ by the well-known Hamilton–Jacobi structure

$$\Lambda_\sigma^T[\mathbf{J},\mathbf{r},\mathbf{u},\mathbf{w}] = -\mathbf{C}(\mathbf{u})\frac{\partial \mathbf{w}}{\partial t} - H_\sigma(\mathbf{J},\mathbf{r},\mathbf{u}). \tag{2.79}$$

which is essential for direct development of functional canonical equations of the field theory [see Eqs. (2.84), (2.86), (2.96) and (2.98)].

It follows from the definition of L_σ, Eq. (2.73) that the dissipative functions H_σ and L_σ are connected by the Legendre transformation

$$H_\sigma(\mathbf{p}_\sigma,\boldsymbol{\pi}_\sigma,\mathbf{u}) = \frac{\partial L_\sigma}{\partial \mathbf{J}}\mathbf{J} + \frac{\partial L_\sigma}{\partial \mathbf{r}}\mathbf{r} - L_\sigma(\mathbf{J},\mathbf{r},\mathbf{u})$$
$$= \frac{\partial L_\sigma}{\partial \mathbf{v}}\mathbf{v} + \frac{\partial L_\sigma}{\partial \mathbf{r}}\mathbf{r} - L_\sigma(\mathbf{v},\mathbf{r},\mathbf{u}),$$
(2.80)

where the momenta $\mathbf{p}_\sigma$ and $\boldsymbol{\pi}_\sigma$ satisfy Eqs. (2.74) and (2.75), whereas $\mathbf{v}$ is the velocity matrix whose entries $v_{ai} = J_{ai}/C_i$ are transport velocities related to the fluxes $\mathbf{J}_i$. As in mechanics, only rate-type variables are subject of the Legendre transformation, whereas the static state variables, $\mathbf{u}$, play the role of a parameter vector.

However, for variational dynamics of fields, a particular representation of H_σ is essential in which the momenta are eliminated on account of the (gradients of) Lagrange multipliers of balance constraints (Sieniutycz, 1994). This is just the representation suitable to direct development of canonical equations and functional Poissonian brackets. Similar multipliers, which are used in the theory of perfect fluids (in the context of action approaches though), are known as the velocity potentials or Clebsch variables (Seliger and Whitham, 1968; Herivel, 1955). In our case, the pertinent variables of H_σ are the coordinates of the static thermodynamic state ($\mathbf{u}$ or $\mathbf{C}$) and the Lagrange multipliers $\mathbf{w}$. From Eqs. (2.74), (2.75), and (2.78) an extremum Hamiltonian follows

$$H_\sigma = \frac{1}{2}\mathbf{L}(\mathbf{u}):\nabla\mathbf{w}\nabla\mathbf{w} - \frac{1}{2}\mathbf{L}(\mathbf{u}):\nabla\mathbf{u}\nabla\mathbf{u} + \frac{1}{2}\mathbf{R}^{-1}:$$
$$(v'^T\mathbf{w})(v'^T\mathbf{w}) - \frac{1}{2}\mathbf{R}^{-1}:(v'^T\mathbf{u})(v'^T\mathbf{u})$$
(2.81)

In terms of the aforementioned Hamiltonian, the thermodynamic integrals S_σ^T, Eqs. (2.37), (2.71) and (2.82), conform the general structure of the field-theory actions of Eulerian fluids that have the Hamilton–Jacobi expressions as their integrands

$$S_\sigma^T = -\int_{t_1,V}^{t_2}\left[H_\sigma(\mathbf{u},\mathbf{w},\nabla\mathbf{u},\nabla\mathbf{w})+\mathbf{C}(\mathbf{u})\frac{\partial \mathbf{w}}{\partial t}\right]dVdt. \quad (2.82)$$

Let us note that, in the state of the local thermal equilibrium, that is, when $\mathbf{u} = \mathbf{w}$ and $H_\sigma = 0$, the quantity S_σ^T describes the ratio of the grand potential and temperature, as a suitable Legendre transform of the entropy of the system.

For the heat and mass transfer theory, Eq. (2.82) and its particular form Eq. (2.83) have the meaning analogous to that which the Hamilton–Jacobi equation has in mechanics of particle motion. For a thermally inhomogeneous, chemically reacting fluid field, a working form of the thermodynamic integral (2.82) is

$$\begin{aligned}S_\sigma^T = \int_{t_1,V}^{t_2}&\left[\mathbf{C}(\mathbf{u})\frac{\partial \mathbf{w}}{\partial t}+\frac{1}{2}\mathbf{L}(\mathbf{u}):\nabla\mathbf{w}\nabla\mathbf{w}-\frac{1}{2}\mathbf{L}(\mathbf{u}):\nabla\mathbf{u}\nabla\mathbf{u}\right.\\ &\left.+\frac{1}{2}\mathbf{R}^{-1}:(\mathbf{v}'^T\mathbf{w})(\mathbf{v}'^T\mathbf{w})-\frac{1}{2}\mathbf{R}^{-1}:(\mathbf{v}'^T\mathbf{u})(\mathbf{v}'^T\mathbf{u})\right]dVdt\end{aligned} \quad (2.83)$$

With the Hamiltonian (2.81), the first canonical equation is obtained by taking the variational (Volterra) derivative for action integral (2.83) with respect to the intensity vector $\mathbf{u}$

$$-\mathbf{a}(\mathbf{u})\frac{\partial \mathbf{w}}{\partial t} = -\frac{\delta H_\sigma}{\delta \mathbf{u}} = -\nabla(\mathbf{L}\nabla\mathbf{u})+\mathbf{v}'\mathbf{R}^{-1}\mathbf{A}^s(\mathbf{u}). \quad (2.84)$$

This is the vector representation of all equations of change in terms of vectors $\mathbf{u}$ and $\mathbf{w}$. The last term on the right hand side describes the chemical sources in terms of the affinity $\mathbf{A}^s$ and chemical resistance $\mathbf{R}$.

In the special case of pure heat transfer in nonreacting continuum, one can introduce an energy diffusivity $D = La^{-1}$. Since $L = kT^2$ and $a = \rho c_v T^2$, in the local equilibrium limit the quantity D is the standard heat diffusivity; $D = k(\rho c_v)^{-1}$. Moreover, since the variable $\mathbf{w}$ approximates T^{-1}, the partial derivatives obey $\partial \mathbf{w}/\partial t \cong -T^{-2}\partial T/\partial t$ and $\nabla^2\mathbf{w} \cong -T^{-2}\nabla^2 T$. Thus, in the local equilibrium

$$\frac{\partial T}{\partial t} = D\nabla^2 T, \quad (2.85)$$

which is the second Fourier law. Note that the local equilibrium postulate was necessary again because the variational principle yields an inherently nonequilibrium description, with $\mathbf{w}$ and $\mathbf{u}$ as two independent variables. For unstable situations, with diverging $\mathbf{w}$ and $\mathbf{u}$, such a postulate (the equality $\mathbf{w} = \mathbf{u}$) may be unacceptable.

The second canonical equation or the stationarity condition of the integral S_σ^T with respect to the vector **w** is the set of conservation equations for energy, mass and electric charge

$$\frac{\partial \mathbf{C(u)}}{\partial t} = -\frac{\delta H_\sigma}{\delta \mathbf{w}} = -\nabla(\mathbf{L}\nabla\mathbf{w}) + \mathbf{v}'\mathbf{R}^{-1}(\mathbf{v}'^T\mathbf{u}) = -\nabla\mathbf{J} + \mathbf{v}'\mathbf{r}. \qquad (2.86)$$

where Eqs. (2.74) and (2.75) have been exploited.

The presence of the state dependent properties in L_σ lub H_σ gives rise to nonlinear behavior. Note that the two canonical Eqs. (2.84) and (2.86) coincide at the local equilibrium, and the resulting equation can be written (in terms of the state variables **u** only) as the following nonlinear equation of change for vector **u**

$$\mathbf{a(u)}\frac{\partial \mathbf{u}}{\partial t} = \nabla(\mathbf{L}\nabla\mathbf{u}) - \mathbf{v}'\mathbf{R}^{-1}\mathbf{A}^s(\mathbf{u}) \qquad (2.87)$$

It incorporates the Fourier–Onsager transports, Ohm's electric conductivity and nonlinear chemical kinetics of Guldberg and Waage. Thus it describes the classical reaction–diffusion systems.

In effect, for H_σ defined by Eq. (2.81), the process investigated satisfies a bracket formalism with (quasicanonical) Poissonian brackets as the sole sort of brackets

$$-\mathbf{a(u)}\frac{\partial \mathbf{w}}{\partial t} = \{\mathbf{w}, H_\sigma\} = -\frac{\delta H_\sigma}{\delta \mathbf{u}} \qquad (2.88)$$

and

$$\frac{\partial \mathbf{C(u)}}{\partial t} = \{\mathbf{u}, H_\sigma\} = \frac{\delta H_\sigma}{\delta \mathbf{w}}. \qquad (2.89)$$

Consider Eq. (2.82) to conclude that the exact canonical form does require the use of densities C_k rather than intensities u_k.

Through transformations of the Hamiltonian matrix the theory of Poissonian brackets allows transformations of the aforementioned field equations into various equivalent sets governed by noncanonical Poissonian brackets (Kupershmidt, 1990a).

2.9 Extension of Theory to Multiphase Systems

From the viewpoint of the considered variational theory, thermodynamics of multiphase systems can be formulated in two ways. In the first approach one assumes that in the investigated system all phases and interphase surfaces are known (or we are able to determine them), and, by this, within each phase and for each interface a separate analysis can be performed. Such an analysis uses, for each phase, methods stemming from thermodynamics of single-phase systems, and supplements the results obtained by those stemming from analyses of the phenomena occuring at interfaces. Examples of such approaches are provided by various authors (Standard, 1964; Slattery, 1992; Sieniutycz, 2000h).

Frequently, however, an extremely large number of phases is present in the system (eg, in dispersed or boiling systems), and the interfaces may vary in the process. The matter can also be complicated by phase transformations and surface reactions. The first approach is then fruitless. In the second (more effective) approach it is assumed that the multiphase system constitutes an isotropic mixture that has definite operational properties, thermodynamic and transport (Bilicki, 1996; Badur et al., 1997). One deals then with the model of fluid with miscrostructure where thermodynamic and transport properties are described by such operational quantities as: operational specific heat, operational first and second viscosity, operational heat conductivity and so forth. Effectiveness of the approach based on operational quantities was shown in series of works on flow boiling and condensation. Therefore, in the context of the variational approach used, we extend the idea of entropy generation that is usually applied to single-phase systems. We may illustrate the approach of this sort on example of the so-called flashing, that is, boiling of expanding water in the nozzle, where intense evaporation occurs caused by a rapid pressure drop and the increase of the flow velocity.

Conditions of the vanishing first variation for the entropy functional (2.29) with respect to the Lagrange multipliers recover balance equations that are built in this function. On the other hand, the varying of this functional (with respect to state coordinates **u**, fluxes **j** and reaction rates **r**) yields kinetic equations. Because of the volume restriction and due to the problem complexity we shall briefly discuss here only an equation describing the kinetics of phase of new phase creation (vapor) during flashing, that is, boiling of expanding water in a nozzle in which an intense evaporation occurs caused by the large pressure drop and consistent increase of the velocity.

Production of vapor in the flashing zone and its later disappearance in the "shock wave" are described by a relaxation–diffusion model (Badur et al., 1997). Assuming that the disequilibrium caused by the creation of vapor bubbles is defined as the difference in the dryness fraction, $x^e - x$, the vapor production can be treated as a chemical reaction described by the resistance equation (the special case of Eq. (2.4))

$$R(x) = R\frac{\ln[k^f(1-x)/(k^b x)]}{k^f(1-x) - k^b x} \qquad (2.90)$$

Let us note that after introducing the equilibrium dryness fraction, $x^e = k^f/(k^f + k^b)$, the kinetics of the phase change during flashing or condensation becomes described by the expression $r = (k^f + k^b)(x^e - x)$ which constitutes a well-known relaxation formula with the relaxation time $\theta_x = (k^f + k^b)^{-1}$; see, for example, (Bilicki, 1996; Badur et al., 1997). This kinetics is, in general, neither isothermal nor isobaric because the rate constants k^f and k^b can be referred to two different temperatures that depend on the process course at a given instant. If the variables used to describe relaxation are, for example, x, p, and s, then the difference $x^e - x$ may be evaluated as the linear combination of the differences $(x^e - x)_{s,p}$, $(p^e - p)_{s,x}$ and $(s^e - s)_{p,x}$ multiplied by related coefficients. This is consistent with the general structure of the relaxation equations of this process. The variational principle that applies the resistance (2.4) in the particular form (2.90) provides therefore a way to describe the flashing process as an interphase chemical reaction. In the regimes of flashing and condensation, that is, where sources of new phase are important, the fluid model is necessarily to model a viscous fluid that conducts the heat. It is also the model of the fluid with microstructure, in fact, a new phase, where transfer processes are described by operational coefficients of viscosity and heat conductivity. Yet, a true difficulty in this example consists in the derivation of a complete set of kinetic and balance equations, along with equations describing pressure, velocity field, and various modes of momentum transport. This, in fact, could be broken down to the variational derivation of the equation of change for momentum (Navier–Stokes equation). Equations of motions are in this case inevitable if we don't want to distort our modeling far beyond an admissible accuracy.

2.10 Final Remarks

Problems that were considered in this chapter introduced variational formulations which involved both original and new state coordinates. In the present review two-bracket methods are omitted (Grmela, 1985; Beris and Edwards, 1994; Grmela and Öttinger, 1997; Öttinger and Grmela, 1997), which (while offering a formalism similar to the Hamiltonian and thus somehow related to variational approaches) do not assure extremum value for a definite functional. Nonetheless, two-bracket methods, in particular GENERIC (Grmela and Öttinger, 1997; Öttinger and Grmela, 1997), and also a close "matrix model" proposed by Jongschaap (Edwards et al., 1997), have turned out to be a powerful tool in modeling reologically complex fluids. The reader is referred to the literature for reviews and comparisons of these formalisms.

Variational formulations obtained in the context of potential theories suggest that the theory of a limiting reversible process need to serve as a basis and indicator when one has to make choice of the kinetic potential. The results seem suggest that thermodynamic irreversibility may not change neither kinetic potential L nor action functional A; it may only complicate potential representations of physical fields in comparison with those describing the reversible evolution. The original "reversible" structure of the kinetic potential L should remain unchanged when an irreversible evolution is described, and, in view of many successful results (Markus and Gambar, 1991, 2004, 2005; Gambar and Markus, 1993), the main difficulties caused by the irreversibility need to be defined as those stemming from the necessity of treating balance constrains with sources. Unnecessity of adjoining conservation laws for energy and momentum in the limiting reversible process could, perhaps, be interpreted as the consequence of the fact that in the course of a reversible evolution in which the physical information is not lost, conservation equations should (as in mechanics) follow uniquely from dynamical equations of motions determined for a given Lagrangian, or from symmetry principles with respect to translations of space or phase coordinates, applied to the extremum action functional. In view of absence of clear intepretations of factors causing the success of potential theories, it should be explained that they transfer the problem to spaces of different variables, thus omitting a variational formulation in the space of original physical variables that does not exist in this particular space if the symmetry of Frechet derivatives is not ensured (Sieniutycz, 1994; Finlayson, 1972; Van and Nyiri, 1999).

In this chapter we limit our variational analyses to classical nonrelativistic systems. The supplementary chapter: A Causal Approach to Hydrodynamics and Heat Transfer offers a least action principle that preserves heat flux, viscous stress, and thermal inertia in conservation laws for energy and momentum of the relativistic one-component system.

CHAPTER 3

Wave Equations of Heat and Mass Transfer

3.1 Introduction

This chapter discusses the so-called paradox of infinite speed of propagation of thermal and concentration disturbances. The paradox has influenced the construction and development of the important thermodynamic theory now called extended irreversible thermodynamics (EIT). See, for example, Jou et al. (1988, 2001) and vols. 6 and 7 of Advances in Thermodynamics series for a presentation of developments in that theory (Sieniutycz and Salamon, 1992a, b, c, d). To acquaint the reader with the historical development of attempts to eliminate the paradox, properties of solutions of parabolic and hyperbolic partial differential equation of heat are compared. A plausible concept of relaxation time is proposed; a simple evaluation of relaxation times for heat, mass, and momentum in ideal gases with a single propagation speed is presented. Extending these ideas to imperfect systems, the simplest extended theory of coupled heat and mass transfer is constructed based on a disequilibrium entropy. The entropy source analysis yields a matrix formula for the relaxation matrix, τ. This matrix can be applied to the coupled transfer and corresponds to the well-known values of relaxation times of pure heat transfer and isothermal diffusion. Various simple forms of wave equations for coupled heat and mass transfer are discussed. Simple disequilibrium extensions of the second differential of the entropy and the excess entropy production are given. They allow one to prove the stability of the coupled heat and mass transfer equations by the second method of Lapunov. Dissipative variational formulations are presented, leading to approximate solutions by a direct variational method. Applications of the hyperbolic equations to a description of short-time effects and high-frequency behavior are outlined. Some related ideas dealing with variational formulations for hyperbolic equations of heat can be found elsewhere (Vujanovic, 1971, 1992; Vujanovic and Djukic, 1972; Vujanovic and Jones, 1988; Lebon, 1976, 1986; and chapter: Variational Approaches to Nonequilibrium Thermodynamics).

When describing heat, mass, and momentum transfer it is customary to typically use one of Fourier's, Fick's, or Newton's constitutive equations. These equations relate irreversible (diffusional) fluxes with respective gradients. In this case the application of conservation laws leads to parabolic equations of change. However, all such standard equations of change (with parabolic terms) have an absurd physical property: a disturbance (thermal, concentrational, etc.) at any point in the medium is felt instantly at every other point; that is, the velocity of propagation of disturbances is infinite. This paradox is clearly seen certain routine solutions of parabolic equations; for instance, in the case of heat conduction in a semiinfinite solid on the surface of which the temperature may suddenly increase from, for example, $T = T_0$ to $T = T_w$ or T_{wall}. The solution (Baumeister and Hamil, 1968), which is based on the error integral, provides $T = T_0$ for the time $t = 0$, but for the sufficiently short time t and arbitrarily large distances x from the wall one has a nonvanishing $T(x, t)$ in the whole space implying infinitely fast propagation of the disturbance (Fig 3.1).

Baumeister and Hamil (1968) determined the effect of the propagation velocity of heat on the temperature and heat-flux distribution in a semiinfinite body due to a step change in temperature at the surface. Their solution yields a maximum but finite heat flux under the conditions of a step change. This is contrary to the infinite value predicted by the error function solution to the Fourier transient conduction equation. Also, assuming that convection is conduction limited, an upper limit for convective heat transfer coefficients is postulated (Fig 3.2).

Flumerfelt and Slattery (1969) performed an experimental study of the validity of Fourier's law. Experiments aimed at the evaluation of the effect of steep temperature gradient on the heat conductivity showed that for the two ceramic cements tested and for the absolute value of the temperature gradient $\nabla T < 10^2 °C/cm$, thermal conductivity is nearly independent of ∇T and Fourier's law represents an excellent approximation. Yet, one should expect effects such as these should be dependent upon molecular structure; it is possible that for other materials more pronounced deviations from Fourier's law will be revealed.

Sandler and Dahler (1964) gave a simple reasoning to prove that the telegrapher's equation provides a more rigorous description of diffusion than does the familiar diffusion equation.

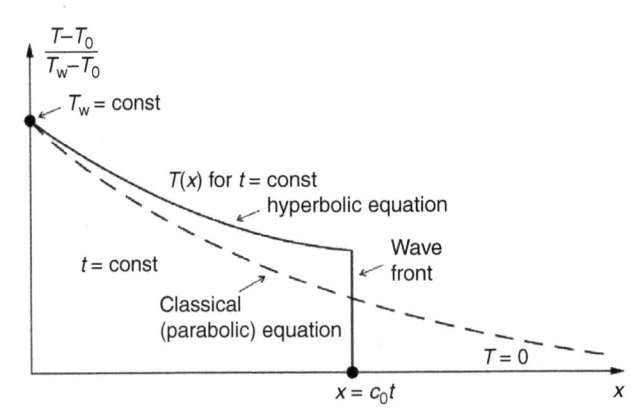

Figure 3.1: Comparison of solutions of the heat problem described by hyperbolic (*solid line*) and parabolic (*dashed line*) equation when the temperature of a semiinfinite solid was suddenly increased at the wall wall (Baumeister and Hamil, 1968; Kao, 1977; Sadd and Didlake, 1977).

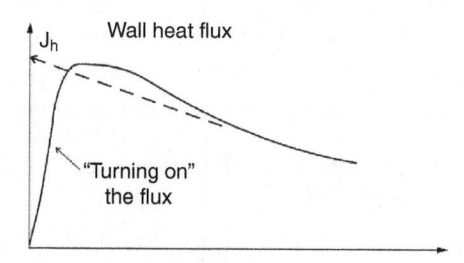

Figure 3.2: Heat flux density at the wall for Fourier model (*dashed line*) and wave model (*solid line*), when the wall temperature was suddenly increased at the wall of a semiinfinite solid (Baumeister and Hamil, 1968).

Solutions of these two equations were compared and it was shown that they are numerically indistinguishable for most cases of interest. They obtained the signal velocity of a concentration pulse to be $(p/\rho)^{1/2}$ with p the pressure and ρ the fluid density. They found that in special circumstances reflection and/or dispersion of concentration waves can occur.

Chester (1963) calculated a critical frequency for thermal fluctuations above which heat transport proceeds by wave propagation rather than by diffusion. This transport phenomenon should occur in some dielectric solids. The wave heat transfer in a solid is an analog of the second sound in helium II. See Lemaitre and Chaboche (1988) for basic mechanics of solid materials and Keesom and Sarris (1940), Keesom (1949), and Putterman (1974) for the information about the heat conductivity and other properties of liquid helium. Chester (1963) applied macroscopic viewpoint which relies upon a modification in the Fourier heat equation. Some quantitative results were obtained on the basis of this modification. Guyer and Krumhansl (1964) derived a dispersion relation for second sound. The starting point of the analysis was a Boltzmann equation for a phonon gas undergoing a temperature perturbation $\delta T_0 \exp[i(\mathbf{k}\cdot\mathbf{x} - \omega t)]$, and Callaway's approximation to the collision term. A dispersion relation obtained explicitly exhibits the need for a "window" in the relaxation time spectrum. Further, the dispersion relation shows that measurement of the attenuation of second sound as a function of frequency is a direct measurement of the normal process and umklapp process relaxation times. Prochovsky and Krumhansl (1964) have examined the conditions necessary for the occurrence of second sound in solids. Their results indicate that second sound can propagate at frequencies greater than the reciprocal umklapp relaxation time and smaller than the reciprocal normal relaxation time. At frequencies less than the reciprocal umklapp relaxation time, the solutions are the same as those for normal thermal conductivity. The frequency range and damping of second sound have been computed using relaxation times determined for sodium fluoride. Macroscopic equations for energy density and energy flux show their relation to the macroscopic equations with which Chester (1963) has treated the second sound. Wiggert (1977) analyzed early-time transient heat conduction by method of characteristics. Shnaid (2003) developed a thermodynamically consistent model based on a thermal mediator which describes the heat conduction with finite propagation speed. The author compared transient temperature distributions determined from the derived governing equation, Fourier equation and hyperbolic Cattaneo (1958) equation. He concluded that governing equations can be derived directly from the classical thermodynamics.

Applying the nonequilibrium thermodynamic theory and some specific results of kinetic and statistical theories, Luikov (1966a) achieved a systematic description of transport phenomena: heat conduction with a finite propagation speed, relaxation of stresses in viscoelastic bodies, motion of moisture in capillary-porous bodies, and also processes of turbulent transfer. He also gave some solutions of hyperbolic equations of mass transfer in porous bodies. Luikov et al. (1976) and Predvolitelev (1981) described mathematical aspects of wave solutions of the heat conduction equation. Berkovsky and Bashtovoi (1972) analyzed the effect of the finite velocity of heat propagation from the viewpoint of the kinetic theory. Their shape of moving wave front associated with the finite velocity of heat propagation does not correspond with the shock wave front of a finite height, but rather with the vanishing height of the limiting surface of contact with the resting medium. Yet, Sieniutycz (1981a, b) described a coupled heat, mass, and momentum transfer in a shock "second sound" wave. Martynenko and Khramtsov (2004) analyzed nonstationary evaporation of liquid on the basis of the phenomenological theory of thermodynamics of irreversible processes. Relaxation of vapor generation was observed for a sharp change of heat flux. Povstenko (2005, 2007) formulated a theory of diffusive stresses based on the diffusion-wave equation with time-fractional derivative. Because the obtained equation interpolates the parabolic equation and the wave equation, the proposed theory interpolates the classical theory of diffusive stresses and that without energy dissipation.

3.2 Relaxation Theory of Heat Flux

The nonphysical behavior, described earlier for the Fourier model, has been pointed out by many authors (Cattaneo, 1958; Vernotte, 1958; Luikov, 1966a, b, 1969; Chester, 1963; Kaliski, 1965; Müller, 1966, 1967, 1971a, b, 1985; Baumeister and Hamil, 1968; Kazimi and Erdman, 1973; Bubnov, 1976a, b; Lebon, 1978; Sieniutycz, 1977, 1979, 1992a, 1996; Nonnenmacher, 1980 and many others). The dilemma was resolved by the acceptance of the hypothesis of heat flux relaxation. The link between the hypothesis and certain results of the nonequilibrium statistical mechanics (Burnett, 1935; Waldmann, 1958), especially Grad's solution of the Boltzmann kinetic equation (Grad, 1958), was discovered, either in the context of phenomenological equations (Lebon, 1978; Lebon et al., 1980; Jou et al., 1988), or conservation laws (Sieniutycz and Berry, 1989, 1991; Sieniutycz, 1994). Important results were also obtained in the context of viscoelastic relaxation functions compatible with thermodynamics of complex fluids (Fabrizio and Morro, 1988; Fabrizio et al., 1989; Huilgol and Phan-Thien, 1986; Edwards and Beris, 2004).

Leonov (1976), Lebon and Rubi (1980), Lebon et al. (1986, 1988), Lebon and Boukary (1988) and Lebon (1992) developed the thermodynamics of viscoelastic polymers and other rheological materials in the context of the extended irreversible theory, aimed at modeling of nonisothermal viscous fluids in nonequilibrium. They showed that EIT has the capability of a general unifying formalism which is able to produce a broad spectrum of viscoelastic constitutive equations. In particular, linear viscoelasticity is easily interpreted within EIT. It may be shown how the classical rheological models of Maxwell, Kelvin–Voight, Poynting–Thompson and Jeffrey's are obtained as special cases of the formalism developed. A general Gibbs equation for viscoelastic bodies is given and its special cases, pertaining to the particular models specified earlier, are discussed. Then the author poses the question: what are the consequences of introducing a whole spectrum of relaxation modes for the pressure tensor, instead of working with one single mode? For that purpose the Rouse–Zimm relaxation models (Rouse, 1953; Zimm, 1956) are analyzed. These models are based on a whole relaxation spectrum; kinetic models frequently predict such spectra. These models have proved useful for describing dilute polymeric solutions. It is shown that EIT is capable of coping with the relaxation spectrum of the Rouse–Zimm model. Finally, the analysis is extended to fluids described by nonlinear constitutive equations (non-Newtonian). Lebon et al. (1986) propose a thermodynamic description of viscoelastic materials in which the thermodynamic state space is widened by including the stress tensor and the heat flux vector among the set of independent state variables. These extra variables satisfy evolution equations taking the form of first order time differential equations. The resulting formalism encompasses classical rheological models of Maxwell, Kelvin–Voight, Poynting–Thompson as well as a thermoelastic body and the Newtonian fluid. The virtue of the formalism is the description without appealing to an artificial decomposition of the strain and stress tensors into elastic and inelastic parts. In fact, Lebon et al. (1986, 1988, 1990, 1993) shows that extended irreversible thermodynamics provides a natural scheme for describing viscoelastic bodies. The suitable description is achieved by introducing the inelastic stress tensor as variable in complement of the standard variables. The Poynting–Thomson, Maxwell, and Kelvin–Voigt models are recovered as particular cases of the formalism. Nonlinear and more complicated models, like Jeffreys' model, are also suggested. Propagation of plane harmonic waves and the consequences of the application of external sinusoidal solicitations are investigated. Finally, a comparison with some other theories is made. Finally, applying EIT, Lebon et al. (1993) examine the influence of mechanical stresses on polymer solutions and chemical reactions. They show that the presence of a flow raises the Gibbs free energy. As a consequence, one observes in polymer solutions a shift of the critical point with respect to quiescent situations and in chemical reactions, a modification of the value of the chemical constant. In contrast with other treatments of viscoelastic materials, no hidden variables are introduced. These EIT-based approaches to viscoelasticity may be compared with other thermodynamic approaches, based on other theories: classical (Meixner, 1949, 1961; Kluitenberg, 1966; Kestin and Bataille, 1980) or rational (Rivlin and Ericksen, 1955; Koh and Eringen, 1963; Huilgol and Phan-Thien, 1986).

Morro (1985, 1992) presented the thermodynamics of linear viscoelasticity in the framework of rational thermodynamics and extremum principles. First, a general scheme for thermodynamics of simple materials is outlined. This scheme is applicable to materials with internal variables and those with fading memory (Maugin, 1975, 1979, 1987; Muschik, 1990b). Rigorous definitions are given for states, processes, cycles and thermodynamic laws. His thermodynamic approach is then applied to viscoelastic solids and fluids (Nettleton, 1960, 1969, 1986, 1987a). The approach is essentially entropy-free in the sense that the dissipative nature of the theory is assured by the negative definiteness of the half-range Fourier sine transform of a (Boltzmann) relaxation kernel. The fact that such negative definiteness yields all conditions derived so far by various procedures is a new result. In addition, such a condition has been proved to be sufficient for the validity of the second law. Necessary and sufficient conditions for the validity of the second law are derived for fluids. Evolution equations are analyzed. The Navier–Stokes model of a viscous fluid follows as a limit when relaxation functions are equal to delta-functions. The extremum principles are shown to be related to thermodynamic restrictions on the relaxation functions. This treatment is quite involved from the mathematical viewpoint, but the reader gets to reexamine basic features of the rational theory in a concise and precise way. In this context, the reader is also referred to the work by Altenberger and Dahler (1992) which discusses applications of convolution integrals to the entropyless statistical mechanical theory of multicomponent fluids.

The relaxation hypothesis is based on the position that Fourier's law (as well as Fick's and Newton's) is an approximation to a more exact equation, called Maxwell–Cattaneo equation,

$$J_h = -\lambda \nabla T - \tau_h \frac{\partial J_h}{\partial t} \qquad (\mathbf{v} = 0) \tag{3.1}$$

where τ_h is the relaxation time of heat flux.

The analogous equations for the irreversible fluxes of mass and momentum have also been found. The important observations are:

1. For the mass diffusion, an equation identical with Eq. (3.1) results from the nonstationary version of the Maxwell–Stefan equation under certain broad conditions (Sandler and Dahler, 1964).
2. For the momentum diffusion, equation similar to Eq. (3.1) is obtained, which is the Maxwell's equation of a viscoelastic fluid, describing the relation between shear stress and velocity gradient.
3. Experiments in acoustic dispersion and absorption (Carrassi and Morro, 1972, 1973) show the superiority of the generalized (wavelike) Navier–Stokes equation with respect to the classical Navier–Stokes equation in the high-frequency regime.

Eq. (3.1) combined with the simplest conservation law for the classical thermal energy (the case of a rigid solid) leads to an equation of change

$$\rho C_p \frac{\partial T}{\partial t} = \lambda \left(\nabla^2 T - \frac{\partial^2 T}{c_0^2 \partial t^2} \right), \tag{3.2}$$

where

$$c_0 = \sqrt{a_h \tau_h^{-1}} = \sqrt{\lambda \rho^{-1} C_p^{-1} \tau_h^{-1}}, \tag{3.3}$$

is called the propagation speed of the thermal wave. When c_0 approaches infinity, Eq. (3.2) simplifies to the well-known parabolic equation of heat. On the other hand Eq. (3.2) is of hyperbolic type. Its solution (Baumeister and Hamil, 1968; Luikov, 1969) for the above-mentioned case of a semiinfinite solid has the following property: temperature $T(x, t)$ has a jump at the distance $x = c_0 t$ from the source of heat placed at $x = 0$ (heating by an ideally mixed fluid). Therefore two regions exist in a solid (Fig. 3.1): the first, in which the heat transfer takes place (disturbed region), and the second, where no transfer is present (undisturbed region). In contrast, as mentioned earlier, Fourier theory predicts the appearance of the disturbances everywhere, even for distances x_c greater than ct (c is the light speed) which is of course unphysical behavior.

For these hyperbolic systems, an interesting effect appears for the wall heat flux ($x = 0$) when the "driving force is turned on." Namely, the wall heat flux, $J_h(0)$, does not start instantaneously, but rather grows gradually (Baumeister and Hamil, 1968) with the rate which depends on the relaxation time τ_h. This is just relaxation in the current and not in the state, the latter being known better than the former one. The acoustic and chemical reaction phenomena may serve as examples of the state relaxation, whereas the relaxations of the heat flux and the viscous stress (and also the current relaxation in an electric circuit, associated with a change in its magnetic energy) are the examples of current relaxation. After some time, the wall heat flux arrives at a maximum and then it decreases in time, similarly as in Fourier case. This decrease is a classical effect and it occurs since the temperature gradient at the wall decreases in time in the course of heating of our solid. Consequently the Fourier and Fick theories are inappropriate for description of the short-time effects, and although relaxation times are typically very small, such effects can still have theoretical importance.

Examples of relaxation time data are (τ_h, τ_d, and τ_m, mean the relaxation time for heat, mass diffusion, and momentum diffusion, respectively):

Metals $\tau_h = 10^{-12}$ s
Gases $\tau_h = 10^{-9}$ s
Typical liquids $\tau_h = 10^{-11} - 10^{-13}$ s

Relaxation times can be much greater in rarefied gases, viscoelastic liquids, capillary porous bodies, dispersed systems, Brownian systems, helium II: for capillary porous bodies, for example, Luikov (1966a, b) evaluated effective values of $\tau_d = 10^{-4}$ s. Calculations exploiting some drying experiments (Mitura, 1981) seem to confirm the order of magnitude of these evaluations (Sumińska, 1983; Sieniutycz, 1989a).

For other special systems one has the following values of τ.

Brownian diffusion in the diameter range $10^{-7} - 10^{-3}$ m:

$$\tau_d = 10^{-8} \div 3 \, s$$

(Sieniutycz, 1984), based on use of some experimental data (Davies, 1966). These evaluations were based on Stokes term of the equation of motion, confirmed experimentally.

Liquid helium:

$$\tau_h = 4.7 \times 10^{-3} \, s$$

(Peshkov, 1944; Keesom, 1949; Bubnov, 1976a, b, 1982).

Turbulent flows:

$$\tau = 10^{-3} - 10^3 \, s$$

(Luikov, 1966a, b; Monin and Yaglom, 1967; Builties, 1977; Petty, 1975).

Kinetic theory (Natanson, 1896; Uhlenbeck, 1949; Wang Cheng and Uhlenbeck, 1951; Resibois and Leener, 1977; Sieniutycz, 1977) predicts the following upper limit velocities for the propagation of disturbances (propagation speeds):

$$c_h \equiv \sqrt{D_h \tau_h^{-1}} = c_d \equiv \sqrt{D_d \tau_d^{-1}} = c_m \equiv \sqrt{\nu \tau_m^{-1}} = \sqrt{P/\rho} \qquad (3.4)$$

where ν is the kinematic viscosity, or the diffusivity of the momentum. Hence one can determine the relaxation times as

$$\tau_h = \frac{\rho D_h}{P} \quad \tau_d = \frac{\rho D_d}{P} \quad \tau_m = \frac{\rho \nu}{P} = \frac{\eta}{P} \qquad (3.5)$$

where η is the dynamic viscosity. For liquids the shear modulus G appears instead of pressure in the formula for the propagation speed

$$c_0 = \sqrt{G/\rho} \qquad (3.6)$$

Boven and Chen (1973) consider thermodynamic restriction on the initial slope of the stress relaxation function. Day (1971) derives restrictions on relaxation functions in linear viscoelasticity. Huilgol and Phan-Thien (1986) review advances in the continuum mechanics of viscoelastic liquids.

At present no exact general criterion is available when it is necessary to include relaxation terms in equations of change. Usually one assumes that these terms are essential when a frequency of the fast variable transients is comparable (or greater) to the reciprocal of a longest relaxation time. The above-mentioned cases are usually cited when these terms should be practically significant for: viscoelastic fluids, capillary porous bodies, dispersed systems, rarefied gases, and helium II for high-rate, unsteady state processes.

Experiments confirming the wave nature of heat are available (Peshkov, 1944; Pellam, 1961; Ackerman et al., 1966; Ascroft, 1976; Ziman, 1960). In their pioneer measurements, Peshkov (1944) and Pellam (1961) have shown the interference of thermal waves. Ackerman et al. (1966) observed onset of the ballistic flow of phonons using temperature pulse techniques below 0.2 K in solid ^{4}He crystals grown from the superfluid. First-sound velocities obtained from the measurements compare well with ultrasonic measurements. See also Jackson et al. (1970) who have studied propagation of heat pulses in NaF to high temperatures in a pure crystal. At the highest temperatures the second-sound velocity fails to level off at the theoretically predicted limiting value. Fox et al. (1972) observed the ballistic flow of phonons using temperature pulse techniques below 0.2 K in solid ^{4}He crystals grown from the superfluid. First-sound velocities obtained from the measurements compare well with ultrasonic measurements. Among the solid-state physicists an opinion prevails that underestimation of the so-called "ballistic" regime of the phonon transfer, which is in fact the wave regime, lead researches to unusual number of invalid statements about the heat transfer in solids (Ascroft, 1976).

3.3 Extended Thermodynamics of Coupled Heat and Mass Transfer

Simultaneous heat and mass transport occurs in many important industrial applications, and plays a basic role in various physical, chemical, and biological processes; hence a vast amount of work on the subject is available in the literature (Demirel and Sandler, 2001, 2002a). Coupled heat and mass transfer occurs in separation by absorption, distillation, and extraction, in evaporation–condensation, in drying, in crystallization and melting, and in natural convection. Assuming local equilibrium, Demirel and Sandler (2001) have used the phenomenological equations relating the conjugated flows and forces defined by the dissipation function of the irreversible transport and rate process.

Various formulations and methodologies have been suggested for describing combined heat and mass transport problems. As pointed out earlier, the nonstationary equation of diffusive motion, that is, Maxwell–Stefan equation of diffusion, leads to the presence of extra (relaxation) terms in the phenomenological equations. Further we will show that the relaxation terms are also substantiated by irreversible thermodynamics. However, the classical expression for entropy and entropy source do not apply in the present (relaxation) case as every relaxation phenomenon is a consequence of the local nonequilibrium in the macroscopic medium. Both phenomenological approaches (Sieniutycz, 1981 a, b, c, d, e) and statistical theories based on Enskog and Grad's iteration methods (Lebon, 1978; Jou et al., 1988; Sieniutycz and Berry, 1989), lead to the conclusion that the entropy of a locally nonequilibrium medium differs from the static (ie, equilibrium) entropy, and the difference depends on all diffusing fluxes $\mathbf{J}_1, \mathbf{J}_2, ...\mathbf{J}_{n-1}, \mathbf{J}_q$. This is the objective of the extended thermodynamics (Nettleton, 1960, 1969, 1986, 1987a; Mc Lennan, 1974; Gyarmati, 1977; Nonnenmacher, 1980; Lebon et al., 1980; Jou et al., 1988; Garcia-Colin, 1987; Eu, 1982, 1986, 1987; Sieniutycz and Berry, 1989, 1991), and many others. This effort culminated in a basic book by Jou, Casas-Vazquez, and Lebon (Jou et al., 2001). A hidden variable approach was developed by Bampi and Morro (1984).

Banach and Piekarski (1992) developed a coordinate free description of nonequilibrium thermodynamics. They attempted to give the meaning of temperature and pressure beyond the local equilibrium and obtain the theory of disequilibrium thermodynamic potentials. To achieve these goals they applied the following maximum principle: among all states having the same values of conserved variables, the equilibrium state gives the specific entropy its greatest value. In association with the coordinate free description of the thermodynamic space the entropy maximum principle allows one to draw upon results from the critical point theory. Namely, it is possible to find a coordinate system for the thermodynamic space such that the entropy can be written as a sum of two physically different terms: the thermostatic entropy depending on the conserved variables and the second term given by a quadratic form depending only on disequilibrium variables. This representation of the specific entropy allows one to arrive at the definition of disequilibrium temperature, pressure, and thermodynamic potential for systems not infinitesimally near to equilibrium.

Kremer (1985, 1986, 1987, 1989, 1992) extends the EIT theory to a broader class of processes. He investigates EIT of ideal and real monatomic gases, molecular gases and mixtures of ideal gases. For the ideal and real gases he develops the formalism of 14 fields, where the 13 fields of mass density, pressure tensor, and heat flux are supplemented by an additional scalar field associated with nonvanishing volume viscosities. Coefficients of shear and volume viscosities, thermal conductivity, three relaxation times, and two thermal-viscous coefficients are identified in the theory. The theory of ideal gas results as a limiting case from that of the real gas. Explicit results are given for classical ideal gases and for degenerate Bose and Fermi gases. Then, an extended theory of molecular gases is formulated which contains 17 fields of density, velocity, pressure tensor, heat flux, intrinsic energy, and intrinsic heat flux. The intrinsic energy pertains to the rotational or vibrational energy of the molecules. Coefficients of thermal and caloric equations of state as well as transport or rate coefficients are identified (shear viscosity, thermal conductivity, self-diffusion, absorption and dispersion of sound waves). Finally, the EIT theory for mixtures of monatomic ideal gases is described. Generalized phenomenological equations follow corresponding to the classical laws of Fick, Fourier, and Navier–Stokes. Onsager reciprocity follows without statistical arguments. An unusual breadth of EIT theory and its capability to synthesize static, transport, and rate properties of disequilibrium systems in one unifying approach is shown.

Perez-Garcia and Jou (1986, 1992) present EIT theory as a suitable macroscopic framework for the interpretation of generalized transport coefficients. Classical thermodynamic theory holds only in the hydrodynamic limit of small frequency and small-wave vector (vanishing ω and k) for any disturbances. A generalized hydrodynamics (GH) is postulated which replaces dissipative transport coefficients by general memory functions justified by some microscopic models. [See the paper by Altenberger and Dahler (1992) and that by Morro (1992) which also deals with memory effects.] A phenomenological theory involving higher-order hydrodynamic fluxes is given for the description of GH processes. Additional variables playing role of currents of the traceless pressure tensor and viscous pressure are introduced into the extended entropy and balance equations. The entropy flux contains the most general vector that can be constructed with all the dissipative variables. As the specific applications, the velocity autocorrelation functions are calculated around equilibrium, and ultrafast thermometry is analyzed. Parallelism of the macroscopic theory with the projection–operator techniques is outlined. Memory functions and continued fraction expansions for transport coefficients are interpreted in terms of the generalized thermohydrodynamics.

Despite of various differences in diverse approaches the basic procedure leading to relaxation terms can be characterized by the representative scheme outlined later. For brevity the momentum transfer is neglected. The difference between the true local entropy of a nonequilibrium state, s', and the thermostatic entropy, s (evaluated for the same internal energy, e), called relaxation entropy, is associated with the tendency of every element of a continuum to recover thermodynamic equilibrium during the vanishing of diffusive fluxes of heat and mass, or the relaxation of the viscous stresses (maximum s' equals to s for $\mathbf{J} = 0$ and for $\Pi = 0$). Since the relaxation is an irreversible process, the relaxation entropy, $s'-s$, corresponding with a stable equilibrium, is always the negative quantity, the consequence of the concavity of the entropy around the macroscopically stable equilibrium. The role of the relaxation entropy in the total entropy expression can be described by the formula (Sieniutycz, 1981a, b, c, d)

$$ds' = T^{-1}de + PT^{-1}d\rho^{-1} + \sum_{1}^{n-1}(\mu_n - \mu_i)T^{-1}dy^i + \sum_{1}^{n}\sum_{1}^{n} g_{ik}\mathbf{J}_i\mathbf{J}_k \tag{3.7}$$

This is a generalized Gibbs equation which in the linear case can be transformed to the concise integrated matrix form

$$s'(\mathbf{z},\mathbf{J}) = s(z) + \frac{1}{2\rho G}\mathbf{J}^T\mathbf{C}^{-1}\mathbf{J} \tag{3.8}$$

In the aforementioned equations:

$$\mathbf{z} = \text{col}(y_1, y_2, \ldots y_{n-1}, e) \quad \text{– state matrix} \quad (3.9)$$

$$\mathbf{J} = \text{col}(\mathbf{J}_1, \mathbf{J}_2, \ldots \mathbf{J}_{n-1}, \mathbf{J}_e) \quad \text{– flux matrix} \quad (3.10)$$

$$\mathbf{u} = \text{col}\left(\frac{\mu_n - \mu_1}{T}, \frac{\mu_n - \mu_2}{T}, \ldots + \frac{\mu_n - \mu_{n-1}}{T}, \frac{1}{T}\right) \quad \text{– transfer potential matrix} \quad (3.11)$$

$$\mathbf{C} = \frac{\partial \mathbf{z}}{\partial \mathbf{u}} = \frac{\partial^2 s}{\partial z_i \, \partial z_j} \leq 0 \quad \text{– entropy capacity matrix} \quad (3.12)$$

$$g_{ik} \equiv \mathbf{C}/\mathbf{G} = \mathbf{C}/\rho c_0^2 \quad \text{– inertial matrix} \quad (3.13)$$

One may ask about the entropy source form which corresponds to the Gibbs equation (3.7) or with its integrated counterpart, Eq. (3.8). The answer is found analogously, as in classical thermodynamics, by combining the Gibbs equation with the equations describing the conservation laws for mass and energy:

$$\rho \frac{dy_i}{dt} = -\nabla \cdot \mathbf{J} \quad (3.14)$$

$$\rho \frac{de}{dt} = -\nabla \cdot \mathbf{J}_e \quad (3.15)$$

As a result, the following entropy balance is obtained

$$\rho \frac{ds'}{dt} = -T^{-1} \nabla \cdot \mathbf{J}_e - \sum_1^{n-1} (\mu_n - \mu_i) T^{-1} \nabla \cdot \mathbf{J}_k + \mathbf{J} \cdot \frac{\mathbf{C}^{-1}}{\mathbf{G}} \cdot \frac{d\mathbf{J}}{dt} \quad (3.16)$$

This equation can be split into the sum of divergence and source terms

$$\rho \frac{ds'}{dt} = -\nabla \cdot \frac{\mathbf{J}_e - \sum_1^n (\mu_k \mathbf{J}_k)}{T} - \mathbf{J}_e \cdot \nabla T^{-1} + \sum_1^{n-1} \mathbf{J}_k \cdot \nabla\left(\frac{\mu_n - \mu_k}{T}\right) + \mathbf{J} \cdot \frac{\mathbf{C}^{-1}}{\mathbf{G}} \cdot \frac{d\mathbf{J}}{dt} \quad (3.17)$$

or more concisely

$$\rho \frac{ds'}{dt} = -\nabla \cdot \mathbf{J}_s - \mathbf{J} \cdot \left(\nabla \mathbf{u} + \frac{\mathbf{C}^{-1}}{\mathbf{G}} \cdot \frac{d\mathbf{J}}{dt}\right) \quad (3.18)$$

where the diffusive entropy flux $\mathbf{J}_s$ has been defined as

$$\mathbf{J}_s \equiv \frac{\mathbf{J}_e - \sum_1^n (\mu_k \mathbf{J}_k)}{T} \qquad (3.19)$$

The condition of nonnegativeness of the entropy source in Eq. (3.18) leads to a phenomenological equation. In the standard matrix notation, the equation has the form

$$\mathbf{J} = \mathbf{L} \cdot \left(\nabla \mathbf{u} + \frac{\mathbf{C}^{-1}}{G} \cdot \frac{d\mathbf{J}}{dt} \right) \qquad (3.20)$$

This formula simplifies to the well-known Onsager's relationship

$$\mathbf{J} = \mathbf{L} \cdot \nabla \mathbf{u} \qquad (3.20a)$$

when G and c_0 tend to infinity. The result obtained, Eq. (3.20), can be written in the form of the following two equations

$$\mathbf{J} + \boldsymbol{\tau} \cdot \frac{d\mathbf{J}}{dt} = \mathbf{L} \cdot \nabla \mathbf{u} \qquad (3.21)$$

$$\boldsymbol{\tau} = -\mathbf{L} \cdot \frac{\mathbf{C}^{-1}}{G} \quad \text{and} \quad G = \rho c_0^2, \qquad (3.22)$$

where $G^{\text{ideal*gas}} = \rho RT/M$ which constitute the phenomenological equation and the relaxation time definition, respectively. Thus we have found the matrix generalization of the Maxwell–Cattaneo equation, Eq. (3.20), describing coupled heat and mass diffusion with the finite wave speed, as well as the supplementary formula, Eq. (3.22), which serves to calculate the elements of the relaxation matrix. One may see that the matrix $\boldsymbol{\tau}$ can be computed on the basis of the two important thermodynamic matrices, the capacity matrix (the reverse of entropy hessian in its natural frame), and Onsager's matrix $\mathbf{L}$, Eq. (3.20a). The latter is usually found from an experiment. Since the product $-\mathbf{L} \cdot \mathbf{C}^{-1} \rho^{-1}$ represents the general tensor of diffusion $\mathbf{D}$ defined in a classical manner (De Groot and Mazur, 1984), which includes the thermal diffusion terms, Eq. (3.22) can be written in the simple alternative form as

$$\boldsymbol{\tau} = \frac{\mathbf{D}}{c_0^2} \qquad (3.23)$$

The special case of this result, for the pure heat transfer, was previously known (Luikov, 1969; Chester, 1963).

3.4 Various Forms of Wave Equations for Coupled Heat and Mass Transfer

The conservation equations

$$\rho \frac{d\mathbf{z}}{dt} = -\nabla \mathbf{J} \tag{3.24}$$

constitute the matrix form of the conservation laws for mass and energy, Eqs. (3.14) and (3.15). Taking the divergence of Eq. (3.24) and using phenomenological relationships, Eqs. (3.21) through (3.23), the matrix system of the wave equations is obtained,

$$\frac{d\mathbf{z}}{dt} = \mathbf{D}\left(\nabla^2 \mathbf{z} - \frac{d^2\mathbf{z}}{c_0^2 dt^2}\right) \tag{3.25}$$

If, instead of the state variables, z, Eq. (3.19), the transport potentials, Eq. (3.11), are used an alternative form is found

$$\rho \mathbf{C} \frac{d\mathbf{u}}{dt} = -\mathbf{L}\left(\nabla^2 \mathbf{u} - \frac{d^2\mathbf{u}}{c_0^2 dt^2}\right) \tag{3.26}$$

In resting systems substantial derivatives simplify to the partial derivatives, and Eqs. (3.25) and (3.26) contain on their right sides d'Alembert (wave) operators instead of usual Laplacians. The form of wave equations representing resting media is useful for finding relativistic generalization of the wave equations (3.25) and (3.26). To obtain this generalization it is sufficient to perform the Lorentz transformation in every equation describing resting system (Sieniutycz, 1979). Of course, when Gallilean transformations are used, the set, Eqs. (3.25) and (3.26) is recovered. The relativistic counterparts of Eqs. (3.25) and (3.26) are,

$$\gamma \frac{d\mathbf{z}}{dt} = \mathbf{D}\left(\nabla^2 \mathbf{z} - \frac{d^2\mathbf{z}}{c_0^2 dt^2}\right) - \left(c_0^{-2} - c^{-2}\right)\gamma^2 \mathbf{D} \frac{d^2\mathbf{u}}{dt^2} \tag{3.27}$$

$$\rho \gamma \mathbf{C} \frac{d\mathbf{u}}{dt} = -\mathbf{L}\left(\nabla^2 \mathbf{u} - \frac{d^2\mathbf{u}}{c_0^2 dt^2}\right) - \left(c_0^{-2} - c^{-2}\right)\gamma^2 \frac{d^2\mathbf{u}}{dt^2} \tag{3.28}$$

where,

$$\gamma = \left(\sqrt{1 - v^2 c^{-2}}\right)^{-1}$$

and c is the light speed. See references (Kranys, 1966; Sieniutycz, 1979; Pavon et al., 1980) for more information.

The physical structure of Eq. (3.25) is obscured by its matrix form. We operate here with the not so common quantities, $\mathbf{z}$ and $\mathbf{u}$. Therefore another approach has been made (Sieniutycz, 1981a, b, c, d) where the fluxes, forces and diffusivities were transformed to the quantities related to the pure heat flux rather than to the energy flux [present in Eqs. (3.21) and (3.25)]. After applying such transformations the following wave system is obtained,

$$\frac{d\mathbf{y}}{dt} = \mathbf{D}_m\left(\nabla^2\mathbf{y} - \frac{d^2\mathbf{y}}{c_0^2 dt^2}\right) + \frac{\mathbf{D}_T}{T}\left(\nabla^2 T - \frac{d^2 T}{c_0^2 dt^2}\right) \qquad (3.29)$$

$$\rho C_p \frac{dT}{dt} = -\rho T \mathbf{D}_T^T \mathbf{C}_m^{-1}\left(\nabla^2\mathbf{y} - \frac{d^2\mathbf{y}}{c_0^2 dt^2}\right) + \lambda\left(\nabla^2 T - \frac{d^2 T}{c_0^2 dt^2}\right), \qquad (3.30)$$

with

$$\mathbf{C} = \mathrm{diag}(\mathbf{C}_m^{-1}, -C_p^{-1} T^2) \text{ and } \mathbf{D}_T = (\rho T)^{-1}\mathbf{L}_T,$$

where $\mathbf{L}_T$ is the part of Onsager's matrix related to the thermal diffusion. This set operates with the most common variables, temperature T and concentrations y_i and simplifies to the classical set for c_0 approaching infinity. It is expected that Eqs. (3.29) and (3.30) describe heat and mass transport better than the classical ones, especially in the case of highly nonstationary cases, for example, during the travel of sound or electromagnetic waves through a medium when thermal diffusion is intensified, separation of isotopes (London, 1961), or when the ultrasonic or dielectric drying of solids solutions takes place.

Analogously, the inclusion of relaxation terms into the momentum transport equation can be performed. Clearly the Maxwell equation of viscoelastic body is then obtained as the phenomenological equation. It is seen that the relaxation theory of heat and mass transfer exhibits a certain connection to the rheological concepts. An extension of these concepts was also observed in the so-called power models of diffusion

$$\sigma_{ik}^0 = -\tilde{a}_m(\nabla^0 \mathbf{v})^p \qquad (3.31)$$

$$\mathbf{J}_h = -\tilde{a}_h(\nabla T)^n \qquad (3.32)$$

$$\mathbf{J}_d = -\tilde{a}_d(\nabla y)^m \qquad (3.33)$$

These equations also lead to the finite propagation of disturbances as the simplest equations of the wave theory,

$$\sigma_{ik}^0 = -2\eta\nabla^0 \mathbf{v} - \tau_m \frac{d\sigma_{ik}^0}{dt} \qquad (3.34)$$

$$\mathbf{J}_h = -\lambda \nabla T - \tau_h \frac{d\mathbf{J}_h}{dt} \tag{3.35}$$

$$\mathbf{J}_d = -\rho D \nabla y - \tau_d \frac{d\mathbf{J}_d}{dt} \tag{3.36}$$

which can be derived from the memory concepts (Nunziato, 1971; Berkovsky and Bashtovoi, 1972). Shter (1973) showed how to generalize Onsager's relations and Eq. (3.36) by using convolution integrals to link thermodynamic fluxes and forces.

Since $\tau = D/c_0^2$, the knowledge of the propagation velocity c_0 is essential. The phonon gas theory has been applied (Chester, 1963; Guyer and Krumhansl, 1964; Prohofsky and Krumhansl, 1964) to evaluate c_0. The random walk model (Goldstein, 1953) also leads to data of c_0. The use of sound dispersion and absorption experiments for the computation of c_0 have also been proposed (Bubnov, 1976a, b).

Link of Eqs. (3.34)–(3.36) with rheological concepts and modeling of rheological systems is obvious. Aubert and Tirrel (1980) investigate the behavior of macromolecules in nonhomogeneous velocity gradient fields. They have obtained some kinetic theory results for bead–spring–type macromolecular models (linearly elastic dumbbell and Rouse) in nonuniform velocity gradient fields. Interestingly, they have shown that contrary to the homogeneous flow result, in general, a macromolecular solute does not move with the local center-of-mass solvent velocity in a nonhomogeneous flow. This phenomenon results essentially from a unique coupling between segmental Brownian motion of the macromolecule in the flow field and the center-of-mass translational diffusion possible only in nonhomogeneous flow. Aubert and Tirrel (1980) consider Poiseuille flow between parallel plates and circular Couette flow; both problems are solved for the dumbbell and some results are obtained for the Rouse model. Migration along streamlines occurs in the Poiseuille flow and migration across streamlines occurs in the Couette flow. Additional energy dissipation results for the Rouse model in Poiseuille flow relative to the simple shear flow case. Nonuniform concentration profiles are calculated in Couette flow by taking the appropriate average of the exact equation for the dumbbell configuration space distribution function. See also: Lebon et al. (1986, 1988, 1990, 1993), Drouot and Maugin (1983, 1985, 1987, 1988), Drouot et al. (1987).

Baranowski (1991, 1992) applies nonequilibrium thermodynamics to membrane transport and diffusion in elastic media with stress fields. A treatment of active transport in terms of linear thermodynamics is given. Bird and Öttinger (1992) review the molecular theory of disequilibrium properties of polymeric liquids: rheological properties, translational diffusivity, and thermal conductivity. (Electrical and optical properties are excluded.) The authors are primarily concerned with work that has appeared since the publication of the treatise by Bird, Amstrong, and Hassager (Bird et al., 1977) which generalizes the classical theory of transport phenomena (Bird et al., 1960) to the rheological systems. Sluckin (1981) shows the influence of flow on the isotropic-nematic transition in polymer solutions.

Ionescu et al. (2008) examined thermodynamic aspects of polymeric liquids subjected to nonisothermal flow from the complementary perspectives of theory, experiment, and simulation. In particular, they refer to the energetic effects, in addition to the entropic ones that occur under conditions of extreme deformation. Comparison of experimental measurements of the temperature rise generated under elongation flow at high-strain rates with macroscopic finite element simulations offer clear evidence of the persistence and importance of energetic effects under severe deformation. The authors evaluated performance of various forms of the temperature equation with regard to experiment, and concluded that the standard form of this evolution equation, arising from the concept of purely entropic elasticity, is inadequate for describing nonisothermal flow processes of polymeric liquids under high deformation. In their opinion, complete temperature equations, in the sense that they possess a direct and explicit dependence on the energetics of the microstructure of the material, provide very good agreement with experimental data.

Applying statistical mechanics Bearman and Kirkwood (1958) derived equations of hydrodynamics including the equations of continuity, equations of motion of the individual components as well as the overall equation of motion, and the energy transport equation. Introduction of perturbations to the singlet and pair space distribution functions linearized in the temperature gradient, diffusion velocities, and the local rate of shear lead in the stationary case to the linear relations of irreversible thermodynamics between the gradients of temperature and chemical potential and the fluxes of heat and matter. Moreover, Bearman (1959) derived the phenomenological equations, including inertial and viscous terms, from statistical mechanical theory. The influence of the additional terms has been evaluated with reference to their effect on the operational significance of the phenomenological coefficients and on the interpretation of diffusion data. Bearman–Kirkwood generalized transport equations (Bearman and Kirkwood, 1958; Bearman, 1959) introduced a new quantity into the transport theory, namely, a coupling between diffusion and both inertia and local stresses. It seems interesting to also investigate the role of the coupling in barodiffusion and migration processes. Bartoszkiewicz and Miekisz (1989) have investigated generalized thermodynamics founded by Bearman–Kirkwood equations. They have shown that the Prigogine theorem on the entropy production invariance also holds when inertia and local stresses contribute to the diffusion–dependent part of the dissipation function. They have compared their generalized theory with other approaches and obtained conditions which reduce this generalized theory to the classical phenomenological approach for the case of mechanical equilibrium. They have also discussed the role of coupling between diffusion and viscous flow in barodiffusion processes and estimated Bearman–Kirkwood partial salt viscosities for the low concentration region. The generalized theory predicts that in the presence of viscous effects and in the low concentration region, electrolyte barodiffusion depends mainly on local stresses. As a consequence, under the mentioned conditions, electrolyte barodiffusion is directed "against" the motion of the mass center. Thus, for some porous membranes some "practical" coefficient $L_{\pi P}$ may be interpreted as the manifestation of the diffusion–viscous flow coupling. Bartoszkiewicz and Miekisz (1989) discuss the possibility of a measurement of the Bearman–Kirkwood ionic partial viscosities. They also investigate diffusion-viscous flow coupling and global transport coefficients symmetry in matter transport through porous membranes (Bartoszkiewicz and Miekisz, 1990). If the statistical mechanical theory of Bearman and Kirkwood is applied to describe isothermal permeation of an incompressible solution through a porous membrane at mechanical equilibrium and not far from thermodynamic equilibrium and if the pores are homogeneous and cylindrical, the global transport coefficients of the system are symmetric if and only if switching off the short-range membrane-solution interactions leads to an uncoupling of diffusion and viscous flow (Bartoszkiewicz and Miekisz, 1990).

3.5 Stability of Dissipative Wave Systems

Although the classical (parabolic) equations and wave equations have the same steady state solutions, the stability depends on the transient behavior which is different in both cases. It is well known (Glansdorff and Prigogine, 1971) that the stability of local equilibrium transients with respect to small perturbations described by parabolic equations can be proved by considering the second differential of entropy. An analogous method was used for wave equations of heat (Lebon and Casas-Vazquez, 1976; Lebon and Boukary, 1982; Sieniutycz, 1981a, b, c, d, e; Sieniutycz and Berry, 1991). Equations of hydrodynamics were included in the last reference cited. Taking into account that in the wave case the entropy expression contains the flux term,

$$\delta^2 S = \int_V \rho \left(\delta T^{-1}\delta e + \delta(PT^{-1})\delta\tilde{v} + \sum_1^{n-1} \delta(\mu_i T^{-1})\delta y^i + \sum_1^{n-1} \delta\zeta_i\, \delta \mathbf{J}_i \right) dV \qquad (3.37)$$

where ξ_i is the adjoint variable of $\mathbf{J}_i$ in the expression describing perfect differential of entropy. Using this expression as the Lapunov functional for the disturbances of e, y_i and $\mathbf{J}_i$ related by the conservation and phenomenological equations (3.14), (3.15), and (3.20) it was found (Sieniutycz, 1981c, d) that if

$$\mathbf{C} \leq 0 \qquad (3.38)$$

the following holds

$$\delta^2 S \leq 0 \quad \text{and} \quad \frac{\partial \delta^2 S}{\partial t} \equiv \int_V (\delta \mathbf{J} \mathbf{L}^{-1} \delta \mathbf{J})\, dV \geq 0 \qquad (3.39)$$

for the constant potentials at the boundaries of the system. This means that the negative sign of the entropy capacity $\mathbf{C}$ (well known in classical thermodynamics) is the sufficient stability condition. Thus, by the application of the second Lapunov method (Sieniutycz, 1981c, d) shows that every trajectory of the distributed process of simultaneous heat and mass transfer in an nonreacting fluid will approach the stationary state for t approaching infinity if the well-known thermodynamic matrix of capacities, **C**, is negative, even if the inertial (ie, relaxation) terms are important.

Alternatively, a functional generalizing the excess entropy production was tested as the Lapunov criterion (Sieniutycz, 1980a)

$$\tilde{V} = \frac{1}{2} \int_V \left(\sum_{i=1}^n \sum_{k=1}^n L_{ik} \nabla \delta u_i \nabla \delta u_k + \sum_{i=1}^n \sum_{k=1}^n L_{ik} \frac{\partial \delta u_i}{c_0 \partial t} \frac{\partial \delta u_k}{c_0 \partial t} \right) dV \geq 0 \tag{3.40}$$

For transients of the perturbations governed by Eq. (3.26) it was obtained that

$$\frac{\partial \tilde{V}}{\partial t} = \int_V \left(\sum_{i=1}^n \sum_{k=1}^n \rho C_{ik} \frac{\partial \delta u_i}{c_0 \partial t} \frac{\partial \delta u_k}{c_0 \partial t} \right) dV \geq 0 \tag{3.41}$$

which again proves that the negative sign of **C** is the sufficient condition of stability. Thus similarly as in the classical situation of infinite $c0$ it has been concluded that the trajectories of simultaneous heat and mass transfer processes are always asymptotically stable if they are described by our wave model. Broadly speaking this means that the two arbitrary solutions **u**(x,t) and **u**₀(x, t) will approach one another despite the essential inertial terms existing for wave equations. This also means that one should not expect principally new results concerning stability of transients around equilibrium (where the inequality **C** < 0 always holds) when the wave equations are used.

3.6 Variational Principles

An intriguing fact is (Sieniutycz, 1977, 1979, 1980, 1981a, b, c, d, 1982, 1983, 1985, 1987, 1989a), (Sieniutycz and Berry, 1989, 1991) that the set of the wave equations of heat and mass transfer can be derived from a certain modification of Hamilton's principle of least action in the form containing the relaxation time τ. The action structure was achieved by applying Bateman's structure of the Lagrangian with its usual exponential term (Bateman, 1929, 1931). When, for example, pure heat transfer is considered, an action functional of the resting solid has the form

$$\tilde{S} = \int_{Vxt} \varepsilon^{-1} \left(\frac{1}{2} J_h^2 c_0^{-2} - \frac{1}{2} \rho_h^2 + J_h \frac{\partial H_h}{c_0^2 \partial t} + \rho_h \nabla H_h \right) \exp\left(\frac{t}{\tau} \right) dVdt \tag{3.42}$$

where ε is the equilibrium energy of an undisturbed state, $\mathbf{J}_h$ is the pure heat flux, $\mathbf{H}_h$ is the heat (Biot's) vector, $\rho_h = \rho C_v(T - T_{eq})$ is the "volumetric charge of the nonequilibrium heat energy" and ρ is the mass density. The stationarity conditions for the action (3.42) are

$$\mathbf{J}_h = \frac{\partial \mathbf{H}_h}{\partial t} \tag{3.43}$$

$$\rho_h = -\nabla \mathbf{H}_h \tag{3.44}$$

$$\frac{\partial \rho_h}{\partial t} + \nabla \mathbf{J}_h = 0 \tag{3.45}$$

$$\frac{\partial \mathbf{J}_h}{\partial t} + \frac{\mathbf{J}_h}{\tau} = -\nabla \rho_h = -\rho C_v \nabla T \tag{3.46}$$

the last result being an equation of heat conduction equivalent to the Maxwell–Cattaneo equation. From Eqs. (3.45) and (3.46) wave equation (3.2) is obtained.

Consequently, a description of the thermal field is obtained from a modified principle of least action with the relaxation exponential term. In a similar manner many new variational formulations leading to coupled wave equations of heat, mass, and electric charge transport have been found (Vujanovic, 1971; Vujanovic and Djukic, 1972; Vujanovic and Jones, 1988; Lebon, 1976, 1986; Bhattacharya, 1982; Sieniutycz, 1978, 1979, 1980, 1982, 1983, 1984, 1985, 1988; Mornev and Aliev, 1995) On the statistical ground a variational method appeared for approximating distribution functions with a "Liouville operator" Eyink (1996).

A functional which leads to the coupled wave equations (3.25) or (3.26) is of the form

$$\tilde{S} = \int_{V_{xt}} \frac{1}{2} \left(\exp(\mathbf{CL}^{-1} \rho c_0^2 t) \right) \mathbf{L} : \left(\nabla u \nabla u - c_0^{-2} \dot{u} \dot{u} \right) dV dt \tag{3.47}$$

which holds for a resting medium, $\mathbf{v} = 0$. Classical and relativistic generalizations of this functional for the moving systems were found (Sieniutycz, 1979). Certain relation to the minimum entropy production theorem should be noted. Namely, for $\mathbf{C}$ tending to zero, that is, when the process becomes purely dissipative and for the stationary state, the principle of minimum entropy production

$$\tilde{S} = \delta \int_V (\mathbf{L} : \nabla u \nabla u) \, dV = 0 \tag{3.48}$$

is recovered from Eq. (3.47). Thus the variational approach based on the modified principle of least action leads to a counterpart of the minimum entropy theorem for the highly nonstationary processes in which both dissipative and dynamical terms are essential. The Lagrangian representation of diffusion (replacing Euler representation, used here) has also been used in the least action type formulations (Sieniutycz, 1983).

Vast applications of variational methods in the context of wave equations are only briefly characterized here. For example, one can study a drying process and a moisture content distribution $Y(x, t)$ in a resting infinite plate the sides of which are maintained at a constant concentration (ie, $Y = Y_{eq}$ and Biot's number is tending to infinity) and the initial moisture content is parabolic (Sieniutycz, 1985). Thus, the initial and boundary conditions pertain to the onset of the so-called second drying period, that is,

$$Y = Y_e + (Y_{max} - Y_e)[1 - (x/X)^2] \quad \text{at } t = 0, \quad \text{for } -X < x < X \tag{3.49}$$

$$Y = Y_e \quad \text{at} \quad x = +X, \text{ and } x = -X \text{ for } t > 0 \text{ if } Bi^{-1} = 0 \tag{3.50}$$

One has to determine approximate solution of the equation

$$\frac{\partial Y}{\partial t} = D\left(\nabla^2 Y - c_0^{-2} \frac{\partial^2 Y}{\partial t^2}\right) \tag{3.51}$$

by minimizing the functional

$$\tilde{S} = \frac{1}{2}\rho D \int_0^{t_f} \int_{-X}^{X} \left(c_0^{-2}\left(\frac{\partial Y}{\partial t}\right)^2 - \left(\frac{\partial Y}{\partial x}\right)^2 \right) \exp\left(\frac{t}{\tau}\right) dx\, dt \tag{3.52}$$

the stationarity condition of which is the wave equation (3.51). Applying the Kantorovich method, one may assume the following form for the moisture content field

$$Y(x,t) = Y_e + (Y_{max} - Y_e)[1 - (x/X)^2]h(t) \quad h(0) = 1 \tag{3.53}$$

where $h(t)$ is an unknown function of time which must extremize the functional (3.52). Applying Eq. (3.53) into Eq. (3.52) one finds

$$\tilde{S} = \frac{8\rho D(Y_{max} - Y_e)^2 X}{15} \int_0^{t_f} (c_0^2 \dot{h}^2(t) - 2.5 X^{-2} h^2(t)) \exp\left(\frac{t}{\tau}\right) dx\, dt \tag{3.54}$$

The best approximating function $h(t)$ must obey the Euler equation for the integral (3.54) which is

$$\tau \ddot{h} + \dot{h} = -2.5 h D X^{-2} \tag{3.55}$$

Hence, the moisture content field is determined by Eqs. (3.53) and (3.55). Averaging procedure applied to Eqs. (3.53) and (3.55) gives

$$h(t) = \frac{\bar{Y}(t) - Y_e}{\bar{Y}(0) - Y_e}, \quad \text{with} \quad \bar{Y} = \frac{1}{2X} \int_{-X}^{X} Y(x,t)\, dx \tag{3.56}$$

and hence Eq. (3.55) assumes the form containing the average moisture content

$$\tau \frac{d^2\overline{Y}}{dt^2} + \frac{d\overline{Y}}{dt} = -K(\overline{Y} - Y_e) \tag{3.57}$$

This is, in fact, a generalized drying equation for the second period of drying. It takes into account the relaxation of the drying rate (τ term), with the well-known drying coefficient $K = 2.5DX^2$. Obviously, for $\tau = 0$ the conventional drying equation is recovered.

Eq. (3.57) was used for experimental evaluation of the drying relaxation times from the drying kinetics curves (Mitura, 1981; Sumińska, 1983). Effective values of τ are of the order of magnitude 10^{-4} seconds in agreement with the Luikov's evaluations (Luikov, 1966a, b).

A comprehensive investigation of the frequency dependent properties of Eq. (3.57) was made for a second drying period (Sieniutycz, 1989a). As for its classical counterpart the equation describes the time changes in the space-averaged free moisture content in solids. When the equilibrium state oscillates around a certain constant absolute moisture content, the RDE takes its most general form, namely that of an oscillator with an external force. The overdamped nature of the drying system (lack of free oscillations) is pointed out. Effective relaxation times and drying coefficients were evaluated from the experimental drying data describing the average solid moisture versus time. The details of the evaluation method, based on a squared error minimization, were discussed. The agreement of the experimental relaxation times obtained with Luikov's (1966a, 1975) and Luikov et al. (1976) computational relaxation times was shown. Quasi-steady-state drying—moisturizing cycles caused by periodic changes in the gas temperature (steady forced moisture content oscillations in solid) were analyzed.

Many other applications of the variational principles with exponential relaxation terms, in particular to fluid mechanics and boundary layer theory, are given in a book (Vujanovic and Jones, 1988).

With a limited success, the variational approaches involving containing the relaxation time τ exponentially can be applied to resting chemically reacting systems, for which a coupled set of equations holds for a fluid with a non-Fickian diffusion and chemical reaction described by a non-Gulberg and Waage kinetics (Garcia-Colin et al., 1986; Sieniutycz, 1987b; Garcia-Colin, 1988, and chapter: Thermodynamic Controls in Chemical Reactors). However, an extension of the variational approach using the relaxation time τ to the case of moving fluids with an arbitrary variable velocity $\mathbf{v}(\mathbf{x}, t)$ is probably impossible. In fact, considerable difficulties have been observed to make this approach consistent with exact, nontruncated forms of the balance laws for the energy, momentum, mass, and entropy. Therefore, it seems that the use of the relaxation-time terms in the functional like Eqs. (3.42) or (3.47) is more mathematical than physical concept, and a physical variational structure taking into account irreversibility has yet to come. Some results in the variational theory of conservation laws (Sieniutycz, 1988, 1989a, 1990, 1994) proves the power of the entirely new approach in variational description of disequilibrium processes. Its physical origin is based on the independent transfer of the entropy and mass. This approach distinguishes the matter and the entropy as the two fundamental entities with the material and thermal degrees of freedom, indicates the appropriateness of the momenta carried by either of these degrees of freedom, and allows to construct a new general structure of disequilibrium dynamics using the material and thermal momenta as well as corresponding quantum phases as the natural variables of the Gibbs fundamental equation out of equilibrium (Sieniutycz, 1990, 1994).

3.7 Other Applications

Other applications of heat wave equations involve:

> Analytical description of thermal laser shocks (Domański, 1978a, b, 1990).
> Impulsive radiative heating of solids (Grigoriew, 1979).
> Short-time heat conduction in thin surface layers (Kao, 1977).
> Short-time contacts and collisions (Kazimi and Erdman, 1973).
> Fast-phase changes and melting (Sadd and Didlake, 1977).
> Short-time heat conduction transients (Wiggert, 1977).
> Experiments in solid helium and ionic crystals, to confirm heat waves (Ascroft, 1976).
> Interference and diffraction of heat waves (Peshkov, 1944; Pellam, 1961).
> Showing wave behavior of atmospheric isotherms in Shuleikin's works (Bubnov, 1976a, b).
> Rate relaxation in Brownian systems (Davies, 1966; Sieniutycz, 1984).
> Acoustic dispersion in high-frequency regime (Carrasi and Morro, 1972, 1973).
> Chemical transients and extended affinities (Garcia-Colin, 1992; Sieniutycz, 1987b).
> "Hyperbolic" reactors (Holderith and Reti, 1979; Reti, 1980).
> Multiphase chemical reactions (Reti and Holderith, 1980).

Determining the thermodynamic curvature of nonequilibrium gases (Casas-Vazquez and Jou, 1985).
Phonon gas hydrodynamics based on the maximum entropy principle and the extended field theory of a rigid conductor of heat (Larecki and Piekarski, 1991).
Polymer diffusion in nonhomogeneous velocity gradient flows (Drouot and Maugin, 1983).

3.8 High-Frequency Behavior of Thermodynamic Systems

One may note that the drying equation (3.57) has the form typical of a damped oscillator. The notion that the drying (dissipative) system can be described by an equation of this kind is rather surprising because such an equation may often imply damped or undamped oscillations. However an analysis of the material constant magnitude (Sieniutycz, 1989a) indicates that the physical nature of the drying is such that only forced oscillations are possible, no free oscillations can occur, which is, of course, in agreement with our experience. Clearly, the wave drying equation (3.57) does not imply the free oscillations for the real material constants. Consequently, although our thermodynamic system is described in the same manner as mechanical or electrical system, the qualitative behavior of this system is quite different; it is over-damped. In general the over-damping effect prevails, thus, as an analysis of the steady forced oscillation shows, the maximum of the rate amplitude occurs for the frequency $\omega_0 = (K/\tau)^{1/2}$ in the form of plateau which extends in the vast range of frequencies ω between K and τ (Sieniutycz, 1989a). When $\tau = 0$ (conventional theory) the plateau extends to infinity. Therefore, the discrepancies between the two theories, wave and conventional, start at the frequencies of the order of τ^{-1}. For lower ω an excellent agreement between the two theories is obtained, that is, each theory is acceptable. It can be shown that the efficiency of energy transmission due to fast disturbances is practically equal to zero in the classical regime $K < \omega < \tau^{-1}$. In the nonclassical regime, described by the relaxation equation, this efficiency can approach even unity in the high-frequency range due to the predominant role of the inertial terms. This fact serves as a justification of a growing interest in the investigation of high-frequency phenomena with a finite wave speed.

Thermodynamics of these phenomena can be investigated either in the context of the kinetic theory (Grad, 1958; Chapman and Cowling, 1973; Cercignani, 1975; Klimontovich, 1982; Karkheck, 1986, 1989, 1992; Kreuzer, 1981; Lifshitz, and Pitajevsky, 1984; Eu, 1980, 1986, 1987, 1989; Jou et al., 1988), or in the framework of thermohydrodynamic principle of least action, which takes into account an additional kinetic term in the kinetic potential (Sieniutycz, 1988, 1990). The origin of this term can also be found in the kinetic theory (Lebon, 1978; Sieniutycz and Berry, 1989, 1991).

3.9 Further Work

Packed bed reactors offer the good opportunity to test virtues and disadvantages of dispersion effects. Considering the description of phenomena in packed bed reactors, Iordanidis et al. (2003) present a critical comparison between the wave model and the standard dispersion model. Capabilities of wave model are tested on the basis of three industrially important reactions. In case of moderate reaction rates very good agreement is found between predictions of both models and the experimental data. However, for highly exothermic reactions with steep temperature and concentration profiles the standard dispersion model fails to describe the experiments, whereas the wave model gives a good agreement with the experiments London (1961).

Karkheck (1992) develops the Boltzmann equation and other kinetic equations in the context of the maximum entropy principle and the laws of hydrodynamics. He uses a general approach based on the maximum entropy principle, and provides a specific example with a constraint imposed on the one-particle distribution function. Hydrodynamics emerges from the observation that a vast amount of detailed microscopic information is macroscopically irrelevant. His interpretation of minimum information (maximum entropy) is not the conventional interpretation of information which is missing but rather information which is irrelevant to a chosen level of observing the behavior of a system. His approach yields generalizations of Boltzmann and Enskog theories. The generalizations are applicable to real fluids and dense gases. An interparticle potential is used which can be adapted to real fluids by using perturbation theory of equilibrium statistical mechanics (Mansoori and Canfield, 1969). A sequence of kinetic equations is generated in the maximum entropy approach by imposing different constraints to yield different forms of closure. Examples of closed kinetic theories for simple liquids are given (Schmidt et al., 1981), and a correspondence with Boltzmann and Vlasov ensembles is shown when the interparticle potential is neglected. Then, from the moment approach, the conservation laws of mass, momentum and energy follow, with very general flux expressions. By expansion of the fluxes in powers of the gradients of density, velocity and temperature, the transport coefficients are obtained. Existence of two distinct temperatures out of equilibrium is shown. A modification of the Chapman–Enskog procedure (Chapman and Cowling, 1973; Lopez de Haro and Cohen, 1984) is proposed for solving the kinetic equations in order to include extra degrees of freedom. A successful comparison with molecular dynamics is presented. This paper should be read by everyone who is interested in the origin and nature of irreversibility.

In the complementary action approaches, which are believed to be physical approaches, Bateman's exponential relaxation terms do not enter (Bateman, 1929, 1931), and the primary role is played by a generalized kinetic energy of diffusion involving heat terms and quantum phases (Sieniutycz, 1990, 1994). The partial derivative of the kinetic potential density, L, with respect to the velocity of entropy transfer, having the significance of a thermal momentum, is an important physical variable of the extended Gibb's equation involving heat terms (Sieniutycz and Berry, 1989, 1991; Sieniutycz, 1990).

The idea of thermal momentum is the physical idea emerging from the fact that any heat flux, an ordered quantity, causes a decrease of the system entropy from its isoenergy equilibrium value [Eq. (3.8), for instance], or the corresponding increase of the energy taken at given entropy (an increase of the availability of the system). Since the nonequilibrium increase of the internal energy Δe is well defined in terms of Grad's solution of Boltzmann kinetic equation (Sieniutycz and Berry, 1989) one can expect to find the corresponding momenta as every Lagrangian approach requires. In the frame moving with the fluid, the thermal momenta are internal momenta similar to the diffusional momenta of mass. All internal momenta compensate in the fluid frame, giving rise to the vanishing hydrodynamic velocity. They are also essential for evolution equations, preserving the finite propagation of heat and diffusion. An apparent redefinition of the hydrodynamic velocity is required when taking into account the contribution of the thermal momenta (Sieniutycz, 1990). Neglect of thermal momenta leads inevitably to local equilibrium descriptions, excluding any deviation of internal energy or entropy from the equilibrium Gibbs surface, in disagreement with the results of the kinetic theory. In fact, meaning of the thermal momentum is an implicit consequence of the kinetic theory, but it may also be viewed independently as a basic starting point of the Lagrangian approach with the kinetic potential involving the material and thermal coordinates, velocities, momenta, and quantum phases. All dissipative phenomena, that is, heat flow, mass diffusion, and viscous motion emerge from relative motions of the entropy and components, and the dissipation can be regarded as an effect of the motion of the various entities with respect to mass bulk of the system. Importantly, the extended theory changes essentially the structure of the energy-momentum tensor and conservation laws in the system far from equilibrium preserving simultaneously its correspondence with the well-known structure of these laws for close-to-equilibrium and nondissipative equilibrium fluids (perfect fluids).

Sieniutycz and Berry (1997) have introduced concepts of thermal inertia and thermomass as tools that offer means to describe transport of heat in disequilibrium fluids. However there are options regarding how to separate the part of the total flow that carries entropy from the "mechanical," nonentropy-bearing part. In their paper some hypotheses are examined and compared for constructing such a field theory of thermal mass in the energy representation. A global intrinsic symmetry and a finite thermal momentum imply that any formulation which hypothesizes a constant ratio θ of thermal mass to the entropy (Veinik, 1966, 1973) must tie the thermal mass to the so-called bare mass of particles, to preserve the global conservation of matter. However, in any formulation consistent with the Grad–Boltzmann theory, where θ must be variable, the thermal mass behaves as a separate variable governed by the entropy and the second law. Nonetheless, in this case θ has a reasonably broad plateau of values within which entropy is a measure of the thermal mass associated with changes of state. Nonlinear transformations linking usual thermodynamic variables with those of the thermal mass frame preserve the components of the tensor of matter, including Noether's energy and pressure. A formula is given for the fraction of the observed mass assignable as thermal mass, in accordance with Grad's solution of the Boltzmann equation.

Guo and Cao (2008) and Guo and Hou (2010) link thermal waves with existence of the thermomass effect. They argue that in times comparable to the characteristic time of the energy carriers, Fourier's law of heat conduction breaks down and heat may propagate as waves. Based on the concept of thermomass, which is defined as the equivalent mass of phonon gas in dielectrics, according to the Einstein's mass–energy relation, the phonon gas in the dielectrics is described as a weighty, compressible fluid. Newton mechanics is applied to establish the equation of state and the equation of motion for the phonon gas as in fluid mechanics because the drift velocity of a phonon gas is normally much less than the speed of light. The propagation velocity of the thermal wave in the phonon gas is derived directly from the equation of state for the phonon gas, rather than from the relaxation time in the Cattaneo–Vernotte model (CV model: Cattaneo, 1958; Vernotte, 1958). The equation of motion for the phonon gas gives rise to the thermomass model which depicts the general relation between the temperature gradient and heat flux. The linearized conservation equations for the phonon gas lead to a damped wave equation, which is similar to the CV-wave equation, but with different characteristic time. The lagging time in the resulting thermal wave equation is related to the wave velocity in the phonon gas, which is approximately two orders of magnitude larger than the relaxation time adopted in the CV-wave model for the lattices. A numerical example for fast transient heat conduction in a silicon film is presented to show that the temperature peaks resulting from the thermomass model are much higher than those resulting from the CV-wave model. Due to the slower thermal wave velocity in the phonon gas, by as much as one order of magnitude, the damage due to temperature overshooting may be more severe than that expected from the CV-wave model.

Thermal transport at small length scales has become of great technological importance in understanding the thermal behavior of systems such as microelectronics, MEMS, and nanoscale systems. In microelectronics, the process of Joule heating and the formation of hotspots often occur on length scales that are on the order of or smaller than the mean free path of the dominant heat carriers (phonons). The transport of heat has been shown to result in temperatures higher than those predicted by diffusive heat transport models. Donmezer and Graham (2014) describe microscopic mechanism of heat transport in these systems and present a coupled phonon Boltzmann Transport Equation (BTE). They also present a heat equation solver to model the heat transport. The coupled method results are validated both in diffuse and ballistic limits. The solver captures the ballistic and diffusive transport effects in large domains. The proper size of the BTE domain is impacted by the phonon mean free path. The proper size of the BTE domain is impacted by phonon boundary scattering.

The modeling of heat transport for small-length scales requires accounting for the physics of heat carrier transport such as the propagation, scattering, and relaxation of phonons. However, the models used to describe such phenomena are often too computationally expensive to model large domains where relevant system boundaries are far removed from the region where ballistic–diffusive phonon transport is dominant. To address the response of such systems to its far field thermal boundary conditions, a multiscale thermal model is needed to efficiently account for transport phenomena in each domain. Donmezer and Graham (2014) present a multiscale thermal modeling approach, where the transport of phonons is treated with a Finite Volume Discrete Ordinates Model (FVDOM) solution to the phonon Boltzmann Transport Equation (BTE) which is embedded in a region treated by diffusive thermal transport. The approach shows that it is possible to create a coupled multiscale model where the boundary conditions of the FVDOM model are derived from the diffusive transport model. The limitations of the method and the key parameters for accurate modeling are investigated and discussed. The final model is used to explore the impact of the size of heat generation regions with respect to the overall domain size on the peak temperature. The results show that by using such a model it is possible to capture the ballistic–diffusive transport in pertinent areas of a domain that is mostly dominated by diffusive transport.

Sellitto and Cimmelli (2012) have developed a continuum approach to the thermomass theory for nonlinear heat transport, aimed at testing its compatibility with the general framework of continuum thermodynamics. The authors analyze propagation of heat waves along a nonequilibrium steady state. The heat flux is supposed to depend on the absolute temperature together with a vector internal variable that is proportional to the drift velocity of the heat carriers. A generalized heat-transport equation follows, which is capable to bring Fourier, Maxwell–Cattaneo–Vernotte and thermomass-theory equations as special cases.

Nanosystems show electronic, photochemical, electrochemical, optical, magnetic, mechanical, and catalytic properties which differ significantly from those of macroscopic systems. There are several reasons for these different behaviors, such as quantum-size effects, or memory and nonlocal effect. Jou et al. (2011) obtained the dispersion relation of heat waves along nanowires, displaying the influence of the roughness of the walls. This information may be useful for the development of new experimental techniques based on heat waves, complementary to current steady-state measurements, for the exploration of phonon–wall collisions in smooth and rough walls. Zanette (1999) shows that Tsallis's generalized statistics provides a natural frame for the statistical-thermodynamical description of anomalous diffusion. Within this generalized theory, a maximum-entropy formalism makes it possible to derive a mathematical formulation for the mechanisms that underlie Lévy-like superdiffusion, and for solving the nonlinear Fokker–Planck equation.

Lebon's (2014) reviews a description of heat transport at micro and nanoscales from the viewpoint of extended irreversible thermodynamics. After a short survey of the classical heat conduction laws, like those of Fourier, Cattaneo, and Guyer–Krumhansl, he briefly overviews the hypotheses and objectives of Extended Irreversible Thermodynamics, which is particularly well suited to cope with processes at short length and time scales. A selection of some important items typical of nanomaterials and technology are reviewed. Particular attention is brought to the notion of effective thermal conductivity: its expression in terms of size and frequency dependence is formulated, its significant increase in nanofluids formed by a matrix and nanoparticles is discussed, the problem of pore-size dependence and its incidence on thermal rectifiers is also investigated. Transient heat conduction through thin one-dimensional films receives special treatment. The results obtained from a generalized Fourier law are compared with those provided by a more sophisticated ballistic-diffusion model. The survey ends with general considerations on boundary conditions whose role is of fundamental importance in nanomaterials.

Quantum theories, such as de Broglie microthermodynamics (de Broglie, 1964), may accommodate EIT effects (Sieniutycz, 1992b, 1994). An approach toward the Hamiltonian formulation as a basis of quantized heat process (Markus, 2005) has proved to be very fruitful. The approach is based on a generalized Hamilton–Jacobi equation for dissipative heat processes (Markus and Gambar, 2004) as the basic component of the variational theory developed by these authors (see chapter: Variational Approaches to Nonequilibrium Thermodynamics). The intriguing results are thermal quasiparticles (Veinik, 1966, 1973; Markus and Gambar, 2005; Vazquez et al., 2009), resembling phonons of quantized acoustic fields.

As predicted by the theory of thermal mass, the invariant nonequilibrium temperature of the system, defined in terms of the kinetic potential density or energy density is

$$T = -\left(\frac{\partial L}{\partial S}\right)_{v_s} = \left(\frac{\partial E}{\partial S}\right)_{i_s} \quad (3.58)$$

where v_s is the entropy transfer velocity and i_s is the corresponding thermal momentum. T obeys the requirements of the canonical formalism and it is, simultaneously, the appropriate coefficient linking the heat flux and diffusional entropy flux (Sieniutycz, 1988, 1990, 1992b; Sieniutycz and Berry, 1989, 1991). This fact allows one to construct a consistent theory of the nonequlibrium potentials remaining in agreement with the modern concepts of some "microscopic" thermodynamics (de Broglie, 1964). These ideas are presented in the papers and a book (Sieniutycz, 1990, 1992b, 1994). The basic role played by quantum phases in the fundamental Gibbs equation of physical systems in disequilibrium (Sieniutycz, 1990), and the link between the invariance of the action, second law, and chemical stoichiometry, prove that the thermal inertia effects may have a quantum origin (de Broglie, 1964; Vazquez et al., 2009).

Research by Guo et al. (2016) investigates the flux-limited behaviors in existing nonlinear models of heat transport in terms of their theoretical foundations. Different saturation heat fluxes are obtained, whereas the same qualitative variation trend of heat flux versus temperature gradient is found in diverse nonlinear models. The phonon hydrodynamic model follows as a standard to evaluate heat flux limiters characterized by rigorous physical foundation.

CHAPTER 4

Classical and Anomalous Diffusion

4.1 Introduction

In this chapter basic equations of classical and anomalous diffusion are simultaneously derived from the two generalized theories of diffusion which apply Tsallis's (1988) generalization of statistical mechanics, the theory of Compte and Jou (1996) and the theory of Zanette (1999). The classical diffusion occurs for the unit value of nonextensivity parameter, $q = 1$. Following this introduction, in the first part of the chapter we consider the link between nonequilibrium thermodynamics and anomalous diffusion as described by Compte and Jou (1996). In the second part we apply the new thermodynamic framework for the description of anomalous diffusion (Tsallis, 1988) as explored in the complementary Zanette's (1999) theory.

We begin with a discussion of basic physical aspects of irreversibility of diffusion using the didactic examples worked out by Hofman (2002). Let us explain the basic difference between the description of a simple mechanical system, for example, a simple molecule, and a complex system composed of very large number (eg, 1 mole) of molecules. A good example is the simplest macroscopic system represented by a gas closed in a container having a barrier (Hofman, 2002). At the beginning of the experiment whole gas is located in one half of the container; this can be accomplished by the barrier (closure). Next the barrier is removed, and the effect is shown in Fig. 4.1. Now assume that this experiment is repeated for the system containing only one particle, Fig. 4.2 (Hofman, 2002).

Let us analyze how each of considered systems evolves in time. In the first system, Fig. 4.1, we observe an irreversible process which consists in filling up the empty part of the container and the simultaneous equalizing of the density. However, in the second system, Fig. 4.2, it is difficult to recognize any direction of the process. One can only observe chaotic motions of our particle. Making pictures of the particle locations in the second system, it is impossible to decide which is the state before the process and which is the state after the process.

This observation leads to a very important conclusion: purely mechanical systems, Fig. 4.2, do not possess distinguished direction of time, that is, they are reversible in time. On the other hand, macroscopic systems, as shown in Fig. 4.1, characterize themselves by the existence of the "time arrow." Processes undergoing in these systems are irreversible. One may unsuccessfully wait until the gas particles spontaneously lump together in the left hand part of the container (Hofman, 2002). When the system's boundary conditions are homogeneous, irreversible processes terminate at a stable state of equilibrium. Hofman (2002) also considers quantitative computational aspects of his probabilistic considerations.

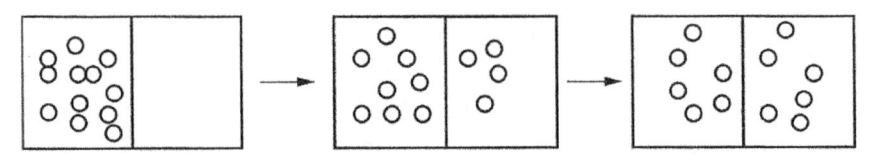

Figure 4.1: Diffusion of gas particles located initially in the left half of container (Hofman, 2002).

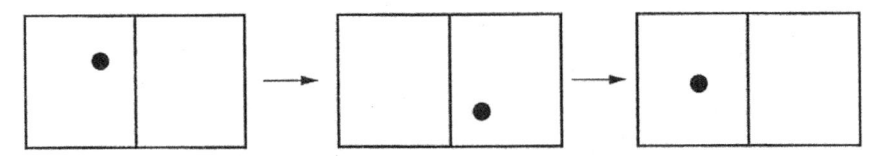

Figure 4.2: "Diffusion" of single particle (Hofman, 2002).

4.2 Classical Picture of Diffusion

Here is some historical information on diffusion. One of its first observations was made by the botanist Robert Brown in 1828. He noticed that pollen particles dispersed in water exhibit a very irregular swarming "motion." In 1905 Einstein conjectured that this motion is due to the interaction of pollen with the water molecules and, in fact, proved that microscopic particles suspended in a liquid "perform movements of such magnitude that they can be easily observed in a microscope, on account of the molecular motions of heat" (Einstein, 1905). Since then, Brownian motion is used as a synonym of diffusion (Zanette, 1999). It is worth stressing that the understanding of Brownian motion is necessary to develop an effective theory of deposition from moving aerosols (Fuchs, 1955; Davies, 1966). In this chapter basic equations of classical diffusion are not derived separately, rather they are recovered at the limit of $q = 1$ from a generalized theory of anomalous diffusion, based on Tsallis's (1988) generalization of statistical mechanics. This is both to fit the brevity of the text and to show the correspondence between the anomalous and classical diffusion.

Among the elementary processes underlying natural phenomena, diffusion is certainly one of the most ubiquitous. In an ensemble of moving elements—atoms, molecules, chemicals, cells, or animals—each element usually performs, at a mesoscopic level, a random path with sudden changes of direction and velocity. As a result of this highly irregular individual motion, which is microscopically driven by the interaction of the elements with the medium and of the elements with each other, the ensemble spreads out. At a macroscopic level, this collective behavior is—in contrast with the individual microscopic motion—extremely regular, and follows very well-defined, deterministic dynamical laws. It is precisely this smooth macroscopic spreading of an ensemble of randomly moving elements that we associate with diffusion (Zanette, 1999).

A very suitable and useful mathematical model for Brownian motion is provided by random walks (Goldstein, 1953). In its simplest version, a random walker is a point particle that moves on a line at discrete time steps Δt. At each step, the walker chooses to jump to the left or to the right with equal probability, and then moves a fixed distance x. This stochastic process can be readily generalized, first, by allowing the walker to move in a many-dimensional space. In addition, time can be made continuous by associating a random duration with each jump or by introducing random waiting times between jumps. Finally, the length of each jump can be also chosen at random from a continuous set with a prescribed probability distribution.

Observing the random course of the particle suspended in the fluid led to the first accurate measurement of the mass of the atom (Lavenda, 1985b). In the hands of Einstein, Brownian motion was turned into a conclusive, observational method for confirming the atomic theory of matter. Brownian motion also contributed to better theoretical understanding of thermodynamic principles. Brownian motion now serves as a mathematical model for random processes. The recognition that Brownian motion is a manifestation of the statistical fluctuations among the microstates of a thermodynamic system had even greater consequences for the study of nonequilibrium systems than it had for systems in equilibrium At equilibrium the order in which events take place is immaterial: the fluctuations among microstates hardly ever give rise to a distinguishable macrostate. For disequilibrium systems, however, the temporal order of events becomes important. Lavenda (1985a, b) requires that a statistical version of the phenomenological second law must explain how nonequilibrium systems evolve to the state of equilibrium, that is, how ordered macrostates spontaneously give rise to disordered ones. Only over a short time can one detect the fluctuations superposed on the smooth, average drift that is generally observed. Lavenda's (1985a) book on nonequilibrium statistical thermodynamics develops the nonlinear generalization of the Onsager and Machlup (1953) formulation of nonequilibrium fluctuations which was restricted to linear (Gaussian) processes. Just as equilibrium is characterized by the state of maximum entropy, corresponding to maximum probability, nonequilibrium states are characterized by the principle of least dissipation of energy, corresponding to the maximum probability of a transition between nonequilibrium states that are not well separated in time. This principle can be generalized to non-Gaussian fluctuations in the limit of small thermal noise and constitutes a kinetic analog to Boltzmann's principle.

Working in stochastic context, Lindenberg (1990) presents a mesoscopic analysis of the dynamic behavior of a diffusion system interacting with a heat bath in both classical and quantum formulations. Lindenberg (1999) complements some earlier works by extending the parameter regime of analysis and confirms her earlier results as well as her new ones against numerical simulations. Some earlier works are also discussed in Van Kampen's (2001) book. Working along this line Romero-Salazar and Velasco (1997) propose a microscopic model to construct a quantum Langevin equation for a system in a heat bath with temperature varying in time.

Kuiken (1994) provides a comprehensive and systematic treatment of phenomenological aspects of classical diffusion; his considerations are from the viewpoint of fluxes, driving forces and irreversible thermodynamics. Krishna and Standart (1979) and Krishna and Taylor (1986) review engineering aspects of basic problems of mass and energy transfer in multicomponent systems. At the frontier between classical and nonclassical diffusion, Wiśniewski (1984) formulates the balance equations of extensive quantities for a multicomponent fluid taking into account diffusion stresses. Baranowski (1992) explores the influence of stress fields on diffusion in isotropic elastic media and develops the thermodynamics of solids with stress fields. He also gives an extension of the chemical potential to stress fields under the experimentally proved assumption neglecting the role of off-diagonal elements in the stress tensor. His attention is directed to stresses developed by a diffusing component, causing displacements, and resulting stresses in the solid. An ideal elastic medium is assumed, that is, plastic deformations are neglected. Diffusion equations are derived in a one-dimensional case. The resulting equation of change for concentration is a nonlinear integro-differential equation. Nonlocal diffusion effects are verified by an original experiment. Experimental results of hydrogen diffusion in a Pd–Pt alloy are compared with the theory. Importance of the stress field to diffusion in solids is shown. Effect of stresses on the entropy production is discussed with the conclusion that a stress-free steady-state obeys the minimum entropy production principle as a stationarity criterion.

Following an earlier approach on the stress-induced diffusion of macromolecules (Tirrell and Malone, 1977), nonequilibrium thermodynamics is applied to membrane transport and diffusion in elastic media with stress fields. Baranowski (1991, 1992) traces the way of derivation of the phenomenological equations and characterizes the framework of linear nonequilibrium thermodynamics (LNET). Applications to membrane transport phenomena are presented in several approaches. A treatment of active transport in terms of LNET is given. Also the importance of individual balance equations for the coupling of diffusion and viscous phenomena is shown. The invariance of the diffusional entropy production in the Prigogine's expression is proven (see also Baranowski, 1974). Possible nonlocal phenomena due to the stress fields are indicated and generalized advantages and disadvantages of the thermodynamic treatment are summarized. As an example of a more general approach to membrane transport, named extended nonequilibrium thermodynamics (ENET) is outlined.

Bearman–Kirkwood statistical mechanical theory of transport equations (Bearman and Kirkwood, 1958; Bearman, 1959) yields phenomenological equations containing inertial and viscous terms and a new quantity introduced into the theory, namely, a coupling between diffusion and both inertia and local stresses. Bartoszkiewicz and Miekisz (1989) have investigated the generalized thermodynamics founded by Bearman–Kirkwood equations. They have shown that the Prigogine's theorem on the entropy production invariance also holds when inertia and local stresses contribute to the diffusion. Their generalized theory predicts that in the presence of viscous effects and low concentrations, the electrolyte barodiffusion depends mainly on local stresses; it is directed "against" the motion of the mass center. Thus, for some porous membranes certain "practical" coefficient $L_{\pi P}$ may be interpreted as the manifestation of the diffusion–viscous flow coupling. See also Bartoszkiewicz and Miekisz (1990) for an uncoupling of diffusion and viscous flow.

Nonisothermal reaction–diffusion (RD) systems with coupled heat and mass transfer control the behavior of many transport and rate processes in physical, chemical and biological systems (Demirel, 2006a, b). A considerable work has been published on mathematically coupled nonlinear differential equations of RD systems with neglecting the possible thermodynamic couplings among heat and mass fluxes and reaction rates. In Demirel's (2006b) work, the thermodynamic coupling sustains the flux which occurs without its primary driving force, which may be gradient of temperature, or gradient of chemical potential, or reaction affinity. His study presents the modeling equations of nonisothermal RD systems with coupled heat and mass fluxes excluding the coupling of chemical reactions using the LNET approach. For a slab catalyst pellet this study shows the dynamic behavior of composition and temperature profiles obtained from the numerical solutions of nonlinear partial differential equations.

Kincaid et al. (1983, 1987) developed the Enskog theory for mutual diffusion and thermal diffusion in multicomponent mixtures. Lopez de Haro et al. (1984) constructed a similar theory which takes into account transport properties of dense binary mixtures with one tracer component. Maugin and Drouot (1983a, 1992) developed an internal-variable-based thermodynamics of macromolecules solutions. The relevant internal variables of state are the components of the conformation tensor. This quantity models some properties of macromolecules by treating them as "material" particles undergoing deformations. The dissipation inequality is obtained in the form of the Clausius–Duhem inequality involving the free energy. Two different types of dissipative processes are distinguished. The first type involves only the observable variables of the thermo-electro-mechanics. The second are the relaxation processes (of the type encountered in chemical kinetics) involving electric and conformational relaxation. Processes of plasticity type do not appear. Evolution equations for conformation, electric polarization and chemical processes are given. An important effect for gyroscopic type reversible processes is developed. Applications to equilibrium conformations, flow-induced and electrically induced conformational phase transitions, electric polarization, and mechano-chemical effects are discussed. Possible generalizations accounting for diffusion and interaction between vortex fields and conformations are given. This paper contributes mainly to the theory of complex fluids such as liquid crystals and emulsions.

In their study of self-diffusivity and viscosity of liquids, Harada et al. (1974) propose a relaxation model describing both the dissipative and oscillatory modes of motion. The resultant velocity autocorrelation function is of the damped oscillators pattern. Self-diffusivity and shear viscosity of a simple saturated liquid can be predicted from the equations derived from the model with the configuration internal energy and a parametric density which is characteristics to the substance. The density determines the probability of dissipative mode of motion. A simple relation of corresponding states holds for several liquids by employing the reduced density defined as the ratio of liquid density to the parametric density.

Transport phenomena in polymers and polymeric liquid mixtures have received considerable attention both in theoretical and applied thermodynamics. Fujita (1961) reviews diffusion processes in polymer-diluent systems from the viewpoint of topics such as: Fundamentals, Fickian sorption, non-Fickian sorption, permeation, and interpretation of diffusion coefficient data. Methods of examining whether sorption is Fickian or not are particularly useful when the experimental sorption curve has a shape equivalent to the true Fickian curve. In particular, concentration dependence of the thermodynamic diffusion coefficients, and uncertainties of using free volume theory of the viscosity for concentrated polymer solutions are discussed. Curtiss and Bird (1996) gave a summary of the kinetic theory of flexible macromolecules, represented by bead-spring model of arbitrary connectivity. Applications involve polydisperse systems, multicomponent mixtures, dilute or concentrated solutions and fluids with concentration, temperature and/or velocity gradients. Fundamental changes have taken place in the description of diffusion in polymers (Neogi, 1996). Diffusivity is no longer a phenomenological coefficient and firm validation from molecular theories now exists for Fick's law. High-speed computers have become available to calculate these diffusivities. Vrentas and Duda (1979) review methods for estimating mutual diffusion coefficients for polymer-solvent systems. They recommend procedures for the determination of the temperature, concentration, and molecular weight dependences of diffusivities both for dilute and concentrated solutions. They recommend a method for anticipating anomalous diffusion phenomena and for determining when classical diffusion theory can be used to analyze diffusional processes involving macromolecules.

Coupling between diffusion and hydrodynamics is seen, for example, in polymer solutions. The diffusion is, in general, non-Fickian. Considering a mixture of two simple or one simple and one complex fluids, Elafif et al. (1999) treat systematically diffusion and rheology in the framework of GENERIC modeling (Grmela and Öttinger, 1997; Öttinger and Grmela, 1997). Diffusion with inertia in simple and complex fluids is in particular considered. The main result is a set of intrinsically consistent equations governing the time evolution of the overall mass density, overall momentum density, relative concentration of one fluid, diffusion flux, and the microstructure of the complex fluid. (The momentum equation involves explicit formulae for the scalar hydrodynamic pressure and the extra stress tensor.) In the process of solving these equations the researchers recover familiar results describing the influence of stresses on diffusion and obtain new results concerning the influence of diffusion on stresses.

Noyola (1999) presents an application of linear irreversible thermodynamics to the problem of diffusion among miscible incompressible fluids. First, he assumes that the mixture density is a function of the volume fractions. This hypothesis allows him to reformulate the interface diffusion problem in terms of transport equations for the concentrations or the mass–volume fractions, the momentum balance equations for the diffusion flux and the energy balance equations for the relative internal energies, respectively. The transport equations so obtained are generalized equations of the classical diffusion equations obtained by means of fluctuating hydrodynamics and the H model of critical dynamics. Afterwards he applies linear irreversible thermodynamics in order to get the entropy production and then assumes a linear relation between fluxes and forces to obtain the constitutive equations for the diffusion flux, the heat flux, and the stress tensor of each component. Finally, he compares his results with those originated from fluctuating hydrodynamics and continuum theory. He shows that the constitutive equation for the diffusion flux is a Maxwell–Cattaneo's type. The constitutive equation for the stress tensor generalizes the Korteweg de Vries stress tensor equations.

Lhuillier et al. (2003) develop a comparative study of the coupling of flow with non-Fickean thermodiffusion. They focus on relation of the entropy source to the definite irreversible processes. Usually, the conservation laws of mass, momentum and energy are of utmost importance. But, sometimes, besides these conserved quantities, the evolution of some extra variables is also needed. These extra variables are connected either to some long-range order in the material, or to some slowly relaxing kinetic energy or structural process. In the former case one speaks of the new variable as an order parameter, while in the latter case one speaks of it as an internal variable. This work is mainly concerned with the description of anomalous diffusion and thermodiffusion in mixtures and suspensions with the help of internal variables. It also makes a short incursion into ordered materials, as the anomalous heat propagation is best understood in the case of superfluids. Classical thermodynamics of irreversible processes is extended to describe nonequilibrium processes which are not directly involved in the conservation laws of mass, momentum and energy. These nonequilibrium processes can be understood as the evolution in time of some internal variables. Thermodynamics with internal describes unusual transport phenomena taking place in binary fluid mixtures and suspensions of particles. Lhuillier et al. (2003) also revisit the unusual heat transport of superfluids to compare the description of these unusual phenomena given by Extended Irreversible Thermodynamics and the GENERIC framework.

Hydrodynamic theory of diffusion in its classical form takes into account only the motion resistance of a single spherical particle which diffuses in the solvent continuum. It does not include effects of interactions between the moving particles and their influence on the frictional force in the Nernst–Einstein equation, which links the particle mobility with the diffusion coefficient. Considering this status of affair Sęk (1996) proposes a new hydrodynamic model for prediction and quantitative description of diffusion phenomena in liquids. To derive diffusion coefficients in liquids he compares the force acting on a diffusing particle which results from the Nernst–Einstein equation with the same force resulting from the classical equation of Kozeny–Karman for the fluid flow through the porous bed. The calculated diffusion coefficients agree with the experimental data with the accuracy better than 4%. Literature of diffusion contains numerous expressions applicable for estimation of values of diffusion coefficients in liquids and gases. It may be noted that equations serving for prediction of diffusion coefficients in gases are derived on the basis of the kinetic theory of gases. On the other hand, models appropriate for determining of diffusion coefficients in liquids have their origin in the so-called hydrodynamic theory of diffusion (Bird et al., 1960). In that theory diffusion of particles is treated as a motion of spherical solids in a viscous continuum. Sęk's (2003) idea is that such a concept of two-phase fluid can be helpful in determining diffusion coefficients in all fluids, that is, in both liquids and gases. The aim of his research is the validation of his own model (Sęk, 1996) when determining numerical values of diffusion coefficients in liquids and gases using the same assumptions. Introductory results show that indeed the model assuming the flow through a porous bed can be used to predict diffusion coefficients in gases. These data may be compared with those in Reid et al. (1987).

Theories of fluctuating hydrodynamics (Landau and Lifshitz (1974a, 1987) have been developed in connection with steady-state nonequilibrium fluctuations. Experimental data verifying these theories have been presented (Ronis and Procaccia, 1982). Cussler (1984) describes how in the absence of hydrodynamic motion, macroscopic concentration gradients relax toward equilibrium by diffusion. Farkas (1997) presents two models of treating cross diffusion. In his model A substance u flows from places where its density is high toward places where the density is low. At the same time, substance v has an attracting or repelling effect on u, so that u flows toward high (resp. low) densities of v. In the model B, there is another possibility to express the attracting (resp. repelling) effect of u on v. This version involves the assumption that the diffusivity of u is a decreasing function of v which means that u is reluctantly diffusing away from the places where the density of v is high. Associated reaction–diffusion models may follow.

Gaveau et al. (1999a) consider the disequilibrium thermodynamics of a reaction–diffusion system at a given temperature, using the Master equation. The information potential is defined as the logarithm of the stationary state. They compare the approximations, given by the Fokker–Planck equation and the Wentzel–Kramers–Brillouin method directly applied to the Master equation, and prove that they all lead to very different results. Finally, they show that the information potential satisfies a Hamilton–Jacobi equation and deduce general properties of this potential, valid for any reaction–diffusion system, as well as a unicity result for the regular solution of the Hamilton–Jacobi equation. In a second article of this series (Gaveau et al., 1999b) consider path integrals, large fluctuations, and an estimation of rate constants in reaction–diffusion systems that can be out of equilibrium. They show that using this path integral formulation it is possible to calculate rate constants for the transition from one well to another well of the information potential and to give estimates of mean exit times. Gaveau et al. (2001a) continue the development of the formalism of disequilibrium thermodynamics in a variational form. They prove that in the framework of progress variables, the Hamilton–Jacobi equation has always a simple solution, and also prove that this solution yields a state function if and only if there is a thermodynamic equilibrium for the system. They study an inequality between the dissipation of energy and of information, and discuss the notion of relative entropy. Finally, they also study the case of a system with one chemical species, where all the previous quantities can be calculated explicitly.

The publications by Gaveau et al. (1999a, b, 2001a) showed that for a chemical system maintained far from equilibrium a new thermodynamic potential, the information potential, can be defined by using the stochastic formalism of the Master Equation. Gaveau et al. (2001b) extend their previous theory to a general discrete stochastic system. For this purpose, they use the concept of extropy, which is defined in classical thermodynamics as the total entropy produced in a system and in the reservoirs during their equilibration. They show that the statistical equivalent of the thermodynamic extropy is the relative information. If a coarse-grained description by means of the thermodynamic extensive variables is available, the coarse-grained statistical extropy allows one to define the information potential in the thermodynamic limit. Using this potential Gaveau et al. (2001b) study the evolution of such systems toward a nonequilibrium stationary state. They derive a general thermodynamic inequality between energy dissipation and information dissipation, which sharpens the law of the maximum work available from nonequilibrium systems.

Diffusion of particles or molecules in a fluid is a natural manifestation of its thermal energy. It is observed in the Brownian motion of dust particles in a fluid or a gas, and it ensures the mixing of different molecules in the fluid. Therefore mixing, at the shortest length scales, follows from diffusion rather than convection. Weitz (1997) illustrates how large fluctuations can occur in diffusion. He also comments on Vailati and Gigglio's (1997) report on giant fluctuations in a free diffusion processes. Weitz points out that while the diffusion relaxes the concentration gradient it also creates enormous fluctuations in the direction perpendicular to the gradient, which are large enough to be seen from the variations they cause in optical phase change and in the pattern of the scattered light. Where do these fluctuations come from? Some authors propose that naturally occurring velocity fluctuations transport small volumes of fluid along the gradient, which then relax most quickly by diffusion, leading to large-scale spatial fluctuations in the concentration perpendicular to the gradient (Weitz's Fig. 1a). If the volume transported by the velocity fluctuation is too large, then it is relaxed by convection due to buoyancy, rather than diffusion—the displaced volume returns to its original location without contributing to the spatial fluctuations (Weitz's Fig. 1b). Traditional measurements do not have this unusual perspective, explaining why these fluctuations are typically not observed in diffusion coefficient measurements. More generally, the reason why these fluctuations escaped our observation until now is probably that mixing is controlled by diffusion only on short length scales. On large scales, either convection or stirring usually dominates. See also Vailati and Giglio (1997) on giant fluctuations in a free diffusion.

In the realm of statistical mechanical theory of diffusion it is worth pointing out Altenberger's and Dahler's (1992) approach in which the researchers have proposed the statistical theory of diffusion and heat conduction based on an entropy source-free method. The approach allows one to answer unequivocally questions concerning the symmetry of the kinetic coefficients. See chapter: Contemporary Thermodynamics for Engineering Systems for more information.

Tsirlin and Leskov (2007) developed general approaches to optimization of practical diffusion systems, based on finite time thermodynamics.

4.3 Introducing Anomalous Diffusion

We shall now abandon the classical diffusion systems and pass to systems exhibiting the anomalous diffusion. The anomalous diffusion is treated in depth, between others, by Compte and Jou (1996) whose motivation is inspired in two different aspects: on the one hand, some extensions of nonequilibrium thermodynamics to generalized transport equations and, on the other hand, some attempts to exploit a nonadditive entropy for multifractal systems. Our aim here is to follow their study of the thermodynamic aspects of anomalous diffusion from these two different, complementary, viewpoints. In this context the importance of Mandelbrot's statement on fractals should be recalled: "Clouds are not spheres, mountains are not cones, coastlines are not circles, and bark is not smooth, nor travel in a straight line" (Aharony and Feder, 1989). Fractals occurs vastly in nature. Berry (1989) calculated the rate at which a D-dimensional cluster of N smoke spherules falls through the air. The cluster may be large or small in comparison with the mean free path of air molecules. As his examples prove, a cluster with $N = 1000$ falls 10 times slower with $D = 1.8$ than when compacted into a sphere of $D = 3$. This shows that fractality is important, and it should be taken into account in future modeling.

Anomalous diffusion occurs in a multitude of physical or other phenomena (Li et al., 2003). It is characterized microscopically by a time-dependent distribution of particles in space where the distance, r, a particle has moved in time, t, increases algebraically with t but with an exponent other than 0.5. Such deviation from the normal diffusion behavior has been observed, for instance, in hydrogen diffusion in amorphous metals, diffusion of water in biological tissues, or in disordered systems more generally. It has also been discussed in conjunction with the growth of thin films on a solid surface and with the important class of diffusion processes on fractal structures (Aharony and Feder, 1989). Fractals in nature originate from self-organized critical dynamical processes (Bak and Chen, 1989). This is a topic where observation was for years far ahead of theory. Much new research can be done ranging from the basic mathematics to direct practical computations (Li et al., 2003).

Zanette (1999) and Naudts and Czachor (2001) show, among others, that Tsallis generalized statistics (Tsallis, 1988, 2011; Salinas and Tsallis, 1999) provides a natural frame for the statistical-thermodynamical description of anomalous diffusion. Within this generalized theory, a maximum-entropy formalism makes it possible to derive a mathematical formulation for the mechanisms that underlie Lévy-like superdiffusion, and for solving the nonlinear Fokker–Planck equation. In mathematical treatments Schulzky et al. (2000) analyze the similarity group and anomalous diffusion equations, whereas Davison and Essex (2001) study the initial value problems for fractional differential equations.

Hoffmann et al. (1998) calculated the entropy production rate for fractional diffusion processes. Their results shows an apparently counterintuitive increase of the entropy production with the transition from dissipative diffusion behavior to reversible wave propagation. This property is deduced directly from invariant and noninvariant factors of the probability density function arising from a group method applied to the fractional differential equation existing between the pure wave and diffusion equations. However, the counterintuitive increase of the entropy production rate within the transition turns out to be a consequence of increasing quickness of processes as the wave case is approached. When this aspect is removed the entropy shows the expected decrease with the approach to the reversible wave limit. In the presence of strong external fields the Einstein relation between diffusion coefficient and friction coefficient must be modified, and it becomes dependent on the fields. Zakari and Jou (1998) investigated how this relation should be generalized far from equilibrium. They showed that the introduction of saturation effects in the diffusion flux yields a modification of the Einstein transport equation. Saturation effects in the mobility also yield corrections to the Einstein equation.

Li et al. (2003) analyzed the link between the fractional diffusion, irreversibility and entropy. They studied three types of equations linking the diffusion equation and the wave equation: the time fractional diffusion equation, the space fractional diffusion equation and the telegrapher's equation. For each type, the entropy production was calculated and compared. It was found that the two fractional diffusions, considered as linking bridges between reversible and irreversible processes, possess counterintuitive properties: as the equation becomes more reversible, the entropy production increases. The telegrapher's equation does not have the same counterintuitive behavior. It was suggested that the different behaviors of these equations might be related to the velocities of the corresponding random walkers. The two fractional diffusion equations show completely counterintuitive behavior, that is, the entropy production decrease along the path, or equivalently, its increase toward the wave end of the bridging regime. In the case of the telegrapher's equation, on the path from the wave equation to the diffusion equation the entropy production increases at the wave end, but decreases at the diffusion end.

Large classes of stochastic processes where fractal scaling properties play an important part can be described by probability distributions with unusually long tails (Li et al., 2003). Fractals have been extensively studied as models for geometrically disordered systems (Aharony and Feder, 1989). They are known to display anomalous diffusion which is characterized by the sublinear, dispersive behavior of the time evolution of the mean-squared displacement determined by the fractal dimension and the spectral dimension. Another quantity of interest is the (ensemble averaged) autocorrelation function $p(0, t)$, the probability to be at the origin at time t having started at the origin at $t = 0$. Saied (2000) presented analytical calculations for the propagator $p(r, t)$, the probability to reach a distance r at time t having started at the origin $t = 0$, on a generalization of the Euclidean diffusion equation for diffusion on fractals. The results were confronted with exact analytical expressions. It was shown that the autocorrelation function $P(0, t)$ scales according to $p(0, t)$ proportional to $t^{-d_s/2}$. The results extend the previously derived expressions. Progress has been made in this direction by introducing the symmetries of the generalized diffusion equation.

A very suitable and useful mathematical model for Brownian motion is provided by random walks (Goldstein, 1953). In its simplest version, a random walker is a point particle that moves on a line at discrete time steps Δt. At each step, the walker chooses to jump to the left or to the right with equal probability, and then moves a fixed distance x. This stochastic process can be readily generalized, first, by allowing the walker to move in a many-dimensional space. In addition, time can be made continuous by associating a random duration with each jump or by introducing random waiting times between jumps. Finally, the length of each jump can be also chosen at random from a continuous set with a prescribed probability distribution.

As a stochastic process, random walk admits a probabilistic description in terms of probability distributions for the relevant quantities (Van Kampen, 2001). In particular, one is interested at studying the probability of finding the walker in a certain neighborhood $d\mathbf{r}$ of point $\mathbf{r}$—in general, in a d-dimensional space—at time t, $P(\mathbf{r}, t) \, d\mathbf{r}$. Note that, besides its interpretation as a probability distribution, $P(\mathbf{r}, t)$ can be related to the density in an ensemble of noninteracting identical random walkers. In fact, if the ensemble contains N walkers, $n(\mathbf{r}, t) = N P(\mathbf{r}, t)$ stands for the space density of walkers (Zanette, 1999).

Suppose that the random walk is defined in continuous time and space, with a waiting time probability distribution $\psi(t)$ and such that the probability that the walker jumps from any point $\mathbf{r}$ to $\mathbf{r} + \mathbf{x}$ is $p(\mathbf{x}) \, d\mathbf{x}$. The normalization of probabilities imposes

$$\int_0^\infty \psi(\tau) d\tau = 1 \qquad \int p(\mathbf{x}) d\mathbf{x} = 1 \qquad (4.1)$$

If, moreover, $\psi(\tau)$ and $p(\mathbf{x})$ satisfy

$$\langle \tau \rangle = \int_0^\infty \tau \psi(\tau) d\tau < \infty \qquad \langle x^2 \rangle = \int x^2 p(\mathbf{x}) d\mathbf{x} < \infty, \qquad (4.2)$$

($\mathbf{x} \equiv |\mathbf{x}|$) it can be proven that $P(\mathbf{r}, t)$ obeys the diffusion equation

$$\frac{\partial P}{\partial t} = D \nabla_\mathbf{r}^2 P \qquad (4.3)$$

where $D \propto \langle x^2 \rangle / \langle \tau \rangle$ is the diffusion constant, or diffusivity. This equation has to be solved for a given initial condition $P(\mathbf{r}, 0)$ with suitable boundary constraints. The density $n(\mathbf{r}, t)$ of an ensemble of noninteracting diffusing particles obeys the same equation.

A typical solution to the diffusion equation describes a density profile that, as time elapses, is smoothed out and broadens. In fact, it can be straightforwardly shown from the general solution that the width of the spatial distribution grows with time as

$$\langle r^2 \rangle = \int r^2 p(\mathbf{r}, t) d\mathbf{r} = 2 dDt \qquad (4.4)$$

where d is the space dimension (Zanette, 1999). Correspondingly, it can be shown that the mean square distance between the present position and the initial position of a random walk that satisfies Eq. (4.2) is proportional to time. This proportionality between mean square displacement and time is the fingerprint of diffusion, as it can be used experimentally, numerically, and theoretically to detect this kind of transport mechanism in a given natural process (Zanette, 1999).

Though, being a form of transport, diffusion is inherently a disequilibrium process, the large-time asymptotic dynamics of an ensemble of diffusing particles can be described in the frame of equilibrium statistical mechanics. In fact, it is expected that, for very large times, the system reaches a state of thermodynamic equilibrium with the medium and between the particles, if they interact. In such state, the diffusing particles and the medium participate of a balanced interchange of momentum and energy, which maintains the particles in their characteristic irregular motion. Once this situation is reached, a connection between the parameters that characterize thermodynamic equilibrium and particle dynamics should exist. Einstein investigated this problem and concluded that diffusivity D and temperature T are proportional:

$$D = \mu k_B T \qquad (4.5)$$

(Einstein, 1905). Here k_B is Boltzmann constant, and μ is the mobility. The mobility is defined as the inverse of the friction coefficient, in the present case, of the diffusing particles in the medium (Kubo, 1988). Einstein's relation, Eq. (4.5), provides the connection between diffusion and thermodynamic equilibrium (Zanette, 1999).

4.4 Compte's and Jou's (1996) Treatment
4.4.1 Application of Tsallis Theory

Classically, the relation between irreversible thermodynamics and transport laws (Fick's, Fourier's laws, for instance) is well established (De Groot and Mazur, 1984; Kuiken, 1994). These laws have a wide range of validity and they are very useful for applications. However, there are situations in which they must be generalized taking into account, for instance, memory effects, nonlocal effects or nonlinear effects (Compte and Jou, 1996). The generalization of such transport laws has stimulated in recent years a corresponding generalization of the underlying nonequilibrium thermodynamics (Keizer, 1987a; Jou et al., 1988, 2001; Müller, 1985, 1992; Sieniutycz, 1994). The basic idea of these developments is that dynamics and thermodynamics must be dealt with in parallel, rather than assuming that there is an a priori thermodynamics (namely, the local equilibrium thermodynamics), which restricts the corresponding admissible dynamics (Compte and Jou, 1996). Thus, whereas some dynamic behaviors are incompatible with the local equilibrium hypothesis, it is possible to introduce nonclassical entropies which are compatible with such dynamic behaviors (Keizer, 1987a; Jou et al., 1988, 2001; Compte and Jou, 1996).

In parallel for the past 20 years, another trend in modern thermodynamics has focused on possible generalizations of the definition of entropy, which conserve their interpretation as lack of information but do not retain certain properties hitherto considered essential, such as extensiveness (Tsallis, 1999a, b; Salinas and Tsallis, 1999). Tsallis (1999a, b) discusses the domain of validity of standard thermodynamics and Boltzmann–Gibbs statistical mechanics (Schmidt et al., 1981; Lifshitz and Pitajevsky, 1984) and then formally enlarges this domain in order to cover a variety of anomalous systems. The generalization involves nonextensive thermodynamic systems. These new definitions apparently may apply to very complex systems with long-range interactions, persistent memory or systems evolving in a fractal space, where the intuitive reasons which justified extensiveness do not apply any longer (Compte and Jou, 1996). Furthermore, in these complex systems (for instance, magnetic systems, random walks, etc.), the usual thermodynamic formalisms fail whenever the relevant thermostatistic quantities are computed, which turn out to diverge. Compte and Jou (1996) consider the definition proposed by Tsallis in 1988 which has been seen to be a consistent generalization both of thermodynamics and of statistical mechanics (Salinas and Tsallis, 1999; Curado, 1999; Zanette, 1999; Beck, 2002b)

$$S_q = -k_B \frac{1 - \sum_i p_i^q}{1 - q} \qquad (4.6)$$

where k_B is the Boltzmann's constant, p_i the probability of microstate i, and q a constant parameter. This generalized entropy reduces to the classical Boltzmann entropy in the limit $q \rightarrow 1$.

Instances where this entropy has been successfully applied include stellar polytropes, with the parameter q being a function of the polytrope index, and Lévy walks (Lévy, 1937), where q is related to the dimension of the resulting fractal trajectory (Compte and Jou, 1996).

Wilk and Włodarczyk (2000) show that nonextensivity parameter q occurring in some of the applications of Tsallis statistics (known also as index of the corresponding Lévy distribution) can be determined, in the $0 \geq q > 1$ case, entirely by the fluctuations of the parameters of the usual exponential distribution. There is a very large variety of physical phenomena described most economically (by introducing only one new parameter, q) and adequately by the nonextensive statistics of Tsallis (1999a, b). They include all situations characterized by long-range interactions, long-range microscopic memories, and space–time (multi)fractal structure of the process (see numerous papers by Tsallis and his coworkers). The high-energy physics applications of nonextensive statistics are quite recent but are already numerous and still growing (Kaniadakis, 2002). All adduced examples have one thing in common: the central formula employed is the following power-like distribution:

$$G_q(x) = C_q \left(1 - (1-q)\frac{x}{\lambda}\right)^{1/1-q}, \tag{4.7}$$

which is just a one parameter generalization of the Boltzmann–Gibbs exponential formula to which it converges for $q = 1$:

$$G_{q=1} = g(x) = c \exp\left(-\frac{x}{\lambda}\right). \tag{4.8}$$

When $G_q(x)$ is used as probability distribution of the variable x (Lévy, 1937 distribution), the parameter q is limited to $1 \leq q < 2$. Also, as demonstrated by Wilk and Włodarczyk (2000) in their earlier paper, Lévy distribution $G_q(x)$ emerges from the fluctuations of the parameter $1/\lambda$ of the original exponential distribution $G_{q=1}$, and the parameters of the distribution $f(1/\lambda)$ define parameter q in a unique way.

Compte and Jou (1996) deal with the thermodynamic aspects of non-Fickian diffusion. In contrast to classical diffusion, where the characteristic value r of the displacement (defined, for instance, as the root mean square of the displacement or as the radius of the sphere which contains 90% of the total number of particles) is proportional to $t^{1/2}$ in the long-time limit, in non-Fickian diffusion, or anomalous diffusion, it behaves as

$$r \sim t^{\sigma/2} \tag{4.9}$$

with $\sigma \neq 1$. If $\sigma = 1$ one recovers the usual diffusion, and for $\sigma < 1$, or $\sigma > 1$ one has subdiffusive or superdiffusive behavior, respectively. Such non-Fickian aspects of diffusion have received much attention in the past decade, both because of their theoretical interest as well as for their applications (diffusion in fractal spaces, transport in ion-conducting materials, flow in rocks, particle diffusion in fluctuating magnetic fields, transport in chaotic dynamics, and so forth (Compte and Jou, 1996). However, whereas the dynamic aspects of diffusion leading to Eq. (4.9) have deserved much interest, the thermodynamic aspects have not been studied in detail. Only recently, Zanette (1999) and some others have focused on the thermodynamic description of Lévy-like anomalous diffusion, even though most of the vast domain of anomalous diffusive phenomena remains unexplored from the thermodynamic viewpoint.

Compte and Jou (1996) try to improve this situation by studying the convenience of the fractal entropy for the thermodynamic description of a kind of correlated anomalous diffusion, namely that arising from a nonlinear diffusion equation:

$$\frac{\partial u}{\partial t} = \nabla^2 u^\gamma \tag{4.10}$$

This kind of equation has been seen to arise naturally when one considers the flow of homogeneous fluids through porous media (Muskat, 1937), particle diffusion across magnetic fields or the spatial spread of biological populations. However, the main aim of the approach by Compte and Jou (1996) is to establish whether a new thermodynamic frame is actually needed for this dynamics as far as nonequilibrium thermodynamics is concerned.

Following the results of Compte and Jou (1996) the further text is organized as follows: the Section 4.4.2 contains a short overview of the standard thermodynamic derivation of the diffusion equation which then serves as a model to obtain the diffusion equation associated with the fractal entropy. In parallel, the diffusion transport coefficient is allowed to depend on the particle density, thus introducing some nonthermodynamic contributions to the anomaly in diffusion. The resulting equation is then solved and the consequences of the second principle on this dynamics are eventually discussed. In Section 4.4.3 three particular cases are studied, and we finally present the conclusions in Section 4.4.4.

4.4.2 Fractal Entropy and Generalized Diffusion Equation

The classical approach to diffusion from thermodynamics, as may be found in classical references such as De Groot and Mazur (1984) or Prigogine (1947), starts from the usual definition for entropy as applied to a continuous distribution $P(\mathbf{x}, t)$:

$$S(t) = -k_B \int P(\mathbf{x}, t) \ln P(\mathbf{x}, t) d^N \mathbf{x} \tag{4.11}$$

By taking the continuity equation

$$\frac{\partial P}{\partial t} = -\nabla \cdot \mathbf{J} \tag{4.12}$$

and assuming a linear relation between the flux and the thermodynamic force $\nabla(\delta S/\delta P)$ in the form

$$J = L\nabla \frac{\delta S}{\delta P} \qquad (4.13)$$

the following diffusion equation emerges

$$\frac{\partial P}{\partial t} = k_B \nabla (L\nabla \ln P) \qquad (4.14)$$

Then, it is generally assumed that the transport coefficient L in Eq. (4.13) is a constant times the distribution function P:

$$L = \frac{D}{k_B} P \qquad (4.15)$$

By using this relation Compte and Jou (1996) are finally able to cast equation (4.14) into the more familiar diffusion equation

$$\frac{\partial P}{\partial t} = D\nabla^2 P \qquad (4.16)$$

Compte and Jou (1996) now turn to a nonstandard thermodynamics and study whether a nonclassical definition of entropy can be related to nonclassical diffusion. Instead of the usual Boltzmann definition for the entropy, namely

$$S[p] = -k_B \sum_i p_i \ln p_i \qquad (4.17)$$

where k_B is Boltzmann's constant and p_i the probability of microstate i, Compte and Jou (1996) consider Eq. (4.6) proposed by Tsallis (Salinas and Tsallis, 1999). The motivation of Eq. (4.6) is to be found in measures arising in analysis of multifractals, and it has been used in several contexts by a number of authors in the past years (Curado, 1999; Zanette, 1999, and many others). Making use of the fact that $\sum_i p_i = 1$, one may rewrite entropy (4.6) as

$$S_q = -\frac{k_B}{1-q} \sum_i p_i (1 - p_i^{q-1}) \qquad (4.18)$$

which clearly shows its positivity. A thorough analysis of the mathematical properties of (4.18) may be found in Salinas and Tsallis (1999). Next, Compte and Jou (1996) generalize (4.18) to a continuous distribution function $P(x, t)$ as

$$S_q(t) = -\frac{k_B}{1-q} \int P(\mathbf{x}, t)[1 - P^{q-1}(x, t)] d^N \mathbf{x} \qquad (4.19)$$

where x is the position in an N-dimensional space and t is the time. By means of the formal structure of classical irreversible thermodynamics (De Groot and Mazur, 1984) one may identify the flux of probability in this generalized thermodynamic formalism as

$$J = L\nabla \frac{\delta S_q}{\delta P} = \frac{k_B q}{1-q} L\nabla P^{q-1} \qquad (4.20)$$

where L is a transport coefficient. When (4.20) is combined with the conservation of probability equation (4.12) one is led to the following diffusion equation for processes evolving under this thermodynamic frame:

$$\frac{\partial P}{\partial t} = -\frac{k_B q}{1-q} \nabla (L\nabla P^{q-1}). \qquad (4.21)$$

It is worth noting that this kind of dynamics has a positive entropy production, as may be seen by evaluating the time derivative of S_q, namely

$$\frac{\partial S_q}{\partial t} = -\frac{k_B q}{1-q} \int P^{q-1} \frac{\partial P}{\partial t} d^N x \qquad (4.22)$$

By introducing (4.21), integrating once by parts and assuming that the surface term vanishes one obtains

$$\frac{\partial S_q}{\partial t} = \left(\frac{k_B q}{q-1}\right)^2 \int L(\nabla P^{q-1})^2 d^N x \qquad (4.23)$$

which, provided $L > 0$, is always positive or zero. This is only so, however, as long as the surface term in the derivation of (4.23) vanishes, namely when

$$LP^{q-1}\nabla P^{q-1}\big|_\Sigma = 0 \qquad (4.24)$$

where Σ stands for the boundary of the volume of space available for diffusion (from now on we will take this volume to be unbounded). This relation will prove useful later in this paper to restrict the range of the possible values for the parameters of the scheme.

At this point Compte and Jou (1996) note that, since in classical irreversible thermodynamics one has $L = DP/k_B$ with D a constant, we are here generally led to accept a dependence for L such as $L = DP^\alpha/k_B$ with D and α constants, and k_B the Boltzmann constant. In the case $\alpha = 1$ one recovers the classical picture and for $\alpha \neq 1$ anomalous diffusive effects of a dynamic nature are introduced into the scheme. A concentration dependent diffusivity of this type has been proposed for fractal diffusion [Stephenson (1995), and references therein], for particle diffusion across magnetic fields and for the motion of a polytropic gas in a porous medium (Muskat, 1937).

Stephenson (1995) obtained some scaling solutions of a class of radially symmetric nonlinear diffusion equations in an arbitrary dimension d. His case (1) pertains to an initial point source with a fixed total amount of material, and (2) for a radial flux of material through a hyperspherical surface. In his macroscopic model the flux density depends on powers of the concentration and its radial gradient. He analyzed the dimensional dependence of these solutions and compared them with scaling solutions of the corresponding linear equations for fractal diffusion. The nonlinear equations contain arbitrary exponents which can be related to an effective fractal dimension of the underlying diffusion process.

Following Compte and Jou (1996) one may further study the effects of the parameters q and α in the description of anomalous diffusion and thus evaluate the convenience of a thermodynamic ($\alpha = 1$) or a dynamic ($\alpha \neq 1$) description of the phenomenon, respectively.

Introducing the relation $L = DP^\alpha/k_B$ into Eq. (4.21) and exploiting the equality

$$P^\alpha \nabla P^{q-1} = (q-1)P^{\alpha+q-2}\nabla P = (q-1)(\alpha+q-1)^{-1}\nabla P^{\alpha+q-1} \qquad (4.25)$$

Compte and Jou (1996) obtained the following generalized diffusion equation,

$$\frac{\partial P}{\partial t} = D\frac{q}{\alpha+q-1}\nabla^2 P^{\alpha+q-1} \qquad (4.26)$$

which is a nonlinear equation of the form advanced by Müller and Ruggeri (1993), elsewhere referred to as the "porous media equation." This equation has been thoroughly studied in several mathematical works. Fractional equations of this sort are known in the theory of porous media (De Pablo et al., 2011). In their discussion Compte and Jou (1996) quote the Barenblatt–Pattle solution (Barenblatt, 2003). This is the physical solution for an initial delta distribution $P(\mathbf{x}, t) = \delta^N(\mathbf{x})$ that has the basic property of being the asymptotic ($t \to \infty$) limit distribution of the solutions of (4.26) with a localized initial distribution $P(\mathbf{x}, 0)$ (Peletier, 1981). Therefore, this solution plays the same role in the dynamics (4.26) as the Gaussian does in the standard linear diffusion.

Assuming $q > 0$, the solution of Eq. (4.26) is

$$P(\mathbf{x}, t) = bt^{-\mu}\{[a^2 - x^2 t^{-2\mu/N}]_+\}^{1/(q+\alpha-2)} \quad \text{for} \quad q+\alpha > 2 \qquad (4.27)$$

$$P(\mathbf{x}, t) = bt^{-\mu}\left(\frac{1}{a^2 + x^2 t^{-2\mu/N}}\right)^{1/(2-q+\alpha)} \quad \text{for} \quad 2\frac{N-1}{N} < q+\alpha < 2 \qquad (4.28)$$

where

$$b = \left(\frac{|q+\alpha-2|}{2Dq[N(q+\alpha-2)+2]}\right)^{1/(q+\alpha-2)} \qquad \mu = \frac{N}{N(q+\alpha-2)+2} \qquad (4.29)$$

$[.]_+ = \max(., 0)$ and a is a constant (depending on q and α) to be determined by normalization. Compte and Jou (1996) now advance the proposition that these solutions correspond to the subdiffusion and superdiffusion cases, respectively, as we shall substantiate in the following paragraph. The solutions with $q + \alpha < 2(N-1)/N$ or $q < 0$ correspond to nonphysical probability distributions (nonnormalizable) and they are, therefore, ignored altogether in further developments. A similar restriction, namely $q > 0$, may be imposed in on the grounds of purely mathematical arguments (Compte and Jou, 1996).

An important feature of these solutions is that they are of the form

$$P(\mathbf{x}, t) = t^{-\mu} f(\mathbf{x} t^{-\mu/N}) \tag{4.30}$$

and, therefore, their characteristic scaling may be easily seen to correspond to anomalous diffusive behavior: if one takes a measure $r(t)$ of the spread of $P(x, t)$ to be the radius of the sphere which contains a fraction $\beta < 1$ of the total probability

$$\int_0^{r(t_0)} P(|x|, t_0) \Omega_N |x|^{N-1} \, d|x| = \beta \tag{4.31}$$

where Ω_N is the surface of the N-dimensional sphere of unitary radius and then consider (4.31) again for a different time t, the change of variable $x' = (t/t_0)^{-\mu/N(x)|x|}$ leads one to

$$\int^{(t/t_0)^{-\mu/Nr(t)}} P(x', t_0) \Omega_N x'^{N-1} \, dx' = \beta \tag{4.32}$$

A simple comparison of (4.31) and (4.32) leads to the conclusion that

$$r(t) = \left(\frac{t}{t_0}\right)^{\mu/N} r(t_0) \tag{4.33}$$

independently of β, whence one recognizes the typical anomalous diffusive scaling

$$r \sim t^{\mu/N} = t^{1/(N(q+\alpha-2)+2)} \tag{4.34}$$

From Eq. (2.34) it is easy to identify the coefficient of anomalous diffusion as the quantity

$$\sigma = 1/(N(q+\alpha-2)+2) \tag{4.35}$$

which satisfies $\sigma > 1$ (superdiffusion) as long as $q + \alpha < 2$ and, conversely, $\sigma < 1$ if $q + \alpha > 2$.

Compte and Jou (1996) next study under what conditions solutions (4.27) and (4.28) are acceptable in the scheme presented, in the sense that they lead to a finite S_q and that they satisfy the assumption on which the positivity of the entropy production relies, namely condition (4.24). On the one hand, solution (4.27) is bounded and vanishes identically at infinity so that both the convergence of S_q and condition (4.24) are trivially satisfied. On the other hand, the asymptotic behavior of (4.28) is

$$P \sim |x|^{2/(q+\alpha-2)} \tag{4.36}$$

whereas for S_q to be finite one must require that $P(x, t)$ fades away faster than $|x|^{-N/q}$ N/q for $|x| \to \infty$ and the fulfillment of condition (4.24) demands that

$$P \sim |x|^{(1-\gamma)/(2q+\alpha-2)} \tag{4.37}$$

with $\gamma > 0$ in this same limit. Consequently, after elementary calculations Compte and Jou (1996) find that the finiteness of the entropy and its positivity in the superdiffusive case ($q + \alpha < 2$) impose the following condition relating the values of q and α

$$\max 2\frac{N-1}{N}, 2\frac{N-q}{N}, 2(1-q) < q+\alpha < 2 \text{ and } q > 0. \tag{4.38}$$

For clarity, Compte and Jou (1996) represent in Fig. 4.3 (their Fig. 1) the domain of the plane (q, α) where the physical values of the parameters lie according to condition (4.38).

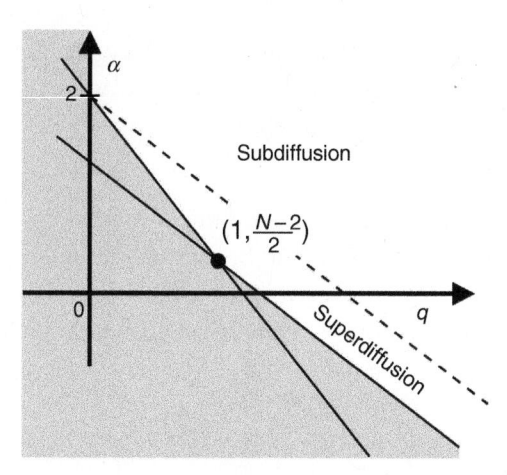

Figure 4.3: Distribution in the (q, α) plane of the domain of nonphysical situations (*shaded area*), superdiffusion, subdiffusion, and standard diffusion (*dashed line*) for a space of N dimensions. From Compte and Jou, 1996.

4.4.3 Some Special Cases

We shall now consider two limiting cases. First, let us suppose that all the anomaly in diffusion is to be accounted for by dynamical mechanisms, so we then have that in the considered scheme $q = 1$ (standard thermodynamics) and $\alpha \neq 1$. It may be shown that in this situation, by choosing α according to $\alpha = 1 + (2/N)(1/\sigma - 1)$, one can reproduce any arbitrary anomalous diffusive behavior:

$$r \sim t^{\sigma/2} \qquad 0 < \sigma < \infty \qquad (4.39)$$

This is easily seen by imagining the function $\sigma = 2/(N(q + \alpha - 2) + 2)$ represented on the z-axis of the domain in Fig. 4.3 and slicing the resulting surface by the plane $q = 1$ (Compte and Jou, 1996).

The other limiting case which is interesting to explore is the situation in which anomalous diffusion is to be considered exclusively as a thermodynamic phenomenon. In this case $\alpha = 1$ (standard dynamics) and $q \neq 1$ with $q = 1 + (2/N)(1 - \sigma/\alpha)$. When we do this, we are led to a restriction on the range over which the coefficient of anomalous diffusion is allowed to vary. This can be seen graphically by slicing the representation of $\sigma = 2/(N(q + \alpha - 2) + 2)$ over the domain in Fig. 4.3 along the plane $\alpha = 1$.

As a result, we see that if anomalous diffusion is to be accounted for purely on account of a fractal thermodynamics, we find that $\sigma < (N + 2)/2$. Therefore, a description of turbulence when N 6 3 (as long as it is described with the kind of correlated dynamics considered), where one typically has $r \sim t^{3/2}$ (Monin and Yaglom, 1967), should demand at least a dynamic contribution ($\alpha \neq 1$) to the nonclassical thermodynamics.

Essentially, no other substantial differences arise in this scheme regarding the distribution functions or the dynamical equations when one considers a pure dynamic or thermodynamic origin of the anomaly in diffusion, since the parameters α and q enter almost everywhere in this scheme symmetrically as $q + \alpha$.

Therefore, it seems more natural to allow a coupling of both dynamic and thermodynamic aspects of diffusion, so that α and q are not to be considered independent of each other. They could, for instance, be related linearly as $\alpha = Cq + 1 - C$ which ensures that when $q = 1$ the standard picture of diffusion ($\alpha = q = 1$) is encountered. Furthermore, if we assume that $C < -1$, then it is easily proved that $q < 1$ corresponds to superdiffusion and $q > 1$ to subdiffusion. If, in contrast, we have $C < -1$, we get $q < 1$ ($q > 1$) related to subdiffusion (superdiffusion).

As done in the previous cases, one can now slice the surface $\sigma = 2/(N(q + \alpha - 2) + 2)$ defined over the domain of physical parameters (q, α), Fig. 4.3, along the plane $\alpha = Cq + 1 - C$ for each of the above-mentioned cases, namely $C > -1$ and $C < -1$. One then finds that for

$C > -1$ one can have $0 < \sigma < 1 + (N(C + 1)/2)$ which means a restriction on superdiffusion, whereas for $C < -1$ one can reproduce dynamics subject to $2/(2 - N(C + 1)) < \sigma < \infty$ and, therefore, only a limitation on subdiffusion is observed. It is, therefore, easy to see that the proposed combined effect of dynamics and thermodynamics can reproduce the typical behavior of diffusion in fully developed turbulence, where $r \sim t^{3/2}$, as long as $C < -1$ or $C > (4/N) - 1$, and even account for intermittency effects (Ronis and Procaccia, 1982; Vailati and Giglio, 1997).

4.4.4 Conclusions of Compte and Jou (1996)

Compte and Jou (1996) have studied the relation of a nonclassical entropy with anomalous diffusion. The need for a generalization of the entropy to describe some kinds of non-Fickian effects was also pointed out by Jou and Camacho (1991), in the very different context of case 2 diffusion, where relaxational effects must be taken into account in the diffusion flux. The relation pointed out by Compte and Jou (1996) goes farther than in Jou and Camacho (1991), where the relaxation effects implied a correction to the dynamics and the thermodynamics, which was of interest for short times. In the approach of Compte and Jou (1996), the modification is deeper, both in the dynamics because they are important not only in the short-time regime but also for long times, and because it might lead to a redefinition of the entropy rather than to a correction to it. The relation points to a new application of the fractal entropy and reveals the link between dynamics and thermodynamics.

In conclusion, Compte and Jou (1996) have shown that dynamic and thermodynamic aspects of diffusion are dynamically indistinguishable except for limitations on the speed of superdiffusion when an anomalous thermodynamics occurs. An important result of Compte and Jou (1996) has been to show that fractal thermodynamics cannot account alone for the superdiffusive properties of turbulence which must include dynamic arguments, combined or not with an nonstandard thermodynamic frame. To proceed further with the role of thermodynamic effects in anomalous diffusion, it could be advisable to explore the equilibrium properties of a system under such a thermodynamic description. A correlation between equilibrium properties and anomalous diffusion set through the generalized entropy could help elucidate the extent of the thermodynamic contribution to anomalous diffusion.

Compte and Jou (1996) find it interesting to explore if anomalous diffusion in a shear flow leads to differences when one assumes a dynamic or a thermodynamic basis of diffusion. The answer would provide us with an experimental procedure to establish which scheme suits better the description of a particular anomalous diffusive phenomenon and what values of q or α appropriately describe it. Their research points out the need of paying more attention to the thermodynamic aspects of non-Fickian diffusion. As they state, a more detailed knowledge of the entropy could be useful when studying problems related to the dissipation in fractal sets, or the couplings of heat and mass transport.

4.5 Zanette's (1999) Treatment

4.5.1 Introduction

Despite the omnipresence of diffusion as a transport mechanism in natural processes, it is known that a different kind of transport underlies a selected, but ever-growing, class of systems. Due to various motivations most of these systems have recently attracted very much attention. They range from turbulent fluids, to chaotic dynamical systems, to genetic codes (see next section). In these systems, anomalous diffusion—a mechanism closely related to normal diffusion, but with some qualitatively different properties—drives transport processes (Zanette, 1999). Over the past few years, it became more and more clear that anomalous diffusion can be made naturally compatible with equilibrium thermodynamics if the Boltzmann–Gibbs formulation of thermodynamics is replaced by the Tsallis formulation. This compatibility generalizes then the connection between normal diffusion and the usual formulation of thermodynamics. The main aim of (Zanette, 1999) treatment is to analyze this generalization, commenting on some related topics brought to light in recent work.

4.5.2 Anomalous Diffusion and Lévy Flights

Any transport mechanism which, like diffusion, behaves at mesoscopic level as an isotropic random process, but which violates Eq. (4.4), is generally referred to as anomalous diffusion. More specifically, most of the literature on anomalous diffusion has been devoted to processes where the mean square displacement $\langle r^2 \rangle$ varies with time as

$$\langle r^2 \rangle \propto t^{2/z} \tag{4.40}$$

where z ($\neq 2$) is the dynamic exponent, or random walk fractal dimension, of the transport process. Normal diffusion corresponds to $z = 2$. For $z > 2$, the growth rate of the mean square displacement is smaller than in normal diffusion, and transport is consequently said to be subdiffusive. On the other hand, for $z < 2$ the mean square displacement grows relatively faster and transport is thus superdiffusive (Zanette, 1999). In the following, the attention will be mainly focused on this latter case.

4.5.2.1 Anomalous diffusion in nature

As advanced earlier, anomalous diffusion occurs in a wide class of natural systems and processes. In the realm of physics, a paradigmatic example is given by particle transport in disordered media. Consider the motion of particles in a medium containing impurities, defects, or some kind of intrinsic disorder, such as in amorphous materials. Examples are disordered lattices, porous media, and doped conductors and semiconductors. In these heterogeneous substrates, particles are driven by highly irregular forces, which determine a complex variation of the local transport coefficients. This heterogeneity can in fact induce anomalous diffusion (Zanette, 1999). For instance, it has been experimentally shown that in quasi-one-dimensional ionic conductors such as hollandite ($K_{1.54}Mg_{0.77}Ti_{7.23}O_{16}$), where transport is very sensitive to the presence of impurities, the dynamic exponent is $z \approx 1 + 1/\theta$ with q proportional to the temperature (Zanette, 1999).

A reasonable model for transport in heterogeneous media is provided by a random walk in a lattice with quenched disorder (Bouchaud and Georges, 1990). Generally, this waiting time distribution, $\psi(\tau)$, behaves as $\psi(\tau) \sim \tau^{-1-\mu}$ for $\tau \to \infty$. For $\mu > 1$, the first relation in Eq. (4.2) holds, and normal diffusion is observed. On the other hand, for $\mu < 1$ the mean waiting time diverges and diffusion is anomalous. The corresponding dynamic exponent for $0 < \mu < 1$ is

$$z = \frac{2}{\mu} \quad \text{for } d > 2 \qquad z = 2 - d + \frac{d}{\mu} \quad \text{for } d < 2 \tag{4.41}$$

A most important instance of anomalous diffusion in physics occurs in turbulent flows. In fully developed turbulence, fluid particles exhibit very irregular motion over a wide range of space and time scales. Based on empirical motivations, Richardson (1926) proposed that the probability $P(R, t)$ that two fluid particles initially close to one another have a separation R at time t, obeys the equation

$$\frac{\partial P}{\partial t} = \frac{\partial}{\partial R}\left(D(R)\frac{\partial P}{\partial R}\right) \tag{4.42}$$

with $D(r) \sim R^{4/3}$. Comparing with Eq. (4.3), it is obvious that the Richardson's equation is a diffusion equation with space-dependent diffusivity. Its solution immediately implies $\langle R^2 \rangle \propto t^3$, indicating that the relative motion of particles in fully developed turbulence corresponds to anomalous diffusion with a dynamic exponent $z = 2/3$, which is well into the superdiffusive regime (Zanette, 1999).

Richardson's law has to be modified to take into account the intermittent vorticity field in a turbulent flow (Mandelbrot, 1982). This means that vorticity—and, in particular, turbulent activity and dissipation—is concentrated on a relatively small volume in the whole system, which is a fractal set. Experiments of Antonia et al. (1981) suggest that the fractal dimension of this set is $d_f = 2.8 \pm 0.05$. Incorporating this correction, it can be shown (Shlesinger et al., 1987) that $\langle R^2 \rangle \propto t^{12/(1+d_f)}$, or

$$z = \frac{1+d_f}{6} \approx 0.63 \tag{4.43}$$

In the work by Shlesinger et al. (1987) the authors introduce the stochastic Lévy walk which is a random walk with a nonlocal memory coupled in space and in time in a scaling fashion. Lévy walks result in enhanced diffusion, that is, diffusion that grows as $t\alpha$, $\alpha > 1$. When applied to the description of a passive scalar diffusing in a fluctuating fluid flow, the model generalizes Taylor's correlated-walk approach. It yields Richardson's t^3 law for the turbulent diffusion of a passive scalar in a Kolmogorov-(5/3) homogeneous turbulent flow and also gives the deviations from the (5/3) exponent resulting from Mandelbrot's intermittency. The model can be extended to studies of chemical reactions in turbulent flow.

A somewhat more abstract form of anomalous diffusion is present in the evolution of chaotic Hamiltonian dynamical systems in phase space. Hamiltonian systems are characterized by volume conservation in phase space, as stated by the Liouville theorem. The domain occupied by a given set of initial conditions in phase space can be strongly distorted under the effect of evolution, but its volume remains constant. Mechanical processes that preserve energy are instances of Hamiltonian systems, but a huge host of systems—both continuous and discrete in time—is known to belong to the same class (Zanette, 1999). Since a single Hamiltonian system can exhibit both regular and chaotic evolution by simply changing the initial condition, the dynamical geometry of its phase space is usually extremely intricate. Zones of nested regular trajectories which alternate with chaotic regions are typically found at many scales, displaying self-similar structures. In the bulk of chaotic regions trajectories are extremely irregular and resemble random paths. On the other hand, when approaching the boundary with a regular region, the same trajectory can temporarily become much simpler and smoother. Consequently, as it evolves in phase space along a chaotic orbit, a Hamiltonian system alternates intermittently between zones of highly complex behavior and a regime of almost regular dynamics. Globally, this motion can be thought of as a stochastic process, and turns out to have the same statistical properties as anomalous diffusion (Zanette, 1999).

Anomalous diffusion has also been observed in dissipative (non-Hamiltonian) dynamical systems, both in simulations and in experiments. For example, a dynamic exponent $z \approx 1.2$ has been measured in the Taylor–Couette flow. Yet, anomalous diffusion is not restricted to physical systems. This transport has in fact been detected in several biological processes. Some sectors of genomic DNA sequences, for instance, are known to exhibit statistical properties analogous to anomalous diffusion. To stress this correspondence, "DNA walks" have been defined. DNA, which codes genetic information, is a large molecule in the form of a chain of nucleotides. Each nucleotide contains either a purine or a pyramidine base. The information code in DNA is a symbolic chain of four letters: A, C, G, and T. Interestingly, within DNA only a small portion provides code information for protein building (3% in the human genome) whereas other zones are noncoding, their specific role being unknown. A one-dimensional DNA walk is constructed by sequentially running over the chain of nucleotids. Each time a purine is found the walker jumps rightwards, whereas when a pyramidine is found the jump occurs leftwards. In DNA walk, systematic deviations from normal diffusive behavior derive from long-range correlations in the nucleotide sequence. It has been found that the DNA walk is in fact statistically identical to normal diffusion in the zones of the genomic chain that code information. On the other hand, in noncoding sequences the DNA walk is analogous to superdiffusion. An instance of anomalous diffusion in biology appears in the flight patterns of certain birds. It has been suggested that albatrosses specialize in long journeys of random foraging, searching for patchily and unpredictably dispersed prey over several million square kilometers. As discussed in the next section, anomalous diffusion is inherently related with fractal geometry, scale-invariance, and self-similarity, which seem to drive the predator-prey dynamics. Of course, the previous collection of examples of anomalous diffusion in Nature is not at all exhaustive, but only pretends to give a hint on the variety of systems driven by this kind of transport. For a more detailed account, the reader is referred to the review by Bouchaud and Georges (1990). This review is also an excellent reference to the mathematical treatment of anomalous diffusion.

4.5.2.2 Random-walk models of anomalous diffusion

In view of the efficacy of random walks in modeling normal diffusion at a mesoscopic level, it is desirable to find a similar stochastic model describing anomalous diffusion. Introducing a waiting time distribution which, for large waiting times, behaves as $\psi(\tau) \sim \tau^{-1-\mu}$ with $\mu < 1$, produces the anomalous dynamic exponent given in Eq. (4.41). In this case $z > 2$ and, therefore, transport is subdiffusive. Note that the source of anomaly in these random walks is the divergence of the average waiting time $\langle \tau \rangle$. For $\mu < 1$, the long-tailed distribution $\psi(\tau)$ allows for very long waiting times with relatively high probability, violating thus the first relation in Eq. (4.2). These long waiting times produce an overall reduction of efficiency in the transport mechanism with respect to normal diffusion, and leads to subdiffusive behavior. Taking into account the previous argument, it can be expected that the violation of the second of relations (4.2) will lead, on the other hand, to superdiffusion. In fact, having a divergent mean square displacement $\langle x^2 \rangle$ requires the jump probability distribution $p(\mathbf{x})$ to have a long tail for large $\mathbf{x}$. This long-tailed distribution would produce, with relatively high probability, very long jumps. Globally, this transport mechanism should be more efficient than normal diffusion, and superdiffusive behavior is expected.

It can be shown that, in d-dimensional space, the mean square displacement diverges if

$$p(\mathbf{x}) \sim \frac{1}{x^{d+\gamma}} \tag{4.44}$$

for large x, with $\gamma < 2$. Note that, for such distribution, normalization requires $\gamma > 0$. It is therefore expected that a random walk with a jump distribution $p(\mathbf{x}) \sim x^{-d-\gamma}$ with $0 < \gamma < 2$ does not model normal diffusion, but some kind of superdiffusive motion (Zanette, 1999). Of course, a large score of functions satisfy Eq. (4.44), and are thus candidates to play the role of a jump distribution for a superdiffusive random walk. Among them, Lévy distributions have been studied in detail.

Lévy (1937) distributions are defined through their Fourier transform, which reads

$$p(\mathbf{k}) = \int \exp(i\mathbf{k}\cdot\mathbf{x}) p(\mathbf{x}) d\mathbf{x} = \exp(-bk^\gamma), \qquad (4.45)$$

where b is a positive constant, and $k \equiv |\mathbf{k}|$. Although the antitransform $p(\mathbf{x})$ has no analytical expression, it can be shown that it satisfies Eq. (4.44). Moreover, if $\gamma < 2$ the positivity of $p(\mathbf{x})$ is assured. The relatively simple form of this jump distribution in the Fourier representation makes it a very suitable tool for analytical manipulation. However, the main interest of Lévy functions in the mathematical theory of distributions comes from the fact that they are stable. Essentially, this means that two Lévy functions with the same Lévy exponent γ produce, upon convolution, a third Lévy function with the same exponent. This can be readily proven in the Fourier representation, where the convolution transforms into ordinary product:

$$p_1(\mathbf{k}) p_2(\mathbf{k}) = \exp(-b_1 k^\gamma) \exp(-b_2 k^\gamma) = \exp[-(b_1 + b_2) k^\gamma] = p_3(\mathbf{k}) \qquad (4.46)$$

Lévy functions are not the only stable distributions, the Gaussian $p(\mathbf{x}) \propto \exp(-x^2)$ being probably the best-known example. The one-dimensional Cauchy distribution, $p(x) \propto (1 + x^2)^{-1}$, is another instance. Stable distributions play a fundamental role in probability theory since according to the central limit theorem—which is usually stated for the Gaussian function—the addition of random variables tends to one such distribution. In particular, as Lévy demonstrated through his generalization of the Gaussian central limit theorem (Lévy, 1937), adding random variables with a power-law distribution as in (4.44)—whose second moment $\langle x^2 \rangle$ diverges for $\gamma < 2$—leads asymptotically to a Lévy distribution.

Another important property of Lévy distributions, which is reflected in its power-law large-x asymptotic behavior, Eq. (4.44), is the absence of characteristic length scales. This implies that random walks with Lévy jump distributions have self-similar properties. In particular, it can be shown that the set of points visited by this kind of random walk is a fractal of dimension γ. As a consequence, these distributions are ubiquitous in the realm of self-similarity geometry—the geometry of fractals (Zanette, 1999).

A discrete-time random walk whose jump distribution is given by a Lévy function as in Eq. (4.45) is called a Lévy flight. Compte suggested that the Fourier transform $P(\mathbf{k}, t)$ of the probability distribution of the walker at a given point at time t satisfies the evolution equation

$$\frac{\partial P}{\partial t}(\mathbf{k}, t) = -D_\gamma k^\gamma P(\mathbf{k}, t) \qquad (4.47)$$

that generalizes diffusion equation (4.3) in the Fourier representation. A simple dimensionality analysis shows that the anomalous diffusivity is $D_\gamma \propto b/\Delta t$, where Δt is the time step of the random walk. In free space, Eq. (4.47) can be readily solved:

$$P(\mathbf{k}, t) = P(\mathbf{k}, 0) \exp(-D_\gamma k^\gamma t) \qquad (4.48)$$

For a delta-like initial condition, $P(\mathbf{r}, 0) = \delta(\mathbf{r})$, one has $P(\mathbf{k}, 0) = 1$ and, thus,

$$P(\mathbf{r}, t) = (2\pi)^{-d} \int \exp(-i\mathbf{k}\cdot\mathbf{r} - D_\gamma k^\gamma t) d\mathbf{k} \tag{4.49}$$

(Zanette, 1999). This function remains unchanged, except for a constant factor, if both space and time are rescaled, $P(\alpha^{1/\gamma}\mathbf{r}, \alpha t) = \alpha^{-d/\gamma} P(\mathbf{r}, t)$. Using this scale invariance, one can write

$$P(\mathbf{r}, t) = t^{-d/\gamma} \Pi(rt^{-1/\gamma}) \tag{4.50}$$

where Π is a function of a single variable. In turn, this implies

$$\langle r^2 \rangle \propto t^{2/\gamma}, \tag{4.51}$$

for $0 < \gamma < 2$. This result, which has been here derived for a delta-like initial distribution, can be generalized by simple superposition to more general initial conditions. It shows that a Lévy flight with $0 < \gamma < 2$ represents superdiffusion with a dynamic exponent $z = \gamma$. On the other hand, for power-law jump distributions with $\gamma > 2$ the dynamic exponent corresponds to normal diffusion, $z = 2$, Fig. 4.4 (Zanette, 1999).

The fact that in Lévy flights the mean square displacement $\langle x^2 \rangle$ of a single step diverges, implies—in contrast with Eq. (4.51)—that the mean square displacement of the walker after a certain time is, on the average over infinitely many realizations, also infinite. The arguments

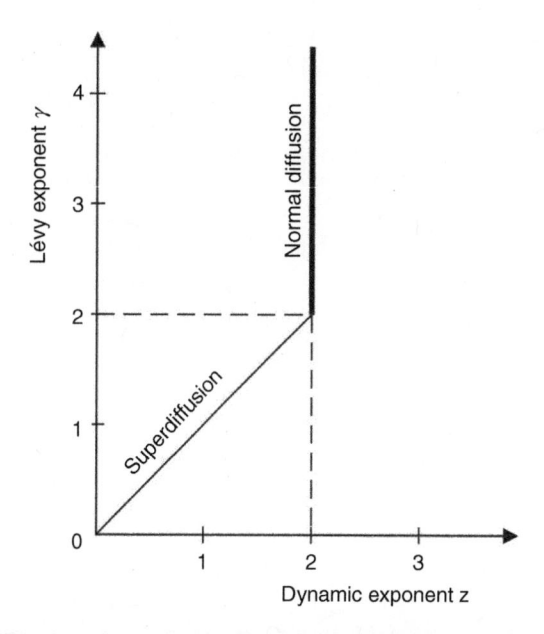

Figure 4.4: The diffusion dynamic exponent z as a function of the Lévy exponent γ. Fig. 1, Zanette, 1999.

used to derive Eq. (4.51) are still not solid (Zanette, 1999). The result (4.51) is expected to be valid for finite times, that is, in a certain portion of the whole random walk, and on averages over a finite number of trajectories. The same result would be valid during a certain time if the jump distribution $p(\mathbf{x})$ is a Lévy function in some (large) range of values of x, but has a cutoff for sufficiently large x (Zanette, 1999). In spite of this drawback, Lévy flights provide a very powerful tool for modeling superdiffusion because of the mathematical properties of the Lévy distributions. They are thus a very satisfactory starting point as a model for studying the statistical mechanics of superdiffusive transport, generalizing the results for normal diffusion (Zanette, 1999).

Before passing to the discussion of superdiffusion in a statistical-mechanical frame, a comment is in order on the numerical simulation of superdiffusive random walks. Due to the cumbersome properties of Lévy distributions in real space (Zanette, 1999), it is not easy in numerical calculations to work directly with these functions. Rather, power-law distributions with the same asymptotic properties as Lévy's, Eq. (4.44), are used. For instance, one can take

$$p(\mathbf{x}) = \frac{N}{(1+x)^{d+\gamma}} \qquad (4.52)$$

with N a normalization constant. The Fourier transform of these distributions behaves like a Lévy distribution for small k, $p(\mathbf{k}) = 1 - bk^\gamma$. Also, they show the same scale invariance for large x, which leads to self-similar properties in the associated random walks. Zanette (1999) illustrates in his Fig. 2 the first 10^4 points visited by a random walk in two dimensions, with the a power-law jump distribution given in Eq. (4.52) and with exponent $\gamma = 1.5$, starting at the center of the main frame. The amplification illustrates the self-similar properties of this process. One observes clustered, fractal-like structure of this set of points.

4.5.3 Maximum-Entropy Formalism for Anomalous Diffusion

Entropy plays a central role in the foundations of equilibrium and nonequilibrium statistical mechanics. It is well known from the work of Boltzmann and others that entropy provides a natural link between disequilibrium processes and their asymptotic states of thermodynamic equilibrium. In addition, the whole theory of equilibrium statistical mechanics can be derived from a variational formalism for the entropy, as follows.

Define the classical entropy S, Eq. (4.17), as a functional of the probability distribution p_i over the states i of a given system,

$$S[p] = -k_B \sum_i p_i \ln p_i \qquad (4.17)$$

where k_B is Boltzmann constant. Find then the values of p_i that maximize $S[p]$, taking into account the normalization constraint, $\sum_i p = 1$, and—if required by particular conditions for the system—any additional constraint on p_i. The value of p_i resulting from this maximization procedure gives the probability of finding the system in state i when thermodynamic equilibrium has been reached. For instance, introducing the energy constraint

$$\sum_i \varepsilon_i p_i = E, \tag{4.53}$$

where ε_i is the energy of state i and E is the thermodynamic energy, the maximization of entropy produces the well-known Boltzmann distribution

$$p_i \propto \exp(-\beta_i \varepsilon_i) \tag{4.54}$$

(Levine and Tribus, 1979; Kubo, 1988; Hofman, 2002).

4.5.3.1 Traditional formalism

As a starting point for including normal diffusion in the frame of equilibrium statistical mechanics, the procedure of entropy maximization has been applied to obtain the jump probability distribution $p(\mathbf{x})$ in a discrete-time random walk (Montroll and Shlesinger, 1983). In this case, entropy is defined as a following generalization of (4.17).

$$S[p] = -k_B \int p(\mathbf{x}) \ln[\sigma^d p(\mathbf{x})] d\mathbf{x} \tag{4.55}$$

Here σ is a characteristic length, whose meaning will become clear soon. The distribution $p(\mathbf{x})$, in fact, has units of length to the power $-d$. The maximization of $S[p]$ is carried out taking into account the normalization of $p(\mathbf{x})$,

$$\int p(\mathbf{x}) d\mathbf{x} = 1 \tag{4.56}$$

and imposing the additional constraint

$$\int x^2 p(\mathbf{x}) d\mathbf{x} = \sigma^2 d, \tag{4.57}$$

which is inspired in the second relation of Eq. (4.2). Except for a dimensionality factor, σ^2 is thus the mean square displacement associated with $p(\mathbf{x})$.

Under these conditions, the maximization of entropy yields

$$p(\mathbf{x}) = (2\pi\sigma^2)^{-d/2} \exp(-x^2/2\sigma^2) \tag{4.58}$$

namely, a Gaussian jump distribution. Since, in view of the constraint (4.57), the mean square displacement associated with $p(\mathbf{x})$ is finite, the maximum-entropy formalism applied as earlier to the jump distribution of a random walk describes normal diffusion.

The question on whether anomalous diffusion can be derived from a variational formalism for the entropy arises now. Montroll and Shlesinger (1983) have shown that this is in fact possible, but requires replacing the constraint (4.56) by a more complex condition on $p(\mathbf{x})$. In particular, Lévy flights, Eq. (4.45), are obtained from the maximization of the entropy (4.55) if the jump distribution satisfies, along with normalization,

$$\int \ln[(2\pi)^{-d} \int \exp(-i\mathbf{k}.\mathbf{x} - bk^{\gamma})] dk] p(\mathbf{x}) dx = \text{constant}. \quad (4.59)$$

However, this is a quite unsatisfactory answer to the previous question. Indeed, besides its complexity, the constraint (4.59) is anything but a natural condition to impose to the jump distribution. In Montroll and Shlesinger's (1983) words, "it is difficult to imagine that anyone in an *a priori* manner would introduce" such a condition for maximizing the entropy with respect to $p(\mathbf{x})$. This remark would at once exclude Lévy flights, and anomalous diffusion with them, from the frame of the maximum-entropy formalism and, therefore, from a natural connection with equilibrium statistics.

In Alemany and Zanette (1994), a different approach has been proposed to tackle the problem of deriving anomalous diffusion from the maximization of entropy. Indeed, random walks in which the set of visited points is a fractal emerge from a maximum entropy formalism applied to the generalized entropies. Since replacing the constraint on the jump distribution implies imposing unconventional, forced conditions on $p(\mathbf{x})$, a possible way out is to replace the form of the entropy instead. In particular, it has been found that the form of the Tsallis entropy (Tsallis, 1988; Curado and Tsallis, 1991) produces, upon maximization under the constraints prescribed by this theory, power-law jump distributions with the asymptotic behavior. Especially Curado and Tsallis (1991) show the way through which the Tsallis (1988) generalization of the Boltzmann–Gibbs statistics becomes consistent with a generalized thermodynamics preserving the Legendre-transformation framework of standard thermodynamics. In addition to that they generalize the Shannon additivity property. As described in the following, random-walk models of anomalous diffusion find a natural statistical-mechanical basis in Tsallis's theory.

4.5.3.2 Generalized formalism

Inspired in the theory of multifractals, Tsallis (1988) proposed to generalize the traditional Boltzmann–Gibbs statistical mechanics by introducing new forms for the entropy and for the constraints to be applied in the maximization procedure. For a system whose ith state is occupied with probability p_i, the generalized entropy reads

$$S_q = -k_B \frac{1 - \sum_i p_i^q}{1 - q} \quad (4.6)$$

where q is a real parameter. For the canonical ensemble, where the energy of the ith state is ε_i and the average energy is E_q, the constraint imposed to p_i which generalizes Eq. (4.53) along with probability normalization, is

$$\sum_i \varepsilon_i p_i^q = E \qquad (4.60)$$

In this generalized formalism, in fact, the average of any observable O is defined as

$$O_q = \sum_i O_i p_i^q \qquad (4.61)$$

This average is usually referred to as the q-expectation value of O. The generalized statistical-mechanical formalism based on Eqs. (4.6) and (4.60) has some remarkable properties. First of all, it reduces to the Boltzmann–Gibbs formulation in the limit $q = 1$. In fact, Eq. (4.17) is recovered from (4.6) in that limit. The constraint (4.53) reduces in turn to the traditional definition of mean energy. The new formalism preserves the full Legendre-transformation structure of thermodynamics for all q (Curado and Tsallis, 1991) leaving invariant in form the main results of statistical thermodynamics, such as the Ehrenfest theorem, the H-theorem, the von Neumann equation, the Bogolyubov inequality, and the Onsager reciprocity theorem. Its seems to be particularly useful in dealing with systems involving long-range correlations and nonextensivity, as the formalism itself is nonextensive for $q \neq 1$. The Tsallis exponent q has thus been interpreted as a measure of nonextensivity. Since its introduction, Tsallis statistics has found successful applications to a large class of problems of high interest, ranging from gravitational systems, to turbulent flows, to optimization algorithms, and so on.

In order to apply Tsallis statistics to discrete-time random walks in the spirit outlined in the previous section, Eqs. (4.55) and (4.57) have to be generalized according to (4.6) and (4.60), respectively. As a function of the jump probability, the generalized entropy can be written as (Alemany and Zanette, 1994)

$$S_q[p] = -\frac{1}{1-q}\left\{1 - \sigma^{-d}\int [\sigma^d p(\mathbf{x})]^q d\mathbf{x}\right\}, \qquad (4.62)$$

whereas the canonical constraint transforms into a condition on the q-expectation value of x^2:

$$\langle x^2 \rangle_q = \sigma^{-d}\int x^2 [\sigma p(\mathbf{x})]^q d\mathbf{x} = \sigma^2, \qquad (4.63)$$

Here, σ preserves its identification as a typical length associated with the jump probability. However, for $q \neq 1$, σ^2 does not coincide with the mean square length of the jumps. For simplicity, the dimensionality factor in the right-hand side of Eq. (4.57) has now been absorbed by σ.

It is shown in the following that the maximization of $S[p]$ as defined in Eq. (4.62) with the constraints (4.56) and (4.63), which, as in the case of the traditional formalism, is carried out by the standard method of Lagrange multipliers (Tsallis, 1988; Curado and Tsallis, 1991) and produces a power-law jump distribution. The exponent of the power-law depends on the space dimension and on the Tsallis exponent q. This jump distribution is not a Lévy distribution like (4.45), but has the same type of asymptotic behavior, Eq. (4.44). For suitable values of d and q, this form of $p(\mathbf{x})$ define a random walk with anomalous properties. For clarity, the results in one dimension are exemplified in Zanette (1999). The jump distribution resulting from the maximization procedure is, in this case,

$$p(x) = Z_q^{-1}[1 - \beta(1-q)x^2]^{1/(1-q)} \qquad (4.64)$$

where the partition function $\mathbf{Z}_q$ is given by

$$Z_q = \int_{-\infty}^{+\infty} [1 - \beta(1-q)x^2]^{1/(1-q)}\, dx \qquad (4.65)$$

The positive constant β is one of the Lagrange multipliers, which can be expressed as a function of σ using the constraint (4.63), as shown later. In the generalized formulation of statistical mechanics β is related to the temperature T in the standard form, as $\beta \propto 1/T$. Zanette (1994) gives the calculation results of the partition function for $1 < q < 3$, and for $q < 1$ which are omitted here. For $q = 1$, the Gaussian (4.58) is recovered. Note that the Tsallis exponent q is related to the exponent γ in Eq. (4.44) (Alemany and Zanette, 1994). Fig. 4.5 shows the profile of $p(x)$ for several values of q.

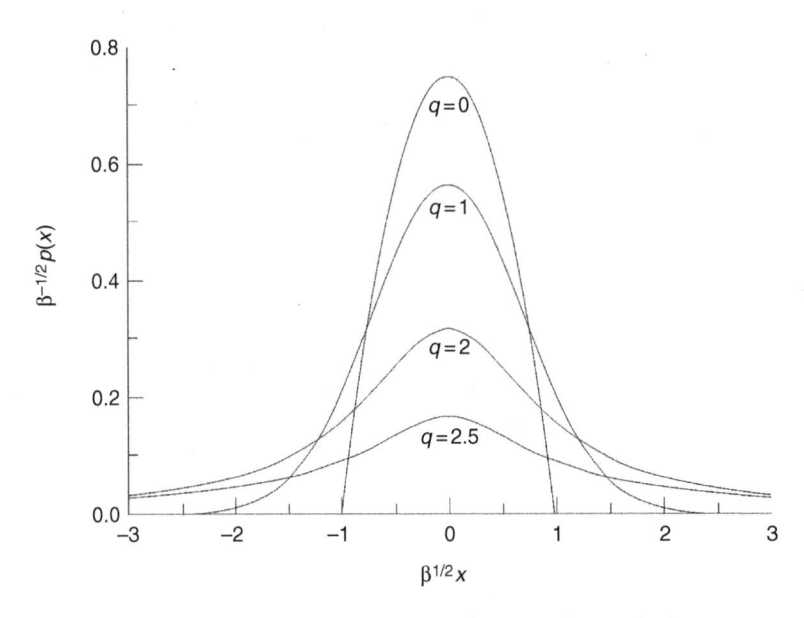

Figure 4.5: Jump probability distribution derived from Tsallis statistics, for some values of the Tsallis exponent q.
For $q = 1$ the standard Gaussian is obtained. Note the cutoff for $q < 1$, and the long-tailed power-law distributions for $q > 1$. Fig. 3, Zanette, 1999, with permission.

Regarding anomalous diffusion, it is shown that the case $q < 1$ is irrelevant. In fact, for such values of the Tsallis exponent $p(x)$ exhibits a cutoff at $|x| = (\sqrt{\beta(1-q)})^{-1}$, and vanishes for larger $|x|$. This implies that the mean square displacement associated with $p(x)$ is finite and the resulting random walk corresponds to normal diffusion. The attention is consequently focused in the following on the case $1 < q < 3$. In this case, for $1 < q < 5/3 \approx 1.67$ the mean square displacement is still finite, and the random walk corresponds to normal diffusion. On the other hand, anomalous superdiffusion is obtained for $5/3 \leq q < 3$. Zanette (1999) also calculates and discusses the q-expectation values of x^2 which, in the frame of Tsallis statistics, replace the mean square displacement of the standard formulation.

4.5.4 Conclusions of Zanette (1999)

Though normal diffusion is ubiquitous in Nature, a large and still growing class of real systems is driven by a different kind of transport processes, namely, by anomalous diffusion. In view of the current importance of many of these systems, which range from turbulent flows to disordered media and chaotic dynamics to flight patterns in birds, it is of high interest having a formulation able to place anomalous diffusion in a statistical-thermodynamic frame, generalizing thus Einstein's theory for normal diffusion. However, within Boltzmann–Gibbs statistics "the wonderful world of clusters and intermittencies and bursts that is associated with Lévy distributions would be hidden from us if we depended on a maximum entropy formalism that employed simple traditional auxiliary conditions" (Montroll and Shlesinger, 1983). Instead, Tsallis generalized that statistics is a strong candidate to successfully yield such a formulation.

Tsallis statistics provides a frame for the mathematical foundations of anomalous diffusion in two forms. In the first place, jump distributions of random-walk models for Lévy-like superdiffusion can be derived from a maximum-entropy principle within the generalized theory. Such distributions exhibit power-law long tails, which are an essential feature in the results of the theory. At once, Tsallis statistics furnishes a nice explanation for the Lévy distributions in other natural phenomena, as the result of the superposition of random variables with long-tailed distributions. In the second place, the functional form of the distributions resulting from Tsallis's formalism successfully suggests the solution to the nonlinear Fokker–Planck equation, which describes both subdiffusion and superdiffusion.

4.6 Discussion of Selected Works
4.6.1 General Statistical Mechanics and Stochastic Systems

Statistical mechanics (SM) is the subject that models and analyzes the macroscopic properties of matter through a statistical representation of its microscopic dynamics. It relies on the identification of macroscopic variables with the mean values of their respective microscopic analogs (random variables). Working in the statistical mechanics context Almeida (2005) adopts a Hamiltonian approach to derive Tsallis's generalized thermostatistics (Salinas and Tsallis, 1999). Following the usual routine of classical statistical mechanics, Almeida starts with a Hamiltonian system composed by weakly interacting elementary subsystems and writes its microcanonical and canonical distributions in terms of its structure function in the phase space. Exponential and the power-law canonical distributions emerge naturally from a unique differential relation of the structure function. When the heat capacity of the heat bath is infinite the canonical distribution is exponential, while if it is constant and finite the canonical density follows a power-law behavior. Almeida then derives the microscopic analogue of the main thermodynamics quantities (temperature and entropy) and shows the validity of the thermodynamics relations. The physical entropy for finite dimensional heat baths turns out to be a version of the generalized entropy of Tsallis. In the thermodynamics limit he recovers the traditional Boltzmann–Gibbs statistics. Furthermore, he shows that the classical thermodynamics relations are independent of the explicit form of the canonical distribution. Finally, he discusses the issue of additivity of the physical entropy and presents theoretical and numerical examples of Hamiltonian systems obeying the generalized statistics.

The recent development of the statistical mechanics based on the non-Gibbsian entropy, or Tsallis's nonextensive statistical mechanics has intensified interests in search for a possible extension of Shannon information theory. This is mainly due to the similarity of the form of the entropy function involved. One can develop the non-Shannon (but includes Shannon's as a special case) information theory to some extent from a formal point of view. As a possible generalization of Shannon's information theory, Yamano (2001) reviews the formalism based on the nonlogarithmic information content parametrized by a real number q, which exhibits nonadditivity of the associated uncertainty. Importance of the generalization resulting from the establishment of the mutual information concept is discussed.

Antoniouk et al. (2014) introduce a notion of generalized stochastic flows on manifolds that extends to the viscous case the one defined by Brenier for perfect fluids. Their kinetic energy extends the classical kinetic energy to Brownian flows, defined as the L^2 norm of their drift. They prove that there exists a generalized flow which ensures the infimum of the kinetic energy among all generalized flows with prescribed initial and final configuration. They also construct generalized flows with prescribed drift and kinetic energy smaller than the L^2 norm of the drift. The results are actually presented for general Lq norms, thus including not only the Navier–Stokes equations but also other equations such as those the porous media.

Assaf (2010) investigates in his PhD thesis the theory of large fluctuations in stochastic populations. Large deviation becomes of importance when the discrete (discontinuous) nature of a population of "particles" (molecules, bacteria, cells, animals, or even humans) drives it to extinction. A standard way of taking the discreteness of particles into account is the master equation which describes the evolution of the probability of having a certain number of particles of each type at time t. The master equation can seldom be solved analytically, hence various approximations are used. One is the Fokker-Planck equation which usually gives accurate results in the regions around the peaks of the probability distribution, but fails in its description of large deviations—that is, the distribution tails. Not much is known beyond the Fokker–Planck regime. In some cases, as for single-step birth–death processes, complete statistics, including large deviations, were determined from various approximations made directly to the pertinent master equation. A different group of approaches employs the generating function formalism. In Assaf's (2010) work the master equation is transformed into a linear partial differential equation for the generating function, and this PDE is analyzed and solved by various techniques such as the method of second quantization or the more recent time-dependent WKB approximation. Assaf (2010) applies the spectral method to the paradigmatic problem of branching A + X→2X and annihilation X + X→E, where A and E are fixed. This multistep single-species birth–death process describes, for example, chemical oxidation reactions. If the branching rate is much higher than the annihilation rate a long-lived metastable, or quasistationary, state exists where the two processes almost balance each other. Still, this long-lived state slowly decays with time because a sufficiently large fluctuation ultimately brings the system into the absorbing state of no particles from which there is a zero probability of exiting. This work dealt with extinction of an isolated long-lived stochastic population describable by a continuous-time Markov process. In his impressive work Assaf (2010) has identified two extinction scenarios, A and B, based on the stability properties of the fixed points of the rate equation for population size dynamics. He supplemented the WKB theory by two additional approximations in the regions where it breaks down.

4.6.2 Diffusion

One of the great paradoxes of statistical physics and diffusion theory has been the so-called Loschmidt's paradox—how does irreversible behavior emerges from a particle system that obey reversible equations of motion? In a paper on irreversibility, diffusion and multifractal measures Dettmann et al. (1997) show how the irreversibility arises from reversible microdynamics in a number of diverse and important cases (Lorentz gas in an external field, color diffusion and planar Couette flow in N particle system). Macroscopic irreversibility is not only shown to be consistent with reversible equations of motion, but follows directly without any ad hoc assumptions. Argument in which time-reversed motions have much lower probabilities than the original motion is not necessarily an extra postulate, as it might emerge as a natural consequence of some reversible microscopic dynamics. The irreversibility examples are limited to models containing thermostats.

Diffusion in disordered systems such as random fractal structure does not follow the classical laws which describe transport in ordered crystalline medium, and this leads to many anomalous physical properties. Some typical examples include transport properties in fractured, porous rocks and the anomalous density of states in randomly diluted magnetic systems. The problem of diffusion in disordered media is a part of the general problem of the transport in disordered media. El-Wakil and Zahran (2000) apply two different classes of fractional differential operators, Liouville-Riemann and Nishimoto, to represent the fractal differential operators in time. Their main purpose is to introduce the fractional Fokker–Planck equation in which the time derivative is replaced with a derivative of fractional order. They obtain a fractional Fokker–Planck equation in the domain of fractal time evolution with a critical exponent α, where $0 < \alpha \leq 1$. This fractional equation can be used to describe the anomalous behavior of a transport process through the fractal medium (El-Wakil and Zahran, 2000; El-Wakil et al., 2001). By applying the technique of eigenfunction expansion to get the solution of the fractional Fokker–Planck equation and some further results, they are finally led to the relation between fractal time evolution and the theory of continuous tine random walk. From the structure of maximum entropy, the structure of the distribution function follows. See also El-Nabulsi (2011) work on the fractional Boltzmann transport equation.

Giona (1996) discusses the renormalization of transport equations and the exact solution of linear and nonlinear transport equations on fractals. He also presents exact results for sorption on fractals and for first-order reaction-diffusion kinetics. Artuso (1997) reviews how anomalous diffusion for classical deterministic dynamical systems is handled within the framework of cycle expansions. Using zeta function, he considers how marginally stable orbits influence deterministic diffusion. He remarks that within this framework normal and anomalous diffusion are dealt with in the same context: the essential role is played by the analytic properties of the zeta function near its first zero. When long time tails are present, probabilistic approximations for the zeta function work well, even for nontrivial examples.

Mainardi (1996) presents an analytical treatment of fractional relaxation–oscillation and fractional diffusion-wave phenomena. After giving essentials of the Riemann–Liouville fractional calculus, he adopts the definition of the fractional derivative in terms of Euler gamma function

$$D_b^\alpha f(t) := D^n I^{n-\alpha} f(t) = \frac{1}{\Gamma(n-\alpha)} \int_0^t \frac{f^{(n)}(\tau)}{(t-\tau)^{\alpha+1-n}} \tag{4.66}$$

Then he considers the fractional relaxation–oscillation equation

$$\frac{d^\alpha u}{dt^\alpha} + \omega^\alpha u(t;\alpha) = 0, \tag{4.67}$$

where ω is a positive constant with the dimensions of a frequency, and the field variable $u(t;\alpha)$ is assumed to be a function of time. According to the cases

$$0 < \alpha \leq 1 \quad \text{and} \quad 1 < \alpha \leq 2 \tag{4.68}$$

Eq. (4.67) can be referred to as the fractional relaxation or the fractional oscillation equation, respectively.

The formalism also leads to the fractional diffusion-wave equation

$$\frac{\partial^{2\beta} u}{dt^{2\beta}} = D \frac{\partial^2 u}{\partial x^2} \qquad (4.69)$$

where D is a positive constant and the field variable $u(t; \alpha)$ is assumed to be causal function of time. We have to distinguish two cases, namely

$$0 < \beta \leq 1/2 \quad \text{and} \quad 1/2 < \beta \leq 1 \qquad (4.70)$$

for which the previous fractional diffusion-wave equation can be referred to as the fractional diffusion or the fractional wave-propagation equation, respectively. The solution of these equations can be obtained by applying the Laplace transform. Mainardi (1996) presents relevant examples along with the discussion of some qualitative properties of solutions.

Chaos is the topic which has attracted interest in many fields but its most essential applications have to do with predictability. Long-term prediction remains an unrealized goal in many contexts from weather forecasing and population biology to stock market forecasting. The aim of a research is to explore the mechanisms that lead to failure in predictability associated with dynamical processes and to consider what may be done about it. Gaspard's (1998) book titled *Chaos, Scattering and Statistical Mechanics* describes recent advances in the application of chaos theory to classical scattering and disequilibrium statistical mechanics generally, and to transport by deterministic diffusion in particular. The author presents the basic tools of dynamical systems theory, such as dynamical instability, topological analysis, periodic-orbit methods, Liouvillian dynamics, dynamical randomness and large-deviation formalism. These tools are applied to chaotic scattering and to transport in systems near equilibrium and maintained out of equilibrium. The book is useful for researchers interested in chaos, dynamical systems, chaotic scattering and statistical mechanics in theoretical, computational and mathematical physics and also in theoretical chemistry. See also Gaspard et al. (2001) on fractality of the hydrodynamic modes of diffusion.

Hwa and Kardar (1989) study fractal structures arising from dissipative transport in open system. To describe fluctuation around of a steady state in a flowing "sandpile" they construct a simple continuum equation. The authors explicitly demonstrate how self-organized criticality emerges in their example from height conservation. Formation of fractals is thus understood in terms of a conservation law in dynamics. A dynamic renormalization group calculation allows to determine fractal dimensions and various critical exponents exactly in all dimensions. They conclude by describing other processes and universality classes exhibiting "self-organized criticality." Different university classes are shown to depend on possible choice of current and on the spectrum of applied noise.

Lapasin et al. (1996) pursue a fractal approach to rheological modeling of aggregates suspensions. They propose a rheological model describing the dependence of the shear viscosity on the shear stress and the disperse phase volume fraction. When aggregate structures are sufficiently large and self-similar, they can be appropriately described in terms of fractal dimensionality D, whose value is a measure of the compactness of the aggregate structure. Flow curves calculated from the model, both for dense and fractal aggregates, closely resemble those observed for real colloidal and noncolloidal suspensions. The experimental data of some real systems, such as high solids coatings and mill base suspensions, are examined in the light of the model. Kantor (1989) compares two complementary (energetic and entropic) elasticity models which exhibit a power law scaling of the elastic constants.

Pfeifer and Avnir (1983a) laid the foundations for appreciation of chemical surfaces as D-dimensional objects where $2 \leq D < 3$. The concept of fractal dimension D of surfaces, advanced as natural measure of surface irregularity in Pfeifer and Avnir (1983a), is shown by Pfeifer and Avnir (1983b), to apply to a remarkable variety of adsorbents: graphites, fume silica, faujasite, crushed glass, charcoals, and silicagel. See more in chapter: Power Limits in Thermochemical Units, Fuel Cells, and Separation Systems.

West and Deering (1994) review fractal physiology for physicists in the context of Lévy statistics. According to their opinion biomedical applications of fractal concepts have led to a wealth of new insights in biology and physiology, including a new formulation of the concept of health. They review how over the past decade the complexity inherent in physiological structures and processes has been described by random fractals. In particular they argue that the scaling observed in these biological data sets is a consequence of the Lévy statistics of the underlying processes. This review interleaves arguments from renormalization group theory, fractal scaling and Lévy stable distributions in various physiological contexts including the mammalian lung, the beating of the heart, DNA sequencing, the dendritic branching of neurons and blood vessels, the dynamics of proteins, ion channel gating and radioactive clearance curves from the body in order to reveal an underlying unity to physiological processes. Wiebel promotes the use of fractal geometry as a design principle for living organisms. However West and Deering (1994) suggest that Lévy stable statistics in being more inclusive, may be an even more useful, design principle.

CHAPTER 5

Thermodynamic Lapunov Functions and Stability

5.1 Introduction

Stability problems stretch from equilibrium to highly disequilibrium situations. This chapter considers the stability for two kinds of problems.

The first kind deals with the stability of singular points and behavior of surrounding paths. The problem is considered in the context of two-phase flow systems occupied by fluids with dispersed solids. Simultaneous heat and mass exchange is the typical process between the gaseous and the solid phase. In this part we focus on qualitative properties and stability of process paths as subsumed by Lapunov functions derived from constant-sign rates of dissipated exergy rather than from thermodynamic potentials (whose sign properties are in general indefinite). In principle this part considers stability problems in both chemically reacting and chemically nonreacting systems. Our special interest are gas–solid processes in dryers, adsorbers, catalytic reactors and regenerators. Typical systems include batch fluidization, steady continuous fluidization, fluidizing horizontal exchanger, countercurrent exchangers (continuous and multistage), and cocurrent continuous exchanger, Fig. 5.1.

The second kind of problems deals with the stability of fields with disequilibrium fluids. Some attention is also given to associated problems, such as: fluctuations, pattern formation, turbulence, and so forth. It is possible to link the path problems and their stability with spatiotemporal patterns in systems driven away from equilibrium. Cross and Hohenberg (1993) give a review of pattern formation in systems outside of equilibrium. Books by Chandrasekhar (1961); Joseph (1976); Glansdorff and Prigogine (1971); Denn (1982), and others, treat stability of distributed steady states with chemical reactions and transport processes. Assuming the reader's familiarity with basic mathematical issues such as formulation of partial differential equations, perturbed forms, and so forth, we limit ourselves to several examples of corresponding eigenvalue problems and brief discussions of solution properties. While our review of these issues here must necessarily be scarce, we hope to give the reader at least beginning of the information about the reach field, encouraging him to study the suitable literature.

Stability problems are described in many books, reviews and research papers. Many can be found in books of theoretical physics; they arise in problems of discrete and continuum mechanics (Landau and Lifshitz, 1974a, 1976, 1987), electrodynamics (Landau and

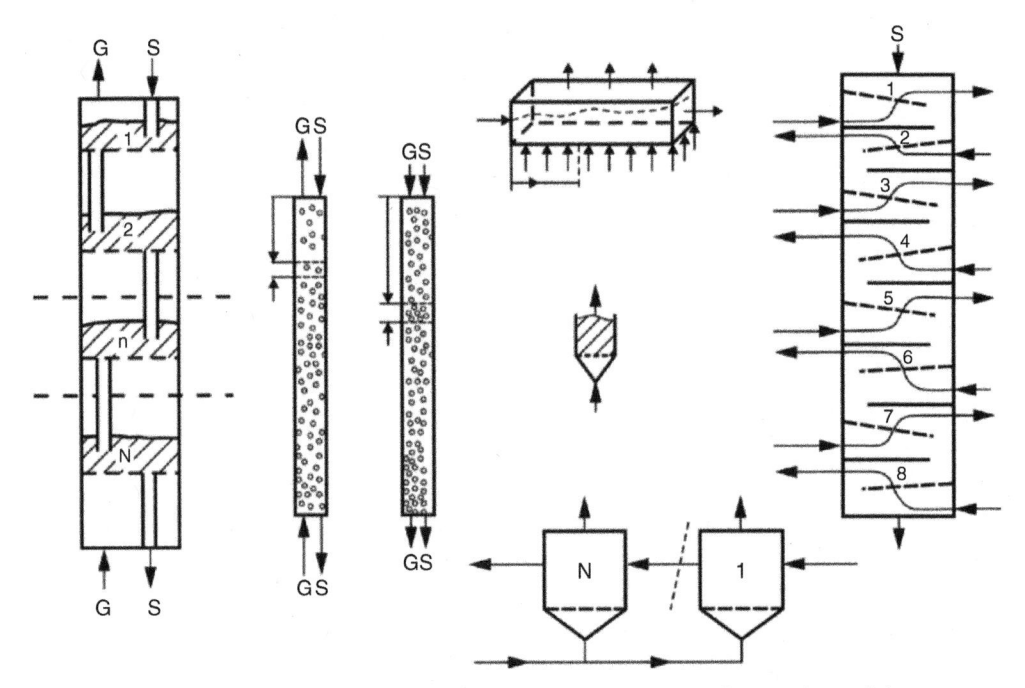

Figure 5.1: Various drying processes between gas and granular solid.
From left to right: multistage countercurrent drying; countercurrent and cocurrent drying of particles freely falling in air; fluidized drying in a horizontal exchanger; batch fluidized drying; and two cases of multistage cross-current drying.

Lifshitz, 1967), theory of fields (Landau and Lifshitz, 1974b), statistical physics (Landau and Lifshitz, 1975), physical kinetics (Lifshitz and Pitajevsky, 1984), and others. Reviews of results obtained in various research groups are available. Associated research fields grow, and the new information is gained for neighbors systems. For example, Asegehegn et al. (2011) review simulation of dense gas–solid multiphase flows. Brauner (1996) reviews the role of interfacial shear modeling in stability of stratified two-phase flows. Using classical and nonequilibrium approaches researchers underline the link between complex behavior and stability of systems. Demirel (2005) treats the stability of equilibrium and disequilibrium systems of transport and rate processes with some case studies. Haltchev and Denier (2003) and Demirel (2005) formulate an eigenvalue problem for stability of boundary-layer fluxes with diffusion and interfacial coupling. In a porous medium, instabilities are related with the temperature dependence of the viscosity, thus with the pressure drop for Darcy flow in the porous combustor. Captained (2003); Albers (2003) and Sureshkumar (2001) apply linear stability analyses with perturbation equations to test mechanisms of thermal instabilities.

Both diffusion and chemical instabilities are driven by the differences of chemical potentials. Baierlein (2001) offers some qualitative understanding of the chemical potential, a topic that the readers frequently find difficult. Three "meanings" for the chemical potential are stated and then supported by analytical development. Two substantial applications—depression of the melting point and batteries—illustrate the chemical potential in action.

The prediction of equilibrium in liquid phases is of great importance in the chemical and process industries. Equilibrium problems of thermodynamic stability can be treated by using thermodynamic potentials. When liquids of different chemical structure are mixed they either form an homogeneous solution or they separate into two phases. This partial solubility is in general a complex function of temperature and composition. In the equilibrium thermodynamics relations which enable one to predict phase stability and critical mixing points are of basic importance. Wisniak (1983) reviews classical thermodynamic conditions for phase stability using concept of virtual displacements. Simultaneous phenomena of critical mixing and azeotropy are discussed. This author also analyzes the ability of thermodynamic models to predict critical mixing. Hua et al. (1996) stress a key step in calculations of phase equilibria, that is, determining if, in fact, multiple phases appear. These authors present a general computational method for the van der Waals equation of state to determine stability for many systems. In fact, they describe and test interval methods for phase stability in binary mixtures. They show that all singular points can be found with a great certainty, and thus the phase stability problem can be solved exactly.

Dynamical approaches formulate stability problems in terms of differential equations for perturbations around a singular point (or a singular subspace) and Lapunov functions (functionals), V. Solution $X(t)$ of the differential equation is called stable in the sense of Lapunov for $t_0 \leq t < +\infty$ if for arbitrary $\varepsilon > 0$ exists $\delta = \delta(\varepsilon, t_0) > 0$ such that for arbitrary solution $Y(t)$ the inequality $\|Y(t_0) - X(t_0)\| < \delta$ implies the inequality $\|Y(t) - X(t)\| < \varepsilon$ for $t_0 \leq t < +\infty$. Roughly this means that two solutions do not diverge with time.

Let us consider a perturbed solution around a singular point $\mathbf{x}_s$, that is, $\mathbf{a}(t) = \mathbf{x} - \mathbf{x}_s$. Symbol $\mathbf{x}$ refers to an actual state which may be regarded as that resulting from a perturbation, whereas $\mathbf{x}_s$ pertains to a singular point, which may be an equilibrium point or a steady state point. Whenever a function $V(\mathbf{a})$ exists which satisfies the conditions $V(\mathbf{0}) = 0$ and $V(\mathbf{a}) > 0$ for $a \neq 0$, and the sign of its total time derivative with respect to the perturbed equations $V_a(\mathbf{a}) \dot{\mathbf{a}}(\mathbf{a})$ is opposite to the sign of the function V, the perturbed solution $\mathbf{a}(t) = \mathbf{x}(t) - \mathbf{x}_s$ is stable in the Lapunov's sense. This is, in brief, the content of the second Lapunov's theorem. In the spatially distributed systems the aforementioned statements are generalized to Lapunov's functionals.

A didactical outline of stability analyzes by Lapunov's direct method is given by La Salle and Lefschetz (1961) who present a readable account of the stability problems treated by Lapunov's direct method. Smith (1966) formulates mathematical principles of matrix calculations for Lapunov quadratic forms. Sancho (1999) analyzes the relation between local and global thermodynamic stability. His examples show the existence of dynamic structures in the considered system, and illustrate how the global stability can be understood in terms of the structures existent in the local dynamics. The dynamics of the stability are generally nonlinear. Sancho states that, probably, in more complex systems, it will also be possible to derive the global behavior of the stability in terms of the local dynamics. For associated considerations, see Lavenda and Santamato (1982).

Keizer (1987a) stresses the versatility of the second Lapunov's theorem as not restricted to using the Euclidean distance. However, he limits his considerations to positive and quadratic V. He points out that the symmetry and positiveness of quadratic V imply the existence of an orthogonal transformation for the perturbation coordinates $\mathbf{a} = \mathbf{x} - \mathbf{x}_s$, which diagonalizes the quadratic form V. The transformation, called the principal axes transformation, rotates the coordinate axes of the elliptical surface $V =$ constant. As a result of this transformation one obtains the function V expressed as the sum of products of new (transformed) coordinates $\mathbf{a}'$ and corresponding eigenvalues λ_i. A system is stable if all eigenvalues are negative. Diagonalization of V and axis transformation are, however, unnecessary; the original system is sufficient to test stability or instability.

When the singular point coincides with the thermodynamic equilibrium point, the Lapunov function of an isolated system can be represented by the excess entropy, whereas its time derivative by the rate of irreversible entropy production. Considering an equilibrium thermodynamic system, Ishida and Matsumoto (1975) and Tarbell (1982), apply the negative excess entropy $W = S^e - S$ as a Lapunov functional in which the entropy density ρ_s is the same function of the thermodynamic state as prescribed by classical thermodynamics (local equilibrium assumption). Invoking the local form of the second law $\sigma > 0$ they obtained in the gradient-less case the conditions $dW/dt < 0$ which proves the stability of all system trajectories approaching the equilibrium singular point. Thus, classical thermodynamics predicts the global stability of the thermodynamic equilibrium state.

Disequilibrium systems provide much more possibilities for instabilities than equilibrium ones. Inhomogeneous boundary conditions imposed on the system set its distance from the equilibrium in terms of the magnitude of a gradient or other thermodynamic force, and influence the type of the stability or instability. After achieving certain critical distance on the thermodynamic branch, systems may enhance perturbations, and evolve to organized states, often called the dissipative structures (Glansdorff and Prigogine, 1971). While the coefficients of transport and rate processes are the effect of short-range interactions, physical and chemical instabilities may lead to long-range order and coherent time behavior, such as a chemical clock, known as the Hopf bifurcation (Demirel, 2005). Stability analyses for linear and nonlinear modes in stationary systems help understand the formation of organized structures. For oscillating reactions in two variable systems organized structures were studied by Ishida and Matsumoto (1975) with emphasis on the direction of rotation of the process trajectory.

Stability analyses of disequilibrium fields refer to the complicated cases in natural systems. More than 60 years ago Turing's treatment of instabilities demonstrated that even simple reaction–diffusion systems might lead to spatial order and differentiation, while the Rayleigh-Bénard instability showed that the maintenance of disequilibrium might be the source of order in fluids subjected to a thermodynamic force above a critical value (Demirel, 2005). There are many examples of problems extending the earlier two to the different situations. Ross and Vlad (1999) review progress in the nonlinear chemical kinetics and complex reaction mechanisms. Ross (2008) synthesizes the results of his research group on thermodynamic fluctuations far from equilibrium. Of special importance are their classification of chemical oscillators and formulation of kinetic schemes by deduction from experiments.

The distance from equilibrium measured in terms of an imposed thermodynamic force constitutes a stability constraint from one side and the organizational stimulus from another side. Systems may enhance perturbations and evolve to organized states called dissipative structures after crossing a critical distance on the thermodynamic branch. While kinetic and transport coefficients summarize short-range interactions, instabilities may lead to long-range order and coherent time behavior as in Hopf bifurcation. Stability analyses for stationary systems are useful in understanding the formation of organized structures.

Cross and Hohenberg (1993) give a comprehensive review of pattern formation in systems outside of equilibrium. While the scope of their paper is broader than the scope of this chapter, they practically omit gas–solid systems, which are of main interest to us at least in a part of the present chapter. Nonetheless there is an important property that links their treatment with ours: the emphasis on the link between theory and quantitative experiments. Yet, their examples are different than ours; they treat mainly single phase systems. They include hydrodynamic patterns such as thermal convection in fluids and mixtures, Taylor–Couette flow, wave instabilities, as well as patterns in interphase fronts, oscillatory reactions and excitable biological media. The theoretical starting point is a set of deterministic equations of motion. The theory has to describe solutions of the deterministic equations starting from typical initial conditions and persisting long times. For linear instabilities of a homogeneous state the theory leads to a classification of patterns in terms of the characteristic wave vector and frequency of the instability.

In the stability problem of a singular point our interest are qualitative properties and behavior of process paths. However, an effective description of gas–solid or gas-catalyst interfaces requires transfer coefficients or resistances for mass and heat transfer. In the disequilibrium thermodynamics equilibrium profiles are sufficient for determining interfacial resistances. Working in D. Bedeaux research group, Johannessen et al. (2011) applied the density functional theory to find equilibrium profiles of vapor–liquid interfaces. The classical density functional theory (DFT) is based on a theorem stating, in the realm of classical fluids that for a system in equilibrium at given temperature T, volume V, and chemical potentials μ_i, an external potential uniquely relates to a distribution in single particle density (Evans, 1992). On this basis integral relations were developed for resistances. In fact, Johannessen et al. (2011) related the interfacial resistances (for heat transfer, mass transfer, and for coupled heat and mass transfer) to only one local thermal resistivity. All data were described in their temperature behavior. These data may be useful in path stability problem which is considered in detail in Sections 5.2–5.5.

5.2 Qualitative Properties of Paths Around Equilibrium and Pseudoequilibrium Points

Ordinary differential equations are sufficient to determine the stability and quantitative properties of paths when a process is described by a lumped parameter model. The paths are then analyzed around a singular point which is characterized by the vanishing process rate. The thermodynamic nature of the singular point pertains to equilibrium, quasiequilibrium or disequilibrium state. The approach discussed here deals with those flow systems in which the singular state represents the thermodynamic pseudoequilibrium between two phases (the name "pseudoequilibrium" state is attributed to a small amount of the dissipated heat caused by the slip between two flowing phases).

We shall exemplify the approach on example of selected systems of grain drying and sorption (Fig. 5.1). If trajectories of the considered process are stable, then a slight change in their initial conditions will result in small diverging of trajectories in time. In an unstable situation, a minor change in their initial conditions causes that these variations will grow indefinitely with the residence or holdup time, t. Because of conservation laws, the nature of the processes investigated is such that the original set of their equations, of certain dimensionality, say r, is constrained to a manifold of dimensionality $k = r - n$ where n is the dimensionality of a hyperspace on which reside the trajectories attributed to a definite singular point.

The singular point may be achieved only via a subclass of trajectories characterized by the prescribed values of the process invariants, the components of a vector $\mathbf{C} = (C_{n+1}, C_{n+2}, \ldots C_r)$. For steady cocurrent and countercurrent flow processes these invariants represent the conserved energy flux and mass flux of the two-phase flow. The state vector, composed of irreducible coordinates, is $\mathbf{x} = (x_1, x_2, \ldots x_n)$. The number of the invariants is $k = r - n$, that is, equals the dimensionality of the equilibrium (pseudoequilibrium) surface. When $k = r - n$ invariants are present, the number of nontrivial coordinates can be reduced to n. Consequently, the set of the process equations can be written in the following vector form

$$\frac{d\mathbf{C}}{d\tau} = 0 \quad (\dim \mathbf{C} = k) \tag{5.1}$$

$$\frac{d\mathbf{x}}{d\tau} = \mathbf{g}(\mathbf{x}, \mathbf{C}) \quad (\dim \mathbf{x} = n) \tag{5.2}$$

The second set describes the irreducible kinetic equations. Assumption of definite values for parameters $C_{n+1}, \ldots, C_r$ in these equations isolates from the family of the original set a subclass attributed to the unique singular point $\mathbf{x} = \mathbf{x}^0$ around which the trajectory behavior is examined. The location of a singular point is found by prescribing $C_{n+1}, \ldots, C_r$ at the level $C_{n+1}^0 \ldots C_r^0$ and by solving the set of irreducible equations for zero rates. Lapunov's stability theorem and Chetaev's instability theorem are then applied around $\mathbf{x} = \mathbf{x}^0$. Lapunov function

$V(\mathbf{x})$ is a potential whose sign in a stable process is opposite to the sign of its total time derivative with respect to the kinetic equations (5.2), $V_x(\mathbf{x})\mathbf{g}(\mathbf{x}, \mathbf{C})$. This difference of signs is the sufficient condition that the trajectories approach the singular point for $t => +\infty$.

Using thermodynamic approaches to Lapunov functions, Sieniutycz and Komorowska-Kulik (1974, 1975, 1978) studied qualitative properties of trajectories of simultaneous heat and mass transfer processes in various gas–solid flow systems. They applied the second method of Lapunov in which the Lapunov functions had forms of excess thermodynamic potentials with respect to an equlibrium or pseudoequilibrium singular point. Eqs. (5.1) and (5.2) and Lapunov criteria were used to study qualitative properties of the cocurrent, countercurrent and crosscurrent gas–solid systems shown in Fig. 5.1. In these systems only the cocurrent case always exhibits stable trajectories achieving the focus-type equilibrium point, whereas other cases show unstable behavior in relation to the singular point. As shown by Fig. 5.2, trajectories of the countercurrent case are unstable around the pseudoequilibrium singular point (calculated for the constant energy of two-phase flux).

The considered problem defines the qualitative behavior of *steady paths* around a singular point $\mathbf{x} = \mathbf{x}^0$, at which the process rates vanish. This *path-analyzing approach* has nothing in common with the stability problem stated for *steady thermal fields* which are described by PD equations for energy and mass balances [Eqs. (5.4)–(5.9)].

Note also the path-analyzing approach of Luna and Martinez (1999), which is applied to the stability analysis of multicomponent liquid mixtures. It deals with ordinary differential equations of gas-phase-controlled evaporation during drying and leads to prediction of process paths and final state of a system. Albers (2003) develops a related analysis of relaxation and linear stability versus adsorption in porous media. He presents the stability properties in dependence on the two most important model parameters, namely the bulk and surface permeability coefficients.

Discontinuous contours of constant Lapunov function $V(\text{- - - -})$ describe a saddle surface in the countercurrent case and a convex surface with minimum of V in the cocurrent case. Closed continuous contours of constant derivative $\dot{V}$ describe in both cases surfaces of definite sign in accordance with the second law of thermodynamics. Only in the cocurrent case signs of V and $\dot{V}$ are always opposite (Sieniutycz, 1967).

5.3 Stability of Steady States Close to Equilibrium

The thermodynamic study of nonequilibrium systems requires concepts which go beyond those traditionally applied by equilibrium thermodynamics. One fundamental difference between the equilibrium thermodynamics and nonequilibrium thermodynamics consists in the emergence of inhomogeneous systems that require to study properties of fluxes or rates which are not considered in equilibrium thermodynamics.

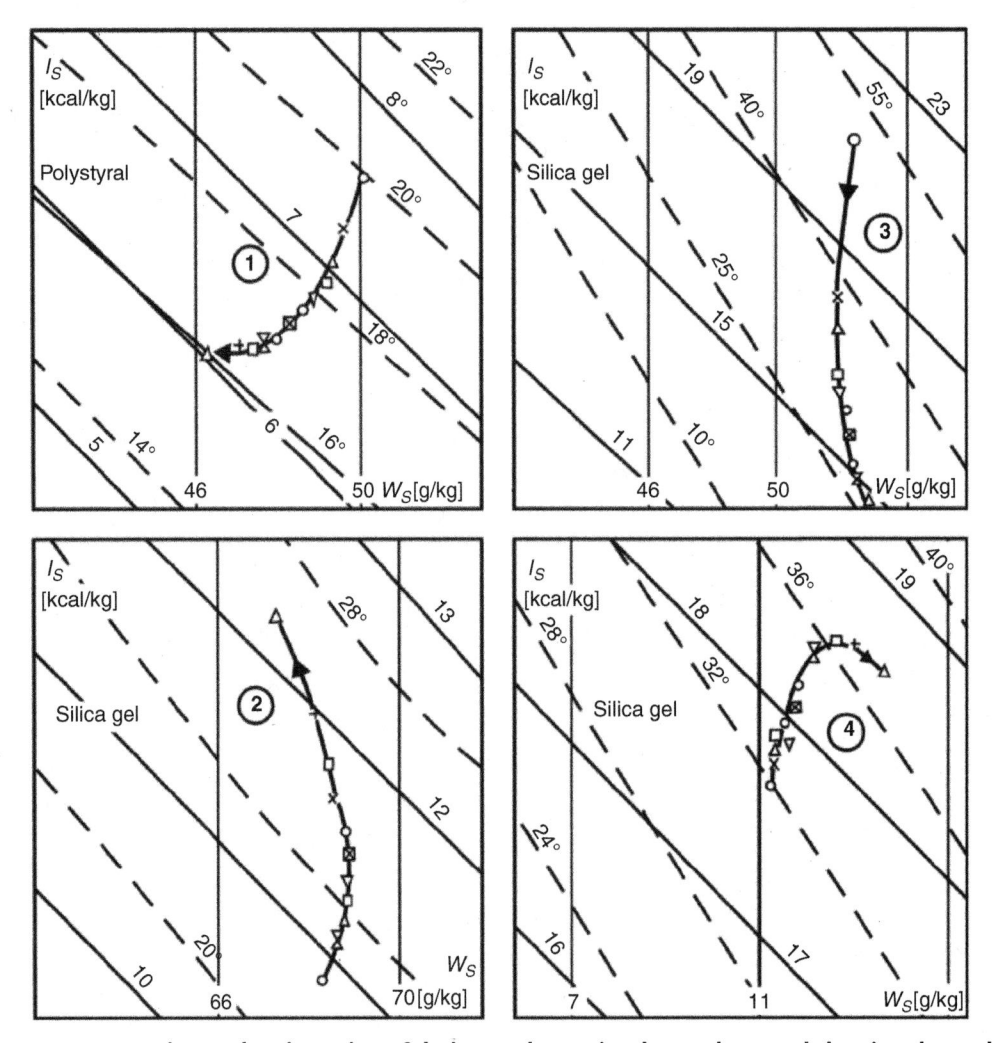

Figure 5.2: Experimental trajectories of drying and sorption located around the singular saddle point in the process of countercurrent contacting of gas and granular solid (Sieniutycz, 1967). The saddle point of trajectories lies in the origin of the coordinate system (*the center point of the graph*).

Thermodynamics of irreversible processes can be included into Lagrange-formalism (Anthony, 1988). Along the line of the related variational formalism, a methodological unification of thermodynamics, continuum mechanics and electromagnetism of matter may be obtained. In his earlier paper (Anthony, 1988) and in Chapter I.2 of a book on variational and extremum principles (Sieniutycz and Farkas, 2005) Anthony reviews the Lagrange formalism for irreversible thermodynamics and the principle of least entropy production. In the classical irreversible thermodynamics the principle is restricted to stationary systems and quite special (linear, or quasilinear) equations.

Anthony claims that, associated with the entropy concept, a more general principle of least entropy production holds within the Lagrange formalism. This entropy is implicitly defined by an inhomogeneous balance equation expressing the second law of thermodynamics. His second law formulation results from the universality of Lagrange formalism and the Lagrangian invariance with respect to a group of common gauge transformations applied to the basic field variables of the system. In Anthony's theory irreversible thermodynamics is included into the framework of Lagrange formalism, which offers a unified method for reversible and irreversible processes. Together with a particular Lie-group of invariance requirements, the relevant set of *observables* of the system is defined by means of Noether's theorem. The observables are implicitly derived from the Lagrangian as density functions, flux density functions, and production rate densities, all entering balance equations (Noether balances). The important quantity is the observable of energy and its conservation associated with time translational invariance. Both laws of thermodynamics are included into the Lagrange-formalism. However, the entropy balance is related to the properties of the second variation rather than the first; it also refers to system's stability and Jacobi sufficient condition of extremum. This role of entropy S differs from that known for perfect fluids, where it is a basic field variable satisfying a sourceless balance. Yet, the different role of the entropy constraint in action principles of irreversible processes seems to be implied by other approaches (Sieniutycz and Farkas, 2005).

Anthony's approach to stability derived from the Hamilton principle, Lagrangian invariance and system symmetries differs substantially from Keizer's (1987a) approach to stability which is based on his extension of covariance matrix [Section 5.5, the text below Eq. (5.13)].

Although the classical (parabolic) and wave equations of change have the same steady state solutions, the stability depends on the transient behavior which is different in both cases. Nonetheless, Glansdorff's and Prigogine's (1971) work implied that the stability of the local equilibrium transients with respect to small perturbations described by parabolic or hyperbolic equations can be proved by considering the second differential of entropy. The criteria based on the second differential of entropy were used to show the stability of disequilibrium transients (Lebon and Casas-Vazquez, 1976; Sieniutycz, 1981c; Sieniutycz and Berry, 1991). Equations of hydrodynamics were included in the last reference cited. Taking into account that, in the wave case, the entropy expression contains the flux term, one has

$$\delta^2 S = \int_V \rho \left(\delta T^{-1} \delta e + \sum_{i=1}^{n-1} \delta(T^{-1}\mu_i)\delta y_i + \sum_{i=1}^{n-1} \delta\zeta_i . \delta \mathbf{J}_i \right) dV \qquad (5.3)$$

where ξi is the adjoint variable of $\mathbf{J}_i$ in the expression describing perfect differential of entropy.

The aforementioned expression was applied as the Lapunov functional for the disturbances of yk, e, and $\mathbf{J}_i$, where $\mathbf{y}$ is the vector of independent mass concentrations and e is the energy density. The static state vector is $\mathbf{z} = (\mathbf{y}, e)$. Correspondingly, "the overall matrix of fluxes," or the matrix of all independent fluxes, is $\mathbf{J} = (\mathbf{J}_1, \mathbf{J}_2, ...\mathbf{J}_{n-1}, \mathbf{J}_e)$. The vector of transfer potentials whose gradients are classical driving forces is $\mathbf{u} = (u_1, u_2, \ldots u_{n-1}, T^{-1})$, where u_k refer to Planck's chemical potentials of species $k = 1, 2, \ldots n-1$ relative to the potential of nth component, that is, to quantities $\tilde{\mu}_k \equiv T^{-1}(\mu_n - \mu_k)$.

The perturbed conservation equations in the vector-matrix notation for the state $\mathbf{z} = (\mathbf{y}, e)$ are

$$\rho \frac{d\delta \mathbf{z}}{dt} = -\nabla \cdot \delta \mathbf{J}, \tag{5.4}$$

whereas the perturbed phenomenological equations are contained in the formula

$$\delta \mathbf{J} + \boldsymbol{\tau} \frac{d\delta \mathbf{J}}{dt} = \mathbf{L}(\nabla \delta \mathbf{u}) \tag{5.5}$$

where the relaxation matrix is defined as $\boldsymbol{\tau} \equiv -\rho^{-1} c_0^{-2} \mathbf{L} \mathbf{C}^{-1}$. Since the classical matrix of diffusion, $\mathbf{D}$, is linked with Onsager's matrix $\mathbf{L}$ and entropy capacitance $\mathbf{C}$ by the expression $\mathbf{D} = -\rho^{-1} \mathbf{L} \mathbf{C}^{-1}$ (De Groot and Mazur, 1984), the definition of $\boldsymbol{\tau}$ in Eq. (5.5) satisfies the formula $\boldsymbol{\tau} \equiv c_0^{-2} \mathbf{D}$, that is, constitutes a matrix generalization of the well-known formula linking the scalar relaxation time with the scalar of heat diffusivity in processes of pure heat transfer, $\tau = c_0^{-2} D$ (Sieniutycz, 1981c).

Applying the previous formulae in Eq. (5.3) it was found that if

$$\det \mathbf{C} \leq 0 \tag{5.6}$$

the following holds

$$\delta^2 S \leq 0 \tag{5.7}$$

and

$$\frac{\partial \delta^2 S}{\partial t} \equiv \int_V (\delta \mathbf{J}^T \mathbf{L}^{-1} \delta \mathbf{J}) dV \geq 0 \tag{5.8}$$

for the constant potentials at the boundaries of the system (Sieniutycz, 1981c). This means that the negativeness of the entropy capacity, $\mathbf{C}$, characterizing the stable equilibrium of every macroscopic system, and the positiveness of Onsager's kinetic matrix, $\mathbf{L}$, provide the sufficient stability condition. The second Lapunov's method thus proves that every trajectory of the process of simultaneous heat and mass transfer in an nonreacting fluid will approach the stationary state for t approaching infinity if the well-known thermodynamic matrix of capacities, $\mathbf{C}$, is negative, even if the inertial (ie, relaxation) terms are important.

Currently many generalized wave models exist, which we ignore here. For example, Atanackovic et al. (2007) proposed a generalized wave equation with fractional derivatives of different order.

Using Lapunov's procedure, other stability conditions for the heat conduction problem with finite wave speed were derived. Lebon and Casas-Vazquez (1976) and Lebon and Boukary (1982) constructed various Lapunov's functions in extended irreversible thermodynamics. Classical conditions requiring the negativeness of the entropy Hessian were shown to be sufficient for stability in the simplest case, Sieniutycz (1980a, 1981c, d, 1984a, 1989a, 1994), Lebon and Boukary (1982) and Sieniutycz and Komorowska-Kulik (1978, 1982). In particular, it was shown that the inequality in Eq. (5.8) can be valid only in a certain vicinity of equilibrium. This was consistent with earlier results of Keizer and Fox (1974) who raised doubt concerning the range of validity of the stability criterion for nonequilibrium states proposed by Glansdorff and Prigogine.

Prigogine won the Nobel Prize in Chemistry in 1977 in recognition of his advances in disequilibrium thermodynamics. His book with Glansdorff on stability and fluctuations (Glansdorff and Prigogine, 1971) is still a very inspiring volume. Prigogine is best appreciated for transferring the second law of thermodynamics to systems that are far from equilibrium, and demonstrating that ordered structures could exist under such conditions. Prigogine called these forms "dissipative structures" stressing that they cannot exist independently of their environment. In agreement with the second law, ordered structures disintegrate into disordered ones. Yet, Prigogine claimed that the emergence of dissipative structures allows order to be created from disorder in disequilibrium systems. Prigogine's passion on dispelling the second law, his attention to time's arrow and the uniqueness of living systems in their environments, on complexity and the emergent order inspired many researchers in the fields of physics, chemistry, information, biology, and related sciences. See the attractively titled Landsberg's (1972) review of the Glansdorff's and Prigogine's (1971) book: The fourth law of thermodynamics.

However, by considering the stability of the equilibrium state, Keizer and Fox (1974) concluded that the second differential of the entropy, which is at the heart of the Glansdorff-Prigogine criterion, is likely to be relevant for stability statements closely to equilibrium only. Explicit counterexamples which invalidated generality of $\delta^2 S$ as the Lapunov criterion were presented (Keizer, 1976a; Fox, 1980). Other counterexamples were given: Winkiel (1976); Keizer, (1976b); Lavenda and Santamato (1982); and Sieniutycz and Komorowska-Kulik (1978, 1980, 1982). The final conclusion achieved was that $\delta^2 S$ is allowed to make the stability statements only for steady states close to equilibrium.

Alternatively a generalized excess entropy production functional may be tested as a Lapunov functional (Sieniutycz, 1981a)

$$V = \frac{1}{2}\int_V \left(\sum_i \sum_k L_{ik} \nabla . \delta u_i \nabla . \delta u_k + \sum_k \frac{L_{ik}}{c_0^2} \frac{\partial \delta u_i}{\partial t} \frac{\partial \delta u_k}{\partial t} \right) dV \geq 0 \qquad (5.9)$$

For the transients of the perturbations governed by Eqs. (5.4) and (5.5) it was obtained that

$$\frac{\partial V}{\partial t} = \int_V \left(\sum_i \sum_k \rho C_{ik} \frac{\partial \delta u_i}{\partial t} \frac{\partial \delta u_k}{\partial t} \right) dV \leq 0 \qquad (5.10)$$

which again proves that the negative det **C** is the sufficient condition of stability.

Thus, similarly as in the classical situation of an infinite c_0 it has been concluded that, under definite conditions, trajectories of simultaneous heat and mass transfer processes are always asymptotically stable if they are described by the wave model. Broadly speaking this means that the two arbitrary solutions $u(\mathbf{x}, t)$ and $u_0(\mathbf{x}, t)$ will approach one another despite the essential inertial terms residing in wave equations. This also means that one should not expect principally new results concerning stability of transients around equilibrium (where the inequality **C** < 0 always holds) when wave equations are used instead of classical ones.

5.4 Chemically Reacting Systems, Fluctuations, and Turbulent Flows

Literature for chemical systems is abundant. In a landmark theoretical publication, Turing (1952) showed that, in response to small perturbations, a system of reacting and diffusing species may spontaneously evolve from an initially uniform state to spatially heterogeneous patterns. He investigated a system of two chemicals, with one as an activator and the other an inhibitor. The activator, stimulated and enhanced the yield of the other chemical, which, in turn, depleted or inhibited the activator formation. Turing demonstrated that if the diffusivity of the inhibitor was greater than that of the activator, then diffusion-driven instability could arise. This is counterintuitive, as we usually expect a homogenizing effect of diffusion. In fact, he demonstrated that the diffusion originates instability, leading to spatial pattern where no earlier pattern existed.

Li (1958) studied the classical theory of Thomson, Eastman, and Wagner, devoted to the thermodynamics of nonisothermal systems in the light of the Gibbs formulation of heterogeneous equilibrium. He found that an extension of Gibbs's derivation should include the classical theory of nonisothermal equilibrium as a special case. His hypothesis was that Onsager's reciprocal relations can be obtained from macroscopic considerations, which also give a relation between the second–order phenomenological coefficients. He also introduced the concept of second–order independence, which provides a way of finding the actual processes. Li (1962a) applied Carathéodory's principle of the integrability of Pfaffian differential equation to nonequilibrium systems in order to establish a thermokinetic potential for the steady state. Such a potential can only decrease in time in all natural processes. Examples of viscous flow, heterogeneous heat conduction, heterogeneous diffusion, consecutive reactions, and a radiation heating are used to demonstrate that it is this potential, not the rate of entropy production, which becomes a minimum at the steady state with adjustable variables. Also it was shown how a small perturbation can cause the system to spiral into the steady state, consistent with the kinetic potential.

Another paper by Li (1962b) considers the link between stable steady states and the thermokinetic potential. It applies the mathematical postulate of Carathéodory for the existence of the entropy function. Such postulate is shown to be a consequence of the physical postulate of the existence of a stable equilibrium for a given set of state variables. Similarly, the existence of thermokinetic potential is shown to be a direct consequence of the physical postulate of the existence of a stable steady state, for any given set of nonequilibrium state variables.

In view of great diversity of complex chemistries, Spalding pointed out the thermodynamic similarity between various chemically reacting systems. In his early work, Spalding (1959a) aims at unifying and clarifying a branch of reactor theory in which the phenomena of extinction and stable or unstable burning are essential. The stirred homogeneous reactor, the solid catalytic surface and the solid fuel surface are shown to be governed by the same equations for energy and mass conservation. A relation is derived between the temperature of the reaction region and two dimensionless control parameters, one representing the chemical loading, the other representing heat loss. The well-known phenomena of extinction, stable and unstable burning equilibria, and the existence of an upper permissible limit to the heat loss are then demonstrated.

Spalding's papers revisited the idea of the Reynold's flux and developed conserved property diagrams for the rate-process calculations (Spalding, 1959b, 1961). His model of mass transfer through an interface involves the Reynolds flux g interpreted as the mass transfer coefficient and the ratio $g/\dot{G}$—as the nondimensional Stanton number. As a basic example, he analyzed graphically heat and mass transfer between the gaseous and liquid phases of a binary mixture (Spalding, 1961), and generalized ideas developed therein in the two books. The first book deals with the convective mass transfer (Spalding, 1963) and, second, with the combustion mass transfer (Spalding, 1979). These two books show the effectiveness of his approach in practically all applications involving the kinetic and balance equations of the two-phase systems with multicomponent phases. Some examples related to his approach can be found in the literature.

Exact mathematical conditions of chemical reactor stability were determined quite early; see Bilous and Amundson (1955) and Aris and Amundson (1958). Berger and Perlmutter (1965) generalized the results on stability conditions for well-stirred flow reactors in their work with several Lapunov functions. The benefit of their work was extended region of asymptotic stability. Numerical examples have shown the conservative nature of this analysis after comparing the condition for local stability with the criteria arising from a linearized treatment. Considering stability criteria for chemical reactions Aris (1969) applied the analogy between the stirred tank reactor and the catalyst particle to introduce some techniques of stability analysis. He developed the relationship between various classes of criteria for uniqueness and stability. The methods presented were next used in the analysis of other stability problems. In his readable book, Aris (1975) gave a comprehensive mathematical theory for diffusion and chemical reactions in permeable catalysts. Razón and Schmitz (1987) reviewed chemical multiplicities and instabilities in chemically reacting systems with a special attention to related experiments.

Considering the chemical reaction with the stages

$$\alpha A_i \to \sum_{j=1}^{n} \beta_j A_j \qquad (5.11)$$

Gorban (1984) and Gorban et al. (1986) considered the reaction scheme (5.11), where A_1, $A_2,\ldots, A_n$, are symbols of the substances, α and β_i are stoichiometric coefficients (integers, $\alpha > 0$, $\beta_j \geq 0$, $j = 1,2,\ldots n$), and n is the number of reagents. There is one initial substance in each elementary reaction (5.11), though $a > 1$ is also possible. Linear reaction mechanisms are typical examples. However, nonlinear reaction mechanisms without interaction of the various substances are also not seldom (eg, CO oxidation on Pt). The concentration of A_i (in mol/m^3 for homogeneous reactions and in mol/m^2 for heterogeneous reactions) is denoted by c_i, and the concentration vector by c. The heterogeneous case is especially essential. Aimed to find Lapunov's functions, this work provided a kinetic model implying steady-state uniqueness and global stability in the reaction polyhedron. The kinetic law generalizing Marcelin–de Donder kinetics was applied for a separate stage. Lapunov functions apply for various conditions of the reaction proceeding in closed systems. Absence of damped oscillations was shown as the equilibrium is approached.

Further work (Gorban, 2014) led to the following conclusion: (1) For every reaction mechanism there exists an infinite family of Lapunov functions for reaction kinetics that depend on the equilibrium but not in the rate constants, (2) These functions are produced by the operations of conditional minimization and maximization from the Boltzmann–Gibbs–Shannon (BGS) relative entropy. This author also proposed Lapunov functions for nonclasical systems governed by the Tsallis relative entropy. For more information see Gorban in arXiv and references therein.

Catalytic gas–solid systems provide many attractive examples of spatiotemporal patterns and instabilities. Luss (1997) reviews current understanding and open questions for formation and motion of temperature fronts and patterns in packed-bed reactors and on single catalytic pellets. Luss and his coworkers show numerous applications of dynamical models and stability results in catalytic processes occurring in regeneration of Diesel particulate filters. Khinast et al. (1998) show how loci of singular points can be used to construct maps of parameter regions with qualitatively different steady and dynamic bifurcation diagrams. The maps describe desirable operation regions and show potential stability problems in other regions. Luss's and Sheintuch's (2005) experiments for various oxidation reactions reveal high-T domains whose boundaries are either stationary, oscillating, moving, or rotating. These phenomena are observed on catalytic wires, rings, cylindrical pellets and thin catalytic beds. Such evolutions are different from the classical Turing mechanism which explains many of the pattern formations in R-D systems.

Bedeaux et al. (2006) apply disequilibrium thermodynamics to describe systems of heterogeneous catalysis. To begin with they recall common assumptions that the resistance to mass transfer towards a catalytic surface is localized in a diffusion layer in front of the surface and that the temperature and chemical potentials are continuous, while the coupling of a heat flux to the mass fluxes is negligible. In their opinion all these assumptions are questionable. Applying nonequilibrium thermodynamics they consider how to integrate the coupling between heat and mass fluxes in the description of the film. Furthermore, following Gibbs, they introduce the surface as a separate subsystem where the coupling between the vector heat flux and the scalar reaction rate is allowed and can be significant in heterogeneous catalysis. Disequilibrium thermodynamics of surfaces allows one to find suitable rate equations. The theory of Bedeaux et al. (2006) allows for a complete description of mass and heat transfer through the film and subsequently from the film to the surface where the reaction occurs. In heterogeneous catalysis fast endothermic or exothermic surface reactions may cause steep temperature gradients between a catalyst surface and the media. Ignoring these gradients will lead to incorrect evaluations of many basic properties of catalysts, including their activity, and selectivity. This paper sketches how to systematically set up the complete description, in which the film and the surface "sum up" to one effective surface.

Yang and Epstein (2003) investigated oscillatory Turing's (1952) patterns in reaction–diffusion systems with two-coupled layers. They proposed a model that mimics a R-D system with two coupled layers to study pattern formation. A novel type of oscillatory Turing pattern was formed from interaction between a Turing mode and an oscillatory mode in the two layers. Depending on the wavelength and the structure of the Turing pattern, Yang and Epstein (2003) found twinkling eyes, localized spiral or localized concentric waves, or pinwheels. Experimental oscillatory Turing patterns may be obtained in the chlorite–iodide–malonic acid reaction. Wave instability and standing waves arose by coupling an oscillatory layer with a stationary layer of the BZ reaction.

Horsthemke and Moore (2004) investigated open diffusively coupled reactors from the viewpoint of competition between the Turing instability to steady patterns and the Hopf instability to oscillations. They applied rigorous formal criteria for the appearance of these instabilities in arrays of coupled reactors. Applying a computational software, they found conditions for the Turing and Hopf bifurcations in small arrays of reactors with an inhomogeneous profile of the substrate concentration. Their basic result was the evaluation of a critical concentration profile, above which the Turing instability occurs before the Hopf instability. This result lead to a condition for stationary Turing patterns to be experimentally detected in arrays of coupled reactors.

Keeping in mind huge material related to stability of chemically reacting systems, our discussion needs to accommodate the insufficiently explored field of thermodynamic Lapunov functions. Tarbel (1977) described a thermodynamic framework within which

Lapunov functions may be generated, and constructed a Lapunov function for the continuous stirred tank reactor (CSTR) with a steady state near equilibrium. His Lapunov function resembles the thermodynamic entropy production function, and the asymptotic stability principle is reminiscent of the theorem of minimum entropy production. Yet, the author showed that the theorem of minimum entropy production does not apply to the near equilibrium CSTR steady state due to the presence of convective material exchange between the CSTR and its surroundings. See also Tarbell (1982).

Working in a maximum-entropy-type formalism Bernstein et al. (1993) presented two Lapunov functions each ensuring the unconditional stability and robust performance of a system with a damped natural frequency. Each Lapunov function involves the sum of two matrices, the first being the solution to the so-called maximum-entropy equation and the second being a constant auxiliary portion. Their significant feature is that the guaranteed stability region is independent of the weighting matrix, while the performance bounds are relatively tight if compared to alternative approaches.

Lavenda's doctoral thesis "Kinetic analysis and thermodynamic interpretation of nonequilibrium unstable transitions in open systems," (Lavenda, 1972), showed a number of properties of nonlinear chemical reactions far from equilibrium. Within the thermodynamic branch (which governs an extended law of mass action), the reactions become unstable and make transitions to kinetic branches with entropy production lower than that specific to the thermodynamic branch. This result was initially contested by Prigogine who expected from the theory of hydrodynamic instabilities (like the Rayleigh–Benard instability) a larger entropy production beyond the critical point, arguing that an increase of dissipation should help to maintain emerging spatial structures. Prigogine considered these spatial structures as unstable chemically diffusing systems described by Turing's morphological models. He called them "dissipative structures," and received the Nobel Prize in Chemistry in 1977 for their discovery. Prigogine later acknowledged that transitions to lower states of entropy production were possible since no spatial structural changes were involved, and incorporated Lavenda's work into a chapter of his book coauthored with P. Glansdorff (Glansdorff and Prigogine, 1971). After receiving his doctorate from the Universite Libre de Bruxelles, Lavenda published a note (Lavenda, 1972) criticizing the Glansdorff–Prigogine universal criterion of evolution, which attributes an inequality to a potential expressed as a function only of intensive variables: the forces. He pointed out that no such thermodynamic potential could exist for it would be devoid of all information regarding the size of the system or the number of particles contained in it. The inequality would be a criterion of stability, but, on account of the assumption of local (stable) equilibrium of the components that the system is broken up into, the sum of stable components can hardly become unstable. The note would probably have gone unnoticed, were it not for Landsberg's citation in his Nature review of the Glansdorff–Prigogine book, where he stated that "the occasional lack of lucidity in the book may give rise to some discussion within the next few years" (Landsberg, 1972).

In a quite broad-ranging review that includes many kinetic and thermodynamic aspects of chemical oscillations, Nicolis and Portnow (1973) discuss the issues such as: mathematical backgrounds, impossibility of oscillations in the linear region of irreversible processes, oscillations past an instability, and physical prerequisites for occurrence of chemical oscillations. Also, experimental results are analyzed for homogeneous oscillations, heterogeneous and inhomogeneous oscillations, thermochemical oscillations, electrochemical, and biochemical oscillations. Biochemical oscillations are considered in the diverse contexts of genetic aspects, organism metabolism, mitochondria, membranes, and muscles. Mathematical equations are derived for conservative oscillations, autocatalytic and cross-catalytic models, end product inhibition and/or substrate inhibition, and activation by the product. Physical significance of oscillations is in the context of problems such as: rhythmic activity of nervous system, circadian clocks, and competing populations. Molecular aspects of chemical oscillations are also discussed.

Costa and Trevissoi (1973) investigated thermodynamic stability of chemical reactors. They made reference to the criteria for thermodynamic stability of stationary states far from equilibrium in order to find the global sufficient conditions for the continuous well-stirred tank reactor and for the catalytic particle. They state that when the thermodynamic conditions are obtained in a form which permits their being symmetrically compared with the well-known necessary and sufficient kinetic conditions, it appears that the former are in fact more restrictive, but they may be led back to the kinetic conditions by isolating some particular features of the theory of normal modes.

Taking into account that the inverse process of the dissipation of the macroscopic motion can occur through fluctuation and including the effect macroscopic motion represented by the hydrodynamic velocity and total energy into the formulation of the second differential of entropy, Oono (1976) showed that $\delta^2 z$ of Glansdorff and Prigogine (1971) is, in fact, $\delta^2 s$ itself. However, Keizer and Fox (1974) raised doubt concerning the range of validity of the stability criterion for nonequilibrium states proposed by Glansdorff and Prigogine (1971). For a particular autocatalytic reaction, the stability analysis of Glansdorff and Prigogine and of Eigen and by Katchalsky in their reviews of this problem does not agree with Keizer and Fox analysis, which is based upon exact solution of relevant rate equations. Keizer and Fox (1974) also found disagreement between the analysis based upon the Glansdorff–Prigogine criterion and their own analysis of an example which involves nonequilibrium steady states. They concluded that the second differential of the entropy, which is at the heart of the Glansdorff–Prigogine criterion, is likely to be relevant for stability questions close to equilibrium only. Conclusions regarding stability may also be changed due to different recognition of relations between fluxes and forces. Analyzing nonlocal relations between thermodynamic fluxes and forces in continuous systems, Aono (1975) showed that the nonlocal vector flux couples with the scalar force even in an isotropic system. This has application to active transport in living organisms and to thermonuclear fusion research (Fig. 5.3).

Yang (1974) has derived a simple binary model from a proposed kinetic scheme for the oxidation of carbon monoxide. The model, incorporating a free radical mechanism, has

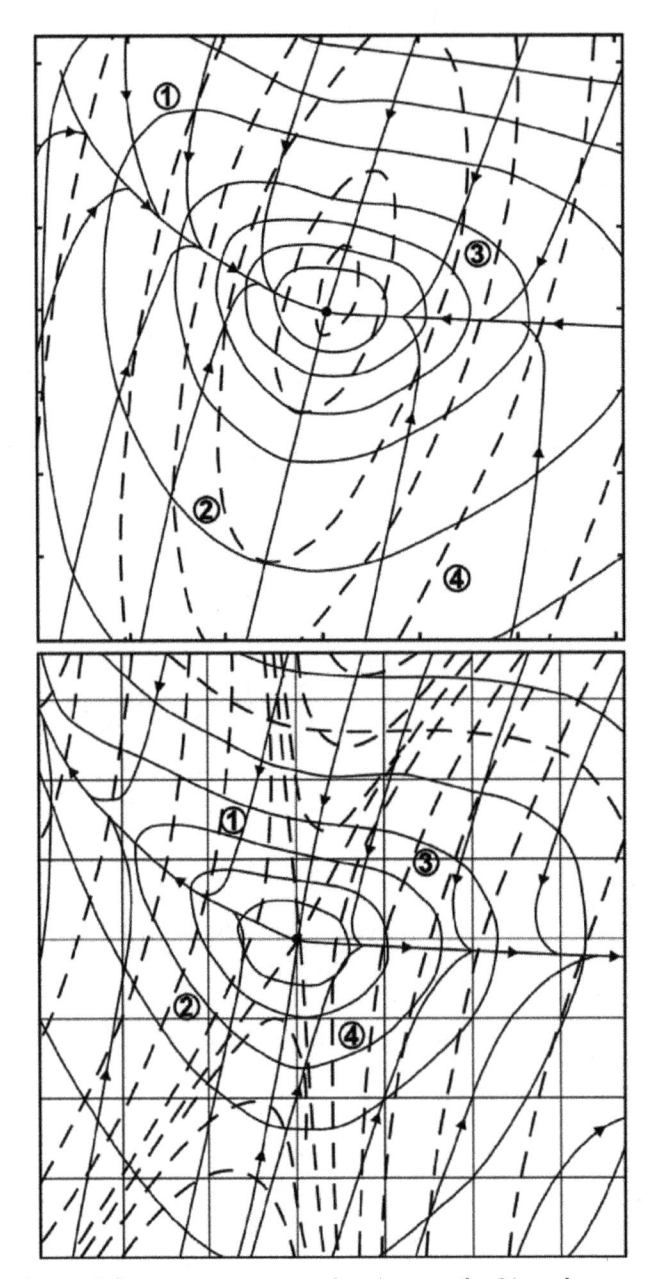

Figure 5.3: Comparison of the cocurrent operation (upper draft) and countercurrent operation (lower draft) between gas and granular solid (Sieniutycz, 1967).

In the state space of temperature T and moisture concentration W, process trajectories, that is, *continuous lines with arrows*, approach, in the cocurrent case, the singular point of focus type, whereas, in the countercurrent case, they leave the singular point of the saddle type along two separatrices. Discontinuous contours of constant Lapunov function V (*dashed lines*) describe a saddle surface in the countercurrent case and a convex surface with minimum of V in the cocurrent case. Closed continuous contours of constant derivative $\dot{V}$ describe in both cases surfaces of definite sign in accordance with the second law of thermodynamics. Only in the cocurrent case signs of V and $\dot{V}$ are always opposite (Sieniutycz, 1967).

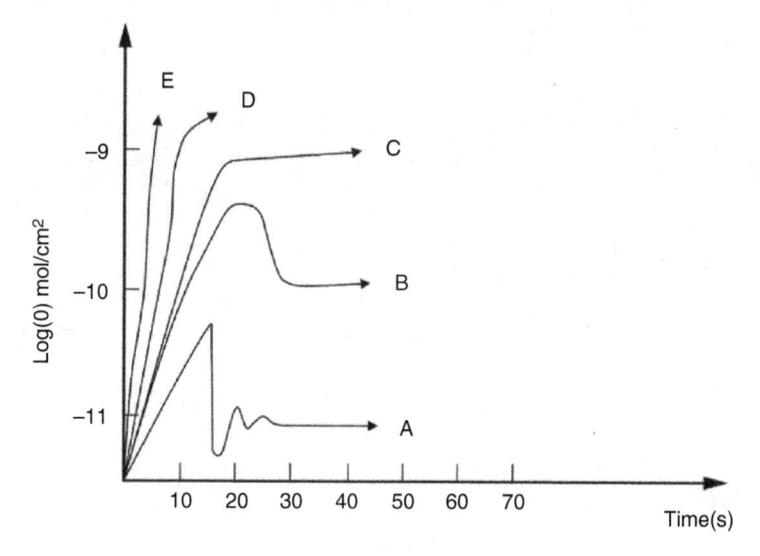

Figure 5.4: Possible trajectories in the process of CO oxidation (Yang, 1974).

been analyzed in the phase plane, where the three distinctive kinetic phenomena, explosion, glow, and oscillation, observed experimentally, were depicted by its solutions, Fig. 5.4. Sieniutycz and Komorowska-Kulik (1980) used Yang's model to derive the dynamical state equations and study thermodynamic criteria of stability of CO oxidation in a fluidized cracking regenerator. Oscillatory effects were similar to those described by Lotka (1920). Dynamical equations, based on the free radical mechanism of Yang (1974), were further generalized to flow systems (Jurkowska, 1982) to show the origin of quenching effects caused by a convective flow of the reacting mixture by the system. Obtained charts showed the stabilizing effect caused by the flow, that is, quenching of instabilities.

Proper understanding of the system behavior during carbon monoxide oxidation is important in many combustion systems and especially in regenerators of catalytic cracking. Sieniutycz and Komorowska-Kulik (1980) investigated the thermodynamic criteria of stability of CO oxidation in a cracking regenerator. They used the dynamical equations that follow from a free radical mechanism proposed by Yang (1974). Ertl (1991) explained the creation of concentration distributions of the adsorbed species on the surface in terms of spatio-temporal patterns. In Yang's equations two basic dynamical variables are the concentration of atomic oxygen $x = [O]$ and that of excited carbon dioxide $[CO_2^*]$. Analyzing this model it may be shown that the second differential of entropy $\delta^2 S$ (a negatively defined function), is not a valid stability criterion in the situation when the singular point is located far from the thermodynamic equilibrium point. In fact, $\delta^2 S$ does not determine neither stability nor instability in the majority of tested cases. When the second Lapunov theorem is applied, only different signs of $\delta^2 S$ and its time derivative would constitute sufficient stability criterion for the process. However, the calculations (Sieniutycz and Komorowska-Kulik, 1980;

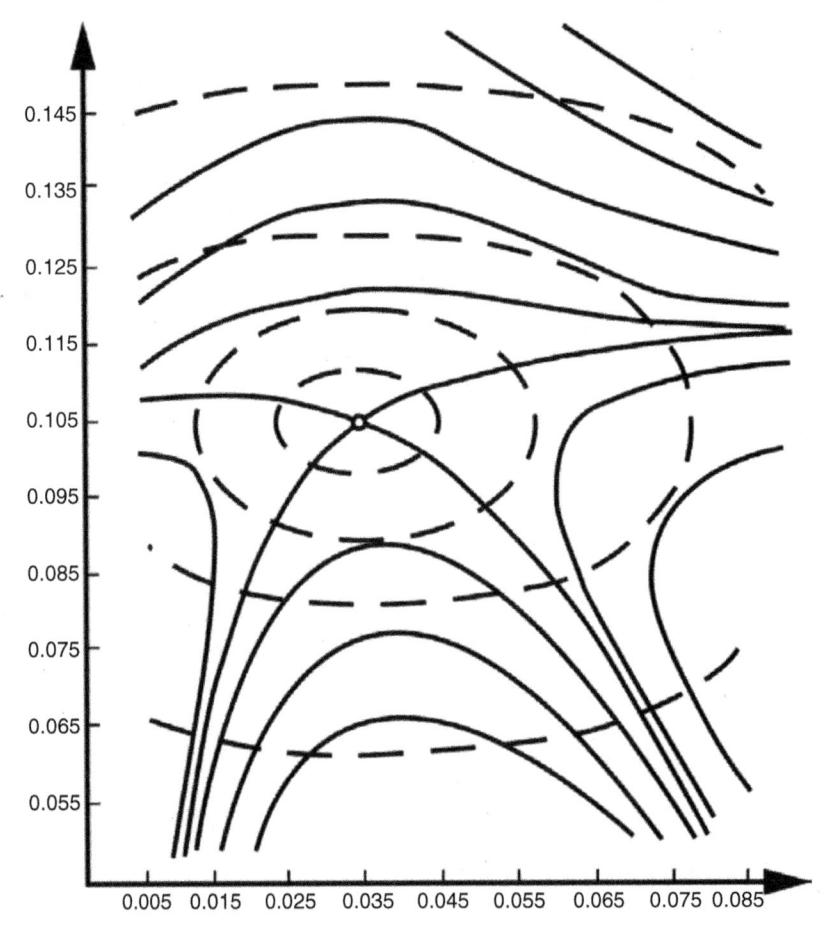

Figure 5.5: An example of inapplicability of the second differential of the entropy $V = \delta^2 S$ (*dashed lines*) and its time derivative (*solid lines*).

The nature of the derivative dV/dt in the state space of $x = [O]$ and $y = [CO_2^*]$ show explicitly the impossibility of proving stability of instability by applying the second differential of the entropy $V = S$ as the Lapunov function (Winkiel, 1976). Yet, a nonclassical function V_s consisting of the sum of S and a kinetic term (quadratic in process rates) provides, in general, a more efficient criterion of stability or instability.

Winkiel, 1976), showed that the time derivative dV/dt with respect to the process equations for $V = \delta^2 S$ is not a function of a constant sign. A suitable example is provided in Fig. 5.5 (Winkiel, 1976). Diagrams which prove the saddle nature of dV/dt were obtained even for stable singular points, showing again the inappropriateness of $V = \delta^2 S$.

Ertl (1991) considered many aspects of oscillatory kinetics and spatio-temporal self-organization in reactions at solid surfaces including CO oxidation on a platinum single-crystal surface. Chemical reactions far from equilibrium on solid surfaces may exhibit typical phenomena of nonlinear dynamics, as exemplified by the catalytic oxidation of CO. Depending on the external parameters (temperature and partial pressures of the reactants), the temporal variation of the reaction rate may become oscillatory or even chaotic. In a parallel way, the concentration distributions of the adsorbed species on the surface form spatio-temporal patterns including propagating and standing waves, rotating spirals, as well as irregular and rapidly changing structures denoted "chemical turbulence." Ertl (2008) gave a brief account of some of the work that had been considered to be the basis for the awarding him the Nobel Prize for his research in the field.

Zhu and Li (2002) performed a linear stability analysis of a reaction–diffusion model for solid-phase combustion. Using the linear stability analysis they investigated the dynamic behavior of a reaction–diffusion model in which the diffusion coefficients of the oxygen gas and the vapor of the combustible solid (Mg) were taken as two controlling parameters in the analysis. Their bifurcation map obtained showed three dynamic regions. Region I only showed stable combustion. Regions II and III both showed stable combustion and oscillatory combustion depending on the ratio of the two diffusion coefficients. Interestingly, region II also showed a small range of a bistable state consisting of a stable focus and an oscillating state, which is like the critical phenomena in phase transitions. These results indicated that the occurrence of oscillating combustion requires that the value of the diffusion coefficient of the Mg vapor should be comparable to or less than that of the oxygen gas at the same temperature.

As stated by Keizer (1976a) in his paper on the kinetic meaning of the second law of thermodynamics, the second law of thermodynamics has never been justified on the basis of mechanical laws, in particular, things which are termed "impossible" by the second law are in reality only highly improbable, and so in principle might occur. This point of view, represented since Boltzmann and Smoluchowski, c.f. Maxwell (2003a, b), has been widely accepted and is central in the modern theory of statistical thermodynamics. To examine the second law of thermodynamics, Keizer (1976a) used a kinetic–molecular theory which connects dissipation and fluctuations. Keizer's discussion of the second law is carried out in two parts. The first part concerns the average aspects of the second law and uses the canonical form of the transport equations. Considerations are restricted to systems with stable equilibrium states and are based on a conservation condition satisfied by transport processes which obey microscopic reversibility. This leads to a restriction on the dissipative behavior of extensive variables which is comparable to Caratheodory statement of the second law. Using a generalization of the Rayleigh–Onsager dissipation function, a kinetic proof is given of the Clausius inequalities for the entropy. Insofar as the transport of heat into a system is the only process which violates microscopic reversibility, this statement is equivalent to the second law. Keizer's (1976a) treatment also gives a kinetic proof of the Clausius inequalities

$$T_R dS/dt \geq dQ/dt \tag{5.12}$$

and

$$dS/dt \geq 0 \tag{5.13}$$

for the entropy. Using statistical aspects of the fluctuation–dissipation postulates, he defines a class of state functions related to the equilibrium statistical distribution, and verifies that the entropy is one of these functions. The second part of his paper extends these considerations by describing the time dependent probability distribution of the extensive variables. The fluctuation–dissipation principles show that at an asymptotically stable equilibrium, this probability distribution is uniquely defined by the equilibrium values of the extensive variables. This leads to a class of state functions related to the statistical properties, and to Einstein's relationship between the entropy and the equilibrium distribution. A discussion is given of how to extend these results to systems with multiple phases or at disequilibrium steady states. Although is not clear that the entropy will be a relevant function for a disequilibrium state, it is still possible to discuss the nature of the disequilibrium statistical distribution at a steady state.

For an asymptotically stable steady state, arguments identical to those used for equilibrium yield a unique Gaussian distribution at the steady state. Since the covariance matrix is symmetric, it may be considered as the matrix of second partial derivatives of a state function evaluated at the steady state and be used in this way to produce state functions related to the statistical distribution. This prepared the ground for future research and a book of this author, which deal with the thermodynamics of disequilibrium steady states (Keizer, 1976b; 1987a).

Contrary to Nicolis's and Prigogine's (1979) reassertion that the "excess entropy" around nonequilibrium steady states, $\delta^2 S_{ss}$, is a Lapunov function, simple, explicit counterexamples which invalidate this claim have been presented, for example, Winkiel (1976) and Fox (1980). Other counterexamples were given by Winkiel, (1976); Keizer (1976a, b); Lavenda and Santamato (1982); and Sieniutycz and Komorowska-Kulik (1978, 1980, 1982). Particularly Winkiel (1976) worked out a systematic thermodynamic analysis of stability problems for dryers and reactors. Also, the existence of an alternative theory possessing a proper Lapunov function for steady states has been reviewed by a number of authors. Plau and Tarbell (1980) worked out an approach to the stability of the CSTR by correlation of Lapunov functions. See also Favache and Dochain (2009). Kawczyński's (1990) book gave a rigorous and concise treatment which analyzes chemical reactions stretching from equilibrium through dissipative structures to chaos.

Assuming that a linearization of the dynamical equations around a selected singular state (equilibrium or steady) is performed, the perturbation *a* of this singular state satisfies a linear set of equations and there exists a region surrounding in which the stability or instability can be proved. While the second differential of the entropy $\delta^2 S$ tested as a Lapunov's function *V* does not determine in general stability or instability for a nonequilibrium singular state (an indefinite sign of dV/dt) a suitable Lapunov function can still be found (Winkiel, 1976; Sieniutycz and Komorowska-Kulik, 1980; Keizer, 1987b). All these approaches are similar from a formal viewpoint as they differ only by the way of how the second derivative of entropy is treated (it is either included into a new *V* or kept apart as an additive component). An original improvement of these approaches is proposed in Section 5.7 entitled "A new approach to Lapunov's functions and functionals".

Farkas and Noszticzius (1992) presented a unified treatment of chemical oscillators treated as thermodynamic systems far from equilibrium. They start with the theory of the Belousov–Zhabotynski (BZ) reaction and point out the role of positive and negative feedback. Then, they show that consecutive autocatalytic reaction systems of the Lotka–Volterra type play an essential role in the BZ reaction and other oscillatory reactions. Consequently, they introduce and then analyze certain (two-dimensional) generalized Lotka–Volterra models. They show that these models can be conservative, dissipative, or explosive, depending of the value of a parameter. The transition from dissipative to explosive behavior occurs via a critical Hopf bifurcation. The system is conservative at certain critical values of the parameter. The stability can be examined by testing the sign of a Lapunov function selected as an integral of the conservative system. By adding a limiting reaction to an explosive network, a limit cycle oscillator surrounding an unstable singular point is obtained. Some general thermodynamic aspects are discussed pointing out the distinct role of oscillatory dynamics as an inherently far-from-equilibrium phenomenon. The existence of the attractors is recognized as a form of the second law. This text gives a pedagogical perspective on the very broad class of problems involving the time order in homogeneous chemical systems.

In the strict sense of the term "reaction–diffusion" refers to systems involving constituents locally transformed into each other by chemical reactions and transported in space by diffusion (Nicolis and De Wit, 2007). They arise, quite naturally, in chemistry and chemical engineering but also serve as a reference for the study of a wide range of phenomena encountered beyond the strict realm of chemical science such as environmental and life sciences. At the present time one can speak about the phenomenology of reaction–diffusion systems. Reaction–diffusion systems in a closed vessel and in the absence of external forces evolve eventually to the state of chemical equilibrium, whereby the constituents involved are distributed uniformly in space and each elementary reactive step is counteracted by its inverse (Nicolis and De Wit, 2007). It has long been realized that the approach to equilibrium can be both in the form of a simple exponential decay, or more involved transient behaviors associated with damped oscillations or nontrivial space dependencies including wavelike patterns. The development of irreversible thermodynamics in the 1950s and onwards provided an explanation of the origin of these two kinds of behaviors by linking them to time evolutions starting close to and far from equilibrium, respectively. Experiment and modeling on laboratory scale reactive systems such as the Belousov–Zhabotinsky reaction (Zhabotinsky, 1967, 1974) confirmed this view. Still, as long as reaction–diffusion processes were carried out in a closed thermostated reactor there was no way to analyze systematically what was going on, since, by virtue of the second law of thermodynamics, the system was bound to reach sooner or later the state of equilibrium (Nicolis and De Wit, 2007). A major development that opened new horizons in the experimental study of reaction–diffusion systems and stimulated, in parallel, important theoretical developments has been the systematic use of open reactors, whereby the system is maintained in a nonequilibrium state as long as desired through the pumping of fresh reactants (the rate of which determines the distance from equilibrium) and the outflow of used products (Nicolis and De Wit, 2007).

When spatial homogeneity is maintained within the reactor through stirring in CSTRs (continuously stirred tank reactors), a rich phenomenology was revealed as key parameters were varied, in addition to the steady state extrapolating the familiar equilibrium like behavior: simple periodic, multiperiodic and chaotic oscillations; multistability, that is, the coexistence of more than one simultaneously stable states; and excitability, whereby once perturbed a system performs an extended excursion before settling back to its original stable state (Nicolis and De Wit, 2007). A most exciting set of behaviors pertains to space patterning, made possible when stirring is not imposed within the reactor as is the case in many real world situations in chemistry, engineering, and biology. Propagating wave fronts, or stabilized ones (like a flame separating fresh and burnt reactants in combustion for instance), are familiar examples. A still different form of spatial organization is the formation of regular steady state patterns arising from the spontaneous symmetry breaking of a spatially uniform state (Nicolis and De Wit, 2007).

Liu et al. (1996) shows that the numerical study under the guidance of symbolic dynamics may provide a suitable method to gain global knowledge on ODEs, which is difficult to obtain either by purely analytical or by completely numerical means. On the example of the periodically forced Brusselator they demonstrate how this approach works in practice; they study the forced Brusselator from the viewpoint of the dissipative map. The second system of investigation is the Lorenz model. In particular, they explore the transition from annular type dynamics to interval dynamics in terms of symbolic dynamics. Their results are instructive for the study of other nonlinear systems with competing frequencies, as the method is topological in nature, but complemented with numerical details. They also discuss the general implication of the approach.

Favache and Dochain (2009) have performed the stability analysis of the continuous stirred tank reactor (CSTR) by considering arguments based on thermodynamics. Different candidates for Lapunov function of thermodynamic type—more specifically those based on the entropy, entropy production, and internal energy—were considered. These provide new insight and physical interpretation in the stability/instability of the equilibrium points of the CSTR. This includes in particular extension of results for less restrictive conditions on system's dynamics and thermodynamics, and invariants sets for the stable equilibrium points of the CSTR by considering an internal energy based Lapunov function.

Hoang et al. (2013) consider thermodynamic stability analysis and its use for nonlinear stabilization of the nonisothermal CSTR. They show how an availability function derived from the concave entropy can be used for the stability analysis and derivation of control strategies for a CSTR. First they perform an overview of related thermodynamic concepts, and then show how the availability function within the thermal domain is the suitable Lapunov function. Numerical simulations illustrate the application of the theory.

Fluctuation theory provides efficient tools to understand disequilibrium processes. Studying fluctuation and relaxation of macrovariables, Kubo et al. (1973) assume that a macrovariable follows a Markovian process. The extensive property of its probability distribution is shown to propagate, which is a generalization of the Gaussian properties of the equilibrium distribution to disequilibrium nonstationary processes. It is basically a WKB-like asymptotic evaluation in the inverse of the size of the microsystem. The authors follow evolution of the variable along the most probable path, and the fluctuation properties around the path are considered with emphasis on the relation of nonlinearity of evolution and the fluctuation. In a general treatment for the asymptotic properties of relaxation eigenmodes path integral formulation and the Hamilton–Jacobi method are applied. Anomalous fluctuations are linked with unstable, critical, or marginal states.

Stability problems are inherently connected with the fluctuation theory. In Chapter 12 of his book on analytical thermodynamics Soo (1962) gives a readable description of fluctuation and relaxation phenomena in the irreversible thermodynamics. Some information contained therein was next used in his later book on direct energy conversion (Soo, 1969). Jou and Llebot (1980) introduce electric current fluctuations in extended irreversible thermodynamics. In his PhD thesis on nonlinear hydrodynamic fluctuations Van Saarloos (1982) discusses such topics as: variational hydrodynamics of ideal fluid, nonlinear hydrodynamic fluctuations, the Fokker–Planck equation, the form of dissipative currents, and the canonical transformation from their Eulerian to Lagrangian description. Nonlinear hydrodynamic fluctuations are limited to those around equilibrium. De la Selva and Garcia-Colin (1986) treat hydrodynamic fluctuations in a chemically reacting fluid, and show how the (nonlocal in time) second moments of these fluctuations can be predicted by the extended irreversible thermodynamics. Their results imply that the fluctuating reaction rate and diffusive flux yield a non-Gaussian noise and that the classical results follow from linear fluctuating hydrodynamics. For the thermostatic second moment in the fluctuations of the energy new terms arise if fast variables intervene for their description. Jou (1989) initiates works on the fluctuation theory in extended irreversible thermodynamics. For later treatments, see Ross (2008).

Keizer (1976b) considered fluctuations, stability, and generalized state functions at nonequilibrium steady states. Fluctuation–dissipation postulates, which describe the kinetic effects of molecular processes, were used to characterize nonequilibrium steady states. Attention is restricted to stable, noncritical states which develop in systems with inputs that are time independent. For these systems it is shown that the steady state distribution is Gaussian, which provides a generalization of the well-known Einstein formula for equilibrium states. For certain systems it is shown that the time dependence of the covariance matrix of the extensive variables gives a necessary and sufficient condition for the stability of a noncritical state. These considerations are illustrated for the steady states accompanying diffusion, heat transport, chemical reactions with linear coupling, and certain nonlinear chemical reactions. These examples show that the covariance matrix is not necessarily related to the local equilibrium entropy. When the covariance matrix is invertible, it can be used to construct generalized state functions which reduce to familiar thermodynamic functions at equilibrium. The generalization of the entropy, called the σ function, is related to stability, the probability density, and generalized "thermodynamic forces" in the same way as the entropy is at equilibrium. Generalized chemical potentials and chemical affinities are defined for steady states of some chemical reactions.

Grabert and Green (1979) reexamined the link between fluctuations and nonlinear irreversible processes. Deterministic equations for nonlinear irreversible processes were derived from a minimum principle, which leads to a set of variables η canonically conjugate to the microvariables, a. Using an action integral, the minimum principle, and the conjugate variables η the authors construct a covariant expression for the conditional probability of fluctuations for a small interval of time. The short-time conditional probability leads to the conditional probability for finite times in the form of a path integral, a generalization of a corresponding integral of Onsager and Machlup (1953). The conditional probability satisfies a Fokker–Planck equation, which has the form derived from statistical mechanics.

Grabert et al. (1980) form the second part of a study which reexamines the relationship between fluctuations and nonlinear irreversible processes and include macroscopic variables which transform odd under time reversal. The fluxes of some of these variables may be purely reversible so that the diffusion matrix may be singular. The deterministic equations for nonlinear processes are again derived from a minimum principle. The fluctuations of the macroscopic variables are treated by a Fokker–Planck equation derived from the statistical mechanics. The conditional probability of the fluctuations is constructed as a path integral. The connection between the deterministic and the stochastic descriptions of macroscopic dynamics is formulated in the way independent of the coordinate frame (a covariant form). The metric tensor in the state space is particularly simple in the frames where microvariables are sums of molecular variables. Also, Sieniutycz and Salamon (1990b) present a phenomenological approach to the problem of the disequilibrium metric. Their approach allows one to compute the length of a disequilibrium path, bound the related dissipation, and determine stability conditions for given differential equations describing transient modes.

Moreover, working in the context of fluctuations, Lavenda and Santamato (1982) discuss controversies associated with the Lapunov function approach to the stability of equilibrium and disequilibrium stationary states. Their analysis rests on the formulations offered by Glansdorff and Prigogine (1971); Keizer (1976a, b), and Fox (1980). It is their contention that these formulations neglect completely the effect of random thermal fluctuations, and, at most, reduce to the approaches involving deterministic Lapunov functions. As they state, the neglect of thermal fluctuations is witnessed by the fact that neither of the approaches discussed earlier leads to a correct asymptotic form of the excess entropy about the equilibrium or a stable disequilibrium steady state. Motivated by these circumstances, Lavenda and Santamato

(1982) define a stochastic excess entropy and stochastic stability conditions. In comparison with their deterministic counterparts, there appears an additional term due to the presence of random thermal fluctuations. They stress that: (1) a general property of linear stochastic processes is that the random fluctuations tend to work against the process stability, and (2) a stochastic inequality subsuming the effect of random thermal fluctuations depends on the choice of initial conditions. Further development of these approaches can be found in a book (Lavenda, 1985a).

On a deterministic level, chemical reactions are described by well-established kinetic law of mass action (KMAL), where chemical kinetics have the form of power laws. On a more refined level, which includes fluctuations of concentrations, the reaction is described in terms of a stochastic process. Onsager's (1931a, b) theory gives the solution of the problem of mutual connection between the deterministic and the stochastic levels of description in the linear regime near equilibrium. Yet, for chemical reactions the linear theory provides only a poor approximation (De Groot and Mazur, 1984). The paper by Grabert et al. (1983) proposes an approach to fluctuations of reversible chemical reactions in closed systems. The deterministic rate laws are cast into the form of nonlinear, Onsager's type, closed laws. The nonlinear chemical resistances (conductances) are in a complete agreement with the experimentally confirmed kinetic law of mass action (Grabert et al., 1983; Shiner, 1987; Sieniutycz, 1987b).

By means of the nonlinear transport theory a Fokker–Planck equation for the stochastic process of concentration fluctuations is obtained. Grabert et al. (1983) and Gaveau et al. (2005) show that the stochastic formulation reduces to the deterministic rate laws in the thermodynamic limit with fixed concentrations. They give examples of reactions in ideal mixtures, and compare their results with those of the usual approach by means of a birth and death type master equation. They show that both approaches lead to the same stationary probability and natural boundaries reflecting the fact of a restricted state space. The Fokker–Planck equation proposed by (Grabert et al., 1983) is different than the Fokker–Planck equation obtained from the master equation by truncating its Kramers–Moyal expansion. However, the two equations have identical Fokker–Planck coefficients in the vicinity of the equilibrium. Compared with the usual master-equation approach the stochastic modeling of chemical reactions has the advantage of allowing for straightforward extension to nonideal mixtures.

The notion of entropy, which was introduced by Clausius in 1865, is central in physics of fluctuations. Originally developed only for systems in thermodynamic equilibrium, entropy was applied to disequilibrium processes since the pioneering work of Onsager, Meixner, De Groot, Mazur, Prigogine, and many others. Ebeling and Engel-Herbert (1986) devoted their investigation to the question of the miscroscopic meaning of entropy for stationary nonequilibrium states. They have begun with the discussion of several entropy concepts (phenomenological Clausius entropy, Shannon and Boltzmann entropy, Kolmogorov entropy) when showing the significance of these concepts in many driven systems.

The concepts of entropy and entropy lowering in comparison to the equilibrium state are good indicators of disequilibrium. Ebeling and Engel-Herbert (1986) show that entropy lowering is connected with the contraction of the occupied part of the phase space due to the formation of attractors. Excitation of oscillations in solids and turbulence in liquid flows are analyzed as concrete examples. The entropy statement formulated by Klimontovich in the turbulence context is discussed by Ebeling et al. (1986). Two examples are investigated: (1) self-oscillations of nonlinear oscillators, and (2) laminar and turbulent flows in tubes. The authors show that the entropy is decreasing with excitation under the condition of fixed energy. Engel-Herbert and Ebeling (1988a) study the Brownian motion of nonlinear oscillators imbedded in a heat bath. For the excitation of sustained oscillations, realized by van der Pol oscillators, the entropy is calculated in the limit of weak dissipation and weak noise. At fixed average energy this entropy decreases monotonically if the feedback strength increases.

The structure of a turbulent flow may be expressed by the collective mechanisms for the transfer of momentum, heat, and mass and by the entropy lowering determined relative to an equilibrium system of the same energy. One can then evaluate the variation of the entropy during the transition of a flow from laminar to turbulent motion. Engel-Herbert and Ebeling (1988b) have calculated the entropy lowering from the one-particle distribution function f using the Boltzmann formula applied in the laminar as well as in the turbulent regime. They assumed that deviations δf from the local Maxwell distributions are small. Under these assumptions they showed that the entropy lowering of the incompressible flow in a tube with increasing Reynolds number increases monotonically if the energy of the flow is kept fixed. The transition from laminar to turbulent motion is a nonequilibrium transition to a more ordered state. Therefore one can regard the development of the turbulent flow beyond the critical Reynolds number as a disequilibrium transition. The Reynolds shear stress may be an order parameter of the transition, and the degree of order may be characterized by the entropy and by the entropy production.

As the Reynolds number grows, a certain portion of the energy of the flow is transferred to the collective macroscopic motions of the molecules, which are reflected, for example, by the hydrodynamic velocity, by the turbulent pulsations, and the randomness, which, apparent from a causal observation, is not without some order (Sreenivasan, 1999). Fixing the energy as Re varies, this portion of the transferred energy is taken from the microscopic thermal motion. But energy contained in the thermal degrees of freedom possesses a higher value of entropy. Accordingly (Engel-Herbert and Ebeling, 1988b) the transformation of energy from the microscopic scale of thermal motion to macroscopic scales of collective molecular motion at fixed energy lowers the entropy. This agrees with the conclusion of Engel-Herbert and Schumann (1987) who have studied the behavior of the entropy during a nonequilibrium phase transition. They have shown that for the generation of sustained oscillations the entropy decreases monotonously if the average oscillator energy remains fixed.

For a turbulence practitioner a main goal is the prediction and control of turbulent effects, their suppression or enhancement depending on actual needs (Sreenivasan, 1999). Even with the smoothest and simplest boundaries, flowing fluids assume the irregular state of turbulence. This feature, while not understood well enough at the present time, bears a connection with the dynamical chaos, the phenomenon that causes appropriate corrections to the turbulence mechanism (Landau and Lifshitz, 1987). Important are thermodynamic aspects of turbulence elucidated by Klimontovich (1982, 1986) and Ebeling (1983), and also by coworkers of these researchers who compared with experiments of Klimontovich's hydrodynamic theory.

Klimontovich (1986, 1991) compares the entropies of laminar and turbulent motion for the same values of the average kinetic energy with respect to the velocity of the laminar flow and the velocity of the averaged turbulent flow. He shows that when transition occurs from laminar flow to turbulent flow, the entropy production and entropy itself decrease. This indicates that the disequilibrium phase transition from laminar flow to turbulent flow is a transition to a more ordered state. He also proves that the turbulent motion should be thought as having a lower temperature than laminar motion. See also Ebeling and Klimontovich (1984) and Klimontovich (1999). Ebeling (1983) compares with experiments the theory of hydrodynamic turbulence of Klimontovich (1982, 1986), which yields effective turbulent viscosity as a linear function of the Reynolds number. This theory seems to be able to describe the observed effective viscosity of vortices and, after modifications, the fluid flow in tubes up to very high Reynolds numbers. Klimontovich (1991) examines the link between the turbulent motion and the structure of chaos.

Ruppeiner (1990) reviews the thermodynamic metric in the context of fluctuation theory. He proposes a covariant theory extending the classical theory of thermodynamic fluctuations beyond the Gaussian approximation and presents a path integral formulation. He shows that the scalar Riemannian curvature plays a special role as the indicator where the classical theory of fluctuations breaks down. Physically, this refers to the correlation volume at the critical point. Examples illustrate the thermodynamic curvature for fluid and magnetic systems. The magnitude of the curvature is related to interactions in nonideal systems.

Using Master equation Gaveau et al. (1999a) consider the nonequilibrium thermodynamics of a reaction–diffusion system at a given temperature. Gaveau et al. (1999b) treat the reaction–diffusion systems out of equilibrium. Based on the earlier-derived path integral formulation of the Hamilton–Jacobi approximation for the Master equation they show that with this path of integral formulation it is possible to calculate rate constants for the transition from one well to another well of an information potential, and to give estimates of mean exit times. Grassia (2001) considers a large particle moving through a sea of small particles. On the micro scale, all particle collisions are elastic. However, on the macro scale, where only the large particle is properly resolved, dissipative forces and fluctuating random forces are observed. These forces are connected by a fluctuation–dissipation theorem. The latter is proved by Grassia in two different ways, first via statistical mechanics, and second by classical mechanical principles of momentum and energy conservation. With simple ideas of classical mechanics, the dissipative forces are related to the properties of small particles on the microscale. The microscale collisions are entirely elastic, yet they lead to dissipation when viewed on larger scales. The classical proof of the fluctuation–dissipation theorem elucidates the physical meaning of the result by demonstrating that fluctuation and dissipation both ultimately arise as effect of collisions.

Theories of fluctuating hydrodynamics (Landau and Lifshitz, 1974a, 1987) have been developed for steady-state nonequilibrium fluctuations. Experimental data verifying these theories are available (Ronis and Procaccia, 1982). In the absence of hydrodynamic motion macroscopic concentration gradients in physical systems relax towards equilibrium by diffusion; see Cussler (1984). This is normally regarded as a spatially homogeneous mixing process. Vailati and Giglio (1997) show, however, that unexpectedly large spatial fluctuations in concentrations can occur during a free diffusion process. Large fluctuations encode information on the fine-scale dynamics. Large deviation relations, known as fluctuation theorems, also capture basic disequilibrium properties. Eyink's (1990) fluctuation hypothesis leads to nonlinear version of Onsager–Machlup (1953) theory. See also Gaveau et al. (2005).

The shadowgraph images and low-angle light scattering show evidence for large concentration fluctuations, orders of magnitude larger in amplitude than those observed in the equilibrium state. They show that the strong inhomogeneity is caused by a coupling between velocity and concentration fluctuations in the disequilibrium state. Gravity cuts off these fluctuations above a certain wavelength, and the amplitude of the fluctuations at longer wavelengths does not depend on any relevant thermodynamic property of the fluid. Consequently, these giant fluctuations should be observed in any mixture undergoing mixing by diffusion. In a related comment Weitz (1997) describes cases when the disequilibrium fluctuations dominate because they cannot relax with a sufficiently high rate. Dykman et al. (1999) also treat large fluctuations and irreversibility in nonequilibrium systems. A fluctuating system typically spends most of its time in a close vicinity of a stable state. Occasionally, however, it will undergo a much larger departure before returning again or perhaps, in some cases making a transition to the vicinity of a different stable state (Dykman et al., 1999). Despite their rarity, large fluctuations are important in diverse circumstances, such as: nucleation at phase transitions, chemical reactions, mutations of DNA sequences, protein transport in bio-cells, and failures of electronic and photonic devices. In many cases of practical interest, the fluctuating system is far from thermal equilibrium. Dykman et al. (1999) study large rare fluctuations in a disequilibrium system, theoretically and by analogue electronic experiment. They specify examples of such systems, which include lasers, pattern-forming structures, trapped electrons which exhibit bistability and switching in a periodic field, and Brownian ratchets which can support a unidirectional current under disequilibrium conditions. They emphasize that the optimal paths calculated via the eikonal approximation of the Fokker–Planck equation can be identified with the locus of the ridges of the prehistory probability distributions which can be calculated and measured experimentally for paths terminating at a given final point in configuration space. The pattern of optimal paths and singularities, such as caustics, cusps, and switching lines is calculated and measured experimentally for a periodically driven overdamped oscillator; the results are in good agreement with each other.

The approach to the analysis of large fluctuations is through the concept of the optimal path, the path that the system is predicted to follow with overwhelming probability during the course of the fluctuation. The optimal path is calculated as a trajectory of an auxiliary Hamiltonian system, yet, for many years it remained unclear how the optimal path is related to the behavior of real fluctuating system. As stated by Dykman et al. (1999), the large fluctuations problem is now solved after introducing the prehistory probability distribution, and, it has been shown that optimal paths are physical observables experimentally measured for both equilibrium and disequilibrium systems. The authors review these results and stress the related advantages. Their theory is in principle a theory of the Hamiltonian system for which, in the range of small noise intensities, the optimal path is defined by the minimum condition for an action functional subject appropriate boundary conditions. The optimal action $S(q, t)$, is a smooth single-valued function of position on a Lagrangian manifold, and, simultaneously, it is a Lapunov function. Therefore $S(q, t)$ may be regarded as a generalized disequilibrium potential for a fluctuating dynamical system. Optimal paths which ensure the minimum action are visualized experimentally by the prehistory probability distribution. For more information on the theory of large fluctuations in stochastic populations see a PhD Thesis by Assaf (2010).

Ortiz de Zárate et al. (2007) discuss fluctuations in a homogeneous no-flow simple reaction-diffusion system, a binary fluid mixture with an association-dissociation reaction. The non-equilibrium concentration fluctuations are spatially long ranged, with intensity depending on the wave number q. The intensity exhibits a crossover depending on whether the wavelength is smaller or larger than the penetration depth of the mixture. This distinguishes between diffusion or activation-controlled regimes by measuring the fluctuations. In a related work Bedeaux et al. (2011) find that fluctuations with wavelengths q smaller than the penetration depth are not affected by the chemical reaction. Yet, the intensity of fluctuations with wavelengths q larger than the penetration depth varies as q^{-2} caused by the presence of the chemical reaction.

Markus and Gambar (1996) presented another possible description of fluctuations and developed quantization procedure in the field theory of thermodynamics (especially heat conduction). This description handles thermodynamic variables as operators, and is suitable to express the magnitude of energy fluctuation in heat conduction. As they point out, one can examine a nonequilibrium process, and such formulation of fluctuations is not in contradiction with the statistical mechanics. Also, one can discuss in such a treatment whether it would be possible to measure the temperature around the absolute zero temperature.

Turbulence problems may be approached from the viewpoint of their link with the fluctuation theory (Monin and Yaglom, 1967). Memory effects in turbulent flows are treated by Builtjes (1977). Durbin (1993) offers a near-wall turbulence model based on k, ϵ, and $\sqrt{2}$ equations. The model predicts flow and heat transfer in two-dimensional channels and boundary layers. It leads to the good agreement with data on skin friction, Stanton number, mean velocity, and turbulent intensities. Solutions to the model show correct Reynolds number dependence without building it into any of the coefficients. Calculations of zero and adverse pressure gradient show that in both cases the results agree well with an experiment.

Eyink (1997) derives a fluctuation law of the energy in freely decaying, homogeneous, and isotropic turbulence within standard closure hypotheses for the three-dimensional incompressible flow. In particular, he derives a fluctuation–dissipation relation which relates the strength of a stochastic backscatter term in the energy decay equation to the mean of the energy dissipation rate. The theory is based on an "effective action" of the energy history and illustrates the Rayleigh–Ritz method developed to evaluate approximately the effective action within the probability density-function (PDF) closures. Eyink (1997) states that these effective actions generalize the Onsager–Machlup (1953) action of disequilibrium statistical mechanics to turbulent flow. They yield detailed predictions for fluctuations, such as multitime correlation functions of arbitrary order, which cannot be obtained by direct PDF approaches.

Eyink (2010) formulates a stochastic least-action principle for the incompressible Navier–Stokes equation. This stochastic principle simplifies to Hamilton's principle for the incompressible Euler equation in the case of zero viscosity. He uses his principle to give a new derivation of a stochastic Kelvin theorem for the Navier–Stokes equation, recently established by Constantin and Iyer (2008), which shows that this stochastic conservation law arises from particle-relabeling symmetry of action. In the framework of his stochastic variational principle Eyink (2010) discusses irreversibility, energy dissipation, and the inviscid limit of Navier–Stokes solutions. In particular, he discusses the connection of the stochastic Kelvin Theorem with his previous "martingale hypothesis" for fluid circulations in turbulent solutions of the incompressible Euler equation. He appreciates another variational principle for fluid equations, Onsager's principle of least dissipation (Onsager, 1931b; Onsager and Machlup, 1953). It quantifies the probability of fluctuations away from hydrodynamic behavior in terms of the dissipation required to produce them. Eyink (1990) gives a derivation of Onsager's principle for incompressible Navier–Stokes equation in a microscopic lattice-gas model. It remains a still open question if any relation exists between the least-action and least-dissipation principles. See also Gomes (2005).

Paraphrasing von Karman, Eyink (2002) states: "In the turbulent flow we see not only randomness but a tendency toward some order." Eyink concludes that turbulent flow is statistically stable: the same averages result for a large domain of initial data. Likewise, small perturbations from the invariant measure relax back to zero. The H-theorem for the relative entropy is a formal expression of this stability property. Fluctuations from average quantities show the same statistical tendency to relax as initial perturbations. Thus, it is posssible that most probable fluctuations exhibit the least dissipation. This fluctuation–dissipation relation is implied by the Rayleigh–Ritz method but likely remains valid beyond it. Eyink (2007) investigates material lines and passive vectors in turbulent flows at infinite-Reynolds numbers, the ensemble of velocities (white-noise in time and roughly continuous in space). In his view "a spontaneous stochasticity" generalizes to material lines and the conservation of circulations generalizes to "a martingale property" of the stochastic process of lines.

For a turbulence practitioner a main goal is the prediction and control of turbulent effects, their suppression or enhancement depending on actual needs (Sreenivasan, 1999). Even with the smoothest and simplest boundaries, flowing fluids yield the irregular state of turbulence. This feature, while not understood well enough at the present time, bears a connection with the dynamical chaos, the phenomenon that causes appropriate corrections to the turbulence mechanism (Landau and Lifshitz, 1987). Important are thermodynamic aspects of turbulence elucidated by Klimontovich (1982, 1986) and Ebeling (1983), and also by coworkers of these researchers who confirmed Klimontovich's hydrodynamic theory by experiments.

Bałdyga and Pohorecki (1995) review turbulent micromixing in chemical reactors. First, they outline the idea of micromixing, its definition and measures. Next, concepts of mixing environments and mixing earliness are presented. Chemical reaction is a molecular-level process and only mixing on the molecular scale can affect its course. The authors focus on the effects of turbulent mixing of incompressible fluids in single-phase systems on the chemical reactions. Processes of turbulent micromixing are discussed in detail. The fluid–mechanical interpretation of turbulent micromixing (effect of fluid element deformation on the acceleration of molecular diffusion, engulfing of environment, inertial-convective disintegration of large eddies, and local intermittency) is presented. The authors conclude that stretching of material elements and vortices accompanied by molecular diffusion results in the growth of the mixing zones. The growth of the zone mixed on the molecular scale is a characteristic feature of micromixing, and should be included in micromixing modeling. The characteristic time constants for the consecutive stages of mixing are compared with the characteristic time for chemical reaction; the numerical criteria are outlined. Two approaches, Eulerian and Lagrangian, require different description methods, and generate specific problems (closure problem, problem of environment). Applications are in the important fields of industrial practice, such as complex reactions, precipitations, and polymerizations.

The ideas, methods, results, and experiments that worked earlier for fluctuations and turbulence in single-phase systems were transferred soon to practical two-phase flows, which are of particular interest to us. Simonin et al. (1993) worked out a theoretical scheme for the numerical prediction of turbulent dispersed two-phase flows by using the two-fluid model approach. Their predictions turned out to be strongly dependent on the closure assumptions applied to calculate the statistical characteristics of the fluid turbulent motion felt by the particles. These authors showed that the interaction of fluctuating motions in both phases is directly related to the fluid/particle velocity covariance tensor. Therefore, their predictions were made using the fluid/particle covariance tensor given by large-eddy simulations (LES). Their closure model obtained for the fluid/particle characteristics was validated by comparison with LES of a particle cloud suspended in an homogeneous and isotropic fluid turbulence. Another virtue of this work is the new approach for the derivation of the fluid/particle turbulent moments applied in the practical closure models.

Two-phase media exhibit a strongly dispersive nature, that is, the strong dependence of the propagation velocity of disturbances on the frequency (Bilicki and Badur, 2003). Their main result is a fruitful model of flowing two-phase systems, in which the basic role is played by operative (effective) quantities such as: operative viscosity, operative thermal conductivity and operative heat capacity. The model is composed of the averaged balance equations for the balances of mass, energy and momentum, and it is supplemented by an equation describing the interphase mass exchange. For systems with liquid boiling, the modeling can predict dispersive shock waves, pseudocritical phenomena, wave propagation, and the differences between the hydrodynamic pressure and thermodynamic pressure. To improve the agreement between the numerical results and the experiments, the authors extended the original version of their model towards the so-called relaxation–diffusive version, which allows one to take into account the phenomena of thermal diffusion and relaxation. The improved model was obtained by using approaches based on the nonequilibrium thermodynamics (thermodynamics of hidden variables and extended irreversible thermodynamics). It contains the relaxation time and the operative heat conductivity, and it shows better agreement with the experimental data than the original model. It represents the extension of the theory of the turbulent transport of heat, mass and momentum in two-phase systems regarded as a pseudohomogeneous phase. This extension leads to the discovery of generalized constitutive relationships in systems with the microstructure, in which scalar and tensor generalizations of the chemical affinity are important. While there is still uncertainty that all important variables were taken into account, the results show the basic improvement in the concepts of modeling of turbulent dispersive media. Especially they show the necessity of rejection of frequent trials towards the use of the theory of a turbulent single-phase medium in turbulent multiphase systems. Such trials are ineffective, especially in turbulent systems with phase changes and microstructure.

The early research in the field of two-phase flow which started from the beginning of the 20th century, created an interest concerning the phenomenon of the thermally induced flow-instability in two-phase systems. Kakaç and Bon (2008) sum up the experimental and theoretical work in two-phase flow instabilities carried out by investigators over a period of several years, demonstrating and explaining three main instability modes of two-phase instabilities (density-wave type, pressure-drop type, and thermal oscillations) that are encountered in various boiling flow channel systems. Typical experimental investigations of these instabilities in tube-boiling systems are indicated, and the most popular models for predicting two-phase dynamic instabilities (the homogenous flow model and the drift-flux models) are clarified. Solution examples and the validation results are also provided.

We discuss here only a broad field of stability problems with composite solids. Advanced composites are increasingly used in structural applications due to their favorable and objectively enhanced properties. They are ideal for applications (eg, aircrafts and automobiles) where high stiffness-to-weight and strength-to-weight ratios are needed (Qiao and Shan, 2005). Because of their excellent properties—such as lightweight, corrosive resistance, nonmagnetic, and nonconductive—the structures made of composite materials have also shown efficient and economical applications in civil engineering structures, such as bridges and piers, retaining walls, airport facilities, storage structures exposed to salts and chemicals, and others. In addition, composite structures exhibit excellent energy absorption characteristics, suitable for seismic response; high strength, fatigue life, and durability; competitive costs based on load capacity per unit weight; and ease of handling, transportation, and installation. These structures are usually in thin-walled configurations, and the fibers (eg, carbon, glass, and aramid) are used to reinforce the polymer matrix (eg, epoxy, polyester and deck panels are typical composite structures commonly used in civil infrastructures (Qiao and Shan, 2005). FRP (fiber-reinforced polymer) structural shapes are primarily made of E-glass fiber and either polyester or vinylester resins. The manufacturing processes include pultrusion, filament winding, vacuum-assisted resin transfer molding (VARTM), hand lay-up, and so forth; while the pultrusion process, a continuous manufacturing process capable of delivering one to five feet per minute of prismatic thin-walled members, is the most prevalent one in fabricating the FRP shapes due to its continuous and massive production capabilities. Due to geometric (that is, thin walled shapes) and material (ie,, relatively low stiffness of polymer and high fiber strength) properties, FRP composites structures usually undergo large deformation and are vulnerable to global and local buckling before reaching the material strength failure under service loads. Thus, structural stability is one of the most likely modes of failure for thin-walled FRP structures. Since buckling can lead to a catastrophic consequence, it must be taken into account in design and analysis of FRP composite structures. Because of the complexity of composite structures (eg, material anisotropy and unique geometric shapes), common analytical and design tools developed for conventional materials cannot always be readily applied to composites. On the other hand, numerical methods, such as finite elements, are often difficult to use in the realm of composites, which requiring specialized training and are not always accessible to design engineers. Therefore, to expand the applications of composite structures. Rather an engineering design approach for FRP shapes is developed. Such a tool, which includes stability analyses, should allow designers to evaluate perform stability analysis of customized shapes as well as to optimize innovative sections. Variational principle as a viable method is often used to develop analytical solutions for stability of composite structures. Variational principle-based formulations form a powerful basis for obtaining approximate solutions to structural stability of FRP shapes. Qiao and Shan (2005) describe the application of variational principles in stability analysis of composite structures, in particular local and global buckling of FRP structural shapes, from which explicit solutions and equations for design and analysis are developed.

5.5 Periodic States, Oscillatory Systems, and Chaotic Solutions

In externally controlled mass transfer systems without chemical reactions only forced oscillations are possible. It may be shown that the space averaging of a wave PD diffusion equation,

$$\frac{\partial Y}{\partial t} = D\left(\nabla^2 Y - c_0^{-2}\frac{\partial^2 Y}{\partial t^2}\right), \tag{5.14}$$

obtained by combining the Cattaneo equation for the diffusive flux density

$$J = -\rho D \nabla Y - \tau \frac{\partial J}{\partial t} \tag{5.15}$$

with the unperturbed form of mass balance (5.4) for $i = 1$ leads to the wave generalization of popular drying equation

$$\tau \frac{d^2 \bar{Y}}{dt^2} + \frac{d\bar{Y}}{dt} = -K(\bar{Y} - Y_e). \tag{5.16}$$

The latter equation contains the space-averaged moisture content, $\bar{Y}$. This is none other than the equation of the second drying period, which takes into account a relaxation effect (τ term). The equation contains the well-known drying coefficient $K = 2.5DX^2$. Obviously, for $\tau = 0$ the conventional drying equation is recovered. Eq. (5.14) was used for experimental evaluation of drying relaxation times from the kinetic curves (Mitura, 1981; Sumińska, 1983). Effective values of τ are of the order of magnitude 10^{-4} s (Luikov, 1966a).

Results of investigation of the frequency dependent properties of Eq. (5.14) are available (Sieniutycz, 1989a).

One may note that the drying Eq. (5.16) has the form typical of a damped oscillator. The notion that the drying (dissipative) system can be described by an equation of this kind is rather surprising because such formula may often imply damped or undamped oscillations. However a comprehensive analysis of the material constant magnitude (Sieniutycz, 1989a) indicates that the physical nature of the drying process is such that only forced oscillations are possible, that is, no free oscillations can occur, which, of course, is in agreement with our experience. Clearly the wave-drying equation does not imply the free oscillations for the real material constants. Consequently, although this thermodynamic system is described in the same manner as the mechanical or electrical system, the qualitative behavior of this system is quite different; it is overdamped. In general an overdamping effect prevails, thus, as an analysis of the steady forced oscillation shows, the maximum of the rate amplitude occurs for the frequency $\omega_0 = (K/\tau)^{1/2}$ in the form of plateau which extends in the vast range of frequencies ω between K and τ (Sieniutycz, 1989a). When $\tau = 0$ (conventional theory) the plateau extends to infinity. Therefore the discrepancies between the two theories, wave and conventional, start at the frequencies of the order of τ^{-1}. For lower ω an excellent agreement between the two theories is obtained that is, each theory is acceptable. It can be shown that the efficiency of energy transmission due to fast disturbances is practically equal to zero in the classical regime $K < \omega < \tau^{-1}$. In the nonclassical regime described by the relaxation equation, the energy efficiency can approach even unity in high frequency range due to predominant role of inertial terms. This fact serves as a justification of an interest in the investigation of high frequency phenomena with a finite wave speed.

However, oscillations in chemical and biological systems are most important. Chemists are aware that certain chemical reactions can freely oscillate in time or space. Prior to about 1920 most chemists believed that oscillations in closed homogeneous systems were impossible. The earliest scientific evidence that such reactions can oscillate was met with extreme scepticism. The reader can study the early history and the debate surrounding oscillating reactions in numerous references [Glansdorff and Prigogine (1971); Nicolis and Portnow (1973); Gray and Scott (1975, 1990); Nicolis and Prigogine (1977, 1979); Kawczyński et al. (1992); Nicolis and Rouvas-Nicolis (2007a, b), and many others]. Gray and Scott (1975) give an account of past and contemporary research and achievements of individual researchers in the field. The most famous oscillating chemical reaction is the Belousov–Zhabotinsky (BZ) reaction (Zhabotinsky, 1967, 1974). This is also the first chemical reaction to be found that exhibits spatial and temporal oscillations. The reader can demonstrate and carry out experiments on this reaction by recipes available in the literature. Theoretical models of oscillating reactions have been studied by chemists, physicists, and mathematicians. The simplest one may be the Lotka–Volterra model (Lotka, 1920). Some other models are the Brusselator, proposed by Prigogine and his collaborators at the Free University of Brussels

(Glansdorff and Prigogine, 1971), and the Oregonator, where the chemical mechanism for the oscillatory Belousov reaction has been described by a general model composed of five steps involving three independent chemical intermediates (Field and Noyes, 1974). This latter model, which shows a limit cycle behavior in a real system, was designed to simulate the Belousov–Zabotinsky reaction (Zhabotinsky, 1967, 1974).

Winfree (1972, 1973, 1974) was the first who launched two different experimental efforts—to reset the circadian (24 h) rhythms of fruit-flies using short light flashes, and to describe the patterns of an oscillating chemical reaction, the BZ reaction. Although these appear to be in completely different scientific domains, Winfree demonstrated theoretical links that led to testable predictions. To understand the connection, first think of a rotating disc, for example, an old-fashioned long-playing record. At the very center of the disc lies a fixed point that does not move at all. Winfree recognized that, in any mathematical equation describing a biological or chemical oscillator, there must also be a fixed point that does not lead to oscillation. But how could this fixed point be observed experimentally? In fact, there are lots of ways. For instance, if stimuli are delivered to the oscillator at different phases of the cycle, and the magnitude of resetting of the oscillator cycle is plotted as a function of the timing of the stimuli, the graph will have different topological properties for small and large stimuli. Winfree demonstrated these features in studies of the circadian rhythms of fruit flies in the 1970s, using light pulses as the stimuli. For chemical reactions, if the fixed point could be trapped in a thin layer of a solution, spiral waves of chemical activity would be observed. In a landmark 1972 paper, Winfree rediscovered spiral waves in the BZ reaction (Zhabotinsky, 1974). Winfree's lectures on this work were enlivened by his distribution of the BZ reaction on filter papers. Winfree extended this work by showing how a stimulus delivered to the heart would lead to the initiation of fibrillation. But to analyze the geometry of wave propagation in the pumping chambers of the heart, it would be necessary to consider waves in three dimensions. Winfree introduced the concept of scroll waves to describe the three-dimensional rotating waves that, in cross-section, appear as a spiral. An information about the chemical waves in confined regions, related experiments and Hamilton–Jacobi–Bellman theory can be found (Sieniutycz and Farkas, 1997).

Chemical and biological wave phenomena became well known and are widely studied in the recent years. While these waves show many similarities with the classical mechanical and electromagnetic waves there are important differences as well. For example, when electromagnetic or mechanical waves travel through a dissipative medium their amplitude decreases inevitably due to energy losses. An amplifier should be applied if we want to restore the original amplitude of a damped wave. Even in the absence of any dissipation, only planar electromagnetic or mechanical waves can propagate with a constant amplitude; their amplitude changes in all other cases. Biological and chemical waves are different. They propagate in a so-called excitable or active media containing evenly distributed energy sources which can be regarded as local amplifiers. As a result the amplitude of chemical and biological waves are preserved while they are propagating in a homogeneous active medium. Such waves can be described simply by evolving fronts and orthogonal rays and this is the basis of the geometric wave theory. In the paper by Volford et al. (1999) the geometrical wave theory and its application for rotating waves are discussed. The waves are rotating around a circular obstacle which is surrounded by two homogeneous wave conducting regions with different wave velocities. The interface of the inner slow and the outer fast region is also a circle but the two circles (the obstacle and interface) are not concentric. The various asymmetric cases are classified and described theoretically. In the second experimental part chemical waves rotating in a so-called moderately asymmetric reactor are studied. A piecewise homogeneous wave conducting medium is created applying a novel reactor design. All the three theoretical cases of the moderately asymmetric arrangement are realized experimentally and qualitative and quantitative comparison of these results with the theoretical predictions show a good agreement.

The most common formulation of the Fermat principle states: the light propagates in such a way that the propagation time is minimal. The article by Farkas et al. (2005) outlines the historical background as well as the different formulations of the Fermat principle. This principle is applied to rotating chemical waves in confined spaces (chemical pinwheels) designed by Winfree (1972). Theory and experiments of rotating chemical waves was further developed by many researchers, for example, Volford et al. (1999). The Fermat principle is valid for any traveling wave, not only for light. In optics, the underlying equation is the wave equation, which describes all details of propagation. Geometrical optics is a limiting case when wavelength tends to zero. The opposite limiting case—when wavelength tends to infinity—is the field of geometrical wave theory. That theory is based on Fermat principle as well as on the dual concepts of rays and fronts. A context to analyze wave fronts can also be provided by the dynamic programming method (Sieniutycz and Farkas, 1997). Recent literature on chemical pinwheels can be found in Volford et al. (1999), Yang and Epstein (2003), and Farkas et al. (2005).

Chemical waves can be described in details by reaction–diffusion equations. These equations may have traveling wave solutions. The essential features of the evolution of chemical wave fronts can also be derived from the geometric theory of waves. Some recent formulations of the Fermat principle require only stationarity of the extremals instead of the more strict requirement of minimal propagation time. This problem is discussed, and it is concluded that for chemical waves only the requirement of minimum is relevant: maximum and "inflexion-type" local stationarity does not play any role (Figs. 5.6 and 5.7).

The chemical waves and its prairie fire picture underlines that only the quickest rays have effects. Nevertheless, there are singularities, when there are several paths of rays with the same propagation time. For these cases the related fields of singularities and caustics are relevant. Special attention is paid to aplanatic surfaces, where all the reflected or refracted

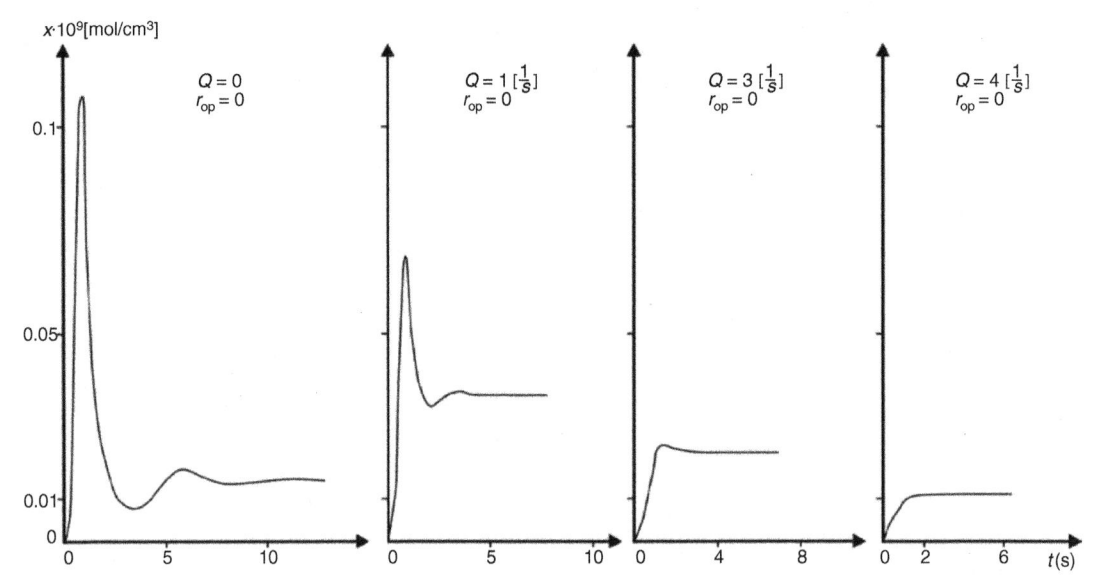

Figure 5.6: Various trajectories showing the quenching effect of flow Q in the process of CO oxidation (Jurkowska, 1982).

paths require the same time, that is, stationarity holds globally. Nonaplanatic refraction is discussed, too: the special example of refracting sphere is treated analytically. Finally, the shape of a chemical lens is derived: this "chemical lens" is able to perfect image formation with chemical waves; that is, circular fronts will be also circular after refraction (Fig. 5.8).

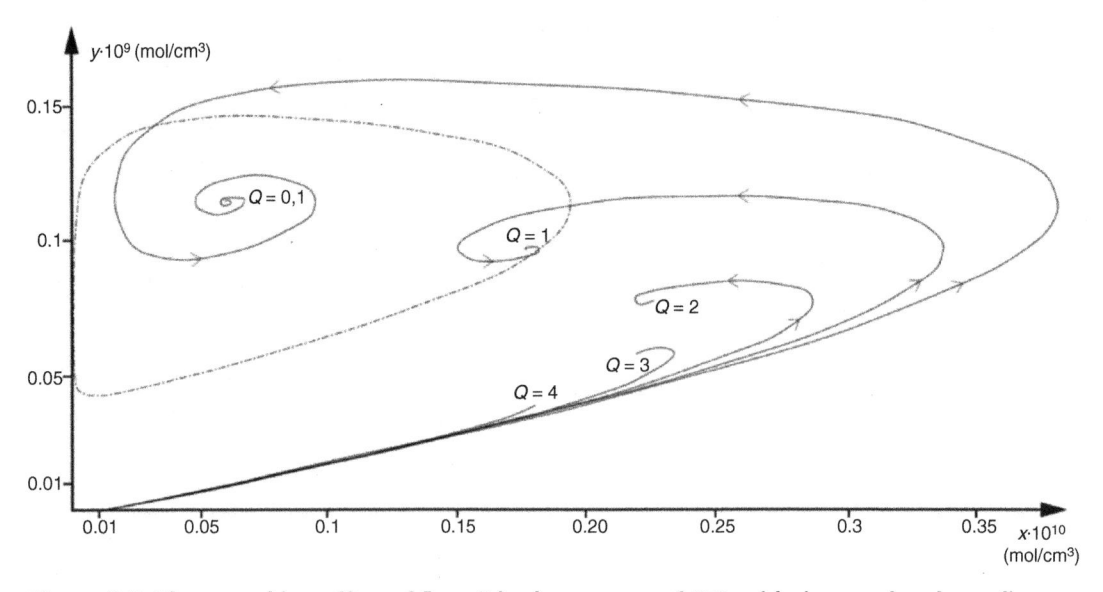

Figure 5.7: The quenching effect of flow Q in the process of CO oxidation on the phase diagram (Jurkowska, 1982).

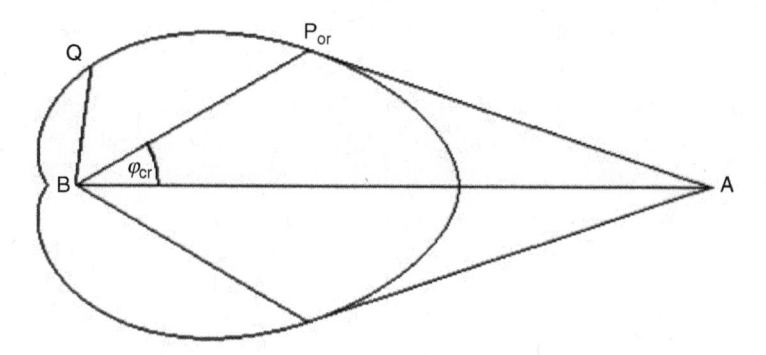

Figure 5.8: Chemical lens (Farkas et al., 2005).

Månsson and Lindgren (1990) described the role of the transformed information in creation processes by showing how the destructive processes of entropy production are related to the creative processes of formation. The role of information for the theory of thermodynamic equilibrium and relative (Kullback, 1959) information for the definition of the exergy were reviewed. Superiority of the information theoretic approach over traditional thermodynamics has been advocated for analyzing systems with emergent spatial structures, for example, spatially self-organizing chemical systems (Yates, 1987). Using examples of oscillatory reactions, these authors have shown that the average entropy production in an oscillating system is not necessarily lower than the entropy production of an (unstable) steady state. They concluded that a general principle of minimum entropy production cannot exist for oscillatory systems (Fig. 5.9).

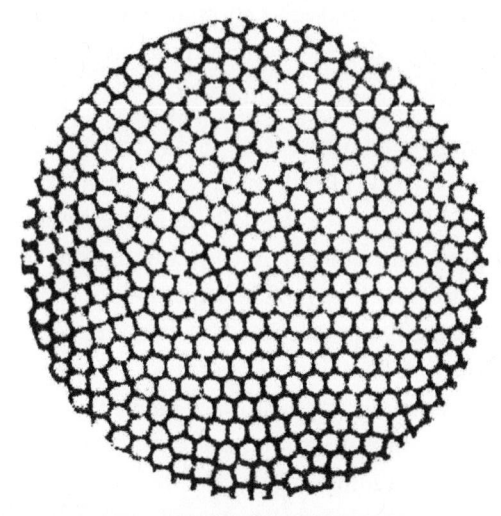

Figure 5.9: The spontaneous organization of water due to convection: once convection begins and the dissipative structure forms a pattern of hexagonal Bénard cells appear (from Capra, 1996 p. 87).

Chaos is detected very frequently in the chemical world. Chaos theory received its universal form since Feigenbaum's (1979) discovery showing the role of functional equations to describe the exact local structure of bifurcated attractors of $x_{n+1} = \lambda f(x_n)$ independent of specific f. In a paper on the transition to aperiodic behavior in turbulent systems Feigenbaum (1980) presented a recursive method that allows the explicit computation of the aperiodic temporal behavior achieved thorough the production of successive subharmonics. The results, which have universal character, resulted in appropriate corrections to the turbulence theory in the books on fluid mechanics (Landau and Lifshitz, 1987). Vulpiani (1989) revised the classical theory of Bachelor which gives k^{-1} law for the power spectrum of a passive scalar at wavenumbers k, for which molecular diffusion is unimportant and much smaller than the fluid viscosity. Applying some ideas from the dynamical systems theory he shows that this power law is related to the chaotic system of marker particles (Lagrangian chaos) and to the incompressibility constraint. This approach also shows that k^{-1} regime is present in nonturbulent fluids and is valid for all dimensions d not smaller than two. In their book on fractals, Bak and Chen (1989) focus on physical aspects of fractals. As they state, fractals originate in nature from self-organized critical dynamical processes.

The study of nonlinear dynamics associated with the motion of atoms or molecules inside matter is a relatively new topic of nonequilibrium statistical mechanics (Mori, 1965; Gaspard and Nicolis, 1990). At this microscopic level, the nonlinear dynamics may induce chaotic behavior over the time scale of the collisions between the particles. Relationships have been established between the characteristic quantities of chaos and the transport coefficients, in particular, thanks to the escape-rate theory. These relationships are based on the observation that the microscopic chaos induces fractal structures in the phase space of out-of-equilibrium systems and in the modes of exponential relaxation toward the thermodynamic equilibrium. These fractal structures are the manifestation of a subtle order in the deterministic chaotic motion of interacting particles.

For dynamical systems of large spatial extension giving rise to transport phenomena, like the Lorentz gas, Gaspard and Nicolis (1990) have established a relationship between the transport coefficient and the difference between the positive Lapunov exponent and the Kolmogorov–Sinai entropy per unit time, characterizing the fractal and chaotic repeller of trapped trajectories. They also discussed the consequences of their results for nonequilibrium statistical mechanics and thermodynamics. See, for example, "complex systems" in Wikipedia, the free encyclopedia, en.wikipedia.org/wiki/Complex system. Gaspard's (1998) book describes recent advances in the application of chaos theory to classical scattering and nonequilibrium statistical mechanics generally, and to transport by deterministic diffusion in particular. The objective of fluctuation theory and theory of stochastic processes is to analyze the response of nonlinear dynamical systems to internally generated fluctuations and to stochastic forcing of external origin (Gaspard, 1998). Special emphasis is placed in the stochastic dynamics of nonequilibrium systems (breakdown of detailed balance) and in the connections between stochastic behavior and the underlying deterministic dynamics. Tools include analytic and numerical methods for solving master and Langevin equations, Markov processes, and large deviation theory. Research on this topic has led to the discovery of the laws governing the fluctuations around nonequilibrium states, including their critical behavior around bifurcation points and their transient behavior in systems involving multiple time scales. Noise induced transitions have also been identified in the response of dynamical systems to external stochastic forcings. Other topics of interest are: Langevin equation description of chemical and hydrodynamic systems (critical fluctuations, persistence of collective behavior, transient phenomena); escape processes from metastable states; recurrence time and extreme value distributions in autonomous and nonautonomous deterministic and stochastic dynamical systems; effect of stochasticity in the vicinity of instabilities and bifurcation points; stochastic limit cycles; microscale thermodynamics; and so forth. Moreover, Gaspard et al. (2001) investigated the fractality of the hydrodynamic modes of diffusion. Gaspard (2004) formulated a fluctuation theorem for nonequilibrium reactions. Zalewski's (2005) thesis on chaos and oscillation in heterogeneous chemical reactions provides a number of examples, in particular those describing the influence of catalyst deactivation on the location of chaotic regions (Zalewski and Szwast, 2006). Chaos and oscillations in biological arrangements of the prey–predator type are also discussed (Zalewski and Szwast, 2007).

A concise, synthesizing book on engineering application of chaos is available (Kapitaniak, 1998). In Chapters 1 and 2, basic concepts and methods used in nonlinear dynamics are introduced. Chapter 3 deals with discrete dynamical systems. Chapter 4 introduces fractals, the Hausdorff dimension and fractal basins of attraction. Chapter 5 is devoted to routes to chaos. In Chapter 6 the author concentrates on chaos role in mechanical, chemical, electrical, and structural engineering. The final chapter is concerned with a new and promising field of controlling chaos. See El Naschie's (2000) review of this book.

Bag and Ray (2000) investigate the fluctuation–dissipation relationship in chaotic dynamics. They consider a general N-degree-of-freedom dissipative system that exhibits chaotic behavior. Based on a Fokker–Planck description associated with the dynamics, they establish that the drift and the diffusion coefficients can be related through a set of stochastic parameters which characterize the steady state of the dynamical system in a way similar to the fluctuation–dissipation relation in nonequilibrium statistical mechanics. The proposed relationship is verified by numerical experiments on a driven double-well system. The main conclusions of their study are the following: a class of dynamical stochastic parameters that attain their steady-state values in the long-time limit of the dynamical system may be used to characterize the dynamical steady state of the system. The first one, first proposed years ago by Casartelli et al. (1976) as a measure of the chaoticity of the system, is closely related to Kolmogorov entropy. Bag and Ray (2000), however, have established a connection between the drift and the diffusion coefficients of the Fokker–Planck equation and the dynamical stochastic parameters in the spirit of fluctuation dissipation relation. The realization of this relation in chaotic dynamics therefore carries the message that although comprising a few degrees of freedom, a chaotic system may behave as a statistical-mechanical system although in a somewhat different sense. Many chaotic systems with algebraically simple representations were described by Sprott (2000). They involve a single third-order autonomous ordinary differential equation (jerk equation) with various nonlinearities. Piecewise linear functions permit electronic implementation with diodes and operational amplifiers. Several new simple and robust chaotic electrical circuits were evaluated.

Periodic solutions in porous catalyst pellets attract the chemical engineers. Datsevich (2003) showed oscillations in pores of a catalyst particle in exothermic liquid (liquid–gas) reactions. According to his model, if the released heat in the pore exceeds certain critical value, the alternating motion of the liquid driven by the formation of the bubble takes place in the pore. Burghardt's and Berezowski's (2003) analysis and numerical simulations have revealed the possibility of the generation of homoclinic orbits as a result of homoclinic bifurcation in the model which describes transport phenomena and chemical reaction. They proposed a related method for the development of a special type of diagrams—the so-called bifurcation diagrams. These diagrams comprise the locus of homoclinic orbits together with the lines of limit points bounding the region of multiple steady states as well as the locus of the points of Hopf bifurcation. Thus, they define a parameter domain for which homoclinic bifurcation can take place. They also facilitate determining of conditions under which homoclinic orbits arise. Two kinds of homoclinic orbits have been observed: semistable and unstable orbits. The nature of the homoclinic orbit depends on the stability features of the limit cycle which is linked with the saddle point. Interesting dynamic phenomena are linked with the two kinds of homoclinic orbits; these phenomena were illustrated in the solution diagrams and phase diagrams.

Unstable homoclinic orbit appears as a result of the merger between an unstable limit cycle and a saddle point and surrounds stable stationary point. This orbit plays therefore the role of a closed separatrix, dividing the region of state variables in the phase diagram into two subregions with two attractors. Two stable stationary points or a stable stationary point inside the orbit and a stable limit cycle outside the homoclinic orbit can be these attractors. On the other hand, merger of a stable limit cycle and a saddle point leads to the formation of a semistable homoclinic orbit which is stable on the inside and unstable on the outside. The internal stability of the orbit leads to the fact that the trajectories originating from the region surrounded by the semistable homoclinic orbit tend to this orbit and, consequently, to the unstable saddle point (obviously, after an infnitely long time). A region of state variables is thus created with a special property: the trajectories issuing from this region can reach the unstable stationary point—a phenomenon impossible without the existence of the homoclinic orbit. In the following research the authors develop stability analysis of steady-state solutions for porous catalytic pellets in which they test the role of the shape of the pellet.

Two books may attract the reader's attention. Epstein's (1998) book, which covers oscillations, waves, patterns, and chaos in the realm of chemical reactions, provides many relevant examples. It includes numerous practical suggestions on reactor design, data analysis, and computer simulations. Throughout the authors emphasize the chemical mechanistic basis for self-organization. Goldbeater's (1996) book covers the field of biochemical oscillations and cellular rhythms. His approach rests on the analysis of theoretical models related to experiments. Among the main rhythms considered are glycolytic oscillations observed in yeast and muscle.

5.6 Stability of Thermal Fields in Resting and Flow Systems Far From Equilibrium

The problem of thermoconvective instabilities in liquid layers has become a classical subject in fluid mechanics with the pioneering works of Benard (1900) and Rayleigh (1916). A comprehensive review can be found in Koschmieder's (1993) book. Rayleigh (1916) first described convection currents in a horizontal layer of fluid, when the higher temperature is on the underside. In his famous book on hydrodynamic and hydromagnetic stability, Chandrasekhar (1961) treats, between others, the thermal instability of a layer of fluid heated from below. This problem is well suited to illustrate many aspects of the general theory of hydrodynamic stability. If the problem is enlarged to include effects of rotation and magnetic field, conflicting tendencies to which a fluid can be subject may be revealed. However, in Chapter 2 of his book, the author considers the simplest problem in the absence of rotation and magnetic field. The temperature gradient in the layer of fluid heated from below is qualified as adverse since, on account of thermal expansion, the fluid at the bottom will be lighter than the fluid at the top, the situation which is potentially unstable. Thus, there will be a tendency to redistribute, the tendency inhibited by the fluid's viscosity. One can expect that the maintained adverse gradient of temperature must exceed a certain value before the instability can manifest itself. Although the thermal convection had been recognized earlier, the earliest experiments which definitively demonstrated the onset of thermal instability in fluids are those of Benard (1900). A detailed description of these experiments is available (Benard, 1900; Chandrasekhar, 1961; Koschmieder, 1993). In all investigated cases, Benard found that when the temperature of the lower surface was gradually increased, at a certain instant, the layer became recirculated and revealed its dissection into cells. He also noticed that there were motions inside the cells: of ascension at the center, and of descension at the boundaries with the adjoining cells. Further searches showed that in all investigated liquids instability sets in at the experimental Rayleigh number 1700 ± 51 which is in quite good accord with the theoretical value 1708.

Busse (1967) discusses the stability of finite amplitude cellular convection and its relation to an extremum principle. The cellular convection flow in a layer heated from below occurs for Rayleigh number R close to the critical value R_c. It is shown that in this region the stable stationary solution is determined by a minimum of the integral

$$\int_0^{H_0} R(H)\,dH, \qquad (5.17)$$

where $R(H)$ is a functional of arbitrary convective velocity fields which satisfy the boundary conditions. For the stationary solutions $R(H)$ is equal to the Rayleigh number. H_0 is a given value of the convective heat transport. In a second part of the paper explicit results are derived for the convection problem with deviations from the Boussinesq approximation owing to the temperature dependence of the material properties.

Plows (1968) reports of calculations of the Nusselt number for steady two-dimensional laminar convection in a thin fluid layer between parallel conducting plates. He uses an iterative numerical technique to calculate the Nusselt number and the structure of the roll for Rayleigh numbers. He concludes that various theoretical determinations of Nu in terms of Ra and Pr in the range $Ra < 22{,}000$ and $0.5 < Pr <$ infinity, are agreeably consistent for the special geometry of the two-dimensional roll. It remains to be seen how applicable this precision is to the laboratory or to the nature. Since the methods of solving flow stability problems involve modeling and/or simplifying assumptions, it is desirable that the analytical results obtained for the natural convection and fluid stability are subjected to the ultimate test of experiment. Assuming that the disturbance velocity is two dimensional, that is, the secondary flow consists of simple Taylor vortices, Snyder (1968) considers the stability of rotating Couette flow. In order to check some recent numerical results he performs measurements of the onset of instability and compared available data with the computations of different investigators. With a few exceptions the agreement was found to be excellent. He also presented a simple formula for calculating the instability onset value when only the inner cylinder rotates.

Gage (1972) states the viscous stability theory for thermally stratified flows: discontinuous jets and shear layers. Considering the influence of thermal diffusion on the convective stability of a horizontal layer of a two-component fluid heated from below, Schechter et al. (1972) found that contrary to intuitive notions, the exceedingly small thermal diffusion separations give rise to large modifications of the usual Benard instability. The small separations, for example, can give rise to an oscillatory instability rather than the usual nonoscillatory marginal states observed when a plate is heated from below. That behavior should be observable in liquids when the less dense fluid migrates to the upper cooled wall. If the more dense fluid migrates to the upper plate, a stability occurs at Rayleigh numbers much below the usual Benard point. Quite interestingly, an instability can exist even if the overall density gradient is not adverse for systems heated from above provided the more dense component of the binary mixture migrates to the upper boundary. Platten and Chavepeyer (1973) observed oscillatory motion in Bénard cell due to the Soret effect.

Investigating the influence of heat transfer on the stability of plane Poiseuille flow, Potter and Graber (1972) include additional viscosity terms in a modified Orr–Sommerfeld equation. They show that it is the presence of these extra terms which leads to a prediction of more unstable flows, for without the inclusion of these additional terms, the flow is stabilized. The results show that the temperature difference between the walls is always destabilizing and leads to a significant reduction of the critical Reynolds number. For example, a temperature difference between the walls of 100°F causes the reduction in the critical Reynolds number from 7500 to 4600. Gage (1972) treats viscous stability theory for thermally stratifies flow, in discontinuous jets and shear layers. For viscous, incompressible fluid the stability of small disturbances is governed by the Orr–Sommerfeld equation

$$(D^2 - \alpha^2)^2 \phi(z) = i\alpha R[(U-c)(D^2 - \alpha^2)\phi(z) - (D^2 U)(\phi(z))] \tag{5.18}$$

(Gage, 1972; Glansdorf and Prigogine, 1971; Joseph, 1976). In this equation $U(z)$ is the basic flow velocity, α is the disturbance wavenumber, c is the disturbance wave speed, and $\phi(z)$ is the z dependent part of the disturbance stream function. The x coordinate is taken in the direction of flow. The vertical coordinate, z, is directed across of the flow, and D represents d/dz. The stability analysis of the Orr–Sommerfeld equation involves the formulation of a characteristic value problem in which the solutions of the equation are required to satisfy certain boundary conditions. Eigenvalues and eigensolutions obtained from the solution of the characteristic value problem depend upon the basic velocity profile and the boundary conditions. The main difficulty in obtaining the solution of the characteristic value problem for a stability analysis is often obtaining solutions to the Orr–Sommerfeld equation. Several approaches are used, yet each method has its limitations. Using the theory of hydrodynamic stability, Gage (1972) exemplifies these approaches in his study of the effect of thermal stratification on the stability of unbounded jets and shear layers. The stabilizing effect of the thermal stratification is seen especially for small wave-number (large-scale) disturbances. As stated by Villers and Platten (1984), the critical Raleigh number in Benard problem depends on type of boundary conditions adopted (rigid or free boundaries). The amplitude of the convective motion is proportional to $(Ra - Ra^{crit})^n$, where $n = 1/2$ near the threshold of convection. For layers of liquid mixtures, thermal diffusion (also called Soret effect) must be taken into account; this effect induces a molecular separation due to the thermal gradient, and the resulting mass fraction gradient drastically changes the stability. In mixtures, when the Soret effect is operative, the Raleigh number and the heat flux are not always monotonically related, and this can lead to the situation where a stable stationary solution does not exist for all values of heat flux, even when there are stable stationary solutions for all Raleigh numbers. As shown by Villers and Platten (1984) heating curves in the two-component Benard problem present hysteresis loops, and, when the temperature gradient is imposed, the obtained period of transient oscillations is in agreement with the computed value from the (linear) theory, and the hysteresis loop is qualitatively identical to the theoretical curve obtained for free and pervious boundaries. If, however, heat flux is imposed, the hysteresis loop is different and sustained oscillations arise, in agreement with experiments.

It is not unusual for a Soret coefficient to change sign with temperature. Mojtabi et al. (2002) develop the theory for the onset of convection in such systems, heated from below or from above, provided that the mean temperature is precisely that at which the change of sign occurs. They also consider the realistic case of rigid, conducting, impervious boundaries for later comparison with laboratory experiments. The authors conclude that a non-Boussinesq Soret effect with a zero mean Soret coefficient always destabilizes a two-component liquid layer, simply because there is in some part of the layer an unstable concentration stratification, independent of whether heating is from below or from above. This effect has been quantified using a linear hydrodynamic stability theory approach, and experiments are shown to be, if not easily achievable, at least possible.

Several numerical and experimental methods have been developed to investigate enclosures with and without obstacle because these geometries have great practical engineering applications. In Jami et al. (2007) a numerical investigation of laminar convective flows in a differentially heated, square enclosure with a heat-conducting cylinder at its center, is carried out. The flow and the temperature are computed using respectively the lattice Boltzmann equation and finite difference with suitable coupling to take natural convection into account. The investigation is performed for $Pr = 0.71$, Rayleigh numbers of $Ra = 103–106$ and temperature-difference ratio of $\Delta T^* = 0–50$. The average hot and cold walls Nusselt numbers, the flow and temperature fields are presented and discussed. For a constant Ra, the average Nusselt number at the hot and cold walls (Nu_h and Nu_c) vary linearly with ΔT^*: Nu_h decreases with ΔT^* while Nu_c increases with ΔT^*. Lattice Boltzmann method applied to the laminar natural convection in an enclosure with a heat-generating cylinder conducting body.

Most investigations have dealt with fluid layers steadily heated from below, with a static temperature gradient and without evaporation. Zhang et al. (2013) reported experimental results on the convective instability of a transient evaporating thin liquid layer. Evaporation is identified as an agent causing Rayleigh–Benard convection and/or Marangoni–Benard convection. Convective flow occurs in the evaporating liquid layer as long as the evaporation is strong enough, regardless of whether the layer is heated or cooled from below. The wavelength of the cells maintains a preference value in steady evaporation. When an evaporating thin layer is strongly cooled from below, both the nonlinear temperature profile of the layer and the flow pattern change rapidly during the transient evaporation process. The wavelength of convection cells increases with time and tends towards the preference value with the approach of a steady evaporation stage. A modified Marangoni number and a modified Rayleigh number serve as the dimensionless control parameters for this system.

In Barletta's (2015) paper viscous dissipation as a source of thermoconvective instability is studied. Thermal instability is not necessarily induced by an external thermal forcing due to the temperature boundary conditions. The author expresses the opinion that the effect of viscous dissipation may be the sole cause of a thermoconvective instability either in a fluid clear of solid material or in a fluid-saturated porous medium. Several recent investigations have contributed to illustrate this result under different flow and thermal conditions. The elementary physical nature of the dissipation-induced instability is the same as that of the Rayleigh–Bénard instability, namely the onset of a secondary buoyant flow taking place when the effect of basic temperature gradient becomes sufficiently intense. The model of Oberbeck–Boussinesq convection with viscous dissipation is described (Barletta, 2015). The essential difference is that the dissipation instability is not induced by an external thermal forcing due to the temperature boundary conditions, as it happens for the Rayleigh–Bénard instability. On the other hand, the cause of the dissipation instability is the basic flow rate itself, acting thermally as a heat generation mechanism. Thus, in this work, the governing parameter determining the transition from stability to instability is not the Rayleigh number, as in the classical Rayleigh–Bénard problem, but the product between the Gebhart number and the square Péclet number. Dissipation-instability in a porous channel with an open boundary is investigated. Literature is surveyed relative to clear fluids and to porous media.

Influence of the relaxation time and viscoelastic properties of fluids on the solution of the Benard problem have attracted many researchers. Many important industrial processes involve (steady) flows of polymeric liquids which exhibit complex non-Newtonian rheological behavior. Advanced, contemporary treatise of simple and complex models of non-Newtonian fluids is available (Bird et al., 1977). Production rates in these processes are often limited by the onset of flow instabilities. Two well-known examples are film casting and fiber spinning, where a periodic variation in film thickness or fiber diameter is observed when a critical take-up speed is exceeded. Numerous attempts have, therefore, been made to understand polymer-processing instabilities by theoretical stability analyzes for model viscoelastic fluids. For mathematical modeling of these systems, see Beris and Edwards (1994) and Beris and Öttinger (2008). Early analyses of viscoelastic fluids showed that, in such media, instability might appear in the form of oscillating convection cells under certain conditions. This is in contrast to the usual Newtonian solution where steady secondary flow is observed. Sokolov and Tanner (1972) considered Maxwell fluid with free boundaries and with rigid boundaries, and worked out a method of analyzing the overstable Benard problem for a general viscoelastic fluid. They have solved the Benard problem for an incompressible viscoelastic case and investigated conditions under which oscillatory motion is the first mode of instability. Their basic equation describing the onset of instability in terms of disturbed quantities may be written in the form

$$(D^2 - a^{*2})[\sigma^* - (D^2 - a^{*2})][\sigma^* - q(\sigma)\Pr(D^2 - a^{*2})]w = -\Pr Ra^{*2} w \qquad (5.19)$$

Compare this result with other Orr–Sommerfeld equations (Gage, 1972; Glansdorff and Prigogine, 1971; Luikov and Berkovsky, 1974; Palm 1975; Joseph, 1976). It was of most interest here to investigate overstability, that is, the situation when the boundary between stable and unstable states is not $\sigma^* = 0$ but $\sigma^* = iw^*$, or $\sigma = iw$, where w is a real number. Sokolov and Tanner (1972) found that the oscillating mode of instability will occur practically in fluids with large ratios of relaxation tine to thermal relaxation coefficient where the thermal relaxation coefficient is the ratio of the squared fluid slab thickness d^2 to the thermal conductivity. Their theoretical results have been applied to a Maxwellian fluid and to some real viscoelastic solutions. The numerical results for the latter suggest that although oscillations of Benard cells is possible, very high temperature gradients or high gravitational fields would be required before the oscillating cells could be observed in common polymeric solutions of moderate viscosity. Joseph's (1976) book gives a comprehensive presentation of the theory of hydrodynamic stability, including the stability of motion of viscoelastic fluids.

Lagnado et al. (1985) have performed a linear stability analysis for an Oldroyd-type fluid undergoing steady two-dimensional flows in which the velocity field is a linear function of position throughout an unbounded region. This class of basic flows is characterized by a parameter λ which ranges from $\lambda = 0$ from simple shear flow to $\lambda = 1$ for pure extensional flow. The time derivatives in the constitutive equation can be continuously varied from corotational to codeformational as a parameter β varies from zero to unity. The linearized disturbance equations are analyzed to determine an asymptotic behavior as time $t \to \infty$ of a spatially periodic initial disturbance. The authors have found that unbounded flows in the range of λ between 0 and 1 are unconditionally unstable with respect to periodic initial disturbances which have lines of constant phase parallel to the initial streamline in the plane of the basic flow. For sufficiently small Weissenberg numbers only disturbances with sufficiently small wavenumber in the direction normal to the basic flow plane are unstable. Yet, for certain values of β, Weissenberg numbers exist above which flows are unstable for all values of the wavenumber α. Stabilizing effect of the retardation time is discovered which stabilizes the basic flow by increasing the critical value of the Weissenberg number.

Captained (2003) studied the onset of the unstable temperature distribution appearing in a plane of a catalytic burner. He assumed an instability mechanism governed by the temperature dependence of the viscosity. An area of lower temperature (dark zone) characterized by a smaller value of viscosity increased the local mass flow, velocity, while the pressure drop remained constant over burner's cross-section. The increased mass velocity produced enhanced cooling of the area, whereby the heat conduction from the hotter area tends to restore a homogeneous temperature distribution. A linear analysis of this mechanism yielded a simple solution for the neutral stability.

5.7 New Approach to Lapunov Functions and Functionals

As in Section 5.2, the conclusion regarding the stability or instability of disequilibrium singular submanifolds can be judged solely on the basis of the properties of function or functional V. This approach is the consequence of assuming that the derivative function

[dV/dt with respect to the kinetic equations or $\dot{V}(\mathbf{a})$] of a constant, for example, negative, sign is the basis in the procedure searching for V. The proposed reorientation with respect to the traditional approach (traditionally one assumes $V(\mathbf{a})$ rather than $\dot{V}(\mathbf{a})$) follows from numerous observations which show that the derivative functions $\dot{V}(\mathbf{a})$ around disequilibrium singular states behave in the state space in the way qualitatively similar to those known for equilibrium or pseudoequilibrium systems (ie, behave as those in Section 5.2). However, the form of potential V in the state space is in general different in disequilibrium situations, which implies different properties of stability or instability.

Below we outline the mathematics involved in the construction method for the Lapunov function, V, which, we believe, is "more physical" and more effective than the methods applied in earlier approaches.

Writing the linear set of perturbed equations around the selected singular state in the form

$$\frac{d\mathbf{a}}{dt}(\mathbf{Aa}) \tag{5.20}$$

($\mathbf{a} = \delta X$ is a perturbed state) one may search for a suitable quadratic form

$$V(\mathbf{a}) \equiv \frac{1}{2}\mathbf{a}^T \mathbf{D} \mathbf{a} \tag{5.21}$$

where $\mathbf{D}$ is an unknown symmetrical matrix. Differentiating this equation and using Eq. (5.20) yields

$$\dot{V}(\mathbf{a}) = \frac{1}{2}\mathbf{a}^T(\mathbf{A}^T\mathbf{D} + \mathbf{D}\mathbf{A})\mathbf{a} \tag{5.22}$$

We shall also use a function $\dot{V}(\mathbf{a})$ which, by definition, satisfies a dissipation-like formula

$$\dot{V}(\mathbf{a}) \equiv a\dot{\mathbf{a}}^T \mathbf{B} \dot{\mathbf{a}} \tag{5.23}$$

where a is a constant of suitable units. Substitution of Eq. (5.20) into previous equation yields

$$\dot{V}(\mathbf{a}) \equiv a\mathbf{a}^T \mathbf{A}^T \mathbf{B} \mathbf{A} \mathbf{a} \tag{5.24}$$

After comparing Eqs. (5.22) and (5.24) we obtain the following matrix equation

$$a\mathbf{A}^T \mathbf{B} \mathbf{A} = \mathbf{A}^T \mathbf{D} + \mathbf{D}\mathbf{A} \tag{5.25}$$

Its solution leads to matrix $\mathbf{D}$ of unknown function V in terms of given matrices $\mathbf{A}$ and $\mathbf{B}$. As opposed to Keizer (1987a), we did not assume definite signs of V and $\mathbf{D}$ but rather imposed the requirement that the time derivative dV/dt with respect to the kinetic equations, that is, $\dot{V}(\mathbf{a})$, is a given function of a definite sign (negative in the energy representation).

This result may be seen as a stability-related second-law formulation for a general system which may open or closed, and which admits an indefinite form (and sign properties) of the thermodynamic potential, but always requires a definite sign of the dissipation rate. Callen's (1988) book on thermodynamics is perhaps the sole text where the second law issues are understood in the similar way although they are not directly described in the form of statements of Lapunov stability theory.

Interestingly, this approach can easily be extended to fields which are spatially inhomogeneous disequilibrium systems described by partial differential equations. Let us recall that considering the stability of coupled heat and mass transfer described by the perturbed wave model in the form of perturbed conservation equations, Eq. (5.4), and perturbed phenomenological equations Eq. (5.5), we obtained the entropy production expression in the form of the identically satisfied second law inequality (5.8) or

$$\dot{V} \equiv \frac{\partial \delta^2 S}{\partial t} = \int_V (\delta \mathbf{J}^T \mathbf{L}^{-1} \delta \mathbf{J}) dV \geq 0. \tag{5.26}$$

Taking the inequality on the right-hand side of Eq. (5.8) or (5.26) as the starting point, we conclude that the Lapunov functional V coincides with $\delta^2 S$ and, in view of the identically nonnegative $\dot{V}$ of Eq. (5.8), stability and instability conditions of simple wave systems are contained in the properties of the second differential of the entropy. This analysis also shows that the classical stability conditions are violated for a negative specific heat (and for change of sign in other classical inequalities).

Similarly, another inequality, Eq. (5.10), was obtained for the transients of the perturbations governed by Eqs. (5.4) and (5.5). In this case, as Eq. (5.10) shows, the time derivative $\dot{V}$ is of nonpositive sign. Therefore the system described by functional V of Eq. (5.9) is unstable whenever the semipositive sign property of the integral (5.9) is violated. This could occur, for example, if the system is no longer characterized by a positively defined det $\mathbf{L}$ or the eigenvalues l_i are not all positive (degenerated dissipation function).

Summing up, the present approach assures that for systems in which the singular state is an equilibrium or quasiequilibrium one, the "purely thermodynamic" Lapunov function, V, which also may be related to the exergy, its balance and its negative source, follows as an efficient criterion of stability (or instability). Also, for a stationary nonequilibrium singular state, a nonclassical function V exists (whose time derivative is positive by the second law), which is suitable for predicting the qualitative properties and stability of trajectories.

Indefinite sign of the function V allows one to deduce the trajectory properties and their stability or instability by testing geometrical properties of V, that is, to proceed for fields in the same way as for equilibrium or pseudoequilibrium singular points (Sieniutycz and Komorowska-Kulik, 1980). Eq. (5.19) stresses that the negative sign of the time derivative function is required by the second law, secured by the negative sign of the matrix $\mathbf{B}$ in the energy representation. The solution of the stability problem is thus broken down to the investigation of the properties of function V or matrix $\mathbf{D}$. For an equilibrium singular point $\mathbf{B}$ is the classical dissipation function in the state space (rather than in the space of rates) whereas matrix $\mathbf{D}$ simplifies to the thermodynamic matrix of second derivatives of entropy or another classical potential (in general, with the accuracy to a multiplicative constant). See Izüs et al. (1995) for a similar approach and the history of previous efforts along this line.

It is interesting to extend this approach to construction of Lapunov's functionals describing disequilibrium fields in terms of partial differential equations for perturbations.

5.8 Further Work

Bhalekar and Burande (2005) identified a new positive definite Lapunov function, L_s, which is the excess rate of entropy production (per unit volume) as opposed to the standard excess entropy production of Glansdorff and Prigogine (1971). A positive definite thermodynamic Lapunov function, L_s, has been defined as the excess rate of entropy production. The total time derivative of L_s in thermodynamic perturbation space then determines the stability or instability of a process. The thermodynamic stability of deformation in viscoelastic fluids has been investigated using the proposed thermodynamic Lapunov function and Lapunov's direct method. To construct the thermodynamic perturbation space, the thermodynamic space based on extended irreversible thermodynamics (EIT) has been used. Three different models of deformation in viscoelastic fluids have been tested for the stability that establishes the regions of asymptotic stability and the stability under small disturbances in each case. The case of instability also surfaces out.

Burande and Bhalekar (2005) investigated thermodynamic stability of a few representative elementary chemical reactions using their recently proposed thermodynamic Lapunov function and Lapunov's second method. The Lapunov function, L_s, used by them is the excess rate of entropy production in the thermodynamic perturbation space, which thereby inherits the dictates of the second law of thermodynamics. This function is not the same as the excess entropy rate that one encounters in the literature. The chemical conversions studied in this presentation are $A + B \rightarrow vxX$ and $A + B \rightleftharpoons vxX$. For simplicity, the thermal effects of chemical reactions have been considered as not adding to the perturbation. The domains of thermodynamic stability under the constantly acting small disturbances, thermodynamic asymptotic stability and thermodynamic instability in these model systems get established.

Tangde et al. (2007) studied the thermodynamic stability of the industrially important reaction of sulfur trioxide synthesis using a framework of the so-called comprehensive thermodynamic theory of stability of irreversible processes (CTTSIP). The authors constructed an expression of the excess rate of entropy production, used the constitutive equations of the perturbation coordinates, and then established the sign of the time rate of Lapunov function. The latter has been defined using an appropriate irreversible thermodynamic expression for the rate of entropy production, that includes the entropy production from the existence of heat flux and the chemical conversion.

Some associated work may be quoted. Lebon and Rubi (1980) present a generalized theory of thermoviscosity aimed at modeling of nonisothermal viscous fluids in nonequilibrium. Perez-Garcia (1981) discusses various physical mechanisms of hydrodynamic instabilities. Upadhyay et al. (2000) consider complexity in ecological systems and indicate that the structural complexity is not necessary for the dynamical complexity to exist. Very simple ecosystems can display dynamical behavior unpredictable in certain situations. In certain cases, when riddled basins are found, even qualitative predictability is denied. Jami et al. (2007) apply the lattice Boltzmann method to the natural convection in an enclosure with a heat-generating cylinder conducting body.

Maini et al. (1997) considered the formation of spatial pattern in biological systems. The authors focused on one of the central issues in developmental biology: the formation of spatial pattern in the embryo. A number of theories have been proposed to account for this phenomenon. The most widely studied is the reaction–diffusion theory, which proposes that a chemical prepattern is first set up due to a system of reacting and diffusing chemicals, and cells respond to this prepattern by differentiating accordingly.

Vlad et al. (2004) investigated response experiments for nonlinear systems with application to reaction kinetics and genetics. They showed that the method involving the linearization of the kinetic equations leads to unpredictable results; because of the interference between measurement and linearization errors, either error compensation or error amplification occurs. Although their approach does not eliminate effects of measurement errors, it leads to more consistent results. For a broad range of inputs no error amplification or compensation occurs, and the error range for the rate coefficients is about the same as the error range of the measurements.

Volpert and Petrovskii (2009) reviewed reaction–diffusion waves in biology. Biological theory of reaction–diffusion waves begun in the 1930s with works in population dynamics, combustion theory and chemical kinetics. Now, it is a developed area which includes qualitative properties of traveling waves for a scalar reaction–diffusion equation and for system of equations, complex nonlinear dynamics, numerous applications in physics, chemistry, biology, medicine.

Some attention should be devoted to interfacial instabilities in fluid–solid systems. Liquid films covering solid surfaces create specific fluid–solid systems whose stability properties exhibit many unusual effects. The phenomenon of liquid flowing along an interface from places with low surface tension to places with a higher surface tension is named the Marangoni effect (Marangoni, 1865). He found that a liquid A spreads on a liquid B when the sum of the interfacial tension and the surface tension of A is lower than the surface tension of B. However, the spreading phenomenon is only one aspect of the Marangoni effect.

When gradients in surface tension arise due to concentration differences within one fluid, flow arises as well, and this phenomenon was first outlined qualitatively by James Thomson (1822–92). Therefore, the Marangoni effect is sometimes referred to as the Thomson effect. The term "Marangoni effect" is also used frequently for a phenomenon that, rather than inducing movement at the interface, retards interfacial motion. Interfacial instabilities can occur. Interfaces often contain traces of surface active substances that reduce the surface tension. In general, surface tension lowering solutes adsorb preferentially in the interface (Gibbs adsorption). The Marangoni effect can occur in a liquid–gas and in a liquid–liquid system. It can be the result of concentration and/or temperature gradients. In the case of concentration gradients, the effect is called the solutal Marangoni effect.

Concentration gradients in a liquid can be the direct result of mass transfer between phases, but can also be the indirect result of buoyancy or another type of convection, forced flow, heat transfer or temperature gradients in general. When temperature gradients are responsible for the Marangoni effect, the effect is frequently called thermocapillarity. Temperature gradients can be the result of heat transfer processes as well as mass transfer processes (involving enthalpy changes) or any type of natural or forced flow. Electric and magnetic fields can also influence flow at an interface by their influence on the surface tension, although this is rarely labeled Marangoni convection. Electric fields can initiate and interact with a nonuniformity of electric conductivity and dielectric constant. Numerous papers are devoted to this electrocapillarity and to the interaction of electric and magnetic fields and Marangoni and Rayleigh convection. In particular, research effort has been devoted to the reduction of thermocapillarity during crystallization from the melt in microgravity, using magnetic fields. Strong nonuniform electric fields have been used to induce the electrocapillary-related spraying effect or electrohydrodynamic effect, resulting in strongly enhanced mass or heat transfer.

Haltchev and Denier (2003) considered the effect of interfacial coupling on the stability of a boundary-layer fluxes over a permeable surface under conditions of intense interfacial mass transfer. The stability of the flow is governed by an eigenvalue problem of Orr–Sommerfeld type coupled to a second-order differential equation for the concentration disturbance field through a flux boundary condition at the permeable surface. Previous studies on this problem have ignored the effect on the stability of the flow of this coupling. Curves of neutral stability and the critical Reynolds number for the flow are obtained. These results show that the fully coupled system produces critical Reynolds numbers and wave-numbers that, in some cases, differ significantly from those obtained when the disturbance coupling is ignored.

The instability mechanism of rupture for thin liquid films is of importance for the understanding of flotation, of foams and emulsions, of coalescence of bubbles and droplets, of vapor condensation on a solid surface, and so forth. Applying basic physics of wetted layers (Deryagin and Tzuraev, 1984) and the hydrodynamic stability theory, Ruckenstein and Jain (1974) studied the rupture of a liquid film on a solid surface and of a free liquid film.

They established range of wavelengths of the perturbation for which instability occurs and evaluate the time of rupture. Film rupture occurs by instabilities, that is, when the perturbations grow in time. Jain and Ruckenstein (1976) theoretical framework determines the stability of liquid films on a solid surface immersed in a viscous fluid. Mandarelli et al. (1980) determine the stability conditions of symmetric and asymmetric thin liquid films to short- and long-wavelength perturbations.

They examine especially influence of surface active agents, of surface viscosity, and of the viscosity of the semiinfinite fluid, on the growth of the perturbation and on the rupture time. They next analyze their unified theoretical framework and study the stability of stagnant films having a wide range of thicknesses and located on a solid surface immersed in a viscous fluid. The stability of films having thicknesses smaller than the distance over which the London dispersion forces are effective is determined by an "effective" surface tension σ_{eff} (the sum of the surface tension and a term due to the difference between the forces acting upon an element of volume at the liquid–fluid interface of the thin film and of the corresponding bulk liquid). The films are unstable and will rupture when $\sigma_{eff} < 0$. The time for rupture is a function of the surface viscosity, the surface elasticity, and the viscosity ratio. The derived equations are also valid for large film thicknesses. In the latter cases, the effective surface tension is equal to the surface tension and the films are always stable.

Dynamic and thermodynamic stability conditions of nonextensive systems are different than those found for classical systems. Naudts and Czachor (2001) prove uniqueness of the equilibrium states of q-thermodynamics for $1 < q \le 2$, using both the unnormalized and the recently introduced normalized energy functional. The proof follows from thermodynamic stability of equilibrium states in case of the unnormalized energy. Dynamic stability conditions are also discussed. The authors expected to find links between dynamic stability conditions of certain nonlinear dynamics and thermodynamic stability conditions with appropriately defined free energy. However, these conditions seem independent.

In this paper we restricted ourselves to nonrelativistic systems. For the stability problems in relativistic dissipative fluids, see, for example, Hiscock and Lindblom (1983) and Pavon et al. (1983). Of course, there are many other references to stability problems in relativistic dissipative fluids. Finally, there remains uncovered the field of the quantum instabilities in the irregular and effectively random fluctuations in the fluid motion, in the causal interpretation of the quantum theory (Bohm and Vigier, 1954).

5.9 Concluding Remarks

One can ask: What are the proper conclusions which reflect the basic findings in this chapter? A basic conclusion is that in spite of great deal of papers searching for Lapunov functions V a relatively small progress was made in finding such forms of V which would reflect well qualitative properties of process paths in the vicinity of the singular point. The reason for this pessimistic statement is associated with the principal assumption present in a majority of works (of both mathematicians and physicists) that the functions V should be of definite signs and the conclusion about the qualitative properties of process paths should be drawn from the properties of time derivatives of V with respect to the dynamical equations for perturbations.

However, as shown here, the imposition of limitations for V to quadratic forms of definite (eg, positive) sign seems to preclude effective testing of stability and the recognition of qualitative properties of process paths. The constant-sign derivatives of dissipated exergy (or produced entropy) proposed as the starting point do a better job; after applying suitable mathematics, for example, Eq. (5.25), they lead to (new generations of) Lapunov functions of indefinite sign, which reflect well the qualitative properties of paths in the vicinity of the singular point.

Consequently, effective testing of paths properties should begin with the selection of constant-sign potentials representing assumed derivative functions $\dot{V}(\mathbf{a})$. For an arbitrary (equilibrium or disequilibrium) singular point $\mathbf{X} = \mathbf{X}_0$ ($\mathbf{a} = 0$), a constant-sign function $\dot{V}(\mathbf{a})$ follows from the second law of thermodynamics, and may be attributed to the intensity of produced entropy or dissipated exergy around this point. These constant-sign source/sink values cannot be mixed with the time derivatives of system's entropy or exergy which, in general, are not constant-sign functions. They are rather sources of sinks of these quantities. This explains why excess entropy production functions (functionals), such as Eq. (5.10), which represent the derivatives of certain Lapunov's functions (functionals), such as Eq. (5.9), constitute useful tools in proving the stability (instability) of these systems.

CHAPTER 6

Analyzing Drying Operations in Thermodynamic Diagrams

6.1 Introduction

Drying is the extraction of moisture from a solid or liquid product. It is an operation involving simultaneous heat and mass transfer (Chen, 2007a). Research in drying is important since drying is an energy-consuming process, in which a large amount of heat needs to be supplied for evaporating moisture. Drying significantly affects the product quality. A reliable drying model should assist process design, simulation, and optimization. In process design, the model can be used to explore new and innovative aspects to evaluate performance of the existing dryer as well as estimate its driving energy supply. For maintaining product quality, a reliable model can be applied to optimize the existing process and explore new processes in order to achieve high-quality products.

Processes of evaporation and condensation play an important role in engineering, vapor-generating systems, contact heat exchangers of industrial power equipment, vacuum metallurgy, and obtaining ultra-pure materials and specimens with special properties in microelectronics and different chemical technologies. They also appear in fuel systems of internal combustion engines, cryogenic engines of rockets, and different technological processes of pharmacy industry. From the point of view of exploitation safety, processes of nonstationary evaporation are of special importance in engineering calculations of transient models for vapor-generating systems of powerful electric power plants.

Fig. 5.1 of the previous chapter depicts various processes between gas and granular solid, with evaporation and condensation of moisture, which will be considered in the present chapter from the viewpoint of graphical methods applied to predict the process performance.

Till date, many thermodynamic problems with evaporation and condensation of moisture have extensively been treated; some of the related publications are cited and described in this chapter. However, purely analytical or numerical presentation often obscures important thermodynamic properties and hampers their understanding. To improve the situation, Sieniutycz (1967) and Ciborowski and Sieniutycz (1969a, b) attempted to visualize solving methods and solutions to some representative drying problems by using thermodynamic diagrams of enthalpy–composition. Suitable graphical representations for experiments follow from the use of graphical methods in thermodynamic charts. These methods are analyzed in the present chapter. Its volume limitation does not allow the inclusion of all derivations to make this chapter self-contained, thus the reader may need to turn to some previous publications cited here for the purpose.

Krischer (1956); Luikov, 1966a, 1975; Mujumdar (2007); Strumillo and Kudra (1987); Kowalski (2003); Chen (2007a), and many others made a significant contribution to scientific basis of the drying theory. In his pioneering books and papers, Luikov (1975) introduced new integral transformations to solve the system of differential equations of heat and mass transfer in capillary-porous bodies, summarized in his book with Mikhailov (Luikov and Mikhailov, 1963). Luikov (1966a) applied nonequilibrium thermodynamics and some results of kinetic and statistical theories to improve the formal theory of evaporating systems. He also obtained a systematic description of various transport phenomena including heat and mass transfer with a finite propagation speed (Luikov, 1966a). Mujumdar (2007) systematically reviewed innovations in industrial drying operations and contributed much to dissemination of novel valuable results in the growing field.

Thermodynamic and kinetic properties (transfer coefficients) are of importance in both analytical and graphical approaches. Kovac (1977) applied the theory of nonequilibrium thermodynamics to multicomponent system containing an interface, using a method developed by Bedeaux et al., 1975. Evaporation is slow, if the resulting velocity field in the vapor is characterized by a low Reynolds number, as shown by Bedeaux and Kjelstrup (1999). Kinetic theory predicts a temperature jump across the surface during slow evaporation. Disequilibrium thermodynamics of interfaces similarly predicts a jump. While kinetic theory gives an explicit size of the jump, disequilibrium thermodynamics, without constitutive coefficients, does not. Bedeaux and Kjelstrup (1999) expected that disequilibrium thermodynamics reproduces the results found from kinetic theory. The reason for their investigation was to show that kinetic theory, together with the boundary conditions, is not in conflict with the second law, a fact, which has often been questioned. A measurement of the temperature profile by Fang (1998) and Fang and Ward (1999) has verified the existence of a temperature jump, and found it to be substantial. This measurement proved that the equilibrium condition for the temperature, $T^g = T^l$, like the equilibrium condition for the chemical potential, $\mu^g = \mu^l$, is not fulfilled during slow evaporation. The temperature was found to change almost up to 10 K from one side of the surface to another, while the chemical potential was found to change up to about 50J. Bedeaux and Kjelstrup (1999) performed an analysis of the experimental results by Fang and Ward (1999) who found a temperature jump up to almost 10 K across the evaporating surface of water, octane, and methyl-cyclohexane. Bedeaux and Kjelstrup (1999) used nonequilibrium thermodynamics to obtain appropriate boundary conditions at the surface. Interfacial transfer coefficients, appearing in these boundary conditions, were determined from the experiments. The authors presented them in a form useful for engineering calculations.

Bedeaux and Kjelstrup (1999) compared their own results with the predictions for the transfer coefficients stemming from kinetic theory. For three chosen materials, the kinetic theory values were found to be 30–100 times larger than those found from the experiment. The authors explained why this finding predicts a liquid surface, which is colder than the adjacent vapor, contrary to the prediction by the kinetic theory. The relative magnitude that they found for interfacial transfer coefficients, suggests that the condensation coefficient of the kinetic theory decreases with increasing internal degrees of freedom in a molecule. Yet, a lower value of this coefficient is insufficient to explain the small value of the transfer coefficients. As a possible explanation, the researchers forward the hypothesis that the single-particle collision model of the kinetic theory should be modified to account for multiparticle events. These issues need to be considered in future analytical and graphical approaches.

In this chapter graphical methods in drying of pulverized solids are described and implemented in an enthalpy chart for a gas–moisture–solid system. Methods solving kinetic equations in enthalpy graphs are of particular interest. Models of single particle drying, fluidized drying, and drying in countercurrent systems are developed, and their graphical solutions in the enthalpy diagram for gas–moisture–solid are outlined. Drying time is found by integration along process path in the enthalpy chart. Components of drying theory, such as the establishment of wet bulb temperature, are interpreted in the chart. Consistent classifications and representations of graphical solutions and experiments are achieved. Advantages of graphical methods are made explicit for balance–kinetic equations with state dependent coefficients, complex gas–solid equilibria, and variable gas states. Application of graphical methods is recommended particularly in those cases when developing analytical solution or effective computer program is very difficult or practically impossible to achieve. Both computer programming and graphical methods can help to realize similar goals and overcome similar difficulties. Graphical approaches, however, show clearly and explicitly the interplay between thermodynamic and kinetic properties of considered systems.

The goal of this chapter is to stress the advantages of graphical methods in solving balance–kinetic equations: cases of state dependent thermodynamic and kinetic coefficients, the case of gas–solid equilibria described by complex equations or their approximations, the case of complex variability of state of drying agent in time, and so forth. The chapter analyzes two basic groups of graphical methods: (1) valuable graphical methods not commonly known (either unpublished or published in local sources not readily available and not in the English language), and (2) newly developed graphical methods to be used in drying practice.

The chapter also presents experiments linked with the graphical classification of solid paths, obtained on the basis of graphical solutions. Most of the experiments described in the chapter are original.

The structure of the present chapter is as follows. Section 6.2 focuses on modeling as a common procedure in analytical, numerical, and graphical approaches. This section also describes main available papers on current modeling. Section 6.3 focuses on graphical methods for drying of single grain and batch drying in steady and dynamical fluidized beds. Section 6.4 develops graphical methods for grain drying in countercurrent, crosscurrent, and concurrent gas flows and moving beds. Section 6.5 imposes a classification of drying paths in enthalpy diagrams, resulting from experiments and from theory. Section 6.6 presents briefly concluding remarks. Appendix discusses some related problems and associated articles, which treat drying problems by analytical and numerical approaches. However, all these problems could equally be treated by graphical methods developed in the present chapter.

6.2 Modeling of Moisture Extraction in Drying Systems

Modeling is an important activity, preceding analysis, synthesis, and optimization of a drying operation. A reliable drying model should be sufficiently simple, accurate, and able to capture the process physics and require minimum sets of experiments to generate the unknown parameters. Drying operations satisfy a general scheme of heat-driven binary separation process shown in Fig. 6.1 (Orlov and Berry, 1991).

Chen and coworkers (Chen and Xie, 1997 and Chen and Putranto, 2013) proposed the so-called reaction engineering approach (REA), which has successfully been used to model some drying processes, mainly thin-layer drying and drying of small particulates of food materials. Physics of drying is captured by the relative activation energy, which represents the level of difficulty to "extract" moisture of drying in addition to evaporating free water. Initially, the relative activation energy may be zero near the start of drying high-moisture product and keeps on increasing as the operation progresses. When a low equilibrium moisture content is reached, the relative activation energy becomes one. The relative activation energy of the same materials can be used to model other drying processes with similar initial moisture content. The REA framework offers an effective way of obtaining the

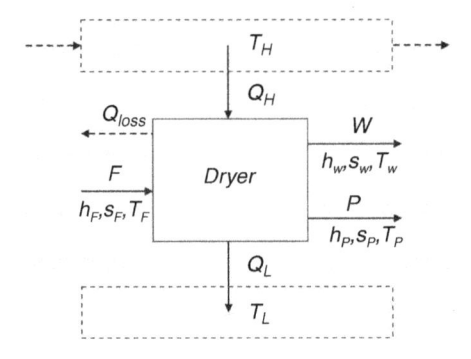

Figure 6.1: Schematic diagram of heat-driven binary separation process.

necessary parameters. Due to the efficiency of the REA, it is worthwhile to develop it further and to implement it into more complex scenarios. REA, which captures the basic drying physics, represents an effective mathematical modeling applied to diverse drying processes. The intrinsic "fingerprint" of the drying phenomena can in principle be obtained through a single accurate drying experiment.

The book by Chen and Putranto (2013), a comprehensive summary of the state-of-the-art and the ideas behind the REA to drying processes, is a valuable resource for researchers, academics and industry practitioners. Starting with the formulation, modeling and applications of the lumped-REA, it goes on to detail the use of REA to describe local evaporation and condensation, and its coupling with equations of conservation of heat and mass transfer, called the spatial-REA, to model nonequilibrium multiphase drying. Finally, the volume summarizes other established drying models, discussing their features, limitations, and comparisons with the REA. Principles of REA are consistent with CFD modeling applied to various heat and mass transfer processes. These principles are also consistent with general principles of graphical methods, in particular those applied in the present chapter.

General equations for simultaneous heat and mass transfer with free enthalpy taken as a universal potential for the moisture transport and the phase transitions have been proposed by Kowalski and coworkers (Kowalski, 2003 and Kowalski and Strumiłło, 1997) who discussed properties of these equations for the boundary conditions characteristic in convective drying of capillary-porous materials. Some applications in the field of drying thermomechanics are given in Kowalski (2003) and Kowalski et al. (2000). In the paper by Kowalski et al. (2000) self-fracturing of fluid-saturated porous materials during drying processes is considered, and a physical model of this widespread phenomenon in drying problems is proposed. The main analysis deals with dispergate systems like ceramics, which after drying and calcination, become porous bodies with relatively high strength. The emphasis is on two problems: an increase of the cohesive force at the initial stage of drying and decohesion of a structure caused by drying-induced stresses.

Some important technological processes in chemical reactors are preceded or followed by drying operations. The use of combustion gases and high-T processes may lead to thermal transformations of previously dried solids. Amar et al. (1989) derived the shape of first–stage sinters produced within a framework, in which a solid skeleton is wetted by a mobile "liquid" pool. Several lattices constitute possible solid skeletons, corresponding to various degrees of surface versus grain boundary and volume transport. The effect of a nonzero dihedral angle at the interface between crystal faces is also treated. In some cases, significant differences exist between shapes calculated by minimizing the free energy and assumed shapes. In conclusion the development of effective drying modeling seldom rests on classical thermodynamics, thus the developments based on disequilibrium theories and concepts are necessary.

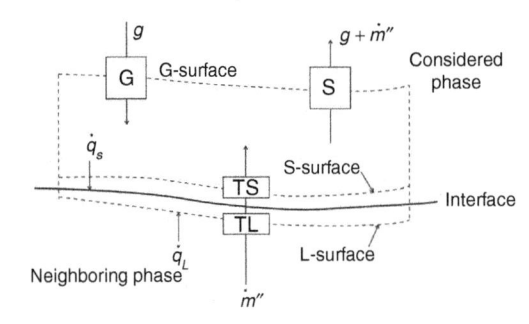

Figure 6.2: In Spalding's model of mass transfer through interface, Reynolds flux g is interpreted as the mass transfer coefficient and the ratio g/G as the nondimensional Stanton number.

Investigating conserved property diagrams for rate-process calculations and considering the heat and mass transfer through an interface between two phases, L and G, Spalding (1959b, 1961, 1963) presented a derivation of a generalized driving force B, and interpreted the mass transfer coefficient in terms of Reynolds flux, g. He also recovered the definition of the nondimensional Stanton number as the ratio g/G.

In the notation used in the book by Spalding (1963), mass transfer through the interface satisfies the relationship of the Ohm's law type,

$$\dot{m}'' = gB, \tag{6.1}$$

where the driving force B, following the conservation law for ith substance and the transfer property $m_{i,T}$, is obtained in the form

$$B = \frac{m_{i,G} - m_{i,S}}{m_{i,S} - m_{i,T}}. \tag{6.2}$$

Correspondingly, the Reynolds flux g is interpreted by Spalding (1963) in terms of the mass transfer coefficient and the ratio g/G as the nondimensional Stanton number, Fig. 6.2.

Modeling, analysis, and some optimizations of drying systems can be accomplished with the help of thermodynamic diagrams. In their books, Bosniakovic (1965) and Ciborowski (1976) have presented diverse analyses of technical processes on thermodynamic charts. Sugar solutions, which represent a special class of salt solutions, are processed on a large scale in sugar factories. The enthalpy–concentration diagram (i–x), which can be used for salt solutions is particularly useful for sugar. The abscissa is chosen to be the water content x that is the water quantity per unit mass of sugar in the solution. For pure sugar, $x = 0$, for pure water x approaches an infinity. Diagrams i–x for sugar serve to interpret the water vaporization from sugar solutions; similar applications may involve moist coal, sorbents, catalysts, and diverse dried materials, in particular various granular solids in fluidizing systems (Bosniakovic, 1965; Ciborowski, 1976).

Drying operations are particularly good examples of irreversible processes, associated with appreciable losses. As a consequence, the heat consumption in a drying operation is usually high, thus the task of applied thermodynamics consists of making a drying process closer to its reversible counterpart. In the entropy diagrams, individual irreversibilities can be found and quantified, so that certain understanding of the installation effectiveness is obtained, and a component in which the largest losses occur can be marked for improvement. Analyses of Bosniakovic (1965) involve not only conventional dryers with heat drying agents, but also units removing moisture from the dryer air by the application of work. Diagrams of enthalpy or entropy versus the moisture content serve to show changes of thermodynamic states and to interpret the influence of process components on the overall thermodynamic efficiency of the operation.

6.3 Graphical Approach to Drying of Single Grain and Drying in Fluidized Beds

A large literature currently exists which is related to the drying of single droplets, single granules, and granular materials. A relatively new modeling of droplet drying is provided by Farid (2003). He derived a model which predicts the decrease in the droplet mass and temperature when it is immersed in hot air, as in the spray drying of solutions with dissolved solids. He assumes that the droplet experiences first the rapid sensible heating without change of mass, and then the droplet experiences some shrinkage, without temperature change but with the rapid mass decrease. This period is followed by a crust generation period with a remarkable change in the droplet temperature weight, and, finally, a period of nearly pure heating of the nearly dry particle. Unlike previous models, Farid's model takes into account shrinkage and temperature distribution within the droplet. The model also ensures a reasonable estimate for the variation in droplet mass and temperature for a number of experimental measurements available in the literature.

6.3.1 Drying of Single Grain by Gas Whose State is a Function of Time

We begin our studies on graphical approaches with a mathematical description of drying of a single grain in a gas and derivation of an equation, which describes the derivative dI/dW in the chart of enthalpy-moisture content for solid phase. For simplicity and brevity of analysis we assume small Biot's numbers (ignoring distribution of temperature and moisture content within the grain). We also assume the occurrence of the evaporation process on the geometric surface of the grain. Under these assumptions equations describing the drying dynamics of the grain have the following form:

$$\frac{m_s dI}{F_s dt} = \alpha\left(t_g - t_s[I,W]\right) + i_{ps} K'_g \left(X_g - X_s[I,W]\right) \tag{6.3}$$

$$\frac{m_s dW}{F_s dt} = K'_g \left(X_g - X_s[I,W] \right) \tag{6.4}$$

The first describes energy balance and the second mass balance. Square brackets refer to functional relationships, which describe the gas–solid equilibrium.

In the problems of simultaneous heat and mass transfer a basic role is played by the Lewis factor

$$Le = \frac{C_g K'_g}{\alpha} \tag{6.5}$$

Assumption of Le = 1 which can be made for unbounded water evaporation leads to the simplest equations in which the gas enthalpy differences $i_g - i_s$ appear as the sole driving force for the energy transfer.

$$\frac{m_s dI}{F_s dt} = K'_g C_g \left(t_g - t_s[I,W] \right) + i_{ps} K'_g \left(X_g - X_s[I,W] \right)$$
$$= K'_g \left(i_g - i_s[I,W] \right) \tag{6.6}$$

See the literature (Sieniutycz, 1967; Ciborowski and Sieniutycz, 1969a, b) for more details and some related subtleties.

For Le different than one the single driving force in the enthalpy equation can be achieved by using an effective enthalpy $i'_s[I,W]$ defined in terms of enthalpy i_{ts} of a gas with temperature $t = t_g$ and equilibrium humidity X_s. The following formula holds (Sieniutycz, 1967; Ciborowski and Sieniutycz, 1969a, e)

$$i'_s = i_s Le^{-1} + (1 - Le^{-1}) i_{ts}. \tag{6.7}$$

In terms of the enthalpy i'_s defined previously, enthalpy formula Eq. (6.3) takes the form

$$\frac{m_s dI}{F_s dt} = K'_g \left(i_g - i'_s[I,W] \right) \tag{6.8}$$

Division of Eq. (6.8) by (6.4) leads to the slope dI/dW in terms of inlet gas parameters i_g, X_g, and parameters of gas in equilibrium with solid, i'_s, X_s.

$$\frac{dI}{dW} = \frac{i_g - i'_s[I,W]}{X_g - X_s[I,W]}. \tag{6.9}$$

The effective equilibrium enthalpy of gas i'_s becomes equal to the true equilibrium gas enthalpy i_s in the case when the Lewis factor Le is equal to one. The simplicity of the slope formula for $Le = 1$ follows from mutual reduction of the effect of heat and mass transfer coefficients. The remarkable property of Eq. (6.9) is its applicability to various drying periods, provided, of course, that the assumptions done while setting Eqs. (6.3) and (6.4) are satisfied. The formal simplicity of Eq. (6.9) is not necessary for development of graphical methods but it is convenient for the brevity of analysis.

In fact, simple details of graphical solving of Eq. (6.9) are not as important as showing that equations of the same structure can be derived for other, more complex, drying processes, such as fluidized drying, packed bed drying, countercurrent drying of falling particles, drying in moving bed systems, and so forth. Most important, Eq. (6.9) is a valid formula not only for drying of a single grain, but also for a large majority of practical drying systems. In particular, as shown later, Eq. (6.9) is valid for drying processes in fluidized beds. For operations in which gas parameters change along the process path, such as those in countercurrent and cocurrent dryers, Eq. (6.9) is still valid but it must be supplemented by balance equations linking changes of state parameters in gas and solid phases. In all these systems parameters of gas contacting with solid are no longer constant but change along the process path, in a different way dependent on the gas–solid contacting type. In these systems assumption of the constancy of gas parameters must be replaced by a rule of changing of gas state coordinates along the process path, which follows from the enthalpy and mass balances in the gaseous phase.

6.3.2 Fluidized Drying of Solid by a Gas with Parameters Variable in Time

We can now pass on to drying processes in fluidized beds. While a number of papers appeared to date related to fluidized drying, our interest is focused on those of direct relation to graphical methods. Considering drying models from a single particle to fluidized bed drying, Tsotsas (1994) has introduced a normalized single particle drying curve integrated into a generic, heterogeneous fluidized bed model to describe batch fluid bed drying. Ciborowski and Sieniutycz (1969c) initiated graphical methods of analysis for basic problems of fluidized heat and mass exchange in thermodynamic diagrams of enthalpy–composition. Case studies which we shall discuss first as extension of their examples illustrate the processes of the steady-state fluidized drying with an external heat, shown in Fig. 6.3 and a batch fluidization in Fig. 6.4. They involve both balance and kinetic aspects, and they use Eq. (6.9). The left part of Fig. 6.3 schematizes the steady nonadiabatic operation, whereas the right one illustrates the two ways of fluidization modeling, first as a pseudo-homogeneous model and the second as a two-phase bubble model. It is the batch fluidization operation for which the application of Eq. (6.9) will be

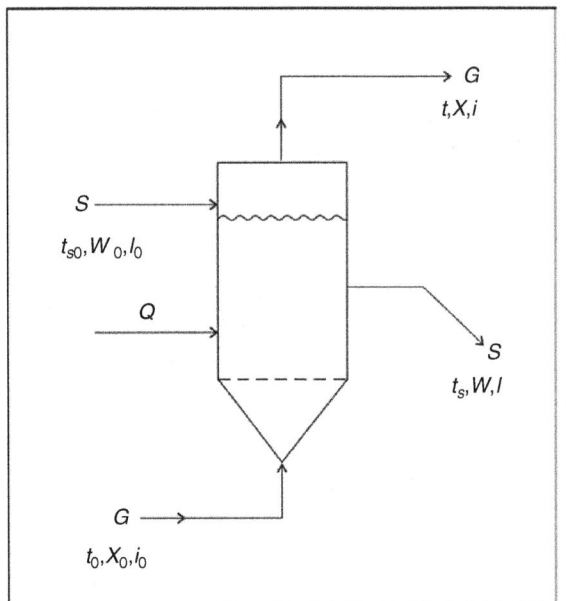

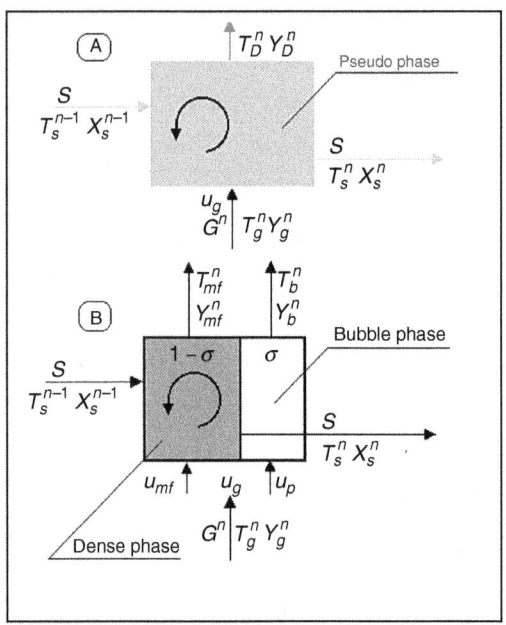

Figure 6.3: Left: Steady-state operation of nonadiabatic fluidized drying of solid particles. Right: Two ways of modeling of fluidized drying.
Case A: Perfect mixing of a single pseudo-phase composed of gas and solid particles (Sieniutycz, 1978, 1991; Szwast, 1990); Case B: Perfect mixing of a dense pseudo-phase containing gas and solid particles and plug flow of bubbles, as in Kunii-Levenspiel model (Kunii and Levenspiel, 1991; Poświata, 2004, 2005; Poświata and Szwast, 2003, 2006; Sieniutycz and Jeżowski, 2013). The scheme allows the simultaneous treatment of the batch fluidization and fluidization in a horizontal exchanger.

considered in Fig. 6.4. In further fluidization examples, as shown in Figs. 6.5–6.7, kinetic issues can be ignored, and purely thermodynamic approaches are not only sufficient but also important and useful.

Chronologically, two ways of modeling of fluidized drying can be distinguished. Case A assumes a model of single mixed pseudo-phase composed of gas and solid particles (Sieniutycz, 1978, 1991; Szwast, 1990). Case B uses a model composed of a dense, mixed pseudo-phase and a dilute phase with plug flow of bubbles (Poświata, 2004, 2005; Poświata and Szwast, 2003, 2006), as described by Kunii and Levenspiel (1991) in their famous book. All these models allow for a simultaneous treatment of the batch fluidization and fluidization in a horizontal exchanger as a continuous limit of a cascade. Since the models of Kunii–Levenspiel type are quite voluminous and numerous details of their application to fluidized dryers are available in the contemporary literature (Kunii and Levenspiel, 1991; Sieniutycz and Jeżowski, 2013) we shall not consider them here.

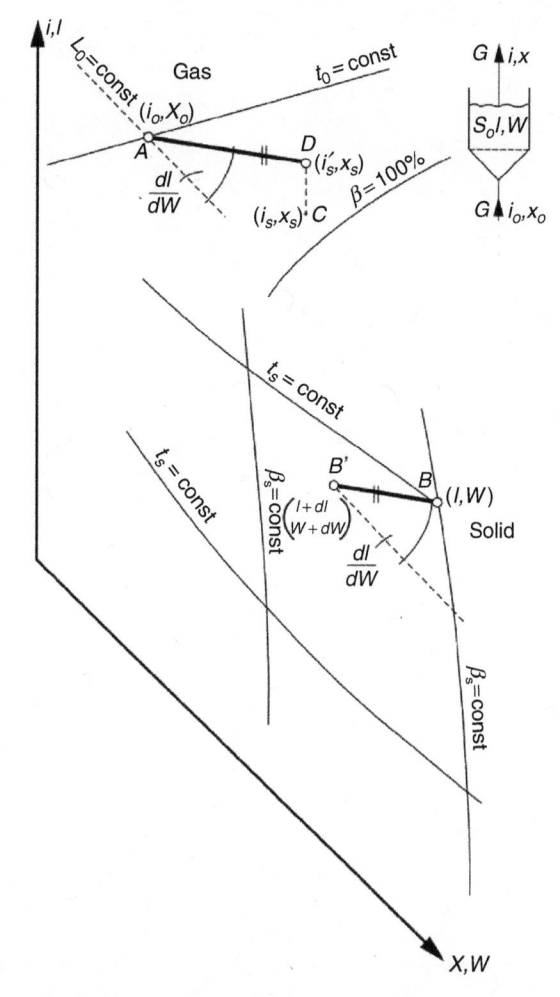

Figure 6.4: Enthalpy–composition chart for gas-water-solid system.
Determining of the local slope dI/dW for the purpose of finding the new solid state in the process of batch fluidization. *Sieniutycz, 1967; Ciborowski and Sieniutycz, 1969c, 1969d, 1969e.*

Accepting brevity as a main criterion, we derive dynamical equations for the batch-fluidized dryer described by the model of a mixed pseudo-phase, which is sufficient for our purposes here. Yet we extend this model by including a set of equations for a steady operation in a horizontal fluidized exchanger. These equations are derived under the assumption that the gas–solid system in the vertical cross-section of the exchanger constitutes a well-mixed disequilibrium pseudo-phase (Sieniutycz, 1991; Sieniutycz and Jeżowski, 2013). As in the batch fluidization, this is again the case when the outlet gas state is in disequilibrium with solid (I, W). This disequilibrium is associated with a finite value of the *number of mass transfer units*, N_g.

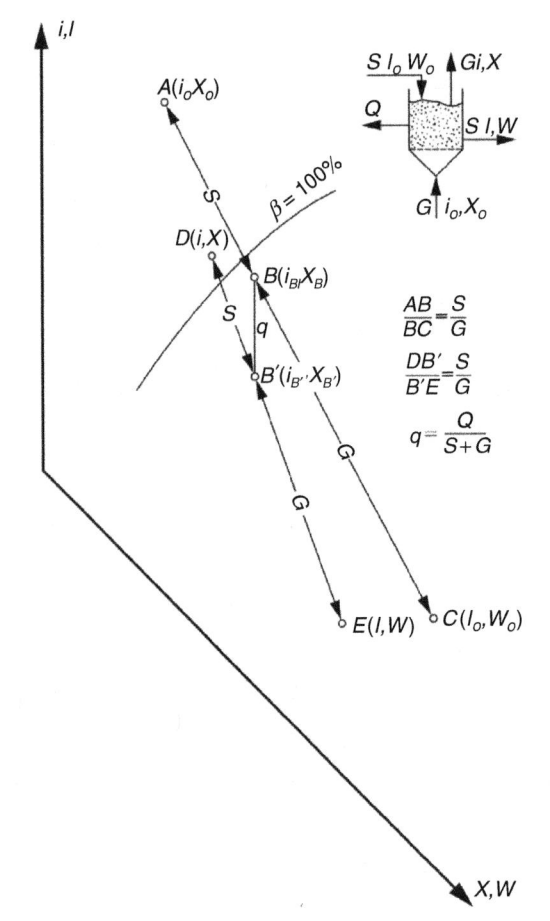

Figure 6.5: Steady nonadiabatic process of fluidized drying in the enthalpy diagram.
Effect of heat losses or external cooling. Sieniutycz, 1967; Ciborowski and Sieniutycz, 1969c, 1969d, 1969e.

Assuming an ideal mixing in the fluidizing pseudo-phase of gas and solid, the balances of the energy and mass are described by a set of two equations

$$S_0 \frac{dI}{dt} = G(i_g - i) = K'_g (i - i'_s) ah \tag{6.10}$$

$$S_0 \frac{dW}{dt} = G(X_g - X) = K'_g (X - X_s) ah \tag{6.11}$$

Index g refers to the inlet gas, whereas the variables without index refer to the unknown state within the dryer. Eliminating parameters of this state from the set of Eqs. (6.10) and (6.11) the state equations of a batch fluidization dryer follow in the form

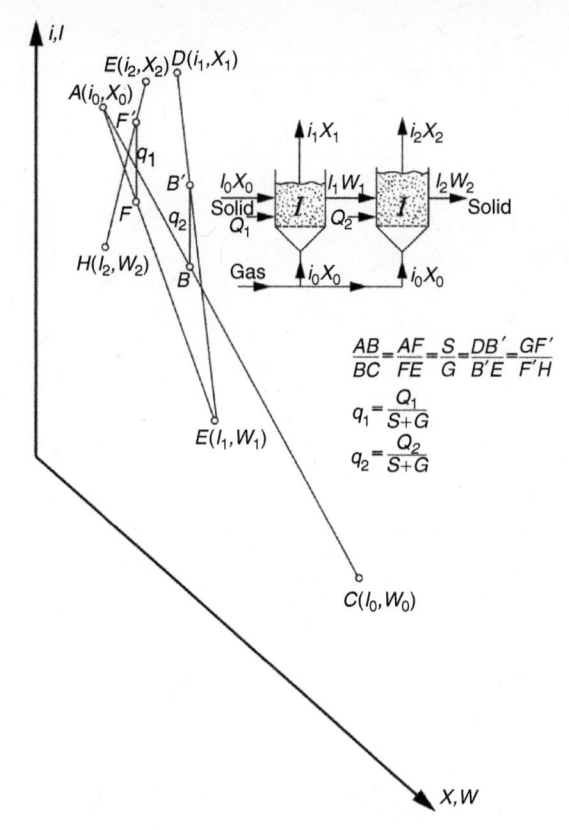

Figure 6.6: Two-stage, steady fluidized drying with external heating at the stages. *Sieniutycz, 1967; Ciborowski and Sieniutycz, 1969c, 1969d, 1969e.*

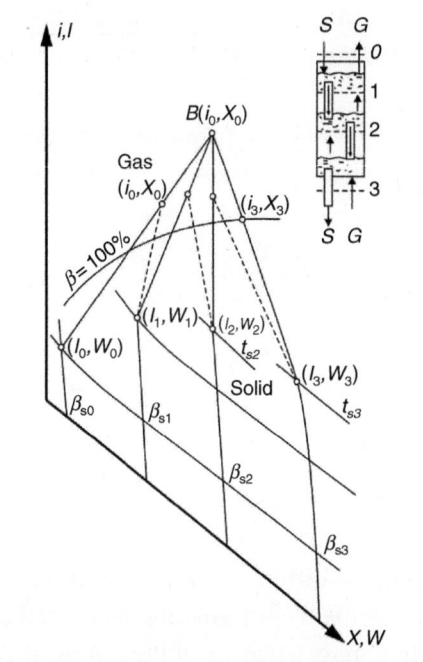

Figure 6.7: Adiabatic sorption in the countercurrent multistage fluidized bed. *Sieniutycz, 1967; Ciborowski and Sieniutycz, 1969c, 1969d, 1969e.*

$$\frac{S_0}{G}\frac{dI}{dt}\left(1+\frac{HTU_g}{h}\right) = i_g - i'_s \tag{6.12}$$

$$\frac{S_0}{G}\frac{dW}{dt}\left(1+\frac{HTU_g}{h}\right) = X_g - X_s \tag{6.13}$$

These are dynamical equations of batch fluidization whose nondimensional form is represented by Eqs. (6.17) and (6.18) later.

Modification of these equations in the form

$$\frac{SdI}{dG}\left(1+\frac{HTU_g}{h}\right) = i - i'_s \tag{6.14}$$

$$\frac{SdW}{dG}\left(1+\frac{HTU_g}{h}\right) = X - X_s \tag{6.15}$$

describes the drying process in a horizontal fluidizing exchanger, under the assumption of a perfect mixing of both phases in the vertical cross-section perpendicular to solid flow.

In terms of the nondimensional quantities defined as follows.

$$\tau = \frac{Gt/S_0 \text{ for batch fluidization process}}{G/S \text{ for the process in the horizontal exchanger}} \tag{6.16}$$

(Gt is the cumulative amount of gas supplied after time t to the batch dryer) the dynamical equations of two considered fluidization processes assume a common form

$$\frac{dI}{d\tau} = \frac{N_g\left[i_g - i'_s(I,W)\right]}{1+N_g} \tag{6.17}$$

$$\frac{dW}{d\tau} = \frac{N_g\left[X_g - X_s(I,W)\right]}{1+N_g} \tag{6.18}$$

where $N_g = h/HTU_g$ is the number of mass transfer units. Summing up, the modeling may include the drying process in a horizontal exchanger for which a nondimensional time τ is simply G/S. Since S is constant, the differential $d\tau = dG/S$. We stress that the differential equations (6.17) and (6.18) combine the information contained in the process kinetics and the one contained in the energy and mass balances of the fluidized systems.

With the collaboration of others Sopornonnarit et al. (1996) designed, fabricated, and tested a cross-flow fluidized bed paddy dryer. Sopornonnarit et al. (1997) also studied corn drying in batch fluidized bed dryer. Drying characteristics of corn were investigated. The experimental results indicated that moisture transfer inside a corn kernel was controlled by internal diffusion. His dryer is, in fact, the same unit as the horizontal exchanger considered previously. Therefore approximate model of Eqs. (6.17) and (6.18) can be applied to the evaluation of the performance of this dryer.

Dividing Eq.(6.17) by (6.18) yields the slope coefficient dI/dW for the considered processes of fluidized drying in the solid enthalpy diagram. The result is again in the form of Eq. (6.9). Note that, in case of fluidization, Eq. (6.9) follows exclusively from balance equations of drying when the thermodynamic equilibrium between the outlet gas and outlet solid can be assumed. For $Le = 1$, effective enthalpy i'_s coincides with the true equilibrium enthalpy of gas, that is, $i'_s = i_s$. In a more general case of an arbitrary factor Le, effective enthalpy i'_s does in fluidization the same job as in the single particle description. Since Eq. (6.9) may be applied to variable inlet gas states, it is not unreasonable to surmise that it holds also for countercurrent and concurrent systems. An analysis in Section 6.4 confirms this expectation (Sieniutycz, 1967).

In conclusion, the modeling of fluidized systems has shown that the slope formula, Eq. (6.9), describes the slope dI/dW in terms of inlet gas parameters i_g, X_g, and parameters of gas in equilibrium with batch of grains, i_s, X_s. Still, as in models of a single grain, effective equilibrium enthalpy of gas i'_s becomes equal to the true equilibrium gas enthalpy i_s when Lewis factor Le is equal to one.

The single pseudo-phase model constitutes an approximate, often insufficient description of fluidization. However it may be used with admissible accuracy (ca. ± 12%) in situations where the bed arrangements (baffles, mixers, etc.) cause a significant mixing of fluidized bed, which makes the dispersion contribution to the plug flow dominant. In the opposite limiting case, that of the negligible dispersion in the gas phase, that is, when an ideal plug flow of gas can be assumed, the two phase model of fluidization formulated by Kunii and Levenspiel (1991) is operative. Dynamical state equations of this two-phase model are in the most general case implicit.

However, for a majority of real fluidizing apparatuses, where the bed height is of order 0.5 m, practical thermodynamic equilibrium is observed between outlet gas and solid due to the very large specific surface of the solid and very small height of the transfer units (HTU). In this case, the number of transfer units $Ng \to \infty$, and limiting state Eqs. (6.17) and (6.18), or any other formulas of this sort, are independent of kinetic phenomena, that is, they describe exclusively equilibrium balances. Needless to say that equilibrium relations in gas–moisture–solid systems may be very complex. Graphical methods often use tables, which describe these equilibria in a vast range of state parameters (Sieniutycz, 1973b).

Poświata and coworkers (Poświata, 2005; Poświata and Szwast, 2006) performed an optimization of drying processes with fine solids in bubbling fluidized beds described by the bubble model of Kunii and Levenspiel. The reader is referred to a book (Sieniutycz and Jeżowski, 2013), where the bubble two-phase model and results of Poświata and coworkers are described and discussed.

In a synthesizing paper, Zahed et al. (1995) modeled numerically the performance of batch and continuous fluidized dryers with diffusion moisture transport in the dense phase particles and interstitial gas-to-particle mass transfer within the dense phase. Their model includes the interphase exchange resistance between gas bubbles and the dense phase. Changes of the bed temperature and product moisture content in the bed with time are predicted. This model can be used for both homogeneous and bubbling fluidized bed drying of cereal grains and granular synthetic polymers.

Ziegler and Richter (2000) analyzed deep-bed drying in enthalpy–water diagrams for air and grain. With regard to the mean moisture content of the grain, the mass transfer driving forces within the drying zone were computed from the diagram and then related to the measured time course of temperature respectively to the drying rate of the grain. The method led to a thin-layer equation for wheat drying based exactly on the thermodynamic relationships between the air and the grain that appear in the drying zone at certain drying conditions. The thin-layer equation was integrated into a mathematical model derived to analyze solar-assisted drying of hygroscopic bulk materials and short-time storage of low-temperature heat. The deep-bed drying model was validated for wheat with a good agreement between experimental and simulation results at varied inlet air conditions. The approach of Ziegler and Richter (2000), in which the enthalpy–water content diagram for moist wheat was combined with the enthalpy–water content diagram for humid air, is not unlike the earlier graphical approach of Ciborowski and Sieniutycz (1969a, b, e). It can be said that they rediscovered the generalized diagram gas–moisture–solid.

When solid fluxes S in Fig. 6.3 are reduced to zero but the apparatus still contains a batch of solid particles, drying can be conducted as an unsteady process shown in the right upper part of Fig. 6.4. This figure outlines basic graphical constructions for the process of batch fluidization in accordance with Eq. (6.9). Provided that thermodynamic data and the enthalpy diagram for the system of gas–moisture–solid are available (Bosniakovic, 1965; Sieniutycz, 1973b; Spalding, 1963; Ziegler and Richter, 2000) important drying or moistening properties can be derived by graphical constructions, which are in principle self-explanatory.

The first stage of the multiple use of Eq. (6.9) is shown in Fig. 6.4, where the differential slope dI/dW is replaced by its difference counterpart $\Delta I/\Delta W$. Successive iterations lead to a discrete approximation of the solid path $I(W)$ in the enthalpy chart. For a solid state (I, W) located in the solid regime of the enthalpy diagram the corresponding equilibrium state

(i_s, X_s) in the gas regime of the diagram. Next the effective enthalpy $i'_s[I,W]$ is determined from Eq. (6.7). This allows calculation of the ratio on the right-hand side of Eq. (6.9), which determines the segment $B'B$ with the slope dI/dW. B' is a new state of the moist solid, for which the procedure is repeated. This procedure leads to the solid path $I(W)$ in the solid part of the enthalpy diagram.

A qualitative picture of drying paths on the solid part of the enthalpy diagram is illustrated in Fig. 6.11 further in the text.

Summing up, it follows from Eqs. (6.8) and (6.9) that for the drying of a single granule or a dispersed solid batch dried by a gas with constant parameters, the construction of the solid path in the I–W chart may be performed by sequential determining of local derivatives dI/dW which satisfy the slope equation in the form given by Eq. (6.9).

The reader can refer the literature of Ciborowski and Sieniutycz (1969a) to consider the generalization of the graphical approach in which the balance and kinetic constructions take into account the coefficient of phase changes ε within the solid phase (see Fig. 6.12 for further information).

6.3.3 Fluidizing Systems with Thermodynamic Equilibrium in the Outlet Stream

When kinetic effects can be ignored, only balance equations are sufficient in graphical constructions. Corresponding examples, of purely thermodynamic nature, are presented in Figs. 6.5–6.7.

Fig. 6.5 describes constructions aimed at determining outlet states of gas and solid in a steady nonadiabatic process of fluidized drying. The example takes into account the effect of heat losses or external cooling (Sieniutycz, 1967; Ciborowski and Sieniutycz, 1969e, c). To find outlet states of both phases, it is only necessary to realize that the adiabatic mixing of inlet humid gas $G(i_0, X_0)$, point A, with the inlet moist solid $S(I_0, W_0)$, point C, occurs in the enthalpy diagram along the straight line ABC, where point B represents the "adiabatic" two-phase mixture of gas and solid. Since however there are usually some heat losses Q, corresponding to unit losses $q = Q(S + G)^{-1}$, the actual state of the mixture is represented by point B', not B. At point B' gas–solid mixture separates along the two-phase equilibrium straight line DE which leads to the outlet humid gas $D(i, X)$ and outlet moist solid $E(I,W)$.

The moisture content W of outlet solid is lower than the initial W_0. Clearly, this drying effect can be enhanced in the two-stage or multistage process, in which an extra gas flux is directed to the second (next) dryer, to dry the solid leaving the first (previous) dryer. The considered two-stage scheme might be realized in the two-stage cascade shown in the right hand side of Fig. 6.6. However, the graphical constructions shown in the enthalpy diagram of Fig. 6.6 refer to dryers with external heating rather than to those with external losses of heat. Such dryers ensure lower final moisture contents of solids than the dryer shown in Fig. 6.5.

6.4 Grain Drying in Countercurrent, Crosscurrent, and Concurrent Gas Flows

An important family of countercurrent operations of drying and sorption (moistening) includes both continuous and multistage systems. One of these countercurrent operations is depicted in Fig. 6.7, which describes the case of the multistage fluidized bed with adiabatic sorption. This stage-wise operation involves the dense-phase fluidizations at stages in which outlet gas and outlet solid are in practical equilibrium. As in rectification columns treated by the Ponchon–Savarite graphical method, point B represents the pole of the system, or the state of fictitious mixture that gives rise to the family of operational lines in the form of straight rays launched from point B. Each operational line connects states of gas and solid in the same cross-section of the apparatus. The graphical construction exploits two-phase equilibrium lines, represented in Fig. 6.7 by straight discontinuous lines. This graphical construction is used to determine the number of theoretical (equilibrium) stages, necessary to achieve a required final concentration of the adsorbed moisture. Dividing this quantity by the mean efficiency leads to the number of real stages when an idealized assumption of perfect stages is unacceptable.

When kinetic issues are essential, graphical methods for countercurrent (and concurrent) systems include Eq. (6.9). However, since the ratio of solid to gas flows are finite, the method should take into account changes of the gas state in the same cross-section of the dryer. These changes are described by the following differential balances

$$\frac{di}{dI} = \pm \frac{S}{G} \tag{6.19}$$

$$\frac{dX}{dW} = \pm \frac{S}{G} \tag{6.20}$$

where the upper sign refers to the countercurrent flow, whereas the lower sign – to the concurrent flow. Eqs. (6.19), (6.20), and (6.9) constitute a complete set describing continuous concurrent and countercurrent systems. This set determines paths of drying or adsorption processes. By assigning a location coordinate h to path parameters (eg, Fig. 6.8), process trajectories can be obtained for all necessary parameters in terms of h. Among a number of possible formulas for h in countercurrent and concurrent systems we specify a few later:

$$h = -\int_{T_1}^{T_0} \frac{GC_g}{\alpha a} \frac{dT}{(T-T_s)} = -\int_{X_0}^{X_1} \frac{G}{K_g'a} \frac{dX}{(X_s-X)} = -\int_{W_0}^{W_1} \frac{S}{K_g'a} \frac{dW}{(X_s-X)} \tag{6.21}$$

Each integral proves that, in the case of constant coefficients, the length of the dryer is the product of the height of the transfer unit, H_{TU}, and the number of transfer units N_{TU}. The right hand part of

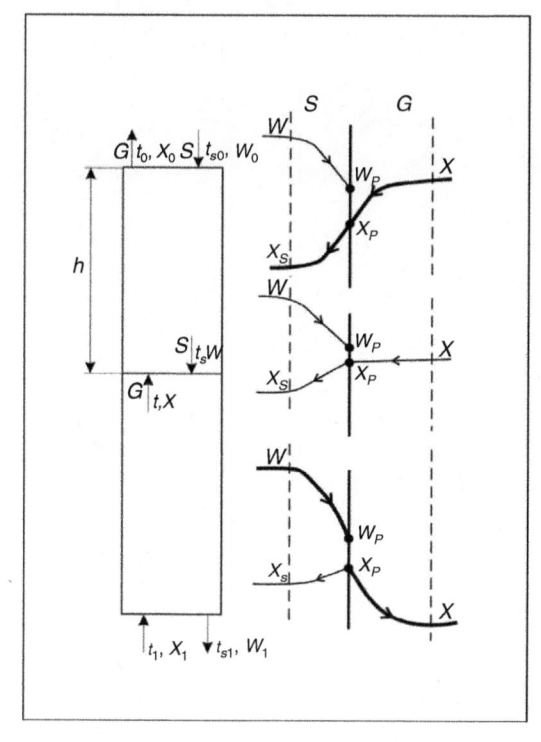

Figure 6.8: Driving forces in gas G and solid S in the process of countercurrent drying of falling solid particles. *(Sieniutycz, 1967; Ciborowski and Sieniutycz, 1969a, 1969b, 1969d).*

Fig. 6.8 shows three possible distributions of gas humidity and solid moisture content in gaseous and solid boundary layers, where especially the latter distribution is of somewhat arbitrary nature (is of rather nontypical nature; see Ciborowski and Sieniutycz, 1969a, b, c, e; Sieniutycz, 1967) for an alternative model assuming the main transfer resistances in solids as well as use of the so-called drying coefficient K in place of the overall gas coefficient K'_g.

Distributions of concentrations W, X, and so forth, in the vicinity of the interface with parameters W_p, X_p, and so forth, may be quite diverse. Some of these distributions may cause the appearance of maxima or minima of moisture content, gas humidity, and temperature within each phase, depending on inlet thermodynamic states of solid and gas, total length of the dryer, and values of transfer coefficients. Many nontypical properties were confirmed by experiments; the results of some of them are shown in Fig. 6.9.

Fig. 6.9 implies a classification for experimental trajectories of drying (upper chart) and sorption (lower chart) for the countercurrent and cocurrent contacting of gas and pulverized solid. This classification is described in the later Section 6.5.

Solid to gas flow ratio, S/G, is one of the important factors influencing shapes of gas and solid paths in the diagram. The constancy of the flow ratio S/G along the path in association with

310   Chapter 6

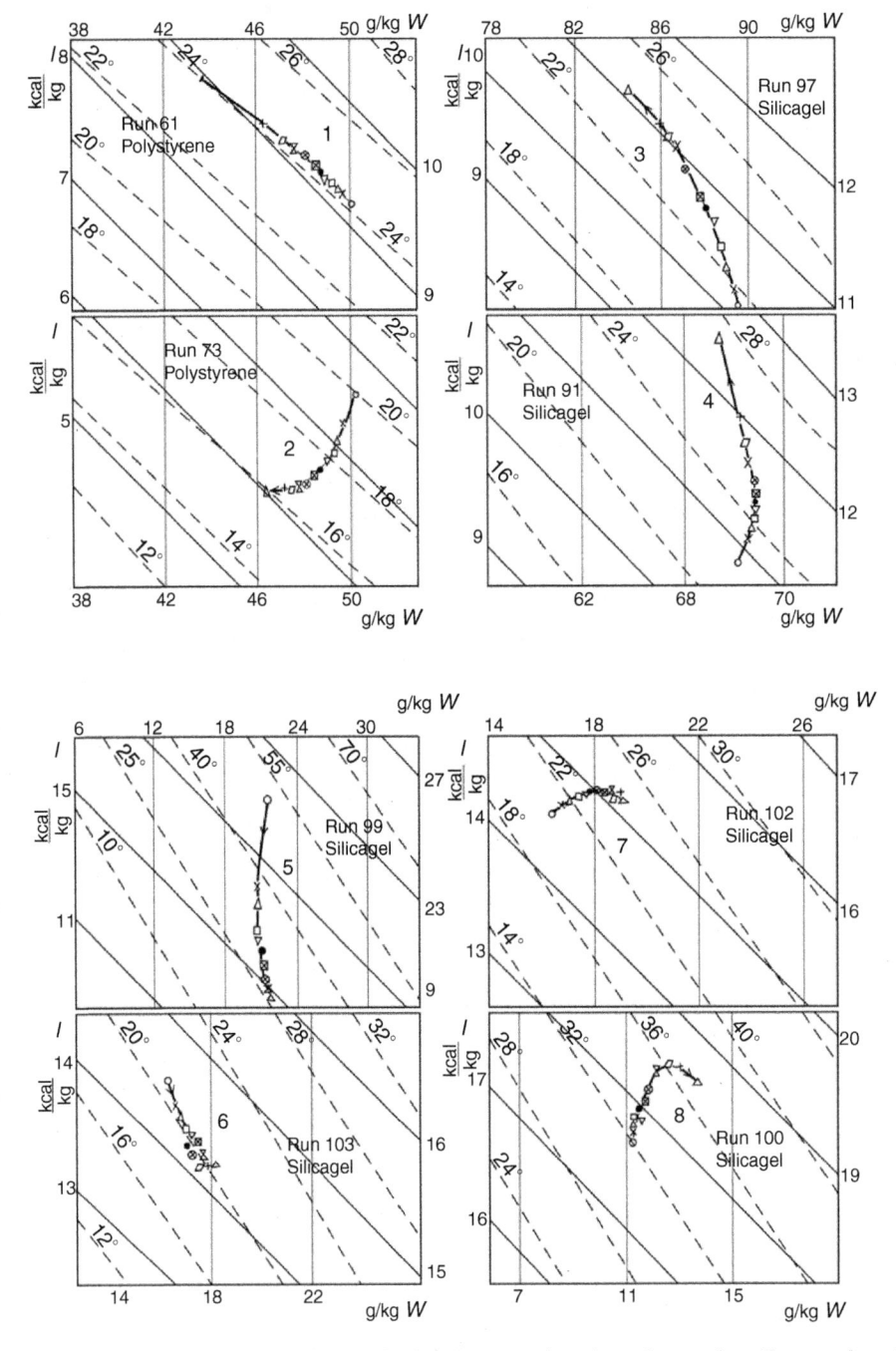

Figure 6.9: Experimental trajectories of drying (upper chart) and sorption (lower chart) in the countercurrent flow of gas and granular solid. *Sieniutycz, 1967.*

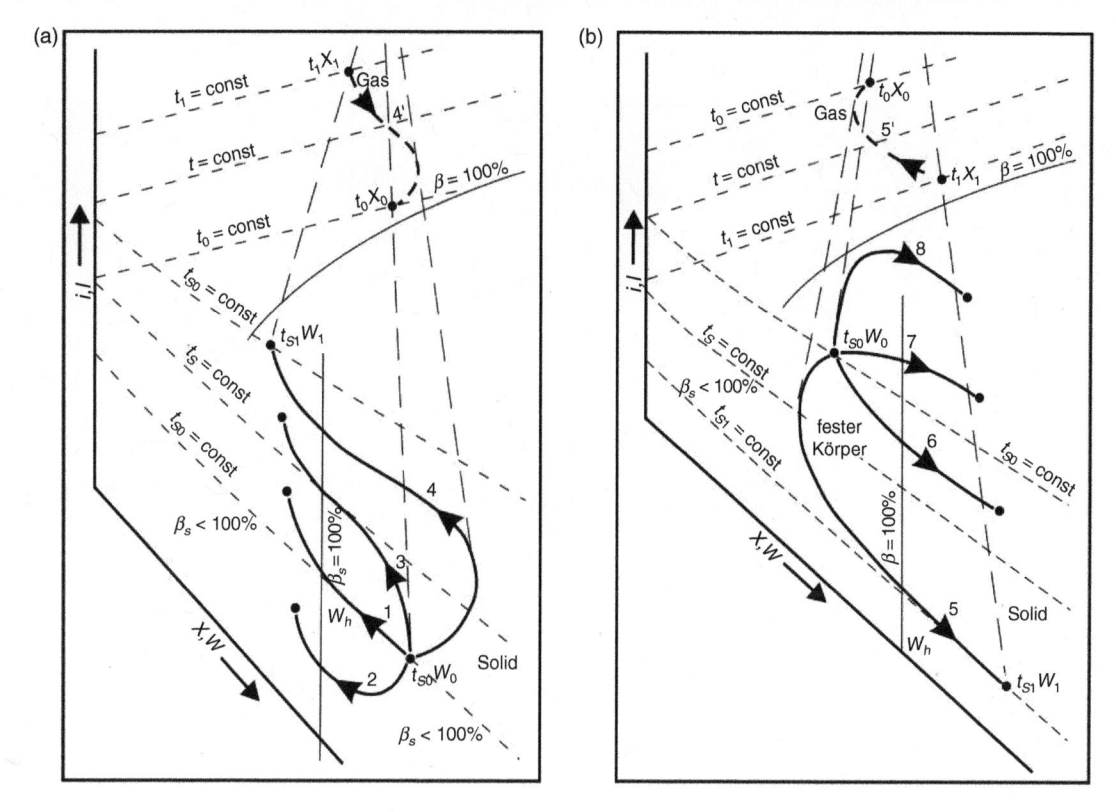

Figure 6.10: Geometric similarities of gas and solid paths in the enthalpy diagrams for drying in countercurrent system (a) and cocurrent system (b).

energy and mass balances yields the condition of geometric similarity of gas and solid paths in enthalpy diagrams. Fig. 6.10 illustrates these geometric similarities for countercurrent processes of drying (left draft) and sorption (right draft). However, note that in cocurrent systems the geometric similarity is accompanied with the simultaneous reflection of the gaseous curve.

6.5 Graphical Classification of Experimental Data

This section links graphical representations with some experimental results and their classification in enthalpy diagrams. Some limitations of the approach are also discussed.

Fig. 6.9 is a basis to classify experimental trajectories of drying (upper chart) and sorption (lower chart) for the countercurrent contacting of gas and pulverized solid. Note that the classification in Fig. 6.9 is for the same pole for all paths.

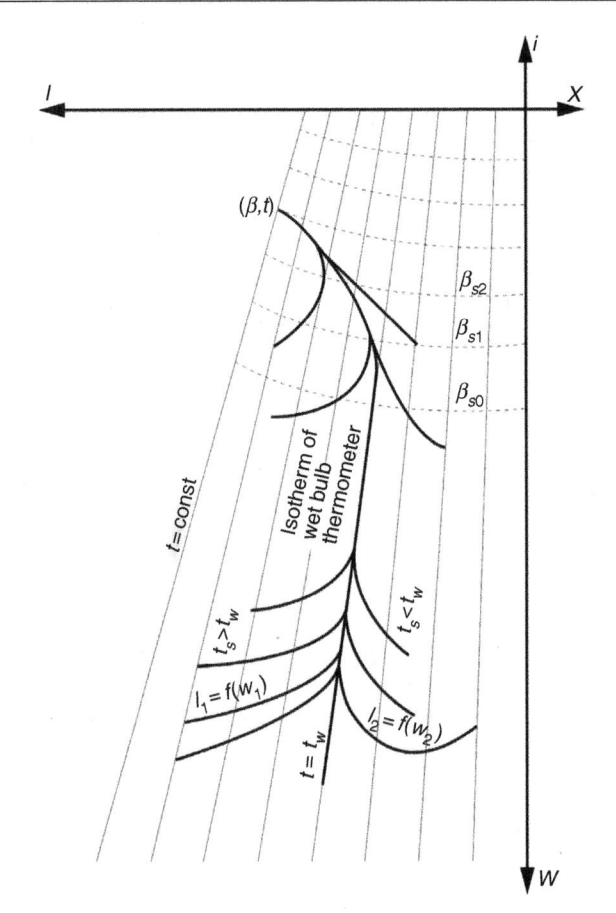

Figure 6.11: Paths of dried solid in the enthalpy–concentration chart for moist solid.
All paths achieve assymptotically the wet bulb temperature or its counterpart attributed to the second drying period.

Experimental data presented therein show that there are only four typical paths for drying and four more paths for sorption. This limitation is the consequence of constraints imposed by balance equations for gas and solid phases, which define the geometric similarity of gas and solid paths in drying systems, Fig. 6.10. The left chart pertains for countercurrent system and the right one for a cocurrent system. In concurrent systems geometric similarity occurs with the simultaneous reflection.

Fig. 6.11 depicts qualitatively drying paths of a moist solid in the enthalpy chart and shows the asymptotic nature of wet-bulb temperature. Qualitatively, this picture may be referred to drying of a single particle or fluidized drying of a solid batch. In the first drying period transient paths of solid fall after a time period into the isotherm of the wet bulb temperature or its counterpart attributed to the second drying period. Souninen (1986) discusses effectiveness of psychrometric charts in their application to drying calculations (Fig. 6.12).

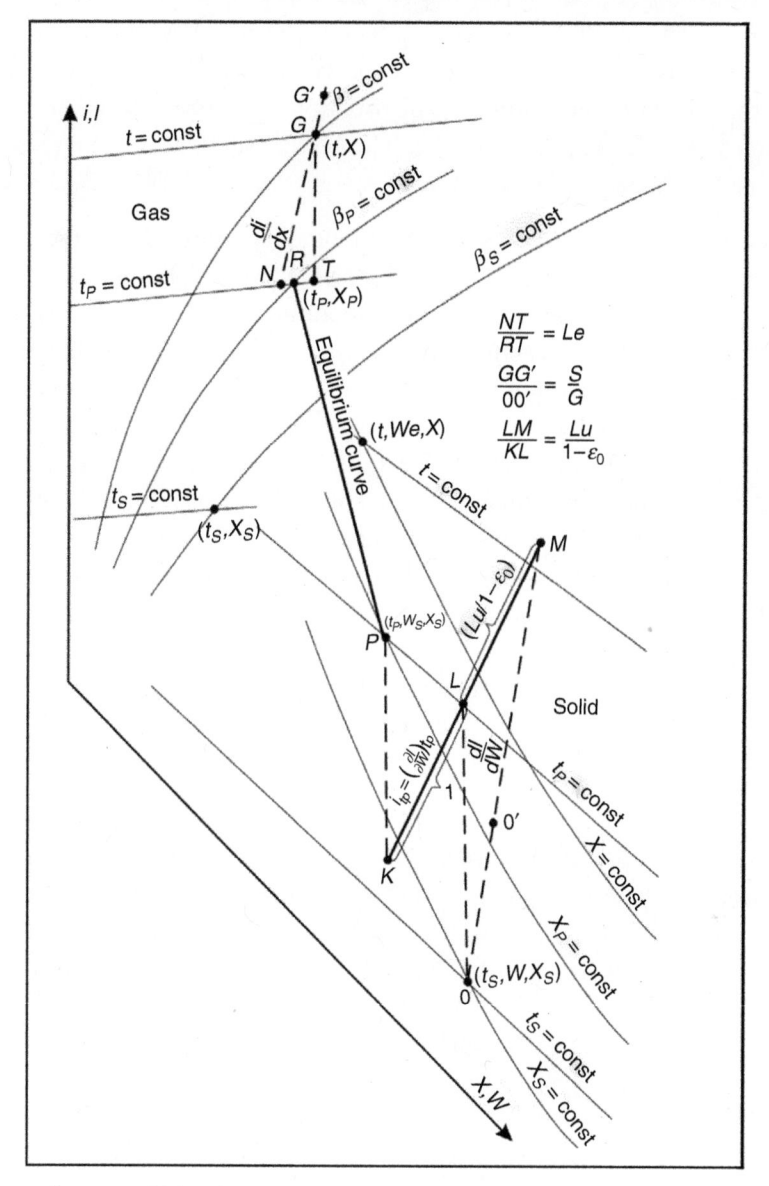

Figure 6.12: Balance and kinetic constructions in the gas–water–solid chart which take into account the coefficient of phase changes ε within the solid phase. *Ciborowski and Sieniutycz, 1969a.*

Limitations associated with the graphical modeling of drying should not be ignored. In fact, any reasonable graphical approach requires some kind of approximation, such as a single averaged temperature and moisture context of the grain. As confirmed by the countercurrent drying experiments (Ciborowski and Sieniutycz, 1969b), Farid's model (Farid, 2003) shows that the temperature distribution within the droplet cannot be neglected even for a small diameter droplet of 200 μm. This corresponds with the results of some earlier works dealing with granular materials, which state that effects of the temperature distribution within the granules with the inhomogeneous moisture distribution cannot be ignored even for small Biot numbers (Ciborowski and Sieniutycz, 1969a, b, c; Farid, 2003). Moreover, there are some other unexplained issues, as discussed later.

Rates of liquid evaporation based on statistical rate theory approach and their applications are analyzed by Fang (1998). According to Fang and Ward (1999), measurements of the temperature profile across the interface of an evaporating liquid are in strong disagreement with predictions from classical kinetic theory or nonequilibrium thermodynamics. However, previous measurements in the vapor were made within a minimum of 27 mean free paths of the interface. Since classical kinetic theory indicates that sharp changes in the temperature can occur near the interface of an evaporating liquid, a series of experiments were made to determine whether the disagreement could be resolved by measurements of the temperature closer to the interface. The measurements reported were performed as close as one mean free path of the interface of the evaporating liquid. The results indicate that it is the higher-energy molecules that escape the liquid during evaporation. Their temperature is greater than that in the liquid phase at the interface and, as a result, there is a discontinuity in temperature across the interface that is much larger in magnitude (up to 7.8°C in experiments) and in the opposite direction to that predicted by classical kinetic theory or disequilibrium thermodynamics. The measurements reported therein support the previous data (Fang, 1998).

Transferring graphical methods to the realm of partial differential equations, Sieniutycz (1970) developed principles of graphical approaches to simultaneous heat and mass transfer in packed beds with stationary packing. All constructions were performed on the two-phase diagram for moist gas and moist solid. The method involving equations of equilibria, conservation, and transfer, holds for an arbitrary period of drying or adsorption. It predicts extrema of temperature and absolute moisture content in solid form, which moves in the direction of the gas flow. Some properties are not anticipated by oversimplified, isothermal analytic theories of packed beds. The new effects are particularly pronounced in the case of large thermal effects accompanying processes of drying or adsorption.

Reader can refer the literature for further examples, in particular for the balance and kinetic constructions in the gas–water–solid chart, which take into account the coefficient of phase changes ε within the solid phase (Ciborowski and Sieniutycz, 1969a).

6.6 Concluding Remarks

In the present chapter, balance and kinetic equations for drying and sorption operations are analyzed and solved in the enthalpy diagram for a gas–moisture–solid system, and principles of thermodynamic classification of qualitative properties of (uncontrolled) trajectories are obtained for various processes of fluidized drying, and concurrent and countercurrent drying and adsorption (Ciborowski and Sieniutycz, 1969a, b). In this classification four typical trajectories pertaining to drying processes and other four trajectories referred to adsorption are obtained. Both graphical and experimental data show, under appropriate conditions, extrema (minima or maxima) of temperature and moisture content in the solid phase. All these extrema are predicted by graphical methods and confirmed by the experiments. Graphical methods of analysis in enthalpy–composition diagrams involve the majority of basic problems of heat and mass exchange in fluidized beds.

Importantly, the knowledge and understanding of graphical approaches considerably facilitates transformation of related mathematical models into computer programs. This is fortunate in view of the common use of computers at the present time. With recent advancements in mathematical techniques and computer hardware, CFD has been found to be successful in predicting the drying phenomenon in various types of industrial dryers. As pointed out by Pakowski and Mujumdar (2006), we have been witnessing a revolution in methods of engineering calculations. See a review by Jamaleddine and Ray (2010) on application of CFD for simulation of drying processes. Computer tools are now easily available. A new discipline of computer-aided progress design (CAPD) has emerged. Today even simple problems are solved using specialized computer software. General computing tools including Excel, MathCAD, MATLAB, and Mathematica are easily available in any engineering company. Bearing this in mind, Pakowski and Mujumdar (2006) have promoted a more computer-oriented calculation methodology and simulation methods to substitute graphical and shortcut methods. Nonetheless they state, "this does not mean that the computer will relieve one from thinking," but rather that the older methods and rules are still valid "and provide a simple commonsense tool for verifying computer-generated results." In this context one should also see the future role and applicability of graphical methods.

For selection of other drying processes in which the use of graphical methods could be fruitful, book and reviews are recommended (Mujumdar, 2007; Chua et al., 2002; Barrozo et al., 2014; Strumillo and Kudra, 1987). The following Appendix briefly reviews several possible drying systems.

6.7 Appendix: Some Associated Drying Problems

Mathematical models and numerical techniques for simulation of steady grain drying in countercurrent and concurrent gas flows have been proposed and tested by Valença and Massarani (2000). To overcome numerical difficulties they developed an approach based on simulation starting from initial transient conditions. They also built a laboratory unit to test mathematical models and numerical techniques through the comparison between calculated results and experimental data. However the majority of research in countercurrent and concurrent systems refers to the moving bed dryers. Some of these works are discussed later.

Jaakko and Saastamoinen (2005) compared moving bed dryers of solids operating in parallel and counterflow modes. He developed a method used to simulate drying in parallel and counterflow moving beds. He also gave an integral solution showing the relation between time or location in the dryer and degree of drying. The method allows for a rapid calculation of the moisture, vapor mass fraction, and temperature distributions along the dryer in drying with moist air or steam. The method applies to various substances to be dried when the dependence of the drying rate on moisture and ambient temperature and humidity (thin-layer drying rate) is known. For the same required degree of drying, unfavorable distribution of the evaporation temperature in the parallel mode increases significantly the size of the dryer in comparison to the counterflow dryer. In steam drying, the difference in the size is not so great, since the evaporation occurs approximately at constant temperature.

Theory of moving beds sheds some extra light on modeling of countercurrent systems with falling grains. The aim of work by Barrozo et al. (1998) was to analyze simultaneous heat and mass transfer between air and soybean seeds in countercurrent and concurrent moving bed dryers by simulation. The equation of drying kinetics was obtained by means of a thin-layer study. The profiles for temperature and humidity of the fluid and the temperature and moisture of the seeds were found by numerical solution of a model consisting of ordinary differential equations. For further work along this line, see papers by (Felipe and Barrozo, 2003; Barrozo et al., 2006a, b). For a summary on quality of seeds undergoing moving bed drying in countercurrent and crosscurrent flows, see especially a review by Barrozo, Mujumdar, and Freire (Barrozo et al., 2014). This article presents a review of the seed air-drying process, including differential equations models derived from mass and energy balances for seeds and air in fixed and moving bed dryers. The article concludes with an overview of several potential drying technologies that can be applied to seeds.

In the paper by Djaeni et al. (2008), two-dimensional computational fluid dynamics (CFD) calculations for multistage zeolite drying (adsorbent drying) were performed for two dryer configurations: (1) a continuous moving bed zeolite dryer and (2) a discrete bed zeolite dryer. The calculations concern drying of tarragon as a herbal product. The results show the profiles of water, vapor, and temperature in dryer, adsorber, and regenerator. The performance of a continuous moving bed zeolite dryer is the best. Residence time of air, product, and zeolite are in accordance to other drying systems. In a latter work of Djaeni, group effects of multistage adsorption drying are evaluated (Djaeni et al., 2009). In a multistage dryer, the product is dried in succeeding stages while the air leaving a stage is fed to the next stage after dehumidification by an adsorbent. Energy efficiency of the drying system is evaluated for low-temperature drying (10–50°C) and compared with conventional condenser drying. Efficiency of the multistage adsorption dryers increases with the number of stages.

Tórrez et al. (1998) studied theoretically and experimentally drying of grain in a cross-flow moving bed type. In Section 6.3 a counterpart of this scheme in fluidization is called the horizontal exchanger). The authors analyzed two different dryer configurations, a dryer with central air distribution and another with multiple air duels. Simulations based on a mathematical model show that distance between inlet air and outlet devices, air-to-solid-flow ratio and dryer height-to-cross-section ratio have a remarkable influence on the process. This conclusion is also valid for countercurrent dryers with falling particles (Ciborowski and Sieniutycz, 1969b).

As stated by Alvarez et al. (2006) product damage, high-energy use, and nonhomogeneous product properties occur in typical drying operations such as pneumatic drying, fluidized-bed dryers, and upward circulating fluidized-bed dryers. The authors constructed a test for an experimental downflow dryer and modeling of the operation. Predictions of pressure, gas velocity, solid holdup, and temperature agreed with experiments.

Jaakko and Saastamoinen (2005) compared the drying rates in moving bed dryers. The gas and the solids to be dried are in parallel flow or counterflow. Their model is based on the solution of arbitrary experimental or theoretical drying rate. Equations of single solid particles (or thin-layer drying rate equation) coupled with heat and mass conservation equations of the dryer. The solution is presented in an integral form of the drying equation showing the relation between time or location in the dryer and degree of drying. The method allows rapid calculation of the moisture, vapor mass fraction, and temperature distributions along the dryer in drying with moist air or steam. The size of the dryer needed to reach the same degree of drying operating in the parallel mode is much greater than that of counterflow type.

Kemp et al. (1994) developed a one-dimensional model and experimental studies of vertical pneumatic conveying dryers. The model takes time increments along the tube; a computer program is required for its implementation. The approach may be used either for initial design or for scale-up from pilot plant data.

Thakur and Gupta (2006) examined experimentally drying of high-moisture paddy under stationary and fluidized bed with and without intervening rest periods. Introduction of a rest period between first and second stage of drying improved drying rate and lowered the energy requirement. Fluidization further improved the overall drying process. During the period of rest stage, paddy grain released a considerable amount of moisture as an effect of residual grain temperature. A considerable amount of energy (21–44%) can be saved by providing a rest period from 30–120 min between the two stages of drying. Fluidization further reduces ($\approx$ 50% against continuous drying under stationary bed) the energy requirement.

Mujumdar and Law (2010) discuss trends and applications in postharvest processing in the context of drying technology. Thermal-drying technologies have attracted significant R&D efforts owing to the rising demand for improved product quality and reduced operating cost as well as diminished environmental impact. Drying materials may appear in the form of wet solid, liquid, suspension, or paste, which require drying to extend the period of storage, ease of transportation, and for downstream processing to produce value-added products. Most of these materials are heat-sensitive and require careful drying; conventional hot air-drying can be detrimental to bioactive ingredients. This article summarizes some of the emerging drying methods and selected recent developments applicable to postharvest processing.

Nicolin et al. (2014) applied the Stefan problem approach to the mathematical modeling of the diffusion process that occurs during hydration of grains. This technique assumes that boundaries of the problem are unknown, and finding them is a task, which is a part of the problem. The problem has boundaries, which move along time. The governing differential equations are numerically solved for the case of transient diffusion, and an analytical solution is obtained for the hydration fronts by considering the pseudo-steady state hypothesis for the diffusion equation. The behavior of hydration fronts is analyzed. The fronts are defined by the moving boundaries and their velocities for both cases. The main differences are analyzed in terms of the transient term of the diffusion equation. The results show the main differences in the behavior of hydration fronts that arise when the transient term is taken into account or the pseudo-steady-state hypothesis is considered and compare the obtained behaviors with experimental data.

Currently, energy and exergy analyses are conducted for the purpose of optimizing operating conditions and quality of the products (Syahrul et al., 2001). In this context, energy and exergy models are developed for drying of granular materials in fluidized beds, packed beds and other systems with granular solids, in order to evaluate energy and exergy efficiencies. The models and their predictions are subsequently verified against the experimental data. The effects of inlet air temperature, fluidization velocity, slip between phases, and initial moisture content on both energy and exergy efficiencies are studied. Also, the hydrodynamic properties, for example, the bed holdup, are determined. The results show that exergy efficiencies are lower than energy efficiencies due to irreversibilities which are not taken into consideration in energy analysis, and that both energy and exergy efficiencies decrease with increasing drying time.

It is an open question whether the flux relaxation phenomena of extended irreversible thermodynamics should be taken into account in graphical methods since the effects are usually not substantial. Mitura (1981) and Martynenko and Khramtsov (2004) analyzed nonstationary evaporation of liquid with the help of the phenomenological thermodynamics of irreversible processes. They stressed the importance of evaporation and condensation processes and operations for the contemporary engineering. These processes and operations are present in contact heat exchangers of industrial equipment, vapor-generation systems, vacuum metallurgy, during production of ultra-pure materials and specimens with original properties in microelectronics and diverse chemical technologies. They are also applied in fuel systems of internal combustion engines, rocket cryogenic engines, and in different technological processes of the pharmaceutical industry.

From the viewpoint of safety of exploitation, nonstationary evaporation processes are particularly important in engineering calculations of transient models of vapor-generating systems of powerful electric power plants. At present, only the problem of a stationary evaporation–condensation process has been extensively investigated. Mitura (1981) and Martynenko and Khramtsov (2004) analyzed unsteady-state evaporation of liquid by promoting thermodynamics of irreversible processes as the governing, basic theory.

Application of the method of the thermodynamics of irreversible processes in the phenomenological analysis of nonstationary evaporation and the results of experimental study of pressure relaxation in thermostated volume with a sharp change in the vaporization flow indicate the considerable effect of inertial properties of the thermodynamic system in the dynamics of development of a nonstationary process. Thus, in calculation of nonstationary modes of operation of fuel-power equipment we should admit such phenomena as additional vaporization and inductive pressure drop with a sharp change in the vaporization flow. Analytical treatment of these effects is difficult, whereas a graphical approach can work relatively easily.

CHAPTER 7

Frictional Fluid Flow Through Inhomogeneous Porous Bed

7.1 Introduction

Since porous systems are suitable arrangements to intensify transfer phenomena and catalytic reactions and to enhance effects of fluid flow, a considerable number of books and research papers on porous systems is available (Muskat, 1937; Bear, 1972; Bear et al., 1999; Dagan, 1989; Dagan and Neuman, 1997; Du Plessis and Masliyah, 1991; Hesse, 1974, 1977a, b), and many others. Lenormand (1989) summarizes the different models of fluid transport in porous media including applications of fractal concepts in petroleum engineering. Porous media have attracted the attention of scientists from various fields, first because they constitute a challenge to basic research (complex systems with complex properties), and, second, they have a genuine and broad field of applications in the chemical and petroleum industry. De Pablo et al. (2011) summarize the information about a fractional porous medium equation. Vidales and Miranda (1996) introduce fractal porous media. Through mathematical considerations they establish relationships between their macroscopic properties (permeability, form factor, and porosity) and check these relationships with experimental data. The good agreement proves the efficiency and validity of the approach.

Yet, for our purposes, the thermodynamically oriented research is of a special interest. Working with a porous medium, Miguel (2000) presented theoretical and experimental research on mass transfer through porous media and used a thermodynamic model to predict the change of mass inside a porous medium and to describe the fluid flow through porous skeleton. The experiments were carried out to provide a thermodynamic chart of a porous medium and to infer the parameters required by the theory. He also investigated the effect of the moisture content and temperature on the fluid transport properties. Packed beds with simultaneous heat and mass transfer in drying and moisturizing were analyzed with the help of enthalpy diagrams (Sieniutycz, 1970). Moving porous systems were investigated in the many papers of Poświata and Szwast (2006), who, in particular, optimized various modes of solid drying in bubbling fluidized beds (vertical and horizontal).

Applying nonequilibrium thermodynamics and modern integral transformations to solve PD equations of transport in porous systems, Luikov and his coworkers developed contemporary mathematical theory of transport and flow phenomena in capillary porous systems, in particular nonisothermal drying systems (Luikov, 1966a, b, 1975; Luikov and Mikhailov, 1963). The theory was also applied to formulate and solve wave equations of the heat conduction (Luikov et al., 1976) and extended to handle the problems of natural convection and thermal waves (Luikov and Berkovsky, 1974). Later, Lorente (2007) and Lorente and Bejan (2006) offered a different (constructal) view of transfers through porous media following Bejan's constructal theory (Bejan, 2000).

Darcy's Law is the basic constitutive relationship that describes fluid motion in porous media and that helps us to understand the movement of fluids in the Earth's crust. Darcy's Law states that the rate of fluid flow through a porous medium is proportional to the gradient of hydraulic head within that fluid. Hydraulic head is interpreted in Fig. 7.1, which shows how water level in the wells measures the head level in the aquifer.

The constant of proportionality is the hydraulic conductivity; it is a property of both the porous medium and the fluid moving through this medium. When this conductivity is a known nonlinear function of a length coordinate x (set here purposely in the direction parallel to the conductivity gradient) a Fermat-like principle can be formulated for steady frictional flows in isotropic media, simple and composite. In contemporary modeling of fluid's seepage, acceleration terms and corresponding transients can be included in a generalized form of Darcy's law, yet they are neglected in the presented variational description for the fluid transport in an inhomogeneous porous medium. Still, there appears the need to take into account the effect of the state changes in the coefficients of the (steady) variational formulation. (The notions of system state and state space are used here in agreement with the dynamic system theory and optimal control.) In Darcy's law, the state-dependent hydraulic conductivity is such a parameter sensitive for the state changes. In this work, we consider a steady, frictional flow of the fluid in an inhomogeneous porous medium. We assume that the hydraulic conductivity is a known prescribed function of the spatial coordinate x. Although in contemporary modeling of fluid's seepage acceleration terms and corresponding transients can be included in generalized form of Darcy's law, they are neglected in the presented

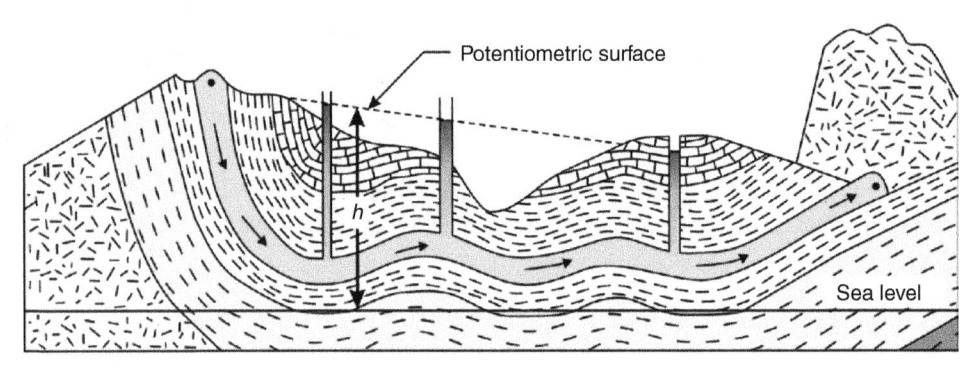

Figure 7.1: Hydraulic head.
Water level in the wells measures the head level in the aquifer.

variational description for the fluid transport. The basic purpose of the present research follows from the observation that, for various reasons, packing can be inhomogeneous (causing location dependent hydraulic resistances). Transport of the fluid through the porous medium is then associated with curved fluid's streamlines, hence the rationale for a variational principle that could treat the curvature effects quantitatively. According to the principle, the shape of a steady trajectory of nonlinear frictional flow is the result of the maximum flux or shortest transition time of the fluid through the medium. This property leads to the prediction of shapes of related "rays" or paths of the fluid flow. This also leads to the process description in terms of wave-fronts and related Hamilton–Jacobi theory which is derived in this chapter from the optimization algorithm of the dynamic programming method. In a part, our approach transfers to the realm of nonlinear Darcy's flows some results obtained in earlier treatments of fluid transfer in the energy and entropy representation (Tan and Holland, 1990; Sieniutycz, 2000b).

Here, however, instead of performing the analysis in the realm of paths, we concentrate on the wave-front description of the fluid flow and the corresponding Hamilton-Jacobi theory. The relevant physical picture refers to tracing of fluid propagation in Lagrange coordinates, where the fluid's flow through the porous skeleton is attributed to motion of the same fluid particles rather than to the fluid's passage through a fixed region of the physical space. Applications of functional equations and the Hamilton–Jacobi–Bellman equation are effective when Bellman's method of dynamic programming (Aris, 1964) is applied to wavefronts. That method involves an alternative view: the observation of mass conduction in porous medium in terms of wave fronts rather than fluid paths. In this method potential functions describing the mimimum resistance or the minimum transition time are obtained in an explicit way by analytical or numerical approaches.

Following an earlier work (Tan and Holland, 1990), but working in the entropy representation as the only correct one that allows the inclusion of conserved energy transfer (Sieniutycz, 2000b), one derives a sort of law of refraction for fluid's frictional transfer. The derivation originates from the boundary conditions that express the continuity of the transfer potential (hydraulic head) and the mass flux continuity through the interface. In this approach we assume a steady state frictional flow of a fluid through a smooth interface separating two regions of different hydraulic conductivity without interfacial resistance and with no sources or sinks of the fluid. (The assumption of the interface transport is only for technical reasons as explained at the end of this section.) Under these conditions hydraulic head $h = P(\rho_m g)^{-1} + z$ is continuous across the interface and the path of the frictional flux (the Darcy's ray) at each point is uniquely determined by the direction of the gradient of hydraulic head $h = P(\rho_m g)^{-1} + z$. Then one shows that a passage of a conserved entity, such as the fluid's mass flux, through an interface is governed by a so-called tangent law of refraction, whereas flows of other quantities, associated with nonconserved fluxes, are governed by laws deviating from the tangent law. In particular, flows of photons are described by sine law of refraction. Thus, we analyze here the nature of conserved flows in both discrete and continuous cases—see Sections 7.2 and 7.3, respectively. We also derive, in Sections 7.4 and 7.5, the variational setting and Hamilton–Jacobi theory for wavefronts. Section 7.6 describes a relation between the present formalism and those other ones describing different physical systems (eg, those with nonconserved flows), and Section 7.7 provides an example for a porous system transport. The concluding Section 7.8 specifies conclusions regarding properties of optimal paths in inhomogeneous porous systems. The results imply applications of the theory to various practical systems with mass transfer (eg, to porous drying systems) in which a nonuniform distribution of fluid's mass flux is observed, caused by local variations of porosity and hydraulic resistance (imperfect systems).

It should be understood that the discrete model of mass transport through an interface between two homogeneous media, is, in fact, only a discrete tool enabling a suitable treatment of an inhomogeneous porous system. The physically significant entity is the continuous inhomogeneous limit of the discrete model, that corresponds to an infinite number of infinitesimal discrete steps, each step involving two neighboring, infinitesimally narrow, vertical homogeneous layers of a porous system, such as in Fig. 7.2. Both neighboring layers differ slightly in their physical properties (eg, in their specific hydraulic conductivity), so that a real porous system is treated as an arrangement composed of an infinite number of infinitesimally narrow vertical layers. The trajectory bending is just the effect of change of the porous system specific conductivity with fluid's residence time. The results refer to flows following curved paths caused by weak nonlinear terms, these terms representing only the effect of state variations (in coefficients), not variations in derivatives (quasilinear models).

While many mathematical results considered here appear in textbooks on variational principles they are new in the context of purely dissipative processes like Darcy's flows,

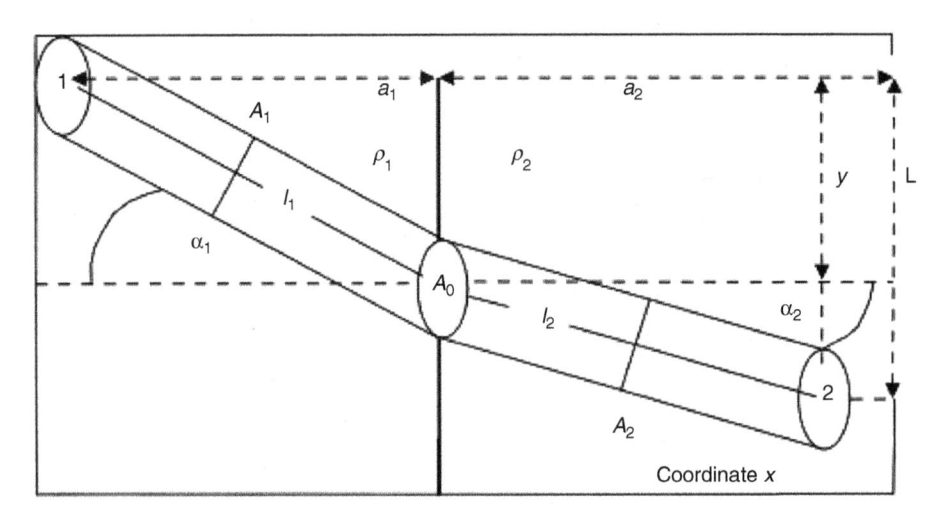

Figure 7.2: Illustration of a Fermat-type principle and the principle of notation.
Bending effect for a Darcy's path in fluid's frictional transfer.

Fick diffusion or Fourier heat transfer. For these processes the Hamiltonian forms of the Lagrange integrand do not apply and the process Lagrangian must be determined prior to the application of the variational procedure.

7.2 A Discrete Model Leading to a Bending Law

Fig. 7.2 depicts a frictional flow of the fluid through a vertical interface separating two regions of different hydraulic conductivity. α_1 and α_2 are respectively the angles of incidence and refraction, whereas ρ_1 and ρ_2 are corresponding specific resistances of the two porous media. By definition, a specific hydraulic resistance is the reciprocal of hydraulic conductivity (ie, ρ_1 and ρ_2 in Fig. 7.2 are reciprocals of hydraulic conductivities k_1 and k_2). By Darcy's law the conductivities (or resistances) relate fluid flows to gradients of the hydraulic head. Assume an interface characterized by good contact between media or the continuity of the quantity h across the interface. The boundary conditions describing the mass conservation in a porous medium whose hydraulic conductivity exhibits a jump through an interface can be written in the following form (IaI refers to the absolute value of a)

$$|\nabla h|_1 \sin \alpha_1 = |\nabla h|_2 \sin \alpha_2 \tag{7.1}$$

for the tangent component of and

$$k_1 |\nabla h|_1 \cos \alpha_1 = k_2 |\nabla h|_2 \cos \alpha_2 \tag{7.2}$$

for the normal component in absence of sources and sinks of mass. Here we have assumed the absence of a reflected flow. The law of the refraction for the conduction rays follows as ratio of these equations

$$\rho_1 \tan \alpha_1 = \rho_2 \tan \alpha_2 \tag{7.3}$$

where, by definition, $\rho = 1/k$ is the specific hydraulic resistance of the porous system. This equation is analogous to Snell's law with the tangent replacing the sine and the specific resistance replacing the refractive index. This law was originally formulated for the heat flow and reciprocal of the usual thermal conductivity λ (Tan and Holland, 1990) which we however attributed (Sieniutycz, 2000b) to the reciprocal of Onsagerian conductivity k as the more proper quantity than λ because of the basic link of k with the entropy production. Eq. (7.3) shows that the bending principle is only of "Fermat type," but it is not really Fermat because the tangent law replaces the sine law.

Bejan (2003) considers the connection between the 'Fermat-type' principle (Sieniutycz, 2000b) and his constructal principle of how geometric form (optimized complexity, organization) is generated in macroscopic flow systems. The consideration of this connection shows the strength of the constructal theory (Bejan, 1997a, c, 2000). Further, Lorente and Bejan (2006) and

Bejan et al. (2005) treat heterogeneous porous media as multiscale structures for maximum flow access. Natural porous structures are heterogeneous with multiple scales that are distributed nonuniformly. Few large pores (fissures, channels, and cracks) are accompanied by numerous finer channels. Bejan and Lorente (2006b) ask: Can this type of flow architecture be attributed to a principle of maximization of global flow access? And they answer: "Features similar to those of multiscale porous structures are exhibited by tree-shaped flow structures. Trees have been deduced from the maximization of flow access between a point and a volume, a point and an area, and a point and a curve (eg, circle)." Invoking the same principle they consider the question of how to bathe with minimal flow resistance a microchannel structure that globally behaves as a porous medium. They develop completely multiscale configurations that guide the flow from one side of the porous structure to the other (line to line and plane to plane) and show analytically the advantages of tree structures over the usual stacks of parallel microchannels. The "porous medium" that has tree-shaped labyrinths is heterogeneous, with multiple scales that are distributed nonuniformly. These features justify comparisons with the design of natural porous structures. The book by Bejan and Lorente (2008a) focuses on constructal theory to design strategies for more effective flow and transport. Constructal theory is the view that the generation of flow configuration is covered by a principle (the constructal law): "For a finite-size flow system to persist in time (to live) it must evolve such that it provides greater and greater access to the currents that flow through it" (Bejan and Lorente, 2008a). This principle has been used to account for many features of design in nature, from the tree-shaped flows of lungs and river basins, to the scaling laws of animal design, river basin design, and social dynamics. The best configurations that connect one component with very many components are tree-shaped, and this is why dendritic flow architectures occupy a central position in the field. For applications in porous and complex flow structures, see collective work by Bejan et al. (2005).

Let us return to the bending principle (7.3). Examining the deviation of the refracted ray from the incident ray, it is convenient to introduce the relative hydraulic resistance of the second porous layer with respect to the first: $\beta = \rho_2/\rho_1$. In terms of β and the angle of incidence, α_1, the angle of refraction is

$$\alpha_2 = \arctan(\tan \alpha_1/\beta) \tag{7.4}$$

whereas the angle of deviation of the reflected ray, $\Delta = \alpha_2 - \alpha_1$ equals

$$\Delta \equiv \alpha_2 - \alpha_1 = \arctan(\tan \alpha_1/\beta) - \alpha_1. \tag{7.5}$$

Figures were obtained (Tan and Holland, 1990) that depict the angle of refraction, α_2, and deviation, Δ, as functions of the incidence angle α_1 and the resistance ratio, $\beta = \rho_2/\rho_1$. The case $\beta > 1$ is the optical analogue of refraction from the rarer to the denser medium, and inversely. In contrast to Snell's refraction law in optics, where the deviation Δ decreases monotonically with the incidence angle α_1, in the tangent law case the deviation attains an extremum and approaches zero for both normal and grazing incidences, $\alpha_1 = 0$ and $\alpha_1 = 90°$, respectively. In the limit as β tends to zero, the incidence angle α_1 approaches zero which means that the Darcy's flux entering a perfect hydraulic conductor is normal to the "interface" (understood as a geometric surface separating two regions with different hydraulic conductivities). Otherwise, as β tends to infinity, the refraction angle α_2 becomes zero, which means that the Darcy's flux leaving a perfect conductor is normal to the "interface."

In the flow tube in Fig. 7.2 the fluid is transferred along the length dl by the cross-section perpendicular to the mass flux. The perpendicular cross-section has the area A which may change with l; the volume differential $dV = Adl$. As distinguished from more standard treatments, we integrate here over the volume V "moving with the fluid"; in this case x and y are Lagrange coordinates and the fluid's flow is attributed to motion of the same portion of the fluid rather than to flow through a fixed area in the space. We introduce the fluid current $I = dQ/dt$ as the amount (mass or volume) of the fluid Q transferred by the system per unit time t. The current I is the conserved property, that is, $I_1 = I_2 = I$. The related density of fluid flux is dQ/Adt or $J_q = I/A$. We also use here the resistance ρ which is the reciprocal of the hydraulic conductivity. The corresponding total resistances: $R_1 = \rho_1 l_1/A_1$ and $R_2 = \rho_2 l_1/A_2$ play also a role as we shall see soon. Here A_1 and A_2 are the cross-sectional areas of a "tube" of the Darcy's flux in porous media 1 and 2. The products $R_1 I$ and $R_2 I$ of resistances and current I have units of the hydraulic head, consistent with the fact that they are determined by the difference of h.

The Darcy's ray travels between two fixed points, 1 and 2. Fig. 7.2 shows the change of the area A perpendicular to a Darcy's ray or a steady streamline of the fluid when passing from the medium with the specific resistance ρ_1 to the medium with the specific resistance ρ_2 (hydraulic conductivities k_1 and k_2). Note that, for example, $A_1 < A_2$ for $\rho_1 < \rho_2$. Tan and Holland (1990) have shown the essential role of the dependence of the total flow resistance on the cross-sectional area A for conserved fluxes in media whose specific resistance exhibits a jump through an interface. The cross-sectional areas of the flux tubes in two porous media can be evaluated as

$$A_1 = A_0 \cos\alpha_1, \qquad A_2 = A_0 \cos\alpha_2 \tag{7.6}$$

where A_0 is the constant area of a flux tube intercepted by the interface (the constant area of projection of the mass flux tube cross-sectional area on the surface of constant resistance). Thus the total resistance for the fluid's flow between the points 1 and 2 is

$$R^{1,2} = \frac{\rho_1 l_1}{A_1} + \frac{\rho_2 l_2}{A_2} = \frac{1}{A_0}\left(\frac{\rho_1 l_1}{\cos\alpha_1} + \frac{\rho_2 l_2}{\cos\alpha_2}\right). \tag{7.7}$$

Substituting to this equation the values of $\cos\alpha_1$ and $\cos\alpha_2$ from Fig. 7.2 as

$$\cos\alpha_1 = \frac{a_1}{\sqrt{a_1^2 + y^2}}, \qquad \cos\alpha_2 = \frac{a_2}{\sqrt{a_2^2 + (L-y)^2}} \tag{7.8}$$

and the lengths l_1 and l_2 as

$$l_1 = \sqrt{a_1^2 + y^2}, \qquad l_2 = \sqrt{a_2^2 + (L-y)^2} \tag{7.9}$$

one finds

$$R^{1,2} = \frac{1}{A_0}\left\{\frac{\rho_1(a_1^2 + y^2)}{a_1} + \frac{\rho_2[a_2^2 + (L-y)^2]}{a_2}\right\}. \tag{7.10}$$

We stress that it is the vertical coordinate y of the intersection point with the interface which is allowed to vary; the horizontal coordinate x of that point is always constant and equal a_1. Since $y/a_1 = \tan\alpha_1$ and $(L-y)/a_2 = \tan\alpha_2$, the condition requiring the first derivative $dR_{1,2}/dy$ to vanish

$$\frac{dR^{1,2}}{dy} = \frac{2}{A_0}\left[\frac{\rho_1 y}{a_1} - \frac{\rho_2(L-y)}{a_2}\right] = 0 \tag{7.11}$$

is equivalent with the requirement that the tangent law, Eq. (7.3), is satisfied. Differentiating Eq. (7.11) with respect to y once again, one obtains

$$\frac{d^2 R^{1,2}}{dy^2} = \frac{2}{A_0}\left(\frac{\rho_1}{a_1} + \frac{\rho_2}{a_2}\right) > 0 \tag{7.12}$$

which proves the minimum property of $R^{1,2}$ at the stationary point. Consequently, the postulate that the fluid follows the trajectory of least resistance is the correct physical principle that leads to the tangent law (7.3), implied by boundary conditions (7.1) and (7.2). Eqs. (7.1) and (7.2) may be seen as the consequence of the mass conservation and the principle of least resistance. (As the latter is incorporated in the theorem of the least entropy production, one may regard Eqs. (7.1) and (7.2) as the consequence of the minimum entropy production applied to any conserved mass flux in a purely frictional flow.) In what follows we shall describe an extension of the aforementioned (discrete) problem to the continuous situation where an inhomogeneous porous material with a variable local resistance $\rho(x)$ is regarded as a sequence of many vertical infinitesimal layers, and the frictional flow of the fluid occurs through a porous material whose hydraulic conductivity varies continuously with x.

7.3 Bending of Fluid Paths in Inhomogeneous Porous System

In a corresponding continuous description that describes the fluid flow within an inhomogeneous porous structure we use the reference frame (x, y) in which the local resistance of the fluid's frictional flow changes along the axis x (Fig. 7.2), and the axis y is tangent to a surface of constant specific resistance ρ. The Darcy's ray is the steady streamline of the fluid $y(x)$, whose shape should be determined. Within the macroscopic description involving averaged continuous paths we define the derivative $u = dy/dx$, formally a control variable, that coincides with the local direction of the gradient of hydraulic head h. As in the previous discrete analysis, in the continuous system we assume the layered structure of the material with many thin layers arranged vertically. To derive optimality conditions for the nonlinear Darcy's flows in the multilayer system composed of N layers, the optimization theory (Bellman's optimality principle: Aris, 1964; Findeisen et al., 1980; Pontryagin et al., 1962; Leitman, 1966) admits that one can consider two arbitrary layers under assumption of their fixed states at the left and right boundaries, whereas the state between the considered layers must be free, to allow an optimization. Keeping in mind the analogy between thermal and frictional problems, the shape of Darcy's rays can be described as an optimal control problem for a minimum of the total resistance (Sieniutycz, 2000b). Certain specific coordinates x and y are usually applied to describe the problem locally; the axis y is parallel to the surfaces of constant resistance. In this case a general structure of the variational principle is represented by an equation

$$\delta J = \delta \int_{l_i}^{l_f} p\, dl = \delta \int_{x_i}^{x_f} p(x, y, u)(\sqrt{1+u^2})\, dx = 0 \qquad (7.13)$$

where $u = dy/dx$ is the slope of the tangent to the path and the indices i and f refer to the initial and final points of the path. In this reference frame the local resistance of the fluid flow changes along the axis x, the axis y is tangent to a surface of constant resistivity $\rho = C$ and $u = dy/dx$ is the local direction of the gradient of hydraulic head. The comparison of Eq. (7.13) and (7.17) in the further text shows that the function p of Darcy's rays corresponds with the function

$$p(x, y, u) = A_0^{-1} \rho(x)\sqrt{(1+u^2)} \qquad (7.14)$$

Eq. (7.13) with function $p(x, y, u)$ defined by Eq. (7.14) should then be optimized with respect to the control $u = dy/dx$ within each infinitesimal layer. In Eq. (7.14) A_0 is the constant area of projection of the mass flux tube cross-sectional area on the surface of constant resistance. The reader is referred to Section 7.5 and the literature (Sieniutycz, 2000a, b) for details regarding numerical studies of Eq. (7.13) by the dynamic programming method.

To derive the tangent law of bending for continuous Darcy's rays, let us regard the porous system as a pseudocontinuum and assume a variable hydraulic resistance ρ. In our coordinate system the planes of constant resistance ρ are perpendicular to the axis x, that is, the flow resistance is a continuous function of x only. Let us imagine that we rotate the system about the axis x until the gradient of the hydraulic head is parallel to the x–y plane. A set of flux tubes with the flowing fluid matter can be defined as in the discrete problem analyzed previously. While in the discrete problem the total resistances are $R_1 = \rho_1 l_1/A_1$ and $R_2 = \rho_2 l_2/A_2$, in the continuous problem these relations are represented jointly by the single local relationship, $dR = \rho A^{-1} dl$, in which A is the variable cross-sectional area of a "tube" of mass flux perpendicular to the flow in a inhomogeneous medium. The products $R_1 l$ and $R_2 l$ have the units of length and describe the differences of the hydraulic head.

As in the discrete problem, we postulate that the path of the fluid in porous medium flowing between two fixed points 1 and 2 is that along which the total resistance is a minimum. Similarly to the discrete problem, $A = A_0 \cos\alpha$, where A_0 is the constant (x-independent) area of the flux tubes intercepted by the interface. The variable cross-sectional area of the flux tube in the medium is described by a continuous counterpart of Eq. (7.6)

$$A = A_0 \cos\alpha \tag{7.15}$$

(see Fig. 7.2) whereas the incidence angle varies with x according to the formula

$$\cos\alpha = \frac{dx}{dl} = \frac{dx}{\sqrt{dx^2 + dy^2}} \tag{7.16}$$

In the aforementioned equations α is the angle between the gradient of the hydraulic head (or the Darcy's ray) and the normal to the planes of constant resistance ρ. Eqs. (7.4), (7.13), and (7.15) then imply the formulation

$$\delta\int_{l_i}^{l_f}\frac{\rho}{A}dl = \delta\int_{l_i}^{l_f}\frac{\rho(x)}{A_0\cos\alpha}dl = \delta\int_{x_i}^{x_f}\frac{\rho(x)\sqrt{dx^2+dy^2}}{A_0 dx(\sqrt{dx^2+dy^2})^{-1}} = \delta\int_{x_i}^{x_f}A_0^{-1}\rho(x)(1+(dy/dx)^2)dx = 0 \tag{7.17}$$

which describes the vanishing first variation for the functional of total resistance defined as

$$R^{i,f} \equiv \int_{x_i}^{x_f} A_0^{-1}\rho(x)[1+(dy/dx)^2]dx \tag{7.18}$$

Comparison of Eqs. (7.13) and (7.17) proves that in Darcy's flows the function $p(x, y, u)$ has the form consistent with Eq. (7.14). The dependence of p on u is caused by the change of the area A perpendicular to the flow with x, as shown in Fig. 7.2.

An important conclusion should be kept in mind when stressing differences between propagation of Darcy's and optical rays: While the simplest optical rays are described by Euclidean and Riemannian geometry, it is rather Finslerian geometry (Ingarden, 1996) that is valid for Darcy's rays. The essence of this geometry is an explicit dependence of the metric tensor on directions, a property that enables one to include nonisotropic systems into considerations. Finslerian geometric formalism deserves mention as it provides geometry for the variational formalism of Euler–Lagrange. Yet, the mathematics of Finslerian geometry is more involved than that of Riemannian geometry, and, therefore, one usually applies classical variational calculus instead of Finslerian geometry.

7.4 Variational Approach to Nonlinear Darcy's Flow

When a pressure field in an isotropic porous medium is imposed along with fixing the hydraulic head gradient, the flow of the fluid in a permeable porous skeleton can be described in terms of "rays," or paths of fluid flow in direction of hydraulic head gradient. Their deviation from straight lines results from a variable conductivity. In fact, frictional rays travel along paths satisfying the principle of minimum of entropy production (Sieniutycz, 2000b), which seems at first glance quite different from the well-known Fermat principle of minimum time for optical rays. However, the principle assures the minimum resistance of the path, which, in a dual problem, causes the maximum of fluid flux through the porous medium and makes the residence time of the fluid in this medium as short as possible. This is quite similar to Fermat principle for propagation of light (Landau and Lifschitz, 1967). Our purpose is to investigate this phenomenon by methods of optimization. We use a particular reference frame (x, y) in which the specific resistance of fluid flow changes along the axis x (Fig. 7.2), the axis y is tangent to a surface of constant specific resistance ρ, and $u = dy/dx$ is the local direction of the gradient of the hydraulic head. A family of frictional paths entering at various angles α_1 is considered corresponding with various gradients of the hydraulic head. To determine each path as an extremal is the goal of the variational approach applied. The approach can also be extended to handle the energy flow in layered composites (Sieniutycz, 2000b). The shape of Darcy's ray caused by nonuniform resistance $\rho(x)$ can be described as an optimal control problem for a minimum of total resistance

$$J = \int_{x_i}^{x_f} A_0^{-1} \rho(x)(1+u^2)dx \qquad (7.19)$$

In the frictional Darcy's flow the specific resistance ρ is the reciprocal of the conductivity related to the gradient of the hydraulic head (grad T in the case of energy flow; Sieniutycz, 2000b). A is a variable area perpendicular to fluid flow, and A_0 is a constant transfer area projected on axis y. A path or flow or "ray" bends depending on the ratio of ρ_2/ρ_1; the bending constant for a ray equals c. Eq. (7.22) defines c in terms of a function R describing the minimum total resistance.

The total resistance of an *optimal* path linking two arbitrary points 1 and 2 is described by the minimum resistance function $R(x_1, y_1, x_2, y_2)$. Eq. (7.19) should be optimized with respect to the slope $u = dy/dx$ as a control variable within each differential layer dx. For a discussion of the role of functional (7.19) in energy transfer problems the reader is referred to literature (Tan and Holland, 1990; Sieniutycz, 2000b). Essential for the functional simplicity is the suitable frame (x, y) in which the specific resistance of the fluid flow changes locally along the axis x whereas diverse frictional paths may enter the investigated region at various angles α_1, (and various controls $u = \tan\alpha = dy/dx$), where each angle corresponds to a different gradient of the hydraulic head and hence to a different flux density of the flowing fluid.

Summing up, we have shown that the approach proposed embeds Bear's refraction law of streamlines (Bear, 1972; Fig. 7.2 and Section 7.2) in the context of variational calculus and optimization techniques. In fact, the method works effectively in association with the dynamic programming algorithm, common in the discrete optimization, as shown in Section 7.5.

7.5 Use of Hamilton–Jacobi–Bellman Theory

For a porous system treated as a set of infinitesimal layers, in which the specific resistance ρ varies along the axis x, a minimum resistance function of the problem is defined,

$$R(x_1, y_1, x_2, y_2) \equiv \min \int_{x_1}^{x_2} A_0^{-1} \rho(x)(1+u^2)dx, \qquad (7.20)$$

where the integrand describes the total derivative of cumulative resistance with respect to horizontal coordinate x. The state or potential function R satisfies an important equation of the optimization theory called the Hamilton–Jacobi–Bellman equation (HJB equation) (Findeisen et al., 1980; Pontryagin et al., 1962; Leitman, 1966). A general HJB equation is a quasilinear partial differential equation with the extremization sign with respect to the optimal control u. The function R appears in the HJB equations explicitly, whereas it is only implicit in the ordinary canonical equations of the Pontryagin's maximum principle (Pontryagin et al., 1962). In the latter case, R is hidden in the so-called adjoint variables (components of the gradient vector of R). In thermodynamics, economics or some other disciplines approaches involving HJB equations may be superior with respect to the Pontryagin's method whenever the principal function R has to be determined.

Usually HJB equations are derived in an exact and systematic way by the dynamic programming method (Findeisen et al., 1980; Pontryagin et al., 1962; Leitman, 1966). However, in this chapter a simple and brief while insightful method based on Caratheodory's lines of reasoning is outlined (Rund, 1966). From the definition of the minimum resistance function R in functional (7.20) we obtain

$$\max_{u(t)} \left\{ R(x_1, y_1, x_2, y_2) - \int_{x_1}^{x_2} A_0^{-1}\rho(x)(1+u^2)dx \right\} = 0. \qquad (7.21)$$

After introducing the total derivative of optimal characteristic function R with respect to a variable upper limit of integration

$$\max_{u(t)} \left\{ \int_{x_1}^{x_2} \left[\frac{dR(x_2, y_2)}{dx} - A_0^{-1}\rho(x)(1+u^2) \right] dx \right\} = 0 \qquad (7.22)$$

This describes a vanishing maximum of the negative resistance (negative integrand of Eq. (7.20)) gauged by the total derivative of optimal function, dR/dx, at the state 2. Expanding the total time derivative of R in terms of the partial derivatives, $\partial R/\partial x$ and $\partial R/\partial t$, the derivative of R with respect to time can be taken off this equation and index 2 can be omitted for variable final states. This procedure shows that the optimal function R with satisfies a quasilinear partial differential equation

$$\frac{\partial R}{\partial x} + \max_u \left[\frac{\partial R}{\partial y} u - A_0^{-1} \rho(x)(1+u^2) \right] = 0. \tag{7.23}$$

Eq. (7.23) is called the Hamilton–Jacobi–Bellman equation (HJB equation), of the problem. The formalism represented by Eqs. (7.20) and (7.23) can also be derived from the continuous algorithm of the dynamic programming method applied to the performance criterion (7.19), the derivations are then slightly more complicated and longer (Findeisen et al., 1980).

For Eq. (7.23) the resistance ρ is independent of y, and the HBJ equation is linear rather than quasilinear. Extremizing the Hamiltonian expression in Eq. (7.23) yields an optimality condition for the fluid flow within each infinitesimal layer. This optimality condition is written as follows in the form of the tangent *law of bending* for a Darcy's ray

$$\rho(x)\frac{dy}{dx} = \frac{A_0}{2}\frac{\partial R}{\partial y} \equiv c, \tag{7.24}$$

where c is a constant which may be both positive or negative. The constancy of the partial derivative $\partial R/\partial y$ follows from an explicit independence of density ρ in Eq. (7.19) with respect to y. A suitable integral formula for bending constant in terms of deviation $y - y^0$ follows in the form

$$c = (y - y^0)\left(\int_{x^0}^{x} \rho^{-1}(x')dx' \right)^{-1}. \tag{7.25}$$

The optimal control u is the solution (7.24) of the HJB Eq. (7.23) in terms of $p = \partial R/dy$. Substituting this solution into Eq. (7.23) yields the Hamilton–Jacobi equation for the frictional fluid flow in an arbitrary layer

$$\frac{\partial R}{\partial x} + A_0^{-1} \rho(x)\left[\left(\frac{A_0}{2\rho(x)} \frac{\partial R}{\partial y} \right)^2 - 1 \right] = 0, \tag{7.26}$$

where the second term of the left hand side expression is the *optimal* Hamiltonian. The solution to the aforementioned equation can always be broken down to quadratures, see Eq. (7.28). However, if the function of specific resistance $\rho(x)$ is too complicated, the integrals cannot be evaluated analytically. This difficulty calls for a discrete approach that solves numerically the associated Bellman's recurrence equation of the problem. Under the thin layer assumption this equation has the form

$$R^n(y^n, x^n) \equiv \min_{u^n, \theta^n} \{A_0^{-1} \rho(x^n)[1 + (u^n)^2]\theta^n + R^{n-1}(y^n - u^n\theta^n, x^n - \theta^n)\} \quad (7.27)$$

where $\theta^n = x^n - x^{n-1}$. This still cannot be analytically solved for an arbitrary $\rho(x)$, thus a dynamic programming sequence of R^n must be generated numerically. Yet, in the limit of an infinite number of stages, an analysis shows that the potential function satisfying Eq. (7.23) takes the limiting form

$$R(x, y) = \int_{x_0}^{x} A_0^{-1} \rho(x') dx' + A_0^{-1}(y - y^0)^2 \left(\int_{x_0}^{x} \rho^{-1}(x') dx' \right)^{-1}. \quad (7.28)$$

This function satisfies both HJB Eq. (7.23) and Hamilton–Jacobi Eq. (7.26). In fact, the analytical solution of the continuous problem is given quite generally by Eq. (7.28), that is valid for an arbitrary function $\rho(x)$ and that includes, of course, the case of the linear ρ as well. In an example of Section 7.7 the case of exponential ρ is discussed in some detail.

Brief information should be given on the numerical solution of Eq. (7.27). In fact, Eq. (7.27) is a typical recurrence equation of dynamic programming although the symbols it uses follow from the description of the physical problem rather than are typical to those used in the literature of optimal control. A typical dynamic programming (DP) equation at stage n is described in terms of a state vector $\mathbf{X}^n$ and a control vector $\mathbf{U}^n$, state transformation functions $\mathbf{X}^{n-1}(\mathbf{X}^n, \mathbf{U}^n)$ and a "cost intensity function" L_0^n. For the Darcy's path problem the state vector is represented in Eq. (7.27) by vertical coordinate y^n and horizontal coordinate x^n. The control vector $\mathbf{U}^n$ consists of slope $u^n = dy^n/dx^n$ and horizontal coordinate interval $\theta^n = x^n - x^{n-1}$. The "cost intensity" or Lagrangian function L_0^n of standard DP algorithms is represented in Darcy's path problem by the resistance per unit length x. With the suitable changes in notation the dynamic programming Eq. (7.27) takes the standardized form

$$R^n(\mathbf{X}^n) \equiv \min_{\mathbf{U}^n} \{L_0^n(\mathbf{X}^n, \mathbf{U}^n)\theta^n + R^{n-1}[\mathbf{T}^{n-1}(\mathbf{X}^n, \mathbf{U}^n)]\}. \quad (7.29)$$

where the vector transformation function $\mathbf{T}^{n-1}(\mathbf{X}^n, \mathbf{U}^n) \equiv (y^n - u^n\theta^n, x^n - \theta^n)$. Fig. 7.3 shows a block scheme of dynamic programming calculations of family of Darcy's rays associated with the aforementioned equation. The scheme refers to the multistage control described by the forward algorithm of the dynamic programming method. Ellipse-shaped balance areas pertain to sequential subprocesses of the numerical procedure which grow by inclusion of subsequent stages.

Numerical aspects of DP algorithms are only briefly mentioned here as there is a vast literature available that discusses these issues (Bellman, 1957; Aris, 1964; Berry et al., 2000;

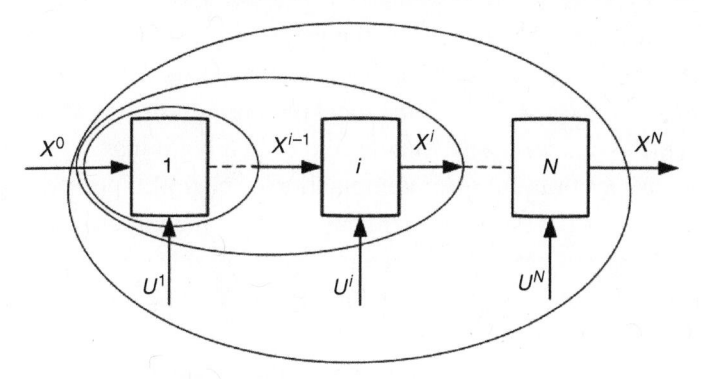

Figure 7.3: Block scheme illustrating the dynamic programming calculation of family of rays satisfying Darcy's law for the fluid motion.

Sieniutycz, 2000a). Starting with $R^0 = 0$ data of optimal resistance functions $R^1, \ldots R^n, \ldots R^N$ are generated recursively over subprocesses composed subsequently of: stage 1, stages 1 and 2, ... stages 1, 2 ... n ... and, finally, stages 1, 2 ... N. Organization of computations requires a grid in nodes of which data of optimal functions and optimal controls are computed and stored. A total number of stages, N, has to be assumed. The numerical DP algorithm generates the resistance potential $R^n(\mathbf{X}^n)$ from the function $R^n(\mathbf{X}^{n-1})$ of the $n-1$ stage subprocess and the state transformations. According to Bellman's optimality principle (Bellman, 1957, Aris, 1964), the forward algorithm maximizes the sum of the optimal cost of all previous $n-1$ stages (the optimal function R^{n-1}) and the nonoptimal cost at the stage n. To determine R^n exactly for a definite n, one would have to numerically determine values of this function for every point $\mathbf{X}^n$, an impossible task. Therefore, one must decide to determine these values on a discrete subset of variables $\mathbf{X}$, and use the data in the way that makes possible evaluation of $R^n(\mathbf{X}^n)$ in a required domain. "Curse of dimensionality," interpolation in tables of results of previous subprocesses and problems of grid expansion are usually main difficulties (Aris, 1964; Berry et al., 2000). Accuracy of 3-D solutions is, of course, much worse than 2-D ones.

7.6 Extensions to Other Systems

Both similarities and differences exist between frictional Darcy's flows and nondissipative flows. While the latter are physically distant from Darcy's flows, it is interesting to learn that the Fermat type principle can be formulated in diverse cases. Here we shall consider extensions including nondissipative flows in various systems, for example, optical and mechanical ones. When A is the area of the cross-section perpendicular to the flow and coordinates x and y are applied the general form of the variational principle is given by Eq. (7.30)

$$\delta \int_{l_1}^{l_2} p \, dl = \delta \int_{x_1}^{x_2} p(x, y, u)\sqrt{1+u^2} \, dx = 0, \qquad (7.30)$$

where $u = dy/dx$ is the slope of the tangent to the path. In this reference frame the local resistance of the flow ρ changes along the axis x, the axis y is tangent to a surface of constant specific resistance $\rho = X$ and $u = dy/dx$ is the local direction of the gradient of certain potential, say Π, which not necessarily is hydraulic head h but it may represent chemical potential of certain species or even linear combination of chemical potentials in the form known for substrate or product terms of chemical affinity (Sieniutycz, 2004a). A set of flux tubes with the flowing matter can be defined again. In a discrete problem total quantities p_k in two neighbor subsystems are $p_1 = I_1^2 \rho_1/A_1$ and $p_2 = I_2^2 \rho_2/A_2$, where both symbols I_k designate corresponding currents. In the continuous case these relations are represented jointly by the single local relationship, $p(x) = I^2 \rho A^{-1}$, in which A is the variable cross-sectional area of a "tube" perpendicular to the flux passing an inhomogeneous medium. It may be shown that products $p_1 l_1$ and $p_2 l_2$ are related to the entropy production in both subsystems. It appears that for nonconserved flows the principle of minimum resistance ceases to hold as an exact law, yet the principle of minimum of entropy production is still valid.

We test the postulate that the path of a generalized (say, chemical) current I flowing between two fixed points 1 and 2 is that along which the total entropy production is a minimum. In the general case the chemical Lagrangian L corresponding with Eq. (7.27) has the form

$$L(x, y, u) \equiv p(x, y, u)\sqrt{1+u^2} = I^2(x, y, u)A^{-1}(u)\rho(x)\sqrt{1+u^2} \qquad (7.31)$$

where $p(x, y, u)$ is the formal analogue of the refraction coefficient of the optical Fermat problem. The optical case and its Hamilton–Jacobi theory are well known (Rund, 1966). Yet, the case of constant I is that of a conservative flux with $A^{-1}(u) = A_0^{-1}\sqrt{1+u^2}$. In this case $p(x, y, u) \equiv I^2 A_0^{-1} \rho(x)\sqrt{1+u^2}$ and the optimization problem is that of the minimum resistance. The dependence of p on u is caused by the change of the area A perpendicular to the flow with x, Fig. 7.2. It is Finslerian rather than Riemannian geometry which describes problems of this sort. In fact, the Finslerian geometry refers to "dissipative" rather than to "reversible" Lagrangians L (Ingarden, 1996).

The Euler–Lagrange equation of the general problem described by Eq. (7.31) is

$$\sqrt{1+u^2}\frac{\partial p}{\partial y} - \frac{d}{dx}\left(\frac{pu}{\sqrt{1+u^2}} + \frac{\partial p}{\partial u}\sqrt{1+u^2}\right) = 0 \qquad (7.32)$$

This equation admits a simple and important interpretation. If the tangent to the extremal makes the angle α with the axis x, then

$$\frac{dy}{dl} = \frac{dy}{\sqrt{dx^2+dy^2}} = \frac{u}{\sqrt{1+u^2}} = \sin\alpha. \qquad (7.33)$$

In the case of a conserved flux (constant I), we apply $p \equiv I^2 A_0^{-1} \rho(x)\sqrt{(1+u^2)}$ and then

$$\sqrt{(1+u^2)}\frac{\partial p}{\partial y} - \frac{d}{dx}\left[\frac{pu}{\sqrt{(1+u^2)}} + I^2 A_0^{-1}\rho(x)u\right] = 0 \qquad (7.34)$$

or, since $u\left(\sqrt{(1+u^2)}\right)^{-1}$ is the sine of α

$$(\sqrt{(1+u^2)}\frac{\partial p}{\partial y} - \frac{d}{dx}\{p\sin\alpha + I^2 A_0^{-1}\rho(x)\tan\alpha\} = \sqrt{(1+u^2)}\frac{\partial p}{\partial y} - \frac{d}{dx}\{2I^2 A_0^{-1}\rho(x)\tan\alpha\} = 0 \qquad (7.35)$$

In view of the y-independent L and p the flux I is conserved and Eq. (7.30) simplifies to

$$\frac{d}{dx}[2\rho(x)(dy/dx)] = 0 \qquad (7.36)$$

or, since $dy/dx = \tan\alpha$

$$\rho(x)\tan\alpha(x) = c \qquad (7.37)$$

where c is a constant. Here we have recovered the tangent law of bending for an inhomogeneous porous medium in which the resistance is a function of x. Eqs. (7.24) and (7.37) transfer the discrete result (7.3) to continuous systems.

On the other hand, in the simple optical case when $p = n$ (the refraction coefficient), p is independent of u and only $L = p\sqrt{(1+u^2)}$ depends on u

$$\sqrt{(1+u^2)}\frac{\partial p}{\partial y} - \frac{d}{dx}\left(\frac{pu}{\sqrt{(1+u^2)}}\right) = 0. \qquad (7.38)$$

Eq. (7.38) is then reduced to

$$\frac{\partial p}{\partial y} - \frac{d}{dl}(p\sin\alpha) = 0. \qquad (7.39)$$

This equation describes a geometrical property of the extremal path, which is independent of the choice of axes. Therefore its interpretation can use any convenient system of axes. As p and n depend on x and y only, the equation n = constant is that of the plane curve and by varying the constant we obtain a family of the curves, the so-called level curves. If O is a point on the extremal, let us take it as the origin and the normal and tangent to the level curve through O as the x and y axes respectively. Now at O the tangent to the curve n = constant is perpendicular to the axis x, hence $(\partial n/\partial x)/(\partial n/\partial y)$ must be infinite and so $\partial n/\partial y = \partial p/\partial y = 0$. In this particular frame Eq. (7.39) becomes the Snell (sine) law of refraction for optical systems

$$n\sin\alpha = \text{constant} \qquad (7.40)$$

This result holds for all points along the extremal. Eq. (7.40) is not restricted to optical case. it is valid also in the mechanical case where the level curves are given by $(h - V)^{1/2}$ = constant, and, as this is equivalent to V = constant, the level curves are curves of the constant potential energy. We observe that frictional and chemical cases are in general more complicated than optical. Yet, while we can formulate diverse descriptions that formally include the well-known optical principle, solid physical information is necessary to attribute a definite formulation to a concrete physical system.

7.7 Example

Let us focus again on porous media and consider an example of flow of fluid through a porous system whose hydraulic resistance increases exponentially with the coordinate x. We begin with the Euler–Lagrange equation for the functional (7.19)

$$\frac{d}{dx}[2\rho(x)(dy/dx)] = 0 \qquad (7.41)$$

that coincides, in this problem, with a differential form of bending formula, Eq. (7.24). Since the equation for an extremal is the second order differential equation, its solution depends on two integration constants. When the function $\rho(x)$ is known, variables in Eq. (7.41) can be separated. Integration between a fixed initial point (x^0, y^0) and an arbitrary final point (x, y) yields a general integral

$$y = y^0 + c \int_{x^0}^{x} \frac{dx'}{\rho(x')}, \qquad (7.42)$$

where the integration constants are c and y^0. This equation is a counterpart of formula (7.25) and applies when one wants to evaluate trajectory $y(x)$ for a given c rather than inversely. For the exponential increase of hydraulic resistance with x,

$$\rho(x) = \rho^0 \exp(\gamma x), \qquad (7.43)$$

the integration of Eq. (7.42) between points (x^0, y^0) and (x, y) yields an optimal path

$$y(x) = y^0 + \frac{c}{\rho^0} \int_{x^0}^{x} \frac{dx'}{\exp(\gamma x')} = \frac{c}{\gamma \rho^0} [\exp(-\gamma x^0) - \exp(-\gamma x)]. \qquad (7.44)$$

A corresponding slope of Darcy's ray is described by an equation

$$\frac{dy}{dx} = \frac{c}{\rho^0} \exp(-\gamma x). \qquad (7.45)$$

For practical purposes information about the case of linear change of hydraulic resistance with coordinate x is usually more appropriate

$$\rho(x) = \rho^0(1+\gamma x). \tag{7.46}$$

Then it can easily be found that in vicinity of $x^0 = 0$ an optimal Darcy's ray is a straight line

$$y(x) = y^0 + \frac{c}{\rho^0}\int_{x^0}^{x} \frac{dx'}{1+\gamma x'} = \frac{c}{\gamma\rho^0}\ln\left(\frac{1+\gamma x}{1+\gamma x^0}\right) \underset{x\to x^0}{\cong} \frac{c}{\rho^0}(x-x^0). \tag{7.47}$$

whose slope dy/dx (independent of γ and x) is determined only by c. Yet for larger x the slope in the linear problem is

$$\frac{dy}{dx} = \frac{c}{\rho^0}(1+\gamma x)^{-1} \tag{7.48}$$

that is, it depends on x. In Eqs. (7.44), (7.45), and (7.48) the ratio c/ρ^0 is the initial slope $(dy/dx)^0$ at the point $(x^0, y^0) = (0, 0)$.

Eqs. (7.45) and (7.48) describe the slopes of Darcy's rays decreasing with x, thus turning toward the direction of the resistance gradient. Indeed, in order to minimize the total resistance, the ray spanned between two given points must take the shape that assures that a relatively large part of it resides in the "rarer" region of the medium. This is in agreement with the tangent law of refraction from a rarer to a denser medium in Fig. 7.2. Eq. (7.44) proves that as x tends to infinity and $x^0 = 0$, y approaches the asymptotic value $c/(\gamma\rho^0)$. For an infinite ratio ρ^0/c, or for the vanishing initial slope $(dy/dx)^0$, one obtains $y = 0$ and $dy/dx = 0$ for all x. In this case a Darcy's ray initially in the direction of the hydraulic resistance gradient propagates undeviated. Otherwise, considering the inverted form of Eq. (7.44) for $x^0 = 0$

$$x = -\gamma^{-1}\ln(1-c^{-1}\gamma\rho^0 y) \tag{7.49}$$

and Eq. (7.45) one concludes that if $\rho^0 c^{-1} = 0$ then x and dx/dy are zero for all y. This means that a ray perpendicular to the resistance gradient (tangent to a surface of the constant resistivity) also propagates undeviated. This is consistent with the discrete tangent law of refraction, but, as first pointed out by Tan and Holland (1990), this is alien to Snell's law because in geometrical optics a ray always bends toward the gradient of the index of refraction (Landau and Lifschitz, 1967).

In an opposite case, the hydraulic resistance can be exponentially decreasing function of x

$$\rho(x) = \rho^0 \exp(-\gamma x), \tag{7.50}$$

the corresponding formulae follow from the previous ones when γ is replaced by $-\gamma$. Eqs. (7.44) and (7.45) of exponential problem take respectively the form

$$y(x) = y^0 + \frac{c}{\rho^0} \int_{x^0}^{x} \frac{dx'}{\exp(-\gamma x')} = \frac{c}{\gamma \rho^0}[\exp(\gamma x) - \exp(\gamma x^0)]. \qquad (7.51)$$

and

$$\frac{dy}{dx} = \frac{c}{\rho^0} \exp(\gamma x). \qquad (7.52)$$

Whereas for the problem with the linear decrease of hydraulic resistance

$$\frac{dy}{dx} = \frac{c}{\rho^0}(1 - \gamma x)^{-1} \qquad (7.53)$$

Now, in both problems the slope of Darcy's ray increases with x, bending away from the direction of the initial slope $(dy/dx)^0 = c/\rho^0$. This is in agreement with the tangent law of refraction from a denser to a rarer medium. In this case there is no asymptotic value of y. For a vanishing initial slope c/ρ^0 one obtains $y = 0$ and $dy/dx = 0$ for all x, which means that a Darcy's ray initially in the direction of the resistance gradient propagates undeviated. Otherwise, considering the inverted forms of Eq. (7.51) and Eq. (7.52) one concludes that x and dx/dy are zero for all y if the initial slope is infinite ($\rho^0/c = 0$). This means that a ray perpendicular to the resistance gradient (tangent to a surface of the constant resistance) also propagates undeviated. Again, while this is consistent with the tangent law, it is alien to Snell's law. (In geometrical optics a ray always bends towards the gradient of refraction index.)

7.8 Final Remarks

In conclusion we stress, between others, applications of the theory to various practical systems with mass transfer, for example, to porous drying systems, in which a nonuniform distribution of fluid's mass flux is observed, caused by local variations of porosity and hydraulic resistance (imperfect systems). The present results show that a sort of refraction law governs the fluid conduction through an inhomogeneous porous skeleton, the so-called tangent law of bending. It was first found in studies of boundary conditions for the heat flow through a discontinuity at which the thermal conductivity has a jump (Tan and Holland, 1990; Sieniutycz, 2000b). The tangent law is different from Snell's law of refraction in optics, with the tangents of the angles of incidence and refraction replacing the sines and the conductivity reciprocal taking the place of the refractive index. Such law is known for the electric field intensity at the boundary between two dielectrics (Landau and Lifschitz, 1967), and it also applies to potential fields in general (Tan and Holland, 1990). By considering in Section 7.7 the example when the flow resistance increases exponentially with x (the porous system becoming "denser" with x) one shows that the slope of the steady Darcy's ray decreases with x, thus turning toward the direction of the specific resistance gradient. Indeed, by minimizing the total resistance, the ray spanned between two given points takes the shape that assures that a relatively large part of it resides in a "rarer" region of the porous medium, that is, a region of its low resistance. In other words, in inhomogeneous porous system, the fluid path bends into the direction that ensures its shape corresponding with the longest residence time of the fluid in regions of low resistance. This makes it possible for one to predict shapes of Darcy's rays or paths of the fluid flow. This also leads to a description of the fluid flow in terms of wave fronts and corresponding Hamilton–Jacobi theory (Rund, 1966). It can also be shown that a Darcy's ray initially in the direction of the resistivity gradient propagates undeviated. However, since the tangent law holds, a ray perpendicular to the resistivity gradient (tangent to a layer of the constant resistivity) also propagates undeviated. This is consistent with the tangent law of bending but this is not in agreement with Snell's law where a ray always bends toward the gradient of the index of refraction. These results represent basic characteristics of steady nonlinear transfer of fluids in frictional porous media.

The constructal theory (Bejan, 2000; Bejan and Ledezma, 1998) allows the imbedding of these results into a broader framework of systems with optimal geometry assuring quickest access between a finite-size volume and one point, their link with the present approach can be considered (Bejan, 2003). Frictional generalizations of Euler equation of fluid motion (Wagner, 2005) imply that Darcy's law is valid not only to unidirectional flows (for which it was experimentally established) but also to three-dimensional flows; consistent with the results obtained here. Yet there are limitations of the method presented. First, any extension of the variational methodology to unsteady flows is very difficult and requires the use of entirely different methods (Sieniutycz and Farkas, 2005). Second, nonlinearities in derivatives of state variables and abrupt changes of state are excluded. Third, only very slow, linearly damped laminar flows (creeping flows) are correctly described (Darcy Reynolds Number <10). In fact, Darcy's law has been found to be invalid for high values of Reynolds number (turbulent flows). Darcy's law has also been found to be invalid either at very low values of hydraulic gradient in some very low-permeability materials, such as clays (Manning, 1997; Freeze and Cherry, 1979), and in some other media, such as ceramic foams (Philipse and Schram, 1991). A review of these issues is available (Innocentini et al., 1999).

On the other hand, some generalized equations for transport in porous media, such as the Dupuit–Forchheimer equation, Darcy–Weisbach equation or Ergun's equation (Ruth and Ma, 1992; Bear, 1972; Charbeneau, 2000; Du Plessis and Masliyah, 1991; Ingebritsen and Sanford, 1998), can describe both laminar or turbulent flows through the bed. Extension of the present approach to these generalized physical situations is a challenging issue, even for steady systems.

Serious limitations on validity of Darcy's law appear in systems affected by uncertainty, that is, in geologic media. They are heterogeneous and show both discrete and continuous spatial variations on a multiplicity of scales. In traditional predictive models, state coordinates (parameters) have been treated as well-defined local quantities that can be assigned unique values at each point in physical space-time and can be found by experiments. This has promoted deterministic attitude towards analysis and modeling and overconfidence in the predictive powers of models for subsurface flow and transport. However, experience shows that hydrologic models require frequent recalibration as flow and transport regimes and the database vary. Also, their predictability deteriorates as space-time hydrologic scales increase and process complexity increases. In practice, state variables can be determined experimentally only at selected locations where their values depend on the scale and mode of measurement. All this leads to uncertainty of aquifers parameters and associated heads, resistances, fluxes, and so forth. It has become common in groundwater models to quantify this uncertainty by treating all parameters and variable as random fields (Dagan, 1989; Gelhar, 1993; Dagan and Neuman, 1997; Cvetkovic, 1997; Rubin, 1997). Therefore, the problem of solving flow quantities becomes that of determining their probability distribution. Many applications involve numerical (conditional) Monte Carlo simulation, which is conceptually simple and applicable to a broad range of flow and transport problems. Outputs include mean system behavior as an optimum predictor (unbiased, with minimum variance) of the actual behavior and second moments, which serve as measures of prediction errors. An alternative simulation is with conditional moment equations, which yield corresponding predictions of flow and transport deterministically, along with quantification of prediction errors. These equations include nonlocal parameters that depend on more than one point in space-time and depend not only on local medium properties but also on the information one has about these properties (location, scale, quantity, and quality of data). Darcy's law and Fick's analogy are generally not obeyed by the predictors except in special cases or as localized approximations. Traditional deterministic equations emerge as approximate (localized) versions of nonlocal flow and transport equations. Like their conditional mean counterparts, deterministic models yield at best good estimates. Whereas nonlocal equations contain information about predictive uncertainty, localized ones do not assure this property (their predictions are of undetermined quality). This corresponds to the practice of taking the deterministic form of Darcy's law for granted, associating it with an effective or equivalent parameter, and estimating its spatial distribution by model calibration against measured heads and fluxes.

Transforming the present variational theory of Darcy's flows to the theoretical framework of *stochastic* Fermat principle and the comparison of the latter with the Lagrange formulation of flow and transport in porous media under uncertainty (Cvetkovic, 1997; Rubin, 1997) could be the subject of a further effort.

The present results imply applications to various practical systems with mass transfer where nonuniform distribution of fluid's mass flux is caused by local variations of porosity and permeability (imperfect systems). Examples are various groundwater flows, drinking wells, porous drying systems, petroleum ground systems, and so forth, where most problems involve flow and solute transport. A special situation arises in coastal aquifers, where excess pumping may cause seawater intrusion, thus threatening the use of the aquifer as a source of fresh water. Modeling flow and pollution of groundwater and seawater intrusion in coastal aquifers are recent, attractive applications (Bear and Verruijt, 1987; Bear et al., 1999).

Let us now explain what effect our results could have on the practical formulation of the aforementioned problems concerning flows of fluids in inhomogeneous media satisfying Darcy's law. Effective management of water resources requires the ability to forecast the response of the managed system, for example, an aquifer, to a collection of operations, such as pumping, recharging, and control of operational variables at aquifer boundaries. Any planning of mitigation, cleanup, or control, requires forecasting the trajectory and history of the fluid and contaminants in both the unsaturated zone and the aquifer. The prediction tool is the numerical model that simulates the flow and pollution motion and transformation. The construction of suitable models should be based on good understanding of the phenomena within the modeled domain and on the domain's boundaries, including chemical and biological processes, on the ability to express this information in the compact form of a well-posed formal, possibly analytical model, and, eventually, in the form of its numerical counterpart contained in the related computer program. By applying suitable numerical techniques, a computer leads to the solution describing predictions of the response of the investigated domain to planned activities. For flow and solute transport in the subsurface many suitable programs are now available. Whenever a porous system is inhomogeneous and its resistance ρ (or associated permeability) is a given function of the space variables x and y the computer program should contain the recurrence equation of dynamic programming, Eq. (7.27) of the present chapter, which solves the problem of the fluid's flow trajectory governed by nonlinear Darcy's law and replaces Darcy's equation itself in the computational program. Let us also note that in problems with undetermined resistance functions that are set spontaneously in the process (such as, for example, for Darcy's flows in fluidized beds) Eq. (7.27) should be solved simultaneously by a trial and error procedure along with balance laws for a fluid in a porous medium. This is, in fact, a general feature in many complex problems where the determination of Darcy's trajectories constitutes merely a component in the complete solution describing a practical process in a inhomogeneous porous system, the other components being: the balance and kinetic equations for energy exchange, equations of balance and movement for the contaminants and equations of pollution transformation by chemical reactions.

Working in the context of Bejan's constructal theory, Ordonez et al. (2003) investigate "designed porous media" or optimally nonuniform flow structures connecting one point with more points. They show both analytically and numerically how an originally uniform flow structure transforms itself into a nonuniform one when the objective is to minimize global flow losses. The flow connects one point (source, sink) to a number of points (sinks, sources) distributed uniformly over a two-dimensional domain. In the first part of their work, the flow between neighboring points is modeled as fully developed through round tubes. Ordonez et al. (2003) have shown that flow "maldistribution" and the abandonment of symmetry are necessary for the development of flow structures with minimal resistance. The search for better flow structures can be accelerated: tubes that show a tendency of shrinking during the search can be assumed absent in future steps. In the second part of their work, Ordonez et al. (2003) show that the flow medium is continuous and permeated by Darcy flow. The development of flow structures (channels) is modeled as a mechanism of erosion, where elements of the original medium are removed one by one, and are replaced with a more permeable medium. The elements selected for removal are identified based on two criteria: maximum pressure integrated over the element boundary, and maximum pressure gradient. The flow structures generated based on the pressure gradient criterion have smaller flow resistances. As flow systems become smaller and more compact, the flow systems themselves become "designed porous media." These design optimization trends are generally applicable in constructal design, that is, where miniaturization, global performance, compactness and complexity rule the design. In conclusion, Ordonez et al. (2003) documented the emergence of nonuniform flow structures when the objective is the minimization of the resistance to flow between one point and many points.

The flow structures developed by Ordonez et al. (2003) illustrate the progress from simple structures to more complex structures. The mechanism that generates flow structures with minimal resistance, is applicable in many fields where designed porous structures are needed. The trend for smaller and more compact heat generating packages, which characterizes everything from heat exchangers to electronics, calls for flow structures that are effective in accessing all the points of a volume. In this direction of miniaturization and increased compactness and complexity, the flow systems themselves become designed porous media.

Other approaches could also be related to the porous systems. Losses due to the flow through conduit components in a pipe system can be accounted for by head-loss coefficients (Schmandt et al., 2014). These losses correspond to the dissipation in the flow field or, in a more general sense, to the entropy generation due to the conduit component under consideration. When only one single mass flow rate is involved, an entropy-based approach is straightforward since the flow rate can be used as a general reference quantity. If, however, one mass flow rate is split or two partial flow rates come together like in junctions, a new aspects appears: There is an energy transfer between the single branches that has to be accounted for. This energy transfer changes the total head in each flow branch in addition to the loss of total head due to entropy generation. As an example, the method is applied to laminar flows. Head change coefficients for dividing and combining flows in a T-shape micro-junction are determined for both branches and discussed with respect to their physical meaning. The newly defined coefficients can be used for the design of a flow network.

Turbulent flow through porous media is of considerable theoretical and practical interest in various engineering disciplines such as chemical, mechanical, nuclear, geological, environmental, petroleum, and so forth. Kundu et al. (2014) develop the numerical modeling of turbulent flow through isotropic porous media. The porous media was considered to be made of a spatially periodic array of infinite square cylinders. A turbulent model was employed for the description of the flow in the porous media. The transport equations were solved through the flow domains having different porosities. The macroscopic pressure variation across the flow domain was investigated and the predictions of the pressure gradient were found to be in good agreement with the Forchheimer-extended Darcy law. The analysis of fluid flowing through porous media has applications in a number of industries such as chemical, mechanical, nuclear, petroleum, geological, environmental and especially in engineering systems based on fluidized bed combustion, enhanced oil reservoir recovery, combustion in an inert porous matrix, underground percolation of chemical waste and chemical catalytic reactors, and so forth. Reservoir engineering deals with the flow of fluids and fluid mixtures through porous media since the advent of oil production, particularly in oil recovery processes.

A variety of natural and engineering systems can be characterized as porous structure through which a fluid flows in turbulent regime (Kundu et al., 2014). In petroleum extraction, the flow accelerates toward the pumping well while crossing the regions of variable porosity. During the fluid transport, the flow regime gradually shifts from laminar to transition and eventually to turbulent regime. This affects the overall pressure drop and well performance. The fluidized-bed combustors and chemical catalytic reactors also experience pressure loss variation due to changes in the flow regime inside the pores. Also, the turbulence within the porous zone plays a significant role in the thermal processes. The wind transport of pollutants through forests, croplands, and ventilated structures is also modeled as flow through porous media. The geometric structures of the aforementioned areas and the wind speed induce turbulence generation. Turbulent boundary layers over forests and vegetation are examples of environmental flows which can benefit from appropriate mathematical modeling. In all cases, better understanding of turbulence through adequate modeling can, more realistically, simulate real-world environmental/natural and engineering flows (Kundu et al., 2014).

A number of turbulence models, such as Reynolds-averaged Navier–Stokes (RANS) and LES-based simulation models, are available for studying the turbulence phenomena. Besides, the Navier–Stokes equations are numerically solved by using direct numerical simulation (DNS) in computational fluid dynamics encompassing the entire flow domain. Direct numerical simulation (DNS) is a simulation in computational fluid dynamics, wherein the Navier–Stokes equations are numerically solved all possible degrees of freedom appearing in the flow field without any turbulence model and approximating assumptions. The DNS offers insight into the mechanism of drag reduction, although its application in engineering flows is generally restricted to the Reynolds numbers $Re < 10,000$. As the Reynolds number Re

increases, the number of grid points required for the simulation increases drastically. Thus, the memory and the computational speed of the computing system restrict the range of Re for the application of DNS. The LES model uses filtered three-dimensional unsteady Navier–Stokes equations separating the large scale eddies from the small scale ones. The main flow structure is then resolved directly, and the small scale eddies are modeled by a subgrid scale model. However, the DNS and LES models require large memory and CPU time. Due to these limitations, the RANS model is preferred for many engineering problems. The low Re k–ε model assures stable converged solution readily as compared to DNS and LES. This model also requires relatively low computational effort. See also Minier and Pozorski (1997) who applied classical ideas from statistical physics to derive a PDF model for turbulent flows. Their model is built by adopting a Lagrangian point of view and by considering separately the statistical effects of the viscous and of the pressure gradient forces which act on a fluid particle.

It may be shown that although the simplest optical rays can be described by Riemannian geometry, it is rather Finslerian geometry that is valid for thermal rays. Finslerian geometry, sometimes called the geometry of Lagrangians, is the generalization of the Riemannian geometry for the case when the metric tensor depends on both state coordinates and state derivatives. Yajima and Nagahama (2015) study the Darcy's flows through inhomogeneous porous media by the theory of Finslerian geometry.

CHAPTER 8

Thermodynamics and Optimization of Practical Processes

8.1 Introduction

Engineering processes are very often far from equilibrium processes, they must go to the desired extent of completion within a finite time, and they must produce at least a minimum amount of product (Sieniutycz and Shiner, 1994). Furthermore, an efficiency based solely on physical grounds is more often than not an insufficient criterion for the performance of a chemical engineering process. Economic and ecological considerations also play important roles. All this means that classical equilibrium thermodynamics can at most place upper or lower limits on chemical engineering processes and that one must turn to irreversible thermodynamics for more information. All classical measures of thermodynamic perfection, such as the Carnot efficiency, exergy efficiencies, or dissipated exergy, have one characteristic in common: They all take the reversible process as their reference. Therefore, one can ask whether any real process operates close enough to the reversible limit for the traditional measures to be useful or even relevant. If not, is it possible to extend the concepts of reversible thermodynamics to provide limits for the performance of processes constrained to operate in finite time or at nonvanishing rates? If yes, what are the optimal paths related to the finite-time transitions? These questions arose from the need to evaluate how effectively real devices and real industrial processes use energy, from the viewpoint of their own limits rather than the reversible process limits. Clearly, in order to utilize in full power of thermodynamic analyses in engineering one must often go beyond conventional second law analyses. Sciubba and Wall (2007) presented the history of exergy from the beginnings to 2004, and Tsatsaronis (2007) commented this review. Majority of optimization problems for chemical processes and operations involve the minimization of costs, consumption of resources or maximization of profit or production size, and so forth.

A typical distillation column in Fig. 8.1 (Sieniutycz and Jeżowski, 2009) is an example of the engineering system whose optimization has attracted a number of active researchers. The distillation column is an energy consuming system because it requires a large heat power input to ensure the separation of components.

We may be interested in bounds established under the condition that, in any circumstances, the process will run with a minimum required intensity, yet yielding a desired product. This requirement usually yields bounds that are orders of magnitude higher than those classical

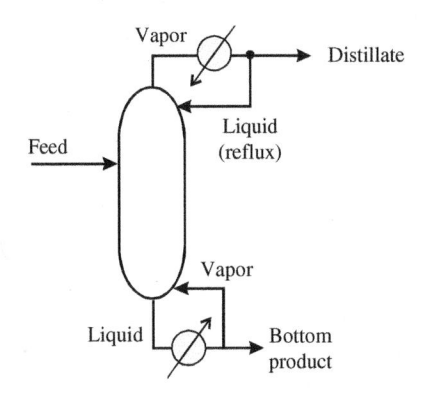

Figure 8.1: Scheme of a simple distillation column (Sieniutycz and Jeżowski, 2009).

known from the textbooks. Consider, for example, the heat consumption in the distillation column in Fig. 8.1 and evaluate a realistic bound that corresponds with the lower limit of heat associated with the use of the theoretical plates instead of the real plates of certain unknown efficiency (lower than unity), in the same column. This bound is usually two to three times larger than another (lower) bound, the heat consumed at minimum reflux conditions (already a quite artificial quantity as it pertains to a column with infinite number of theoretical stages). Next, the heat at minimum reflux is usually several times larger than that of the reversible evaporation heat. In effect, a design engineer must expect that the heat consumption he should dispose of should be at least an order of magnitude higher than the evaporation heat. In complex separation processes as those involving cycles, losses, and nonlinearities, such evaluations are highly nontrivial. Complex optimization techniques must be used to obtain dynamic limits for various unsteady processes, including those in separation systems.

Before optimizations, a minimum work requirement for distillation processes should be known. Cerci (2001) proposes to analyze a typical ideal distillation process by using the first and second laws of thermodynamics with particular attention to the minimum work requirement for individual subprocesses. The distillation column consists of an evaporator, a condenser, a heat exchanger, and a number of heaters and coolers. Several Carnot engines are also employed to perform heat interactions of the distillation process with the surroundings and determine the minimum work requirement for processes. These Carnot engines ensure the maximum possible work output or the minimum work input associated with the processes, and therefore the net result of these inputs and outputs leads to the minimum work for the entire distillation process. The minimum work relation for the distillation is the same as the minimum work input relation for an incomplete separation of incoming saline water, and depends only on properties of incoming saline water and outgoing pure water and brine.

Converse and Gross (1963) and Fan (1966) gave a nice account of optimal distillate-rate policy in batch distillation. Carrera-Patino and Berry (1991) worked out a model for coupling between heat and mass transfer in batch distillation. Rivero and Garcia (2001) presented exergy analysis of the reactive distillation system. Shen et al. (2013a, b) analyzed the batch and continuous extractive distillations. Some of these papers are depicted subsequently in more details (Sec. 8.6).

An example of the system with energy production is the well-known endoreversible thermal engine analyzed by Novikov, Chambadal, Curzon, and Albhorn. Curzon and Ahlborn (1975) showed that the efficiency, η, at maximum power output is given by the expression $\eta = 1 - (T_2/T_1)^{1/2}$ where T_1 and T_2 are the respective temperatures of the heat source and heat sink. De Vos (1992) has shown that the efficiency of existing engines is well described by the aforementioned result. Yet, Gyftopoulos (2002, 2003) criticized the Curzon–Ahlborn efficiency "for its lack of connection to power producing processes," although he recognized a tribute to energy systems by scientists and engineers. Many doubts in this discussion were explained in the reply of Chen et al. (2001a).

Basic characteristics of this system, describing the heat, power output and entropy production in terms of thermal efficiency η were discussed by De Vos (1992). They are depicted in Fig. 8.2. It is an open question if this model can directly be applied to radiation units. In the opinion of Badescu (2013), radiation reservoirs cannot be assimilated to heat reservoirs (chapter: Radiation and Solar Systems).

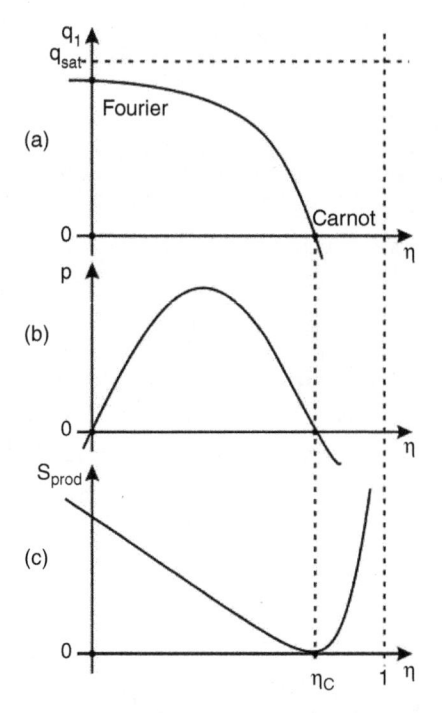

Figure 8.2: De Vos' (1992) book provides all details of derivations for analytical formulas describing driving heat (a), power (b), and efficiency (c), and a discussion of basic theoretical issues and practical conclusions.

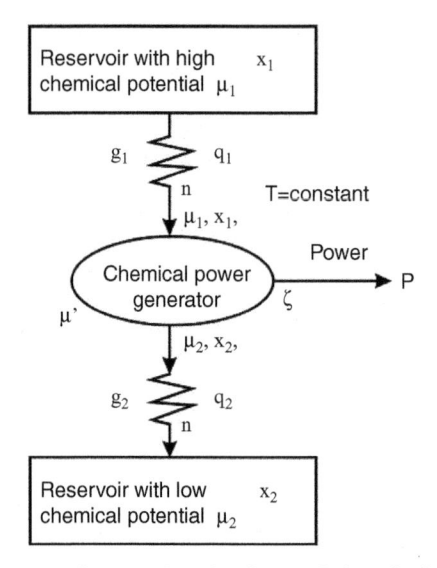

Figure 8.3: Scheme of an isothermal chemical engine.

Another example of the power producing system is an isothermal chemical engine shown in Fig. 8.3. Its characteristics are well known (De Vos, 1992; Sieniutycz and Jeżowski, 2009, 2013). A scheme similar to that in Fig. 8.3 can be attributed to a fuel cell who is an electrochemical engine (chapter: Power Limits in Thermochemical Units, Fuel Cells, and Separation Systems, Fig. 10.1).

In the past five decades increasing number of scientists and engineers have devoted their research to applications of optimal control theory in macroscopic practical systems. This has been true in the context of availability analysis (exergy optimization) and the related optimizations of thermodynamic processes in finite time. Finite-time thermodynamics (FTT) emerged and quickly began to grow.

R.S. Berry's intuition, care, patience, and creativity were crucial to putting forward a great deal of the original research in FTT (Berry, 1979, 1989a, b, 1990; Berry et al., 1978, 2000; Ondrechen et al., 1980a, b; Ondrechen, 1990; Mozurkewich and Berry, 1981, 1982). A group of works by Band, Kafri, and Salamon and Hoffmann led to modeling and optimization external combustion engines (Band et al., 1980, 1981, 1982a, b; Hoffmann et al., 1985, 1997). The research comprised many engineering processes and operations, in particular heat and mass transfer processes, separation operations and chemical conversion systems (Andresen, 1983; Andresen et al., 1984 a, b, c; Salamon and Berry, 1983; Salamon et al., 1980, 1982, 1984, 1990, 2001; Bejan, 1987, 1988; Orlov and Berry, 1990, 1993; Sieniutycz and Szwast, 1982).

In Russia, Tsirlin et al. (1998) worked out the FTT for the class of operations and processes with given rate and minimal entropy production. The general conditions they obey were derived. It was shown how the application of those conditions to particular systems produces a number of known bounds on entropy production (for heat and mass transfer processes and chemical conversion) as well as previously unknown bounds (for throttling, crystallization, and mechanical friction). Applying FTT, Tsirlin and Kazakov (2002a) determined limiting possibilities of irreversible separation processes and defined realizability areas for systems with given productivity (Tsirlin and Kazakov, 2002b). This series of works quantified the limiting performance of diffusion engines and membrane systems (Tsirlin et al., 2005; Tsirlin and Leskov, 2007). Of other authors who contributed to FTT we should also stress Feidt (1996, 1997, 1999, 2006); Feidt et al. (2007), and V. Badescu (2000, 2004a, b, 2008, 2013, 2014). The research of Badescu is depicted in chapter: Radiation and Solar Systems, since it is focused on radiation systems. Here we cite only his approaches bounding the efficiencies for endoreversible conversion of thermal radiation (Badescu, 1999) and optimal paths minimizing lost availability in heat-exchange processes (Badescu, 2004b).

The progress affected also the related field called the thermoeconomics which links economic analyses with thermodynamics (Berry et al., 1978; Clark, 1986; Sieniutycz, 1973a, 1978; Szwast, 1990; Benelmir and Feidt, 1997, 1998; Corning and Kline, 2000; Corning, 2002; Durmayaz et al., 2004; El-Sayed, 2003; Erlach et al., 1999; Le Goff et al., 1990; Månsson, 1985, 1990; Sciubba, 2000, 2001; Sieniutycz and Salamon, 1990c; Tsatsaronis, 1984, 1987, 1993, 1999, 2007; Tsatsaronis and Winhold, 1985a, b; Tsatsaronis et al., 1993; Tsirlin, 1974, 2008; Valero, 1995, 1996; Valero et al., 1993, 1994a, b, 2006) and many others. The results of these analyses have explored in-principle limits to the efficiency of energy conversion in finite time (Salamon et al., 1980; Salamon and Berry, 1983; Nulton et al., 1985), have found ways to improve the power that can be extracted from various engines (Rubin, 1980; Salamon et al., 1982; Salamon and Nitzan, 1981; Band et al., 1980, 1981, 1982a, b; Hoffmann et al., 1985; Mozurkewich and Berry, 1981, 1982; Sieniutycz and Berry, 1991), and have provided ways to improve important industrial processes (Månsson and Andresen, 1985). Associated directions also appeared in the context of "cooling" such quasi-physical systems as simulated annealing problems in which one exploits a physical analogy to solve optimization problems plagued by many local minima (Bass, 1968; Nulton and Salamon, 1988; Harland and Salamon, 1988; Andresen and Gordon, 1994; Sieniutycz and Jeżowski, 2013). An article by Salamon et al. (2001) presents a synthesis of the progress in control thermodynamics by laying out the main results as a sequence of principles. The authors state and discuss a number of general principles for finding bounds on the effectiveness of energy conversion in a finite time.

Optimization results and models of chemical and mechanical engineering processes and operations are reported in the following:

Books (Aris, 1961, 1964; Berry et al., 2000; Beveridge and Schechter, 1970; Dryden, 1975; Brodowicz and Dyakowski, 1990; Mironova et al., 2000; Boyle, 2004; El-Sayed, 2003;

Geworkian, 2007; Feidt, 2006; Fan and Wang, 1964; Fan, 1966; Sieniutycz, 1978, 1991; De Vos, 1992, Sieniutycz and Szwast, 1982; Sieniutycz and Jeżowski, 2009, 2013, Plath, 1989; Tsirlin, 1997; Wu et al., 2000; Zhu, 2014).

Papers (Aghbalou et al., 2006; Alvarez et al., 2006; Amelkin et al., 2005; Assari et al., 2013; Avnir et al., 1989; Bak et al., 2002; Band et al., 1982b; Bedeaux et al., 1999; Bejan, 1996a, b, 2000a; Bejan and Lorente, 2001; Benelmir and Feidt, 1997, 1998; Berry et al., 2000; Bochenek et al., 2011; Bracco and Siri, 2010; Burghardt, 2008; Campanari, 2001; Cenusa et al., 2004; Chato and Damianides, 1986; Chato and Khodadadi, 1984; Chen and Andresen, 1999; Chen and Zhao, 2009; Chen and Sun, 2004; Chen et al., 1997a, b, 2001b, 2004a, b, 2005, 2008, 2010, 2011, 2013; Costea et al., 1999; De Vos, 1985, 1991, 1992, 1993a, b, 1997; Durmayaz et al., 2004; El-Genk and Tournier, 2000; Ensinas et al., 2007; Feidt et al., 2007; Gayubo et al., 1993a, b; Hoffmann et al., 1985; Kalina, 2011; Ganjeh et al., 2013; Lazzaretto et al., 1993a, b; Lazzaretto and Macor, 1995; Le Goff et al., 1990; Lems et al., 2003; Li et al., 2010a; Lozano et al., 1993; Mozurkewich and Berry, 1981, 1982; Ni et al., 2007; Nulton and Salamon, 1988; Orlov and Berry, 1993; Palazzi et al., 2007; Poświata, 2004, 2005; Poświata and Szwast, 2003, 2006; Sahin et al., 1997; Salamon and Nitzan, 1981; Sieniutycz, 1973a, 1975, 1976b, 1978, 1991, 1997a, b, 1998b, 1999, 2000a, c, h, 2001a, c, 2007b, 2008a, 2009a, c, 2010a, 2013a, b; Sieniutycz and Kuran, 2006; Sieniutycz and Poświata, 2012; Sieniutycz and Szwast, 1999; Sieniutycz et al., 2011; Szargut, 2002a; Szargut and Stanek, 2007; Szepe, 1966; Bykov and Yablonskii, 1980; Szwast, 1990; Szwast and Sieniutycz, 1998, 1999, 2001, 2004; Szwast et al., 2002; Tondeur, 1990; Tondeur and Kvaalen, 1987; Tsatsaronis, 1984, 1987; Tsirlin, 1974; Tsirlin and Leskov, 2007; Vargas and Bejan, 2004; Vargas et al., 2005; Velasco et al., 2000; Wang, 1998; Wierzbicki, 2009; Wu et al., 2008; Xia et al., 2010a, b, 2011a, b; Zhu et al., 2001, 2003, 2004; Yan, 1993).

Entropy generation minimization (EGM), an approach very close to FTT in the sense of objectives, goals, methods, and tools, has received a mature status of the applied thermodynamic theory of finite-size devices and finite-time processes (Bejan, 1987, 1988, 1996a, b, c; Bejan and Schultz, 1983, and many others). Constructal theory of shape and structure evolved from the EGM-based and FTT-based considerations of heat devices (Bejan, 1997d) and attempts to predict shape and structure in natural macroscopic systems (Bejan, 1997d, 2000; Bejan and Lorente, 2004, 2005). Dryden's book covers many classical issues associated with the efficient use of energy (Dryden, 1975). Exergy analyses (Szargut and Petela, 1965) were applied for optimization of drying systems with granular solids (Szwast, 1990) and to other chemical engineering operations (Sieniutycz, 1973a, 1974c, 1975, 1976b, 1978, 1991; Sieniutycz and Szwast, 1982). Earlier analyses used classical exergies, and later, applied in the context of FTT, lead to finite-rate exergies (Sieniutycz, 1997b, 1998a, b, 1999, 2000a, c, d, e, i, 2001a, b, e, 2002a, a, 2003b, c, d, 2006b, c, d, 2007b, d, 2008a, b, 2009a, b, c, d, 2010a, b, c, d, 2011a, b, c, d, e,

2012a, b, c, 2013a, b). For FTT theory and further development of finite rate exergies, see: Li et al., 2010b; Lin et al., 2004; Wu et al., 2000; Xia et al., 2009, 2010a, b, c, 2011a, b; Zhu, 2014. For considerations regarding the appropriateness of FTT optimization criteria and ecological performance functions, see: Angulo-Brown, 1991; Ayres, 1978, 1999; Ayres et al., 1996, 1998; Bakshi and Grubb, 2013; Baumgärtner, 2004; Chen et al., 2002, 2004a, b, 2005, 2010; Faber et al., 1996; Hau and Bakshi, 2004; Jorgensen, 2001; Kåberger and Månsson, 2001; Kay and Schneider, 1992b; Odum, 1971, 1988; Sousa, 2007; Szargut, 1986, 1990, 2005; Szargut and Stanek, 2007; Valero, 1995, 1996; Zhu et al., 2003, 2004; Yan, 1993, and others.

The applications stretched from the realm of thermal technology (Szargut, 2001) to chemical, metallurgical, and other technologies (Szargut et al., 1988). Particularly fruitful were ecological applications of exergy for the analysis of cumulative exergy consumption and cumulative exergy losses (Szargut, 1990), determination of a proecological tax to replace the personal taxes (Szargut, 2002b), and the minimization of nonrenewable resources depletion (Szargut, 2005). In book by Hausen (1976) book, classic analytical treatment of heat exchangers working in cocurrent, countercurrent, and crosscurrent contacting modes is given. Second-law-based optimization of heat and mass exchanger networks is now a large field governed by entropy criteria (Chato and Damianides, 1986; Garrard and Fraga, 1998; Chato and Khodadadi, 1984; Chato, 1989). Links can be drawn between the second law analyses and Linnhoff's pinch method (Linnhoff, 1979, 1982; Linnhoff and Ahmad, 1990; Sieniutycz and Jeżowski, 2009, 2013). See also Ziębik (1995) and other works of this author regarding applications of second law analysis in industrial energy and technological systems.

One should stress many applications of finite size thermodynamics for the optimization of heat pumps, engines, energy cogeneration systems, and energy management (Clarke and Morgan, 1983; Feidt, 1996, 1997, 1999, 2006; Feidt et al., 2007; Salamon and Nitzan, 1981; Salamon et al., 1980, 1982, 1990, 2001). Gordon (1991, 1993) and Gordon and Orlov (1993) proposed generalized power-efficiency diagrams as performance characteristics of endoreversible chemical engines. Salamon et al. (2001) synthesized the progress in control thermodynamics by laying out the main results as a sequence of principles; nine general principles were suggested for finding bounds on the effectiveness of energy conversion in a finite time. There are also FTT applications attacking geothermal units (Bodvarsson and Witherspoon, 1989), biotechnology systems (Benelmir and Feidt, 1997, 1998; Costea et al., 1999; Bejan and Mamut, 1999; Berthiaume et al., 2001; Bracco and Siri, 2010), and phase-change heat transfer (Charach and Zemel, 1992). Exergy analyses of "CO_2 zero-emission" high-efficiency plants lead to another group of problems (Calabro et al., 2004). Applying systematically energy and exergy analysis, Ganjeh et al. (2013) have developed an exergoenvironmental optimization of heat-recovery steam generators in a combined-cycle power plant. Thermoeconomics of energy conversions is covered in El-Sayed's (2003) book.

The are many research works which apply the second law analyses and finite-time thermodynamics to heat pumps (HP). Classifications of HPs can be found (Moser and Schnitzer, 1985). Duminil (1976) developed basic thermodynamic principles applied to HPs. Using a cycle model Chen and Andresen (1999) investigate optimal performance of an endoreversible Carnot cycle used as a cooler. The cooling rate was optimized for specific power input and total heat-transfer area of the cooler. The optimal bounds of the primary performance parameters, such as the cooling rate, the power input, and the temperatures of working substance, are found. The results provide a guide for the understanding of the cooler performance and the improvement of the theory of endoreversible Carnot cycles. They stimulated a series of further works, in particular on the field of heat pumps (Reay and Mac Michael, 1979). For example, J. Chen, Bi, and Wu (1999) developed a generalized model of a combined heat pump cycle and its performance. J. Chen et al. (2000) worked out the performance of irreversible combined heat pump cycles. J. Li et al. (2010b) formulated an optimal relation for a generalized Carnot-like heat pump with complex heat transfer. Applications of heat pumps for the purpose of decreasing the energy use in drying are known (Żyłła and Strumiłło, 1981). Fernández-Seara et al. (2012) developed experimental analysis of a direct expansion solar assisted heat pump with integral storage tank for domestic water heated under zero solar radiation conditions. Kuang et al. (2003) conducted experiments on a solar-assisted HP system for heat supply. Dong La et al. (2013) summarized effects of irreversibilities on the thermodynamic performance of open-cycle desiccant cooling cycles. Zhu et al. (2001) investigated optimal performance of a generalized irreversible Carnot heat pump with a generalized heat transfer law. Morosuk et al. (2004) presented a new proposal in the thermal analysis of complex regeneration system. Mujumdar (2007) examined the role of innovation in drying in various industrial sectors, for example, paper, wood, foods, waste management, and so forth, and discussed examples of intensification and innovation in drying designs with the use of HPs.

By adopting a course representative to the approaches of the optimal control, the observer becomes an active participant who interacts with the system. As R.S. Berry says, this is nothing new to engineers, who operate this way regularly, but is novel to other scientists. Optimal control is "an approach that implies we want to do something with our scientific insight besides observe and describe systems" (Berry, 1990).

Finding the strategy which optimally drives a physical system is always a nontrivial task. Accordingly, the mathematical theory of optimal control has become a standard tool for solving these problems. The importance of thermodynamics as a tool in many disciplines endows these problems with an interdisciplinary character which touches the theory of heat engines, solar energy engineering, mechanical and chemical engineering, economics of industrial processes, ecology, and so forth.

Thermodynamics and thermo-economics of energy conversion, transmission, and accumulation constitute the main problems of this chapter. The structure of the chapter is as follows. Section 8.2 exposes the role of optimization theory in thermodynamic and thermoeconomic approaches. Then, after reviewing mathematical methods of optimum seeking in Section 8.3, the application of optimization theory in minimizing of catalysts deactivation is exemplified in Section 8.4. As there are already many comprehensive sources on Bejan's optimizing in his constructal theory, this theory is characterized in Section 8.5 only verbally and from the standpoint of basic literature. Section 8.6 describes selected papers referred to applications, mainly in the fields of chemical, biochemical, and mechanical engineering.

When there is only a finite amount of time available to accomplish an useful engineering task, for example, withdrawal of a definite work from a solar unit, separation in a mixture, or conversion in a reactor, usable energy losses and costs are inevitable, and operations must occur with a finite intensity for given process requirements. As an operation becomes faster the losses of usable energy and corresponding exploitation costs increase. However, with some process conditions minimum usable energy input could be supplied to a separation unit or a thermal machine (engine, heat pump, etc.), but the thermodynamic efficiency would then be much less than that of a related reversible process. One can ask the question: what are the optimality conditions for such minimum consumption of the usable energy or what is economically optimal consumption of this energy corresponding to the minimum of the sum of the exploitation and investment costs, for given process requirements. This sets respectively the problems of thermodynamic and thermoeconomic optimization approached in this project.

The thermodynamic limits of industrial processes are not often considered when economists describe manufacturing systems with production functions. It is therefore worthwhile to testify the relevance of certain thermodynamic measures, limits, and efficiency to economic efficiency. All classical measures of thermodynamic perfection, such as the Carnot efficiency, exergy efficiencies, and so forth take the reversible process as their reference. Therefore, one can ask the question whether any real process operates closely enough to the reversible limit for the traditional measures to be useful or even relevant? If not, is it possible to extend the concepts of reversible thermodynamics to provide limits for the performance of processes constrained to operate in finite time or at nonvanishing rates? These questions arose from the need to evaluate how effectively real devices and real industrial processes use energy, from the viewpoint of their own limits rather than the reversible process limits when the basic requirements and related average intensity are given. These questions have led to the contemporary efforts to find bounds and related thermodynamic and economic optima corresponding to given process demands performed in a finite time.

The two branches of nonequilibrium thermodynamics: finite-time thermodynamics (FTT) and thermoeconomic optimization (TEO) have contributed most to the solution of the aforementioned problems. The goal of FTT is to isolate the temporal characteristics of some basic generic models representing relatively broad classes of irreversible processes which have the same basic irreversibilities in common, irregardless of the individual details.

FTT retains the philosophy of model idealization known from reversible thermodynamics (the Carnot cycle) but uses somewhat more realistic models which have some basic irreversibilities incorporated, for example, thermal resistances between the reservoirs and working fluid (cycle by Curzon and Ahlborn, 1975). While FTT investigates in principle the processes in a given equipment, TEO includes into consideration investment costs and the related economic analysis and economic optima. It is well known that the optimum configuration of a given practical or industrial system will vary depending on such economic factors as equipment and capital costs, sources of capital, method of financing, period of investment, cost of displaced fuel, taxes, resale value. TEO involves naturally a tradeoff between investment cost (which increases with equipment size) and exploitation cost. When the sum of the exploitation and the investment cost is minimized, an optimal size for equipment is obtained. This perspective which combines technical, thermodynamic, and economic approaches and links the principles of economics with thermodynamic analyses of energy or resources based on the first and second law has been named "thermoeconomics."

8.2 Backgrounds for Optimizing in Thermodynamic Systems

Roughly speaking there are the two broad categories of approaches where the methods of the contemporary mathematical theory of optimization can be used for analyzing thermodynamic systems. In the first approach, when one wants to optimi*z*e these systems by changing some external parameters (decisions or controls), both extremal and nonextremal solutions pertain to real physical situations. In this case the governing equations describing the changes of the internal state are known and a control is imposed through the system boundaries. This first approach includes studies which seek in-principle limits to the operation of thermodynamic processes and studies which seek to improve existing engineering systems. In the second approach, one wants to predict the system behavior (usually irreversible) under prescribed conditions and therefore seeks to derive its governing equations from some variational or extremum principles (see chapter: Variational Approaches to Nonequilibrium Thermodynamics). In these systems, extremal controls are set by nature rather than by man. Here, however, we discuss the first approach, that is, the use of thermodynamics in optimization of controlled systems.

R.S. Berry characterizes a book (Sieniutycz and Farkas, 2005) as follows: "the volume deals with *descriptive* variational methods, some with methodology and some with applications. A few of the chapters follow the other kind of use of variational methods, what we may call the *prescriptive*. This direction is perhaps epitomized by optimal control theory. Here we use the power of variational calculus to tell us what we should do in order to achieve some goal. Here, we can sometimes open new directions by recognizing that some variable previously considered beyond our control could, in fact, become a control variable and enable us to construct entirely new kinds of devices."

In a similar spirit, Fung and Tong (2001) state in the revision to the classic engineering text on foundations of solid mechanics: "Engineering is quite different from science. Scientists try *to understand* nature. Engineers try to *make things* that do not exist in nature. Engineers stress invention. To embody an invention the engineer must put his idea in concrete terms, and design something that people can use. That something can be a device, a gadget, a material, a method, a computing program, an innovative experiment, a new solution to a problem, or an improvement on what is existing. Since a design has to be concrete, it must have its geometry, dimensions, and characteristic numbers. Almost all engineers working on new designs find that they do not have all the needed information. Most often, they are limited by insufficient scientific knowledge. Thus they study mathematics, physics, chemistry, biology and mechanics. Often they have to add to the sciences relevant to their profession. Thus engineering sciences are born."

The discussion of the role of thermodynamics in optimization of controlled systems is done here briefly and in general terms since detailed analyses are the subject of a number of papers. While thermodynamics (mostly irreversible) is the dominant ingredient in availability analysis, exergy optimization and the minimization of entropy production, it plays only a partial role in thermoeconomics where the overall problem deals with the balance of value (economic balance). Since the economic objective is always formulated in economic terms (prices), the solution of any thermoeconomic problem is not equivalent to that of the corresponding thermodynamic problem (Durmayaz et al., 2004). It does however reduce to thermodynamic optimizations in the limit when the price of certain thermodynamic quantities such as the availability used or the power produced becomes much larger than the prices of other participating quantities (Andresen et al., 1983). This limit represents an energy theory of value, that is, a value system in which one considers energy as the single valuable commodity (Berry et al., 1978). A second way for economic optima to coincide with thermodynamic optima is if the economic value of the exergy unit is the same for all forms of matter and energy taking part in the process. In this case the thermodynamic problem of minimum availability loss (due to the internal irreversibilities in the system) is equivalent to minimum economic costs. This case is however rather special, as the prices of the exergy units generally differ (Sieniutycz, 1978).

Hence, the general (thermo)economic optimization problem cannot be, as a rule, broken down to the problem of minimum irreversibility. Even the thermodynamic problem of minimizing the driving exergy (available energy of the input flows) is not equivalent to minimizing irreversibility since it does not take into account the exergy of the outgoing flows. The problem of minimum cumulative exergy consumption (a generalization of the previous problem), which is of interest for some complex industrial and ecological systems with many units and interconnections, is also not equivalent to the problem of minimum dissipation for the same reasons. These problems belong rather to ecology and energy management. Extra unconventional terms in the performance index appear in ecological analyses where a cost-type optimization criterion takes into account not only the cumulative consumption of exergy but also the exergy losses resulting from the deleterious impact of the wastes on human activity and health, quality of agricultural crops, forestry, natural resources, and so forth.

Nevertheless, a number of complex economic or ecological problems were born as extensions and generalizations of irreversibility analysis and their connection with thermodynamics is still strong enough to justify a common macroscopic context.

Although various performance criteria can be formulated for various problems, the physical constraints (balance equations, equations of state, thermal equilibria, and kinetic relationships) are the product of thermodynamic analyses. Even if the thermodynamic criteria are replaced by the more appropriate economic ones, or by suitable ecological objective functions, the set of process constraints remains often similar, and the same optimum seeking method can be used in thermodynamic, economic, and ecological cases. Thus most of the methodological experience in the formulation of the objective functions and even the majority of the constraints can be preserved when passing from the thermodynamic to the economic or ecological optimization.

In his early book Bejan (1982) considers technical and physical consequences of entropy generation through heat and fluid flow. Bejan's (1987) review places in perspective of the new direction devoted both to the analysis of the thermodynamic irreversibility of heat and mass transfer components and systems and to the design of these devices on the basis of entropy generation minimization. The review assesses the second law analyses in the heat and mass transfer. It focuses on the mechanisms of the entropy generation in heat and fluid flow and on the design tradeoff of balancing the heat transfer irreversibility against the fluid flow irreversibility. Applications pertain to the design and thermal energy storage of the exchangers. Bejan (1996a) identifies three current optimization approaches to energy systems as in principle identical. As stated by Bejan (1996a) entropy generation minimization, finite-time thermodynamics, or thermodynamic optimization is the method that combines into simple models the most basic concepts of heat transfer, fluid mechanics, and thermodynamics. These simple models are used in the optimization of real (irreversible) devices and processes, subject to finite-size and finite-time constraints. Bejan's (1996a) review traces the development and adoption of the method in several sectors of mainstream thermal engineering and science: cryogenics, heat transfer, education, storage systems, solar power plants, nuclear and fossil-fuel power plants, and refrigerators. Bejan's emphasis is placed on the fundamental and technological importance of the optimization method and its results, the pedagogical merits of the method, and the chronological development of the field. Bejan's (1996b) development of entropy generation minimization, as an equivalent alternative of thermodynamic optimization of finite-size systems and finite-time processes, gives a systematic treatment of refrigeration plants, heat pumps and engines and offers plausible definitions of coefficients of performance and efficiencies of these units.

Many remarks showing a link between second-law analyses, FTT, and thermoeconomics, and the limitations of thermodynamic analyses in the light of the latter, are expressed by Clark (1986) who has pointed out that economic cost analysis introduces a new and different set of variables which must be considered simultaneously with thermodynamic ones to obtain the best technically acceptable design. He has developed evaluation methods which bridge the interface between thermodynamic and economic considerations for the purpose of optimum design. An optimum configuration will vary depending on such economic factors as equipment and capital costs, sources of capital, period of investment, cost of displaced fuel, resale value, and so forth. Other constraints follow from ecology when one takes into account not only the exergy consumption but also the exergy losses resulting from the deleterious impact of wastes on human activity and health, crops, forestry, natural resources, and so forth.

Månsson (1985, 1990) discusses some general aspects of the significance of thermodynamics for economic analyses. The former is to be used to identify and quantify basic constraints resulting from conservation laws as well as losses due to irreversibility. Also, thermodynamics provides clear and quantitative definitions of efficiency. The similarity of the value concept in thermodynamics (exergy) and economics is illustrated along with the conclusions about the inappropriateness of efforts to reduce all aspects of value to a single physical standard. An example of the maximum outgoing exergy obtained from given set of sources is given. A simple macroeconomic growth model is presented to show the role of the thermodynamic limitations.

Bergman et al. (2008) subsume the recommendations of a U.S. National Science Foundation international workshop held in May 2007 at the University of Connecticut. Based upon the final discussion on the topical matter of the workshop, several trends become apparent. A strong interest in sustainable energy is evident. A continued need to understand the coupling between broad length (and time) scales persists, but the emerging need to better understand transport phenomena at the macro/mega scale has evolved. The need to develop new meterology techniques to collect and archive reliable property data persists. Societal sustainability received significan attention. Matters involving innovation, entrepreneurship, and globalization of the engineering profession have emerged. Integration of research thrusts and education activities has been highlighted throughout. Based upon the involvement of, and input from, a significant portion of the international heat- and mass-transfer community, challenges and opportunities have been identified in the areas of sustainable energy systems, biological systems, security, information technology, nanotechnology, and education.

The following was suggested:

1. Recognition of the rapidly evolving changes in research thrusts, education initiatives, and national policies.
2. Energy and other workshops should direct more attention to the impact of globalization made possible by today's capability for instantaneous communication ability among researchers, educators, and policy-makers.

3. Subsequent actions might include a focus on university–industry–government interaction, including aspects of technology transfer and commercialization.
4. The heat transfer community must not only respond to technical needs of the society but also take a proactive stance in setting the national research agenda in areas where the community has demonstrated leadership and has great credibility, such as in sustainable energy.
5. The heat transfer community must take a proactive stance in leading a national discussion including the involvement of the nonengineering community, in understanding the societal implications.
6. Important common interests of researchers working on sustainable energy and clean water developments should intersect and provide a pressing motivation for interdisciplinary actions in the future.

Baumgärtner (2004) states that integrating methods and models from thermodynamics and from economics promises to yield encompassing insights into the nature of economy-environment interactions. The division of labour between thermodynamics and economics seems to be obvious. Thermodynamics should provide a description of human societies' physical environment, while economics should provide an analysis of optimal individual and social choice under environmental scarcities. But the task is difficult. Being a branch of physics, thermodynamics is a natural science. It explains the world in a descriptive and causal, allegedly value-free manner. On the other hand, economics is a social science. While it pursues to a large extent descriptive and causal (so-called 'positive') explanations of social systems, it also has a considerable normative dimension. Valuation is one of its basic premises and purposes. Therefore, bringing together thermodynamics and economics in a common analytical framework raises all kinds of questions, difficulties and pitfalls. Baumgärtner (2004) lays out the rationale, concepts, and caveats for developing and using thermodynamic models in ecological economics. He sketches the historical origins of this endeavor and develops the fundamental rationale of employing thermodynamic concepts and models in ecological economics. He also introduces the elementary concepts and laws of thermodynamics and gives an overview of approaches to incorporating thermodynamic concepts into economic analysis. He assesses their respective potential for ecological economics. He surveys various implications and insights from thermodynamic models in ecological economics. He concludes by assessing the role of thermodynamics for ecological economics and for the discussion of sustainability.

A chapter by Bakshi and Grubb (2013) in a book edited by Cabezas and Diwekar (2013) demonstrates the role that thermodynamics can play in assessing the sustainability of technological activities and in improving their design. Since thermodynamics governs the behavior of all macrosystems, it plays crucial role in understanding physical limits of technologies and for quantifying the contribution of resources. The concept of exergy captures the first and second laws. Since exergy is a common currency that flows and gets transformed in industrial and ecological systems, it allows the joint analysis of industrial and ecological systems. This insight permits accounting of the role of ecosystem goods and services in supporting human activities. As ecosystems are critical to sustainability, accounting for their role must be a part of all methods aimed toward the analysis and design of sustainable systems. Thermodynamics provides a rigorous approach for meeting this challenge. In addition, exergy analysis of industrial processes and life cycles helps in identifying areas of largest resource inefficiency and opportunities for improvement. This approach complements the insight obtained from assessing the impact of emissions. Case studies based on the life cycle of biofuels and nanomanufacturing are used to demonstrate the important role that thermodynamics can play in sustainability engineering.

Dincer and Cengel (2001) discuss the energy, entropy, and exergy concepts and their roles in thermal engineering. Energy, entropy and exergy concepts come from thermodynamics and are applicable to all fields of science and engineering. Therefore, they intend to provide background for better understanding of these concepts and their differences among various classes of life support systems with a diverse coverage. They also cover the basic principles, general definitions and practical applications and implications. Some illustrative examples are presented to highlight the importance of the aspects of energy, entropy, and exergy and their roles in thermal engineering. In conclusion, in a broader perspective (except for the zeroth and third law of thermodynamics), they define thermodynamics as a science of energy, exergy, and entropy. Apparently, the first law of thermodynamics (FLT) refers to the energy analysis which only identifies losses of work and potential improvements or the effective use of resources, in, for example, an adiabatic throttling process. However, the second law of thermodynamics (SLT), that is, exergy analysis takes the entropy portion into consideration by including irreversibilities. During the past decade exergy-related studies have received considerable attention from various disciplines ranging from chemical to mechanical engineering, from environmental engineering to ecology and so on. As a consequence of this, recently, the international exergy community has expanded greatly.

Tsatsaronis (1987, 1993, 1999, 2007) discusses the strengths and limitations of exergy analysis and development, state-of-the-art, and applications of thermoeconomics (exergoeconomics). The latter is a method that combines exergy with conventional concepts from engineering economics to evaluate and optimize the design and performance of energy systems. His papers are written mainly for the person interested in applying thermoeconomics to energy systems. They review the history of exergy analysis and thermoeconomics, the performance evaluation of an energy system from the viewpoints of the second law of thermodynamics and thermoeconomics as well as use of thermoeconomic optimization techniques. Tsatsaronis and Winhold (1985a, b) introduce a new methodology for exergoeconomic analysis and evaluation of energy conversion plants, and exemplify it by analysis of a coal-fired steam power plant. Tsatsaronis et al. (1993) develop an exergy costing theory of exergoeconomics.

Lazzaretto et al. (1993a, b) have developed an exergoeconomic symbolic optimization for energy systems and then optimize the main design variables of a combined cycle plant by using two different performance functions: the specific consumption of the plant and the average unit cost of the product. The researchers suggest that the method is promising for complex thermal systems. As an example of the exergoeconomic symbolic optimization (EXSO) for energy systems (Lazzaretto et al., 1993a) the main design variables of a combined cycle plant have been optimized using two different objective functions: the specific consumption of the plant and the average unit cost of the product. Aplications to more complex systems are predicted. Most of the thermoeconomic accounting and optimization methods for energy systems are based upon a definition of the productive purpose for each component (Lazzaretto and Macor, 1995). On the basis of this definition, a productive structure of the system can be defined in which the interactions among the components are described by their fuel product. The aim of this work is to calculate marginal and average unit costs of the exergy flows starting from their definitions by a direct inspection of the productive structure. As a main result, it is noticed that the only differences between marginal and average unit cost equations are located in the capital cost terms of input-output cost balance equations of the components. Durmayaz et al. (2004) present a lasting account of optimization of thermal systems based on finite-time thermodynamics and thermoeconomics.

Valero et al. (1993) and Lozano and Valero (1993b) formulated the fundamentals of the so-called structural theory of thermoeconomics. Lozano and Valero (1993b) and Lozano et al. (1993) applied this theory to thermoeconomic analysis of gas turbine cogeneration systems. Frangopoulos et al. (2002) presented a review of methods for the design and synthesis optimization of energy systems. The theory is based on four general premises, which permit to obtain the set of characteristic equations or primal of the system. From this set, the costing equations or dual is obtained. The relation primal-dual is bijective and depend on the descriptive function used to build the characteristic equations, that is, enthalpy, energy, or free energy function. When the relative free energy is used the corresponding primal-dual is only sensible to variations of quantities and independent of local variations of the local enthalpy and entropy of the flows. Developing techniques for designing efficient and cost-effective energy systems is one of the foremost challenges energy engineers face. In a world with finite natural resources and increasing energy demand by developing countries, it becomes increasingly important to understand the mechanisms which degrade energy and resources and to develop systematic approaches for improving the design of energy systems and reducing the impact on the environment. The second law of thermodynamics combined with economics represents a very powerful tool for the systematic study and optimization of energy systems. This combination forms the basis of the relatively new field of thermoeconomics (exergoeconomics). During the 1960s, R.B. Evans, Y.M. El-Sayed, R.A. Gaggioli, and M. Tribus, among others, conducted pioneering work in this field. However, the comprehensive effort to apply thermoeconomics systematically to the analysis, optimization and design of energy systems did not start until the eighties. New methodologies have flourished, giving rise to new concepts with their own nomenclature, definitions, and applications. In 1990, a group of concerned specialists in the field (C. Frangopoulos, G. Tsatsaronis, A. Valero, and M. von Spakovsky) decided to compare their methodologies by solving a predefined and simple problem of optimization: the CGAM problem, which was named after the first names of the researchers. Valero et al. (1994a) consider definition and conventional solution of CGAM problem, whereas Valero et al. (1994b) discuss the application of the exergetic cost theory to the CGAM problem. An optimization strategy for complex thermal systems is then presented. The strategy is based on conventional techniques and incorporates assumptions and consequences of the exergetic cost theory (ECT) and of symbolic exergoeconomics. In addition to the results obtained by conventional techniques, the method provides valuable information about the interaction of components. Thermoeconomics is a physico–mathematical background for energy–economic–ecological analyses which are made in different fields of knowledge. To this end the concept of exergetic cost (Valero, 1995; Valero et al., 2006) is employed; this is a concept close to that of embodied energy or cummulative exergy consumption (Szargut et al., 1988). The analysis of exergetic cost focuses in a rigorous and detailed way on the process of cost formation, imposing a rationale on one of the most important problems of the techniques for calculating costs which is that of the costs allocation in bifurcations. In the same way, the problem of the truncation of the calculation of costs is solved by starting from the possibility of calculating the exergy of all natural resources counted on the basis of a rationally chosen reference environment for the whole planet. Elsewhere the author studies the physico–mathematical reasonings which underpin the theory of cost allocation, as well as its discrepancies and analogies with other analyses of use, such as the analysis of embodied energy and the emergy analysis (Odum, 1971, 1988). The author also declares a critique of exergy as a measurement of the quality of products, concluding that the concept of quality is much wider and more complex than that of thermodynamic exergy, which leaves theoretical space remaining for future exploration. Finally, a calculation of exergoecological costs is proposed for all those products and services which our society produces. This would be brought about by means of wide international cooperation in which, first, concepts, methods, and sources of analysis would be fixed and then projects of cost calculation undertaken in the widest possible range of cases. See also: Erlach et al. (1999); Valero et al. (2006), and Geworkian (2007).

Demirel's (2014) book on nonequilibrium thermodynamics, emphasizes the unifying role of thermodynamics in analyzing natural phenomena. The book contains a chapter on stochastic approaches to include the statistical thermodynamics, mesoscopic nonequilibrium thermodynamics, fluctuation theory, information theory, and modeling the coupled biochemical systems. The range of applications of thermodynamics and entropy is constantly expanding and new areas finding a use for the theory are continually emerging. Singh's (2013) volume responds to the need for a book that deals with basic concepts of entropy from a hydrologic and water engineering perspective and that uses such concepts to solve a range of water engineering problems. The applications of concepts and techniques vary across different areas, and the book aims to relate them to practical problems of environmental and water engineering. The book presents and explains the Principle of Maximum Entropy (POME) and the Principle of Minimum Cross Entropy (POMCE), and presents their applications to different types of probability distributions. Entropy-based criteria are important for urban planning. Illustrative examples are given for maximum entropy spectral analysis and minimum cross entropy spectral analysis, which are two powerful techniques for the problems faced by environmental and water researchers.

8.3 Mathematical Methods of Optimization

Mathematically, optimization is seeking of the best solution under given conditions (constraints). Such a general definition of optimization includes most conscious activity pursued by mankind during its long history. Its mathematical pursuit also has a long history, but several major strides have been made during the last three decades. We pause to review several of the most important recent milestones for the development of the discipline. These are: Dantzig's methods for linear programming (Dantzig, 1968; Llewellyn, 1963), the Karush–Kuhn–Tucker conditions for nonlinear programming (Kuhn and Tucker, 1951; Zangwill, 1969), Bellman's dynamic programming (Bellman, 1957, 1961; Bellman and Kalaba, 1965; Aris, 1964), and Pontryagin's maximum principle (Pontryagin et al., 1962; Fan and Wang, 1964; Lee and Marcus, 1967; Leitman, 1981; Sieniutycz, 1978; Findeisen et al., 1980). These methods are complemented by the classical methods of differential calculus, Lagrange multipliers and the calculus of variations (functional optima; Gelfand and Fomin, 1963). Together they make a powerful collection of optimization tools which can solve many difficult practical problems (Beveridge and Schechter, 1970). Of newer optimization techniques, methods of averaged optimization and sliding regimes in optimized problems (Tsirlin, 1974, 1997, 2008, 2009; Berry et al., 2000) received the status of effective tools in applied finite-time thermodynamics. For these developments, progress in computational techniques has also been essential.

The mathematical nature of the optimization problem lends itself naturally to a division between finite and infinite dimensional problems. In the first, called the mathematical programming problem, the optimization criterion (performance index) is the *function* of the decisions (and possibly also other additional variables like constants, parameters and so forth, called collectively the uncontrolled variables). In the second, called the dynamic optimization problem or the variational problem, the optimization criterion is the *functional* of the decisions (as well as state variables and constant parameters). In the first case, one is searching for an optimum in the form of an *n*-tuple *of numbers*. In the second case the optimizer's task is to determine the *functions* describing the dynamic behavior of the optimal control as a function of time or as a function of any time-like independent variable characterizing the evolution of the process (length, holdup time, etc.). These functions constitute the optimal control which characterizes the best external action on the system in time. The optimal control is obtained as the basic ingredient of the optimization solution simultaneously with another vector function which characterizes the best time behavior of the process state, called the optimal trajectory. The stable asymptotic solution of the optimal control problem (if it exists), is associated with a steady-state situation when the process time tends to infinity and the external controls are time independent. These asymptotic solutions are often the solutions to a corresponding static problem. However, this fact is seldom exploited directly, that is, we seldom seek solutions to the static problems as the special (steady-state) solutions of dynamic problems. The static optimization problems (mathematical programming problems) have their own methods which are simpler than those of the dynamic problems. Hence tractable static problems may involve many more components of the control vector than their dynamic counterparts.

The complexity of the optimization problems is as a rule connected with the complexity and variety of the constraints. Constraining equations and inequalities may be algebraic, difference, differential, integral, and integro-differential. For mathematical programming problems (static optimization problems) the constraints are algebraic. For variational problems (dynamic optimization) differential equations are the natural constraints although any other type of constraints specified previously can additionally appear in more involved variational problems. This may result in an extremely complex structure for dynamic optimization problems. This is not to say that static problems of mathematical programming cannot be complex and difficult to solve. These difficulties are due to the nonlinearities that may appear in the constraining equations and/or the performance index.

8.3.1 Static Methods

When both the constraining equations and the performance index are linear, the optimization problem is a linear programming problem (Gass, 1958; Dantzig, 1968; Llewellyn, 1963). The linear programming (LP) problems include transportation problems, distribution from sources to sinks, traveling salesman problems, allocation of resources among activities, management decisions, and so forth. The LP problem is usually too restrictive for thermodynamic applications where at least the objective function is nonlinear by nature.

When one or more of the constraints, or the objective function are nonlinear, the statics problem is a nonlinear programming problem. It can be formulated as follows. Determine values for n variables $\mathbf{y} = (y_1, y_2, \ldots, y_n)$ that optimize the scalar objective function

$$A(y) = A(y_1, y_2, \ldots y_n) \tag{8.1}$$

subject to l equality constraints

$$g_i(\mathbf{y}) \equiv g_i(y_1, y_2, \ldots y_n) = 0 \qquad i = 1, \ldots l \qquad (8.2)$$

and to $m-l$ inequality constraints

$$g_i(\mathbf{y}) \equiv g_i(y_1, y_2, \ldots y_n) \geq 0. \qquad i = l+1, \ldots m \qquad (8.3)$$

Nonlinear programming problems may have widely varying properties, and certain limitations on the forms of the functions appearing therein are necessary if the problems are to be solved. The Karush–Kuhn–Tucker conditions, generalizing the classical Lagrange multipliers to the case involving inequality constraints, is the basic theoretical tool for solving nonlinear programming problems (Greig, 1980; Varaiya, 1972; Kuhn and Tucker, 1951; Zangwill, 1969). The mathematical content of this condition is contained in the equations

$$\begin{aligned} \nabla A(\mathbf{y}) &= \sum_{i=1}^{m} \lambda_i g_i \\ \lambda_i &\geq 0, g_i(\mathbf{y}) \geq 0, \quad i = l+1, \ldots m \\ \lambda_i g_i &= 0, \quad i = 1, \ldots, m \end{aligned} \qquad (8.4)$$

We refer the reader to the many books available on the subject (Mangasarian, 1969; Beveridge and Schechter, 1970; Findeisen et al., 1980; Bracken and McCormick, 1968; Greig, 1980; Varaiya, 1972; Zangwill, 1969; Sieniutycz, 1978, 1991; Tsirlin, 2008, 2009). This list may be supplemented by the book by Canon et al. (1972), which is a mathematically rigorous, comprehensive treatise focused on the systematic derivation of equations of optimal control from the theory of mathematical programming. In particular, it explains subtle properties of discrete maximum principles from the standpoint of mathematical programming. Typical algorithms for the solution of the problem (8.4) proceed by absorbing the constraints, Eqs. (8.2) and (8.3), into an augmented optimization criterion.

Regarding the applications in chemistry, we mention an important special case: the isothermal and isobaric equilibrium state of a thermodynamic system is in a natural way a nonlinear programming problem where the free energy of the system is minimized. For the example of a chemical mixture, the free energy is minimized subject to the linear constraints resulting from the conservation of the atoms of the elements; the minimization determines the concentrations at chemical equilibrium (White et al., 1958; Van Zeggeren and Storey, 1970; Bailie, 1978; Li et al., 2004; Nicolis and De Wit, 2007). Minimizing the entropy production rate in chemical reactors is another example of use static optimization methods (Nummedal et al., 2003). Still another application involves the search for the maximum entropy in nonequilibrium chemically reacting mixtures (Ugarte et al., 2005). A number of static optimization methods may be applied in the design of reactors with multiple reactions (Mann, 2000). Usefullness of the principles of reaction engineering approach (REA) may be demonstrated, including computational fluid dynamics–based modeling, further expanded to include simultaneous heat and mass transfer processes (Chen and Putranto, 2013). One may also analyze a distillation process by using the first and second laws of thermodynamics with particular attention to the minimum work requirement for individual subprocesses (Cerci, 2001). Novel computational techniques with high accuracy and efficiency are being introduced to experimental identification of biochemical processes (Demirel, 2010) and to computational algorithms in medicine (Zarei et al., 2012, 2013).

The Kuhn–Tucker method is very general and hence not always the most effective. Sometimes, when only simple equality constraints are present, a number of variables (equal to the number of the equality constraints) can be eliminated and the problem can be reduced to the problem of extremizing an unconstrained function of the remaining variables. For this case (unconstrained optimization of a multivariable function) a variety of iterative nongradient as well as gradient techniques can be used to find the optimum (Rosenbrock and Storey, 1966; Findeisen et al., 1980). Most frequently they are based on iterative searches for optima along certain directions in the decision space. These searches start from an arbitrary point and terminate close to the optimum. Many reviews of these methods and their applications are available (Beveridge and Schechter, 1970; Findeisen et al., 1980; Sieniutycz and Szwast, 1982) along with the associated proofs of convergence (Zangwill, 1969). The constraints can be taken into account not only by the Lagrange multiplier approach, as in the Kuhn and Tucker (1951) method, but also by introducing various penalty terms. These add terms to the objective function which penalize violations of the constraints by increasing the objective if any constraint is violated. This forces any search procedure to leave this region quickly (Bracken and McCormick, 1968). Differential geometry based continuation algorithms were developed for separation process applications (Riona and Van Brunta, 1990).

8.3.2 Dynamic Methods

In his book on optimization of chemical reactors Aris (1961) described a number of problems associated with maximum conversion for prescribed duration or with a minimum holdup time for preassigned conversion. These methods involve objective functions in the form of functionals rather than functions and can be further subdivided into discrete and continuous. We begin by addressing methods for discrete problems. Such problems are also referred to as multistage decision processes. Perhaps the widest used method for such processes is the discrete version of the dynamic programming method (Bellman, 1957, 1961, 1967; Bellman and Kalaba, 1965). The dynamic programming (DP) is based on the Bellman principle which asserts that at each stage of the process we need compare only the costs of the current decision plus the minimal cost from the state to which this decision moves us. Unfortunately, the effectiveness of numerical algorithms implementing this method work effectively only when the dimensionality of the state vector is low. Hence, for larger dimensional problems, other methods are needed.

Another popular method, free from DP flaws, is a discrete version of the Pontryagin maximum principle (Fan and Wang, 1964; Halkin, 1966; Sieniutycz, 1978). The discrete problem is defined by an objective function in the form of a sum over the costs of single-stages, and constraints as the difference equations (some local algebraic constraints for the decisions and/or state variables at each stage may also appear). The corresponding optimization solution comprises the sequence of the optimal controls and the sequence of the optimal state variables (the discrete trajectory). The book by Canon, Cullun, and Polak (Canon et al., 1972) offers the systematic derivation of equations of optimal control from the theory of mathematical programming. In particular, it explains subtle properties of discrete maximum principles from the standpoint of mathematical programming. The book by Athans and Falb (1966) focuses on Pontryagin's maximum principle for constrained optimal control systems.

When the number of the discrete intervals (number of process stages) and the initial and final states and times of the process remain the same, the discrete problem tends to a limiting continuous problem. This limit is the primary goal in many research approaches. In the continuous case the optimal objective function A^0 (the cumulative value of the optimum cost corresponding to the final state $\mathbf{X}^f$ and the final time t^f) can be viewed as a function of all components of the current state, $\mathbf{X} = (x_0, x_1,...., x_n) = (x_0, \mathbf{x})$, and the current time t. This is consistent with Bellman's optimality principle (Bellman, 1957, 1967).

In the Mayer formulation of the continuous control problem, considered here, the objective function A is the zero-th component of the enlarged state vector, $\mathbf{X} = (x_0, x_1, \ldots, x_n) = (x_0, \mathbf{x})$, at some final time t^f, that is, $A = x_0(t^f)$. The optimization is subject to the following constraints

$$\frac{d\mathbf{X}}{dt} = \mathbf{F}(\mathbf{X}, \mathbf{u}, t)$$
$$\mathbf{x}_k(t^i) = \mathbf{x}_k^i, \quad k \in \mathbf{K}$$
$$\mathbf{x}_j(t^f) = \mathbf{x}_j^f, \quad j \in \mathbf{J}$$
$$\mathbf{u} \in \mathbf{U}$$

(8.5)

with the $n + 1$ dimensional enlarged state vector $\mathbf{X}$, enlarged rate vector $\mathbf{F} = (f_0, f_1, \ldots f_n)$, r-dimensional control vector $\mathbf{u}$, and the time t. The continuous problem, described previously, pertains to the time evolution of a continuous system with an external control, under the specified boundary conditions. These conditions usually prescribe the initial time and some of the coordinates x_k of the state vector at the initial time and another part of the state coordinates x_j at its end. The zero-th component must of course be unspecified as this is the objective to be maximized or minimized. This is the classical problem of optimal control as formulated by Pontryagin. It is best attacked by the very general and powerful optimization method called the maximum principle (Pontryagin et al., 1962; Leitman, 1981;

Fan and Wang, 1964). The name of the method is associated with the following basic property of optimal solutions: in order to make a performance coordinate x_0 maximum (minimum) over the whole path, the Hamiltonian function,

$$H = \mathbf{ZF}(\mathbf{X}, \mathbf{u}, t) \equiv z_0 f_0 + \mathbf{zf}(\mathbf{X}, \mathbf{u}, t) \tag{8.6}$$

must attain the global maximum (minimum) with respect to the admissible control vector **u** at each time instant on the optimal path. The scalar f_0 is the growth rate of a generalized profit (eg, intensity of power yield in engine case, see an example subsequently).

The Hamiltonian H is defined as the scalar product of the rate vector **F** and the so-called adjoint vector, $\mathbf{Z} = (z_0, \mathbf{z})$. Note that the definition of H includes the zero-th component, that is, the rate of change of the cost and its adjoint. The latter can be taken equal to 1 along the whole path when the rates **F** do not contain the variable x_0 explicitly. The origin of the adjoint variables is seen in the Lagrange multiplier representation of the control problem; the adjoint variables are just the Lagrange multipliers associated with the constraints resulting from the state equations. For specific cases, other suitable interpretations of the adjoint variables are possible, for example, as the partial derivatives of the optimal cost function or as counterparts of the momenta in classical mechanics.

The properties of the optimal solution in terms of the Hamiltonian are described in terms of the canonical equations for the state and the adjoint variables:

$$\frac{d\mathbf{Z}}{dt} = -\frac{\partial H}{\partial \mathbf{X}} \tag{8.7}$$

and

$$\frac{d\mathbf{X}}{dt} = \frac{\partial H}{\partial \mathbf{Z}} \equiv \mathbf{F}(\mathbf{X}, \mathbf{u}, t) \tag{8.8}$$

plus the extremum condition on $H(\mathbf{X}, \mathbf{Z}, \mathbf{u}, t)$ with respect to the controls **u**. Along the extremal path, $H^e(\mathbf{X}, \mathbf{Z}, t) = \max_\mathbf{u} H(\mathbf{X}, \mathbf{Z}, \mathbf{u}, t)$. For an interior maximum of H, this condition has the form:

$$\frac{\partial H}{\partial \mathbf{u}} = 0 \tag{8.9}$$

The boundary conditions for the canonical set (8.7)–(8.9) have usually a complex two-point form depending on the specific formulation of the problem; part of the conditions is specified at the beginning and part at the end of the process. For instance, if the initial state of the system is completely specified, the initial conditions specify $\mathbf{X}(t^i)$. The right-end (final) conditions have the simplest form when the final process state is free (unconstrained). This state should then be found as a component of the optimal solution. In this case, the final conditions deal only with the components of the adjoint vector and have the form $z_0(t^f) = 1$, as well as $z_k(t^f) = 0$ for $k = 1, 2, \ldots, n$. However, more complex transversality conditions appear when the final state coordinates are constrained to lie on some final submanifold in the state space (Pontryagin et al., 1962; Leitman, 1966, 1981; Sieniutycz, 1978, 1991).

One consequence of the canonical equations of the continuous problem is an equation describing the time dependence of the optimal Hamiltonian function

$$\frac{dH}{dt} = \frac{\partial H}{\partial t} \qquad (8.10)$$

This equation shows that the Hamiltonian has the constancy property along any optimal path when the system does not involve the time explicitly. This represents, of course, the conservation of energy for conservative systems of mechanics, that is, the existence of a definite first integral for any autonomous physical system. If the final time is left unspecified (free-end time problem), then the final value of the Hamiltonian must vanish, that is, $H(t^f) = 0$.

A discrete principle can be formulated which, for convex systems, is strongly analogous to Pontryagin's principle provided that the time increments

$$\theta^n = t^n - t^{n-1} \qquad (8.11)$$

at the stages of the discrete process are unconstrained except for a possible constraint on their sum, that is on the total time (Sieniutycz, 1978; Sieniutycz and Szwast, 1982). The freedom of choosing θ^n allows one to formulate the discrete optimal control problem in the canonical form governed by the Hamiltonian

$$H^{n-1} = \mathbf{Z}^{n-1}\mathbf{F}^n(\mathbf{X}^n, \mathbf{u}^n, t^n) \qquad (8.12)$$

obeying the following canonical equations (note the presence of both continuous time t and discrete time n in the equations)

$$\frac{\mathbf{Z}^n - \mathbf{Z}^{n-1}}{t^n - t^{n-1}} = -\frac{\partial H^{n-1}}{\partial \mathbf{X}^n} \qquad (8.13)$$

and

$$\frac{\mathbf{X}^n - \mathbf{X}^{n-1}}{t^n - t^{n-1}} = \frac{\partial H^{n-1}}{\partial \mathbf{Z}^{n-1}} \equiv \mathbf{F}^n(\mathbf{X}^n, \mathbf{u}^n, t^n) \qquad (8.14)$$

$$\frac{\partial H^{n-1}}{\partial \mathbf{u}^n} = 0 \qquad (8.15)$$

$$\frac{H^n - H^{n-1}}{t^n - t^{n-1}} = \frac{\partial H^{n-1}}{\partial t^n} \qquad (8.16)$$

The boundary conditions for this discrete set have a form which is similar to the continuous version, discussed previously. As distinguished from the continuous version, the relationship characterizing the change of optimal Hamiltonian, Eq. (8.16), does not result from the canonical Eqs. (8.13) and (8.14) but it constitutes an additional optimality condition associated with the optimal choice of the (unconstrained) time increments $\theta^n = t^n - t^{n-1}$. Eq. (8.16) is here the optimality condition coming from the additional freedom of choosing the time increments, θ^n. The use of this discrete formalism is illustrated on examples in an excellent article by Szwast treating a class of multistage fluidized drying processes (Szwast, 1990).

The maximum principle reduces the global problem of the maximizing the functional to the problem of the maximizing the function H at each instant. In the most frequent case, when an analytical solution is impossible or very difficult, these methods allow one to organize an approximating numerical scheme using a large but finite number of the time instants which give a discrete approximation for the optimal path and the corresponding optimal control.

Changes in the boundary conditions do not influence the form of the Hamiltonian function and the canonical equations. Nonetheless, they may drastically simplify or complicate the associated numerical solution. The two-point boundary conditions appearing as a rule in problems of this kind are quite involved from the computational viewpoint, and require the use various iterative approaches to gain an acceptable match between the computed and the prescribed boundary values. These methods are comprehensively described in the textbooks on optimal control (Lee and Marcus, 1967; Beveridge and Schechter, 1970; Leitman, 1981; Sieniutycz, 1978, 1991). Aris (1964) presents examples of dynamic programming applications in the field of chemical reactors.

Other optimization conditions may be derived from the maximum principle. The continuous version of the dynamic programming leads to a partial differential equation of the Hamilton–Jacobi type with an additional condition of the Erdmann–Weierstrass type which expresses the maximality of H with respect to the control vector $\mathbf{u}$ (Bellman, 1967).

The Euler equations of variational calculus can be recovered by investigating an extremum of the objective $x_0(t^f)$. The canonical Eqs. (8.7) and (8.8) and the use of Eq. (8.9) to determine the stationary momenta yield

$$\frac{d}{dt}\left(\frac{\partial L}{\partial \dot{\mathbf{x}}}\right) = \frac{\partial L}{\partial \mathbf{x}} \qquad (8.17)$$

which are the well-known Euler equations of the variational calculus, written in vector form. They are suitable for describing the reversible microscopic motion of elementary particles or the frictionless motion of macroscopic bodies in inviscid media or in a vacuum. See chapters: Contemporary Thermodynamics for Engineering Systems and Variational Approaches to Nonequilibrium Thermodynamics, regarding attempts to describe some dissipative processes by equations of motion following in the form of the Euler equations of variational calculus or Hamilton's canonical equations. Large effort of many researchers in the theory and application of variational principles for macroscopic systems is summarized in a book (Sieniutycz and Farkas, 2005).

Although we have introduced here maximum principles as alternative methods, apparently independent of dynamic programming, it should be kept in mind that, under differentiability assumptions, maximum principles may be derived from the dynamic programming approaches (Boltyanski, 1971, 1973; Fan, 1966; Findeisen et al., 1980; Sieniutycz, 1978, 1991). Moreover, in terms of the optimal performance function V, defined as a maximum value of the optimized functional, the so-called Hamilton–Jacobi–Bellman (HJB) equation and the Hamilton–Jacobi (HJ) equation can be obtained as the locally sufficient optimality conditions.

In fact, there are various optimization algorithms whose solutions converge to the solutions of continuous HJB equations and HJ equations. However, classical differentiable solutions have their own analytical theory. This theory deals with differentiable value function $V(\mathbf{x}, t)$ or function $R(\mathbf{x}, t) = -V(\mathbf{x}, t)$, and is outlined here. Both equations may be derived by continuous dynamic programming (Bellman, 1961; Fan, 1966; Findeisen et al., 1980; Sieniutycz, 1991). Yet, for brevity, a simple approach exploiting Caratheodory's idea of potentiality of optimal performance function (Rund, 1966) is applied subsequently.

We shall derive the Hamilton–Jacobi–Bellman equation and exemplify the procedure on a power yield process. The optimization problem in this case is the maximum power delivery from a thermal engine whose characteristics are shown in Fig. 8.2. They describe driving heat q_1, power output p and entropy production, S_{prod}, in terms of thermal efficiency η.

In general terms, the optimization problem is governed by the characteristic function

$$V(x^i, t^i, x^f, t^f) \equiv \max_{u(t)} \int_{t^i}^{t^f} f_0(\mathbf{x}, \mathbf{u}, t) \, dt \tag{8.18}$$

which contains state vector, $\mathbf{x}$, control vector, $\mathbf{u}$, and time t.

In the power example $\mathbf{x} = T$ is a state coordinate representing the variable bulk temperature of the resource fluid, and $\mathbf{u} = T'$ is a control variable called Carnot temperature (Sieniutycz, 2003b, 2011c). In terms of the thermal efficiency η, the Carnot temperature in any irreversible case can be calculated as $T' = (1-\eta)\Phi T^e$, where is T^e a constant temperature of the environment or of a lower reservoir, and Φ is the internal irreversibility coefficient.

As it follows from the definition of the maximum performance function V

$$\max_{u(t)} \left(\int_{t^i}^{t^f} f_0(\mathbf{x}, \mathbf{u}, t) \, dt - V(x^i, t^i, x^f, t^f) \right) = 0 \tag{8.19}$$

Here $f_0 = -l_0$, is the profit generation rate, or, in the power yield example, the intensity of power production. Differentiation of Eq. (8.19) with respect to the upper limit of the integral, t^f, yields

$$\max_{u(t)} \left(\int_{t^i}^{t^f} f_0(\mathbf{x},\mathbf{u},t) - \frac{dV(x^i,t^i,x^f,t^f)}{dt^f} \right) = \max_{u(t)} \left(\int_{t^i}^{t^f} f_0^f(\mathbf{x},\mathbf{u},t) - \frac{\partial V}{\partial t^f} - \frac{\partial V}{\partial x^f} \mathbf{f}^f(\mathbf{x},\mathbf{u},t) \right) = 0 \quad (8.20)$$

Observe that all rates (f_0 and $\mathbf{f}$) and partial derivatives of V are evaluated at the final state (the so-called 'forward equation'). In the second expression, total time derivative is expanded in terms of all rates. Now, partial derivative of the characteristic function V with respect to time can be taken off this equation and the superfluous index f can be omitted (variable final states). Then, after rejecting superscript f, replacing V by $-R$, and introducing the Lagrangian $l_0 = -f_0$, a Hamilton–Jacobi–Bellman equation (HJB equation) is obtained

$$\frac{\partial R}{\partial t} + \max_{u(t)} \left(\frac{\partial R}{\partial \mathbf{x}} \mathbf{f}(\mathbf{x},\mathbf{u},t) - l_0(\mathbf{x},\mathbf{u},t) \right) = 0 \quad (8.21)$$

It is consistent with Eq. (8.20) and incorporates the definitions, $l_0 = -f_0$ and $R = -V$.

For the power yield example, when a driving resource relaxes with linear kinetics (Newtonian heat exchange), the Lagrangian $l_0 = c(1-T^eT'^{-1})(T'-T)$ contains Carnot temperature control T' (Sieniutycz, 2003b, 2011c; Sieniutycz and Jeżowski, 2009, 2013). The kinetics of heat transfer refers to a nondimensional time τ rather than to the usual time t. Eq. (8.21) then yields

$$\frac{\partial R}{\partial \tau} + \max_{T'(t)} \left\{ \left(\frac{\partial R}{\partial T} - c\left[1 - \frac{T^e}{T'}\right] \right)(T'-T) \right\} = 0 \quad (8.22)$$

The solution of this equation implies an exponential decrease of the resource temperature T in time, caused by the finiteness of the resource fluid. Also the Carnot temperature T' decreases exponentially in time in this case. The details are available elsewhere (Sieniutycz and Jeżowski, 2009, 2013).

We may exploit the Hamiltonian formula, Eq. (8.6), and the definition of state adjoints in terms of the partial derivatives value function, $z_0 = \partial R/\partial x_0 = 1$ and $\mathbf{z} = \partial R/\partial \mathbf{x}$, to express Eq. (8.6) in the form

$$H(z,x,u,t) = \sum_{i=1}^{s} \left(\frac{\partial R}{\partial x_i} \right) f_i(x,u,t) - l_0(x,u,t) \quad (8.23)$$

that contains state vector $\mathbf{x}$, control vector $\mathbf{u}$, rates f_0 and f_i and state adjoints $z_i = \partial R/\partial x_i$. Eq. (8.23) has the structure of the energy of a mechanical system. Nonoptimal Hamiltonian

H should therefore become in optimal case an energy-type quantity which is constant along an optimal path of an autonomous system. A quantity associated with H is the state adjoint z_k that may be interpreted as a momentum-type variable. For k-th coordinate x_k

$$z_k \equiv \frac{\partial R}{\partial x_k}. \tag{8.24}$$

In terms of $\mathbf{z} = \partial R/\partial \mathbf{x}$, $\mathbf{x}$, and t in function H, a general Hamilton–Jacobi–Bellman equation can be written in the form

$$\frac{\partial R}{\partial t} + \max_{\mathbf{u}(t)} H\left(\frac{\partial R}{\partial \mathbf{x}}, \mathbf{x}, \mathbf{u}, t\right) = 0 \tag{8.25}$$

In view of the linearity of Eqs. (8.23) and (8.25) with respect to $\partial R/\partial \mathbf{x}$, Hamilton–Jacobi–Bellman equation is a quasilinear partial differential equation with a maximum operation.

Maximizing the Hamiltonian with respect to vector $\mathbf{u}$ yields the optimal control $\mathbf{u}$ in terms of the variables $\partial R/\partial \mathbf{x}$, $\mathbf{x}$, and t

$$\mathbf{u}\left(\mathbf{x}, \frac{\partial R}{\partial \mathbf{x}}, t\right) \equiv \arg\max_{\mathbf{u}(t)} H\left(\frac{\partial R}{\partial \mathbf{x}}, \mathbf{x}, \mathbf{u}, t\right) \tag{8.26}$$

Eq. (8.26) means that the optimal control $\mathbf{u}$ is an argument of the Hamiltonian maximization with respect to $\mathbf{u}$. When the feedback control function (8.26) is substituted into the Hamilton–Jacobi–Bellman Eq. (8.25), an equation known as the Hamilton-Jacobi equation is obtained (Benton, 1977)

$$\frac{\partial R}{\partial t} + H\left(\frac{\partial R}{\partial \mathbf{x}}, \mathbf{x}, t\right) = 0 \tag{8.27}$$

This equation is known from analytical mechanics; it does not contain controls $\mathbf{u}$. It is a nonlinear partial differential equation whose solution leads to the optimal cost function $R(\mathbf{x}, t)$.

For discrete models, the theory of Hamilton–Jacobi–Bellman equations can be restated and live with its own life in the realm of difference equations, sums, recurrence relations, two-stage criteria, and so forth, often achieving a form quite dissimilar to the original HJB theory (Sieniutycz, 2000a). Numerical methods apply discrete optimization models for given rates f_0 (or l_0) and $\mathbf{f}$. These models are discrete by nature when they represent a real system with a few finite-size stages, or constitute discrete approximations of real continuous systems, arranged for computational purposes. Examples are especially provided from the FTT engines theory (Sieniutycz, 1999, 2000a). Working mainly in the discrete context, Sieniutycz (2000a) enunciates parallelism for structures of optimization in mechanics and thermodynamics in terms of the duality for thermoeconomic problems of maximizing of production profit and net profit which can be transferred to duality for least action and least abbreviated action which appear in mechanics. With the parallelism in mind, he reviews theory and macroscopic applications of a recently developed discrete formalism of Hamilton–Jacobi type which arise when Bellman's method of dynamic programming is applied to optimize active (work producing) and inactive (entropy generating) multistage energy systems.

Books and monographs on the optimization theory and dynamic programming method describe the link between Hamilton–Jacobi–Bellman Eq. (8.25) and discrete algorithms of dynamic programming (Bellman, 1961; Fan, 1966; Findeisen et al., 1980; Sieniutycz, 1991, 2000a). Consequently equations modeling continuous processes can be solved numerically by the discrete algorithms of dynamic programming which are equivalent with the Hamilton–Jacobi–Bellman Eq. (8.21) or (8.25). For instance, one can apply the DP algorithm which solves the following Bellman's recurrence equation

$$R^n(x^n,t^n) = \min_{u^n,\theta^n}\{l_0^n(x^n,t^n,u^n,\theta^n)\theta^n + R^{n-1}[(x^n - f^n(x^n,t^n,u^n,\theta^n)\theta^n, t^n - \theta^n]\} \quad (8.28)$$

(Sieniutycz and Jeżowski, 2009, 2013). Fig. 8.4 shows an example of a computational block scheme describing the optimization principle for this multistage process by the dynamic programming method (forward algorithm). In accordance with Bellman's optimality principle, ellipse-shaped balance areas pertain to sequential subprocesses which grow by inclusion of proceeding units.

Important issue of convergence conditions of discrete solutions to nonclassical (viscosity) solutions is discussed in detail elsewhere (Sieniutycz and Jeżowski, 2009, 2013, their chapter: Frictional Fluid Flow Through Inhomogeneous Porous Bed). Only a short information is given subsequently.

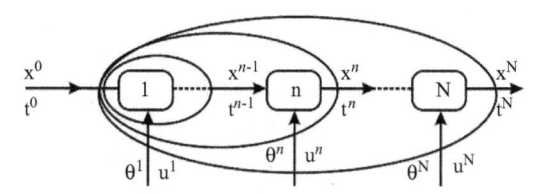

Figure 8.4: Block scheme describing the optimization calculations of a multistage process by the dynamic programming method (forward algorithm).

8.3.3 Viscosity Solutions

Viscosity solutions provide a suitable framework to study those Hamilton–Jacobi–Bellman equations or Hamilton-Jacobi equations which do not admit classical continuous solutions (Crandall and Lions, 1986; Crandall et al., 1992; Bardi and Capuzzo-Dolcetta, 1997; Chen and Su, 2003). The dynamic programming context for the viscosity solutions has been worked out (Capuzzo Dolcetta and Falcone, 1989; Bardi and Capuzzo-Dolcetta, 1997). Many contemporary papers provide techniques for approximating the viscosity solution of a HJB or HJ equation in deterministic control problems involving discrete dynamic programming. Monotonically convergent schemes in computational algorithms were obtained and error estimates were proved. The approximate solutions converge uniformly to the viscosity solution of the original problem. Stochastic control problems (Cadenillas and Karatzas, 1995) can also be treated using the dynamic programming approach. For control Markov diffusion HJB equations are nonlinear PD equations of second order. Viscosity solutions provide a suitable framework to study these HJB equations. Generalized solutions of Hamilton–Jacobi equations were given (Lions, 1982; Crandall and Lions, 1986; Crandall et al., 1992, and many others).

Clarke (1997) presents many contemporary aspects of nonsmooth analysis and control theory. Summaries of approaches to viscosity solutions are in books (Fleming and Soner, 1993; Bardi and Capuzzo-Dolcetta, 1997; Falcone and Makridakis, 2001). The last book addresses numerical methods for viscosity solutions. Bauer et al. (2006) presented an adaptive spline technique to solve HJB equations.

8.4 Sorption Models for Minimal Catalyst Deactivation

In this section we limit ourselves to thermodynamic aspects of catalyst deactivation and a discussion of a few engineering applications. Froment and Bischoff (1979) is the standard text on chemical reaction engineering, beginning with basic definitions and fundamental principles and continuing all the way to practical applications. Levenspiel (1965) and Fogler (2006) give necessary overall information on the role of catalysts in chemical reaction engineering. Butt and Petersen (1988) treat various mechanisms associated with activation, deactivation and poisoning of catalysts.

A basic deactivation mechanism associated with sorption processes is that in which the irreversible occupation of active sites (blockage) occurs through selective adsorption and subsequent degradation to carbonaceous material (coke) of some of the process components. To describe deactivation through coke deposition, mechanistic kinetic models have been derived since 1970s that take into account deactivation as a series of reaction steps together with the main reaction. Beeckman and Froment (1980) describe the deactivation of a catalyst by coke deposition in terms of two mechanisms: site coverage and pore blockage. Their relation between the deactivation function and the coke content is derived for single ended pores, for pores open at both ends and for networks of pores. Hughes (1984) has nicely synthesized early results on deactivation of catalysts. A great deal of original research on catalyst deactivation is contained in proceedings of international symposia (Delmon and Froment, 1980, 1987).

For a batch reactor Szepe (1966) and Szepe and Levenspiel (1968, 1971) obtained important conditions regarding decreasing temperature profiles in catalyst deactivation, interpreted later in terms of necessity of catalyst saving (Szwast, 1994). Bykov and Yablonskii (1980) described simulation and optimization of catalytic processes with varying activity on the basis of a detailed kinetic model. General scheme of optimal control for these processes was proposed. The Pontryagin's maximum principle was chosen as method of optimization. A possibility of critical phenomena related to catalyst activity changes was shown. Corella and Asua (1982) formulated and analyzed kinetic equations of mechanistic type with nonseparable variables for catalyst deactivation by coke (Froment and Bischoff, 1979; Beeckman and Froment, 1980; Corella and Asua, 1982; Hughes, 1984; Butt and Petersen, 1988). Beuther et al. (1980) investigated the mechanism of coke formation in the catalytic processing of heavy oils. Extensions have been made of these models taking into account aspects such as: various deactivation mechanisms (Corella and Monzon, 1988); the existence of the limiting residual activity and a maximum coke content (Corella et al., 1988); the activation of catalyst (Agorreta et al., 1991; Szwast, 1994; Gayubo et al., 1993a, b); selectivity in deactivation of each individual main reaction in complex reaction systems (Corella et al., 1985); nonhomogeneity of active sites Corella and Menéndez, 1986), who also presented application to the deactivation of commercial catalysts in the FCC process. Megiris and Butt (1990a, b) considered effects of poisoning on the dynamics of fixed bed reactors. Agorreta et al. (1991) applied a kinetic model for simultaneous activation and deactivation processes in solid catalysts to kinetic data obtained in isopropyl alcohol dehydrogenation on a Cu/SiO_2 catalyst. They have shown influence of different catalyst pretreatments on the relevant kinetic parameters of the system. A data analysis method for the deactivation of a Ni catalyst, used in the benzene hydrogenation, where the deactivation occurs by a poisoning step that is independent of the main reaction, has been proposed. In this case the deactivation fits to kinetics of separable functions that is first order with respect to the poison (tiophene) concentration. Consequently, it is obvious that in the case of the sorption models of catalyst deactivation the treatment of the reaction and deactivation kinetics requires careful attention being paid to the effect of the concentration of all the reaction components. Morbidelli et al. (2001) consider in their chapter: Classical and Anomalous Diffusion, mechanisms of catalyst deactivation followed by an insightful treatise on optimal distribution of catalyst in pellets, reactors, and membranes as a basis to catalyst design. Sieniutycz (1978, 1991) presented formulations and solutions of several optimization problems with catalyst deactivation and regeneration, including those in moving bed reeactors. Regeneration may be attributed to various processes where a substance (catalyst, sorbent, solvent, protective layer, etc.) undergoes in turn a regeneration and a decay. Regeneration of valuable properties is in general costly, but the deeper regeneration is, the better chemical yield is achieved. Its research involves the mathematical modeling, simulation, and optimization.

Fractal models of catalysts and porous systems appeared. Lenormand (1989) showed that the problems for application of the different models of fluid transport in porous media and of fractal concepts are of three types: no validity of the models due to viscous forces always present in both fluids at large scale, difficulty to use fractal geometry as a model of reservoir and finally the need for analytical or differential equations. Sieniutycz (2009c) investigated power production in complex multireaction systems propelled by either uncoupled or coupled multicomponent mass transfer. The considered system contained two mass reservoirs, one supplying and one taking out the species, and a power-producing reactor undergoing the chemical transformations characterized by multiple (vector) efficiencies. An approach was applied that implements balances of molar flows and reaction invariants to complex chemical systems with power production. Reaction invariants, that is, quantities that take the same values during a reaction, follow by linear transformations of molar flows of the species. Flux balances for the reacting mixture were written down by equating these reaction invariants before and after the reactor. Obtained efficiency formulas were applied for steady-state chemical machines working at the maximum production of power. Total output of produced power was maximized at constraints which take into account the (coupled or uncoupled) mass transport and efficiency of power generation. Special attention was given to nonisothermal power systems, stoichiometric mixtures, and internal dissipation within the chemical reactor. Optimization models lead to optimal functions that describe thermokinetic limits on power production or consumption and extend reversible chemical work W_{rev} to situations in which decrease of chemical efficiencies, caused by finite rates, is essential. The classical thermostatic theory of reversible work is recovered from the present thermokinetic theory in the case of quasistatic rates and vanishing dissipation.

Comparing with nonreacting systems, thermodynamic optimization approaches entered the realm of chemical reactors with a delay. Nevertheless, Ondrechen et al. (1980a) studied FTT of chemical problems and Plath (1989) was very apt with his volume on optimal structures in heterogeneous reaction systems. Ondrechen et al. (1980b); Ondrechen (1990) applied FTT to determine power yield of chemically driven engines. Månsson (1985) determined optimal temperature profile for an ammonia reactor. Nummedal et al. (2003) minimized the entropy production rate of an exothermic reactor. McGlashan and Marquis (2007) put forward availability analysis of postcombustion carbon capture systems to select unavoidable irreversibilties. Mert et al. (2007) worked out the exergoeconomics of a proton exchange membrane (PEM) fuel cell transportation system. The analysis includes the PEM FC stack and system components such as compressor, humidifiers, pressure regulator and the cooling system. The system efficiency increases with temperature and pressure and a decrease in membrane thickness. Assuming natural gas as a reference fuel, Milewski and Lewandowski

(2009) substantiated the use of biogases as fuels for solid oxide fuel cells (SOFC). Milewski et al. (2009) used molten carbonate fuel cell (MCFC) to reduce CO_2 emissions from a coal fired power plant. Moroń et al. (2013) conducted a coal experiment for combustion in oxy-fuel atmosphere and investigation of emissions of NO_x, SO_2, fly ash burnout, and kinetic parameters for nitrogen evolution during combustion in oxy fuel. Lems et al. (2003) specified admissible policies for optimization of energy transfer in chemical and biochemical reaction systems. Rao et al. (2004) selected and analyzed thermodynamic efficiencies of electrochemical systems (chapter: Power Limits in Thermochemical Units, Fuel Cells and Separation Systems). Orlov and Berry (1990, 1993) gave analytical expressions for the upper bounds of power and efficiency of a heat engine and an internal combustion engine by taking into account finite rate of heat exchange and chemical reactions. With these results they suggested recommendations of theoretically possible ways of improving internal combustion engines. Riekert (1979) treats practical aspects of large chemical plants.

Szepe (1966); Szepe and Levenspiel (1968), and Park and Levenspiel (1976a, b) developed a few practical approaches to catalyst deactivation and related optimization problems. A decreasing temperature policy to save catalyst in time, first obtained by Szepe (1966) for a batch reactor, appears now frequently as a component of optimization solutions in more complex systems. Sieniutycz (1978, 1991) show this property in optimization solutions for moving bed reactors. Szwast (1994) worked out the comprehensive treatment for optimization of chemical reactions in tubular reactors with moving catalyst deactivation. Szwast and Sieniutycz (1998, 1999) treated optimization of the reactor-regenerator system with catalytic parallel-consecutive reactions. Systems with moving deactivating catalyst, composed of a cocurrent tubular reactor and a catalyst regenerator with an additional flux of a fresh catalyst, has been investigated (Szwast and Sieniutycz, 1998, 1999). For the temperature dependent catalyst deactivation, the optimization problem has been formulated in which a maximum of a process profit flux is achieved by a best choice of temperature profile along tubular reactor, best catalyst recycle ratio, and best catalyst activity after regeneration. The set of parallel-consecutive reactions, A + B→R and R + B→S, with desired product R has been taken into account. A cocurrent tubular reactor with temperature profile control and recycle of moving deactivating catalyst has been investigated (Szwast and Sieniutycz, 2001). For the temperature dependent catalyst deactivation, the optimization problem has been formulated in which a maximum of a profit flux is achieved by the best choice of temperature profile and residence time of reactants for the set of catalytic reactions A + B→R and R + B→S with desired product R. The dynamic unsteady-state reactor process in a cocurrent tubular reactor with single–run reagents (continuous phase) and with multirun moving catalyst (dispersed phase) has been investigated (Szwast and Sieniutycz, 2004). For the temperature-dependent catalyst deactivation the reactor process has been considered in which after optimal number B of catalyst residences in the reactor the whole amount of catalyst leaves the system; then the fresh catalyst is directed to the reactor and the next B-runs cycle of the optimal process starts. Optimization problem has been formulated in which a maximum of an average (for one cycle) process profit flux is achieved by a best choice of number B of catalyst residences in the reactor and best choice of temperature profiles along tubular reactor for each catalyst run, $b = 1,\ldots B$, respectively. Zalewski and Szwast (2008) considered chaotic behavior in a stirred tank reactor with first-order exothermic chemical reaction $A \rightarrow B$ undergoing in a environment of deactivating catalyst particles. The reactor considered in their paper is a continuous one in respect to reacting mixture and periodic one in respect to catalyst particles. A temperature dependent catalyst deactivation with the Arrhenius–type equation describing catalyst deactivation rate is assumed. The direction of displacement areas in which the chaos is generated is studied. Graphs show that if the catalyst is older the areas in which the chaos is generated move toward higher temperatures.

Gayubo et al. (1993a, b) performed comprehensive parametric studies of temperature vs. time sequences to palliate deactivation in parallel and in series-parallel with the main reaction. In their work, the calculation of temperature vs. time sequences to palliate catalyst deactivation in an integral reactor has been studied either by maintaining constant the conversion at the reactor outlet in a simple reaction or by maintaining constant the concentration of a given component at the outlet in a complex reaction system. The experimental systems studied, which are a simple one (dehydration of 2-ethylhexanol) and a complex one (isomerization of *cis*-butene), have kinetic models of the Langmuir–Hinshelwood–Hougen–Watson type for the main reaction and deactivation, with deactivation by coke dependent on the concentration of the reaction components. In the reaction of dehydration of 2-ethylhexanol deactivation occurs in parallel with the main reaction and in the isomerization of *cis*-butene deactivation occurs in series-parallel with the main reaction. A parametric study has been carried out for both reaction systems. The sequences calculated have been experimentally proven in an automated reaction apparatus.

Gayubo et al. (1993a, b) also optimized temperature-time sequences in reaction-regeneration cycles involving the process of isomerization of *cis*-butene. Various aspects of the practical applications of kinetic models in an integral reactor have been comprehesively analyzed. For example, in the triangular isomerization reaction of n butenes

$$\begin{array}{c} 1-\text{butene} \\ r_{t1} \updownarrow r_{1t} \qquad\qquad r_{1c} \updownarrow r_{c1} \\ \\ r_{tc} \\ t-\text{butene} \;\; \begin{array}{c}\rightarrow\\\leftarrow\end{array} \;\; c-\text{butene} \\ r_{ct} \end{array} \qquad (8.29)$$

the kinetic equation for each one of the elementary steps, characterized by the reaction rates r_{ij} which refer to the formation of the j species from the species i, is represented by the expression

$$(r_{ij})_0 = \frac{k_{ij} K_i P_i}{1 + K_1 P_1 + K_c P_c + K_t P_t} \qquad (8.30)$$

From this equation the formation (production) rate equations for each one of the species that take part in the reaction are

$$(r_1)_0 = -(r_{1t})_0 + (r_{t1})_0 - (r_{1c})_0 + (r_{c1})_0 = \frac{-k_{1t} K_1 (P_1 - P_t/K_{1t}) - k_{1c} K_1 (P_1 - P_c/K_{1c})}{1 + K_1 P_1 + K_c P_c + K_t P_t} \qquad (8.31)$$

$$(r_c)_0 = (r_{1c})_0 - (r_{c1})_0 - (r_{ct})_0 + (r_{tc})_0 = \frac{k_{1c} K_1 (P_1 - P_c/K_{1c}) - k_{ct} K_c (P_c - P_t/K_{ct})}{1 + K_1 P_1 + K_c P_c + K_t P_t} \qquad (8.32)$$

$$(r_t)_0 = (r_{1t})_0 - (r_{t1})_0 + (r_{ct})_0 - (r_{tc})_0 = \frac{k_{1t} K_1 (P_1 - P_c/K_{1t}) - k_{ct} K_c (P_c - P_t/K_{ct})}{1 + K_1 P_1 + K_c P_c + K_t P_t} \qquad (8.33)$$

where K_{1t}, K_{1c} and $K_{ct} = K_{1t}/K_{1c}$ are the equilibrium constants found by the group contribution method as definite functions of the temperature (Gayubo et al., 1993a).

The relevance and applicability of these kinetic models has been proven for a vast range of processes where the deactivation occurs by coke deposition in the isomerization reaction of the n-butenes (Gayubo et al., 1993a, b). The feeding of pure 1-butene in an integral reactor leads to the experimental curves of molar fractions versus time, in which deactivation is hardly noticeable, thus indicating that this component is not origin of the catalyst decay. However, it is shown that the decay becomes noticeable when *cis*-butene and *trans*-butene (but not l-butene) is fed and whenever the composition of one of them is noticeable. The balance of the active sites carried out by following the postulates of the Langmuir–Hinshelwood–Hougen–Watson kinetics (L–H–H–W kinetics), and in agreement with the kinetic model of the main reaction, Eq. (8.30), leads to a kinetic expression of (in general) nonseparable functions, in which *activity evolution with time is function of concentrations of all the reaction components and their adsorption equilibrium constants on the active sites*. For the reaction (8.29), from the active-site balance

$$-\frac{da}{dt} = \frac{k_{dc} K_c P_c + k_{dt} K_t P_t + k_{d1} K_1 P_1}{1 + K_1 P_1 + K_c P_c + K_t P_t} a, \qquad (8.34)$$

where $P_i = C_i RT$, $k_{d1} = 0$ for the reaction (8.29), and the catalyst activity obeys the simultaneous relations

$$a = \frac{r_{ij}}{(r_{ij})_0} = \frac{r_i}{(r_i)_0} = \frac{r_c}{(r_c)_0} = \frac{r_t}{(r_t)_0} \qquad (8.35)$$

The adduced kinetic model of deactivation governed by the sorption mechanism of is then a separable one with a kinetic equation of the L-H-H-W kinetics type for each one of the steps of the main reaction and of the deactivation. An equation having the structure of Eq. (8.34) has also been obtained for other reactions, as for example, dehydratation of 2-ethylhexanol where the kinetic model corresponds to a mechanism with control of the surface chemical reaction step on two active sites (Gayubo et al., 1993a). The resulting deactivation model assumes a sorption equilibrium constant for the adsorbed molecules of reactant which take part in the degradation reaction to coke, causing deactivation. This leads to an identification of the different energetic states in the mechanism of these molecules, adsorbed on strongly acidic sites, with respect to those adsorbed on weak (or moderately weak) acidic sites which only take part in the main reaction. It is seen that the catalyst deactivation described by the sorption models depends inherently on concentrations. The assumption of uniform deactivation of the different individual steps of Eq. (8.29) is supported by the homogeneity of the acidic strength of the catalyst used, which is constituted by sites that are weakly or moderately acidic. For the catalysts which have high levels of very strong acid sites, the limitations of Eq. (8.34) to describe the deactivation of *all* the individual reaction steps have experimentally been shown. A more complex model of the selective deactivation is then necessary, with different activities for each individual reaction step. The dynamic way of the reactor operation by the following sequences of temperature and/or flow programmed in time (reaction products sent periodically to the chromatograph and monitoring of the conversion) sets the control of the reactor response for the purpose of maintaining the constant conversion, measuring the distribution of acidic strength, and so forth (Gayubo et al., 1993a, b).

For a simple reaction system with the L–H–H–W kinetics for the main reaction and deactivation the optimal strategy in the fixed bed reactor is to maintain the constant conversion by corresponding optimal temperature-time strategies which depend on time and location in the reactor. For multireaction system the optimal strategy is to maintain constant the selectivity toward a prescribed product. This is conditioned by the difficulty in the rigorous design of an actual reactor under catalyst decay when the decay depends on concentration of the species at each reactor position. The pseudo steady-state assumption where the deactivation time constant is much larger than the fluid residence time is often operative. The temperature sequences of an isothermal reactor (with assumed spatial thermal homogeneity) maintaining the final conversion constant were calculated and the theoretical predictions of the related concentration profiles and outlet conversions were confirmed by experiments (Gayubo et al., 1993a). For the tubular flow reactor at an exact steady state the optimal strategies always depend on location only. The optimal temperature sequences have been combined with operation strategies in the context of the reaction-regeneration cycles. See reviews by Gayubo et al. (1993b) for fixed bed reactors and Szwast (1994) for tubular flow reactors. When formulationg the general optimization problem it is however enough to assume that the optimal temperature sequence is prescribed, and this observation is exploited in the present work.

A kinetic equation describing the *reactivation* of the catalyst is necessary when studying any reaction-regeneration cycles. A reactivation relationship should express the recuperated catalyst activity during the regeneration as a function of the regeneration stage variables that influence coke elimination. The reaction operating variables that cause the coke deposition can also, in principle, influence the catalyst reactivation. While there are a number of papers modeling the catalyst regeneration in terms of a coke combustion process (Długosz, 1985; Hüttner, 1983), there are fewer papers modeling regeneration processes from the difficult perspective of the catalyst activity recovery. Gayubo et al. (1993a) have studied this problem eperimentally for the kinetics (8.29). Their approximate equation that describes the function $a_0(t, T, t_c)$, relates the activity recovered after each regeneration step (simultaneously an initial activity of the reaction step) with the reaction step time t, its temperature and with the coke combustion time t_c. It has been obtained from a number of series of the cycles where, in each series a constant value of reaction time was taken.

The regeneration consistent of stripping operation with helium during a time period (necessary for the temperature to increase from the reaction step up to the coke combustion temperature, ca. 820 K) and a subsequent combustion of the residual coke with an air axcess for a variable t_c. Using this equation, an optimum of the average production rate over the whole cycle (including the fixed bed-reaction and regeneration steps) implies a partial regeneration of the catalyst (a_0 less than unity). Therefeore, in the design of reaction-regeneration cycles with quickly deactivating catalysts, the regeneration kinetics has a considerable incidence on the optimum operation conditions, as these correspond to the partial recuperation of the catalyst activity.

A single-mode regeneration of the solid catalyst or sorbent is seldom a sufficiently relevant policy. To prevent uncontrolled, too fast combustion of the coke, usually the regeneration of a solid catalyst consists of the combination of two steps: the step of the stripping treatment with an innert gas and the step of the coke combustion. Similarly the regeneration of a sorbent is composed of the desorption of an active substance(s) by an inert fluid and next by high T heating of the sorbent. Drying of solid sorbents and moist catalysts should be regarded as a sort of stripping. It is proven that the most essential recovery of the catalytic potential occurs during the stripping-type operations rather than during the combustion-type operations. Similarly a substantial recovery of the sorption potential occurs during the classical drying, not during any subsequent combustion or firing-type steps.

The analogy between the usual drying of a sorbent and the reactivation of a sorbent (catalyst) goes further. Some experiments (Gayubo et al., 1993a, b) showed that it is the strong acidic sites in the solid skeleton that are initially regenerated, the situation which corresponds to activation of the most active capillars at the beginning of drying of a capillary solid. A substantial hysteresis has been discovered in the relationship between the activity–coke-content in the reaction–regeneration cycles. This hysteresis is analogous to that well known for sorption–desorption equilibria, essential during moisturizing and drying of solids.

A thermodynamic approach treating the sorbents and capillary-porous bodies as pseudosolutions is known that proved its effectiveness when computing thermodynamic functions of solids on the basis of the gas–solid equilibria and heat effects (Hougen et al., 1954; Luikov, 1966b, 1975; Nikitina, 1968; Sieniutycz, 1973b).

A trial is made here to extend this approach to estimate the activity of the catalyst or sorbent on the thermodynamic ground. We assume that this activity is identified with the thermodynamic activity of the solid phase treated as the component of a pseudosolution composed of the solid and the substance that causes the decay of a catalyst or sorbent (coke, moisture, poison, etc.). For brevity, in what follows this substance will be called the coke. Under the condition of the constant temperature and pressure the solid-phase activity changes with the thermodynamic activity of coke according to the Gibbs–Duhem equation which can be written in the form

$$d\mu_s + W_c d\mu_c = RTM_s^{-1} d\ln a_s + W_c RTM_c^{-1} d\ln a_c = 0, \qquad (8.36)$$

where is the specific chemical potential of the solid and is the specific chemical potential of the coke. W_c is the concentration of the coke (moisture, poison) expressed as the mass of coke per unit mass of the pure (fresh) solid. The molar masses of solid and coke are respectively M_s and M_c. Now let us assume that the coke activity and its chemical potential can be measured in terms of the pressure of a volatile substance that remains in the thermodynamic equilibrium with coke. Alternatively the molar concentration of this substance $y = p_c/P$ or the ratio of its pressure p_c to its largest possible pressure (saturation pressure at the same T and P) attained at sufficiently large W_c can be used, $\beta_c = p_c/p_c^{max}(T)$. In the case of the solid drying (desorption) the quantity β_c is called the equilibrium relative humidity of the misture. Eq. (8.36) immediately yields

$$d\ln a_s = -W_c M_s M_c^{-1} d\ln p_c = -W_c M_s M_c^{-1} \left(\frac{\partial \ln p_c}{\partial \ln W_c}\right)_{P,T} dW_c = -W_c M_s M_c^{-1} \left(\frac{\partial \ln \beta_c}{\partial \ln W_c}\right)_{P,T} dW_c \qquad (8.37)$$

As the coke concentration increases its activity increases and the solid activity decreases. The thermodynamic activity of the solid a_s can be obtained by integration of the previous equation between the limits $a_s(W_c = 0) = a_s^0$ and $a_s(W_c)$. This yields an equation expressing a general rule of the decrease of the thermodynamic activity of the solid with the increase of the coke concentration W_c

$$a_s = a_s^0 \exp\left[-\int_{W_c^0=0}^{W_c} M_s M_c^{-1} \left(\frac{\partial \ln \beta_c(W_c, P, T)}{\partial \ln W_c}\right)_{P,T} dW_c\right]. \tag{8.38}$$

Should one assume that in the range of small W_c the (de)sorption equilibrium can be approximated by a straight like $\beta_c = kW_c$ (k is a positive constant slope coefficient) the partial derivative in the aforementioned equation would be equal to the unity, and, in this case, the thermodynamic activity of the solid would decrease exponentially with the coke concentration W_c, for the decay coefficient (M_s/M_c) independent of T and P

$$a_s = a_s^0 \exp[-M_s M_c^{-1} W_c]. \tag{8.39}$$

However, as the nature of the sorption equilibria implies, the assumption $\beta_c = kW_c$ is always very rough, and never holds as a good approximation. Nonetheless the experimental data of deactivation (Gayubo et al., 1993a, b) show an exponential decrease of the rate related catalyst activity ($a = r/r^0$) with the coke concentration W_c. Yet the isobaric experiments show that the curves $a_s(W_c)$ are, in fact, affected by the temperature T, the fact which proves that the oversimplified Eq. (8.39) is too rough an approximation and the general Eq. (8.38) should be more appropriate as it directly implies the relation $a_s(W_c, T, P)$ rather than $a_s(W_c)$.

But even if T, and P are effectively not present in the relation $a_s(W_c, T, P)$ the simple exponential function (8.39) is too approximate. As proved by Eq. (8.38) the form of the relation $a_s(W_c, T, P)$ depends on the shape of the sorption equilibrium curve which contains W_c as the basic independent variable. Since the shape of the function $a_s(W_c, T, P)$ at constant T and P never admits the assumption $\beta_c = kW_c$ (under which Eq. (8.39) holds), the frequently made identification of $-d\ln a/dt$ with the molar sorption rate of the coke, $M_s M_c^{-1} dW_c/dt$, implied by Eq. (8.39), is never well substantiated in the framework of the present approach. This identification is what actually persists, implicitly, in the currently used models of the deactivation (Gayubo et al., 1993a, b). As implied by Eqs. (8.37) or (8.38) the derivative $d\ln a/dt$ should be rather related to the product of the rate $M_s M_c^{-1} dW_c/dt$ and the state function, $\partial \ln \beta_c(W_c, T, P)/\partial \ln W_c$, the latter being determined by the properties of the sorption or desorption isoterm. Consequently a more exact form of the deactivation equations derived from the sorption models of deactivation has the general structure

$$\frac{da_s}{dt} = -\left(\frac{\partial \ln \beta_c(W_c, P, T)}{\partial \ln W_c}\right)_{P,T} r_s a_s, \tag{8.40}$$

where $r_s = M_s M_c^{-1} dW_c/dt$ is the molar rate of the sorption (desorption) of a poison or the molar rate of the coke formation.

The virtue of such a generalized kinetics is that it is capable of describing both the deactivation and reactivation of a catalyst in an improved and unified way which takes into account the sorption-desorption hysteresis. The hysteresis effect manifests through the correcting function

$$\xi = \left(\frac{\partial \ln \beta_c(W_c, P, T)}{\partial \ln W_c} \right)_{P,T} \tag{8.41}$$

which is different for the case of sorption and desorption. Thus, in the present approach, the hysteresis between the catalyst deactivation and reactivation has been linked with the hysteresis between the sorption and desorption. The latter is well known from the numerous experiments in sorption or desorption (stripping, solid drying, etc.) and has found a theoretical justification in terms of the capillary condensation of the sorbate. In the standard deactivation approaches used so far the correcting function ξ has been approximated by the unity (corresponding with the equilibrium $\beta_c = kW_c$) which automatically neglected the hysteresis. The approach can be generalized for the multiple reactions. For example, the deactivation kinetics described by Eq. (8.34) the corrected equation takes the form

$$-\frac{da}{dt} = \frac{k_{dc}\xi_c K_c P_c + k_{dt}\xi_t K_t P_t + k_{d1}\xi_1 K_1 P_1}{1 + K_1 P_1 + K_c P_c + K_t P_t} a, \tag{8.42}$$

where the corection functions ξ_i ($i = c, t, 1$) refer to the all components contributing to the catalyst deactivation. However, many simpler empirical or semiempirical deactivation formulae are also used in optimization Szepe (1966); Szepe and Levenspiel (1968); Park and Levenspiel (1976a, b). Note that in spatially distributed systems, for example, in packed beds with catalyst grains, deactivation leads to activity profiles in space and time and movement of activity fronts (Fig. 8.5). In effect, exit conversion in a packed bed decreases in time (Fig. 8.6). This requires to replace the used catalyst by the fresh one after some finite time.

The temperature dependence of the sorption equilibrium curve should condition the role of the temperature T in the function $a_s(W_c, T, P)$. Since the influence of T on the sorption equilibrium is determined by the differential heat of the sorption r_s (the difference between the partial enthalpy

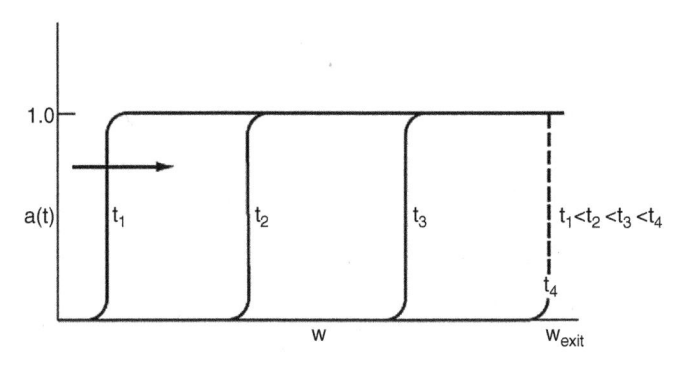

Figure 8.5: Motion of the activity front in a packed bed (Fogler, 2006).

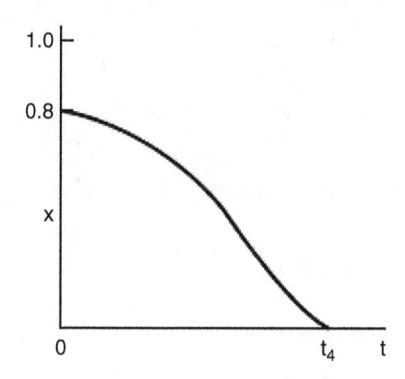

Figure 8.6: Exit conversion in a packed bed (Fogler, 2006).

of the coke in the bounded state in the solid and in the pure condensed state at the same T and P) it follows that the pronounced role of T in the function $a_s(W_c, T, P)$ proves a strong thermal effect of coke bounding through the occupied (deactivated) sites of the solid.

Summing up, the assumption that the thermodynamic activity of the solid a_s can be identified with its kinetic activity a (related to the decrease of the chemical rate) is quite fruitful and deserves more investigations. Its strong feature is that it links the kinetics of the catalyst decay with the particular properties of the (ab)sorbed substance(s). However it is dubious whether the identification of a_s with a can survive in the multiple reaction case on catalyst with inhomogeneous acidic strength since, in that case, the kinetic activity is supposed to be the property of all reactions involved rather than that of the catalytic solid only, and a vector of the various kinetic activities plays supposedly a role in the reaction kinetics (Fig. 8.6).

In chemical process design the real limits on the reactor yield derive from sources other than and more restrictive than the second law. Often they appear as constraints and inefficiencies of available technology; these, however, may have consequences that can be treated by thermodynamic analyses. For example, low overall thermal efficiencies can be a thermodynamic consequence of the lack of a suitable catalyst. The limits related to usual catalysts can, however, be overcome by the use of highly selective catalysts (eg, shape selective zeolite catalysts). They integrate selective transport with catalysis in a way that is not only energy efficient but also cheap. Second-law analyses placed in the context of such problems can play both a more subtle and a more useful role in chemical engineering than they do now. The contemporary design of chemical and energy networks involves real-life objectives which contain both a quantitative part (cost of equipment, energy, and resources) and a qualitative part (safety, operability, controllability, flexibility, etc.). The industrial problem is very complex and involves a combinatorial approach to the match between hot and cold streams, flow configuration choice, equality and inequality constraints on the temperature dependent properties, materials, pressure-drop limitations, and so forth. All this leads to the trend toward complex thermoeconomies and requires abandonment of purely thermodynamic concepts (Durmayaz et al., 2004).

The mission and essence of engineering (especially chemical engineering) is "to come up with processes to make materials wanted by man—new or improved processes to replace older less efficient ones, and processes to make completely new materials." This is now being accomplished by innovations in new catalysts, new reaction pathways or new contacting patterns. In a hydrogenation example (Levenspiel, 1965) the packed bed with the so-called supported liquid catalyst replaces more conventional units (bubble columns, spray columns, tubular reactors, etc.) The packed bed contains ingeniously prepared porous catalyst pellets with an extremely large internal surface. This idea allows one to replace the tons of hot, expensive liquid catalyst in the conventional units (which would cost millions of dollars) with just a few grams of catalytic liquid for the whole operation. This is just one of many examples that illustrate the role of surfaces and thin layers in contemporary chemical engineering. However, the applications of irreversible thermodynamics to such processes are still in an embryonic state. A recent promising treatment uses the irreversible thermodynamics of surfaces to develop a second-law analysis of the phase changes and latent heat storage in a single component system. In general, however, the thermodynamics of surfaces have not been used within second-law analyses. Their realm is now limited to processes where the role of surface effects is negligible or where these effects are purely reversible. Surface dissipation is, however, substantial in many catalytic systems, surfactant layers and electrochemical systems. Here the problems are not only practically ignored in second law analyses but also seldom encountered in standard nonequilibrium theory. This situation must change if thermodynamics is to play a significant role in contemporary chemical engineering.

8.5 An Excursion to Bejan's Constructal Theory

In a series of his papers and books Bejan (1997a, b, c, 2000) developed his original constructal theory of organization in nature. The basic paper (Bejan, 1997c), explains the solution to the fundamental problem of how to collect and "channel" to one point the heat generated volumetrically in a low conductivity volume of given size. The amount of high conductivity material that is available for building channels, that is, high conductivity paths through the volume is fixed. The total heat generation rate is also fixed. The solution is obtained as a sequence of optimization and organization steps. The sequence has a definite time direction, it begins with the smallest building block (elemental system) and proceeds toward larger building blocks (assemblies). Optimized in each assembly are the shape of the assembly and the width of the newest high conductivity path. Bejan shows that the paths form a tree-like network, in which every single geometric detail is determined theoretically. Yet, the tree network cannot be determined theoretically when the time direction is reversed, from large elements toward smaller elements. He also describes far-reaching implications in physics, biology, and mathematics. The paper (Bejan, 1997c) is of the essential significance to the further developments of Bejan's constructal theory of organization in Nature.

Bejan (2003) considers the connection between "Fermat-type" principle (Sieniutycz, 2000b) and his constructal principle which asks how geometric form (optimized complexity, organization) is generated in macroscopic flow systems. The conclusion on this connection shows the strength of the constructal theory (Bejan, 1997a, c, 2000). The constructal principle calls for maximizing global performance (objective, purpose, function) subject to global finiteness constraints, in a flow system the architecture of which is free to change. The result of invoking the constructal principle is the architecture of anything that moves— the very configuration (geometry, design, drawing) of the flow system. The connection is summarized graphically based on the analogy between minimizing travel time, minimizing thermal resistance, and maximizing flow access in general. As stated by Bejan, constructal theory covers all flow systems; the animate, the inanimate and the engineered. Fermat-type invocations of the principle mean that the flow access is maximized between two discrete points. In the earliest papers on constructal theory, the flow access was maximized between one point (source, or sink) and an infinity of points (volume, or area). According to Bejan (2003), however, the principle works in considerably more complex systems.

In the context of constructal theory Bejan and Ledezma (1998) discuss streets tree networks and urban growth as a manifestation of optimal geometry for quickest access between a finite-size volume and one point. Bejan and Lorente (2001, 2004, 2005) review developments in optimization by focusing on the generation of optimal geometric form (shape, structure, topology) in flow systems. The flow configuration is free to vary. The principle that generates geometric form is the pursuit of maximum global performance (eg, minimum flow resistance, minimum irreversibility) subject to global finiteness constraints (volume, weight, time). The resulting structures constructed in this manner have been named constructal designs. The thought that the same objective and constraints principle accounts for the optimally shaped flow paths that occur in natural systems (animate and inanimate) has been named constructal theory. Examples of large classes of applications are drawn from various sectors of mechanical and civil engineering: the distribution of heat transfer area in power plants, optimal sizing and shaping of flow channels and fins, optimal aspect ratios of heat exchanger core structures, aerodynamic and hydrodynamic shapes, tree-shaped assemblies of convective fins, tree-shaped networks for fluid flow and other currents, optimal configurations for streams that undergo bifurcation or pairing, insulated pipe networks for the distribution of hot water and exergy over a fixed territory, and distribution networks for virtually everything that moves in society (goods, currency, information). The principle-based generation of flow geometry unites the thermodynamic optimization developments known in mechanical engineering with lesser-known applications in civil engineering and social organization. As Bejan and Lorente (2001) say, their approach extends thermodynamics, because it shows how thermodynamic principles of design optimization account for the development of optimal configurations in civil engineering and social organization. Bejan and Lorente (2006a) define the constructal theory. According to their text this theory is the view that the generation of flow configuration is a physics phenomenon that can be based on a physics principle first formulated by A. Bejan and called the constructal law: "For a finite-size flow system to persist in time (to survive), its configuration must evolve in such a way that it provides an easier access to the currents that flow through it". This principle predicts natural configuration across the board: river basins, turbulence, animal design (allometry, vascularization, locomotion), cracks in solids, dendritic solidification, Earth climate, droplet impact configuration, and so forth. The same principle yields new designs for electronics, fuel cells, and tree networks for transport of people, goods, and information. This review describes a paradigm that is universally applicable in natural sciences, engineering and social sciences. In conclusion, Bejan and Lorente (2006a), the real world (nature, physics) gains architecture, organization, and pattern. The tissues of energy flow systems such as the fabric of society and all the tissues of biology, are optimized architectures. Not just "any" architectures, as in the black boxes of classical thermodynamics, but the optimal, or the near-optimal architectures.

Climbing to this high podium of performance is the transdisciplinary effort—the balance between seemingly unrelated flows, territories, and disciplines. No flow system is an island. No river exists without its wet plain. No city thrives without its farmland and open spaces. Everything that flowed and lived to this day to "survive" is in an optimal balance with the flows that surround it and sustain it. This balancing act—the optimal distribution of imperfection—generates the very design of the process, power plant, city, geography, and economics. Bejan and Lorente (2006a) review on the basis of constructal theory the physics phenomena of generation of flow configuration in nature, which are now covered by the thermodynamics of nonequilibrium flow systems with configuration. The new and most basic development is this: there are two time arrows in thermodynamics, not one. The old is the time arrow of the second law of thermodynamics, the arrow of irreversibility: everything flows from high to low. The new is the time arrow of the constructal law, the arrow of how every flowing thing acquires architecture. The "how" is condensed in the constructal law: existing configurations assure their survival by morphing in time toward more easily flowing configurations. The constructal time arrow unites physics with biology and engineering. A turbulent flow has "structure" because it is a combination of two flow mechanisms: viscous diffusion and streams eddies). Both mechanisms serve as paths for the flow of momentum. According to the constructal law, the turbulent flow structure is the architecture that provides the most direct path for the flow of momentum from the fast regions of the flow field to the slow regions.

Bejan and Lorente (2006b) ask: Can the flow architecture of multiscale porous structures be attributed to a principle of maximization of global flow access? And they answer: "Features similar to those of multiscale porous structures are exhibited by tree-shaped flow structures. Bejan and Marden (2006) also discuss constructing animal locomotion from the theory. For an earlier, related treatment see Ahlborn (1999) and Ahlborn and Blake (1999). Bejan's and Lorente's (2008a) book focuses on constructal theory to design strategies for more effective flow and transport. Constructal theory is the view that the generation of flow configuration is covered by a principle (the constructal law): "For a finite-size flow system to persist in time (to live) it must evolve such that it provides greater and greater access to the currents that flow through it" (Bejan and Lorente, 2006a, 2008b). This principle accounts for many features of design observed in nature, from the tree-shaped flows of lungs and river basins, to the scaling laws of animal design, river basin design, and social dynamics. This is a fast growing field with contributions from many sources, and with leads in many directions. One is the use of the constructal law to predict and explain the occurrence of natural flow configurations, inanimate and animate. The other is the application of the constructal law as a physics principle in engineering. This philosophy of design as science is the main thread of the book (Bejan and Lorente, 2008a). As stated by the authors, in science the origin (genesis) of the structure of flow systems has been overlooked. Design has been taken for granted at best, it has been attributed to chance, inspiration, talent, and art. The education in the sciences is based on sketches of streams into and out of boxes, to the position that the stream occupies in space and in time (Bejan and Lorente, 2008b). Constructal theory and design is a proposal to change this attitude. The benefits from thinking of design as science are profitable. For example, the march toward small dimensions (micro, nano), the miniaturization revolution, is not about making smaller and smaller components that are to be dumped like sand into a sack. This revolution is about the "living sack," in which every single component is kept alive with flows that connect it to all the other components (Bejan and Lorente, 2008b). Each component is put in the right place, like the neurons in the brain, or the alveoli in the lung, the hill slopes in the river basin, and the mid-size cities on the map. It is the configuration of these extremely numerous components that makes the "whole" perform best. Because natural flow systems have configuration, one treats the emergence of flow configuration as a phenomenon based on a scientific principle. Constructal theory is the mental viewing that the generation of the flow structures that we see everywhere (river basins, lungs, atmospheric circulation, vascularized tissues, etc.) can be reasoned based on a principle of increase of flow access in time.

Geometry or drawing is not a figure that always existed and now is available to look at, or worse, to look through and take for granted. The figure is the persistent movement, struggle, contortion and mechanism by which the flow system achieves global objective under global constraints. (Bejan, 2000). The best configurations that connect one component with very many components are tree-shaped, and for this reason dendritic flow architectures occupy a central position in the field. Trees are flows that make connections between discrete points and continua (Bejan and Lorente, 2008a, b), that is, between one point and infinities of points, namely, between a volume and one point, an area and one point, and a curve and one point. The flow may proceed in either direction, for example, volume-to-point and point-to-volume. Trees are not the only class of multiscale designs to be discovered and used. In the book by Bejan and Lorente (2008a) teach how to develop multiscale spacings that are distributed nonuniformly through a flow package, flow structures with more than one objective, and, especially, structures that must perform both flow and mechanical support functions. Along this route, the authors reveal designs associated with animal design. Every multiscale solid structure that is to be cooled, heated, or serviced by fluid streams must be and will be vascularized. This means trees and spacings and solid walls, with every geometric detail sized and positioned in the right place in the available space. These will be solid-fluid structures with multiple scales that are distributed nonuniformly through the volume. This is like in the prevailing view of animal design, where diversity is mistaken for randomness, when in fact the diversity is the fingerprint of the constructal law (Bejan and Marden, 2006). The future of designs belongs to the vascularized, at all scales (Bejan and Lorente, 2008b). Bejan et al. (2000) extended to economics the constructal theory of generation of shape and structure in natural flow systems that connect one point to a finite size area or volume. It is shown that by invoking the principle of cost minimization in the transport of goods between a point and an area, it is possible to anticipate the dendritic pattern of transport routes that cover the area, and the shapes and numbers of the interstitial areas of the dendrite. It is also shown that by maximizing the revenue in transactions between a point and an area, it is possible to derive not only the dendritic pattern of routes and their interstices, but also the optimal size of the smallest (elemental) interstitial area. Every geometric detail of the dendritic structures is the result of a single (deterministic) generating principle. The refining of the performance of a rough design (eg, rectangles-in-rectangle) pushes the design toward a structure that resembles a theoretically fractal structure (triangle-in-triangle). The concluding section shows that the law of optimal refraction of transport routes is a manifestation of the same principle and can be used to optimize further the dendritic patterns. The chief conclusion is that the constructal law of physics has a powerful and established analog in economics: the law of parsimony. The constructal theory unites the naturally organized flow structures that occur spontaneously over a vast territory, from geophysics to biology and economics.

Technological applications of constructal theory are progressing. In a book by Bejan et al. (2005) the authors treat porous and complex fow structures in technological processes. Vargas and Bejan (2004) propose a methodological structure for the thermodynamic optimization of internal structure in a fuel cell. Vargas et al. (2005) subsume a structured procedure to optimize the internal structure (relative sizes, spacings), single cells thickness, and external shape (aspect ratios) of a polymer electrolyte membrane fuel cell (PEMFC) stack so that net power is maximized. The constructal design starts from the smallest (elemental) level of a fuel cell stack (the single PEMFC), which is modeled as a unidirectional flow system, proceeding to the pressure drops experienced in the headers and gas channels of the single cells in the stack. The polarization curve, total and net power, and efficiencies are obtained as functions of temperature, pressure, geometry, and operating parameters. The optimization is subjected to fixed stack total volume. There are two levels of optimization: (1) the internal structure, which accounts for the relative thicknesses of two reaction and diffusion layers and the membrane space, together with the single cells thickness, and (2) the external shape, which accounts for the external aspect ratios of the PEMFC stack. The flow components are distributed optimally through the available volume so that the PEMFC stack net power is maximized (Vargas et al., 2005). Numerical results show that the optimized single cells internal structure and stack external shape are "robust" with respect to changes in stoichiometric ratios, membrane water content, and total stack volume. The optimized internal structure and single cells thickness, and the stack external shape are results of an optimal balance between electrical power output and pumping power required to supply fuel and oxidant to the fuel cell through the stack headers and single-cell gas channels. It is shown that the twice maximized stack net power increases monotonically with total volume raised to the power 3/4, similarly to metabolic rate and body size in animal design. Using constructal approach Reis (2008) shows that distributions of street lengths and nodes follow inverse-power distribution laws. That means that the smaller the network components, the more numerous they have to be. In addition, street networks show geometrical self-similarities over a range of scales. Based on these features many researchers claim that street networks are fractal in nature (Aharony and Feder, 1989). What is shown here is that both the scaling laws and self-similarity emerge from the underlying dynamics, with the purpose of optimizing flows of people and goods in time, as predicted by the Constructal Law. The results seem to corroborate the prediction that cities' fractal dimension approaches 2 as they develop and become more complex.

Perhaps in view of unequivocal optimization criterion and the absence of its exact analytical formula, constructal theory has been the subject of some criticisms (Kuddusi and Egrican, 2008). Constructal theory applied to the volume to point or point to volume flows aims to decrease global flow resistance by furnishing low resistive flow links in the flow field. It expects to improve the flow performance by increasing the branching of the low-resistive flow links. Fourteen different constructal theory applications involving tree-shaped flow networks were verified with the purpose to check whether an increase in branching (complexity) of tree-shaped flow network leads to an increase in flow performance or not? (Kuddusi and Egrican, 2008). They aimed to answer the question: does the evolution model of constructal theory in increasing the branching (complexity) of tree-shaped flow networks through the sequence of constructal designs, namely, elemental volume, first construct, second construct, and so forth, improve the flow performance? The review supported the conclusion of original statement by Ghodoossi (2004): "constructal theory will not necessarily improve the flow performance if the internal complexity of the flow area is increased. In contrast, the performance will mostly be lowered if the internal complexity of the flow area is increased."

8.6 Discussion of Research Works

We shall now describe some results of the optimization applications in the framework of the first category of approaches. Recall that this category do not try to describe natural evolutions but rather asks for the optimal way to control a process. This summary refers to selected works of researchers who have been involved with the entropy generation minimization, finite-time thermodynamics (FTT), thermoeconomics, and some mixed approaches.

Rivero and Garcia (2001) present application of exergy analysis to the reactive distillation system of a methyl ter-butyl ether (MTBE) production unit of a crude oil refinery. In a refinery, the MTBE is obtained from methanol, and butanes (isobutylenes) are produced in the fluidized catalytic cracking (FCC) unit. The MTBE production process is a relatively new process, which uses reactive distillation columns. Distillation and chemical reaction occur simultaneously in a packed and tray column. This combination shows important advantages over the packed-bed reactor and over the distillation system, including the use of the heat of reaction for the separation of products, a relatively easy control of the temperature profile in the catalytic section, low operation costs due to high reaction yields, and low capital costs due to reduced equipment items. Process simulation studies help in determining the influence of operating parameters such as the column feed location, methanol flowrate, and the reboiler and condenser heat duties. The reactives (isobutylenes) after purification are introduced to the main reaction system, the products of which are sent to the reactive distillation system to complete the reaction. A top rectification section, a bottom stripping section and a medium reaction section, compose the reactive column.

The results of the exergy analysis of the unit indicate that the main exergy losses (about 63 percent) of the MTBE plant occur in the reactive distillation system, particularly in the distillation column itself and in its associated condenser. Gasoline mixtures have been reformulated incorporating ethers such as MTBE, which increases the octane number, and reduces pollutant gases emissions. With the exergy analysis of the simulated process it was possible (1) to establish the optimal operating conditions producing a higher amount of high-quality products, (2) to establish the critical equipment items with the highest exergy losses, and (3) to evaluate the improvement potential of the reactive distillation system.

Shen et al. (2013a) have studied the batch and continuous extractive distillation of minimum- and maximum-boiling azeotropic mixtures with a heavy entrainer. These systems exhibit class 1.0-1a and 1.0-2 ternary diagrams, each with two subcases depending on the location of the univolatility line. The feasible product and feasible ranges of the operating parameters reflux ratio (R) and entrainer/feed flow rate ratio for continuous (FE/F) and batch (FE/V) operation were assessed. Class 1.0-1a processes allow the recovery of only one product because of the location of the univolatility line above a minimum value of the entrainer/feed flow rate ratio for both batch and continuous processes. A minimum reflux ratio R also exists. For an identical target purity, the minimum feed ratio is higher for the continuous process than for the batch process, for the continuous process where stricter feasible conditions arise because the composition profile of the stripping section must intersect that of the extractive section. Class 1.0-2 mixtures allow either A or B to be obtained as a product, depending on the feed location. Then, the univolatility line location sets limiting values for either the maximum or minimum of the feed ratio FE/F. Again, the feasible range of operating parameters for the continuous process is smaller than that for the batch process. Entrainer comparison in terms of minimum reflux ratio and minimum entrainer/feed ratio is enabled by the proposed methodology. Shen et al. (2013b) have studied the continuous extractive distillation of minimum- and maximum-boiling azeotropic mixtures with a light entrainer. The ternary mixtures belong to class 1.0-2 and 1.0-1a diagrams, each with two subcases depending on the location of the univolatility line. The feasible product and feasible ranges of the operating parameters reboil ratio (S) and entrainer/feed flow rate ratio for the continuous process (FE/F) were assessed. Equations were derived for the composition profiles of the stripping, extractive, and rectifying sections in terms of S and FE/F. Class 1.0-1a processes enable the recovery of only one product because of the location of the univolatility line above a minimum value of the entrainer/feed flow rate ratio for both batch and continuous processes. Given a target purity, a minimum reboil ratio S also exists; its value is higher for the continuous process than for the batch process, for the continuous process where stricter feasible conditions arise because the composition profile of the rectifying section must intersect that of the extractive section. Class 1.0-2 mixtures allow either A or B to be obtained as a product, depending on feed location on the composition triangle. Then, the univolatility line location sets limiting values for either the maximum or minimum of the feed ratio FE/F.

Early works involved in thermodynamic calculations of optimal parameters of thermal power stations were summarized by Andriuszczenko (1965). Approaches involving classical exergy analyses, which emerged from the early research of Rant, Grassman, Szargut, and many others, are still popular. They constitute exergy method of thermal and chemical plant analysis, and are well described in several books (Szargut and Petela, 1965; Brodyanskii, 1968, 1973; Kotas, 1985, 1986; Szargut et al., 1988; Szargut, 2005) of which the last one includes ecological applications. Szargut et al. (1988) has developed exergy approaches to chemical and metallurgical processes. Banerjec et al. (1992) performed exergy analysis of kinetic swing adsorption processes (exergy-based comparison of cycles for the production of nitrogen from air). Grassman diagrams have been drawn for typical operating points to show the exergy flows and irreversibilities in the processes. Bosniakovic (1965) and Ciborowski (1965, 1976) have presented in their books many analyses of technical processes on thermodynamic diagrams focusing, however, quite frequently, on processes for which the kinetic information was oversimplified, or of secondary significance. Further works included kinetic information into the analyses in a more systematic way. Le Goff et al. (1990) presented a comparison of exergy-based optimization and economic optimization of industrial processes. Energy, exergy, and value balancing were discussed with various definitions of the profits, costs and efficiencies. Scales of value based on enthalpy, exergy, ecology, and economics were compared. Examples of economic and energy accounting were presented for industrial systems treated as energy converters with various definitions of the energy-equivalent periods and cost-equivalent periods. Operating cost optimizations according to energy, exergy, and monetary scales were given for a heat exchanger, a beet-pulp dryer and some heat transformers which upgrade industrial waste heat. This should be of interest to workers who have to organize a realistic industrial optimization with its characteristic many-compromise property.

Andresen and Gordon (1991, 1992b) derived the optimal heating or cooling strategy that minimizes entropy production for a class of common heat transfer processes that are constrained to proceed in a fixed, finite time. The empirical wisdom embodied in conventional single-pass counterflow heat exchanger design is examined in light of this solution. For judiciously selected system parameters, the counterflow heat exchanger can yield the optimal solution. The solution shows a formal link with the constant thermodynamic speed for minimizing entropy production in thermodynamic processes and simulated annealing (Andresen and Gordon, 1994). See more information in chapter: Multiphase Flow Systems. Ratkje and de Swaan Arons (1995) presented a new way of calculating lost work in a reacting, diffusing mixture. They show that a fast chemical reaction can be treated within the framework of coupled transport equations for heat and mass, after elimination of dependent components. Two approaches to reduction of lost work follow from the equations: reduction of driving forces and utilization of coupling. When Fick's and Fourier's laws apply, losses can be minimized, but not avoided, by reduction of driving forces. When transport processes are coupled, as expressed by the theory of irreversible thermodynamics, loss reductions can be obtained also through the coupling coefficients. Maximum reduction is obtained for strict coupling, that is, unique flux relationships. Coupling implies physical interaction and is discussed for isothermal and nonisothermal systems. The relation between irreversible thermodynamics and exergy analysis implies that low-loss designs are those which prevent irreversibilities. Low-loss designs prevent mixing, promote catalyst specificity or demixing, increase membrane selectivity and enjoy small driving forces. Irreversible thermodynamics shows that strict coupling and large coupling coefficients and steady states also contribute to low-loss designs. Exergy analysis and irreversible thermodynamics may prove their suitability in the search for acceptable ways of exploiting the resources of the world. The derivations show that the concepts used by Forland et al. (1988, 1992) may extend the information obtained by exergy analyses on chemical processes. Their results show that exergy-saving processes are obtained for strict coupling or large values of the phenomenological coefficient l_{kj}. The isomerization reaction gives a simple example of strict coupling: A–B. Minimum lost work is obtained for $-J_A = J_B$; the consumption of A is always equal to the production of B. This can be obtained if B is separated from A as it is produced. One obtains the familiar result that adsorption, membrane separation, or other separation techniques practically preserve the exergy of the system.

Considering extreme performance of heat exchangers of various hydrodynamic models of flows Amelkin et al. (2000) treated the problem of minimization of entropy production for one-pass heat exchangers of various types of description of hydrodynamic characteristics of the flows. Two models of the flows have been considered, namely a model of an ideal mixing and that of an ideal exclusion. The solution of the problem at issue allows one to construct a measure of thermodynamic perfectness of the heat exchanger taking into account the irreversibility of the heat exchange process. In conclusion, by solving the problem of extreme performance of one-pass heat exchangers of different hydrodynamic models of the flows, the obtained results allow one to construct a criterion of thermodynamic perfection of heat exchangers (taking the extreme performance boundary as an ideal regime to compare with). Such a criterion takes into account unremovable losses that is, losses due to irreversibility (as it had been done previously) and due to the hydrodynamics of the flows. This criterion can be used, for example for performance comparison of heat exchangers of either the same size features or the same hydrodynamic characteristics of the flows.

Amelkin et al. (2005) considered power-producing thermo-mechanical systems exchanging heat with several (more than two) heat reservoirs via irreversible heat-transfer processes. The systems were optimized for maximum power output at a given increase of entropy and internal energy of the working fluid for the case of stationary temperatures of the heat reservoirs. The authors found that independently of the number of reservoirs the temperature of the working fluid should attain no more than two different values. These base values are time independent and are attained for a certain fraction of the overall time available for the process. In particular this is the case for cyclic processes. The optimal cycle corresponds to a Carnot cycle of two isotherms and two isentropes/adiabats. A surprising but inevitable consequence of this process structure is that there may exist heat reservoirs that should never connect to the working fluid. These "unused" reservoirs provide heat at temperatures that are between the temperatures of the isotherms. Thus it is more advantageous to ignore the heat available from them than to operate at different temperatures. Bojic (1997) analyzed the annual worth for cogeneration of heat and power of an endoreversible Carnot engine by using the methods of finite-time thermodynamics. He finds that, in order to obtain the maximum annual worth of production of heat and power, in the design of such systems, the heat exchangers on the hot and cold sides of the Carnot engine must have equal products of their size and heat transfer coefficient. Also, for the maximum annual worth, the ratio of the lower and higher temperatures of the Carnot engine should have its optimal value.

In a highly professional review MacLean and Lave (2003) evaluated automobile fuel/propulsion system technologies by examining the life-cycle problems and life-cycle implications of a wide range of fuels and propulsion systems that could power cars and light trucks. L. Chen, Xia, and Sun (2011) determined maximum power output of multistage continuous and discrete isothermal endoreversible chemical engine system with linear mass transfer law. One of these papers is discussed subsequently.

L. Chen et al. (2011) searched for the maximum power output of a multistage isothermal endoreversible chemical engine, a system operating between a finite capacity high-chemical potential reservoir and an infinite capacity low-chemical-potential environment, characterized by the linear mass transfer law. For the fixed initial time and fixed initial concentration of the key component in the high-chemical potential reservoir, the continuous model is optimized by applying Hamilton–Jacobi–Bellman (HJB) theory and Euler-Lagrange equation, respectively. The discrete model is optimized by applying dynamic programming method and discrete maximum principle. Numerical examples are provided for the discrete model with three different boundary conditions. The results are compared with those for the continuous model. The results show that both maximum power output is equal to the difference between the reversible power and a dissipation term linked with entropy production. The relative concentration of the key component in the high chemical potential reservoir decreases with time. The relationships among the maximum power output of the system, the process period and the final concentration of the key component in the high-chemical-potential fluid reservoir are discussed in detail. Similarly, L. Chen et al. (2013) treated a multistage irreversible Carnot heat engine system operating between a finite thermal capacity high-temperature fluid reservoir and an infinite thermal capacity low-temperature environment with a generalized heat transfer law [$q \propto (\Delta(T^n))^m$]. Optimal control theory was applied to derive the continuous HJB equations to determine optimal fluid temperature configurations for maximum power output under the conditions of fixed initial time and fixed initial temperature of the driving fluid. Based on the universality of optimization results, the analytical solution for the case of Newtonian heat transfer law ($m = 1$, $n = 1$) was further obtained. Since there are no analytical solutions for other heat transfer laws, the continuous HJB equations were discretized and the dynamic programming (DP) algorithm was solved to obtain the complete numerical characteristics of the optimization problem. Then the effects of the internal irreversibility and heat transfer laws on the optimization results were analyzed in detail. The results provide theoretical guidelines for the optimal design of energy conversion systems. In another group of researchers chemical engines were treated in a similar fashion (Chen and Sun, 2004; L. Chen et al., 1997a, b). L. Chen et al. (1999b, c) analyzed the performance of an isothermal combined-cycle chemical engine with mass leak. Chen et al. (2005) worked out ecological optimization for generalized irreversible Carnot refrigerators. Chen et al. (2013) presented dynamic programming analyses leading to maximum power in irreversible heat engines under a generalized heat transfer law, and give many numerical examples for multistage engines propelled by heat fluxes described by the linear transfer law. Taking an ecological optimization criterion as the objective, Chen et al. (2010) have derived an optimal policy for the best ecological performance of generalized irreversible chemical engine. This consists of maximizing a function representing the best compromise between the power and entropy production rate of the heat engine. Xia et al. (2010b) treat Hamilton–Jacobi–Bellman equations and dynamic programming for power-optimization of a multistage heat engine with general convective heat transfer law. Also, Xia et al. (2010c) find a finite-time exergy for a finite heat reservoir and generalized radiative heat transfer law. Xia et al. (2011a) depict endoreversible modeling and optimization of multistage heat engine system with a generalized heat transfer law via HJB equations of dynamic programming.

A paper by Xia et al. (2011a) refers to dynamical power-producing system. Accepted is an engine model with the generalized convective transfer law, and considered are effects of losses (irreversibility of external heat resistance and internal dissipation). Optimal control theory serves to derive sufficient conditions for optimum as HJB equations. The authors determine optimal temperature profiles for maximum power output. Yet, the analytical solution is possible only for the Newtonian heat transfer. For complex transfer laws, HJB equations are discretized and Bellman's dynamic programming is adopted to obtain numerical solutions. This makes it possible to discuss relationships between the maximum power output, process duration and the fluid temperature. The main originality of the paper lies in careful discussion of the influence of various heat transfer models on power production and optimal solutions in dynamical systems with power yield. Another multistage endoreversible Carnot heat engine system operating between a finite thermal capacity high-temperature fluid reservoir and an infinite thermal capacity low-temperature environment with a generalized heat transfer law $q \propto (\Delta T)^m$ is investigated by Xia et al. (2011b). Optimal control theory is applied to derive the continuous HJB equations, which determine the optimal fluid temperature configurations for maximum power output under the conditions of fixed initial time and fixed initial temperature of the driving fluid. Based on the general optimization results, the analytical solution for the Newtonian heat transfer law ($m = 1$) is further obtained. Since there are no analytical solutions for the other heat transfer laws, the HJB equations are discretized and a DP algorithm is adopted to get a numerical solution. Then the relationships linking maximum power output, process duration and fluid temperature are determined and discussed.

Kaushik and Kumar (2001) presented an investigation on finite-time thermodynamic evaluation of Ericsson and Stirling heat engines. Finite-time thermodynamics has been applied to optimize the power output and thermal efficiency of both engines, including finite heat capacitance rates of the heat source and sink reservoirs external fluids, finite-time heat transfer, regenerative heat losses and direct heat leak losses from source to sink reservoirs. The presence of these effects causes irreversibilities and affects the performance of the Ericsson and Stirling heat engines using ideal or real gas as the working fluid/substance. Kaushik and Kumar (2001) have found that both cycles with an ideal regenerator [epsilon (R) = 1.00] are as efficient as an endoreversible Carnot heat engine operating at the same conditions, but this is not practical because an ideal regenerator requires either infinite regenerator area or infinite regeneration time. It is also found that the regenerative heat losses, regenerator effectiveness and the direct heat leak losses do not affect the maximum power output of either heat engine.

Angulo-Brown (1991) studied an endoreversible Carnot-type heat engine under usual restrictions: no friction, working substance in internal equilibrium (endoreversibility), no mechanical inertial effects, and under Newton's cooling law for heat transfer between working fluid and heat reservoirs. He has found a monoparametric family of isoefficient straight lines. Along each line the power output divided by the entropy production was constant. From these properties and by using some dissipated quantities, relationships followed between reversible work and finite-time work, and between reversible efficiency and finite-time efficiency. An "ecological" criterion for the best mode of operation consists in maximizing a function representing the best compromise between power and the product of entropy production and the cold reservoir temperature. Engine efficiency results almost equal to the average of Carnot and Curzon and Ahlborn (1975) efficiencies.

Bedeaux et al. (1999) examined the problem of energy-efficient production in an industrial process. By the energy-effcient production they mean minimum entropy production. They use the possibility to redistribute the production in different times or parts of the system for a given total production, and show that a distribution, that equipartitions the derivative of the local entropy production rate with respect to the local production, minimizes the entropy production. Equipartition in time implies stationary state production. Equipartition in space implies production for a given position independent force. The same constant derivative of the local entropy production rate is found if ones optimizes the production for a given total entropy production. Close of equilibrium the equipartition condition is found to reduce to the isoforce principle. Further from equilibrium, this reduction is extended to a whole class of nonlinear flux-force relations. We show that, when one increases the total production, the entropy production per unit produced starts to increase linearly, as a function of this total production. It is shown which process conditions give an optimum path with an equipartition of the entropy production rate. How this relates to the isoforce principle is discussed. In general constraints on process conditions restrict the freedom to optimize, and therefore make it impossible to realize the most favorable conditions. The importance of the Onsager relations for the systematic description of the optimization is discussed. A general optimization criterion is formulated: The derivative of the local entropy production rate at a given time with respect to the local productions at the same time must be constant in order to give the minimum total entropy production.

A commonly used method to dry fine solid particles is drying in a fluidized bed. Poświata and Szwast (2003, 2006) have solved a problem of optimization of exergy consumption in fluidized drying of fine solids. A drying process is conduced in a three-stage cascade of fluidized cross-current dryers. Solid flows from stage to stage, and fresh gas is introduced to each stage of the cascade. The hydrodynamics of bubble fluidized bed and kinetics of heat and mass transfer are taken into account. The bed hydrodynamics is described by the two-phase bubbling model. The drying proceeds in the second period of drying. To optimize this problem a generalized version of a discrete algorithm with constant Hamiltonian is used. In optimization calculations, gas parameters (temperature, humidity and flow rate) minimizing total process cost are sought. The results, presented as graphs, are as follows. The interstage solid humidities for optimum processes are practically independent of fluidized bed hydrodynamics and kinetics of heat and mass transfer. However, the interstage solid temperatures are clearly dependent on the hydrodynamic and the kinetics. Optimum values of temperature and humidity of drying gas decrease, respectively, for each stage of cascade if the value of a coefficient Z_{mf} increases and if the value of a coefficient a_1 increases. Optimum temperature of inlet drying gas increases with increment of stage number of the cascade (independent of values of Z_{mf} and a_1). Recalling that, for the first period of drying, the opposite situation is observed, that is, the optimum temperature decreases as stage number increases. Variation of optimum gas humidity along the cascade is dependent on values of coefficients Z_{mf} and a_1 and is more complicated than dependence on gas temperature.

Sieniutycz (1998a) and Sieniutycz and von Spakovsky (1998) analyzed an irreversible extension of the Carnot problem of maximum mechanical work delivered from a system of two fluids with different temperatures. Considered is the active heat exchange between the fluid of a limited thermal capacity and an ambient or environmental fluid whose thermal capacity is so large that its intensive parameters do not change. In the classical problem the rates vanish due to the reversibility requirement, in the extended problem irreducible irreversibilities occurring in boundary layers are admitted. The thermodynamic modeling is linked with ideas and methods of the optimal control. A variational theory treats an infinite sequence of infinitesimal Curzon–Ahlborn–Novikov (CAN) processes as the pertinent theoretical model leading to a finite-time exergy of the system with a finite exchange area or a finite contact time. This dissipative exergy is discussed in terms of the process intensity and finite duration. An analytical formalism, analogous to the one used in analytical mechanics is effective. A novel approach is worked out based on the HJB equation for the dissipative exergy and related work functionals (HJB theory). The HJB formulation is helpful to deal with work potentials by numerical methods which use Bellman's recurrence equation as practically the sole method of extremum seeking for functionals with constrained rates and states and for complex boundary conditions. It is pointed out that only an irreversible process can be optimal for a finite duration. A link is shown between the process duration, dissipation and optimal intensity measured in terms of a Hamiltonian. Hysteretic properties are effective which cause the difference between the work supplied and delivered, for inverted end states of the process. A decrease of maximal work received from an engine system and an increase of minimal work added to a heat pump system is shown in the high-rate regimes and for short durations. The results prove that bounds of the classical availability should be replaced by stronger bounds obtained for finite rate processes. Sieniutycz (1998b) shows that an extended exergy of the kinetic rather than the static origin can be formulated for finite-time transitions, which simplifies to the classical thermal exergy in the limiting case of infinite duration. The extended exergy can be derived either by gauging a functional of the energy dissipation by the classical exergy or as a finite-time extension of the classical thermodynamic work extracted from a system of a body and its environment. With this exergy he considers optimization applications for various active continuous and cascade processes encountered in the theory of the energy exchange through working fluid of an engine, a refrigerator or a heat pump. They refer, for example, to systems with finite exchange area or with a finite contact time, and they clearly show that a complementary economic theory is possible. The theory is a good example of formulation where nonlinear thermodynamic models are linked with ideas and methods of optimal control. Variational approaches strongly analogous to those in analytical mechanics and the optimal control theory of continuous and discrete systems are effective tools in thermodynamics optimization. Nonlinear equations of dynamics, which follow as combinations of the energy balance and transfer equations, are constraints in the problem of work optimization. Functionals, which describe the maximized work, can effectively be optimized by various optimization methods; in this work variational calculus is used along with some aspect of the canonical transformation theory. In particular one can discuss the role of the finite process intensity and finite duration. The optimality of a definite irreversible process for a finite-time transition of a controlled fluid is pointed out as well as a connection between the process duration, optimal dissipation and the optimal process intensity measured in terms of a Hamiltonian. Discrete processes can be analyzed by methods extending those known for continuous ones. The results show that limits of the classical availability theory should be replaced by better limits which are obtained for the finite-time processes and which are closer to reality. A hysteretic property is discovered for the generalized exergy that describes a decrease of the maximum work received from an engine system and an increase of work added to a heat pump system, the features of which are particularly important in high-rate regimes (for short durations of thermal processes).

Sieniutycz and Poświata (2012) applied optimization methods to determine power generation limits in various energy converters, such as thermal, solar, chemical, and electrochemical engines. Methodological similarity accomanies analysing of power limits in thermal machines and fuel cells treated as electrochemical flow engines. Operative driving forces and voltage are good indicators of imperfect phenomena in converters. The results obtained generalize previous findings for power yield limits in thermal systems with finite rates. While temperatures T_i of participating media were only necessary variables in purely thermal systems, in the FC systems both temperatures and chemical potentials μ_k are essential. This case is associated with engines propelled by fluxes of both energy and substance. In dynamical systems downgrading or upgrading of resources may occur. Energy flux (power) is created in the generator located between the resource fluid ("upper" fluid 1) and the environmental fluid ("lower" fluid, 2). Fluid properties, transfer mechanisms and conductance values of dissipative layers or conductors influence the rate of power production. Numerical approaches to the dynamical solutions are based on the dynamic programming or maximum principle. Downgrading or upgrading of resources may also occur in electrochemical systems of fuel cell type. Yet, the authors restrict themselves to the steady-state fuel cells. An analysis shows that, in linear systems, only at most one-quarter of power dissipated in the natural transfer process can be transformed into the noble form of mechanical power (chapter: Thermodynamic Aspects of Engineering Biosystems).

Cogeneration plants in which power and heat are produced together are now widely used. These cogeneration plants are more advantageous in terms of the energy and exergy efficiencies than plants which produce heat and power separately. Sahin et al. (1997) have performed a performance analysis of coregeneration plants and have been carried exergy optimization for an endoreversible cogeneration cycle using finite-time thermodynamics. The optimum values of the design parameters of the cogeneration cycle at maximum exergy output were determined. Their model is more general than those of the endoreversible power cycle found in the literature. They have investigated effects of design parameters on exergetic performance and have discussed the results obtained. Their results show that the endoreversible power-cycle model may be regarded as a special case of the endoreversible cogeneration cycle. Lior (2002) expressed his thoughts about future power generation systems and the role of exergy analysis in their development. Lior et al. (1991) discussed the combining fuel cells with fuel-fired power plants for improved energy efficiency. For a hybrid fuel cell system that integrates a turbogenerator, Alvarez et al. (2006) have proposed an innovative configuration of fuel cell technology, to overcome the intrinsic limitations of fuel cells in the conventional operation. The analysis by Alvarez et al. (2006) develops the application of molten carbonate FC technology, for the performance assessment of the fuel cell prototype integrated in the hybrid FC system. The approach has been completed with a thermoeconomic analysis of the 100 kW cogeneration fuel cell power plant. Operational results and design limitations were evaluated together with operational limits and thermodynamic inefficiencies (exergy destruction and losses) of the 100 kW fuel cell. This research leads to the design of a hybrid system and demonstrates potential and benefits of the new hybrid structure. The results are quantified through a thermoeconomic analysis in order to get the most cost-effective plant configuration. One promising configuration is the case where the fuel cell in the power plant behaves as a combustor for the turbogenerator. The latter behaves as the balance of plant for the fuel cell. The combined efficiency increases to 57% and NOx emissions are eliminated. The synergy of the fuel cell/turbine hybrids lies mainly in the use of rejected thermal energy and residual fuel from the fuel cell to drive the turbogenerator in a 500 kW hybrid system.

Yet, in their closing text, Alvarez et al. (2006) stated that the analysis of the 100 kW cogeneration fuel cell power plant was mostly based on theoretical grounds, neither with an experimental validation nor an experimental polarization curve defined under specific operating conditions (that is, considering real ohmic, activation, and concentration losses) which in fact determine the behavior of the fuel cell. The following observations have been made:

- The experimental power output obtained came about to be 75 kW as opposed to a theoretical value of 100 kW
- The efficiency obtained experimentally was 30% as opposed to a theoretical value of 47%
- The theoretical operating conditions of gas composition, temperature, and flow rates were not reached, with substantial effects on the behavior and performance of the fuel cell.

Therefore the authors concluded that the theoretical modeling does not guarantee a sound analysis of fuel cell systems. Exact polarization curve and a suitable definition of FC operating conditions are essential in order to obtain fairly reliable values for the performance analyses of existing cell and new prototypes, and for the design of new power systems. Once the operational constraints are known, exergy analysis together with thermoeconomic analysis provides a rational base to optimize the integration of the fuel cell with the corresponding plant, and to evaluate system costs. Fuel cell hybrid power systems have a sufficient potential for attaining efficiency levels well beyond the limits today known for distributed power generation technologies, and quite similar to the last generation of combined cycle power plants operating at power levels about 800 times greater. In terms of electricity cost, this will be a competitive technology, able to challenge existing distributed-generation technologies. It enjoys further features such as modularity, fuel flexibility, high capacity factor, and dispatchability, and will also be competitive with centralized generation technologies, if factors such as costs of transport and distribution and emission costs are taken into account.

Ion exchange processes comprise of the interchange of ions between a solution and an insoluble solid. Water decontamination consists of removal of ionic pollutants or heavy metals discharged in effluent from electroplating plants as well as mining and electronic industries. Bochenek et al. (2011) propose a generic design procedure for a continuous ion exchange process. The procedure is based on the optimized arrangement of parallel batch columns. The process continuity is achieved by proper shifting of the inlet conditions. The authors give a method for the optimization of the process variables. Its concept is demonstrated on the example of ammonia removal from wastewaters. Two flowsheet schemes are considered utilizing fresh or recycled regenerating agent. The superiority of the optimized process over the periodic operation is demonstrated.

The use of biomass for energy generation is getting increasing attention (Chen et al., 2004a, b). At present, gasification of biomass is taken as a popular technical route to produce fuel gas for application in boilers, engine, gas turbine or fuel cell. Up to now, most of researchers have focused their attentions only on fixed-bed gasification and fluidized bed gasification under airblown conditions. In that case, the producer gas is contaminated by high tar contents and particles which could lead to the corrosion and wear of blades of turbines. Furthermore, both the technologies, particularly fixed bed gasification, are not flexible for using multiple biomass-fuel types and also not feasible economically and environmentally for large-scale application up to 10–50 MW_{th}. An innovative circulating fluidized bed concept has been considered in our laboratory for biomass gasification, thereby overcoming these challenges. The concept combines and integrates partial oxidation, fast pyrolysis (with an instantaneous drying), gasification, and tar cracking, as well as a shift reaction, with the purpose of producing a high quality of gas, in terms of low tar level and particulates carried out in the producer gas, and overall emissions reduction associated with the combustion of producer gas. This paper describes an innovative concept and presents some experimental results. The results indicate that the gas yield can be above 1.83 Nm³/kg and the fluctuation of the gas yield during the period of operation is 3.3% at temperature of 750°C. Generally speaking, the results achieved support our concept as a promising alternative to gasify biomass for the generation of electricity.

Thermodynamic equilibrium-based models of gasification process are relatively simple and widely used to predict producer gas characteristics in performance studies of energy conversion plants (Kalina, 2011). However, if an unconstrained calculation of equilibrium is performed, the estimations of product gas yield and heating value are too optimistic. Therefore, reasonable assumptions have to be made in order to correct the results. Kalina (2011) proposes a model of the process that can be used in case of deficiency of information and unavailability of experimental data. The model is based on free energy minimization, material and energy balances of a single zone reactor. The constraint quasiequilibrium calculations are made using approximated amounts of nonequilibrium products, that is, solid char, tar, CH_4, and C_2H_4. The yields of these products are attributed to fuel characteristics and estimated using experimental results published in the literature. A genetic algorithm optimization technique is applied to find unknown parameters of the model that lead to the best match between modeled and experimental characteristics of the product gas. Finally, generic correlations are proposed and quality of modeling results is assessed in the aspect of its usefulness for performance studies of power generation plants. A paper by Li et al. (2004) presents the results from biomass gasification tests in a pilot-scale (6.5-m tall × 0.1-m diameter) air-blown circulating fluidized bed gasifier, and compares them with model predictions. The operating temperature was maintained in the range 700–850°C, while the sawdust feed rate varied from 16 to 45 kg/h. Temperature, air ratio, suspension density, fly ash reinjection and steam injection were found to influence the composition and heating value of the product gas. Tar yield from the biomass gasification decreased exponentially with increasing operating temperature for the range studied. A nonstoichiometric equilibrium model based on direct minimization of Gibbs free energy was developed to predict the performance of the gasifier. Experimental evidence showed that the pilot gasifier deviated from chemical equilibrium due to kinetic limitations. A model adapted from the pure equilibrium model, incorporating experimental results regarding unconverted carbon and methane to account for nonequilibrium factors, predicts product gas compositions, heating value and cold gas efficiency in good agreement with the experimental data.

Extensive research has recently been conducted to reveal the major factors that affect biomass fast pyrolysis. Shrinkage affects the mechanism of pyrolysis in various ways. Biomass properties (porosity, thermal conductivity, specific heat capacity, density, etc.) vary during the thermal degradation, resulting in different temperature gradient inside the particle, due to increase in density as the particle diameter decreases. As a consequence, the product yields are affected by the thinner and hotter char layer formed. In the study by Papadikis et al. (2009) fluid–particle interaction and the impact of shrinkage on pyrolysis of biomass inside a 150 g/h fluidized bed reactor has been modeled. Two 500 μm in diameter biomass particles are injected into the fluidized bed with different shrinkage conditions. The two different conditions consist of (1) shrinkage equal to the volume left by the solid devolatilization, and (2) shrinkage parameters equal to approximately half of particle volume. The effect of shrinkage is analyzed in terms of heat and momentum transfer as well as product yields, pyrolysis time and particle size considering spherical geometries. The Eulerian approach has been used to model the bubbling behavior of the sand, which is treated as a continuum. Heat transfer from the bubbling bed to the discrete biomass particle, as well as biomass reaction kinetics are modeled according to the literature. The particle motion inside the reactor is computed using drag laws, dependent on the local volume fraction of each phase. FLUENT has been used as the modeling framework of the simulations with the whole pyrolysis model incorporated in the form of user defined function (UDF). Papadikis et al. (2009) conclude that shrinkage does not have a significant effect on the momentum transport from the bubbling bed to the discrete biomass particles for small sizes in the order of 500 μm. The effect of shrinkage on momentum transport can be neglected when fluidized beds operate with such small particle sizes. However, the same cannot be stated for shrinkage parameters that shrink the particles close to their total disintegration. The model excluded this extreme condition and studied particles that shrink until half of their initial volume. For fast pyrolysis applications in lab-scale fluidized beds, small particle sizes are necessary (350–500 μm) due to feeding problems. The effect of shrinkage on the pyrolysis of thermally thin particles does not have a significant impact neither on the product yields nor the pyrolysis time. Due to small Biot number, the progress of the reaction is only dependent on the heat transfer inside the particle and the effect of the chemical reaction rate is not significant. The results agree with already developed single particle models in the literature. Computational fluid dynamics models can give important information regarding the overall process of fast pyrolysis. They can efficiently used to derive important conclusions in the industrial sector, regarding the design and optimization of bubbling fluidized bed reactors by deeply understanding the factors that highly influence the process.

Rosen and Dincer (2003) investigate the relation between capital costs and thermodynamic losses for devices in modern coal-fired, oil-fired, and nuclear electrical-generating stations. Thermodynamic loss rate-to-capital cost ratios are used to show that, for station devices and the overall station, a systematic correlation appears to exist between capital cost and exergy loss (total or internal), but not between capital cost and energy loss or external exergy loss. The possible existence is indicated of a correlation between the mean thermodynamic loss rate-to-capital cost ratios for all of the devices in a station and the ratios for the overall station, when the ratio is based on total or internal exergy losses. This correlation may imply that devices in successful electrical generating stations are configured so as to achieve an overall optimal design, by appropriately balancing the thermodynamic (exergy-based) and economic characteristics of the overall station and its devices. The results may (1) provide useful insights into the relations between thermodynamics and economics, both in general and for electrical generating stations, (2) help demonstrate the merits of second law analysis, and (3) extend throughout the electrical utility sector.

Dincer and Rosen's (2007) book deals with the exergy theory and its applications to various energy systems. The application examples are presented as a potential tool for design, analysis, and optimization. The authors enunciate the role exergy in minimizing and/or eliminating environmental impacts and providing sustainable development. Applying an exergy analysis to a FC system, De Groot and Woudstra (1995) showed that heat transfer is the cause of large losses. These losses may greatly be reduced if the amount of heat transfereed in the system is reduced. This may be done, for example, by the application of internal reforming. In the book by Dincer and Rosen (2007) several key topics ranging from the basics of the thermodynamic concepts to advanced exergy analysis techniques in a wide range of applications are covered. These topics are: psychrometric processes, drying operations, heat pump systems, crude oil distillation systems, cryogenic systems, cogeneration-based distric energy systems, crude oil distillation systems, fuel cell systems (PEM fuel cells, SOFC, hybrid SOFC-gas turbine [GT] systems), and aircraft flight systems. Also, basic aspects of exergoeconomics of practical systems are covered. In this frame, see also Bockris et al. (1992) for the electrochemistry of waste removal.

Dijkema and Weijnen (1997) discuss the use of fuel cells in the process industry which, among other purposes, is associated with the utilizing of low-BTU off-gases in the industry by implementing fuel cells on site. The FC systems in industry can be used to convert byproduct hydrogen into electric power and heat. In addition FC stacks can be fully integrated into the process system design to improve the plant's performance. In the latter case, the arrangement of fuel cell stacks is considered one of the unit operations, and the equipment can be shared between the process and the FC system. Finally trigeration systems can be designed that simultaneously produce chemicals, power and heat as equally commercial products. FC applications in the process industry can be identified by a three-step method. The economic feasibility of these applications not only depends on the FC stack lifetime and stack capital cost, but also on the effect on the process performance. As stated by Palazzi et al. (2007), for a steady power generation, fuel cell-based systems are being foreseen as a valuable alternative to thermodynamic cycle-based power plants in small-scale applications. As the technology is not yet established, many aspects of fuel cell development are currently investigated worldwide. Part of the research focuses on integrating the fuel cell in a system that is both efficient and economically attractive. To address this problem, the paper authors present a thermo-economic optimization method that systematically generates the most attractive configurations of an integrated system. In the methodology of Palazzi et al. (2007), the energy flows are computed using conventional process simulation software. The system is integrated using the pinch-based methods that rely on optimization techniques.

This defines the minimum of energy required and sets the basis to design the ideal heat exchanger network. A thermo-economic method is then used to compute the integrated system performances, sizes and costs. This allows performing the optimization of the system with regard to two objectives: minimize the specific cost and maximize the efficiency. A solid oxide fuel cell (SOFC) system of 50 kW integrating a planar SOFC is modeled and optimized leading to designs with efficiencies ranging from 34 to 44%. The multiobjective optimization strategy identifies interesting system configurations and their performance for the developed SOFC system model. The method of Palazzi et al. (2007) proves to be an attractive tool to be used both as an analysis tool and as support to decision makers when designing new systems.

Kazim (2004) presents a comprehensive exergy analysis of a 10 kW PEM fuel cell at variable operating temperatures, pressures, cell voltages and air stoichiometrics. The calculations of the physical and chemical exergies, mass flow rates, and exergetic efficiency are performed at temperature ratios T/T_0 and pressure ratios P/P_0 ranging from 1 to 1.25 and from 1 to 3, respectively. In addition, the analysis is conducted on fuel cell operating voltages of 0.5 and 0.6 V and at air stoichiometrics of 2, 3, and 4 in order to determine their effects on the efficiency of the fuel cell. The calculated results illustrate the significance of the operating temperature, pressure, cell-voltage, and air stoichiometry on the exergetic efficiency of the fuel cell. However, it is recommended that the fuel cell should operate at stoichiometric ratios less than 4 in order to maintain the relative humidity level in the product air and to avoid the membrane drying out at high-operating temperatures. The exergy efficiency of a fuel cell system is determined in terms of the ratio of the power output of the fuel cell over the differences between the total exergies of the reactants and products of the fuel cell electrochemical process. The total exergy of the reactants and the products consist of both physical and chemical exergies calculated for each element in the electrochemical process. From the current results, conclusions could be drawn concerning the fuel cell exergetic efficiency, which can be improved by adopting any or a combination of the four operational measures. Firstly, the exergetic efficiency of a PEM fuel cell can be improved by higher operating pressure. However, a high-pressure difference between the cathode and the anode is recommended in order to enhance the electro-osmotic drag phenomena between the two electrodes. Second, the efficiency of the fuel cell can be increased through increasing the fuel cell operating temperature in spite of the small and low temperature range of a PEM fuel cell as opposed to other types of fuel cells that operate at high temperatures. Third, higher exergetic efficiency could be attained if the fuel cell operates at relatively higher cell voltages that would require less mass flow rates for the reactants and the products to achieve a high electrical output. Fourth, by having a high air stoichiometry, a significant increase in the efficiency of the fuel cell can be achieved, although it would be recommended to have an air stoichiometry in the range between 2 and 4 in order to maintain the relative humidity level in the product air and avoid the fuel cell membrane drying out at high operating temperatures.

Hybrid micro gas turbine fuel cell systems attracted growing attention. Bakalis and Stamatis (2012) deal with the examination of a hybrid system consisting of a precommercially available high temperature solid oxide fuel cell and an existing recuperated microturbine. The irreversibilities and thermodynamic inefficiencies are evaluated after examining the full and partial load exergetic performance and estimating the amount of exergy destruction and the efficiency of each hybrid system component. At full load operation the system achieves an exergetic efficiency of 59.8%, which increases during the partial load operation. Effects of the various performance parameters such as fuel cell stack temperature and fuel utilization factor are assessed. The results show that the components in which chemical reactions occur have the higher exergy destruction rates. The exergetic performance is affected significantly by the stack temperature. Based on the exergetic analysis, suggestions are given for reducing the overall system irreversibility. Finally, the environmental impact of the operation of the hybrid system is evaluated and compared with a similarly rated conventional gas turbine plant. From this comparison it is apparent that the hybrid system has nearly double exergetic efficiency and about half the amount of greenhouse gas emissions compared with the conventional plant.

Calise et al. (2006, 2007) discuss the simulation and exergy analysis of a hybrid solid oxide fuel cell–gas turbine (SOFC–GT) power system. In the SOFC reactor model, by assumption, only hydrogen participates in the electrochemical reaction and the high temperature of the stack pushes the internal steam reforming reaction to completion. The unreacted gases are assumed to be fully oxidized in the combustor downstream of the SOFC stack. Compressors and GTs are modeled on the basis of their isentropic efficiency. For the heat exchangers and the heat recovery steam generator, all characterized by a tube-in-tube counterflow arrangement, the simulation is carried out using the thermal efficiency–NTU approach. Energy and exergy balances are performed not only for the whole plant but also for each component to evaluate the distribution of irreversibility and thermodynamic inefficiencies. Simulations are performed for different values of pressure, fuel utilization factor, fuel-to-air, and steam-to-fuel ratios and current density. Results show that, for a 1.5MW system, an electrical efficiency close to 60% can be achieved using appropriate values of the most important design variables; in particular, the operating pressure and cell current density. When heat loss recovery is also taken into account, a global efficiency of about 70% is achieved. Calise et al. (2007) have constructed a detailed model of a solid oxide fuel cell (SOFC) tube, equipped with a tube-and-shell pre-reformer unit. Both SOFC tube and pre-reformer are discretized along their axes. Energy, mole and mass balances are performed for each slice of the components, allowing the calculation of temperature profiles. Based on this modeling, temperatures, pressures, chemical compositions, and electrical parameters are evaluated for each slice of the two components under investigation. The influence of the most important design parameters on the performance of the system is investigated.

A paper by Ptasinski et al. (2002) concerns a new method of sewage sludge treatment that contributes more than traditional methods to the sustainable technology by achieving a higher rational efficiency of processing. This is obtained by preserving the chemical exergy present in the sludge and transforming it into a chemical one--methanol. The proposed method combines a sludge gasification process and a modified methanol plant. The sludge gasification produces a synthetic gas mainly containing CO and H_2, which gas is next used as a reactant to make methanol. The plant is simulated by a computer model using the flow-sheeting program Aspen Plus. The exergy analysis is performed by Ptasinski et al. (2002) for various operating conditions and their optimal values are found. The irreversibilities of different plant segments: thermal dryer, gasifier, gas cleaning, compressors, methanol reactor, distillation column, and purge gas combustion were evaluated. Ptasinski et al. (2002) show that the rational efficiency of the overall process is much higher than that of traditional plants.

Prins et al. (2003) presented the first and second law analysis in the context of gas–char reactions. Passing from coal to biomass gasification, Prins et al. (2007) studied the effect of fuel contenton the thermodynamic efficiency of gasifiers and gasification system. A chemical equilibrium model was used to describe the gasifier; obviously, the equilibrium model presented the highest gasification efficiency that can be possibly attained for a given fuel. Gasification of fuels with varying composition of organics in terms of O/C and H/C ratio was compared. It was found that exergy losses in gasifying wood (O/C ratio around 0.6) are larger than those for coal (O/C ratio around 0.2). It could thus be attractive to modify the properties of highly oxygenated biofuels prior to gasification, for example, by separation of wood into its components and gasification of the lignin component, thermal pretreatment, and/or mixing with coal in order to enhance the heating value of the gasifier fuel. Ptasinski (2008) reviewed thermodynamic efficiency of biomass gasification and biofuels conversion. The exergy analysis was used to analyze the biomass gasification and conversion of biomass to biofuels. The thermodynamic efficiency of biomass gasification was reviewed for air-blown as well as steam-blown gasifiers. Overall technological chains biomass-to-biofuels were evaluated, including methanol, Fischer-Tropsch hydrocarbons, and hydrogen. The efficiency of biofuels production was compared with that of fossil fuels.

Biomass has great potential as a clean, renewable feedstock for producing modern energy carriers. Ptasinski et al. (2007) focus on the process of biomass gasification, where the synthesis gas may subsequently be used for the production of electricity, fuels, and chemicals. The gasifier is one of the least-efficient unit operations in the whole biomass-to-energy technology chain and an analysis of the efficiency of the gasifier alone can substantially contribute to the efficiency improvement of this chain. The research of Ptasinski et al. (2007) is focused to compare different types of biofuels for their gasification efficiency and benchmark this against gasification of coal. In order to quantify the real value of the gasification process exergy-based efficiencies, defined as the ratio of chemical and physical exergy of the synthesis gas to chemical exergy of a biofuel, are proposed. Biofuels considered include various types of wood, vegetable oil, sludge, and manure. Exergetic efficiencies are evaluated by Ptasinski et al. (2007) for an idealized gasifier in which chemical equilibrium is reached, ashes are ignored, and heat losses are neglected. The gasification efficiencies are evaluated by them at the carbon-boundary point, where exactly enough air is added to avoid carbon formation and achieve complete gasification. The cold-gas efficiency of biofuels was found to be comparable to that of coal. The exergy efficiencies of biofuels are lower than the corresponding energetic efficiencies. For liquid biofuels, such as sludge and manure, gasification at the optimum point is not possible, and exergy efficiency can be improved by drying the biomass using the enthalpy of synthesis gas.

Juraščík et al. (2010) present the results of exergy analysis for a biomass-to-synthetic natural gas (SNG) conversion process. Their study is based on wood gasification, which is analyzed for different gasification conditions like temperature and/or pressure. The analyzed temperature was varied in the range from 650 to 800°C and the pressure range was from 1 to 15 bar. The main process units of biomass-to-SNG conversion technology are gasifier, gas cleaning, synthesis gas compression, CH_4 synthesis and final SNG conditioning. The results shows that the largest exergy losses take place in the biomass gasifier, CH_4 synthesis part and CO_2 capture unit. The overall exergetic efficiency of the biomass-to-SNG process was estimated in the range of about 69.5–71.8%. Biomass is a renewable feedstock for producing energy carriers. Piekarczyk et al. (2013) have performed a thermodynamic evaluation of biomass-to-biofuels production systems. Yet, the usage of biomass is accompanied by possible drawbacks, due to limitation of land and water, and competition with food production. The analysis concerns so-called second-generation biofuels, like Fischer–Tropsch fuels or substitute natural gas which are produced either from wood or from waste biomass. For these biofuels the most promising conversion case involves production of syngas from biomass gasification, followed by the synthesis of biofuels. The thermodynamic efficiency of biofuels production is analyzed and compared using both the direct exergy analysis and the thermo-ecological cost. This analysis leads to the detection of exergy losses in various elements which forms the starting point to the improvement of conversion efficiency. The efficiency of biomass conversion to biofuels is also evaluated for the whole production chain, including biomass cultivation, transportation and conversion. The global effects of natural resources management are investigated using the thermo-ecological cost (Szargut, 1986, 2002b). The energy carriers' utilities such as electricity and heat are generated either from fossil fuels or from biomass. In the former case the production of biofuels not always can be a renewable energy source whereas in the latter one the production of biofuels leads always to the reduction of depletion of nonrenewable resources. Vitasari et al. (2011) worked out an exergy analysis of SNG production via indirect gasification of various biomass feedstock. The process is simulated with a computer model using the flowsheet program Aspen Plus. The analysis is performed for various operating conditions such as gasifier pressure, methanation pressure, and temperature. The largest internal exergy losses occur in the gasifier followed by methanation and SNG conditioning. Exergetic efficiency of biomass-to-SNG process for woody biomass is higher than that for waste biomass.

Periodic adsorption processes have been widely used for industrial applications, mainly because it spends less energy than the usual gas separation processes (Barg et al., 2001). The pressure swing adsorption (PSA) units have been used as high-efficiency and low-cost alternatives to the industrial applications of gas separation. The largest commercial application of periodic adsorption processes is the PSA process of the hydrogen purification. Several works on the numeric simulation of such units have been launched that proposed different mathematical formulations to model the system. Despite of its wide use in the chemical and petrochemical industry, there are limited literature reports about the modeling of a complex commercial unit with multiple adsorbents, multiple beds, and several feed components. The developed mathematical model for a commercial PSA unit for hydrogen purification allows us to study effects of the feed temperature and cycle time. Most of the complex features of the industrial plant are predicted, enabling a realistic approach. In conclusion, when simulating an industrial PSA unit, the process characteristics can be represented well by a nonisothermal equilibrium model with extended Langmuir isotherm and by a linear driving force model to describe the adsorption dynamics. Effects of the feed temperature and cycle time may be studied. The influence of the feed temperature on the product purity depends on operational conditions, and can take the product out of specification. Variations in the cycle time may be done to identify the influence of this operational factor on the product purity and on hydrogen recovery. The product purity can be improved, depending on operational conditions, through a suitable reduction in the cycle time.

Aggregation processes are very common in chemistry and biology. Many aggregation processes, of considerable chemical or biological significance, are facilitated by surfaces. Most real life surface aggregation processes are so complex that it is extremely difficult to model such systems theoretically from fundamental statistical mechanics and reaction rate theories. However, simple model systems can be designed, for which the aggregation rates and equilibrium constants can be calculated theoretically. Nag and Berry (2007) apply a simple model to develop thermodynamics and kinetics for bulk and surface aggregation processes capable of competing with each other. The processes are the stepwise aggregation of monomers in a fluid medium and on an inpenetrable solid-surface bounding the fluid medium, besides the adsorption and desorption of the same species at the solid-fluid interface. Emphasis is on aggregation processes in the high-friction limit. The theoretical model serves to compare the kinetics and thermodynamics of the processes and to infer the conditions in which one process dominates another, in the high-friction limit, such as in a liquid. The motivation of the study is obtaining insight into competition between aggregation in solution and on an adjoining surface, such as a cell membrane. The authors examined the relative importance of bulk solution and of surface as sites for aggregation. Several variables play significant roles in this phenomenon. Three key parameters have been varied: the intermonomer binding energy, the monomer-surface binding energy, and size of the monomers. The calculations show how the rates of binary aggregation and the equilibrium ratios of monomers in solution and on the surface compete and how one or the other can be made to dominate in the high-friction regime. The analysis was extended to the binary association rates of monomers with already-formed dimers and larger aggregates, on the surface and in solution. Either of the adsorption–desorption and solution dimerization equilibrium constants may be made to dominate the other by adjusting the same three parameters. A generalization emerges: the surface equilibrium constant for dimer concentration is always greater than the corresponding solution constant for the physically relevant parameter ranges. Nag and Berry (2007) point out that these results would be a good starting point for comparing the kinetics and thermodynamics of aggregation of globular proteins in solution with that on membranes.

Artificial neural networks are becoming more and more popular. Bator and Sieniutycz (2006) studied the application of artificial neural network (ANN) for predicting suspended particulate concentrations in urban air, taking into account meteorological conditions. Calculations are based on pollution measurements taken in the city of Radom, Poland, in the period 2001–2002. PM10 emission and primary meteorological data were used to train and test the application of network. Two different methods of emission calculation were applied. First, ANN method based on multilayer perceptron with unidirectional information flow was used. Second, a hybrid model based on a modified Gaussian model of Pasquille's type and ANN with radial base function (RBF) was applied. Network architecture and transition function types were described. Statistical assessment of the obtained results was made. In addition, hybrid model results were compared with emission calculations of dust pollution based on the Gaussian model, including various methods of calculation of pollution dispersion coefficients. See also Williams et al. (2001) for the greenhouse gas control technologies and additional information about emissions of pollutants and greenhouse gases.

Ao et al. (2007) have developed gray modeling of heat-pump-surrounding system exergy efficiency. They assume the heat-pump-surrounding (HPS) system as a gray system and makes a gray model, GM (1, N), for the HPS system to assess the system exergy efficiency. It attains the system exergy efficiency laws of the HPS system through some steps including the gray relationship analysis, creating gray number, setting up GM (1, N) model, and so on and also verifies the accuracy of the laws from the gray model. It shows that gray system theory can be used in the HPS system to make gray models and present accuracy system exergy efficiency laws when the HPS as a whole system lacks enough practically tested data. Ao et al. (2007) also suggest the variables linked lower-level heat source parameters should be given obviously in the relation analyzing the HPS system exergy efficiency.

Burghardt (2008) has reviewed future challenges for research in chemical and process engineering (ChPE). In his opinion, the increased demand of specialized materials and active compounds which are of much more complex form than the commodity chemicals has shifted the focus of manufacturing from the process design to product design and the key question is now what product would be manufactured and not how it will be produced. The study presents an analysis of the experimental tools and theoretical methods indispensable in solving such multiscale and interdisciplinary problems. He states that ChPE made recently a great progress in the fiel of research and development of new technologies and now its most important task is to create real connections between all the scales of the manufacturing process and in this way to develop a methodology for the design and optimization of novel production processes based on the multiscale and interdisciplinary approach. More powerful computers and computational techniques as well as new analytical instruments connected with new measurement techniques enable the formulation of problems of product design proceeding from the nanoscale to the macroscale and thus to create the key elements of the future development of new technologies. ChPE, which quantitatively integrates fundamentals of chemical kinetics and catalysis with transport phenomena, followed by modeling, design and optimizations of processes and industrial plants can have a prevailing part in this development. For ChPE, it is now important to shift the gravity center from the traditional sequential solution of problems to the simultaneous global solution methods devoting special attention to the connections between scales. The key need for such an integrated approach is the interdisciplinary procedure of solving problems. This should help to enlarge and modify its core and thus to cope with the future challenges of frontier areas.

However not only engineering but also applied physics uses FTT concepts in its reasoning and analyses. For example, heat flow carried by electrons in a thermoelectric device requires a surprisingly wide "pipe"—a rare case where quantum effects have macroscopic consequences, Whitney (2014). From the theory of finite-time thermodynamics we know that power-producing machines are only Carnot-efficient if they are reversible, but then their power output is vanishingly small. As in a typical FTT work, Whitney (2014) asks, what is the maximum efficiency of an irreversible thermoelectric quantum device with finite power output? He uses a nonlinear scattering theory to answer this question for thermoelectric quantum systems, heat engines, or refrigerators consisting of nanostructures or molecules that exhibit a Peltier effect. He finds that quantum mechanics places an upper bound on both power output and on the efficiency at any finite power. The upper bound on efficiency equals Carnot efficiency at zero power output but decays with increasing power output. It is intrinsically quantum (wavelength dependent), unlike Carnot efficiency. This maximum efficiency occurs when the system lets through all particles in a certain energy window, but none at other energies. A physical implementation of this phenomenon is discussed, as is the suppression of efficiency by a phonon heat flow.

Characteristic equations and average and marginal costs are analyzed by Valero, Serra and their coworkers (Erlach et al., 1999). Structural Theory of Thermoeconomics is proposed as a standard and common mathematical formulation for all thermoeconomic methodologies employing thermoeconomic models that can be expressed by linear equations. In previous works it has been demonstrated that the Exergy Cost Theory (ECT), the AVCO approach, and the Thermoeconomic Functional Analysis (TFA) can be dealt with the Structural Theory. In this paper, it is demonstrated that the last-in-first-out (LIFO) approach, a thermoeconomic cost accounting method, can also be reproduced with the Structural Theory. The LIFO and the Structural Theory are both applied to a combined cycle plant and it is shown that the equation systems obtained from both methods are the same. Moreover, a procedure to develop the productive structure representing the thermoeconomic model of LIFO (ie, from which the same costing equations are obtained as with the LIFO approach) is explained in detail, which provides the tools to reproduce the costs obtained from LIFO with the Structural Theory for complex energy systems. This concludes a series of works where it has been demonstrated that the most developed thermoeconomic optimization and cost accounting methodologies, as all of them employ thermoeconomic models that can easily be linearized, can be dealt with by the mathematical formalism of the Structural Theory. As a consequence the structural theory is a general mathematical formalism either for thermoeconomic cost accounting and/or optimization methods, providing a common basis of comparison among the different methodologies, which could be considered the standard formalism for thermoeconomics. The cost calculation method of the structural theory is based on the rules of mathematical derivation applied to a set of characteristic equations that determine the thermoeconomic model. In this way, as more physical and realistic information is contained in the characteristic equations, more physical significance will be contained in the costs obtained.

Sciubba (2000, 2001) presents an ambitious approach to the evaluation of energy conversion processes and systems, based on an extended representation of their exergy flow diagram. This approach is of thermoeconomic type and represents a systematic attempt to integrate into a unified coherent formalism both Cumulative Exergy Consumption and Thermoeconomics. In fact, the approach constitutes a generalization of both, in that its framework allows for a direct quantitative comparison of nonenergetic quantities like labor and environmental impact. A critical examination of the existing state-of-the-art of energy- and exergy analysis methods and paradigms indicates that an extension of the existing "Design and Optimization" procedures to include explicitly "nonenergetic externalities" is feasible. It appears that it can be successfully argued that some of the issues that are difficult to address with a purely monetary theory of value can be resolved by Extended Exergy Accounting ("EEA" in the following) methods without introducing arbitrary assumptions external to the theory. In this respect, Sciubba's (2001) EEA can be regarded as a natural development of the economic theory of production of commodities, which it extends by properly accounting for the unavoidable energy dissipation in the productive chain. While a systematic description of the EEA theory is discussed in previous work by the author, this paper aims at a more specific target, and presents a formal representation of the application of EEA to a cogenerating plant based on a gas-turbine process. It is shown that the solution indeed leads to an "optimal" design, and that its formalism embeds even extended thermo-economic formulations.

Continuation algorithms ware shown the only assured techniques of solving the highly nonlinear problems: azeotropic distillation, extractive distillation, and liquid extraction (Riona and Van Brunta, 1990). Unfortunately, the robustness of solution is accompanied by increased computation. New procedures that increase the efficiency of these algorithms were documented. Local differential geometry is exploited during the continuation procedure to provide a more accurate prediction of the solution trajectory. A rigorous method was documented for accurate prediction of the unit tangent, principal unit normal, and the curvature of the solution path. The resulting computational procedure is significantly more efficient than other continuation methods. Riona and Van Brunta (1990) show that the effects of increased accuracy of prediction are threefold. The number of (1) continuation steps, (2) corrections to return to the solution path, and (3) trajectory prediction failures are all reduced. See Bennett (1982) for the thermodynamics of computation.

In his book on energy and process optimization for the process industries Zhu (2014) describes techniques to optimize processing energy efficiency in process plants. His book on energy and process optimization for the process industries provides a holistic approach that considers optimizing process conditions, changing process flow schemes, modifying equipment internals, and upgrading process technology that has already been effectively used in a process plant. Tested by operating plants, the book describes technical solutions to reduce energy consumption leading to significant returns on capital. The book provides managers, chemical and mechanical engineers, and plant operators with methods and tools for continuous energy and process improvements.

Tondeur and Kvaalen (1987) and Tondeur (1990) developed a principle of equipartition of entropy production: a practical optimality criterion for transfer and separation processes. They ask a basic and practical question: in what sense can one say that the distribution of the driving forces in one industrial configuration is better than in another. Several realistic chemical engineering examples, involving heat exchangers, plate distillation columns, chromatographic separators, and so forth are analyzed. The duty (useful effect) for each process is defined and the following equipartition of the entropy production is advanced: "for a given duty, the best configuration of the process is that in which the entropy production rate is most uniformly distributed". A generalization of the principle is also proposed for systems with a distributed design variable: "the optimal distribution of an investment is such that the investment in each element is equal to the cost of the energy degradation in this element". It follows that a uniform distribution of the ratio of these two quantities should be preserved. These results show the importance of irreversibility analysis for rational design. There are also associated approaches of this sort. For example, Sieniutycz (1973a) finds that in optimal the higher equipment costs should be concentrated in places where the process is most intense.

Hoffmann (1990) discusses aims and methods of the finite-time thermodynamics from the viewpoint of finite-time bounds, optimally of irreversible engines and applications of simulated annealing. By extending the Salamon and Berry (1983) formula for the bound on the dissipated availability, Hoffmann shows how work deficiency may be bounded by using a geometric quantity: the thermodynamic length. Hoffmann et al. (1997, 2003) discussed optimal process paths for endoreversible systems. Applying FTT, Mozurkewich and Berry (1981) determined the engine performance by optimization of piston motion. Li et al. (2009a, b) determined optimum work in real systems with two finite thermal capacity reservoirs. The results obtained can be compared with those of Sieniutycz (1998b, 2006b). The similarities and differences for these cases are as follows: for fixed initial and final states and duration of processes, if one only controls the external working fluid, whether the second fluid is finite or not and whether the thermal machines are reversible or not, the optimal temperature profiles and the optimal controls of the external working fluid are the same. The optimal work of the cases is different. Kreuzer et al. (1988) formed disequilibrium thermodynamics principles for sublimation, evaporation, and condensation. Poplewski and Jezowski (2003) effectively applied a random search type approach for designing optimal water networks. Sih (1985, 1992) proposed a new approach to nonequilibrium thermomechanics based on the so-called isoenergy density theory. Puncochar and Drahos (2003) found the entropy of fluidized bed as an effective measure of particles mixing.

Sieniutycz and Berry (2000) presented a discrete Hamiltonian analysis of endoreversible cascades of thermal engines. Their limiting solutions for an infinite number of stages represent dynamical evolutions of continuous engines with finite upper reservoirs, whose temperature decreases in time. Sieniutycz and Kubiak (2002) defined dynamical energy limits in traditional and work-driven operations. Sieniutycz and Salamon (1990a, b, c) described the diverse nonequilibrium theories, and their extremum principles, and proposed a phenomenological approach to the problem of the nonequilibrium metric in the field (Eulerian) description of the fluid motion. Considering the state of affair of second-law analyses and finite-time thermodynamics Sieniutycz and Salamon (1990d) and Sieniutycz and Shiner (1994) focused on the relation between thermodynamics of irreversible processes and chemical engineering. Sieniutycz and Szwast (1999) exemplified optimization of multistage thermal machines by a Pontryagin's-like discrete maximum principle. Sieniutycz and Szwast (2003) evaluated work limits in imperfect sequential systems with heat and fluid flow. In a comprehensive report Sieniutycz et al. (2011) developed thermodynamic optimization of chemical and electrochemical energy generators with applications to fuel cells. Sieniutycz et al. (2012) treat power generation in thermochemical and electrochemical engine systems. There are also applications of thermodynamics in electrolysers and membranes. For example, Xu et al. (2011) analyse the performance of a high temperature polymer electrolyte membrane water electrolyser. We do not consider here applications of his sort.

Sieniutycz (2000c) constructs thermodynamic criteria for optimization of sequential work-assisted heating and drying operations which run jointly with "endoreversible" thermal machines. Total power input is minimized with constraints which take into account the dynamics of heat and mass transport. Sieniutycz (2000d) presents thermodynamic analysis of a problem of maximum work produced by an irreversible system with a finite-resource fluid and a bath fluid. Inherent irreversibilities are admitted in fluids' boundary layers. Work functional is derived and the solution of canonical equations is given. Irreversible properties cause the difference between the work supplied and delivered. Sieniutycz (2001a, b) extends to the realm of complex fluids and finite durations the problem of extremum work delivered from (or consumed by) the nonequilibrium system composed of a complex fluid, a perfect thermal machine and the environment or an infinite reservoir. Optimal work functions, which incorporate an inevitable minimum of the entropy production are found as functions of end states, duration, and (in discrete processes) number of stages. Sieniutycz (2003a, b, 2011c) synthesizes thermodynamic models of traditional and work-driven operations with heat and mass exchange. These works provide the principles of construction of the so-called Carnot variables which are particularly suitable for easy determining power limits in various power systems. Applying these controls one can develop a unified analysis showing how various transfer phenomena (heat, mass, and electric-charge transfer) effect power limits in various energy converters, such as thermal, chemical, and radiation engines and fuel cells. Diverse pseudo-Carnot structures govern converters' efficiencies and applied to estimate irreversible power limits in steady state systems. Radiation engines are treated as systems described by Stefan–Boltzmann equations. Both chemical and electrochemical energy generators (fuel cells) satisfy similar principles of modeling and apply similar schemes of power evaluation as thermal machines. The systematic application of Carnot controls as variables which satisfy identically the internal entropy constraint means, in fact, the emergence of a coherent, methodologically novel approach to the family of power systems. Such approach has the virtue of reducing the number of controls and is superior to the traditional approach, which works with an enlarged number of traditional constrained controls. Sieniutycz's (2003c) synthesis reviews problems of thermodynamic limits on production or consumption of mechanical energy in practical and industrial systems including constrained multistage systems described by quasi-canonical structures of optimal control. Sieniutycz's (2006b) development of generalized (rate dependent) availability adds to the earlier formulas corrections resulting from internal irreversibilities (abandoning the endoreversibility). Preserving usual definition of Hamiltonian H as the scalar product of rates and generalized moments Sieniutycz (2006c) developed the optimization theory for basic classes of discrete optimal control processes governed by difference rather than differential equations for the state transformation. Power optimization algorithms in the form of HJB equations lead to work limits and generalized availabilities. Converter's performance functions depend on end thermodynamic coordinates and a process intensity index, h, in fact, the Hamiltonian of the optimization problem. As an example of limiting work from radiation, a finite rate exergy of radiation fluid is estimated in terms of finite rates quantified by Hamiltonian h.

Sieniutycz (2007d, 2008b, 2009b) analyzes steady and dynamic models of chemical engines and the expressions for power and entropy production. Sieniutycz (2009a) describes the application of Lagrange multipliers in the context of optimization of power systems by dynamic programming. In power production problems maximum power and minimum entropy production are inherently connected by the Gouy–Stodola law. Maximum power and/or minimum entropy production are governed by Hamilton–Jacobi–Bellman equations (HJB equations) which describe the potential function of the problem and associated controls. Yet, in many cases the optimal relaxation curve is nonexponential and HJB equations cannot be solved analytically. Systems with nonlinear kinetics (eg, radiation engines) are particularly difficult, thus, discrete counterparts of continuous HJB equations and numerical approaches are recommended. Discrete algorithms of dynamic programming (DP), which lead to work limits and generalized availabilities, are especially effective. Sieniutycz (2009a) considers convergence of discrete algorithms to solutions of HJB equations, solutions by discrete approximations, and role of Lagrange multiplier λ associated with duration constraint. In analytical discrete schemes, the Legendre transformation is a significant tool leading to original work function. Numerical algorithms of dynamic programming and dimensionality reduction in these algorithms are described. Indications showing the method potential for other systems, in particular chemical energy systems, are outlined. Sieniutycz (2009b) presents a thermodynamic approach to simulation and modeling of nonlinear energy converters, in particular radiation engines. Novel results are obtained especially for dynamical engines when the temperature of the propelling medium decreases in time due to a continual decrease of the medium's internal energy caused by the power production. Basic thermodynamic principles determine the converter's efficiency and work limits in terms of the entropy production. The real work is a cumulative effect obtained in a system of a resource fluid, a sequence of engines, and an infinite bath. Nonlinear modeling involves dynamic optimization in which the classical expression for efficiency at maximum power is generalized to endoirreversible machines and nonlinear transfer laws. The primary result is a finite-rate generalization of the classical, reversible work potential (exergy). The generalized work function depends on thermal coordinates and a dissipation index, h, that is, a Hamiltonian of the minimum entropy production problem. This generalized work function implies stronger bounds on work delivered or supplied than the reversible work potential. The role of the nonlinear analyses and dynamic optimization is shown especially for radiation engines. As an example of the kinetic work limit, generalized exergy of radiation fluid is estimated in terms of finite rates, quantified by the Hamiltonian h. Sieniutycz (2009d) investigates an innovative thermodynamic scheme serving to recover some of the large drying energy in the superior form of mechanical work at the cost of a small reduction of drying rate. Sieniutycz (2010d, 2011d, 2012a, 2013a, b) worked out unified power optimization approaches for various energy converters, such like: thermal, solar, chemical, and electrochemical engines. Thermodynamics leads to converter's efficiency and limiting power. Efficiency equations serve to solve problems of upgrading and downgrading of resources. The main novelty of the paper in the energy yield context consists in showing that the generalized heat flux Q (involving the traditional heat flux q plus the product of temperature and the sum products of partial entropies and fluxes of species) plays in complex cases (solar, chemical, and electrochemical) the same role as the traditional heat q in pure heat engines. The presented methodology is also applied to power limits in fuel cells. Examples show power maxima in fuel cells and prove the suitability of the thermal machine theory to chemical and electrochemical systems. The introduction of a reduced Gibbs free energy change between products p and reactants s takes into account the decrease of voltage and power decrease caused by the incomplete conversion of the reaction. This is a synthesizing thermodynamic approach to modeling and power maximization in various energy converters, such as thermal, solar and chemical engines and fuel cells. The efficiency decrease is linked with thermodynamic and electrochemical irreversibilities expressed in terms of polarizations (activation, concentration, and ohmic). Maximum power data provide bounds for SOFC energy generators, which are more exact and informative than reversible bounds for electrochemical transformation.

Applications of thermodynamics in environmental and water engineering are represented well by a book by Singh (2013). The book deals with basic concepts of entropy theory from a hydrologic and water engineering perspective and with applications of these concepts to a range of water engineering problems. Environmental applications of concepts and techniques vary across different subject areas, and the book relates them directly to many practical problems in the field. The book also explains suitable extremum principles for entropy and presents their applications to different types of probability distributions. Techniques of spectral analyses are developed for maximum entropy and minimum cross entropy. These techniques are applied in problems of environmental science and engineering and described with illustrative examples.

CHAPTER 9

Thermodynamic Controls in Chemical Reactors

9.1 Introduction

In this chapter we outline various thermodynamic problems of chemical reactions and chemical reactors, partly in the literature context. These problems are usually linked to the applications of phenomenological and statistical thermodynamics, macroscopic kinetics, and optimal control theory, which are the fields that help formulate static and dynamic models of chemical, electrochemical and biological reactors. The material covered here will also be helpful to formulate the models of fuel cells considered in the next chapter. Obtained mathematical models are frequently used to forecast the reactor's outlet conversion, to design required control arrangements, or to perform various optimizations of chemical systems.

In the industrial reality often the reactor is merely a component of a complex system which, for example, comprises of a preprocessing unit, heating and cooling subsystems, separation train, and so forth, that is all what is encountered on the way between raw materials to products, Fig. 9.1. An engineer who separates the reactor from the whole system, for example, for the analysis or optimization, should at least keep in mind the standard range of values of the process variables under which the reactor runs. Preserving these values in quantitative analyses, numerical calculations, optimization algorithms, and so forth, increases the practical value of the results obtained, by making them closer to everyday practices. However, in free researches, one could go beyond these standard values and test some new regions of possibly better performance.

In the first part of this chapter we analyze nonlinear chemical and thermal transport processes, where transport steps are treated as peculiar chemical reactions described by appropriate affinities. Rate and transport equations contain terms exponential with respect to Planck's potentials and temperature reciprocal. In every elementary step, of transport or rate nature, kinetic mass action law leads to the identification of two competing unidirectional fluxes. While they both are equal in the thermodynamic equilibrium, their difference out of the equilibrium constitutes the observed flux, which represents the resulting rate. A generalized affinity emerges along with the question if it represents the true or at least acceptable a driving force. Correspondence with classical Butler–Vollmer kinetics is secured for electrochemical systems. Near thermodynamic equilibrium the theory converges to the

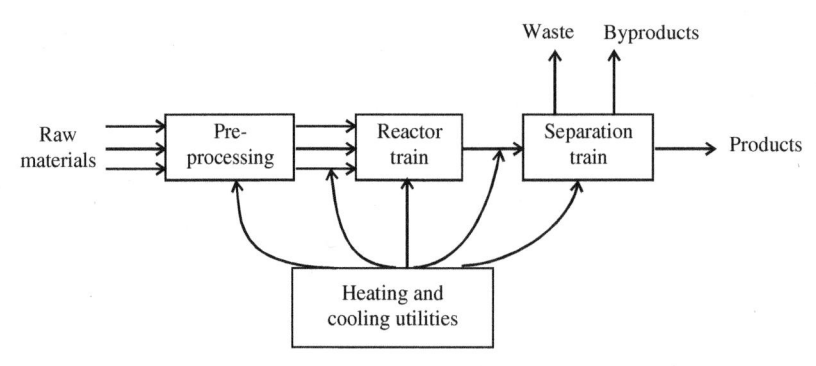

Figure 9.1: Simplified flow sheet of typical system in chemical industry.
Sieniutycz and Jeżowski, 2009.

standard linear kinetics described by Onsager's equations. On the way we discuss the notion of a chemical Ohm's law, a possible extension of the theory to heterogeneous systems, and some problems of stability and limit cycles, supplementary to those considered in chapter: Thermodynamic Lapunov Functions and Stability. Next group of problems refers to chaos and fractals in chemical systems.

In the second part of the chapter we outline the important recent achievements in modeling of chemical power production systems along with corresponding methods of thermodynamic optimization. This text is necessarily brief, first because of chapter's volume limitation and second, because it describes a vast, rapidly developing field linked with many applications of FTT. For more comprehensive studies, the reader is referred to the literature of the subject (de Vos, 1992; Wu et al., 2000; Chen and Sun, 2004; Sieniutycz and Jeżowski, 2009, 2013). In this chapter, the FTT method is aimed at maximum production of power in chemical systems, and after extension to surface reactions, it can efficiently treat complex electrochemical flow systems, in particular fuel cells (see chapter: Power Limits in Thermochemical Units, Fuel Cells, and Separation Systems). With the thermodynamic knowledge and methods of dynamic optimization (especially the dynamic programming), kinetic limits are estimated for optimal work function W_{max} that describes an integrated power output and generalizes familiar reversible work W_{rev} for finite rates. Optimization results lead to power limits, which in the context of fuel cells (next chapter), depend on the overvoltage effects caused by rate processes (electrochemical reactions) and associated transport phenomena.

Many chemical engineers are familiar with the early book of Aris (1961) on optimization of chemical reactors, where the author described a number of problems associated with maximum conversion for prescribed duration or minimum holdup time for prescribed conversion. Dynamic programming was applied as a main method of optimization. Also, an excursion to the maximum principle method of Pontryagin et al. (1962) was made by studying the properties of characteristic equations. Sometime later Aris (1964) gave a basic theoretical treatment on the discrete dynamic programming method, again with examples of maximum yield problems in chemical reactors. In his next, very readable book, Aris (1975) presented a comprehensive mathematical theory of diffusion and chemical reactions in permeable catalysts. Another research direction applied the second law to evaluate efficiencies of chemical processes as in the classical analysis of Denbigh (1956) analysis. The applications to power yield problems came later, and some of these are described in Section 10.

Calculation of chemical equilibria is an important task, which should be done before dealing with chemical kinetics, dynamics and optimization. For a chemical mixture, its free energy may be minimized subject to the constraints resulting from the conservation of atoms of elements. The minimization determines concentrations at chemical equilibrium (White et al., 1958; Van Zeggeren and Storey, 1970; Bailie, 1978; Li et al., 2004; Nicolis and De Wit, 2007). Minimizing the entropy production rate may also be an example of use of static optimization (Nummedal et al., 2003).

Still another approach involves the search for the maximum entropy in the nonequilibrium chemically reacting mixtures (Ugarte et al., 2005). They applied the maximum entropy principle to analyze nonequilibrium chemically reacting mixtures and to model complex chemical reaction processes. The maximum entropy principle was employed within the rate-controlled constrained-equilibrium (RCCE) method to determine concentration of different species in a nonequilibrium combustion process. It was assumed that the system evolves through constrained equilibrium states where the mixture entropy is maximized subject to constraints of the energy, volume, and number of elements. Mixture composition was determined by integrating a set of differential equations of constraints rather than integration of differential equations for species as is done with kinetics techniques. Since the number of constraints is much smaller than the number of species, the number of rate equations required to describe the time evolution is reduced. The method modeled the stoichiometric mixture of the formaldehyde-oxygen combustion process. In calculations 29 species and 139 reactions was used at different sets of pressures and temperatures, ranging from 1 atm to 100 atm, and from 900 K to 1500 K respectively. Three fixed elemental constraints: conservation of elemental carbon, elemental oxygen, and elemental hydrogen and from one to six variable constraints were used. The four to nine rate equations for constraint potentials (Lagrange multipliers conjugate to the constraints) were integrated and RCCE calculations always gave correct equilibria. Only eight constraints were needed for very good agreement with detailed calculations. Ignition delay times and major species concentrations were within 0.5–5% of values predicted by detailed chemistry calculations. Adding constraints improved the accuracy of the mole fractions of minor species at early times, but had only a little effect on the ignition delay times. Rate-controlled constrained-equilibrium calculations reduced the computation time by 50% when using eight constraints.

Continuing the maximum conversion direction, Bak et al. (2002) analyzed the maximum obtainable yield of B, starting from A in the consecutive chemical reactions $A \Leftrightarrow B \Leftrightarrow C$.

They use the temperature as a control variable for a given process duration. They show that all optimal paths start with a branch at infinite temperature. Depending on the energy barriers and collision factors, the optimal temperature may subsequently switch to either a finite temperature (internal optimum) or zero temperature if the "maximum useful time" is exceeded. This is a turnpike structure. The time rate of change of B is positive throughout the process. For certain sets of parameter values the production rate dB/dt will approach zero asymptotically for long durations, for longer it will become zero for finite durations. The authors derived a curve on which switching from this temperature to lower temperatures is possible. The production rate of B at the end of the reaction is equal to the incremental gain of B by increasing the duration when the path is re-optimized for this new duration. For given parameters, there is a unique "maximum useful time" which results in the largest yield of B possible. This duration may be infinite. If a duration longer than this is specified, all reactions should be shut off for that excess amount of time, a situation that makes most optimization routines become unstable. Compare Aris (1961) for earlier results.

Johannessen and Kjelstrup (2005) observed that thousands of numerical solutions of an optimal control problem for plug flow reactors form what they call a "highway in the reactors' state space." The problem was to find the heat transfer strategies, which minimize the entropy production in reactors with fixed chemical conversion. The control variable was always a temperature of the heating/cooling medium along the reactor. The highway represents the most energy efficient way to travel far in state space. Such highways were studied for five reactor systems, endothermic and exothermic ones. Numerical analysis showed that the reactor highway is characterized by approximately constant thermodynamic driving forces/local entropy production for reasonable process intensities. Each solution represents a compromise between the entropy production of reactions, heat transfer, and frictional flow (pressure drop). The solutions enter and leave the highway at different positions depending on how far from the highway their initial and final destinations are. Knowledge about the nature of the highway, for example, when the reactor operates in a reaction mode or a heat transfer mode, may be important for energy efficient reactor design. The theoretical formulation of the optimization problem is valid for plug flow as well as batch reactors. The authors showed that important results in literature like the Spirkl–Ries quantity, the theorems of equipartition of entropy production and equipartition of forces are contained in their general formulation.

9.2 Stoichiometry of General Chemical Reactions

Our present purpose will be the presentation of suitable forms for stoichiometric descriptions of general chemical reactions. Fig. 9.2 depicts a general scheme of steady, multireaction system where all possible species may appear on both sides of reaction. The virtue of such complex description is easy treatment of autocatalytic reactions.

$$n_1, \quad \begin{aligned} v^f_{11}B_1 + v^f_{21}B_2 + \ldots v^f_{i1}B_i \ldots \ldots \ldots + v^f_{m1}B_m &= v^b_{11}B_1 + v^b_{21}B_2 + \ldots v^b_{i1}B_i \ldots \ldots \ldots + v^b_{m1}B_m \\ v^f_{12}B_1 + v^f_{22}B_2 + \ldots v^f_{i2}B_i \ldots \ldots \ldots + v^f_{m2}B_m &= v^b_{12}B_1 + v^b_{22}B_2 + \ldots v^b_{i2}B_i \ldots \ldots \ldots + v^b_{m2}B_m \\ &\ldots \ldots \ldots \ldots \ldots \\ v^f_{1R}B_1 + v^f_{2R}B_2 + \ldots v^f_{iR}B_i \ldots \ldots \ldots + v^f_{mR}B_m &= v^b_{1R}B_1 + v^b_{2R}B_2 + \ldots v^b_{iR}B_i \ldots \ldots \ldots + v^b_{mR}B_m \end{aligned} \quad n_2,$$

Figure 9.2: Stoichiometry scheme of steady, multireaction system.
Flow reactor at a steady state.

This scheme in Fig. 9.2 assumes a set of chemical reactions, $j = 1, 2,\ldots,R$, undergoing between species B_i ($i = 1, 2,\ldots, m$) in the system. Arbitrary reaction j can be written in the form

$$v^f_{1j}B_1 + v^f_{2j}B_2 + v^f_{3j}B_3 + \ldots + v^f_{mj}B_m = v^b_{1j}B_1 + v^b_{2j}B_2 + v^b_{3j}B_3 + \ldots + v^b_{mj}B_m \tag{9.1}$$

where B_i are species participating in R chemical reactions $j = 1, 2,\ldots R$, and $v_{i,j}$ is the *absolute* stoichiometric coefficient of component i in reaction j. The reaction description in terms of Eq. (9.1) involves the reaction stoichiometry in Fig. 9.2.

Instead of the "absolute stoichiometry" of Eq. (9.1) we may introduce traditional stoichiometric coefficients $v_{ij} = v^b_{ij} - v^f_{ij}$ (De Groot and Mazur, 1984). Eq. (9.1) yields then

$$(v^b_{1j} - v^f_{1j})B_1 + (v^b_{2j} - v^f_{2j})B_2 + (v^b_{3j} - v^f_{3j})B_3 + \ldots + (v^b_{mj} - v^f_{mj})B_m = 0 \tag{9.2}$$

or

$$v_{1j}B_1 + v_{2j}B_2 + \ldots v^b_{ij}B_i + v_{i+1,j}B_{i+1} \ldots + v_{mj}B_m = 0 \tag{9.3}$$

Interpretation of the above stoichiometry is illustrated in Fig. 9.3. Negative v_{ij} refer to substrates and positive v_{ij} refer to products. Vanishing v_{ij} refer to inert species.

Considering the general system with arbitrary inlet and outlet molar flows, and assuming stoichiometry Eq. (9.3), the balance of mole numbers for jth chemical reaction yields

$$\frac{(n_{1_1} - n_{1_2})_j}{v_{1j}} = \frac{(n_{2_1} - n_{2_2})_j}{v_{2j}} = \ldots = \frac{(n_{i_1} - n_{i_2})_j}{v_{ij}} = \ldots = \frac{(n_{k_1} - n_{k_2})_j}{v_{kj}} = \frac{(n_{l_1} - n_{l_2})_j}{v_{lj}} = \frac{(n_{m_1} - n_{m_2})_j}{v_{mj}} \equiv -n_j \tag{9.4}$$

While treating flow systems, we here introduced into consideration the chemical flux of jth reaction, n_j defined as the product of the homogeneous reaction rate and the volume. When the reaction is heterogeneous, analogous product of its rate and the surface area is introduced

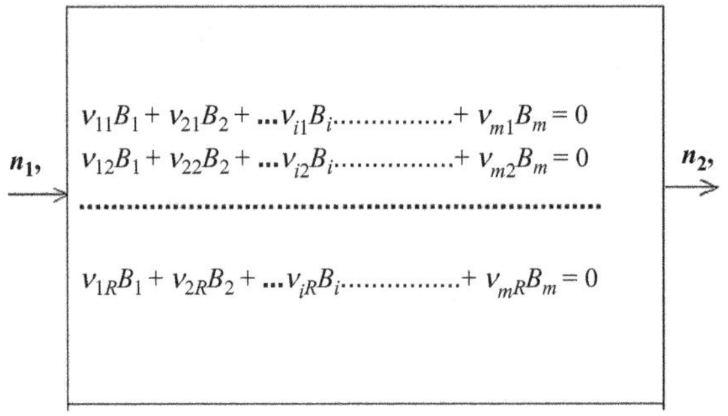

Figure 9.3: A scheme of a multireaction system described by the traditional stoichiometry.
Flow reactor at a steady state.

in the chapter: Power limits in thermo-chemical units, fuel cells and separation systems. For the multireaction system the chemical flux of reaction j is defined as:

$$n_j = \frac{(n_{i2'} - n_{i1'})_j}{v_{i,j}} \tag{9.5}$$

where n_j is the product of reaction rate r_j and the reaction volume, $(n_{i2''} - n_{i1'})_j$ is the molar flux of component i reacting in reaction j and $v_{i,j}$ is the relative stoichiometric coefficient of ith component in reaction j, defined in physical chemistry as $v_{ij} = v_{ij}^b - v_{ij}^f$ (traditional stoichiometric coefficient).

9.3 Driving Forces of Transport and Rate Processes

The aim of this section is a general macrokinetics of transport and rate processes (chemical or electrochemical reactions) and its application to the theory of optimal power yield in chemical or electrochemical systems. Analyzing two competing directions in elementary chemical or transport steps, we investigate equations of nonlinear kinetics of Marcelin–Kohnstamm–de Donder type that contain terms exponential with respect to the Planck potentials and temperature reciprocal. The accepted approach distinguishes in each elementary process, diffusive or chemical, two competing unidirectional fluxes, which are equal at the thermodynamic equilibrium. The observed (resulting) flux, which describes an elementary rate, is the difference of these unidirectional fluxes off equilibrium. We regard the kinetics of this sort as the potential representation of a generalized law of mass action that comprises the effect of rate processes, transfer phenomena, and external fields. Important are physical consequences of these results close and far from equilibrium, which show how diverse processes can be described, and how the basic equation of electrochemical kinetics can be obtained within the general kinetic description. In these considerations we point out the significance of a generalized affinity and correspondence with the standard kinetic mass action law and Onsager's theory near the thermodynamic equilibrium.

We begin with a discussion of the driving forces in generalized kinetic schemes. The general equations of macrokinetics are formulated for a set of chemical reactions, $j = 1, 2, ..., N$

$$\sum_{i=1}^{n} v_{ij}^f B_i^f \leftrightarrow \sum_{i=1}^{n} v_{ij}^b B_i^b \tag{9.6}$$

between species $i = 1, 2,..., n$ and also for the transport (diffusion) processes involving the same species. The absolute stoichiometric coefficients v_{ij} always assume positive values. Both groups of processes are described by the same basic equation; the differences appear in the way the stoichiometric coefficients are treated, so to secure their equality on both sides of the energy barrier for transport processes.

The absolute rates are expressed as functions of potentials F_k, which are components of the vector $\boldsymbol{F} = (T^{-1}, -\mu_1 T^{-1}, ..., -\mu_n T^{-1})$. The classical definition of the chemical affinity is here transferred to the entropy representation (superscript s) and extended so that it can take into account the effect of transport phenomena

$$A_j^s = \sum_{i=0}^{n} (v_{ij}^b F_i^b - v_{ij}^f F_i^f) \tag{9.7}$$

In the kinetic regime a genuine chemical step takes place. In this case the transport thermodynamic force $-X_i = F_i^f - F_i^b = 0$ or $F_i^f = F_i^b$. As the resulting stoichiometric coefficient $v_{ij} = v_{ij}^b - v_{ij}^f$ is nonvanishing for a genuine reaction (it is positive for products and negative for substrates), the direction of an isothermal reaction in the kinetic regime is determined by the classical affinity (De Donder, 1920; de Groot and Mazur, 1984)

$$A_j^s = \sum_{i=1}^{n} (v_{ij}^b - v_{ij}^f) F_i = \sum_{i=1}^{n} v_{ij} F_i = (\boldsymbol{v}^\mathrm{T} \boldsymbol{F})_j \tag{9.8}$$

For a diffusion step $v_{ij}^b = v_{ij}^f = v_{ij}^{f,b}$, that is, the *resulting* stoichiometric coefficient vanishes, or $v_{ij} = 0$. In this case the process is driven by the classical driving forces or the differences $X_i = F_i^b - F_i^f$. Then

$$A_j^s = \sum_{i=1}^{n} (v_{ij}^b F_i^b - v_{ij}^f F_i^f) \rightarrow \sum_{i=1}^{n} v_{ij} (F_i^b - F_i^f) = \sum_{i=1}^{n} v_{ij} X_i. \tag{9.9}$$

Thus in the diffusion regime quantities X_i emerge naturally, which means that the diffusional fluxes are related to Onsager's driving forces. Fig. 9.4 interprets the link between F_i an X_i.

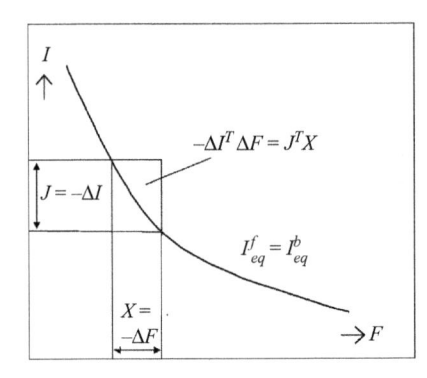

Figure 9.4: Interpretation of absolute fluxes I, potentials F, net fluxes J, and classical thermal forces X.
Oláh, 1997a, b.

For any elementary step the generalized affinity (9.7) can be written down in the form

$$A_k^s = \sum_i v_{ik}^b F_i^b - \sum_i v_{ik}^f F_i^f = R\ln\left(\frac{k_k^f \prod_{i=1}^s c_i^{v_{ik}^f}}{k_k^b \prod_{i=1}^s c_i^{v_{ik}^b}}\right) = R\ln\left(\frac{r_k^f}{r_k^b}\right) \tag{9.10}$$

The quantity $A_k^s = A_k/T$ is the affinity of kth exchange process in the entropy representation. In imperfect systems affinities a_i should replace concentrations c_i. It may show that the rates can be given the form of Ohm's law with appropriately defined resistance (see Section 4).

After calculating net fluxes and forces we observe that in the conventional Onsager's description of irreversible processes one deals with net fluxes and net thermodynamic forces. Traditional rate equations postulated in the Onsager's theory have the following structure

$$\mathbf{J} = \mathbf{f}(\mathbf{X}) \qquad \mathbf{J} = -\Delta\mathbf{I}; \mathbf{X} = \Delta\mathbf{F} \tag{9.11}$$

Here I_k is the vector of "absolute" or unidirectional rates, exemplified by Eq. (9.13) below or a like. In the linear Onsager's theory, the standard notation for first Eq. (9.11) is

$$J_i = \Sigma_k L_{ik} X_k, \tag{9.12}$$

where L_{ik} is the Onsager's phenomenological coefficient. The link of the present formalism with the Onsager's theory can be interpreted graphically (Oláh, 1997b). Each unidirectional flux I_j is a direction-independent function of potentials F_k.

Bykov et al. (1982) first proposed a formula for the macrokinetics of nonisothermal reactions in closed chemical systems in terms of the Marcelin–de-Donder kinetics, which later received a general form and applicative content in the book of Keizer (1987a) book. Bykov et al. (1982)

also suggested some explicit forms of the Lyapunov functions for reacting systems under various conditions. Further, Oláh (1997a, b, 1998) and Sieniutycz (2004a, b, 2005a) described kinetics of nonlinear heat transfer and mass diffusion, treating these processes as peculiar chemical reactions described by their own affinities.

Each approach distinguished in an elementary process, thermal, diffusive, or chemical, two competing (unidirectional) fluxes. They are equal in the thermodynamic equilibrium and their difference off the equilibrium constitutes the observed flux or the resulting process rate. Nonequilibrium thermal and diffusion processes are described by equations containing exponential terms with respect to chemical potentials of Planck and temperature reciprocal that simultaneously are analytical expressions characterizing the transport of substance or energy by the energy barrier. The kinetics of this form are consistent with the law of mass action, and they describe the chemical behavior close and far from equilibrium. One shall also stress the role of a nonlinear resistance and the restrictiveness of the corresponding affinity. Kinetics obtained constitute the potential representation of the generalized law of mass action, which comprises the effect of rate processes, transfer phenomena, and external fields. Remarkable are physical consequences of these kinetics close and far from the equilibrium, and the diversity of processes that can be treated, Keizer (1988).

Garcia-Colin (1988, 1992) commented on the kinetic mass action law from the viewpoint of its compatibility with EIT. The difficulty was in the nonlinearity of the kinetics. He considered a single-step homogeneous chemical reaction with a well-known exponential relation linking the chemical flux and the ratio of the affinity and the temperature. With the postulated disequilibrium entropy depending on the conventional variables and diffusion fluxes (heat flux and chemical flux), expressions for the entropy source and extended phenomenological relations were postulated. Correspondence between classical kinetics and the implications of EIT was possible subject to an ansatz concerning the form of a phenomenological function, P, in the generalized relaxation formula. Fluctuations of the chemically reacting fluid around an equilibrium reference state were discussed in the framework of EIT. Their time decay obeyed an exponential rather than a linear equation due to nonvanishing relaxation times. An open question was the existence of thermodynamic criteria suitable for predicting how a chemical fluctuation evolves in time.

Garcia Colin et al. (1986) investigated the consistency of the kinetic mass action law with thermodynamics within the frame of extended irreversible thermodynamics. Using the standard method developed for this theory they show that it leads to the prediction that the time derivative of the reaction rate is a general function of all scale invariants such as the energy, volume, and so forth, among them the rate of the reaction itself. When the relaxation time for the reaction rate is negligible, a particular form of this function leads to the kinetic law of mass action (KMAL). When simultaneous heat and mass transfer occurs, the divergence of the heat and diffusion fluxes can drive the rate of the reaction. In absence of the heat and diffusion fluxes the usual form of KMAL is recovered. Sieniutycz (1987b) has provided a variational substantiation to these results in the form of the corresponding least action principle, which leads to the mass action law and extended affinity. Sieniutycz (2004d, 2005a) has also presented a theory of nonlinear transport phenomena in the presence of chemical reactions in the form of a unifying thermodynamic framework for nonlinear macrokinetics in reaction–diffusion systems.

Moreover, Shiner (1992) presented an original Lagrangian formulation of chemical dynamics consistent with mass action kinetics, based on the idea of a nonlinear chemical resistance and a corresponding Rayleigh dissipation function. His approach yields a unification of the dynamics of chemical systems with other sorts of processes (eg, mechanical) in the framework of Lagrangian dynamics and yet simplifies the treatment of complex couplings. A major virtue is a physical Lagrangian based on the Gibbs free energy in classical form. The author's formulation stresses a natural algebraic symmetry in the chemical reaction system at stationary states far from equilibrium. This symmetry can be viewed as a more general case of Onsager's reciprocity, and can explain the complex nature of the symmetry properties in chemical systems. The formalism leads to the network representation of chemical systems; this allows a treatment using standard methods of network thermodynamics. When the inertial effects are negligible, the Lagrangian dynamics can be set in terms of a variational formulation, which simplifies near equilibrium to the theory of minimum power dissipation. Shiner and Sieniutycz (1996) show that systems whose equations of motion are derivable from a Lagrangian formulation and subject to constraints expressing conservation of quantities such as mass and charge possess a symmetry property at stationary states far from thermodynamic equilibrium. Examples of such systems are electrical networks, discrete mechanical systems and systems of chemical reactions. The symmetry is in general only an algebraic one, not the differential Onsager reciprocity valid close to equilibrium; the more general property does reduce to the Onsager result in the equilibrium limit, however. No restrictions are placed on the form of the resistances. Thus, the symmetry property is not limited to linear systems. The approach presented leads to the proper set of flows and forces for which the Onsager-like symmetry holds.

9.4 Nonlinear Macrokinetics of Chemical Processes

Consequently, the observed kinetic mechanism follows as the net result of two opposite steps. At the thermodynamic equilibrium both absolute rates I_k and potentials F_k equalize, although they are different at disequilibrium states. In any ith step the quantities I_i^0 ("exchange currents") are the same for both directions. The unidirectional chemical flux in the forward direction (a component of unidirectional vector I^f) satisfies an equation

$$r_j^f = r_j^0 \exp(-\sum_i v_{ij}^f F_i^f), \quad j = 0, 1, \ldots, N. \tag{9.13}$$

The same structure holds for the backward process with only the difference in indices (b instead of f). Yet, there are two competing processes, direct and reverse (or forward and backward). In the discrete model describing exchange processes between two different subsystems the resulting rate of an elementary transport or chemical step subsumes the competition of forward and backward rates

$$r_j = r_j^f - r_j^b = r_j^0 \left(\exp(-\sum_i v_{ji}^f F_i^f) - \exp(-\sum_i v_{ji}^b F_i^b) \right) \quad (9.14)$$

In these equations, potentials $F_i = (1/T, -\mu_i/T)$ are the partial derivatives of the entropy with respect to the extensive variables appearing in the fundamental Gibbs equation for the entropy differential. In the kinetic regime (fast transport; $F_i^f = F_i^b = F_i^{eq}$) classical chemical kinetics is recovered in the Marcelin–Kohnstamm–de Donder form (MKD form). As shown by Eq. (9.14), this is the kinetics in terms of the potential quantities instead of concentrations. Some forms of these equations separate gas constant R as a multiplier, which may have influence on the definition of other constants (see, eg, Eq. (9.26) below). In the diffusive regime (fast reactions) the kinetic system represents a nonlinear diffusion process.

In the framework of mesoscopic nonequilibrium thermodynamics De Miguel (2006a) considered large deviations from the behavior predicted by the Butler–Volmer equation of electrochemistry, which follows from the MKD kinetics. The nonequilibrium thermodynamic hypotheses were extended to include velocity space and cope with imperfect reactant transport leading to departures from Butler–Volmer behavior. This resulted in a modified Butler–Volmer equation in good agreement with experimental data. The distinct advantages of the method and its applicability to analyze other systems were discussed.

De Miguel (2006a) concluded that with disequilibrium thermodynamics a simple expression can be obtained to account for saturation effect in the rates of electrochemical reactions. Imperfect transport of the reacting species to the catalytic surface causes these nonideal effects that are beyond the reach of a simple Arrhenius–Kramers description such as the Butler–Volmer equation. The velocity coordinate has to be brought into the local-state description of the system in order to cope with the possibility of delayed relaxations. While no additional complexity is caused by adding the coordinate, the applicability of the results is valid to a wider range of regimes. This formalism has already been used to account for inertial effects in nucleation and evaporation, and it offers a possibility to analyze other activated processes exhibiting departures from Arrhenius-Kramers behavior, such as viscoplasticity and protein folding. De Miguel (2006b) considers nonisothermal effects in activated processes by means of mesoscopic nonequilibrium thermodynamics. Charge transfer through electrode surfaces is used as a model problem. He shows that, as a generalization of classical nonequilibrium thermodynamics, the theory is capable of incorporating thermal effects into a nonlinear description of activated processes. This leads to a modified law of mass action accounting for nonisothermal conditions. Generalized versions of Nernst and Butler–Volmer equations allowing for thermal gradients follow as a consequence of the modified rate law. He discusses some distinct advantages of the formalism over its classical counterpart.

One may show the correspondence of the present theory with classical kinetics of rate processes governed by the mass action law. We consider the competition of forward and backward chemical steps. Using the usual structure of chemical potentials

$$\mu_i(c,T) = \mu_{i0}(T) + RT \ln c_i \tag{9.15}$$

we substitute concentrations c_i into the classical mass action kinetics

$$r_j = r_j^f - r_j^b = k_j^f(T) \prod_{i=1}^{n} c_i^{v_{ij}^f} - k_j^b(T) \prod_{i=1}^{n} c_i^{v_{ij}^b} \tag{9.16}$$

The result is the chemical rate in the Marcelin–de Donder form

$$r_j(c,T) = r_j^f(c,T) - r_j^b(c,T) = r_j^0(T) \left(\exp \sum_{i=1}^{n} v_{ij}^f \frac{\mu_i}{RT} - \exp \sum_{i=1}^{n} v_{ij}^b \frac{\mu_i}{RT} \right) \tag{9.17}$$

This equation has the typical (yet not the most general) nonlinear structure of macrokinetic equations. Its virtue is a single rate constant, r_j^0, common for both directions, representing the so-called *exchange current*, and the explicit satisfaction of the principle of microscopic reversibility at the equilibrium. The exchange current expressed by rate constants has the form

$$r_j^0(T) \equiv k_j^f(c,T) \left(\exp \sum_{i=1}^{n} -v_{ij}^f \frac{\mu_{i0}}{RT} \right) = k_j^b(c,T) \left(\exp \sum_{i=1}^{n} -v_{ij}^b \frac{\mu_{i0}}{RT} \right) \tag{9.18}$$

In ionic systems the chemical potentials are replaced by electrochemical potentials.

Electrochemistry is in fact, the realm where the notion of the exchange current is best known. In fact, it is the condition of the vanishing electrochemical affinity at equilibrium, the condition that makes it possible to define the universal rate constant r_j^0 or the exchange current of electrochemistry. The related equations are analytical expressions characterizing the transport of substance or energy through the energy barrier. These kinetics may be regarded as potential representations of generalized law of mass action that includes transfer phenomena, external fields, nonlinear symmetries, and generalized affinity. For these representations the correspondence with the classical kinetics near equilibrium may be shown. Generalized affinities provide an alternative to usual driving forces in nonequilibrium transport processes. The field counterpart of the theory implies that diffusion coefficients are not constants but are exponential functions of intensive state coordinates. A remarkable number of experiments may be quoted that confirm such nonlinear behaviors in physical systems.

In the isothermal case, the result of the procedure involving the potential description of kinetics is the Butler–Vollmer equation that describes the electric current J as the difference between the anodic and cathodic currents. Oláh (1997a, b) has given clear interpretation of anodic and cathodic currents as well as of the exchange current in terms of voltage. On his diagrams the abscissa of the crossing point of both currents describes the exchange current. He has also interpreted logarithms of anodic and cathodic currents as well as of the exchange current in terms of electrochemical potential. The potential form of the Butler–Vollmer equation is:

$$J = j^{\text{anod}}(\mathbf{c},T) - j^{\text{cathod}}(\mathbf{c},T) = j^0(T)\left(\exp\sum_{i=1}^{n} v_i^f \frac{\tilde{\mu}_i}{RT} - \exp\sum_{i=1}^{n} v_i^b \frac{\tilde{\mu}_i}{RT}\right) \quad (9.19)$$

Whereas a customary, equivalent form of the Butler–Vollmer equation describes the electric current of a cell in terms of the overvoltage η and universal Faraday's constant F

$$J = j^0\left(\exp\left((1-\alpha)\frac{F\eta}{RT}\right) - \exp\left(\frac{-\alpha F\eta}{RT}\right)\right). \quad (9.20)$$

The coefficient α characterizes the symmetry of the energy barrier and is near to ½. Eq. (9.20) is commonly used to describe various electrochemical systems including fuel cells. Its consequence is the familiar Tafel chart.

We shall now summarize the general nonlinear equations of macrokinetics for homogeneous systems. Characterizing these general equations one may consider coupled transfer of heat (h) and mass (m). Introduced are potentials $F_i = (1/T, -\mu_i/T)$, $i = 0, 1, \ldots, n$, which are the thermodynamic conjugates of the extensive variables in the Gibbs equation for the system's entropy

$$dS = T^{-1}dE - T^{-1}\mu_\alpha dc_\alpha \equiv \sum_{i=0}^{s} F_i dC_i \equiv \mathbf{F}d\mathbf{C} \quad (9.21)$$

The transport kinetics is described by the *general exchange equation* for the net flux J_i

$$J_k = I_k^0\left\{\exp\left(\frac{-\sum_i v_{ik}^* F_i^f}{R}\right) - \exp\left(\frac{-\sum_i v_{ik}^* F_i^b}{R}\right)\right\} \quad (9.22)$$

whose equivalent form in terms of deviations from equilibrium is

$$J_i = I_i^{eq} \Delta \exp\left(\frac{-\sum_i v_{ik}^*(F_k^f - F_k^{eq})}{R}\right) \tag{9.23}$$

with

$$I_i^{eq} \equiv I_i^0 \exp\left(\frac{-\sum v_{ik}^* F_k^{eq}}{R}\right) \tag{9.24}$$

as the common value of the absolute current at equilibrium. We also find

$$\frac{I_k^f}{I_k^b} = \exp\left(\frac{\sum_i (v_{ik}^* F_i^b - v_{ik}^* F_i^f)}{R}\right) \tag{9.25}$$

Eqs. (9.22) and/or (9.25) characterize the thermal kinetics of transfer processes.

For the genuine chemical reactions the formulas analogous to (9.22) and (9.25) are

$$r_j = r_j^f - r_j^b = r_j^0 \left(\exp\left(\frac{-\sum_i v_{ji}^f F_i^f}{R}\right) - \exp\left(\frac{-\sum_i v_{ji}^b F_i^b}{R}\right)\right) \tag{9.26}$$

$$r_j^f = r_j^b \exp\left(\frac{A_j^s}{R}\right) \tag{9.27}$$

The difference between Eqs. (9.26) and (9.14) is the separation of the reciprocity of gas constant R as a suitable multiplier. Eqs. (9.22), (9.25), (9.26), and (9.27) show that the unidirectional (forward and backward) kinetics of kth transport process proceeds between the forward and backward components of the "one-way affinity," that is, the quantity

$$\Pi_k \equiv -\sum_{i=0}^{m}(v_{ik}^* F_i) \tag{9.28}$$

where $F_i = (1/T, -\mu_n/T)$ (Shiner, 1992; Oláh, 1997a; Sieniutycz, 2005a). The quantities Π_k are defined for initial and final states of the process. In the transport kinetics, they play the same role like the temperature does in the heat exchange; each ith elementary transport proceeds from higher Π_i^f to lower Π_i^b.

The same situation takes place for genuine chemical reactions: momentary value of "one-way affinity" of ith reaction, that is, quantity Π_i is the potential of this reaction. The ith reaction proceeds continuously through all intermediate values of Π_i, between high Π_i^f and low Π_i^b. Analogously the charge of heat energy flows from a higher temperature T^f to a lower one, T^b. The basic form Eq. (9.26) is, in fact, a general mass action law in terms of potentials F_i. It is not restricted to the kinetic regime. One may write down Eq. (9.26) in a condensed form

$$r_j = r_j^0 \left(\exp\left(\frac{\Pi_j^f}{R}\right) - \exp\left(\frac{\Pi_j^b}{R}\right) \right). \qquad (9.29)$$

where

$$\Pi_j \equiv -\sum_{i=0}^{n}(v_{ij}F_i) \qquad (9.30)$$

Eq. (9.29) contains two components of one-way affinity Eq. (9.30), in forward and backward directions, respectively. They play the role not only in chemical kinetics Eq. (9.26) or (9.29) but also in the so-called chemical Ohm's law, Eq. (9.31) below.

Consider the jth reaction step characterized by a *constant* specific resistance ρ_j and the (electro) chemical vector $\mathbf{J}_j$ through the surface area A. Eqs. (9.10), (9.21), (9.22), and (9.30) show that the electrochemical or chemical kinetics can be given the form of Ohm's law

$$\mathbf{J}_j = \mathbf{R}_j^{-1}(\Pi_j^f - \Pi_j^b), \qquad (9.31)$$

in which both quantities Π_j are the boundary values of potentials of the reaction j. However, this form of the law is exact only for a constant specific resistance ρ_j. For a complicated motion of the chemical complex through the energy barrier, knowledge of the local resistance formula and its integration are required. This shows the restrictive ingredient of the model of discrete energy barrier which—in spite of many successes—is not capable of avoiding mean quantities characteristic of the whole barrier connected with finite affinities or driving forces, such as chemical resistance $\mathbf{R}_j$

9.5 Heterogeneities, Affinities, and Chemical Ohm's Law

Since the advent of Eyring's absolute rate theory (Glasstone et al., 1941), the idea of reaction coordinate has been a basic principle in studying chemical reactions. The fundamental assumptions of that theory consist of the existence of activated complex or transition state C in the course of the chemical reaction F→B so that the chemical mechanism is F⇔C→B subject to the chemical equilibrium between F and C. For quantitative purposes Fukui and coworkers have introduced a concept of "intrinsic reaction coordinate" (IRC) as the reaction path from the initial F to final B by way of the transition point C on the adiabatic potential surface, and given the dynamic equations of IRC (Tachibana and Fukui, 1979a, b); Fukui, 1981a, b. In the present chapter the notion of IRC and the concept of "chemical resistance" (CR) (Grabert et al., 1983; Shiner, 1987) consistent with the mass action law (Grabert et al., 1983; Shiner, 1987, 1992; Sieniutycz, 1987b; Sieniutycz and Shiner, 1993) serve to describe interphase chemical reactions in two-phase systems or heterogeneous reactions. For the definition of the specific chemical resistance of an elementary step see refs. (Sieniutycz, 1987b, 2000h; Grabert et al., 1983; Shiner, 1987). The macroscopic basis for the derivations are extremum properties of the entropy production and the related Onsagerian framework for chemical reactions (Sieniutycz and Shiner, 1993).

A reasonable postulate is that the union of IRC and CR constitutes a theoretical tool, which allows one quantitative descriptions of complex reactions including catalytic ones. A Fermat-type principle is capable of predicting changes in shapes of reaction paths as "chemical rays" within each phase and through an interface, which separates immiscible media. When one phase differs from another in the value of specific chemical resistance, ρ a sort of refraction law describes the bending of "chemical rays" at the interface. For conserved fluxes the minimum of entropy production can be associated with the minimum resistance of the path. Taking this into account it is easy to conclude that the minimum resistivity at a given flux causes—in the dual picture of the problem—the maximum of flux through the medium along a given path, which ensures that the residence time of the flux-related "charge" in the medium is as short as possible. This makes the principle for travel of physical or chemical entities quite similar to that for propagation of light. To evaluate chemical resistances, findings in the field of heterogeneous kinetics are exploited (Thomas and Thomas, 1967; Bailie, 1978; Yablonskii et al., 1983; Ioffe et al., 1985; Boreskov, 1986; Tabiś, 2000; Molski, 2001).

In a consecutive isothermal chain of several elementary steps (qth chemical chain), the overall chemical affinity of the chain is

$$\mathcal{A}_q^s = \sum_{i=1}^{n}(v_{iq}^f - v_{iq}^b)\mu_i T^{-1} = -\sum_{i=1}^{n}\left(v_{iq}\mu_i T^{-1}\right) = R\left(\ln[k_q^f \prod_{i=1}^{n} a_i^{v_{iq}^f}] - \ln[k_q^b \prod_{i=1}^{n} a_i^{v_{iq}^b}]\right) = R\left(\ln \frac{k_q^f \prod_{i=1}^{n} a_i^{v_{iq}^f}}{k_q^b \prod_{i=1}^{n} a_i^{v_{iq}^b}}\right) \quad (9.32)$$

in the entropy representation (superscript s). Whereas the specific resistance formula for this overall (nonelementary) reaction can be applied in the form corresponding with the Langmuir–Hinshelwood mechanism of kinetics (Boreskov, 1986; Tabiś, 2000; Molski, 2001)

$$\rho_q(\mathbf{a},T) = RL^{-2} \frac{\ln\left(k_q^f \prod_{i=1}^{n} a_i^{v_{iq}^f}\right) - \ln\left(k_q^b \prod_{i=1}^{n} a_i^{v_{iq}^b}\right)}{k_q^f \prod_{i=1}^{n} a_i^{v_{iq}^f} - k_q^b \prod_{i=1}^{n} a_i^{v_{iq}^b}} \left(1 + k_q^f \prod_{i=1}^{n} a_i^{v_{iq}^f} + k_q^b \prod_{i=1}^{n} a_i^{v_{iq}^b}\right)^{\Sigma v_{iq}^f} \quad (9.33)$$

where R is the gas constant and reaction rate constants are effective or operational quantities.

In rough applications activities a_i are replaced by concentrations c_i, and reaction rate constants are effective or operational quantities. The right-hand side structure of Eq. (9.33) is the product of the classical resistance evaluated as in a homogeneous reaction (Grabert et al., 1983; Sieniutycz, 1987, 2001c) and a bracket expression that describes an inhibiting heterogeneous effect typical of the Langmuir–Hinshelwood kinetics (Boreskov, 1986; Tabiś, 2000; Molski, 2001). From the first of these equations the substrate and product part of the chemical affinity, that is, terms Π^f and Π^b in Eq. (9.31) follow as forward and backward contributions in formula (9.32). Equation (9.33) represents a consistency condition for resulting macrokinetics with the Langmuir–Hinshelwood kinetics and also (when taking into account individual separate steps) with the mass action law.

Correspondingly, with Eq. (9.32) the integral of the entropy production for the whole reaction system ($q = 1, 2, ..., N$) is

$$S_s = \sum_q S_q = \sum_q \int R \ln \left[k_q^f \prod_i a_i^{v_{iq}^f} - k_q^b \prod_i a_i^{v_{iq}^b} \right] dQ_q \tag{9.34}$$

Consistently with Eqs. (9.32) and (9.33), the entropy-representation Ohm's law of the qth reaction chain characterized by the *constant* resistance $\mathbf{R}_q$ and chemical flux I_q has the form

$$I_q = \mathbf{R}_q^{-1} (\Pi_q^f - \Pi_q^b) \tag{9.35}$$

which expresses the current through the area A as ratio of the potential difference $\Pi_q^f - \Pi_q^b$ to the total resistance, $\mathbf{R}_q$. The potentials in the above formula are "one-way affinities" that is, quantities Π_q, Eq. (9.30). Eq. (9.35) can be obtained from specific resistance Eq. (9.33) and definition Eqs. (9.36) and (9.37)

$$d\mathbf{R} \equiv \rho \frac{dl}{A} \tag{9.36}$$

and

$$\mathbf{R} \equiv \int_{t_1}^{t_2} \rho \frac{dl}{A} \tag{9.37}$$

for the case of a constant ρ. This is, however, an approximation. In fact, the essential role of the (variable) quantity ρ is to replace the physically insufficient (or even artificial) theory based on the integral-type Eq. (9.37) and rough global formula Eq. (9.35) by the local or differential formula Eq. (9.36) and a variational principle. For more information, see Sieniutycz, 2002e.

Fig. 9.5 depicts the potential-energy lines interpreting the energy barriers and role of surface energy of activation. Increase of optimal cumulative resistance R_m occurs along the reaction

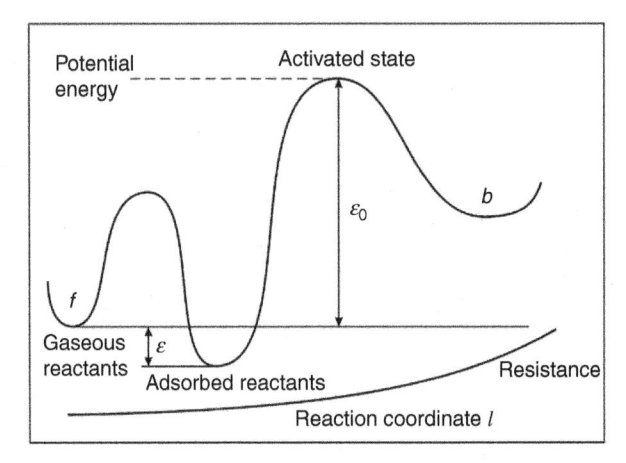

Figure 9.5: States of gaseous and adsorbed reactants, activated state, and state of product b. The low monotonously increasing line describes the chemical resistance R in terms of the reaction coordinate l for a typical heterogeneous reaction.

coordinate. The cumulative resistance of the heterogeneous reaction, optimized in the chemical process, is affected by the kinetic mechanism.

Hesse (1974) derived the transport equations for multicomponent diffusion through capillaries in the transition region between gas diffusion and Knudsen diffusion. Next, Hesse (1977a, b) applied his Hesse (1974) approach to the multicomponent diffusion and chemical reaction of gases in porous catalysts. His results are still helpful in studying of transport phenomena and chemical reactions in heterogeneous systems Shiner (1984) has proposed a dissipative Lagrangian formulation of the network thermodynamics of (pseudo-) first-order reaction-diffusion systems. This is the formulation of the equilibrium and dynamic properties of reaction–diffusion systems, which is analogous to the Lagrangian formulation of dissipative mechanical systems and electrical networks. Since the formulation uses "kinetic" forces for the chemical reactions, not thermodynamic forces, the Lagrangian and dissipation function are only apparent quantities. Nonetheless, the Lagrangian determines equilibrium properties completely, and the dissipation function contains all dissipative effects. The formulation shows that no modulation of the potentials in diffusional pathways is caused by the presence of the chemical reactions, as was previously claimed, if the same scale is used for all reactions and diffusion steps in the system. Furthermore, the formulation renders the system in a network form, which guarantees symmetry in the force–flow relations at the stationary state. This is shown explicitly for the special case of King–Altman–Hill systems. Unfortunately, the Lagrangian approach using kinetic forces seems to be limited to systems with only first-order or pseudo-first-order reactions. To treat higher-order reactions, the thermodynamic forces must be used.

For systems of chemical reactions the phenomenological equations of nonequilibrium thermodynamics display the differential symmetry of Onsager reciprocity in general only for stationary states in the vicinity of thermodynamic equilibrium, as is well known. Shiner (1987) shows, however, that systems of chemical reactions, on the macroscopic, phenomenological level of chemical kinetics, do show a less stringent symmetry at all stationary states arbitrarily far from equilibrium. The key is a reformulation of the kinetic equations for single chemical reactions as resistive laws relating the force driving a reaction to the rate of the reaction through a (nonlinear) generalized chemical resistance; these laws are analogous to the "voltage = current times resistance" laws for resistive elements in electrical networks. The resistive laws for single reactions then lead naturally to stationary state relations for systems of coupled chemical reactions of the form of the phenomenological equations of nonequilibrium thermodynamics, and these relations display algebraic symmetry at stationary states arbitrarily far from equilibrium, although the differential symmetry of Onsager reciprocity is in general not valid. An example illustrates the relation of the symmetry property for chemical reactions to that displayed by RC electrical networks and how the chemical symmetry follows from analogs of Kirchhoff's laws obeyed by systems of chemical reactions.

9.6 Other Important Results Advancing the Field

According his own words, Kramers (1940) considered a particle which is caught in a potential hole and which, through the shuttling action of Brownian motion, can escape over a potential barrier. This approach, in which the effect of the solvent was taken into account in the form of a Markovian dissipation and Gaussian delta-correlated fluctuations, yields a suitable model for elucidating the applicability of the transition state method for calculating the rate of chemical reactions. The classic Kramers' paper stimulated many further treatments. In his study of fluctuation and relaxation of macrovariables Kubo et al. (1973) considered the relation of nonlinearity of evolution and the associated fluctuation. Anomalous fluctuations arose in unstable, critical, or marginal states. Lindenberg (1990) presented a mesoscopic analysis of a dynamical system interacting with a heat bath in both classical and quantum formulations. Many dynamics lead to the same thermal equilibrium and the specific route depended on the detailed nature of the system-bath interactions. For the system and bath as a "universe," bath variables were systematically eliminated from the "universe Hamiltonian" to end up with the system equations of motions containing fluctuating and dissipative forces. The analysis has great value for understanding the limitations of the usual assumptions of statistical mechanics (eg, departures from the Langevin equation). Attention is paid to forms of the fluctuation-dissipation relations, which ensure eventual equilibration of the system with bath; the universality of these relations for classical systems may be violated in quantum systems. The classic work of Kramers (1940), has spawned an enormous and important literature on the results, extensions to wider parameter regimes and to non-Markovian dissipation models. The literature involves generalizations to more complex potentials, many degrees of freedom, analysis of quantum effects, and application to specific experimental systems. One recent direction, deals with the derivation of the rate coefficient so as to obtain not only the asymptotic rate constant but the behavior of the rate coefficient at all times. In the paper of Lindenberg (1999) the author complements many works by extending the parameter regime of analysis and confirms the early and new results against numerical simulations. These advances are described in a synthesizing book on stochastic processes in physics and chemistry (Van Kampen, 2001).

Appreciation should be given to relatively unknown papers of Gyarmati (1961b, 1977) and Lengyel and Gyarmati (1980, 1986), devoted to the thermodynamic studies of homogeneous nonlinear chemical reaction systems. Satisfaction of KLMA (Guldberg and Waage, 1867; Sieniutycz, 1987) is the basic criterion verifying the correctness of a chemical theory. It is often asserted that the linear theory of irreversible thermodynamics, the foundations of which were laid by Onsager, Casimir, Prigogine, and others, does not adequately represent irreversible phenomena in systems far from equilibrium. Such assertions are correct in the linear flux-force constitutive equations of the original Onsager theory, that is, in equations, where the equalities $L_{ik} = L_{ki}$ are the Onsager reciprocal relations (ORR) and L_{ik} are constant coefficients. However, if one takes advantage of the so-called quasi-linear generalization of the thermodynamic theory (Gyarmati, 1977) according to which the conductivity coefficients L_{ik} may depend, for example, on the intensive thermostatic state variables $\Gamma_1, \Gamma_2, ..., \Gamma_j$), then the constitutive equations

$$I_i = \sum_{k=1}^{f} L_{ik}(\Gamma_1, \Gamma_2, ..., \Gamma_f) X_k \qquad (i, k = 1, ... f) \qquad (9.38)$$

with the ORRs in the form

$$L_{ik}(\Gamma_1, \Gamma_2, ..., \Gamma_f) = L_{ki}(\Gamma_1, \Gamma_2, ..., \Gamma_f) \qquad (9.39)$$

and with the Gyarmati supplementary reciprocal relations

$$\frac{\partial L_{ik}(\Gamma_1, \Gamma_2, ..., \Gamma_f)}{\partial \Gamma_j} = \frac{\partial L_{ik}(\Gamma_1, \Gamma_2, ..., \Gamma_f)}{\partial \Gamma_j} \qquad (9.40)$$

ensure good approximation for any example involving transport and rate processes.

Lengyel and Gyarmati (1980) gave a number of examples confirming the validity of this formalism, including the classical problem of heat conduction in anisotropic solids. Moreover, some decades ago Gyarmati and Li laid the foundations of a nonlinear theory, which is based on the generalized reciprocity relations (GRR)

$$\frac{\partial I_i}{\partial X_k} = \frac{\partial I_k}{\partial X_i} \qquad (i, k = 1, \ldots, f) \tag{9.41}$$

Gyarmati (1961a, b) and Li (1958, 1962a, b). In particular, Li found that Onsager's reciprocal relations follow from macroscopic considerations, which also give a relationship between the second-order phenomenological coefficients. He also asserted that the Carathéodory postulate for the existence of the entropy function is a direct consequence of the physical postulate of the existence of a stable equilibrium state. Moreover, the existence of thermokinetic potential follows from the physical postulate of the existence of a stable steady state for any set of disequilibrium variables.

Yet, the question of consistency of chemical kinetics with nonlinear thermodynamics appeared. As explained by Gyarmati, the theory formulated independently by Gyarmati and by Li has been ignored or misinterpreted in publications of other authors. Misinterpretation of the theory led in the case of chemical kinetics to assuming the identity $X_i = A_i$. In the application to chemical kinetics neither Li nor Gyarmati have ever proposed the GRRs in the already accepted linear form. Neither of them has ever stated that affinities are real thermodynamic forces in chemical kinetics in the nonlinear domain. In conclusion, it is to be stressed that, in the nonlinear domain, affinities are not the real forces of chemical reactions, and, therefore, nonlinear thermodynamic theories using reaction rate–affinity constitutive equations must be inconsistent with observed chemical kinetics. The Gyarmati–Li theory allows other (more correct) choices for thermodynamic forces, which make nonlinear thermodynamics consistent. Doubts regarding affinities as correct forces of chemical reactions were also expressed later (Shiner, 1987; Sieniutycz, 1987b; and others). One-way affinities, discussed above, Eq. (9.30), make a much better job than the traditional chemical affinities.

Lengyel (1988) deduced the Guldberg–Waage mass action law from Gyarmati's governing principle of dissipative processes. With the gained theoretical insight, Lengyel (1992) analyses the consistency between chemical kinetics and thermodynamics. He begins with expressions for the entropy production as bilinear forms for the stoichiometric dependent and stoichiometric independent systems of elementary chemical reactions. For the independent reactions, GRR follow, now known as Gyarmati–Li relations (Li, 1958). Dissipation potentials are introduced next, and the nontruncated KLMA is shown to result from the "governing principle of dissipative processes." This seems important since the chemical kinetics is often considered as a field "which only reluctantly obeys the standard formalism of TIP." However, the traditional concept of chemical affinity has to be rejected, although the partial Gibbs free energies of the reagents need to be used. A flux potential is applied as a function of the partial Gibbs free energy of the reactants and of the products. It follows that the theory can solve some difficult problems of chemical kinetics, which were unsolved by other methods. This approach offers several surprises, but its consistency is firmly established by the examples worked out.

9.7 Stability and Fluctuations in Chemical Reactions

Nonisothermal reaction–diffusion systems control the behavior of many transport and rate processes in physical, chemical, and biological systems, such as pattern formation and chemical pumps (Demirel, 2008). Considerable work has been published on mathematically coupled nonlinear differential equations by neglecting thermodynamic coupling between a chemical reaction and transport processes of mass and heat. Study of Demirel (2008) presents the modeling of thermodynamically coupled system of an elementary chemical reaction with molecular heat and mass transport. The thermodynamic coupling refers that a flow occurs without or against its primary thermodynamic driving force, which may be a gradient of temperature or chemical potential or reaction affinity. The modeling (Demirel, 2008) is based on the linear nonequilibrium thermodynamics approach by assuming that the system is in the vicinity of global equilibrium. The modeling equations lead to unique definitions of cross-coefficients between a chemical reaction and heat and mass flows in terms of kinetic parameters, transport coefficients, and degrees of coupling. These newly defined parameters need to be determined to describe some coupled reaction-transport systems. Some methodologies are suggested by Demirel (2008) for the determination of the parameters and some representative numerical solutions for coupled reaction-transport systems are presented. Moreover other important phenomena, such as instabilities and fluctuations can arise in nonisothermal disequilibrium reaction–diffusion systems. Additional information can be found in the paper of Demirel (2009) on thermodynamically coupled heat and mass flows in a reaction-transport system with external resistances. The issues covered in the study of Demirel (2008) are only outlined here because the main treatment of stability problems is presented in the chapter: Thermodynamic Lapunov Functions and Stability.

Keizer and Fox (1974) raised doubt concerning the range of validity of a stability criterion for nonequilibrium states proposed by Glansdorff and Prigogine (1971). For a particular autocatalytic reaction, the stability analysis presented by Glansdorff and Prigogine, and the one by Eigen and by Katchalsky in their review of this problem, does not agree with the analysis of Keizer and Fox based upon exact the solution of the relevant rate equations. Keizer and Fox (1974) also find disagreement between the analysis based upon the Glansdorff–Prigogine criterion and their own analysis of a second example, which involves non equilibrium steady states. The situation is subtle because seemingly innocent approximations (eg, the use of specialized conditions in the autocatalytic reaction $X + Y \rightleftarrows 2X$ may lead to the impression that the scope of validity of the criterion is wider than it actually is. By considering the stability of the equilibrium state, Keizer and Fox (1974) conclude that the second differential of the entropy, which is at the heart of the Glansdorff–Prigogine criterion, is likely to be relevant for stability questions close to equilibrium only.

On a deterministic level chemical reactions are described by the well-established KMAL, where chemical kinetics have the form of power laws. On a more refined level, which includes fluctuations of concentrations, the reaction is described in terms of a stochastic process. Theory of Onsager (1931a, b) theory gives the solution of the problem of mutual connection between the deterministic and the stochastic levels of description in the linear regime near equilibrium. Yet, for chemical reactions linear transport theory provides only a poor approximation (de Groot and Mazur, 1984).

The paper by Grabert et al. (1983) proposes an approach to describe fluctuations of reversible chemical reactions in closed systems. The deterministic rate laws are cast into the form of nonlinear Onsager-type closed laws. The nonlinear chemical resistances (conductances) are found in a complete agreement with the experimentally confirmed KLMA (Grabert et al., 1983; Shiner, 1987; Sieniutycz, 1987b). By means of nonlinear transport theory a Fokker–Planck equation describing the stochastic process of concentration fluctuations is obtained (Grabert et al., 1983). The stochastic formulation reduces to the correct deterministic rate laws in the thermodynamic limit with the concentrations kept fixed. Examples of reactions in ideal mixtures are given and the results of the presented approach are compared with those of the usual approach by means of a birth and death–type master equation. It is shown that both approaches lead to the same stationary probability and exhibit the same natural boundaries reflecting the fact of a restricted state space. The proposed Fokker–Planck equation is different from the Fokker–Planck equation obtained from the master equation by truncating its Kramers–Moyal expansion. However, the two equations are shown to have identical Fokker–Planck coefficients in the vicinity of the deterministic equilibrium state. Compared with the usual master equation approach the proposed stochastic modeling of chemical reactions has the advantage of allowing for a straightforward extension to reactions in nonideal mixtures. See also a related paper by Gaveau et al. (2005).

Escher and Ross (1983) showed multiple ranges of flow rate with bistability and limit cycles for Schlögl mechanism in CSTR. Further, Escher and Ross (1985) investigated the reduction of dissipation in a thermal engine by means of periodic changes of external constraints. They tested the behavior of a thermal engine driven by chemical reactions, which take place in a continuous flow, stirred tank reactor fitted with a movable piston. Work was produced by means of a heat engine coupled to the products and to an external heat bath, and by the piston. The paper by Escher et al. (1985) presents details of specific mechanisms and the occurrence of resonance phenomena.

Sieniutycz and Komorowska-Kulik (1978, 1980, 1982) investigated the second differential of entropy as the possible Glansdorff and Prigogine (1971) criterion for the familiar three singular points, S_1, S_2, and S_3 in CSTR. The results are shown in Figs. 9.6–9.8. The discontinuous lines describe the contours of the tested potential with the positive values of $V \equiv -\delta^2 S/2$, whereas the continuous lines with arrows describe the process trajectories.

Closed continuous lines without arrows describe the contours of the time derivative dV/dt of the potential V with respect to the kinetic equations. The results in Figs. 9.6–9.8 show

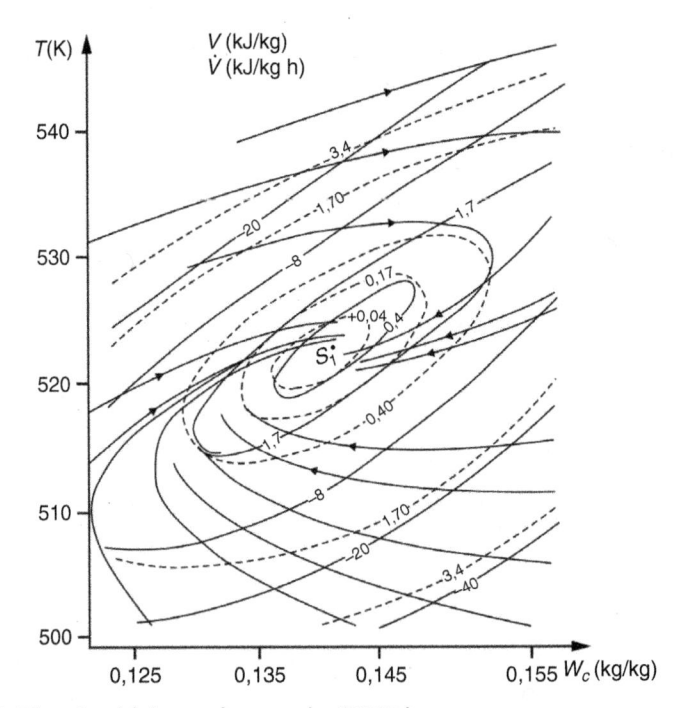

Figure 9.6: First (stable) steady state in CSTR in temperature-concentration chart.

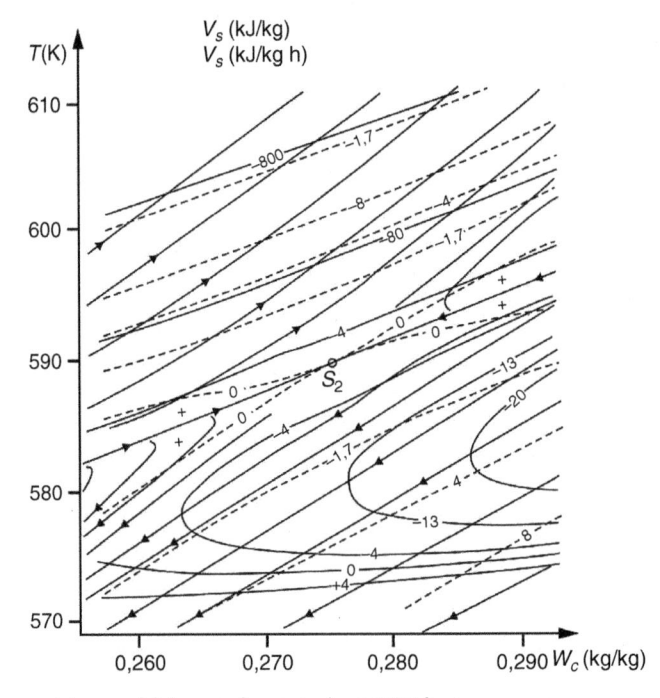

Figure 9.7: Second (unstable) steady state in CSTR in temperature-concentration chart.

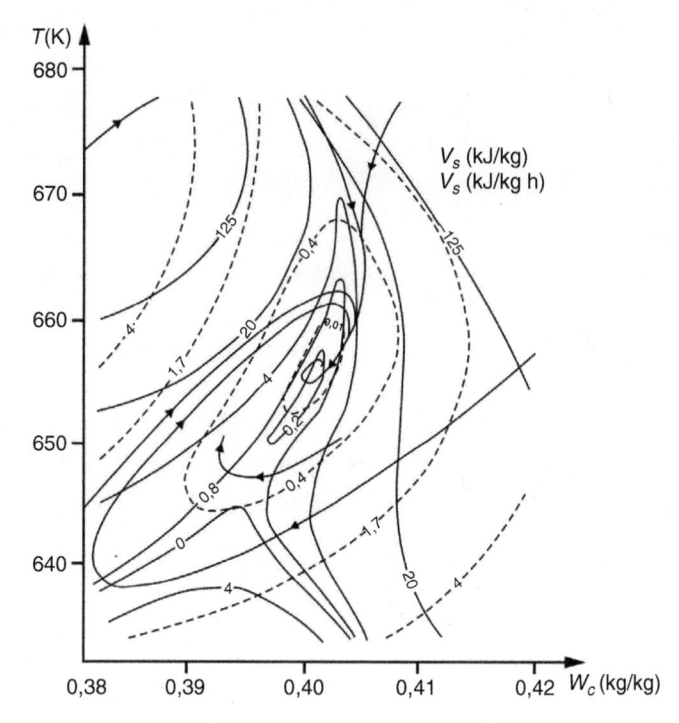

Figure 9.8: Third (stable) steady state in CSTR in temperature-concentration chart.

that this time derivative is, in general, a function of the indefinite sign, which proves that the second differential of the entropy does not provide the generally valid criterion to judge the stability or instability, locally or in large. In the finite regions surrounding the singular point S_1 the inequalities $V > 0$ and $dV/dt < 0$ are satisfied, which proves the local stability of S_1, a correct result. In the regions surrounding point S_2, from the saddle geometry of function V and the inequality $dV/dt < 0$ the instability of S_2 follows. This is also a correct result, known from the theory of CSTR. However, for the most important stable steady-state S_3 in which the reaction rate is the highest, no confirmation of the stability is obtained since the comparison of geometries of V and dV/dt shows that no univocal conclusion is possible. Indeed, whereas the function V surrounding singular point C_3 is everywhere negative, the time derivative is not everywhere positive, which would prove the stability of state S_3 (about which we know it is stable from experience and independent kinetic considerations). Many other results were obtained which show that the second differential of entropy is unable to judge the stability of a disequilibrium steady state if this state is sufficiently far from the equilibrium. On the other hand $V \equiv -\delta^2 S/2$ and any other classical thermodynamic potential can give univocal answer around the thermodynamic equilibrium. Indeed, considering the equilibrium state, Tarbell (1982), applied the negative excess entropy

$W = S^e - S$ as a Lyapunov functional in which the entropy density ρs is the same function of the thermodynamic state as prescribed by classical thermodynamics (local equilibrium assumption). Invoking the local form of the second law $\sigma > 0$ he obtained in the gradientless case the conditions $dW/dt < 0$ which proves the stability of all system trajectories approaching the equilibrium singular point. Thus classical thermodynamics predicts the global stability of thermodynamic equilibrium.

Bahroun et al. (2013) deal with the nonlinear control of a jacketed fed-batch three-phase slurry catalytic reactor to track a reference temperature trajectory. To this end, they define the so-called thermal availability function from the availability function as it has been described in the literature. This thermal availability is used as a Lyapunov function in order to derive a stabilizing control law for the reactor by using the jacket temperature as the manipulated variable. The performance of the control scheme in the presence of measurements noise and parameter uncertainty is illustrated by simulations results.

Analyzing nonlocal phenomenological relation between thermodynamic fluxes and forces in continuous systems, Aono (1975) showed that the vector flux couples with the scalar force even in an isotropic system. This result has some application to active transport in living organisms and to thermonuclear fusion research.

9.8 Instabilities and Limit Cycles

Carbon monoxide oxidation is important in many combustion systems and especially in regenerators of catalytic cracking. Sieniutycz and Komorowska-Kulik (1980) and Jurkowska (1982) investigated stability of CO oxidation in a cracking regenerator where process dynamics follows from a free radical kinetic mechanism proposed by Yang (1974). The results are shown in Figs. 9.9–9.12. The second differential of entropy $\delta^2 S$ was shown to fail as a stability criterion when a singular point was located far from the system's equilibrium point. Inequality of signs of $\delta^2 S$ and its time derivative would constitute a sufficient stability criterion obtained by the second Lyapunov theorem. However, the calculations showed an oscillatory behavior of trajectories of CO oxidation, the regime where the second differential of the entropy $V = \delta^2 S$ does not determine stability because the time derivative dV/dt is not a function of a constant sign. A suitable example is provided in Fig. 5.2 (Winkiel, 1976). Diagrams were obtained proving the saddle nature of dV/dt in the vicinity of stable singular points, showing the inappropriateness of $V = \delta^2 S$. The material in chapter: Thermodynamic Lapunov Functions and Stability supplements the one discussed here.

Biological and chemical waves are different from electromagnetic ones. The first propagate in a so-called excitable or active media containing evenly distributed energy sources, which can be regarded as local amplifiers. As a result the amplitude of chemical and biological

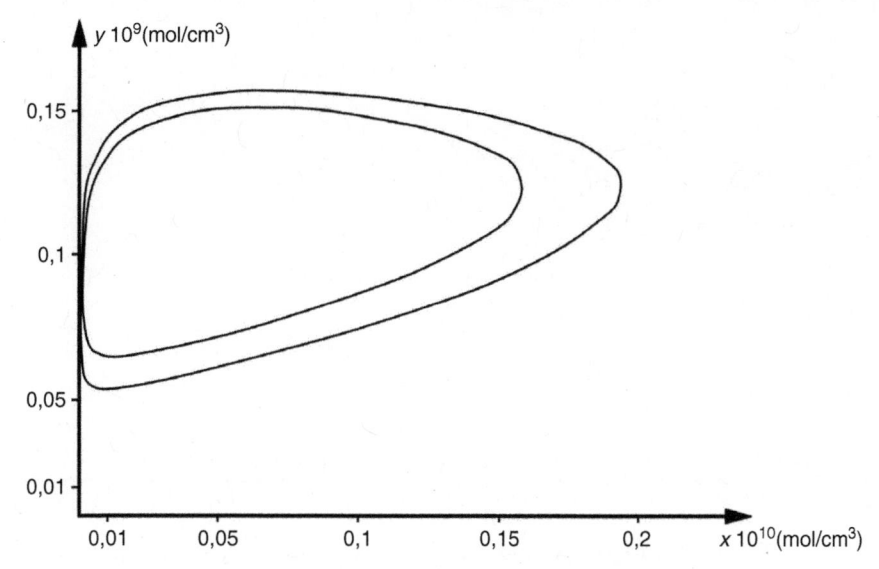

Figure 9.9: Controlling the magnitude of limit cycles.
The area surrounded by the oscillatory trajectories grows with the reduction of convective flow Q in the process of CO oxidation. *Jurkowska, 1982.*

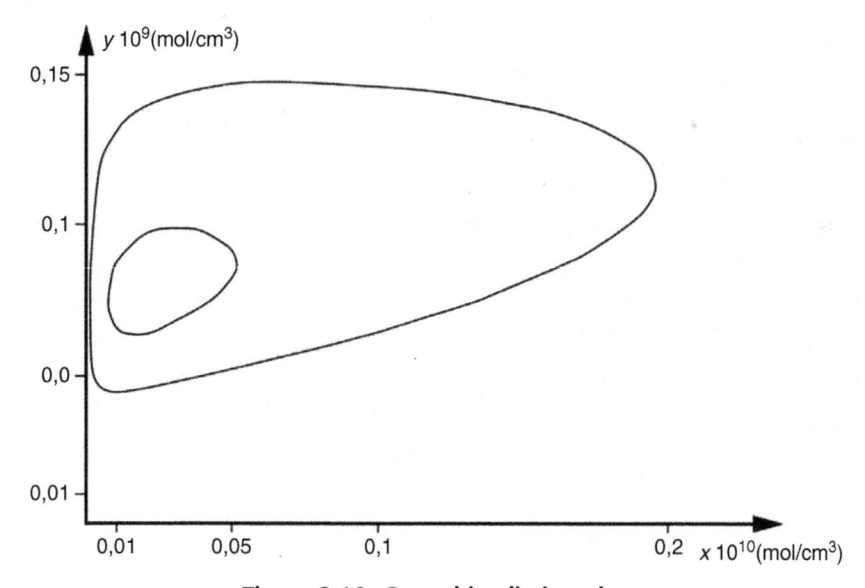

Figure 9.10: Quenching limit cycles.
Oscillatory trajectories reduce the area they surround due to the quenching effect of flow Q in the process of CO oxidation (Jurkowska, 1982). Also see Figs. 5.6 and 5.7 for some nonclosed trajectories showing the quenching effect of flow Q in the process of CO oxidation. *Jurkowska, 1982.*

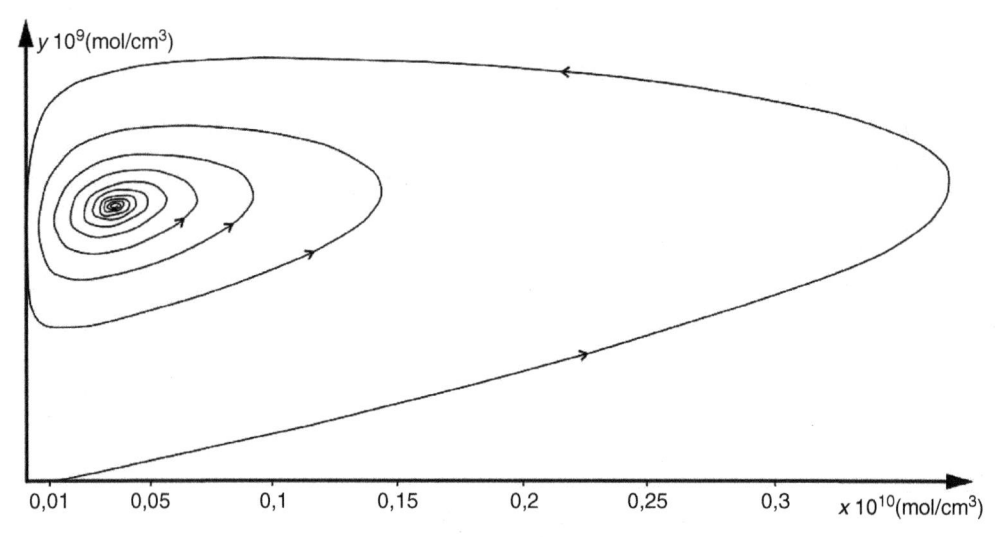

Figure 9.11: All oscillatory trajectories terminate at a stable focus point in the process of CO oxidation.
Jurkowska, 1982.

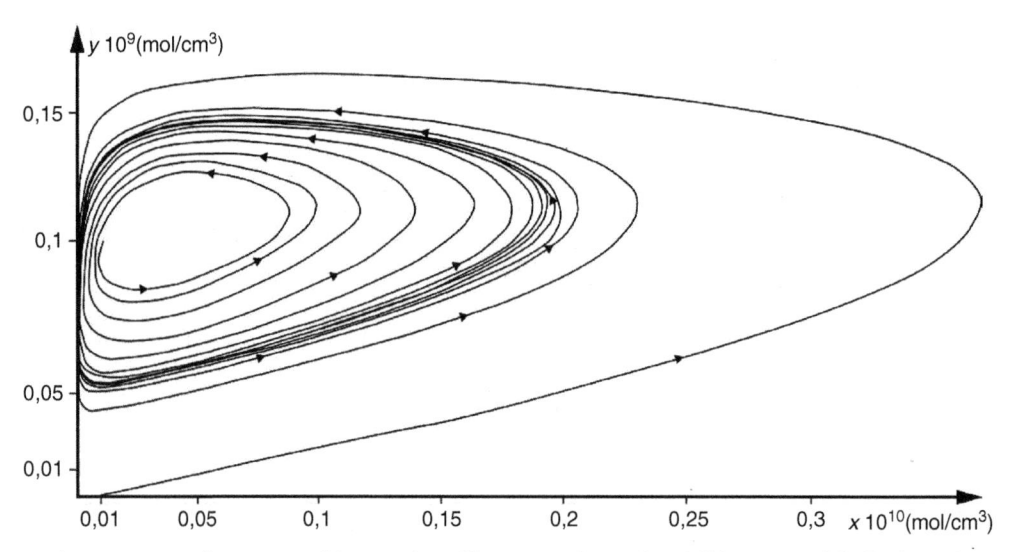

Figure 9.12: All outer and internal oscillatory trajectories fall into a stable limit cycle in the process of CO oxidation.
Jurkowska, 1982.

waves are preserved while they are propagating in a homogeneous active medium. Such waves can be described simply by evolving fronts and orthogonal rays and this is the basis of the geometric wave theory. In the paper by Volford et al. (1999) the geometrical wave theory and its application for rotating waves are described. See more information in the chapter: Thermodynamic Lapunov Functions and Stability.

9.9 Chaos and Fractals in Chemical Word

Classical reaction kinetics has been found to be unsatisfactory when the reactants are spatially constrained on the microscopic level either by walls, phase boundaries, or force fields (Kopelman, 1988). Recently discovered theories of heterogeneous reaction kinetics have dramatic consequences, such as fractal orders for elementary reactions, self-ordering and self-unmixing of reactants, and rate coefficients with temporal "memories." According to the opinion of Kopelman (1988) opinion, the new theories were needed to explain the results of experiments and supercomputer simulations of reactions that were confined to low dimensions or fractal dimensions or both. Among the practical examples of "fractal-like kinetics" are chemical reactions in pores of membranes, excitation trapping in molecular aggregates, exciton fusion in composite materials, and charge recombination in colloids and clouds.

Pfeifer and Avnir (1983a) laid the foundations for appreciation of chemical surfaces as D-dimensional objects where $2 \leq D < 3$. Being a global measure of surface irregularity, this dimension labels an extremely heterogeneous surface by a value far from two. It implies, for example, that any monolayer on such a surface resembles three-dimensional bulk rather than a two-dimensional film because the number of adsorption sites within distance l from any fixed size, grows as l^n. Generally, a particular value of D means that any typical piece of the surface unfolds into m^D similar pieces upon m-fold magnification (self-similarity). The underlying concept of fractal dimension D is reviewed and illustrated in a form adapted to surface-chemical problems. From this the authors derive three major methods to determine D of a given solid surface which establish powerful connections between several surface properties: (1) The surface area A depends on the cross-section area σ of different molecules used for monolayer coverage, according to $A \propto \sigma^{(2-D)/2}$. (2) The surface area of a fixed amount of powdered adsorbent, as measured from monolayer coverage by a fixed adsorbate, relates to the radius of adsorbent particles according to $A \propto R^{D-3}$. (3) If surface heterogeneity comes from pores, then $-dV/d\rho \propto \sigma^{2-D}$ where V is the cumulative volume of pores with radius $\geq \rho$. The authors also discuss statistical mechanical implications.

The concept of fractal dimension D of surfaces advanced as natural measure of surface irregularity in Pfeifer and Avnir (1983a), is shown by Pfeifer and Avnir (1983b), to apply to a remarkable variety of adsorbents: graphites, fume silica, faujasite, crushed glass, charcoals, and silica gel. The values of D found for these examples vary from two to almost three (for smooth and very irregular surfaces, respectively), thus covering the whole possible range. They quantify the intuitive picture that surface inhomogeneities are minor, for example, in graphites, but dominant, for example, in charcoal. The analysis is based on adsorption data, with main focus on adsorbates of varying molecular cross section. They include N_2, alkanes, polycyclic aromatics, a quaternary ammonium salt, and polymers. The straight–line plots so obtained confirm also a number of reported on-surface conformations of specific adsorbates. The converse method to get D from varying the size of adsorbent particles is exemplified for fume silica and crushed glass. The chapter of Avnir et al. (1989) in Plath (1989) book on optimal heterogeneous reaction systems discusses applications of fractal geometry to chemical systems of complex and highly irregular structures. The main topic is optimization of heterogeneous-catalyst structure: simulations and experiments with fractal and nonfractal systems. A group of new mathematical techniques, of particular applications in chromatographic adsorbents, are appropriate to colloidal systems, irregular surfaces, branched polymers, and many other areas of polymer, colloidal, and surface chemistry. In the second chapter of the book of Aharony and Feder (1989) on fractals, Bak and Chen (1989) focus on physical aspects of fractals. Fractals in nature originate from self-organized critical dynamical processes. Their mathematical model explicitly demonstrates how energy, which is injected uniformly can be dissipated on a fractal. The fractal dimension is 2.5 in agreement with analysis of turbulence experiments. The authors suggest that turbulence may be viewed as a sustained "forest fire." Working on reaction kinetics in disordered systems, Zumofen et al. (1989) have considered diffusion-limited reactions of the types $A + A \rightarrow 0$ and $A + B \rightarrow 0$ by modeling the dynamics of a disordered system through random walks. After introducing hierarchical structures, two aspects of disorder followed: (1) temporal and (2) energetic. The temporal disorder was accounted for by continuous time random walks, whose waiting times distribution displays long time tails. The energetic disorder was modeled using ultrametric spaces with hierarchically distributed energy barriers. Interplay between these two disorder aspects was investigated.

Jaroniec et al. (1990) have derived equations for physical adsorption of gases and vapors on fractal surfaces of heterogeneous microporous solids. They studied influence of surface geometry on the gas adsorption on a microporous solid by analyzing the dependence of the differential molar enthalpy ΔH, the immersion enthalpy ΔH_{im}, and the differential molar entropy ΔS on the fractal dimension D, which was used to characterize the surface irregularity. It was shown that $-\Delta H$, $-\Delta H_{im}$, $-\Delta S$, and the average adsorption potential $A-$ increase as the fractal dimension of the surface accessible for adsorption increases. The dependence of these thermodynamic quantities on the fractal dimension D was attributed to the fact that the fraction of small micropores in microporous solids increases as the fractal dimension D increases.

Fractional Fokker–Planck equation was used to describe the anomalous behavior of the transport process through the fractal medium (El-Wakil et al., 2001). Solutions of models of tubular chemical reactors may be of a very complex dynamic character, including chaos. Alhumaizi (2000) has examined a conceptual model of an autocatalytic reaction in a continuous stirred tank reactor in which the autocatalyst undergoes mutation. The model consisted of a set of three ordinary differential equations with seven design parameters. He has shown that self-sustained chaotic behavior can occur in this system. The regions of chaos were entered and exited according to period-doubling cascades or through a process involving intermittency. Calculated Lyapunov exponents confirmed the chaotic behavior.

Berezowski (2003) presented three kinds of fractal solutions of model of recirculation nonadiabatic tubular chemical reactors with mass or thermal feedback. (This feedback is accomplished either by the recycle of mass stream or by external heat exchanger.) The first kind of fractal solutions concerns the structure of Feigenbaum's diagram on the limit of chaos. The second kind and the third one concern the effect of initial conditions on the dynamic solutions of models. In the course of computations two types of recirculation were considered, namely the recirculation of mass (return of a part of products stream) and recirculation of heat (heat exchange in the external heat exchanger). Maps of fractal images were obtained. Zalewski and Szwast (2007) found chaos and oscillations in selected biological arrangements of the prey–predator type. They considered two such models. Dependence of the average concentrations on the parameters of the models has been discussed. The areas of the maximal values of rate constants in particular models were determined. Zalewski and Szwast (2008) considered chaotic behavior in a stirred tank reactor with first-order exothermic chemical reaction ($A \rightarrow B$) undergoing in a reaction mixture with deactivating catalyst particles. Their reactor was continuous with respect to reacting mixture and periodic with respect to catalyst particles. A temperature dependent catalyst deactivation with the Arrhenius–type equation describing catalyst deactivation rate was assumed. The direction of displacement areas in which the chaos is generated was studied. Graphs showed that for older catalysts the areas in which the chaos is generated move toward higher temperatures.

Reviews on micromixing and memory in chemical reactors add more inspiration. Guida et al. (1994) analyze the mathematical description of the behavior of maximum mixedness and maximum segregation reactors in appropriately defined equivalence classes, all members of any such class behaving equivalently in a well-defined sense in both steady and unsteady state. Three important conclusions are reached. First, Guida et al. (1994) establish under which conditions the behavior of at least one member of an equivalence class is, at all times, in the domain of the initial conditions and the reactor has therefore perfect memory. Second, when the reactor has perfect memory the boundary conditions play no role whatsoever, so the discussion regards what are the appropriate ones for a maximum mixedness. Zwietering reactor is shown to be empty (Guida et al., 1994). Third, both the possibility of multiple steady states and of the sustained conduction of a biochemical (or more generally autocatalytic) reaction are shown to be related to the memory of the reactor, and not to backmixing. Another review on turbulent micromixing is by Bałdyga and Pohorecki (1995). They imply that the growth of the zone mixed on the molecular scale is a characteristic feature of micromixing and should be included in micromixing modeling (Fig. 9.13).

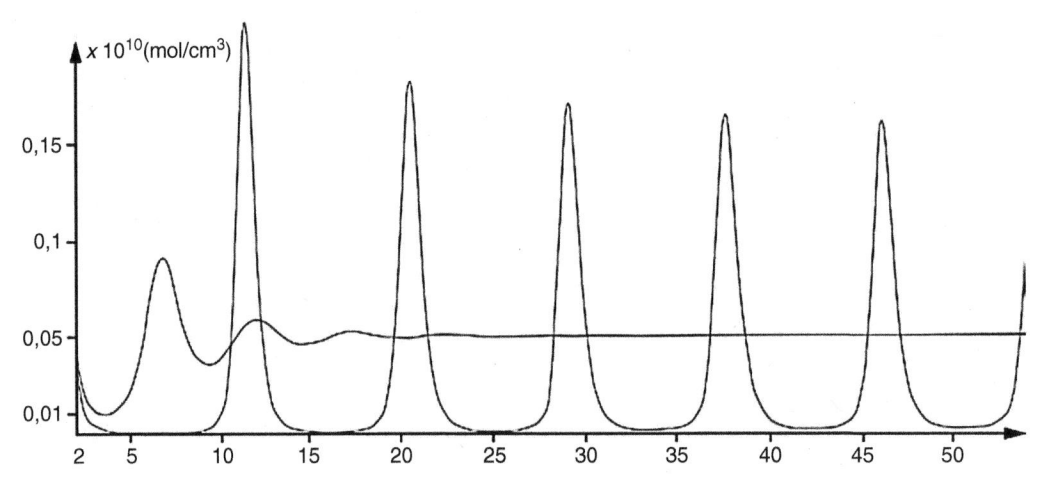

Figure 9.13: Dynamics of oscillatory trajectories terminating at a stable focus point or falling into a stable limit cycle in the process of CO oxidation.
Jurkowska, 1982.

Liu et al. (1996) study the periodically forced Brusselator from the viewpoint of the dissipative standard map. The second system of investigation is the Lorenz model. The authors explore the transition from annular type dynamics to interval dynamics in terms of symbolic dynamics. Finally they discuss the general implication of their approach. See more in chapter: Thermodynamic Lapunov Functions and Stability.

9.10 Power Yield in Chemical Engines
9.10.1 Introduction

Ondrechen et al. (1980b) studied thermodynamics of two chemical problems obeying Arrhenius law kinetics; the studies were performed from the viewpoint of finite time thermodynamics. The first problem consists of the determination of the maximum thermodynamic or chemical efficiency of a synthetic process whose heat input appears as preheating to overcome an activation barrier. In the first problem, an Arrhenius-law synthesis, they show that for typical values of activation energy, there is a temperature to which the reactants may be preheated so as to maximize the reaction rate. The second is the determination of maximum power that can be obtained from an exothermic reaction carried out in a continuous flow system. Power produced by the engine driven by an exothermic Arrhenius-law reaction behaves in some ways like the case of a temperature independent rate, which corresponds to the vanishing activation energy. The more exothermic the reaction, the higher the maximum power and the lesser the time required to achieve the maximum power. The maximum power, achieved with a finite and nonzero flow rate, is a sensitive function of the activation energy of the reaction, a conclusion following form the observation that the power of a chemical engine propelled by a chemical reaction is limited by the activation energy.

Escher and Ross (1983) showed multiple ranges of flow rate with bistability and limit cycles for Schlögl mechanism in CSTR. Further Escher and Ross (1985) investigated the reduction of dissipation in a thermal engine by means of periodic changes of external constraints. Two modes of operation were compared, each with fixed input rate of chemicals: one with periodic variation of an external constraint [mode (b)], in which we vary the external pressure, and one without such variation [mode (a)]. The authors derived equations for the total power output in each of the two modes. The power output in mode (b) can be larger than that of mode (a) for the same chemical throughput and for the same average value of the external pressure. For a particularly simple case the authors have shown that the total power output in mode (b) is larger than that in (a) if work is done by the piston. At the same time the entropy production is decreased and the efficiency is increased. The possibility of an increased power output is due to the proper control of the relative phase of externally varied constraint and its conjugate variable, the external pressure and volume. This control is achieved by the coupling of nonlinear kinetics to the externally varied constraint.

Chen et al. (1997a, b, 1998, 1999a) formulated and solved a group of problems on performance characteristics and maximum power for chemical engines. Chen et al. (1998) determined the performance properties of an isothermal chemical engine with a mass leak. They derived the relations between the optimal power output, the efficiency, the maximum power output and the corresponding efficiency. They also found the maximum efficiency and the power output on the basis of linear (Onsager, 1931a, b) mass-transfer law. As they say, there exists a maximum efficiency bound with nonzero power output besides the maximum power output bound with nonzero efficiency for the engine. This characteristic is different from that known for the chemical engine with mass transfer as the sole irreversibility. Chen et al. (1999a) analyzed the performance of an isothermal combined-cycle chemical engine with mass leak. They derived the relation between the optimal power output and efficiency, and also determined the efficiency at the maximum power output, as well as the relation between the maximum efficiency and the corresponding power output.

Chen et al. (1999b) review the historical background, research development, and the state-of-art of finite time thermodynamic theory and applications from the point of view of both physics and engineering. The emphasis is on the performance optimization of thermodynamic processes and devices subject to the finite-time and/or finite-size constraints (heat engines, refrigerators, heat pumps, chemical reactors, etc.). The following aspects are considered: the systems governed by Newton's law, the effect of heat resistance and other losses on the performance, an analysis of the effect of heat reservoir models on the performance, and applications for real thermodynamic processes and devices.

A cyclic model of a class of chemical engines is set up by Lin et al. (2004), in which not only finite-rate mass transfer and mass leakage but also the internal irreversibility resulting from friction, eddy currents and other loss effects inside the cyclic working fluid are taken into account. The influence of these irreversible properties on the performance of the cycle are revealed. An optimal relation between the power output and the efficiency of the cycle is derived. With the optimal relation, optimal performances, and performance bounds of the cycle are determined and evaluated. For example, the maximum power output and the corresponding efficiency, the maximum efficiency and the corresponding power output, the optimal mass-transfer time, the minimum rate of energy loss, and so on, are calculated and analyzed. The results obtained provide a theoretical guidance for the effective application of energy resources and for the optimal design and development of a class of chemical engines. Some basic results known for the endoreversible engines can be deduced.

Using the finite time thermodynamics Chen et al. (2008) optimize the performance of an endoreversible chemical engine with diffusive mass transfer. They derive analytical relations for power output and efficiency and a relation linking these quantities in the optimal isothermal case. They also study locations and properties of optimal operating regions. Xia et al. (2010a) analyze and then optimize the performance of a generalized irreversible chemical engine with irreversible mass transfer, mass leakage, and internal dissipation. A numerical example shows the effects of chemical potential ratio, mass transfer coefficient ratio, mass leakage coefficient, and internal irreversible parameter on the power output versus the efficiency. The results provide theoretical guidance for the design of engines.

Several remarks should be devoted to the mathematically oriented works. Gadewar et al. (2001) present a systematic method to determine reaction invariants and mole balances for complex chemical systems. Sieniutycz (2009c) investigates power production in complex multireaction systems propelled by uncoupled or coupled multicomponent mass transfer. Evaluation of process invariants and mole balances follows the approach of Gadewar et al. (2001). Tsirlin et al. (1998) derived general optimality conditions for a group of finite-time processes. Application yields many known bounds on the entropy production and chemical conversion and previously unknown bounds (for throttling, crystallization, and mechanical friction). Chen et al. (2011) maximized power outputs of the multistage and continuous models of the chemical engine by applying the Hamilton–Jacobi–Bellman (HJB) theory and the Euler–Lagrange equation. The methods of dynamic programming and discrete maximum principle are used. Numerical results of both models are compared for three different boundary conditions. Each maximum power output is equal to the difference between their classical reversible limits and a dissipation term. Comparing to the previous results (Sieniutycz and Jeżowski, 2009; Sieniutycz, 2008b), the paper provides more detailed data of the maximum power in the chemical engine operating between a finite-capacity high-chemical-potential reservoir and an infinite-capacity low-chemical-potential environment.

9.10.2 Principles of Modeling of Power Yield in Chemical Systems

Power optimization in chemical, power-producing systems brings the difficulty caused by multiple (vector) efficiencies. Steady-state modeling describes a chemical system in which two reservoirs are infinite, whereas an unsteady model treats a dynamic system with finite upper reservoir and gradually decreasing chemical potential of a key fuel component. In the chemical system exemplified below, total power output is maximized at constraints, which take into account mass transport and efficiency of power generation. Dynamic optimization methods, in particular calculus of variations, lead to optimal functions that describe integral power limits and extend reversible chemical work W_{rev} to finite rate situations. Optimization data quantify effects of chemical rates and transport phenomena. When calculating power limits, we search for purely physical extrema with no regard to economic optima.

Below we analyze the performance of a nonisothermal chemical engine in terms of heat and mass fluxes flowing from a fuel reservoir to the power generator. By assumption, a fuel mixture that drives the power generator is composed of an inert and an active component. Efficiency, power yield, and fuel flux are essential variables determining the performance of the chemical system (de Vos, 1992; Sieniutycz, 2000h, 2001c, 2007d). The problem of finite-rate limits, which was treated in many earlier papers for thermal processes (de Vos, 1992; Sieniutycz, 2003a, b, c, 2006c, d), is applied here to chemical systems, steady and unsteady. Enhanced limits caused by finite rates are evaluated for power released from an engine system or added to a power consuming system.

We distinguish two basic models of chemical units producing power. Steady-state model, originated by de Vos (1992); Gordon (1990, 1991, 1993), and Gordon and Orlov (1993), and further worked out by Chen et al., 1997a, b, 1998, 1999a; Lin et al., 2004; Xia et al., 2010a; and others refers to the situation when both reservoirs are infinite, whereas a newer, unsteady model treats a dynamical case with finite upper reservoir and gradually decreasing chemical potential of the active component of fuel (Sieniutycz, 2007d, 2008b, 2012; Li et al. (2010a); L. Chen et al., 2011; Xia et al., 2009, 2010b, c; Sieniutycz and Jeżowski, 2009, 2013). Some of these works extend the original isothermal problem to situations with nonisothermal generation of power. In the dynamical case Lagrangian and Hamiltonian approaches to power functionals and optimization algorithms using canonical equations are effective. Finite-rate models incorporate the effect of entropy production caused by irreversible heat and diffusion phenomena. Modeling a power-assisted chemical operation for the purpose of dynamic limits is a difficult task. Evaluation of dynamic limits requires sequential operations (Berry et al., 2000; Sieniutycz, 1999, 2003c, 2006d, 2007d; Xia et al., 2011a), where the power yield is maximized at constraints, which describe dynamics of energy and mass exchange. The dynamical model can be continuous or discrete; the latter are frequent for computational purposes. The results describe limiting work functions in terms of end states, duration and (in discrete processes) number of stages. Modeling of power generation processes is consistent with general philosophy of optimization (Berry et al., 2000; Beveridge and Schechter, 1970; Chen and Sun, 2004; Sieniutycz and de Vos, 2000; Sieniutycz and Farkas, 2005). Constraints take into account dynamics of mass transport and rates of fuel consumption. Finite-rate, endoreversible modeling includes irreducible losses of classical exergy caused by resistances. Optimal performance functions, which describe extremum power and incorporate a residual minimum entropy production, are determined in terms of end states, duration, and (in discrete processes) number of stages (Sieniutycz, 2003c, 2006c, 2007d).

Similarly as in thermal systems enhanced power limits follow from the constrained optimization of power in chemical engines. Mass transport processes participate in the transformation of differences of chemical potentials into mechanical power. However, if chemical reaction drives the power yield process, a more precise statement refers not to the individual chemical potentials μ_k, but to their linear combination in the form of unidirectional component of chemical affinity, that is, one-way affinity Π_j. This affinity component is, in fact the potential Π_j of the reagents that decreases along the reaction path. The quantity Π_j (which reduces to the chemical potential of a single component only in the case of a simplest isomerisation reaction) plays in chemical engines role analogous to that played by the temperature in heat engines (Sieniutycz, 2004a). Yet, in chemical systems, generalized reservoirs are present, capable of providing heat and substance. Infinite reservoirs are capable of keeping the process potentials (T, μ_k, Π_j, etc.) constant. For such reservoirs problems of extremum power (maximum of power produced and minimum of power consumed) are static optimization problems. On the other hand, for finite upper reservoir, in which the amount of an active reactant is gradually reduced and its chemical potential decreases in time, the considered problems are those of dynamic optimization.

Starting with a steady-state situation we consider a single "endoreversible" chemical machine, an engine or power consumer, in Fig. 9.14. The engine is propelled by a high-μ active reactant of a fluid mixture (fuel mixture) that is supplied to the power generator from an infinitely large reservoir. The fluid mixture is composed of the active reactant and an inert. The flux of this active reactant, designated by n, is a basic quantity determining the intensity of power yield.

Yet, tackling a dynamical situation requires considering a multistage production (consumption) of power in a sequence of infinitesimal chemical engines. This case is associated with a finite upper reservoir and a gradual exhaust of the active reactant in the fuel mixture. In the multistage arrangement the chemically active reactant drives the chemical power generator at each stage from which power is released. In the multistage power consumer the fuel mixture is upgraded in the system to which power is supplied. In each case the second fluid is an infinite reservoir. The fluids are of finite thermal and mass conductivity; hence there are finite resistances in the system. In a multistage engine operation the driving chemical potential decreases at each stage; the whole operation is described by the sequence $\mu^0, \mu^1, ..., \mu^N$. The popular "engine convention" is used: work generated in an engine, W, is positive, and work generated in a power consumer is negative; this implies a positive work $(-W)$ supplied to the power consumer.

The sign of an optimal work function $V^N = \max W^N$ defines working mode for an optimal sequential process as a whole. In engine modes $W > 0$ and $V > 0$. In power consumption modes, $W < 0$ and $V < 0$, therefore working with a function $R^N = -V^N = \min(-W^N)$ is more convenient. Of special attention are two processes: the one which starts with the state $X^0 = X^e$ and terminates at an arbitrary $X^N = X$ and the one which starts at an arbitrary $X^0 = X$ and terminates at X^e. For these processes functions V^N and R^N are counterparts of the classical exergy in state changes with finite durations. The quantity V^N constitutes a generalized potential of extremum work that depends on end states of the fuel mixture and its holdup time in the system (time of fuel consumption). Alternatively, a generalized potential could be expressed in terms of end states of the fuel stream and the optimal Hamiltonian that is a measure of the process intensity along an optimal path.

To find power limits for a finite time degrading of the fuel mixture the system unit must contain a chemically active part or a "reaction zone." Only in "endoreversible" systems this zone is a purely reversible part of the system, that is, its efficiency of energy production is given by a reversible formula. This reversible formula constitutes a chemical counterpart of the familiar Carnot formula $\eta_C = 1 - T_2/T_1$ valid for thermal engines. In chemical engines, however, the reversible formula refers to the chemical affinity, which is the second component of the efficiency vector, as we shall see soon.

Yet, the restriction to external irreversibilities is unnecessary; in fact, thermodynamic models can go beyond "endoreversible limits" that is, they can treat internal irreversibilities as well, see refs. (Berry et al., 2000; Chen et al., 2001a; Sieniutycz, 2006d, 2008a). It is essential, however, that in either of two methodological versions of the approach, of which the first gives up internal irreversibility whereas the second one estimates it from a model, the power limits obtained are stronger than those predicted by the classical exergy. In short, this results from the "process rate penalty" taken into account in every version of the approach. In classical Carnot–like analyses the resource and environment reservoirs are insensitive to the effect of dissipators (boundary layers, resistances, etc.) because the reversible situation requires the spatial homogeneity in each reservoir. In the irreversible analysis, performed here, which admits dissipative transports, inhomogeneity in transport potentials is an essential property.

9.10.3 Thermodynamics of Power Production in Chemical Engines

This section analyzes a single-stage chemical process depicted in Fig. 9.14. To obtain a power yield formula associated with a single isomerization reaction

458 Chapter 9

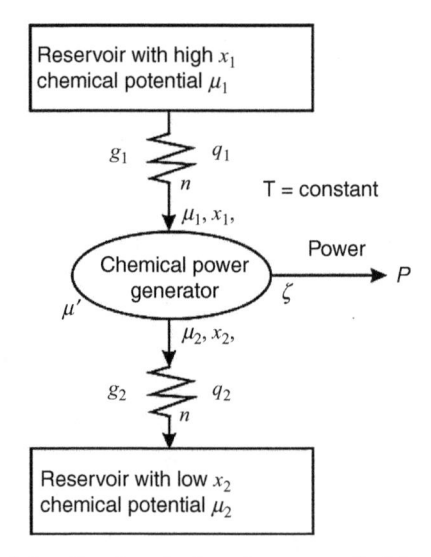

Figure 9.14: Principle of a chemical engine propelled by the mass transfer of an active component through an inert.

$$B_1 \Leftrightarrow B_2 \tag{9.42}$$

we apply balances of energy, entropy and mass. We assume that an active reagent 1 and the sensible heat q_1 are transferred to the reaction zone through an inert gas. As the result of the chemical transformation the reaction zone yields product 2, which is transferred through the same inert gas to the environment. The energy balance

$$\varepsilon_1 = \varepsilon_2 + p \tag{9.43}$$

and the mass balance in terms of molar fluxes

$$n_1 = n_{1'} = n_{2'} = n_2 \tag{9.44}$$

are combined with an equation describing the continuity of the entropy flux

$$\frac{\varepsilon_1 - \mu_{1'} n_{1'}}{T_{1'}} = \frac{\varepsilon_2 - \mu_{2'} n_{2'}}{T_{2'}} \tag{9.45}$$

Equation (9.45) holds in the chemically active, reversible part of the system where the chemical reaction runs.

Equations (9.43)–(9.45) yield the power expression

$$p = \varepsilon_1 - \varepsilon_2 = \left(1 - \frac{T_{2'}}{T_{1'}}\right)\varepsilon_{1'} + T_{2'}\left(\frac{\mu_{1'}}{T_{1'}} - \frac{\mu_{2'}}{T_{2'}}\right) n_{1'} \tag{9.46}$$

The energy flux in the dissipative parts of the system is continuous, for example

$$\varepsilon_1 = q_1 + h_1 n_1 = q_{1'} + h_{1'} n_{1'} = \varepsilon_{1'} \qquad (9.47)$$

Using the primed part of this equation in the power formula (9.46) yields

$$p = \left(1 - \frac{T_{2'}}{T_{1'}}\right) q_{1'} + [(h_{1'} - h_{2'}) - T_{2'}(s_{1'} - s_{2'})] n_{1'}. \qquad (9.48)$$

Thus the combination of the second law and the reversible balance of the entropy leads to power expression (9.48) in which the Carnot efficiency of an endoreversible process

$$\eta = 1 - \frac{T_{2'}}{T_{1'}} \qquad (9.49)$$

is the thermal component of a two-dimensional efficiency vector. The second component of this vector is the exergy-like function of the active component evaluated for the primed state 2 as the reference state

$$\beta' = h_{1'} - h_{2'} - T_{2'}(s_{1'} - s_{2'}). \qquad (9.50)$$

Observe that, in the considered case, where molar flux n is the efficiency basis, the chemical component of efficiency is not nondimensional. Other such cases will be observed, in which efficiencies (in general, nondimensional) coincide with certain *active* driving forces. Surprisingly, therefore, an active driving force always represents efficiency and quantity, which can occasionally be made nondimensional, if necessary.

The primed quantities and equations are often applied in transformed forms expressing all physical quantities in terms of the bulk state variables of both fluids and certain controls. The latter are related to fluxes of heat and matter. Applying the energy flux continuity and mass flux continuity in Eq. (9.48) to eliminate fluxes $q_{1'}$ and $n_{1'}$ on account of q_1 and n_1 yields

$$p = \left(1 - \frac{T_{2'}}{T_{1'}}\right) q_1 + \beta n_1. \qquad (9.51)$$

where coefficient β is defined as

$$\beta = \left[c_p(T_1 - T_{1'}) \left(1 - \frac{T_{2'}}{T_{1'}}\right) + (h_{1'} - h_{2'}) - T_{2'}(s_{1'} - s_{2'}) \right] n_1 \qquad (9.52)$$

Note bilinear power structures in Eqs. (9.48) and (9.51). Both the equations give useful pictures of chemical efficiency, yet they differ in the heat flux (q_1 or $q_{1'}$) accepted as the control variable. The second component of power expressions are Eqs. (9.48) and (9.51) is associated with the work production (consumption) due to the mass transfer. Its interpretation is the product of mass flow n and exergy-like functions β' or β whose structure follows from a combination of the energy and (conservative) entropy balances (Sieniutycz, 1999).

However, we may eliminate exergy-like functions β and β' by passing to the process description in terms of different fluxes. Leaving mass flux n unchanged we introduce a new energy flux, $Q_{1'}$, called total heat flux, which is defined by an equation

$$Q_{1'} \equiv q_{1'} + T_{1'} s_{1'} n_{1'} \qquad (9.53)$$

Clearly, $Q_{1'}$ is the product of temperature $T_{1'}$ and the total entropy flux, the latter being the sum of entropy transferred with heat $q_1/T_{1'}$ and with substance $s_1 n_1$. The virtue of flux $Q_{1'}$ is that the power production $p = \dot{w}$ assumes in terms of $Q_{1'}$ and n_1 an intuitively natural form that contains the Carnot efficiency and chemical potential difference of the active component

$$\dot{w} = p = \left(1 - \frac{T_{2'}}{T_{1'}}\right) Q_{1'} + (\mu_{1'} - \mu_{2'}) n \qquad (9.54)$$

In this case the chemical efficiency component is just the chemical affinity for the single isomerization reaction considered

$$\zeta \equiv \mu_{1'} - \mu_{2'} \qquad (9.55)$$

While this efficiency is not nondimensional, it describes correctly the system, so it will be used further as a convenient quantity.

Alternatively, we may base the efficiency on a certain mass analogue of heat flux $Q_{1'}$ defined as $G_{1'} \equiv \mu_{1'} n_{1'}$. The quantity $G_{1'}$ represents, in fact, the flux of Gibbs free energy of the active component of fuel. The suitability of this quantity, called the Gibbs flux, follows from its capability of measuring the quality of mass flux $n_{1'}$. In fact, flux $G_{1'} = \mu_{1'} n_{1'}$ measures the quality of $n_{1'}$ in the similar way as heat $Q_{1'} = T_{1'} s_{1'}$ measures the quality of entropy flux $s_{1'}$; the associated "quality potentials" are respectively $\mu_{1'}$ and $T_{1'}$. See de Vos (1992) and Sieniutycz and Jeżowski (2013) for various forms of chemical efficiencies and power expressions.

9.10.4 Entropy Generation in Steady Systems

Let us determine an expression for the entropy production from the entropy balance of a steady system. From the entropy balance of an overall system composed of reservoirs and reactor the intensity of the entropy production σ_s follows in terms of the reservoir parameters and system fluxes

$$\sigma_s = \frac{q_2}{T_2} - \frac{q_1}{T_1} + (s_2 - s_1) n \qquad (9.56)$$

Using in this equation the energy conservation law

$$q_1 + h_1 n = q_2 + h_2 n + p \tag{9.57}$$

to eliminate flux q_2 we obtain

$$\sigma_s = \frac{q_1 + h_1 n - h_2 n - p}{T_2} - \frac{q_1}{T_1} + (s_2 - s_1)n. \tag{9.58}$$

Whence, in terms of fluxes q_1 and n

$$\sigma_s = q_1\left(\frac{1}{T_2} - \frac{1}{T_1}\right) + \frac{(\mu_1 - \mu_2)n + (T_1 - T_2)s_1 n}{T_2} - \frac{p}{T_2} = \frac{(\eta_C - \eta)q_1 + (\beta_C - \beta)n}{T_2}, \tag{9.59}$$

Eqs. (9.56) and (9.57) yield in terms of other fluxes and corresponding efficiencies,

$$\sigma_s = Q_1\left(\frac{1}{T_2} - \frac{1}{T_1}\right) + \frac{(\mu_1 - \mu_2)n}{T_2} - \frac{p}{T_2} = \frac{(\eta_C - \eta)Q_1 + (\zeta_C - \zeta)n}{T_2} = \frac{(\eta_C - \eta)Q_1 + (\omega_C - \omega)G_1}{T_2}. \tag{9.60}$$

Eqs. (9.59) and (9.60) generalize to the nonisothermal case an earlier result limited to isothermal situations (Sieniutycz, 2007d). They show that, modulo the multiplier $(T_2)^{-1}$, the isothermal component of the entropy production is the product of the reactant flux and the deviation of chemical efficiency ζ from the corresponding efficiency of the reversible process, ζ_C. Yet, as shown by Eq. (9.60), the entropy production formula of an nonisothermal process contains the thermal component equal to the product of total (ie, mass transfer including) heat flux $Q_1 \equiv q_1 + T_1 s_1 n$ and the deviation of thermal efficiency η from the Carnot efficiency.

The subscript C points out that the reversible efficiencies η_C, ζ_C, and ω_C refer to the Carnot point of the system, also called "open circuit point" (de Vos, 1992). This point is associated with vanishing currents and upper, reversible limits for thermal and chemical efficiencies.

9.10.5 Characteristics of Steady Isothermal Engines

Expressing chemical potentials

$$\mu_k = \mu_{0k} + RT \ln x_k = \mu_{0k} + RT \ln\left(\frac{X_k}{1 + X_k}\right) \tag{9.61}$$

in terms of molar fraction of the active component in "upper" and "lower" part of the system ($k = 1, 2$) we obtain the affinity-related efficiency (9.55) in the form

$$\zeta = \mu_{1'} - \mu_{2'} = \zeta_0 + RT \ln\left(\frac{x_{1'}}{x_{2'}}\right) \equiv \zeta_0 + RT \ln\left(\frac{X_{1'}(1 + X_{2'})}{X_{2'}(1 + X_{1'})}\right), \tag{9.62}$$

where

$$\zeta_0 = \mu_{01'} - \mu_{02'} = \mu_{01} - \mu_{02}. \quad (9.63)$$

The last equation applies the property of chemical passivity for each component in the chemically inactive parts of the system, where only diffusive transport takes place. The equation describes, in fact, the standard Gibbs energy for the isomerization reaction considered. The constant ζ_0 involves chemical potentials of substrate and product in their reference states. The reference value ζ_0 vanishes only if both components are identical. In general, however, the constant ζ_0 is nonvanishing.

When both reservoirs are infinite the process is at the steady state. The mass transfer between each reservoir and the production section of the system is described by certain kinetic equations. For simplicity we assume that these equations are linear. The mass balances for the substance transferred, produced, and consumed are contained in the equations

$$n = g_1(x_1 - x_{1'}) \quad (9.64)$$

and

$$n = g_2(x_{2'} - x_2). \quad (9.65)$$

In order to determine work characteristics of the chemical engine at the steady state (unlimited stock of fuel) one searches for concentrations $x_{1'}$ i $x_{2'}$ expressed in terms of a control variable. For the chemical engine a suitable quantity can be efficiency ζ, Eqs. (9.62) and (9.63). This means that the following system of equations should be solved

$$\zeta = \zeta_0 + RT \ln\left(\frac{x_{1'}}{x_{2'}}\right), \quad (9.66)$$

$$g_1(x_1 - x_{1'}) = g_2(x_{2'} - x_2) = n. \quad (9.67)$$

From the first equation of this set we find

$$\frac{x_{2'}}{x_{1'}} = \exp\left(-\frac{\zeta - \zeta_0}{RT}\right). \quad (9.68)$$

Substituting $x_{2'}$ from this equation to the first equality of Eq. (9.67) written in the form

$$g_1 x_{1'} + g_2 x_{2'} = g_1 x_1 + g_2 x_2 \quad (9.69)$$

one obtains

$$x_{1'} = \frac{g_1 x_1 + g_2 x_2}{g_1 + g_2 \exp\left(-(\zeta - \zeta_0)/RT\right)} \quad (9.70)$$

and

$$x_{2'} = \frac{g_1 x_1 + g_2 x_2}{g_1 + g_2 \exp(-(\zeta-\zeta_0)/RT)} \exp\left(-\frac{\zeta-\zeta_0}{RT}\right) = \frac{g_1 x_1 + g_2 x_2}{g_1(\zeta-\zeta_0)/RT + g_2} \quad (9.71)$$

Each of the last two equations can be used in the balance–kinetic formula (9.67). This leads to an equation for the feed flux of the active component of fuel in terms of chemical efficiency

$$n = g_1 x_1 - g_1 x_{1'} = g_1 x_1 - g_1 \frac{g_1 x_1 + g_2 x_2}{g_1 + g_2 \exp(-(\zeta-\zeta_0)/RT)}. \quad (9.72)$$

This equation can be yet simplified into the form

$$n = g_1(x_1 - x_{1'}) = g_1 g_2 \frac{x_1 \exp(-(\zeta-\zeta_0)RT) - x_2}{g_1 + g_2 \exp(-(\zeta-\zeta_0)/RT)} = g_1 g_2 \frac{x_1 - x_2 \exp((\zeta-\zeta_0)/RT)}{g_1 \exp((\zeta-\zeta_0)/RT) + g_2}. \quad (9.73)$$

Thus, the feed rate of the system by the active reactant can be described by an equation

$$\frac{n}{g_1} = \frac{x_1 - x_2 \exp((\zeta-\zeta_0)/RT)}{1+(g_1/g_2)\exp((\zeta-\zeta_0)/RT)}. \quad (9.74)$$

The function inverse to the above defines the chemical efficiency in terms of the reactant's feed rate n

$$\zeta = \zeta_0 + RT \ln\left(\frac{x_1 - n g_1^{-1}}{n g_1^{-1}(g_1/g_2) + x_2}\right) \quad (9.75)$$

This equation shows that an effective concentration of the reactant in the upper reservoir $x_{1\text{eff}} = x_1 - g_1^{-1} n$ is decreased, whereas an effective concentration of the product in the lower reservoir $x_{2\text{eff}} = x_2 + g_2^{-1} n$ is increased due to the finite mass flux. Consequently, the efficiency ζ decreases nonlinearly with n. When the effect of the resistance g_k^{-1} is ignorable or the flux n is very small, reversible efficiency, ζ_C, is attained. Quite generally the power function described by the product $\zeta(n)n$ exhibits the maximum power for a finite value of the flux n. Eqs. (9.55) and (9.74) yield power in terms of efficiency ζ of an isothermal process

$$p = (\mu_{1'} - \mu_{2'})n = \zeta \frac{x_1 - x_2 \exp((\zeta-\zeta_0)/RT)}{(g_1)^{-1} + (g_2)^{-1} \exp((\zeta-\zeta_0)/RT)}. \quad (9.76)$$

This power function exhibits a maximum for a certain efficiency, ζ, this efficiency being a chemical analogue of the well-known Chambadal–Novikov–Curzon–Ahlborn efficiency (*CNCA* efficiency (de Vos, 1992; Chen et al., 2001a). The maximum power can be considered with respect to control variables n and ζ, for example. Yet, other control variables can be considered.

At the Carnot point (also called "open circuit point of the system"; de Vos, 1992), rates and power vanish, and the system efficiency ζ attains its upper limit. Only then is this efficiency identical with the reversible chemical affinity of the reaction. For the present model

$$\zeta_C = \frac{p}{n} = \mu_1 - \mu_2. \tag{9.77}$$

Putting Eq. (9.73) or (9.74) to zero yields

$$x_1 = x_2 \exp\left(\frac{\zeta_C - \zeta_0}{RT}\right). \tag{9.78}$$

This formula leads to the limiting reversible efficiency in the form determined by the chemical affinity of reaction at the reversible Carnot point

$$\zeta_C = \zeta_0 + RT \ln\left(\frac{x_1}{x_2}\right) \equiv A_C. \tag{9.79}$$

However, the power produced at the "open circuit" state equals zero, corresponding with vanishing feed flux of the driving reactant. Therefore a problem of maximum power or an economic power is the realistic optimization problem in the present context.

Intensity of the entropy generation in the system, Eq. (9.60), was presented in several forms. For the isothermal system considered two forms of σ_s can be discussed. In the first, the controlling quantity is the efficiency ζ

$$\sigma_s = \frac{(\zeta_C - \zeta)n}{T} = \frac{(\zeta_C - \zeta)}{T}\left(\frac{x_1 - x_2 \exp((\zeta - \zeta_0)/RT)}{(g_1)^{-1} + (g_2)^{-1} \exp((\zeta - \zeta_0)/RT)}\right). \tag{9.80}$$

In the second form the control is the fuel flux n; then, from Eq. (9.75),

$$\sigma_s = \frac{(\zeta_C - \zeta)n}{T} = R \ln\left(\left(\frac{x_1}{x_2}\right)\left(\frac{ng_2^{-1} + x_2}{x_1 - ng_1^{-1}}\right)\right). \tag{9.81}$$

In the so-called "short circuit point," there is no power production for any value of ζ_0, despite possible chemical reaction. Only entropy is then produced with intensity

$$\sigma_{szw} = \frac{\zeta_C}{T}\left(\frac{x_1 - x_2 \exp(-\zeta_0/RT)}{(g_1)^{-1} + (g_2)^{-1} \exp(-\zeta_0/RT)}\right) = \frac{(\mu_1 - \mu_2)}{T}\frac{(x_1 - x_2 \exp(-\zeta_0/RT))}{(g_1)^{-1} + (g_2)^{-1} \exp(-\zeta_0/RT)} \tag{9.82}$$

which is the maximum intensity of the engine range. When $\zeta_0 = 0$ the entropy production at the "short circuit point" corresponds with the situation without chemical reaction and energy generation

$$(\sigma_s)_{\zeta=0} = \frac{\zeta_C n_{zw}}{T} = \frac{\zeta_C}{T}\left(\frac{x_1 - x_2}{(g_1)^{-1} + (g_2)^{-1}}\right) = T^{-1}\frac{(\mu_1 - \mu_2)(x_1 - x_2)}{(g_1)^{-1} + (g_2)^{-1}}. \tag{9.83}$$

Clearly, only at the short circuit point and for the associated absence of power the entropy production at this point is proportional to the product of the reaction rate and its chemical affinity. Of course, the proportionality coefficient is the temperature reciprocal, T^{-1}. This is the classical result, which, however, does not hold when the system produces power (ie, belongs to the class of "active systems").

Using inverse function $\zeta(n)$, Eq. (9.75), one may describe power in terms of the fuel flux n

$$p = \zeta(n)n = \zeta_0 n + RTn \ln\left(\frac{x_1 - ng_1^{-1}}{ng_2^{-1} + x_2}\right). \tag{9.84}$$

Equation (9.75) proves that efficiency ζ decreases nonlinearly with feed flux n. The consequence of this property is the maximum of the product $\zeta(n)n$ describing the power yield. The maximum power point is the result of maximizing of the power function with respect to its free control variable, ζ, or n. Analytical methods are seldom effective. However, one can use diagrams describing functions $p(\zeta)$, or $p(n)$ to determine the maximum point graphically. Information on the location of maximum point can also be obtained by a numerical search.

Equation (9.84) shows that a maximum of power p is attained for the fuel flux n satisfying an equation

$$\frac{dp}{dn} = \zeta_0 + RT \ln\left(\frac{x_1 - ng_1^{-1}}{ng_2^{-1} + x_2}\right) - RTn\frac{g_1^{-1}x_2 + g_2^{-1}x_1}{(x_1 - ng_1^{-1})(ng_2^{-1} + x_2)} = 0. \tag{9.85}$$

Its numerical solution generates a function describing the optimal feed of the system by the active reactant

$$n_{mp} = f(\zeta_0, T, x_1, x_2, g_1, g_2). \tag{9.86}$$

Substituting this result into Eq. (9.75), leads to the chemical counterpart of the Chambadal–Novikov–Curzon–Ahlborn (CNCA) efficiency (de Vos, 1992; Chen et al., 2001a). It describes the efficiency ζ_{mp} of the chemical engine at the maximum power point in terms of the system properties

$$\zeta_{mp} = f(\zeta_0, T, x_1, x_2, g_1, g_2). \tag{9.87}$$

The curve of produced power has two zero points. The first is the "short circuit point" (also the point of vanishing efficiency), and, the second is the "open circuit point" or Carnot point at which the feed of the system with the active component is infinitely slow. Equation (9.73) proves that the feed rate of the active reagent at the short circuit point (where $\zeta = 0$) is

$$n_{sc} = \frac{x_1 - x_2 \exp(-\zeta_0/RT)}{(g_1)^{-1} + (g_2)^{-1} \exp(-\zeta_0/RT)}. \tag{9.88}$$

For a nonvanishing $\zeta_0 = \mu_{01'} - \mu_{02'} = \mu_{01} - \mu_{02}$, the short-circuit point corresponds with a particular, "purely dissipative" state of the system at which loss elements predominate so significantly that the chemical reaction does not produce any power, despite of nonvanishing reaction rate and existing finite potential of ζ_0 (but not ζ) for the production of this power. For $\zeta_0 = 0$ the discussed equation describes the situation, in which the reactor does not exist, the fuel stream flows by two sequentially connected conductors, and molar flux of the reactant, n, is determined by the specification of the overall conductance $g = \left[(g_1)^{-1} + (g_2)^{-1}\right]^{-1}$. Consequently, for $\zeta_0 = 0$ and at the short circuit point of the system

$$n_{sc} = \frac{x_1 - x_2}{(g_1)^{-1} + (g_2)^{-1}}. \tag{9.89}$$

which is a form of Ohm's law for the fuel flux transferred by two resistances in series.

9.10.6 Sequential Models for Computation of Dynamic Generators

When resources are limited their quality decreases in time, and dynamical processes arise. Power optimization then requires variational methods to maximize power integrals subject to process differential constraints. Taking into account analytical difficulties we often apply methods of discrete optimization, for example, the discrete maximum principle or dynamic programming method. Regarding computer needs, an extremum problem for a power integral is then broken down into an optimization problem for a cascade with a finite number of steps.

To set a discrete model of the chemical engine, we consider the optimal fuel consumption in a cascade of K engines, with efficiency ζ^k or local feed n^k as possible control variables at kth stage. To describe the fuel consumption we exploit the mass balance. We also introduce a cumulative flux of active substrate over the first k stages of the cascade, $N^k = \Sigma n^l$, where $l = 1, 2 \ldots k$. The sequence of local fluxes n^l (related to "upper" potentials μ_1^k in the kth engine) describes allocations of N^k between stages 1, 2, ..., k. Each local flux n^k of the active reagent at stage k equals the change of cumulative mass flux $N^k - N^{k-1}$.

The mass balance at the stage k,

$$n^k \equiv N^k - N^{k-1} = -I(X_1^k - X_1^{k-1}), \tag{9.90}$$

shows that the local molar flux at the stage k (an interval of the cumulative mass flux N^k) can be evaluated as the (negative) product of molar flux of inert, I, and the change of reactant's concentration, ΔX^k. Then, in the case of an isothermal mass exchange, any change in the cumulative coordinate N can be evaluated in terms of the reactant's chemical potential or its concentration X_1^k.

The reactant's flux n^k can be eliminated on account of difference $X_1^k - X_1^{k-1}$ with the help of Eq. (9.90). Simultaneously it is convenient to have at our disposal a separate expression for efficiency ζ^k in terms of concentrations X_1^k and X_1^{k-1}. Both these needs are satisfied in one stroke below. Mass balance (9.90) and the current-efficiency characteristic (9.74) yield

$$-I(X_1^k - X_1^{k-1}) = N_1^k - N_1^{k-1} = \frac{x_1 - x_2 \exp((\zeta - \zeta_0)/RT)}{1 + (g_1/g_2)\exp((\zeta - \zeta_0)/RT)} \beta_1'^k (A^k - A^{k-1}), \quad (9.91)$$

where $\beta'_1 \equiv \beta_1 y = \beta_1 a_1/(a_1 + a_2)$ and $\beta_1'^k (A^k - A^{k-1}) = g_1^k$. Whence

$$-\frac{I(X_1^k - X_1^{k-1})}{\beta_1'^k (A^k - A^{k-1})} = \frac{n^k}{g_1^k} = \frac{x_1 - x_2 \exp((\zeta - \zeta_0)/RT)}{1 + (g_1/g_2)\exp((\zeta - \zeta_0)/RT)}. \quad (9.92)$$

Since the inversion of Eq. (9.74) or the right-hand side of Eq. (9.92) is (9.75), the chemical efficiency ζ^k in terms of X_1^k and X_1^{k-1} is

$$\zeta^k = \zeta_0^k + RT \ln\left(\frac{X_1^k/(1 + X_1^k) + I(X_1^k - X_1^{k-1})/\beta_1'^k a_v^k F^k (l^k - l^{k-1})}{-(I(X_1^k - X_1^{k-1}))/(\beta_1'^k a_v^k F^k (l^k - l^{k-1}))(g_1/g_2) + x_2} \right). \quad (9.93)$$

In the engine mode the concentration of the active reactant can only decrease along a path, thus the term with the discrete slope $\Delta X^k/\Delta l$ in Eq. (9.93) is negative. Consequently, the efficiency of a stage working in the engine mode is lower than the Carnot efficiency.

The quantity

$$\frac{I}{\beta_1' a_v F} \equiv H_{TU_1} \quad (9.94)$$

has units of length, and is known from the mass transfer theory as the "height of the mass transfer unit" (H_{TU}). In Eq. (9.94) it is referred to partial mass transfer coefficient of active reactant, β_1', although an analogous quantity could be defined for the product. The nondimensional length

$$\tau_1 \equiv \frac{l}{H_{TU_1}} \quad (9.95)$$

is identical with the "number of transfer units" N_{TU} for mass transfer. Since it is proportional to the system's extent l and hence to the contact time of active reactant with mass exchange area, it also plays the role of a nondimensional time, and this is why it is designated by τ_1

$$\frac{p^k}{l} = -\left\{\zeta_0 + RT \ln\left(\frac{\left(X_1^k/(1+X_1^k)\right) + \left((X_1^k - X_1^{k-1})/(\tau_1^k - \tau_1^{k-1})\right)}{x_2 - j\left((X_1^k - X_1^{k-1})/(\tau_1^k - \tau_1^{k-1})\right)}\right)\right\}(X_1^k - X_1^{k-1}) \quad (9.96)$$

The coefficient $j \equiv g_1/g_2$ is the conductance ratio.

The total power delivered from the N-stage process per unit flux of the inert is the sum of contributions at stages. This sum is a discrete functional which is maximized by the suitable choice of the inter-stage concentrations and allocation of time intervals between the stages.

$$W^N \equiv \sum_{k=1}^{N} w^k = -\sum_{k=1}^{N}\left\{\zeta_0 + RT \ln\left(\frac{\left(X_1^k/(1+X_1^k)\right) + \left((X_1^k - X_1^{k-1})/(\tau_1^k - \tau_1^{k-1})\right)}{x_2 - j\left((X_1^k - X_1^{k-1})/(\tau_1^k - \tau_1^{k-1})\right)}\right)\right\}(X_1^k - X_1^{k-1}) \quad (9.97)$$

Results for continuous systems are obtained as limits of results found in discrete systems for a sufficiently large number of stages, to secure reasonable accuracy.

We shall now outline a computational algorithm in which the discrete problem of maximum power is treated as an optimal control problem, and a suitable control variable is chosen to assure a relatively simple model. The simplicity condition is satisfied by a simple link of control with rate change of reactant's concentration in the fuel, $dX/d\tau_1$. In the discrete version this rate is replaced by the difference ratio $\Delta X/\Delta \tau_1$.

We introduce a control variable describing the fuel consumption

$$v^k \equiv -\frac{n^k}{g_1} \quad (9.98)$$

It is negative in engine modes and positive in power consumption modes. For the process investigated we find

$$v \equiv -\frac{n}{g_1} = -\frac{N^k - N^{k-1}}{\gamma_1^k - \gamma_1^{k-1}} = \frac{I(X_1^k - X_1^{k-1})}{\beta_1'^k(A_1^k - A_1^{k-1})} = \frac{X_1^k - X_1^{k-1}}{\tau_1^k - \tau_1^{k-1}}. \quad (9.99)$$

$$u \equiv -\frac{q_1}{g_1} = -\frac{Q^k - Q^{k-1}}{\gamma_1^k - \gamma_1^{k-1}} = \frac{Ic(T_1^k - T_1^{k-1})}{\alpha_1'^k(A_1^k - A_1^{k-1})} = \frac{T_1^k - T_1^{k-1}}{\tau_1^k - \tau_1^{k-1}}. \quad (9.100)$$

The optimization problem searches for a maximum of the performance index

$$W^N = -\sum_{k=1}^{N} \left\{ \zeta_0 + RT^k \ln\left(\frac{X_1^k(1+X_1^k)^{-1} - v^k}{x_2 - jv^k} \right) \right\} v^k \theta^k \qquad (9.101)$$

($j \equiv g_1/g_2$) subject to difference constraints

$$X_1^k - X_1^{k-1} = v^k \theta^k \qquad (9.102)$$

$$\tau^k - \tau^{k-1} = \theta^k \qquad (9.103)$$

In a nonisothermal problem an extra equation has to be included to treat temperature changes

$$T^k - T^{k-1} = u^k \theta^k \qquad (9.104)$$

and the power function has to be enlarged to include the thermal component of power yield.

Change in the sign of performance function Eq. (9.101) transforms the problem into a discrete problem of power minimization. Its numerical solution is described elsewhere (Sieniutycz and Jeżowski (2013).

9.10.7 Further Problems

Power evaluation in practical chemical generators calls for the relaxation of the assumption about reversibility of chemical reaction within the active zone with power yield. This requires introducing internal entropy production within the active zone. Moreover, single chemical efficiency, ζ, defined as power yield per one mole of an invariant molar flux n, is insufficient in nonisothermal generators. In these units, multiple reactions and extra component of power, associated with the temperature difference between the reservoirs, play a role. An example of a multireaction system is illustrated in Fig. 9.2 This scheme assumes a set of chemical reactions, $j = 1, 2, ..., R$, undergoing between species B_i ($i = 1, 2, ..., m$) in the system.

Consider a general while still idealized case of an endoreversible engine with many components and chemical reactions, assuming that the boundary states of the endoreversible zone refer to the cross-sections 1' and 2'. In such an engine fluxes of entropy and matter are continuous through the active (endoreversible) zone, because there is no entropy production within reversible part of the generator, and the mass conservation principle is valid. In terms of the total heat flux Q, entropy flux n_s, and Gibb's fluxes G_j, the continuity equations are

$$\frac{Q_{1'}}{T_{1'}} = \frac{Q_{2'}}{T_{2'}} = n_s \qquad (9.105)$$

$$\frac{G_{j1'}}{\Pi_{j1'}} = \frac{G_{j2'}}{\Pi_{j2'}} = n_j \qquad j = 1, 2, \ldots, R \qquad (9.106)$$

Applying continuity formulas, Eqs. (9.105) and (9.106), in the general power formula of an endoreversible chemical engine driven by heat fluxes and R chemical reactions

$$p = Q_{1'} - Q_{2'} + \sum_{j=1}^{R}(G_{j1'} - G_{j2'}) \qquad (9.107)$$

power expression of an idealized (endoreversible) engine is obtained in several alternative forms contained in the equation

$$\begin{aligned}
p &= Q_{1'} - Q_{2'} + \sum_{j=1}^{R}(G_{j1'} - G_{j2'}) = \left(1 - \frac{Q_{2'}}{Q_{1'}}\right)Q_{1'} + \sum_{j=1}^{R}\left(1 - \frac{G_{j2'}}{G_{j1'}}\right)G_{j1'} \\
&= \left(1 - \frac{T_{2'}}{T_{1'}}\right)Q_{1'} + \sum_{j=1}^{R}\left(1 - \frac{\Pi_{j2'}}{\Pi_{j1'}}\right)G_{j1'} = \left(1 - \frac{T_{2'}}{T_{1'}}\right)T_{1'}n_{s1'} + \sum_{j=1}^{R}\left(1 - \frac{\Pi_{j2'}}{\Pi_{j1'}}\right)\Pi_{j1'}n_{j1'} \quad (9.108) \\
&= \left(1 - \frac{T_{2'}}{T_{1'}}\right)Q_{1'} + \sum_{j=1}^{R}(\Pi_{j1'} - \Pi_{j2'})n_{j1'}
\end{aligned}$$

The last term of this equation contains endoreversible chemical efficiencies expressed as changes of reaction potentials, for all reactions involved, $j = 1, 2, \ldots, R$.

Consider now a general *nonideal case*, characterized by the internal production of entropy and incomplete conversion of substrates into products. Since the identification of active affinity with chemical efficiency is valid only for thermodynamically reversible reactions, we shall explain how internal entropy production and incomplete conversion affect forms of efficiency expressions.

The second law applied to the internal entropy production in the engine mode dictates that $n_{s2'} > n_{s1'}$, or $\Phi > 1$. Moreover, $n_{j2'} < n_{j1'}$, or $\Xi_j = < 1$, for each reaction j, which refers to the reduction of the outgoing chemical flow of product j by incomplete conversion. Thus the following relations hold in an imperfect system

$$\frac{Q_{2'}}{T_{2'}} = \Phi\frac{Q_{1'}}{T_{1'}} = \Phi n_{s1'} \qquad (9.109)$$

$$\frac{G_{j2'}}{\Pi_{j2'}} = \Xi_j \frac{G_{j1'}}{\Pi_{j1'}} = \Xi_j n_{j1'} \qquad (9.110)$$

where imperfection coefficients satisfy the inequalities $\Phi > 1$ and $\Xi_j < 1$. The first inequality follows from the second law, whereas the second is the consequence of the incomplete conversion which dictates that the outlet chemical flux $n_{j2'}$ can only be less that the inlet chemical flux $n_{j1'}$.

With Eqs. (9.109) and (9.110), a nonideal power formula is obtained in the form

$$p = Q_{1'} - Q_{2'} + \sum_{j=1}^{R}(G_{j1'} - G_{j2'}) = \left(1 - \frac{Q_{2'}}{Q_{1'}}\right)Q_{1'} + \sum_{j=1}^{R}\left(1 - \frac{G_{j2'}}{G_{j1'}}\right)G_{j1'}$$

$$= \left(1 - \Phi\frac{T_{2'}}{T_{1'}}\right)Q_{1'} + \sum_{j=1}^{R}\left(1 - \Xi_j\frac{\Pi_{j2'}}{\Pi_{j1'}}\right)\Pi_{j1'}n_{j1'} = \left(1 - \Phi\frac{T_{2'}}{T_{1'}}\right)Q_{1'} + \sum_{j=1}^{R}(\Pi_{j1'} - \Xi_j\Pi_{j2'})n_{j1'} \quad (9.111)$$

When diffusion processes can be ignored, a simple approximation of the above formula for the simple isomerization reaction $a \Leftrightarrow b$ (or $A - B \Leftrightarrow 0$) has the form

$$p \cong \left(1 - \Phi\frac{T_{2'}}{T_{1'}}\right)Q_{1'} + (\mu_{1'} - \Xi\mu_{2'})n_{1'} \quad (9.112)$$

This result contains reduced thermal and chemical efficiencies, and it represents an "imperfect counterpart" of the idealized power formulas Eqs. (9.54) and (9.108) which are limited to reversible conversions.

Power formulae Eq. (9.111) generalize idealized power of Eq. (9.108). Imperfect efficiency vector has two components. The first or thermal component describes power generated by total heat flux $Q_{1'}$ with the lowered (non-Carnot) efficiency, $\eta = 1 - \Phi T_{2'}/T_{1'}$. Whereas the second or chemical component describes power generated with the nonideal chemical efficiency $\xi = \mu_{1'} - \Xi\mu_{2'}$. The fractional nature of coefficient Ξ is consistent with the negative sign of chemical potentials (Baierlein, 2001). In the literature, where this property is not always respected, imperfection coefficients different than Ξ are occasionally applied.

The chemical term in power formulae predominates in systems that work closely to isothermal conditions. In the engine mode, where $\Phi > 1$ and $\Xi < 1$, internally imperfect chemical system behaves as it would work with an increased temperature of lower reservoir ($\Phi T_{2'}$ instead $T_{2'}$) and with a decreased chemical affinity of an effective value $\mu_{1'} - \Xi\mu_{2'}$. Of course, these imperfections reduce power production. An example of the engine power reduction caused by the imperfect chemical efficiency is presented by Lin et al. (2004).

Mann (2000) proposed a new approach to formulate the design of chemical reactors with multiple reactions. The new design methodology is applicable for all reactor configurations, any number of chemical reactions, having any form of rate expressions. Reactor operations are formulated in terms of the smallest set of dimensionless design equations and their solutions provide dimensionless operating curves, indicating progress of the chemical reactions, species compositions and the temperature variations without considering the reactor size or operating time. Reactor sizing represents a "point" on these operating curves and the required volume is readily determined for a specified feed rate and given reaction rates. The design is formulated using the least number of terms in each design equation, thus providing the most robust set of equations for numerical solutions. The methodology also provides new design capabilities for reactor configurations whose designs are not available in the literature. First part of the approach, Mann (2000), presents the basic concepts of the new approach and illustrates how they are applied. However, this theory does not focus on power yield problems in chemical reactors.

Jaworski and Zakrzewska (2008) have presented a general framework of the concept of multiaspect and multiscale approach to reactor modeling. Numerical methods employed in reactor design, based on computational fluid dynamics, computer-aided process engineering, and computational chemistry have been described. The published strategies of integrating those methods and examples of their applications to reactor design have been collected and discussed. Hopefully, a future modification of this theory will include chemical reactors with power production.

CHAPTER 10

Power Limits in Thermochemical Units, Fuel Cells, and Separation Systems

10.1 Introduction

The frontiers in energy research and education are currently attributed to: power systems, biological systems, security, information technology, and nanotechnology (Bergman et al., 2008). Their text stresses the role of various power systems (thermal, chemical, and electrochemical) in contemporary investigations. In recent years, many approaches have been applied that implement methods of thermodynamic optimization to complex energy systems with power production. For instance, Chen et al. (2001) present an advanced discussion on the Curzon–Ahlborn efficiency and its connection with the efficiencies of real heat engines. Sánchez-Orgaz et al. (2013) treat recuperative solar-driven multistep gas-turbine power plants. By applying energy and exergy analyses, Ganjeh et al. (2013) develop exergo-environmental optimization of heat-recovery steam generators in combined cycle power plants. Using different objective functions, Bracco and Siri (2010) propose the exergetic optimization of single level combined gas–steam power plants. Chaisantikulwat et al. (2008) develop an approach to modeling and control of planar anode-supported solid oxide fuel cells (SOFCs). Many papers analyze hybrid combinations of gas turbines (GT) and FC. Pade and Thorsten Schröder (2013) investigate an economic analysis for fuel-cell-based micro-combined heat and power under different policy frameworks. Bakalis and Stamatis (2012) develop an exergetic analysis of a hybrid micro-gas-turbine fuel-cell system based on existing components.

Considering electrolyzers as units inverse to FCs, Xu et al. (2011) analyze the performance of a high temperature polymer electrolyte membrane water electrolyzer. Simultaneously, Faesa et al. (2011) offer an experiment-based design approach applied to reducing and oxidizing tolerance of the anode supported SOFC. El-Genk and Tournier (2011) present a number of experiments showing power maxima in FC systems. These maxima are also displayed in a synthesizing book, based on chemistry and physics of FC systems, Li (2006). Power maxima in FC systems are particularly important because they prove the existence of power limits in real FCs. These limits are excluded in ideal, reversible cells, which, in fact, do not exist in Nature.

The systems considered so far belong to the category of power generators (engines). However, there are also power consumers, such as: refrigerators, heat pumps (HP), and separators (eg, dryers). Working in that context, La et al. (2013) analyze cooling cycles from the viewpoint of the effect of irreversible processes on the thermodynamic performance of open-cycle desiccant cycles. Li et al. (2010) derive a fundamental optimal relation of a generalized irreversible Carnot HP with complex heat transfer law. Appreciating the significance of experimental research, Fernández-Seara et al. (2012) present an experiment-based treatment of the direct expansion solar assisted HP with an integral storage tank for domestic water heating. Also working with experiments, Kuang et al. (2003) study the performance of a solar assisted HP system for heat supply. Celik and Muneer (2013) propose a neural network (NN) based method for conversion of solar radiation data, applicable to solar units and solar supported HPs. In the field of separators, Afriyie et al. (2013) perform mathematical modeling and validation of the drying process in a solar crop dryer. Assari et al. (2013) develop energy and exergy analyses of fluidized bed dryer based on two-fluid modeling. Kaleta et al. (2013) evaluate drying models of apple (Ligol) dried in a fluidized bed dryer. Kavak Akpinar (2010) develops modeling and performance analyses for drying of mint leaves in a solar dryer and under open sun.

In view of the variety of problems, the need for a synthesizing approach to limiting power generated or supplied to various energy systems seems to be both natural and important. The present chapter offers such a synthesizing approach. Caused by systems imperfections, effects of power yield reduction (in engines) or of power consumption increase (in HPs and separators) are evaluated from expressions for entropy production and Gouy–Stodola law. In general power limits are determined by involved optimizations in which total power yield is an extremum with respect to process controls. Classical methods of differential calculus and Lagrange multipliers are usually sufficient to determine power limits in steady systems. Yet, for dynamical systems, which are described by differential optimization models, the determining of power limits is generally a difficult task. Functionals of total power must be extremized (minimized for power consumers) by variational methods.

The present chapter offers a unified approach to power limits in power producing and power consuming systems, in particular in those using renewable resources. As a benchmark system which generates or consumes power, a well-known standardized arrangement is considered in which two different reservoirs are separated by an engine or HP. Either of these units is located between a resource fluid ("upper" fluid 1) and the environmental fluid ("lower" fluid, 2), Fig. 10.1b. Power yield or power consumption is determined in terms of conductances, bulk temperatures and internal irreversibility coefficient, Φ. While reservoir temperatures T_i are only necessary state variables to describe thermal units, in chemical (electrochemical) engines, HPs or separators both temperatures and chemical potentials μ_k are necessary.

Systems producing energy (engines) and those consuming energy are usually treated separately in the literature. An original feature of the present chapter is that it focuses on a

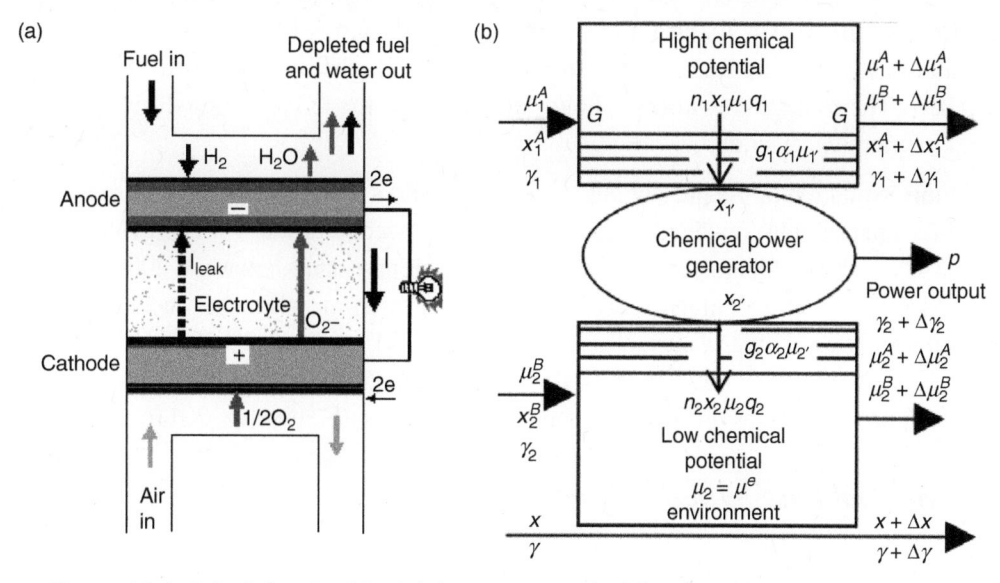

Figure 10.1: Principle of a SOFC (a) as compared with a flow chemical engine (b).

synthesis of limiting properties of energy-production and energy-consumption units, which means that upper limits of power released from engines are confronted with lower limits of power supplied to various "energy consumers" (HPs, separators, electrolyzers, etc.). A novel property is the enunciation of certain, relatively unknown, symmetry in behavior of energy consumers and energy producers (engines).

The development of use of Bellman's dynamic programming (DP) offers for dynamical processes a general and synthesizing treatment seldom applied previously (Kuran, 2006; Kuran and Sieniutycz, 2007; Sieniutycz, 2000a, c, 2003a, 2008b,). Discrete difference equations are derived here from DP as canonical equations for the process Hamiltonian consistent with Bellman's recurrence equation, and solved by numerical methods. When this Hamiltonian and the canonical equations are used in the computational algorithm, the "curse of dimensionality" observed for the DP algorithms is omitted. Pontryagin's type (Hamiltonian-based) approaches (Boltyanski, 1971, 1973; Canon et al., 1972; Findeisen et al., 1980; Halkin, 1966; Poświata and Szwast, 2003, 2006; Sieniutycz and Berry, 2000; Szwast (1988)); are described in Section 10.5. Yet, the DP approaches remain still useful, especially when investigating properties of extremal costs. In fact, the exact derivation of both methods requires determining of locally-sufficient optimality conditions from the so-called Caratheodory–Boltyanski stage optimality criterion that allows the variation of the terminal state which is otherwise fixed in Bellman's method (Boltyanski, 1971; Sieniutycz, 1978; Szwast, 1988). Thus, there is a novelty in the mathematical approach applied, first because of its generalizing formalism, second because of the synthesizing thermodynamic model capable of treating various energy generators (engines) and energy consumers (thermal, radiative, and electrochemical). The inclusion of electrolzers to the family of power consumers, while quite natural, is beneficial to the theory of FC electrolyzers that have recently attracted many researchers. The limitation manifests, however, whenever the FC topology differs from that of thermal or chemical system.

The volume limitation of the present chapter does not allow for inclusion of all derivations to make the chapter self-contained, thus the reader may need to turn to some previous works, (De Vos, 1992; Sieniutycz, 2003a, c, 2008b; Sieniutycz and Jeżowski, 2009, 2013; Kuran, 2006; Kuran and Sieniutycz, 2007). In particular, we must refer the reader to derivations of special while useful control variables (Carnot controls; Sieniutycz, 2003a; Sieniutycz and Jeżowski, 2009, 2013). Carnot controls follow from the identification and selection of free decisions in power systems propelled by transfer phenomena (Sieniutycz, 2011c.)

10.2 Internal Dissipation in Steady Thermal Systems

In this section we recapitulate briefly how the methods and results referring to simplest endoreversible systems are generalized to those with internal dissipation. A great deal of research on power limits published to date includes stationary thermal systems, in which case both reservoirs are infinite. To this case refer steady-state analyses of the Chambadal–Novikov–Curzon–Ahlborn engine (CNCA engine: De Vos, 1992; Sieniutycz, 2003a; Sieniutycz and Jeżowski, 2009, 2013), in which the energy exchange is described by Newtonian law of cooling, or the Stefan–Boltzmann engine, a system with two radiation fluids and the energy exchange governed by the Stefan–Boltzmann law (Badescu, 1999, 2000, 2004, 2008; De Vos, 1992; Landau and Lifshitz, 1975; Landsberg, 1990a, b; Lavenda, 1991; Sieniutycz and Kuran, 2006). Because of their stationarity (caused by the infiniteness of each reservoir), the power maximizing controls are lumped to a fixed point in the state space. For a CNCA engine, the maximum power point may be related to the optimum value of a free (unconstrained) control variable which can be efficiency η or Carnot temperature T'. The later text summarizes basic results in this field obtained for systems with internal dissipation (irreversibility factor Φ).

For a stationary heat operation of CNCA type, with bulk temperatures T_1 and T_2 and internal irreversibility factor Φ, the propelling heat in terms of Carnot temperature is (Sieniutycz, 2003a; Sieniutycz and Jeżowski, 2009, 2013)

$$q_1 = g(T_1, T_2, \Phi)(T_1 - T') \qquad (10.1)$$

where g is an effective overall conductance, which may be function of bulk state and Φ. Hence the power output is

$$p = g(T_1, T_2, \Phi)\left(1 - \frac{\Phi T_2}{T'}\right)(T_1 - T') \qquad (10.2)$$

Setting to zero the partial derivative of p with respect to T', one finds at the maximum power point

$$T'_{opt} = (T_1 \Phi T_2)^{1/2} \quad (10.3)$$

Since the effective environment temperature equals ΦT_2 and the Carnot structure holds for thermal efficiency η in terms of Carnot T', the power-maximizing efficiency follows as

$$\eta_{mp} = 1 - \Phi T_2/T'_{opt} = 1 - \sqrt{\Phi T_2/T_1}. \quad (10.4)$$

This equation represents a generalization of the classical CNCA formula (De Vos, 1992) for the case when internal imperfections (coefficient Φ) persist in the heat power system (Chen et al., 2001a).

An exact expression for the efficiency at the optimal power point cannot be determined analytically for the Stefan–Boltzmann radiation engine. Yet, the temperature can be found graphically from the chart $p = f(T')$. A pseudo-Newtonian model (Kuran, 2006; Sieniutycz and Kuran, 2005, 2006), which treats state dependent energy exchange with coefficient $\alpha(T^3)$, omits to a considerable extent analytical difficulties of the Stefan–Boltzmann equation.

10.3 Selected Results for Dynamical Thermal Systems

The approach outlined previously can be extended to dynamical systems. In the dynamical case power maximization problem requires use of variational methods (to handle extrema of functionals) in place of static optimization methods (which handle extrema of functions). Nonexponential shape of a relaxation curve for the radiation fluid resource is the consequence of nonlinear properties of the radiation fluid as the propelling medium. Nonexponential are also other curves describing the radiation relaxation, for example, those from the exact model based on the Stefan–Boltzmann equation (Kuran, 2006; Sieniutycz and Kuran, 2005, 2006). A discrete model, involving difference equations, is usually used for calculation of a dynamical engine.

As shown elsewhere (Sieniutycz, 2003a), total entropy production of the endoreversible power generation by the simple reaction $A_1 - A_2 = 0$ (isomerisation or phase change of A_1 into A_2), may be written in the form

$$\sigma_s = \varepsilon_1 \left(\frac{1}{T'} - \frac{1}{T_1} \right) + \left(\frac{\mu_1}{T_1} - \frac{\mu'}{T'} \right) n_1 \quad (10.5)$$

where $\varepsilon = q + h_a n_a + h_b n_b \ldots + h_k n_k$ is total energy flux in state 1 (driving the process). Efficiency lowering in endoreversible engines, HPs, and drying separators is the consequence of external irreversibilities quantified by the aforementioned expression.

Dynamical energy yield, may be interpreted as a multistage evolution process driven by a limited amount of a resource fluid. The operation may be viewed as the sequence of small elementary engines of CNCA type, that is, the sequence of small units shown in Fig. 8.4. In a thermal engine this way of power yield means the continual decrease of the resource temperature $T_1 = T(t)$ (approximated here by the sequence of temperatures T^n). The optimization procedure searches for an extremal curve rather than an extremum point.

Power maximizing curve $T(t)$ of a limiting continuous process is accompanied by optimal control $T'(t)$; they both are components of the dynamical solution which satisfies a Hamilton–Jacobi–Bellman equation (HJB equation) for the optimal control (Bellman, 1957, 1961, 1967; Bellman and Kalaba, 1965). Expressions extremized in HJB equations of dynamical power problems are nonoptimal Hamiltonians, H. The optimal temperature T' is obtained in the form of a feedback control, as the quantity maximizing H with respect to Carnot temperature at each point of the path. For example, after using a pseudo-Newtonian model (the one with temperature dependent conductances), we obtain an optimal dynamics of the relaxing thermal radiation in the form of two equations. The first is Eq. (10.6a)

$$\dot{T} = \xi(h_\sigma, T, \Phi)T \tag{10.6a}$$

which describes the optimal trajectory in terms of state T and constant h_σ representing numerical value of H along an optimal path. The second equation describes the optimal Carnot control

$$T' = \big(1 + \xi(h_\sigma, \Phi, T)\big)T \tag{10.6b}$$

Comparing with linear systems, the pseudo-Newtonian relaxation is not exponential. When standard boundary conditions for exergy are used, optimal work functions become generalized (rate dependent) exergies (Sieniutycz, 2003a; Sieniutycz and Jeżowski, 2009, 2013).

A more exact approach to radiation engines, described later, abandons the pseudo-Newtonian approximation and uses the exact Stefan–Boltzmann equations from the beginning.

10.4 Radiation Engines by the Stefan–Boltzmann Equations

In the dynamical radiation engines considered here only one ("upper") reservoir is finite. The power integral involves the product of differential heat and imperfect efficiency

$$\dot{W} = \int -\dot{G}c\left(1 - \frac{\Phi T}{T'}\right)dT \tag{10.7}$$

(Kuran, 2006; Kuran and Sieniutycz, 2007; Sieniutycz, 2003a; Sieniutycz and Jeżowski, 2009, 2013). The integrand of Eq. (10.7) represents power intensity as the intensity of a generalized profit, f_0.

When the propelling medium consists of the radiation fluid, power maximization problem is described by Eq. (10.7) and Eq. (10.8). They describe a *symmetric model* of power yield from radiation (both reservoirs consist of radiation). In the physical space, power exponent $a = 4$ for radiation and $a = 1$ for a linear resource.

In the engine mode, integral (10.7) has to be maximized subject to the dynamical constraint ("state equation")

$$\frac{dT}{dt} = -\beta \frac{T^a - T'^a}{(\Phi'(T'/T^e)^{a-1} + 1)T^{a-1}} \tag{10.8}$$

Eq. (10.8) is derived in several publications (Kuran, 2006; Kuran and Sieniutycz, 2007; Sieniutycz and Kuran, 2005, 2006).

As it follows from the general theory of dynamic optimization, extremum conditions for the problem involving Eqs. (10.7) and (10.8) are contained in the HJB equation of the problem

$$\frac{\partial V}{\partial t} - \max_{T'(t)} \left\{ \left(\dot{G}_c(T) \left(1 - \Phi' \frac{T^e}{T'} \right) + \frac{\partial V}{\partial T} \right) \beta \frac{T^a - T'^a}{(\Phi'(T'/T^e)^{a-1} + 1)T^{a-1}} \right\} = 0 \tag{10.9}$$

where the state function $V = \max \dot{W}$ is often called the value function. In our problem V is simply the maximum power function. We shall also use the minimum power function $R = \min(-\dot{W})$, such that $V = -R$.

As it is impossible to solve Eq. (10.9) analytically, except for the case when $a = 1$, we outline here a way for numerical solving based on Bellman's method of dynamic programming (DP). Taking into account computational needs, discrete difference equations are next derived, and solved by numerical methods. Thus we introduce a related discrete scheme

$$\dot{W}^N = \sum_{k=1}^{N} -\dot{G}_c(T^k)\left(1 - \frac{\Phi T^e}{T'^k}\right)(T^k - T^{k-1}) \tag{10.10}$$

$$T^k - T^{k-1} = \theta^k \beta \frac{T'^{k^a} - T^{k^a}}{(\Phi'(T'^k/T^e)^{a-1} + 1)T^{k^{a-1}}} \tag{10.11}$$

$$t^k - t^{k-1} = \theta^k \tag{10.12}$$

An optimizer searches for maximum of the sum (10.10) subject to discrete constraints (10.11) and (10.12). Applying to this problem the dynamic programming method, the following recurrence equation is obtained for the minimum power function $R = \min(-\dot{W})$

$$R^n(T^n,t^n) = \min_{u^n,\theta^n}\left\{\dot{G}_c(T^n)\left(1-\frac{\Phi T^e}{T'^n}\right)\beta\frac{T^{na}-T'^{na}}{(\Phi'(T'^n/T^e)^{a-1}+1)T^{na-1}}\theta^n \right.$$
$$\left. +R^{n-1}\left((T^n - \frac{\theta^n \beta(T'^{na}-T^{na})}{[\Phi'(T'^n/T^e)^{a-1}+1]T^{na-1}}, t^n - \theta^n\right)\right\} \quad (10.13)$$

While the analytical treatment of Eqs. (10.7) and (10.8) is a tremendous task, it is quite easy to solve recurrence Eq. (10.13) numerically. Fig. 8.4 depicts the multistage computational scheme. Low dimensionality of state vector in Eq. (10.13) assures a good accuracy of DP solution. Moreover, an original accuracy can be improved after performing the so-called dimensionality reduction associated with the elimination of time variable t^n. In the transformed problem, without t^n, accuracy of DP solutions is high (Kuran, 2006).

Decreasing temperatures of radiation relaxing in engine mode and increasing temperature of radiation utilized in HP mode in terms of time, were studied for a constant value of Hamiltonian $H = 1 \times 10^{-8}$ [JK^{-1}m^{-3}] (Kuran, 2006; Kuran and Sieniutycz, 2007). Since the process is nonlinear remarkable deviations of relaxation curves from the exponential decay (characteristic for linear dynamics) was observed.

However, when the number of state variables increases (eg, when several concentrations x_i^n may accompany temperature T^n as in chemical engines), DP algorithms become inefficient and inaccurate. We must then abandon DP approaches and turn into the Pontryagin type (Hamiltonian-based) approaches.

10.5 Hamiltonians and Canonical Equations

For Hamiltonian-based approaches, which apply Pontryagin's canonical equations, problems of large dimensionality of state vector are inessential. The dynamic changes in the discrete state of a multistage system are described by a set of ordinary difference equations called the state transformations. They represent the state from the stage $n-1$ in terms of the state from the stage n and control variables U^n.

The set of discrete state transformations can be written in the following general form

$$x^{n-1} = T^n\left(x^n, t^n, U^n\right) \quad (10.14)$$

where

$$t^n = t^{n-1} + \theta^n, \quad (10.15)$$

and $U^n = (u^n, \theta^n)$ is an enlarged vector of control variables, which includes the discrete interval of time θ^n. The time variable t^n may be identified with any state variable growing monotonically in physical time or space time.

After defining the function

$$\mathbf{f}^n = (\mathbf{x}^n - \mathbf{T}^n(\mathbf{x}^n, t^n, \mathbf{U}^n))/\theta^n \qquad (10.16)$$

the aforementioned state transformations can be given the form of discrete state equations (Poświata, 2005; Sieniutycz, 2006c)

$$\mathbf{x}^{n-1} = \mathbf{x}^n - \mathbf{f}^n(\mathbf{x}^n, t^n, \mathbf{u}^n, \theta^n)\theta^n \qquad (10.17)$$

and

$$t^{n-1} = t^n - \theta^n \qquad (10.18)$$

As they involve the discrete rates (f^n, 1), we call this form the "standard form."

A performance index describing a generalized profit (total power in our case) is defined in this formalism by the following equation

$$P^N \equiv \dot{W}^N = \sum_{n=1}^{N} f_0^n(\mathbf{x}^n, t^n, \mathbf{u}^n, \theta^n)\theta^n \qquad (10.19)$$

where f_0 is the generation rate of the generalized profit (power in the case of energy yield problem).

To solve the optimization problem of extremum W, an enlarged (time-constraint absorbing) Hamiltonian is defined in the following form

$$\tilde{H}^{n-1}(\mathbf{x}^n, t^n, \mathbf{z}^{n-1}, \mathbf{u}^n, \theta^n) \equiv f_0^n(\mathbf{x}^n, t^n, \mathbf{u}^n, \theta^n) + \sum_{i=1}^{s} z_i^{n-1} f_i^n(\mathbf{x}^n, t^n, \mathbf{u}^n, \theta^n) + z_t^{n-1} \qquad (10.20)$$

where z_i are adjoint (Pontryagin's) variables.

In an optimal process the enlarged Hamiltonian $\tilde{H}^{n-1}$ satisfies in the enlarged phase space $x = (\mathbf{x}, t)$ and $z = (\mathbf{z}, z_t)$ the following equations:

$$\frac{x_i^n - x_i^{n-1}}{\theta^n} = \frac{\partial \tilde{H}^{n-1}}{\partial z_i^{n-1}} \qquad (10.21)$$

(state equations) and

$$\frac{z_i^n - z_i^{n-1}}{\theta^n} = -\frac{\partial \tilde{H}^{n-1}}{\partial x_i^n} \qquad (10.22)$$

(adjoint equations),

and also the equations which describe the necessary optimality conditions for decision variables $\mathbf{u}^n$. For example, if the optimal control lies within an interior of admissible control set

$$\partial[\theta^n \tilde{H}^{n-1}(\mathbf{x}^n, z^{n-1}, \theta^n, \mathbf{u}^n)]/\partial \theta^n = 0 \quad (10.23)$$

and

$$\frac{\partial \tilde{H}^{n-1}}{\partial u_j^n} = 0 \quad (10.24)$$

Eq. (10.23) implies vanishing of the enlarged Hamiltonian $\tilde{H}^{n-1}$ along a discrete optimal path whenever discrete rates f_i are independent of θ^n. In addition, the energy-like Hamiltonian (without z_t term) is constant for the process whose rates are independent of time t^n. Under convexity properties for rate functions and constraining sets the optimal controls are defined by the equations

$$\theta^n = \arg\max_{\theta^n} \{\theta^n \tilde{H}^{n-1}(\mathbf{x}^n, z^{n-1}, \mathbf{u}^n, \theta^n)\} \quad (10.25)$$

and

$$\mathbf{u}^n = \arg\max_{\mathbf{u}^n} \{\tilde{H}^{n-1}(\mathbf{x}^n, z^{n-1}, \mathbf{u}^n, \theta^n)\} \quad (10.26)$$

($n = 1,\ldots,N$; $i = 1,\ldots,s+1$ and $j = 1,\ldots,r$

Optimization theory for generalized (θ^n – dependent) costs and rates provides the bridge between constant-H algorithms (Sieniutycz, 1974b, 1991, 2000a) and more conventional ones such as those by Katz (Boltyanskii, 1973; Canon et al., 1972; Fan and Wang, 1964; Halkin, 1966), and many others (Sieniutycz, 2006c; Findeisen et al., 1980). Since, as shown by Eq. (10.23), control θ^n can be included in the Hamiltonian definition, that is, an effective Hamiltonian can be used

$$\mathsf{H}^{n-1} \equiv \tilde{H}^{n-1} \theta^n, \quad (10.27)$$

extremum conditions (10.21)–(10.26) can be written in terms of H^{n-1}. The related canonical set is that of Halkin

$$x_i^n - x_i^{n-1} = \frac{\partial \mathsf{H}^{n-1}}{\partial z_i^{n-1}} \quad (10.28)$$

$$z_i^n - z_i^{n-1} = -\frac{\partial \mathcal{H}^{n-1}}{\partial x_i^n} \tag{10.29}$$

(Canon et al., 1972; Halkin, 1966). Qualitative difference between the role of controls $\mathbf{u}^n$ and θ^n in the optimization algorithm is then lost since they both follow from the same stationarity condition for Hamiltonian H^{n-1} in an optimal process. For example, in the weak maximum principle

$$\frac{\partial H^{n-1}}{\partial \theta^n} = \frac{\partial H^{n-1}}{\partial u_j^n} = 0 \tag{10.30}$$

in agreement with Eqs. (10.23) and (10.24). Moreover, Eqs. (10.23) and (10.30) imply the condition $\tilde{H}^{n-1} = 0$ if $\tilde{H}^{n-1}$ is independent of time interval θ.

Until now Hamiltonian algorithms were used in power systems for models with θ independent discrete rates (Sieniutycz and Berry, 2000). Yet, Poświata and Szwast have shown many applications in exergy optimization of thermal and separation systems, in particular fluidized dryers (Poświata, 2005; Poświata and Szwast, 2003, 2006). Sieniutycz (2006c) has shown some other applications for energy and separation systems and for a minimum time problem. Taking into account the diversity of discrete rates, which may contain explicit intervals θ^n as the consequence of various ways of discretizing, applications of algorithm (10.16)–(10.24) in power or separation systems may be appropriate and useful.

In particular, the algorithm is suitable in the numerical studies of the optimal solutions for the discrete equations of the radiation engine, Eqs. (10.10)–(10.12). For these equations the enlarged Hamiltonian has the form

$$\tilde{H}^{n-1} = \left(z_T^{n-1} - \dot{G}_c(T^n)\right)\left(1 - \frac{\Phi T^e}{T'^n}\right) \times \beta \frac{T^{n^a} - T'^{n^a}}{(\Phi'(T'^n/T^e)^{a-1} + 1)T^{n^{a-1}}} + z_t^{n-1} \tag{10.31}$$

Optimal dynamics corresponding to this function are available (Kuran, 2006; Kuran and Sieniutycz, 2007). A nonlinear relaxation of radiation is observed rather than a simple exponential decay typical of linear systems.

10.6 Simple Chemical and Electrochemical Systems

Thermodynamic approaches can also be applied to chemical and electrochemical engines (De Vos, 1992; Sieniutycz, 2008b) and electrochemical engines, that is, FCs (Wierzbicki, 2009; Zhao et al., 2008). In these engines mass transports drive transformation of chemical energy into mechanical power, Fig. 10.1a. Yet, as opposed to thermal machines, in chemical ones generalized streams or reservoirs are present, capable of providing both heat and substance. Large streams or infinite reservoirs assure constancy of chemical potentials. Problems of maximum of power produced or minimum of power consumed are then the static problems. For a finite "upper stream," however, the amount and chemical potential of an active reactant decrease in time, and considered problems are those of dynamic optimization and variational calculus. Because of the diversity of chemical systems the area of power-producing chemistries is extremely broad.

Power expression and efficiency formula for a chemical system follow from the entropy and energy balances in the power-producing zone of the system ('active part'). The simplest model of power producing chemical engine is that with an isothermal isomerization reaction, $A_1 - A_2 = 0$ (De Vos, 1992; Sieniutycz, 2008b). In an 'endoreversible chemical engine' total entropy flux is continuous through the active zone. When a formula describing this continuity is combined with energy balance we find in isothermal case with a complete conversion

$$p = (\mu_{1'} - \mu_{2'})n_1, \qquad (10.32)$$

where the feed flux n_1 equals to n, an invariant molar flux of reagents. Process efficiency ζ is defined as power yield per molar flux, n. This efficiency is identical with the chemical affinity of reaction in the chemically active part of the system. While ζ is not dimensionless, it describes correctly the system. In terms of Carnot variable, μ', the efficiency satisfies an equation (Sieniutycz, 2008b)

$$\zeta = \mu' - \mu_2. \qquad (10.33)$$

For a steady engine the following function describes chemical Carnot control μ' in terms of fuel flux n_1 and its mole fraction x

$$\mu' = \mu_2 + \zeta_0 + RT \ln\left(\frac{x_1 - n_1 g_1^{-1}}{n_1 g_2^{-1} + x_2}\right) \qquad (10.34)$$

As Eq. (10.33) is valid, Eq. (10.34) also characterizes efficiency control in terms molar flux, n, and fuel fraction x.

Eq. (10.34) shows that an effective concentration of the reactant in upper reservoir $x_{1\text{eff}} = x_1 - g_1^{-1} n$ is decreased, whereas an effective concentration of the product in lower reservoir $x_{2\text{eff}} = x_2 + g_2^{-1} n$ is increased due to the finite mass flux. Therefore efficiency ζ decreases nonlinearly with n. When the effect of resistances is ignorable or flux n is very small, reversible Carnot-like chemical efficiency, ζ_C, is attained. The power function, described by the product $\zeta(n)n$, exhibits a maximum for a finite value of the fuel flux, n.

A related dynamical problem may also be considered (Sieniutycz, 2008b). Application of Eq. (10.34) to the Lagrangian relaxation path leads to a work functional

$$W = -\int_{\tau_1^i}^{\tau_1^f} \left\{ \zeta_0 + RT \ln\left(\frac{X/(1+X) + dX/d\tau_1}{x_2 - jdX/d\tau_1} \right) \right\} \frac{dX}{d\tau_1} d\tau_1 \qquad (10.35)$$

whose maximum describes the dynamical limit of the system. Here $X = x/(1-x)$ and j equals the ratio of upper to lower mass conductance, g_1/g_2. The path optimality condition may be expressed in terms of the constancy of the following Hamiltonian

$$H(X, \dot{X}) = RT\dot{X}^2 \left(\frac{1+X}{X} + \frac{j}{x_2} \right). \qquad (10.36)$$

For low rates and large concentrations X (mole fractions x_1 close to the unity) optimal relaxation rate of the fuel resource is approximately constant. Yet, in an arbitrary situation optimal rates are state dependent so as to preserve constancy of H in Eq. (10.36).

Hjemfelt and Ross (1989) have applied periodic variations to the influxes of oxygen and methane entering a reaction vessel in which takes place a combustion reaction and, in the absence of these variations, the system is in a stable focus. They have measured the reaction of the system to these variations and find a resonant response, and changes of phase relations between the forcing and response of the system, near the autonomous frequency. They have calculated the enthalpy content of the gases and using a simple model of a Carnot engine they studied the power output of the system; they found increases (of around 7% in power) in these quantities near the autonomous frequency. The experimental results have been compared to the predictions of a numerical model specific to their system and to the analytic solution of a linear set of differential equations with a stable focus. Hjemfelt and Ross (1989) found good agreement with both, but, as they say, there is an aspect of the experimental results, which requires additional hypotheses.

Hjemfelt et al. (1991) investigated efficiency of power production in nonlinear electrochemical systems, with a special interest focused on the effect of nonlinear mechanistic steps. Previous work (Hjemfelt and Ross, 1989) has shown that external periodic perturbations of a damped or undamped oscillator may show changes in efficiency. Now, the researchers applied periodic variation of concentration on one of the reactions and analyzed the consequent changes in efficiency. Fig. 10.2 illustrates the scheme of the considered system, which is an electrochemical cell for the reaction mechanism given by the Eqs. (10.37) and (10.38). The electrochemical cell consists of two compartments separated by a semipermeable membrane. On each electrode a reaction occurs:

$$A^- \leftrightarrow X + e_1^- \qquad (10.37)$$

$$X + e_2^- \leftrightarrow B^- \qquad (10.38)$$

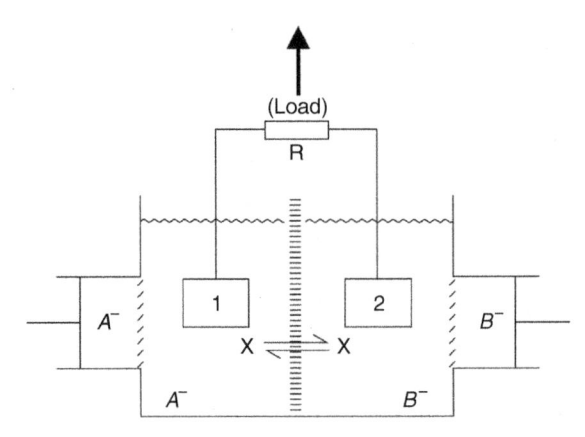

Figure 10.2: Power production in simple nonlinear electrochemical system.
Hjemfelt et al., 1991.

Subscript on the electrons denotes the electrode of the reaction. The net result is the isomerization of A^- to B^-. The flows A^- and B^- are restricted to opposite sides of the cell by a semipermeable membrane, but the uncharged chemical species and the positively charged counter-ion may flow through the membrane. The cell is connected to reservoirs of A^- and B^- which maintain their concentrations as constant. Each compartment is well mixed, and it is assumed that mass flows through the membrane are fast compared to the reaction time (the same concentrations of neutral species and the solution potentials in each compartment). The ionic species react only at the electrodes with the rates given by Butler–Volmer equations.

Fig. 10.2 shows scheme of the electrochemical cell for the reactions described by Eqs. (10.37) and (10.38). Broken lines denote semipermeable membranes (Hjemfelt et al. (1991). The electrochemical model neglects activation–passivation reactions or couplings with mass transport. The instabilities in are due to a coupling of nonlinear reactions in the bulk solution with the Butler–Volmer reaction kinetics on the electrode. Therefore, chemical reactions are the exclusive source of dissipation in the considered model. The authors analyzed three chemical reaction models of power production in the cell. In these models ionic species react on electrodes to form neutral species with rate governed by Butler–Volmer kinetics. The authors allow reactions between the neutral species in the bulk solution. The flow of electrons between the electrodes causes the power production. The ratio of this power yield to the free energy change of the overall reaction is the efficiently of the power yield. The inclusion of one autocatalytic step in the bulk solution leads to bistability. A high-power output, high-efficiency branch connected to the equilibrium point coexists with a low-power output, low-efficiency branch. The inclusion of extra linear mechanistic steps between neutral species leads to damped oscillations and a Hopf bifurcation to undamped, unstable oscillations. Applied periodic variation to the system lead to resonance effects, but the periodic variations have a negligible effect on the power yield efficiency. Calculations, where the concentration of A^- has a sinusoidal component, show the negligible effect on the average power production.

10.7 Power Yield and Power Limits in FCs

Keizer (1987a) developed self-consistent thermodynamic approach describing kinetics of irreversible phenomena as peculiar chemical reactions, including electrochemical ones. The approach is the basis of Keizer's book (Keizer, 1987a). Also Oláh (1997a, b, 1998) investigated nonlinear heat transfer and mass diffusion where steps with energy and mass exchange were treated as some chemical reactions described by appropriate affinities. Any approach of this sort distinguishes in each elementary process, diffusive or chemical, two competing (unidirectional) fluxes. They are equal in the thermodynamic equilibrium and their difference off equilibrium constitutes the observed flux representing the resulting rate of the process. Heat and diffusion processes are described by equations containing exponential terms with respect to chemical potentials of Planck and temperature reciprocal; these equations are analytical expressions characterizing the transport of the substance or energy by the energy barrier. Fig. 9.5 depicts states of gaseous and adsorbed reactants, activated state and state of product. The kinetics of this sort is consistent with the law of mass action and describes its consequences closely and far from equilibrium. One need also stress the role of a nonlinear resistance and the corresponding affinity. With these kinetics, the potential representation of the generalized law of mass action comprises the effect of rate processes, transfer steps, and external fields. Remarkable are physical consequences of it close and far from the equilibrium, and the diversity of phenomena that can be treated. See also Sieniutycz (2004a, 2005a). Fig. 10.3 presents the interpretation by Oláh (1997a, b) of anodic and cathodic currents in electrochemical processes. Abscissa of the crossing point of both currents describes the exchange current.

Bockris and Kainthla (1986) described basic problems in conversion of light and water to hydrogen and electric power. Forland et al. (1988) discussed electrochemical phenomena in the context of irreversible thermodynamics. Fuliński et al. (1998) considered non-Markovian character of ionic current fluctuations in membrane channels. The Smoluchowski–Chapman–Kolmogorov functional equation is the most basic test of Markovian character of finite

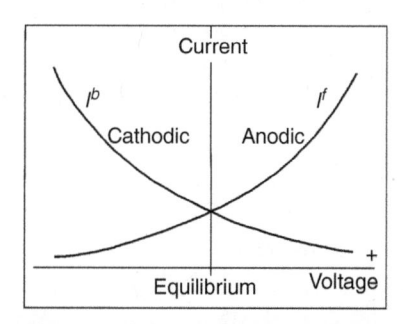

Figure 10.3: Interpretation of Oláh (1997) on anodic and cathodic currents as well as of the exchange current in terms of voltage.

Abscissa of crossing point of both currents describes the exchange current.

stochastic chains, and is simpler in application than other tests. In their work this test was used to analyze the experimental data on potassium current from single BK channel in a biological membrane. The results suggested the non-Markovian character of the analyzed data.

Feudel et al. (1984) investigated reaction–diffusion systems with charged particles. Conditions for the arising of electrical dissipative structures in a compartment system consisting of two boxes separated by a membrane have been derived. The appearance of a polar dissipative structure has been proved for a simple capacitor model in combination with a simple second order chemical kinetics which leads to an analytically solvable problem. Electrical dissipative structures can in principle be considered as nonequilibrium electrical batteries. The theoretical efficiency of such batteries has been estimated.

A two-part paper on disequilibrium electromotive force (EMF), part I: CSTR measurements and part II: theory, Keizer and Chang (1987), Keizer (1987b), refers to predictions, based on statistical thermodynamics, that at disequilibrium steady states the EMF of an electrochemical cell will differ from the local equilibrium value given by the Nernst equation. Using irreversible thermodynamics for bulk and surface systems Kjelstrup and Bedeaux (1997) derived the electric potential profile and the temperature profile across a formation cell. The method was demonstrated with the SOFC. The expression for the cell potential reduces to the classical formula when equilibrium is assumed for polarized oxygen atoms across the electrolyte. Using some literature data, they show how the cell potential is generated at the anode, and how the energy is dissipated throughout the cell. Llebot (1992) developed extended irreversible thermodynamics of electrical phenomena for a rigid solid conductor in the presence of flows of heat and electricity. By combining the balance equations for the internal energy and electron concentration with the disequilibrium entropy an expression for the entropy production was obtained containing the generalized thermodynamic forces with the time derivatives of the energy and the electron flux. Considering nonstationary electrical conduction he showed that only EIT can preserve the positive entropy production σ in general. For sufficiently large frequencies, the classical theory can yield an erroneous expression allowing negative σ during part of the cycle.

Ratkje et al. (1992) presented a nonstandard approach to the thermodynamics of irreversible transport processes in electrolytes and membranes. The description is lumped and preserves many features of the classical Onsagerian formalism. On the other hand, the classical chemical affinity is not applied in the formalism (see also the work of Lengyel, 1992 work giving up the affinity). In such systems scalar-vector couplings can take place. Rubi and Kjelstrup (2003) apply mesoscopic nonequilibrium thermodynamics to derive the Butler–Volmer equation, or the stationary state nonlinear relation between the electric current density and the overpotential of an electrode surface. The equation comes from a linear flux-force relationship at the mesoscopic level for the oxidation of a reactant to its charged components. The surface is defined with excess variables (a Gibbs surface). The Butler–Volmer equation is derived using the assumption of local electrochemical equilibrium in the surface on the mesoscopic level. In conclusion, the authors show that the equation used for reaction-controlled charge transfer or the Butler–Volmer equation, has a general basis in nonequilibrium thermodynamics. In the derivation they take advantage of the systematic procedure offered by mesoscopic nonequilibrium thermodynamics. They arrive at the Butler–Volmer equation using assumptions of local electrochemical equilibrium along the reaction coordinate space, and a linear flux-force relation for this coordinate space. The derivation requires a certain identification of the overpotential with the effective driving force to be true. By setting a more general basis for this electrochemical equation, possibilities become open for other studies when the assumptions do not hold, that is, when there are gradients between the surface and its close surroundings, and when the stationary state condition does not apply.

Llebot (1992) presented extended irreversible thermodynamics of electrical phenomena for a rigid solid conductor in the presence of flows of heat and electricity. By combining the balance equations for the internal energy and electron concentration with the disequilibrium entropy an expression for the entropy production was obtained containing the generalized thermodynamic forces with the time derivatives of the energy and the electron flux. Considering nonstationary electrical conduction he showed that only EIT can preserve the positive entropy production σ in general. For sufficiently large frequencies, the classical theory can yield an erroneous expression allowing negative σ during part of the cycle. Shiner et al. (1996) showed that the equations of motion for systems with electrochemical coupling can be derived from a dissipative Lagrangian formalism and that these systems can also be cast into an equivalent network formulation. These formulations were possible due to the formalisms common to electrical networks and other nonchemical systems. The examples treated are simple electrochemical cells obeying Butler–Vollmer kinetics at the electrodes and mass action kinetics for the other reactions.

FCs have attracted attention by virtue of their inherently clean and reliable performance. Their main advantage as compared to heat engines is that their efficiency is not a major function of device size. A FC, Fig. 10.1, is an electrochemical energy converter which directly and continuously transforms a part of chemical energy into electrical energy by consuming fuel and oxidant. Most FCs currently being developed for use in vehicles and as portable power generators require hydrogen as a fuel. Chemical storage of hydrogen in liquid fuels is considered to be one of the most advantageous options for supplying hydrogen to the cell. In this case a fuel processor (FP) is needed to convert the liquid fuel into a hydrogen-rich stream. FCs are devices that convert the chemical energy stored in a fuel directly into electrical power. The main attractive features of FC systems are quiet operation and low emissions. However, some issues in reducing manufacturing and material costs, improving stack performances, their reliabilities and lifetimes, must be solved to enable commercialization. The solid oxide fuel cell (SOFC) is considered as one of the most promising energy conversion device and as an alternative of existing power generation systems. SOFCs operate at high temperatures from 600 to 1000°C to ensure sufficient ion conductivity through their electrolytes, which are nonconductive to electrons. Main SOFC components include air channel, cathode, electrolyte, anode, fuel channel, and interconnects. Tubular, planar, and monolithic structures are primary SOFC structures.

The contemporary research on FCs is intense, very abundant, and quite unpredictable because its future depends to a large extent on material discoveries. Faghri and Guo (2005) review challenges and opportunities of thermal management issues related to FC technology, modeling development, and application of FC technology. They predict that the development and application of FC technology will be increased significantly through analysis and improvement of the heat transfer in the FC stack and auxiliary components and by implementation of innovative heat transfer schemes. Current technical limitations can be overcome through a combination of novel methodologies, increased efficiency of components, decreased component size and mass, and detailed modeling. Such research also affects the existing FC technologies and developers, FC models, and unresolved issues in FC modeling. In conclusion, FC technology promises to be a highly efficient and environmentally friendly power source with broad applications, including transportation and both portable and stationary power generation. Although tremendous progress in FC stack and system development has been made in past decades, FCs are still relatively far from being fully commercialized. Numerous technical challenges remain, such as performance, cost, system issues, and choice of fuel (Faghri and Guo, 2005).

Thermal management plays an important role in FC design and optimization from single cells to the system level. Manufacturing and operational costs can be reduced by optimizing the efficiency of FCs through detailed analysis of heat and mass transport phenomena taking place within the stack and system. Challenges and opportunities of thermal management issues are related to both low temperature FCs and high temperature FCs (Faghri and Guo, 2005). Thermal management is closely related to water management in PEM FCs, where water is used for both humidification of reactant gases and cooling purposes. Thermal management is also a key tool for system energy balance design. This is particularly true for FC systems that include a reformer, which increases the system complexity dramatically. Heat transfer in auxiliary components such as heat exchangers, blowers, burners, and so forth, also needs careful consideration and implementation, Faghri and Guo (2005). The high operating temperature imposes stringent requirements both on SOFC design and on defining operating conditions. The key to thermal management of SOFC stacks is to satisfy the high temperature reaction requirement but avoid steep temperature gradients. A better understanding of thermal management issues of FC systems will enable the system designer to select a cogeneration turbine, to estimate the size of the auxiliary equipment (pumps, compressor, heat exchangers, etc.), and to estimate the optimal set of stack operating conditions for a higher efficiency and long-term operation.

Faghri and Guo (2005) also provide an overview of FC models developed in the past few years. Though the literature shows that many researchers have focused on development of viable FC models, a number of unresolved issues demand more intensive research. The advanced two-phase models for FC application are needed for better understanding of the behavior of heat and mass transfer issues in flow channel. For example for PEMFC modeling, two-phase flow in the cathode gas channel is not modeled at present, despite the importance of flooding (Faghri and Guo, 2005). Because of the simplicity and reasonable performance within a certain range of applications, the Darcy model is useful for the majority of existing studies on mass flow in porous electrodes. Yet, it has been found that the Darcy model has some limitations, and the second phase should be taken into account when massive bubbles are produced. The loss due to contact resistance is a significant factor in FC operation, and future FC models should address the numerical modeling of contact resistance loss. More accurate models on internal reforming chemistry is needed for SOFC modeling. Advanced stack and system level modeling combined with parallel computing is also required for improved stack and system performance Faghri and Guo (2005).

Because of the volume limitation, we can only mention selected valuable papers in the FC area. Lutz et al. (2002) presented thermodynamic comparison of FCs to the Carnot cycle in the efficiency context. Cheng-Xin Li et al. (2009) investigated the effect of composition of NiO/YSZ anode on the polarization characteristics of SOFC fabricated by atmospheric plasma spraying. P.-W. Li et al. (2011) proposed an analytical model for SOFCs with high power densities, to help optimize the size, spacing, and geometrical shapes of current collectors. The model includes the ohmic, concentration, and activation polarizations, and leads to curves of power density versus current density. It is recommended that 3D pillars be used for the current collectors in SOFCs while incorporating appropriate measures to ensure a uniform flow distribution. Unless limited by the fabrication technologies for the bipolar plates, smaller control areas and current collectors are preferable for SOFCs to have higher power densities. Applying constructal theory, Ordonez et al. (2007) presented a model and a structured procedure to optimize the internal structure of a single SOFC so that the power density is maximized. The key conclusion is that trade-offs exist, and that from them results the internal structure relative sizes and spacings of the flow systems in a single SOFC, that is, constructal design (Bejan, 2000, 2003; Vargas and Bejan, 2004; Vargas et al., 2005). In practice such trade-offs must be pursued based on models that correspond to real applications. In principle, this optimization can be extended on a hierarchical ladder to large and more complex SOFC systems, to explore multiscale packings that use the available volume for maximum performance. Delsman et al. (2006) stressed application potential of FCs as stationary power plants, as power sources in transportation, and as portable power generators for electronic devices. Radulescu et al. (2006b) presented experimental theoretical analysis of the operation of a natural gas (NG) cogeneration system using a polymer exchange membrane FC. Roy-Aikins (2002) worked out principles of thermoeconomic optimization for a hybrid micro-turbine and FC. Rabaey and Verstraete (2005) described microbial FCs as the novel biotechnology for energy generation. Palazzi et al. (2007) presented a thermoeconomic optimization method for thermoeconomic modeling and optimization of SOFC systems that systematically generates the most attractive configurations of an integrated system. In the developed methodology, the energy flows are computed using conventional process simulation software. The system is integrated using the pinch-based methods that rely on optimization techniques. Pramuanjaroenkij et al. (2008) performed mathematical analysis of planar SOFCs. Ołdak (2011) applied thermodynamics in mathematical modeling of hybrid energy systems with SOFCs. Pade and Schröder (2013) performed an economic analysis of FC based micro-combined heat and power under different policy frameworks. Lemański (2007), Lemański and Badur (2005), and Lemański et al. (2004) analyzed various strategies for gas turbine–solid oxide fuel cell hybrid cycles. Lemański (2007) investigated various energy cycles with FC and gas-vapor turbine. Lemański and Badur (2005) treated SOFCs with an internal reforming. Logan (2008) described microbial FCs. Milewski and Lewandowski (2009) presented an analysis of various biogases as fuels for SOFCs. Their results are compared with NG as a reference fuel. The biogases are characterized by both lower efficiency and lower fuel utilization factors in comparison with NG. The governing equations were presented. Milewski et al. (2009) have shown that molten carbonate fuel cell (MCFC) reduce CO_2 emissions from a coal-fired power plant. The MCFC is placed in the flue gas stream of the coal-fired boiler. The main advantages of this solution are: higher total electric power generated by a hybrid system, reduced CO_2 emissions and higher system efficiency. Use of an MCFC could reduce CO_2 emissions by 61%.

Power maximization approaches can be applied to many electrochemical systems, in particular to FCs (Zhao et al., 2008; Wierzbicki, 2009). Power decrease below ideal (Nernst) value and power maxima in engines are caused by the entropy production. The FC modeling should be supported by experiments (Sieniutycz et al., 2011). FCs are expected to play a significant role in the next generation of energy systems and road vehicles for transportation. However, substantial progress is required in reducing manufacturing costs and improving performance. In recent years, considerable progress has been made in modeling to improve the design and performance of SOFC. The SOFC is one of the most promising FCs for direct conversion of chemical energy to electrical energy with the possibility of its use in cogeneration systems because of the high temperature waste heat. The book by Sunden and Faghri (2005) aims to contribute to the understanding of the transport processes in SOFC, proton exchange membrane fuel cells (PEMFC) and direct methanol fuel cells (DMFC), which are of current interest. Kakaç et al. (2007) review the research focused on the numerical modeling of SOFCs. The objective is to summarize the present status of the SOFC modeling and identify evergreen problems. Various mathematical models are analyzed for three geometric configurations (tubular, planar, and monolithic) to solve transport equations coupled with electrochemical processes to describe the reaction kinetics including internal reforming chemistry in SOFCs.

Pramuanjaroenkij et al. (2008) developed a mathematical transport model for a planar SOFC. Its analysis has been performed by the use of an in-house program, which can help developers to understand the effects of various parameters on the performance of the FC. In the model, electrochemical kinetics, gas dynamics, and transport of energy and species are coupled. The model predicts polarization curve, velocity, and temperature fields, species concentration, and current distribution in the cell depending on FC temperatures and electrolyte materials used in the components, such as yttria-stabilized zirconium (YSZ) and gadolinia-doped ceria (CGO). SOFC operating temperatures at 500, 600, 800, and View the MathML source are considered and the modified Nernst equation is used to obtain a reversible cell voltage. It is shown that the anode-supported SOFCs with YSZ electrolyte can be used to obtain a high power density in the higher current density range than the YSZ electrolyte-supported SOFCs when they are operated at View the MathML source. Performance comparisons between two electrolyte materials, YSZ and CGO are made. The YSZ-electrolyte SOFC in this work shows higher power density than the CGO-electrolyte SOFC at the higher temperatures than 750°C.

In a material-oriented research of El-Genk and Tournier (2000) a number of nickel/Haynes-25, vapor anode, multitube alkali metal thermal-to-electric conversion (AMTEC) cell designs, similar to the Pluto/Express series cells, were developed and optimized for improved performance. Optimizations investigated the effects of changing the number and length of the sodium beta-alumina solid electrolyte (BASE) tubes, the cell length and the condenser temperature on the cell's electric power output, conversion efficiency, temperature margin and the temperatures of the BASE brazes and the evaporator. Various options of integrating the high performance nickel/Haynes-25 cells with $238PuO_2$ general purpose heat source (GPHS) modules, into the electric power system's configuration proposed by advanced modular power systems, were explored for satisfying an end of mission (EOM) electric power requirement of 130 We at 28 V DC for 10–15-year space missions. The power systems' integration options explored included changing the number of GPHS modules and the number of AMTEC cells in the generator and using either fresh or aged 238 PuO_2 fuel in the GPHS modules.

Hydrogen is the basic fuel medium in SOFCs. Investigation of the thermodynamic–electrochemical characteristics of hydrogen production by a solid oxide steam electrolyzer (SOSE) plant, energy and exergy analysis has been conducted (Ni et al., 2007). Overpotentials involved in the SOSE cell have been included in the thermodynamic model. The waste heat in the gas stream of the SOSE outlet has been recovered to preheat the H_2O stream by a heat exchanger. The heat production by the SOSE cell due to irreversible losses has been investigated and compared with the SOSE cell's thermal energy demand. It was found that the SOSE cell normally operates in an endothermic mode at a high temperature while it is more likely to operate in an exothermic mode at a low temperature as the heat production due to overpotentials exceeds the thermal energy demand (Ni et al., 2007).

A diagram of energy and exergy flows in the SOSE plant helps to identify the sources and quantify the energy and exergy losses. The energy analysis shows that the energy loss is mainly caused by inefficiency of the heat exchangers. However, the exergy analysis reveals that the SOSE cell is the major source of exergy destruction (Ni et al., 2007). The effects of some important operating parameters, such as temperature, current density, and H_2O flow rate, on the plant efficiency have been studied. Optimization of these parameters can achieve maximum energy and exergy efficiencies. The findings show that the difference between energy efficiency and exergy efficiency is small as the high-temperature thermal energy input is only a small fraction of the total energy input. In addition, the high-temperature waste heat is of high quality and can be recovered. In contrast, for a low-temperature electrolysis plant, the difference between the energy and exergy efficiencies is more apparent because a considerable amount of low-temperature waste heat contains little exergy and cannot be recovered effectively. The study of Ni et al. (2007) provides a better understanding of the energy and exergy flows in SOSE hydrogen production and demonstrates the importance of exergy analysis for identifying and quantifying the exergy destruction. The findings can further be applied to maximize the cost-effectiveness of SOSE hydrogen production.

J. Chen and Zhao (2009) proposed a theoretical modeling approach, which describes the behavior of a FC–heat engine hybrid system in steady-state operating condition based on an existing SOFC model, to select useful design characteristics and potential critical problems. The different sources of irreversible losses, such as the electrochemical reaction, electric resistances, finite-rate heat transfer between the FC and the heat engine, and heat-leak from the FC to the environment were investigated. Energy and entropy analyses were used to indicate the multi-irreversible losses and to assess the work potentials of the hybrid system. Expressions for the power output and efficiency of the hybrid system were derived to determine performance characteristics of the system. The effects of the design parameters and operating conditions on the system performance were studied numerically. It was found that there are certain optimum criteria for some important parameters. The results obtained should provide a theoretical basis for both the optimal design and operation of real FC–heat engine hybrid systems. This approach can be extended to other FC hybrid systems.

Power limits and thermodynamics of FCs were considered in numerous papers (Li et al., 2011; P. -W. Li et al., 2011; Sieniutycz, 2010b, c, d, 2011a, b, d, e, 2012b, Sieniutycz and Poświata, 2011; Sieniutycz et al., 2011; Sunden and Faghri, 2005; Zhao et al., 2008), and many others. Sieniutycz and Poświata (2011) studied thermodynamics of power yield and power limits in electrochemical systems of FC type. Power maximization approach was developed to SOFCs treated as flow engines driven by chemical reagents and electrochemical mechanism of electric current generation. Performance curves of a SOFC system were analyzed. Steady-state model of a high-temperature cell was considered, which refers to constant chemical potentials of incoming hydrogen fuel and oxidant. Lowering of the cell voltage below its reversible value was attributed to polarizations and incomplete conversions of basic reactions. Power formula described the effect of efficiency, transport laws and polarizations. It was shown how reversible electrochemical theory is extended to the case with chemical reactions; this case included systems with incomplete conversions, "reduced affinities" and an idle run voltage. Efficiency lowering was linked with polarizations (activation, concentration, and ohmic). Effect of incomplete conversions was modeled by assuming that substrates can be remained after the reaction and that side reactions may occur. Optimum conditions were found for basic input parameters. Power data differ for power generated and consumed, and depend on system's characteristics, for example, current intensity, number of mass transfer units, polarizations, electrode area, and so forth. They provide bounds for FC generators, which are stronger than reversible bounds for electrochemical transformation.

Voltage lowering in FCs below the reversible value is a reasonable measure of their imperfection. Reversible cell voltage E^0, calculated from the Nernst equation, is often a reference basis. Yet, in more general cases, actual voltage without load must take into account losses of the idle run, which are the effect of flaws in electrode constructions and other imperfections.

In Wierzbicki (2009) research and many other works the operating voltage of a cell is evaluated as the departure from the idle run voltage E_0

$$V = E_0 - V_{int} = E_0 - V_{act} - V_{conc} - V_{ohm} \tag{10.37}$$

Losses, which are called polarization, include three main sources: activation polarization (V_{act}), ohmic polarization (V_{ohm}), and concentration polarization (V_{conc}). Power density is the product of voltage V and current density i. A large number of approaches for calculating polarization losses has been presented in literature, as reviewed by Zhao et al. (2008).

Diagrams of entropy production and power yield in FCs in terms of density of electric current constitute the well-known characteristics (De Vos, 1992; Sieniutycz, 2008b), Fig. 10.4.

Validation of the thermodynamic model of FC is based on the application of the Aspen Plus™ software for simulation purposes and organization of FC power experiments, as described in Wierzbicki's MsD thesis (Wierzbicki, 2009). A complete review of the theory and experiments is presented in the report (Sieniutycz et al., 2011).

For a synthesizing description of FC reaction, the notion of the Carnot reaction potential is useful. It represents the so-called "one-way affinity" or the sum of products of chemical potentials and absolute stoichiometric coefficients of reagents (Sieniutycz and Jeżowski, 2009, 2013). References (Sieniutycz, 2010d, 2011c) explain the construction principle of Carnot reaction potential in complex chemical and electrochemical systems, as illustrated in Fig. 10.5.

Newman (1967, 1973) gives an early synthesis of transport processes in electrolytic solutions. He uses the ionic approach as opposed to the so-called components approach.

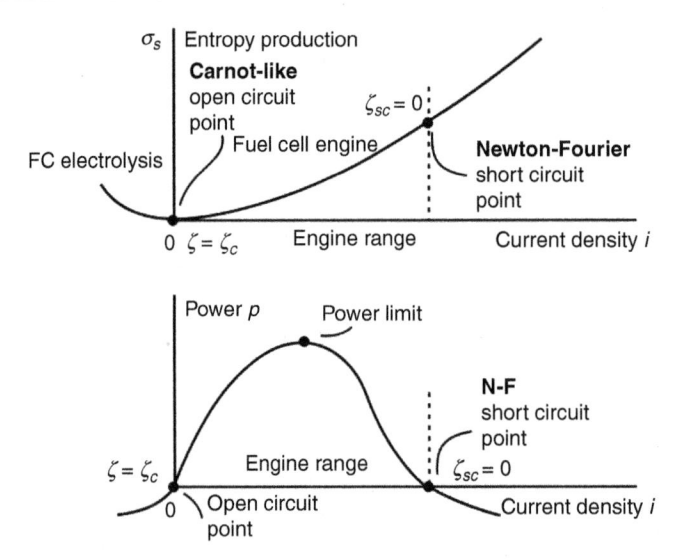

Figure 10.4: Entropy production and power yield in an electrochemical engine or fuel cell in terms of density of electric current.

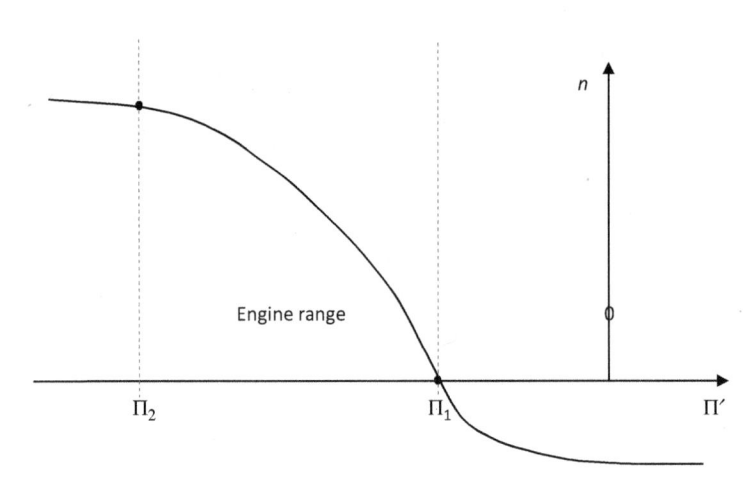

Figure 10.5: Chemical flux n versus Carnot reaction potential Π'. The engine range is between $\Pi' = \Pi_1$ and $\Pi' = \Pi_2$. Chemical efficiency ζ is identical with affinity A, and equals $\zeta \equiv A = \Pi - \Pi_2$. Limiting power is obtained by maximization of product $n(\Pi')(\Pi' - \Pi_2)$ with respect to Carnot control Π'. Sieniutycz, 2011c.

Newman (1967, 1973) discusses transport equations, dilute solutions, boundary conditions, electrode kinetics and polarization or overvoltage issues, mass transfer problems, description of concentrated electrolytes and its link with the dilute–solution theory. The latter approach (Sundheim, 1964; Ekman et al., 1978) was worked out in the context of fused salts and uses solely the inert species present in the system.

To quantify power yield and power limits in FCs we focused on the formalism of inert components (Sundheim, 1964; Ekman et al., 1978) rather than on the ionic description. Below we develop a simple theory of power limits for these systems. Assume, for simplicity, that the active (power producing) driving forces involve only: one temperature difference $T_{1'} - T_{2'}$, single chemical affinity $\mu_{1'} - \mu_{2'}$ and the operating voltage $\phi_{1'} - \phi_{2'}$. Total power production is the sum of thermal, substantial and electric components, that is,

$$\begin{aligned} P &= (T_{1'} - T_{2'})I_s + (\mu_{1'} - \mu_{2'})I_n + (\phi_{1'} - \phi_{2'})I_e \\ &= (T_1 - T_2)I_s + (\mu_1 - \mu_2)I_n + (\phi_1 - \phi_2)I_e \\ &\quad - R_{ss}I_s^2 - R_{nn}I_n^2 - R_{ee}I_e^2 - R_{sn}I_sI_n - R_{se}I_sI_e - R_{ne}I_nI_e \end{aligned} \quad (10.40)$$

Eq. (10.40) applies usually to linear thermo–electro–chemical systems. Linear systems are those with constant, current independent or flux independent resistances or conductances. They satisfy Ohm's type or Onsager's type laws linking thermodynamic fluxes and thermodynamic forces [dissipative driving forces which are represented by products $R_{ik}I_k$ in Eq. (10.40)]. While many FC systems are nonlinear, that is, possess current dependent resistances, the dependence is often weak, so a linear model is a good approximation.

After introducing the enlarged vector of all driving potentials $\tilde{\mu} = (T, \mu, \phi)$, the flux vector $\tilde{\mathbf{I}}$ of all currents and the overall resistance tensor $\tilde{\mathbf{R}}$, Eq. (10.40) can be written in a simple matrix–vector form

$$P = (\tilde{\mu}_1 - \tilde{\mu}_2)\tilde{\mathbf{I}} - \tilde{\mathbf{R}} : \tilde{\mathbf{I}}\tilde{\mathbf{I}} \quad (10.41)$$

Maximum power corresponds with the vanishing of the partial derivative vector (Fig. 10.6)

$$\partial P / \partial \tilde{\mathbf{I}} = \tilde{\mu}_1 - \tilde{\mu}_2 - 2\tilde{\mathbf{R}}\tilde{\mathbf{I}} = 0 \quad (10.42)$$

Activation and concentration polarization occurs at both anode and cathode locations, while the resistive polarization represents ohmic losses throughout the cell. As the voltage losses increase with current, an initially increasing power begins finally to decrease for sufficiently large currents, so that maxima of power are observed (El-Genk and Tournier, 2000; Li, 2006; Zhao et al., 2008; Wierzbicki, 2009). The data include the losses of the idle run attributed to the flaws in electrode constructions and other imperfections.

Experimental lines of power density versus current density obtained by P.-W. Li et al. (2011) (and very many others) confirm the existence of power maxima and subsequent power decrease for sufficiently large densities of electric current, Fig. 10.7.

Consequently, based on Eq. (10.42), the optimal (power-maximizing) vector of currents at the maximum point of the system can be written in the form

$$\tilde{\mathbf{I}}_{mp} = \frac{1}{2}\tilde{\mathbf{R}}^{-1}(\tilde{\mu}_1 - \tilde{\mu}_2) \equiv \frac{1}{2}\tilde{\mathbf{I}}_F \quad (10.43)$$

Characteristics of SOFC at various temperatures

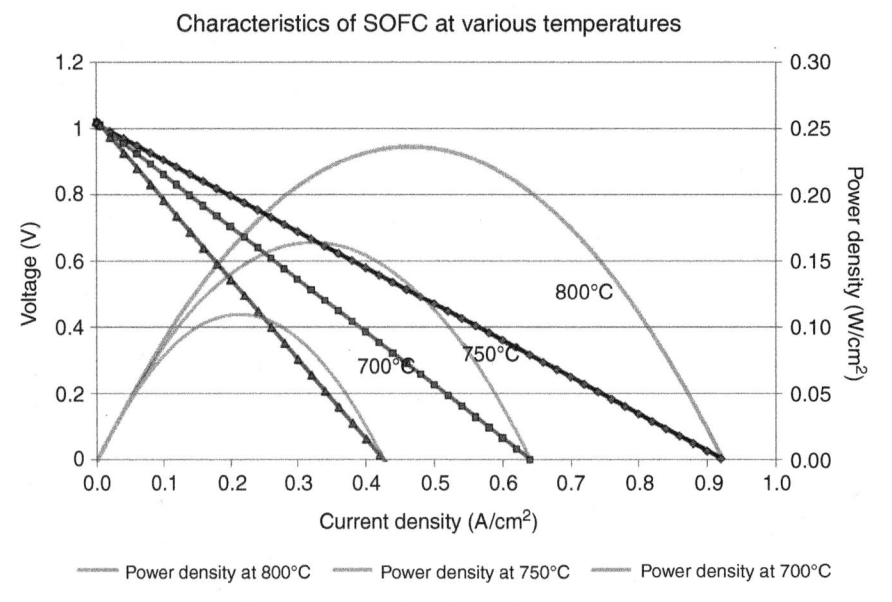

Figure 10.6: Experimental voltage-current density and power-current density characteristics of the SOFC for various temperatures.
Continuous lines represent the Aspen Plus™ calculations testing the model versus the experiments. The lines were obtained in Wierzbicki's MsD thesis supervised by S. Sieniutycz and J. Jewulski (Wierzbicki, 2009). Points refer to experiments of Wierzbicki and Jewulski in Warsaw Institute of Energetics. *Wierzbicki, 2009, and his ref. 18.*

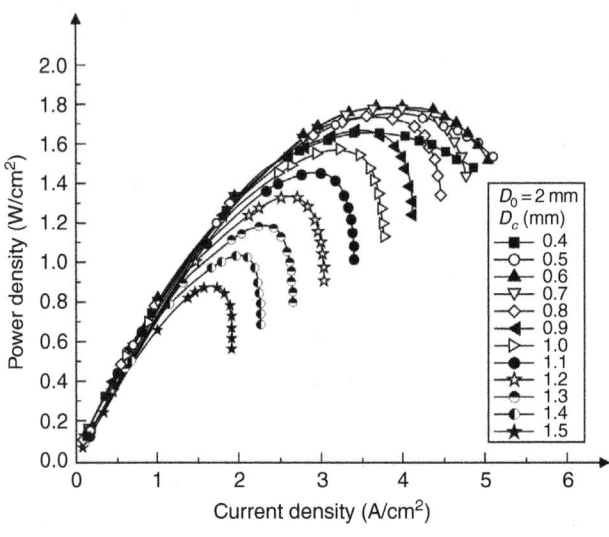

Figure 10.7: Experimental curves of power density versus current density obtained by Li et al. (2011).

This result means that the power-maximizing current vector $\tilde{\mathbf{I}}_{mp}$ in strictly linear systems equals one half of the purely dissipative current at the Fourier–Onsager point, $\tilde{\mathbf{I}}_F$ at which no power production occurs. Moreover, we note that Eqs. (10.41) and (10.43) yield the following result for the maximum power limit of the system

$$P_{mp} = \frac{1}{4}(\tilde{\mu}_1 - \tilde{\mu}_2)\tilde{R}^{-1}(\tilde{\mu}_1 - \tilde{\mu}_2) \qquad (10.44)$$

In terms of the purely dissipative flux vector at the Fourier–Onsager point, $\tilde{\mathbf{I}}_F$ the aforementioned limit of maximum power is represented by an equation

$$P_{mp} = \frac{1}{4}\tilde{\mathbf{R}} : \tilde{\mathbf{I}}_F \tilde{\mathbf{I}}_F \qquad (10.45)$$

Of course, the power dissipated at the Fourier–Onsager point equals

$$P_F = \tilde{\mathbf{R}} : \tilde{\mathbf{I}}_F \tilde{\mathbf{I}}_F \qquad (10.46)$$

Eqs. (10.45) and (10.46) prove that only at most 25% of power (10.46), which is dissipated in the natural transfer process, can be transformed into the noble form of the mechanical power. This result holds for linear systems, and, probably, it cannot be easily generalized to the nonlinear transfer systems where significant deviations from Eq. (10.45) are expected depending on the nature of diverse nonlinearities. Despite the limitation of the result (10.45), its value is significant because it shows explicitly the order of magnitude of thermodynamic limitations in power production systems.

The aforementioned analysis also proves that a link exists between the mathematics of the thermal engines and FCs, and also that the theory of FCs can be unified with the theory of thermal and chemical engines.

A word of explanation of related physical effects is in order. While power ratios involving Eqs. (10.45) and (10.46) can be regarded as some efficiency measures, they should not be mixed with commonly used, popular efficiencies, especially with first law efficiencies. In fact, the considered power ratios represent some specific, second-law efficiencies of the overall thermo–electro–chemical process. The popular FC efficiencies $\eta_{FC} = \Delta G/\Delta H$ and $\eta_{FC} = -P/\Delta H$, commonly applied to many FC systems, can easily achieve numerical values much higher than ¼ [power ratio of Eqs. (10.45) and (10.46)]. They are, however, the first-law efficiencies defined in a different way than the power ratios P_{mp}/P_F satisfying Eqs. (10.45) and (10.46).

There is a number of definitions of FC efficiencies, based on first or second laws, proposed for measuring and comparing the performance of electrochemical processes. Only second-law efficiencies are correct measures which show how close the process approaches a reversible process. Efficiencies based on the first law (often found in the literature) can generate efficiency values greater than 100% for certain systems depending on whether the change in entropy for the overall chemical reaction involved in the process is positive or negative. See, for example, paper by Rao et al. (2004) on various definitions of FC efficiencies.

10.8 Exergy and Second Law Analyses of FC Systems

Applying an exergy analysis to a FC system, De Groot and Woudstra (1995) show that heat transfer causes large losses. These losses may be reduced if the amount of heat transferred in the system is reduced. This may be done by the application of internal reforming. Cownden et al. (2001) report another exergy analysis, of a solid polymer FC system, for transportation applications. The model analyzes all components of the system including the FC stack and the air compression, hydrogen supply, and cooling subsystems. The largest destruction of exergy within the system occurs inside the FC stack. Fryda et al. (2008) applies exergy analysis to SOFC and biomass gasification.

Dijkema and Weijnen (1997) discuss the use of FCs in the industry, associated with the utilizing low-BTU off-gases by implementing FCs on site. These systems can convert byproduct hydrogen into electric power and heat. FC stacks are integrated into the system design to improve the plant's performance. Bedringas et al. (1997) applies an exergy analysis to improve performance of SOFC systems. They include preheating of fuel and air, reforming of methane to hydrogen, and combustion of the remaining fuel in afterburner. An iterative computer program using a sequential-modular approach is applied in the analyses. Simulation of an SOFC system with external reforming leads to first-law and second-law efficiencies of 58 and 56%, respectively, with 600% theoretical air. Heat released from the afterburner serves to reform methane, vaporize water, and preheat air and fuel. The very low exhaust temperature in this case can only be used for heating rooms or water. Because of heat requirements, fuel utilization in the cell is limited to 75%. The remaining fuel is used for preheating and reforming. Reduced excess air leads to reduced heat requirements and allows higher fuel utilization in the FC. Irreversibilities are also reduced, so that efficiencies increase. Recycling fuel and water vapor from the FC results in first-law and second-law efficiencies of 75.5 and 73%, respectively, while 600% theoretical air vaporization of water is avoided and the fuel utilization index increases. Campanari (2001) presents a thermodynamic model and parametric analysis of a tubular SOFC module.

Calise et al. (2006) discuss the simulation and exergy analysis of a hybrid solid oxide fuel cell–gas turbine (SOFC–GT) power system. In the SOFC reactor model it is assumed that only hydrogen participates in the electrochemical reaction and that the high temperature of the stack pushes the internal steam reforming reaction to completion; the unreacted gases are assumed to be fully oxidized in the combustor downstream of the SOFC stack. Compressors and GTs are modeled on the basis of their isentropic efficiency. For the heat exchangers and the heat recovery steam generator, all characterized by a tube-in-tube counterflow arrangement, the simulation is carried out by using the thermal efficiency-NTU approach. Energy and exergy balances are performed for the whole plant and for each component to evaluate the distribution of irreversibility and thermodynamic inefficiencies. Simulations are performed for different values of operating pressure, fuel utilization factor, fuel-to-air and steam-to-fuel ratios, and current density. Results show that, for a 1.5 MW system, an electrical efficiency close to 60% can be achieved for appropriate values of the most important design variables, the operating pressure and the cell current density. When heat loss recovery is also taken into account, a global efficiency of about 70% is achieved. Calise et al. (2007) further construct a detailed model of a SOFC tube, equipped with a tube-and-shell preformer unit. Both SOFC tube and preformer are discretized along their axes. Data of the kinetics of the shift and reforming reactions are introduced. Energy, mole, and mass balances are set for each slice of the component, allowing calculation of temperature profiles. Friction factors and heat exchange coefficients are evaluated from experimental correlations. Detailed modeling serves to evaluate the SOFC overvoltage. Based on this modeling, temperatures, pressures, chemical compositions, and electrical parameters are evaluated, and the influence of the basic design parameters on the system's performance is investigated.

Douvartzides et al. (2004) present the method of exergy analysis for a SOFC power plant involving external steam reforming and fed by methane and ethanol. The parameters for optimal operation of the integrated SOFC plant are determined after minimizing the existing energy and exergy losses. A comparison of methane and ethanol as appropriate fuels for a SOFC-based power plant is performed in terms of exergy efficiency by assuming the minimum allowable (for carbon-free operation) reforming factors for both cases. Then, a parametric analysis provides guidelines for practical design. The analysis employs factors such as the extent of the steam reforming reaction and the hydrogen utilization in the SOFC, and proves that the appropriate adjustment of the plant performance can be achieved by simply interrelating them, to control the energy losses to environment and the effect of combustion processes in the power cycle. Energy losses from the SOFC stack have also a negative impact to plant efficiency, and are minimized through an appropriate thermal management between the mixtures incoming and leaving the stack, to attain the adiabatic regime of SOFC operation. Methane-fed operation is of higher efficiency than that of the ethanol-fed one. In conclusion the exergy calculations locate the losses accurately and the exergy analysis improves the understanding of this complex process.

Delsman et al. (2006) presents a second-law analysis of an integrated FP and FC system. They consider the following primary fuels: methanol, ethanol, octane, ammonia, and methane. The maximum amount of electrical work and corresponding heat effects produced from these fuels are evaluated. The exergy analysis is performed for a methanol processor integrated with a PEMFC, for the use as a portable power generator. The integrated FP–FC system, which can produce 100 W of electricity, is simulated with a computer model using the flow-sheeting program Aspen Plus. The influence of various operating conditions on the system efficiency is investigated, such as the methanol concentration in the feed, the temperature in the reformer and in the FC, as well as the FC efficiency. The calculated overall exergy efficiency of the FP–FC system is higher than that of typical combustion engines and rechargeable batteries. In conclusion, integrated FP–FC systems may have promising applications, particularly as man-portable power supplies in electronic and electric devices as well as in residential and recreational sectors. Hydrogen is the superior fuel for FCs having the highest energy density per unit mass but is difficult to store. Various primary fuels like alcohols and hydrocarbons can be processed to obtain a hydrogen-rich gas. The maximum amount of work produced in a FP–FC system is different for various fuels, but is approximately the same for each fuel when expressed per mole of H_2 generated via steam reforming. An exergy analysis of the methanol processor integrated with a PEM–FC shows that the largest exergy losses occur in the FC, the burner, the vaporizer, and the reformer. The overall exergy efficiency of the system is about 37%. The total exergy loss depends mainly on the methanol concentration in the feed and the FC performance; the temperature in reformer and FC have a much smaller effect. The calculated overall exergy efficiency of the fuel-to-electricity conversion is higher for the FP–FC system than for rechargeable batteries and combustion engines.

Nowadays, the climatic change has addressed the research targets to find renewable energy sources and in order to develop more efficient technologies in a simultaneous way, with the object of promoting a rational use of the energy in the frame of the sustainable development. In this case, an integrated process for sustainable electrical energy production from bioethanol was designed (Hernandez and Kafarov, 2009) which takes advantage of hydrogen fuel as an energy carrier and FCs as efficient and clean devices. The calculated efficiency for this process is better than traditional power cycles, which constitutes a starting point for future developments of this technology.

A considerable group of papers uses NNs to model the FC performance. Cheng-Xin Li et al. (2009) tests the effect of composition of NiO/YSZ anode on the polarization characteristics of SOFC fabricated by atmospheric plasma spraying. Jurado (2003) investigates power supply quality improvement with a SOFC plant by neural-network-based control. Javania et al. (2014) investigate heat transfer with phase change materials (PCMs) in passive thermal management of electric and hybrid electric vehicles where the PCM is integrated with a Li-ion cell. Hattia and Tioursib (2009) develop dynamic NN controller of PEM FC system.

Chaichana et al. (2012) stress the importance of the mathematical model for analysis and design of FC stacks and associated units. In general, a complete description of FCs requires an electrochemical model to predict their electrical characteristics, that is, cell voltage and current density. However, obtaining the electrochemical model is quite a difficult and complicated task as it involves various operational, structural, and electrochemical parameters. In their study, a NN model is first proposed to predict electrochemical characteristics of SOFCs. Various NN structures are trained based on the back-propagation feed-forward approach. The results show that the NN with optimal structure reliably provides a good estimation of FC electrical characteristics. Then, a NN hybrid model of a direct internal reforming SOFC, combining mass conservation equations with the NN model, is developed to determine the distributions of gaseous components in fuel and air channels of SOFC, as well as the performance of the SOFC in terms of power density and FC efficiency. Effects of various key parameters, for example, temperature, pressure, steam to carbon ratio, degree of pre-reforming, and inlet fuel flow rate on the steady SOFC performance under isothermal conditions are also investigated. A combination of the first principle model and NN improves the prediction of the SOFC performance and reduces the computational time.

Care is recommended when assuming minimum entropy production in an arbitrary electrochemical network. Yet Perez (2000) stresses a thermodynamic interpretation of the variational Maxwell theorem for DC circuits based on the minimum of the entropy production. His considerations complete Van Baak (1999) text regarding the use of variational methods on DC circuits. Applying these variational approaches, Van Baak (1999) describes some variational alternatives to Kirchhoff's loop theorem. He recalls that problems involving the flow of direct current in circuits are typically solved through the use of Kirchhoff's rules—the junction theorem and the loop theorem. Since readers typically find the meaning and use of the junction theorem straightforward, but are prone to conceptual and computational difficulties with the loop theorem, he feels motivated to present a little-known alternative method for solving DC circuit problems. In his method, the junction theorem is retained, but the use of the loop theorem is replaced by the minimization of a single global variational quantity for the circuit. The use of this method is illustrated (Van Baak, 1999), and its equivalence to standard methods is proved by vector calculus from the steady-state limit of Maxwell's equations. In summary, he stresses that the foundation for variational methods in DC circuits consists in the minimal entropy production in resistors and generators.

For illustration Perez (2000) presents an example of the minimization the entropy production applied to the simple electric circuit shown in Fig. 10.8. He accepts the power function $P_F = T_e \dot{S}$ with a constant T_e, expressed as

$$P = R(I_1 + I_2)^2 + R_1 I_1^2 + R_2 I_2^2 - 2e_1(I_1 + I_2) - 2e_2 I_2 \qquad (10.47)$$

because the currents pass through generators in the negative sense. The multiplier T_e is the constant temperature of the environment. The circuit equations can be derived from the expressions for vanishing derivatives

$$\left(\frac{\partial P}{\partial I_1}\right)_{I_2} = 2R(I_1 + I_2) + 2R_1 I_1 - 2e_1 = 0 \qquad (10.48)$$

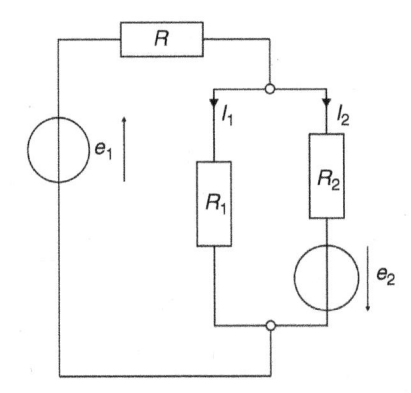

Figure 10.8: An electric network corresponding with the minimization of the function P in a DC circuit with two generators and two unknown variables, the currents I_1 and I_2.
(Perez, 2000).

and

$$\left(\frac{\partial P}{\partial I_2}\right)_{I_1} = 2R(I_1 + I_2) + 2R_2 I_2 - 2e_1 - 2e_2 = 0 \tag{10.49}$$

Therefore, we obtain

$$e_1 = R(I_1 + I_2) + R_1 I_1 \tag{10.50}$$

and

$$e_1 = R(I_1 + I_2) + R_2 I_2 - e_2 \tag{10.51}$$

which is the correct result. This is merely a modest example. According to Meixner (1963) electrical networks furnish many excellent examples for the application of thermodynamics of irreversible processes. In particular, they provide one of the simplest cases where, besides Onsager coefficients, one has Casimir coefficients in the phenomenological equations. Meixner (1963) gives two thermodynamic formulations of the network equations, which closely resemble the Lagrangian and the Hamiltonian formalism, respectively, of classical mechanics.

10.9 Limits on Power Consumption in Thermochemical Systems

Since the power and efficiency formulae derived from analyses of power generators can be simply readjusted to units, which consume power, it is interesting to develop analyses for units such as thermal and solar HPs, drying separators, and electrolyzers. HPs and HP-supported dryers are good representatives of systems driven by a consumed power. A meaningful reduction of drying exergy could be achieved by the optimal control of many dryers; on the average the potential for exergy reduction is more than 20%. Some industrial data are, however, still reported in terms of energy (enthalpy) rather than exergy efficiency, and suffer from the lack of a strong link with well-established kinetic models.

Awareness of limited energy supplies prompts significant effort in developing recovery processes. Possible processes and design modifications involve waste heat recovery from solid, energy recuperation from gas, use of HPs for waste energy upgrading, recycling of exhausted agent, combined mechanical and thermal drying, and use of solar energy.

Application of optimal control principles to processes with consumption of mechanical energy gives certain extra potential for improving their economy. Some of these processes use thermal separators and HPs. The HP is in principle the only device, which would allow exploitation of the low-exergy sources commonly available in nature and industry. HPs increase driving energy by adding low-quality energy taken from a low-exergy source to it to obtain energy of high quality economically. Mathematical analyses of power-assisted processes lead to optimization-determined bounds on the power input and exergy dissipation.

We shall consider an optimal operation of HP-supported drying. The optimization example refers to a multistage drying process in which the gas leaving the previous stage enters the HP and dryer of the next stage; see Fig. 10.9. For relatively efficient stages, it may be assumed that the outlet solid and gas are in the equilibrium due to a large specific solid area. The properties of the HP as the heating device are important in this analysis; the better COP

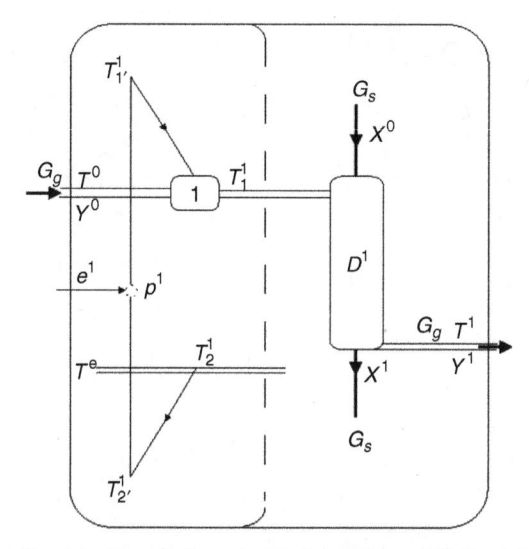

Figure 10.9: A scheme of one-stage drying operation with a HP 1 and a continuous cocurrent dryer D^1 with falling particles.

The multistage idea means, of course, repetition of this single-stage set in the next stages.

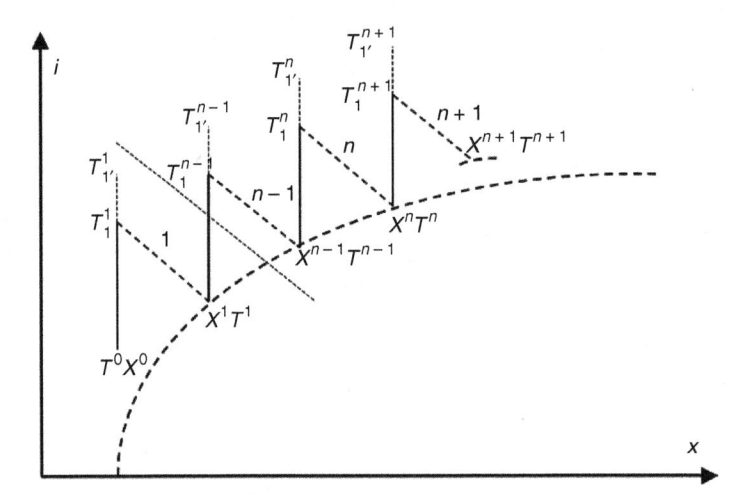

Figure 10.10: Changes of gas states in a multistage work-assisted drying operation.
Primed states refer to temperatures of circulating fluids which heat gases supplied to dryers 1, 2,...,n.

results in more efficient heating. The final effect is a complex process—in fact, a power-assisted drying operation—in which gas states vary as shown in Fig. 10.10.

As shown in Fig. 10.9 a stage of this complex multistep operation often comprises not a single dryer but rather an appropriate group of various units, which is repeated when the process proceeds from one step to another. This is the case of a multistage drying operation, in which gas at each stage is first heated with a HP and then is directed to a dryer (note that only one stage of that operation is shown in Fig. 10.9). In this example a continuous drying process occurs in a cocurrent dryer; yet the countercurrent contacting could alternatively be applied. The purpose is to minimize the work consumption in, say, a two-stage operation by a suitable choice of the intermediate moisture content between the first and the second stage.

One may ask: how many HPs (and stages) are there in the optimal system? The answer is possible by making the economical analysis which leads to determining of operational and investment costs. The optimization calculations for the cascade terminate when the sum of operational and investment cost stops to decrease.

The balances of mass and heat yield

$$\frac{rG_s}{cG_g}(W^0 - W^1) = (T_1^1 - T^1), \tag{10.52}$$

$$T_1^1 - T^1 = -\frac{r}{c}(X_1^1 - X_s(T^1)), \tag{10.53}$$

whereas the power consumed at a single stage per unit flow of gas, e^1, the quantity having the units of a specific work, is described by an expression

$$e^1 \equiv -\frac{p}{G_g} = c\left(1 - \frac{T^e}{T_1^1 + u^1}\right)u^1\theta^1, \qquad (10.54)$$

where u^1 is the energy supply to the drying gas in the condenser of the HP, and $u^1 = -q^1/g > 0$ is a measure of this energy supply in the temperature units. When the effect of internal dissipation in HP is included, the performance coefficient of the supporting HP contains imperfection factor Φ as the multiplicative factor of the bath temperature T^e.

Substituting into Eq. (10.54) the temperature T_1^1 following from Eq. (10.53)

$$T_1^1 = T^1 + \frac{r}{c}(X_s(T^1) - X_1^1) \qquad (10.55)$$

and taking into account that $X_1^1 = X^0$ [also $X_1^n = X^{n-1} = X_s(T^{n-1})$, for $n = 2,...,N$] we find the specific energy consumption at the stage

$$e^1 = c\left(1 - \frac{T^e}{T^1 + (r/c)(X_s(T^1) - X^0) + u^1}\right)u^1\theta^1. \qquad (10.56)$$

Eq. (10.56) is transformed further in view of the link between u^1 and θ^1 (consider difference constraint describing $\Delta T^n = u^n\theta^n$ for $n = 1$)

$$e^1 = c\left(1 - \frac{T^e}{T^1 + rc^{-1}(X_s(T^1) - X^0) + \left((T^1 + rc^{-1}(X_s(T^1) - X^0) - T^0)/\theta^1\right)}\right) \qquad (10.57)$$
$$\times (T^1 + rc^{-1}(X_s(T^1) - X^0) - T^0).$$

In terms of the "adiabatic temperature function"

$$T^a(T^1) \equiv T^1 + rc^{-1}X_s(T^1) \qquad (10.58)$$

the work expression takes the final form

$$e^1 = c\left(1 - \frac{T^e}{T^a(T^1) - rc^{-1}X^0 + \left((T^a(T^1) - rc^{-1}X^0 - T^0)/\theta^1\right)}\right) \times (T^a(T^1) - rc^{-1}X^0) - T^0). \qquad (10.59)$$

An analogous function, but with the shifted superscripts, is valid for the second stage.

$$e^2 = c\left(1 - \frac{T^e}{T^a(T^2) - rc^{-1}X_s(T^1) + \left((T^a(T^2) - rc^{-1}X_s(T^1) - T^1)/\theta^2\right)}\right) \times (T^a(T^2) - rc^{-1}X_s(T^1) - T^1).$$
$$(10.60)$$

The constraint $X^1 = X_s(T^1)$ resulting from the outlet phases equilibrium in the first dryer is incorporated in the second work expression. In terms of the adiabatic temperature at the second stage

$$e^2 = c\left(1 - \frac{T^e}{T^a(T^2) - rc^{-1}X_s(T^1) + \left((T^a(T^2) - T^a(T^1))/\theta^2\right)}\right) \times (T^a(T^2) - T^a(T^1)). \quad (10.61)$$

The sum of both works yields the total work consumed. This total work is the thermodynamic cost that should be minimized. For a fixed holdup time $\tau^2 = \theta^1 + \theta^2$ there are two free controls: θ^1 and T^1. Thus we can accomplish the power minimization (Fig. 10.10).

This example outlined optimizing the multistage operation with drying gases heated by a sequence of HPs. The requirement of sufficiently low final moisture content in solid defines a constraint on the total amount of the evaporated moisture per unit time. The optimal solution assures a minimum of total consumed power. The optimization procedure searched for an optimal inter-stage temperature T^1 and an optimal heat transfer area of the first HP a^1 (measured in terms of the control variable θ^1). The optimal transfer areas are close to each other, whereas the total power supply decreases distinctly with the total transfer area. For more details see Sieniutycz (2000c). Generalization to the N-stage cascade system is obvious.

10.10 Estimate of Minimum Power Supplied to Power Consumers

A general theory of minimum power fed to power consumers can be developed, based on systems' thermodynamic and kinetic properties, as outlined later. Note a symmetry of this approach in relation to maximum power yield in power generators (FCs in Section 10.7).

Consider power consumption in a linear thermochemical system with possible electric fluxes in case of electrolyzers. Assume that a two-reservoir arrangement is sufficient to accomplish a separation process or HP heating. Both these operations need for their running some amount of power supplied to the system. We shall use total resistances of upper and lower parts of two-reservoir system

$$R_s = R_{1s} + R_{2s}, \quad R_n = R_{1n} + R_{2n}, \quad R_{el} = R_{1e} + R_{2e}, \quad (10.62)$$

After considering the coupled transfer of heat, mass, and electricity, the power supply reads

$$\begin{aligned} P &= (T_{1'} - T_{2'})I_s + (\mu_{1'} - \mu_{2'})I_n + (\phi_{1'} - \phi_{2'})I_e \\ &= (T_1 - T_2)I_s + (\mu_1 - \mu_2)I_n + (\phi_1 - \phi_2)I_e \\ &\quad + R_{ss}I_s^2 + R_{nn}I_n^2 + R_{ee}I_e^2 + R_{sn}I_sI_n + R_{se}I_sI_e + R_{ne}I_nI_e \end{aligned} \quad (10.63)$$

where nonprimed quantities refer to bulk states and primed ones to the active, power-consuming part of the system. As shown by this equation, the general thermodynamic setting allows for at least rough assessment of power consumption limits in thermo–electro–chemical systems of the simplest, standardized topology (with no counterflows). This topology corresponds to a power consumption unit (eg, HP) immersed between two reservoirs, one with high potentials and one with low ones, as described many times. The reason why this power assessment can only be rough is explicit in Eq. (10.63), which has ignored information about the topological structure of various flows in the system. Let us also add that electrolyzers are described by the formalism of inert components rather than by the ionic description.

For simplicity Eq. (10.63) assumes that active (power consuming, primed) driving forces involve only: one temperature difference, trivial chemical affinity, and the operating voltage as the difference of electric potentials. Total power consumption is the sum of thermal, substantial, and electric components.

Eq. (10.63) constitutes the simplest account of thermo–electro–chemical separators and HPs; indeed it does not contain any "topology parameter." Complex configurations of contacting such as countercurrent flows that may exist in FC electrolyzers are not taken into account in Eq. (10.63). Linear systems described by this equation are those with constant (current independent or flux independent) resistances or conductances. They satisfy the Ohm or Onsager type laws linking thermodynamic fluxes and thermodynamic forces [dissipative driving forces represented by products $\Sigma R_{ik}I_k$ in Eq. (10.63)]. While many thermal separation systems and fuel-cell electrolyzers are nonlinear, that is, possess current dependent resistances, the dependence is often weak, so a linear model can be a good approximation. Below, by applying Eq. (10.63) we attempt to develop a simple evaluation of power limits for HPs and separation systems under the aforementioned specified assumptions.

After introducing the enlarged flux vector $\mathbf{I} = (I_s, I_n, I_e)$, the enlarged vector of potentials $\tilde{\boldsymbol{\mu}} = (T, \mu, \phi)$, and the related resistance tensor $\tilde{\mathbf{R}}$, Eq. (10.63) can be written in a simple form

$$P = (\tilde{\boldsymbol{\mu}}_1 - \tilde{\boldsymbol{\mu}}_2)\tilde{\mathbf{I}} + \tilde{\mathbf{R}} : \tilde{\mathbf{II}} \tag{10.64}$$

The bulk driving forces $\tilde{\boldsymbol{\mu}}_1 - \tilde{\boldsymbol{\mu}}_2$ are given constants, whence, in systems with constant resistances, we are confronted with a simple minimization problem for a quadratic power consumption function p. While the dimensionality of the potential vector will often be quite large in real systems, the structure of Eq. (10.64) will be preserved whenever the power expression will be considered in the aforementioned matrix–vector notation.

Minimum power corresponds with vanishing partial derivatives of function P. The optimal (power minimizing) vector of currents at the minimum power point can be written in the form

$$\tilde{\mathbf{I}}_{mp} = -\frac{1}{2}\tilde{\mathbf{R}}^{-1}(\tilde{\boldsymbol{\mu}}_1 - \tilde{\boldsymbol{\mu}}_2) \equiv -\frac{1}{2}\tilde{\mathbf{I}}_F \tag{10.65}$$

This result means that power-minimizing current vector in separators and HPs, is equal to the negative of one half of the purely dissipative current at the Fourier–Onsager point. The latter point refers to the system's state at which no power production occurs.

Eqs. (10.64) and (10.65) yield the following result for the minimum power supplied to the HP or thermal separation system

$$P_{mp} = -\frac{1}{4}(\tilde{\mu}_1 - \tilde{\mu}_2)\tilde{\mathbf{R}}^{-1}(\tilde{\mu}_1 - \tilde{\mu}_2) \qquad (10.66)$$

(negative power supply follows because the engine convention was used, in which case the power released is positive).

In terms of the purely dissipative fluxes at the Fourier–Onsager point, where neither power production nor power consumption occurs, the aforementioned limit of minimum power is represented by an equation

$$P_{mp} = -\frac{1}{4}\tilde{\mathbf{R}} : \tilde{\mathbf{I}}_F \tilde{\mathbf{I}}_F \qquad (10.67)$$

The minus sign corresponds with the engine convention, which requires that power supplied to the system is negative.

On the other hand, power dissipated at the Fourier–Onsager point is

$$P_F = \tilde{\mathbf{R}} : \tilde{\mathbf{I}}_F \tilde{\mathbf{I}}_F \qquad (10.68)$$

Comparison of Eqs. (10.67) and (10.68) proves that, in linear thermo–electro–chemical separators with power support, at least 25% of power dissipated in the natural transfer process must be supplied as the power surplus in order to run a power-consuming system with a finite rate. Yet, this result cannot, probably, be sufficiently exact for systems of more complex topology and for those with nonlinearities, where significant deviations may be expected.

In fact, the present result describes the quite disadvantageous power surplus applied to real HPs, dryers, electrolyzers, and other separation systems. Yet, for these systems significant deviations from Eqs. (10.65)–(10.67) are expected depending on nature of nonlinearities and topology variations, and also on topology improvements to include countercurrent contacting. Despite its restrictiveness, the result (10.67) has a certain value since it at least estimates the order of magnitude of thermodynamic limitations in power consumption systems.

The analysis presented proves that a link exists between the mathematics of HPs, separators, and electrolyzers, and also that, possibly, the theory of electrolyzers can be unified with the theory of thermal and chemical separators and HPs. All these systems are power consumers. However, serious topological differences between these systems may occasionally render them quite dissimilar.

The reciprocals of power ratios involving Eqs. (10.67) and (10.68) are the second-law-based performance coefficients, which show how closely a concrete process approaches a reversible process. While these power ratios can be regarded as efficiency measures, they should not be confused with reciprocals of commonly used, popular performance coefficients, especially first law-based coefficients of HPs and some separation systems. There are commonly accepted correct definitions of performance coefficients based on first or second laws, proposed for measuring and comparing performance of HPs and separation processes (Bejan, 1988, 1996b; Berry et al., 2000; Wu et al., 2000; Zhu, 2014).

10.11 Final Remarks

While the modeling and particular numerical results are limited to FCs, continuous cocurrent dryers with falling particles and multistage HPs, our approach can be extended to different and more complex configurations.

The practical application of a given energy source in technology is supported following the theoretical recommendations based on irreversible thermodynamic analysis of the cycles (Mansoori and Patel, 1979; Berry et al., 2000). Upper and lower limits for the coefficient of performance of solar absorption cooling cycles have been derived from the first and second laws (Mansoori and Patel, 1979). These limits depend not only on the environmental temperatures of the cycle components but also on the thermodynamic properties of refrigerants, absorbents and mixtures thereof. Quantitative comparative studies of different refrigerant-absorbent combinations are possible.

The approach of the finite time thermodynamics (FTT) applied in the context of HPs has raised an interesting discussion in which its followers and opponents may find lots in common regarding some issues, but also disagree about the others. For example, a renewed and eloquent criticism of the FTT approach was published (Gyftopoulos, 1999b), in addition to a presence of a stream of FTT publications in the field. Many of the latter papers contain in-depth derivations of the power expression and the Curzon–Ahlborn–Novikov efficiency (CAN efficiency), including thorough explanations of connection between the CAN efficiencies and typical efficiencies of real heat engines (Chen et al., 2001a) and HPs (Li et al., 2010b). In fact, both types of publications quote-experimental data to support their own conclusions and final standpoint (Berry et al., 2000; Chen et al., 2001a; De Vos, 1992, 1985).

The theoretical estimation of minimum work supplied to a HP used in drying technology ought to be compared with the experimental energy consumption criteria. Szwast's exergy optimizations in a class of drying systems with granular solids (Berry et al., 2000; Szwast, 1990) show an agreement of about 25% between the calculated and experimental consumption of the driving exergy. A careful use of the experimental data is given by De Vos who illustrated the practical usefulness of CAN engine theory on the example of the quantitative description of nuclear power plant "Doel 4" in Belgium, and explained the difference between the predicted value of engine's optimal efficiency ($\eta_{CAN} = 0.293$) and the experimental thermal efficiency ($\eta_{exp} = 0.350$). By simple economical considerations De Vos also explained why the actual efficiency of the engine is larger than its CAN efficiency (De Vos, 1992).

This chapter provides data for power production limits, which are enhanced in comparison with those predicted by the classical thermodynamics. In fact, thermostatic limits are often too far from reality to be really useful. Generalized limits, discussed here, are stronger than those predicted by the thermostatic theory. As opposed to classical thermodynamics, generalized limits depend not only on state changes of resources but also on process resistances, process direction and mechanism of heat and mass transfer. Common methodology was developed for thermal, chemical, and electrochemical systems, including FCs (Li, 2006).

In fact, the main methodological novelty of the chapter lies in its synthesizing nature, which includes FCs to the common class of thermodynamic power yield systems. In the thermodynamic optimization, that is, the optimization applying thermodynamic constraints and performance criteria, thermodynamic synthesis means an idea of combining various partial optimization models into a "synthesizing" (not necessarily "generalizing") model from which performances of all the component units can be predicted. With the irreversible thermodynamics we can predict the performance behavior and power limits for quite diverse practical systems.

In comparison to the previous publications, canonical (Hamiltonian) treatment of optimal dynamical processes constitutes a useful approach, which might be contrasted with the dynamic programming approaches (DP approaches) applied somewhat earlier (Sieniutycz, 2000c, 2003a, 2008b; Sieniutycz and Jeżowski, 2009, 2013). Since the analytical treatment of differential optimization models [such as those in Eqs. (10.8) and (10.9) of the present chapter] is most often a very difficult task, solving techniques used previously treated numerically Bellman's recurrence equation of a dynamic programming model (DP model). However, DP approaches are effective only for low dimensionality of the state vector. As the number of state variables in chemical systems is usually large (many concentrations may accompany temperature and catalyst activities), DP algorithms become inefficient and inaccurate in real systems.

Therefore, in this chapter, we limited the DP approaches to special cases of low state dimensionality and stressed the efficiency of Pontryagin type (Hamiltonian-based) approaches, Section 10.5. The development and application of Pontryagin's type (Hamiltonian-based) approaches to optimal power systems is the technical core of the present chapter. As opposed to the dynamic programming algorithms (DP algorithms, applied earlier), Hamiltonian algorithms, which involve discrete or difference equations rather than DP recurrence equations, are particularly effective in power systems with large dimensionality of the state vector. There is also a novelty in the mathematical structure of Hamiltonians, which may admit discrete process rates as entities explicit in time intervals, θ. Until now Hamiltonians were used in power systems quite seldom, and were limited to models with θ-independent discrete rates, the models, which sometimes force a researcher to quite special discretizing (Poświata, 2005; Sieniutycz, 2006c).

Believing that "good old" may be nearly as good as "good new," we adduce here a few parts of A. Schneider's review of the book of Bailie (1978) *Energy Conversion Engineering*, the book which we particularly like, since, in spite of its age, it still is a very useful and interesting textbook.

> This book is intended as a textbook for senior level students in engineering or the physical sciences. As stated by the author, the emphasis on coal is due to his familiarity with this energy source as well as to the widely held belief that coal will be the major fuel for the next century. The chemical engineer's point of view of energy matters prevails throughout the book. A concise historical review and an introduction of simple energy conversion calculations are followed by a brief presentation of the energy demand and supply situation. The treatment of thermodynamics, chemical equilibrium, and chemical reaction kinetics is quite thorough. An excellent survey is given of coal gasification and liquefaction systems, including descriptions of equipment, operation practices, and a review of their current status. ...

> The principles of solar energy are introduced and several collector types (flat plate, focusing, and satellite) are analyzed. Only superficial treatment is given to energy storage, photovoltaic, and biomass conversions. The chapter devoted to environmental considerations, though limited to fossil fuel combustion, is probably the most useful. Pollution sources, dispersion mechanisms, abatement equipment, and cost-benefit tradeoffs are thoroughly treated and many numerical examples are provided. Three pages are devoted to economics, politics, and social implications. ... The text contains many calculation examples, with the solutions presented in tabular form. Each chapter also provides many generally good problems for student assignments. The text is largely devoid of bibliographic references. Omitted from the text are important modes of energy conversion, such as hydroelectric power, internal combustion engines, electric storage batteries, thermoelectric generators, and so forth.

> Energy Conversion Engineering is one of the better texts to have appeared in recent years. As a textbook it will be useful to chemical engineers, but coverage of several important topics may require supplemental texts.

CHAPTER 11

Thermodynamic Aspects of Engineering Biosystems

11.1 Introduction

The development of a rigorous thermodynamic description, for the exceedingly complex world of biotechnology, is one of the major challenges today in establishing the scientific basis for rational, efficient, and fast bioprocess development and design (von Stockar and van der Wielen, 1997). While the equilibrium and disequilibrium aspects in biotechnology and thermodynamics as the main theory underlying them have thus far received certain attention, design and development of biotechnological processes is still carried out today in an relatively empirical way. For this situation to improve, a wider and more systematic use of the thermodynamic methods in biotechnology must be encouraged (von Stockar and van der Wielen, 1997). Thus one of the goals of their overview is the identification of research needs. The development of a rigorous thermodynamic description of the extremely complex world of biotechnology may seem a daunting task, but is also one of the major challenges in establishing the scientific basis for efficient and rapid bioprocess development and design.

Quite a body of knowledge exists already, but a wider use of many branches such as thermodynamics of charged biopolymers, correlative approaches and exposition of open and irreversible systems, needs to be encouraged (von Stockar and van der Wielen, 1997). Further research is needed into many different areas. They include increasing our base of reliable data on phase equilibria and on free energy of biomolecules in their environment, with a particular emphasis on not only proteins but also on DNA and other biopolymers. The emphasis is also on further developing both theoretical and correlative approaches, research into thermodynamic effects in biopolymer stability and function, application of classical and irreversible thermodynamics to cellular systems, large-scale biocalorimetry, energy and free energy loss analysis of bioprocesses, cellular growth processes, and metabolic schemes. The scope for novel research into these and many other related areas is enormous and the results are essential in order to meet the challenges outlined above.

Von Stockar's (2013) book covers the fundamentals of biothermodynamics, showing how thermodynamics can best be applied to applications and processes in biochemical engineering. Its goal is the rigorous and efficient application of thermodynamics in biochemical engineering, to rationalize bioprocess development and obviate a substantial fraction of this need for tedious experimental work. Among many attractive fields covered are the following: Thermodynamics in multiphase catalysis, protein precipitation with salts and polymers, stability of protein solutions, stabilities and melting temperatures of B-form DNA, ion exchange equilibria of weak-electrolyte biomolecules, live cells as nonequilibrium systems, bioprocess monitoring and control. A thermodynamic approach is developed to predict black-box model parameters for microbial growth. Biothermodynamics of live cells is described in the context of energy dissipation and heat generation in cellular cultures. There are sections on thermodynamic analysis of photosynthesis, thermodynamic analysis of dicarboxylic acid production in microorganisms and on thermodynamic analysis of metabolic pathways. This volume can be supplemented by Demirel's (2004) paper on exergy application in bioenergetics.

Haynie (2001) describes in his book many interesting contemporary problems of biophysics and biothermodynamics, including principles of energy transformation, animal energy consumption, energy conservation in living organisms, carbon, energy and life links, biological aspects of third law, irreversibility and life, chemical couplings, photosynthesis, ATP hydrolysis, membrane transport, enzyme–substrate interaction, molecular pharmacology, hemoglobin, DNA, polymeraze chain reaction, transfer free energy of aminoacids, protein stability and dynamics, nonequilibrium thermodynamics of live, and so forth. There is also a section on statistical thermodynamics which includes comments on helix-coil transition theory. Sections on binding equilibria and reaction kinetics include scientific information about the oxygen transport, electron transfer kinetics, enzyme kinetics, proton binding, protein folding, muscle contraction, and molecular motors. The final chapter defines the frontier of biological thermodynamics including research toward: small systems and molecular motors, formation of the first biological macromolecules, bacteria, information and life, biological complexity, and evolution from the viewpoint of the second law.

Contemporary applications of thermodynamics in biosystems are broad. Thermodynamics may serve as a formal and substantive theory in applications involving the analyses of economic and biological systems (Sousa, 2007). In particular, thermodynamics may help to predict selling prices and develop paraeconomic analyses and balances based on unit exergies as measures of economic prices (Szargut and Petela, 1965; Sommerfeld and Lenk, 1970; Szargut, 2005; Sousa, 2007). Thermodynamics and especially exergy theory may also be applied for the determination of the proecological tax replacing actual personal income taxes (Szargut, 2002b).

Thermodynamics of living systems is an introduction to the laws of thermodynamics and their application to biological and environmental systems (Meyer, 2001). Topics involve zeroth, first, second, and third laws, open and closed systems, enthalpy and specific heat, Gibb's free energy and chemical potential for biological and environmental systems. Applications include biochemical potentials, water potential, adsorption, osmosis, radiation, membranes, surface tension, and fugacity. Also of importance are thermodynamic cycles as they apply to living systems. A researcher or a student who will complete this program will be able to

- Apply the first law to energy flow in biological and environmental systems
- Understand the different utility of energy in its different forms

- Analyze the performance of heat engines and heat production in biological systems
- Calculate changes of sensible and latent heat in living organisms and their environment
- Explain water movement in the soil–plant–atmosphere continuum from Gibb's free energy
- Explain movement of materials influenced by membranes and their surroundings

Biological and environmental applications use the laws of classical thermodynamics, but must go beyond thermostatics (Meyer, 2001). This is not entirely a new concept since chemical engineering thermodynamics and physical chemistry probe into their respective application areas. Applications and energy cycles, involving biological and environmental systems have been addressed in many literature sources over the past 50 years. "Thermodynamics of Living Systems" is a required subject for Biological Systems Engineering. Plants and animals require a continual input of free energy or exergy. If sources of free energy (or exergy) are removed, organisms and other related biological processes drive toward equilibrium or consequent cessation of life. In order to understand biological and environmental processes, one needs to apply the Gibb's free energy function, which is the combination of the first and second laws of thermodynamics. Most classical treatments of thermodynamics are better named "thermostatics." However, in most biological and environmental problems, nonequilibrium and dynamic conditions require an understanding of not only thermal, but also chemical potentials, as well.

Konopka (1984) considers some literature data on the information content of the coding regions in DNA in relation to biological evolution. He questions the suggestion that information content of the coding regions in DNA tends to increase with evolution and, therefore, is a suitable indicator of evolutionary progress. In particular, he shows that we should not use the Shannon entropy function for comparison of genes coding for different polypeptides. We can, however, use this entropy function to compare genes coding for the same polypeptide either in the same genome (a multigene family) or in different species. In his formalism of degenerate coding (Konopka, 1985) the proposed theory is presented in a way enabling suitable application to molecular biology. Two kinds of redundancy of a degenerate code are considered (due to the excess in codon length and because of the code degeneracy).

Experiments prove that up to 50% of the DNA consists of repeats. The biological role of repeats is now recognized as very important issue. DNA sequences of higher organisms contain thousands of nearly identical dispersed repetitive sequences. To understand the effect of repeats on word entropies, Herzel et al (1994) constructed a model that can be analyzed analytically. The hypothetical model sequences consist of independent equi-distributed symbols with randomly interspersed repeats. On this basis Herzel et al (1994) predict that the entropy of DNA sequences measuring the information content is much lower than suggested by earlier empirical studies.

11.2 Control of Biological Reactions and Decaying Enzymes

King and Altman's (1956) approach offers a schematic method of deriving the rate laws for enzyme catalyzed reactions. The method pertains to analysis of steady-state enzyme kinetics of very complicated mechanisms. The King–Altman method has been implemented as an interactive web form. The user can type a reaction mechanism in a symbolic form, using the usual chemical notation (eg, $E + S \Leftrightarrow ES \rightarrow E + P$) and obtain the result in the algebraic form. Grubecki (1997) and Grubecki and Wójcik (2000) analyzed optimal temperature policies for a batch reactor subject to an enzyme deactivation. They have compared isothermal and optimal temperature policies for a batch reactor subject to an enzyme deactivation, adopting, in principle, techniques of deactivation modeling developed earlier in chemical systems with catalyst decay.

A cocurrent tubular reactor with temperature profile control and recycle of moving deactivating enzyme or catalyst has been investigated (Szwast and Sieniutycz, 2001). For the temperature dependent enzyme deactivation, the optimization problem has been formulated in which a maximum of a profit flux is achieved by the best choice of temperature profile and residence time of reactants for the set of catalytic reactions $A + B \rightarrow R$ and $R + B \rightarrow S$ with desired product R, the rates of reactions have been described separately for every reagent by expressions containing (temperature dependent) reaction rate constants, concentrations of reagents, catalyst activity, as well as catalyst concentration in the reacting suspension and a measure of the slip between reagents and solid catalyst particles. The algorithms of maximum principle have been used for optimization. The optimal solutions show that a shape of the optimal temperature profile depends on the mutual relations between activation energies of reactions and catalyst deactivation. It has been proved that the optimal temperature profile is a result of the compromise between the overall production rate of desired reagent R (production rate in the first reaction minus disappearance rate in the second one), necessity of saving of reagents residence time (reactor volume) and necessity of saving catalyst; the most important influence on the optimal temperature profile is associated with necessity of saving the catalyst: When catalyst recycle ratio increases (mean number of catalyst particles residing in reactor increases) optimal temperatures save the catalyst, as the optimal profile is shifted in direction of lower temperatures. The same is observed when catalyst slip increases (catalyst residence time in reactor increases). Despite of variation in catalyst concentration the optimal profile is practically the same because the decay rate is affected only by instantaneous activity of catalyst. When reactor unit volume price decreases, catalyst residence time increases, whereas optimal temperature profile is shifted to lower temperatures. When economic value of unit activity of outlet catalyst increases (catalyst with a residual activity still has an economic value), catalyst saving should be more and more intense. As far as possible, the optimal profile is shifted in direction of lower temperatures whereas the optimal residence time is still the same. Then the optimal profile is isothermal at the level of minimum allowable temperature, whereas the catalyst is saved as its residence time in reactor decreases. The system and effects described are illustrated in Figs. 11.1–11.8.

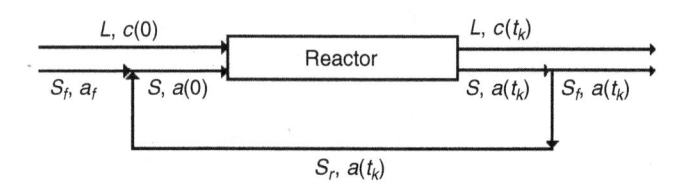

Figure 11.1: Cocurrent tubular reactor with enzyme S_r recycle (Szwast and Sieniutycz, 2001).

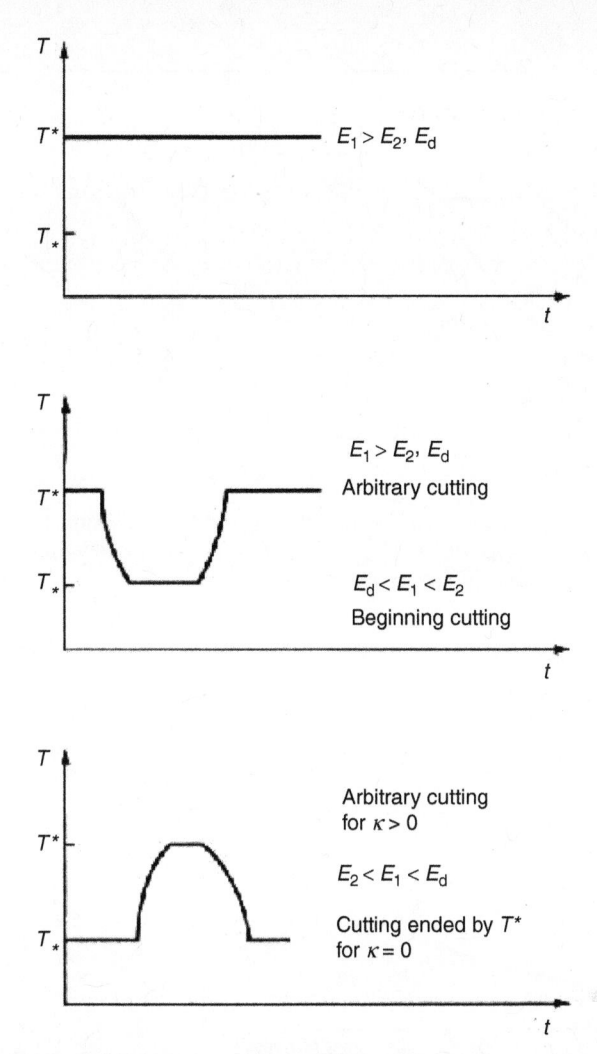

Figure 11.2: Shapes of the optimal temperature profiles for various relations between activation energies of reactions E_1 and E_2, as well as activation energy of enzyme deactivation E_d.

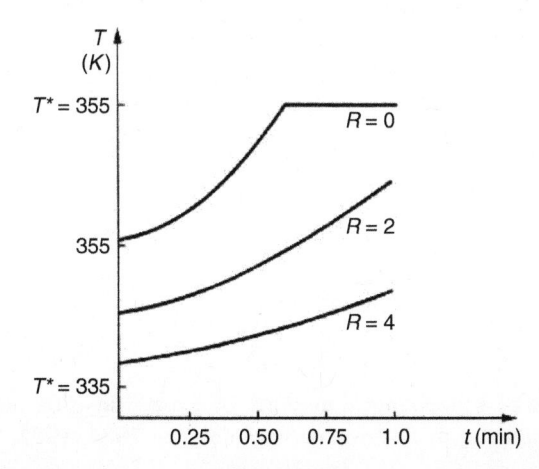

First profile: $E_1 > E_2, E_d$

Second profile: $E_1 > E_2, E_d$ — Arbitrary cutting; $E_d < E_1 < E_2$ — Beginning cutting

Third profile: Arbitrary cutting for $\kappa > 0$; $E_2 < E_1 < E_d$; Cutting ended by T^* for $\kappa = 0$

Figure 11.3: Optimal temperature profiles for various values of catalyst recycle ratio R.

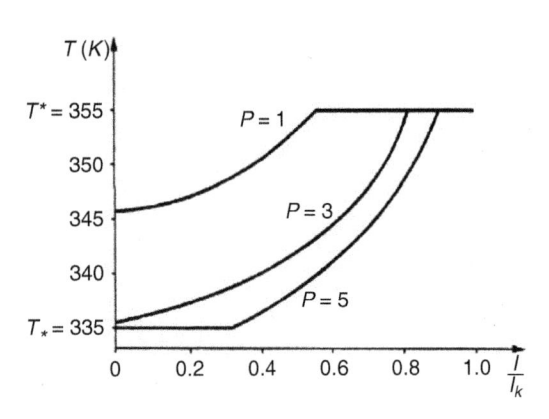

Figure 11.4: Optimal temperature profiles for various values of enzyme slip P.

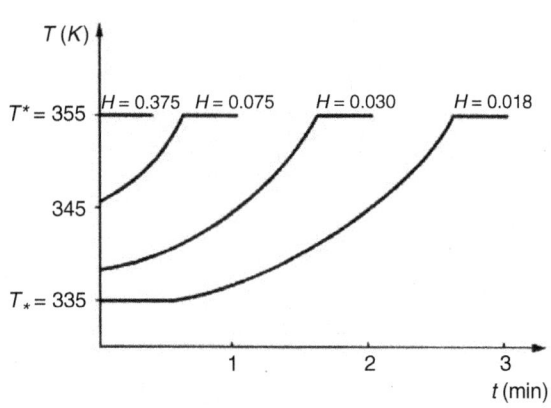

Figure 11.5: Optimal temperature profiles for various values of reactor unit volume price H.

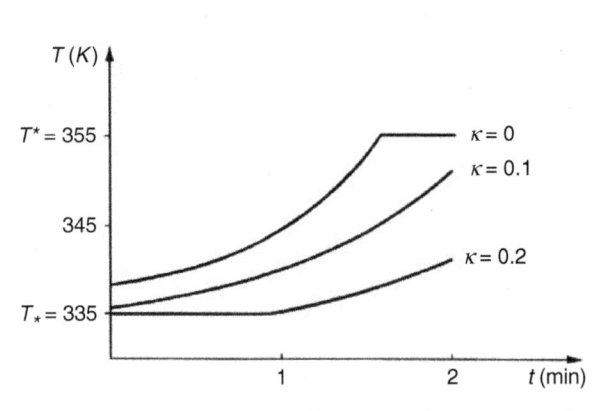

Figure 11.6: Optimal profiles for various economic values of unit activity of outlet catalyst κ.

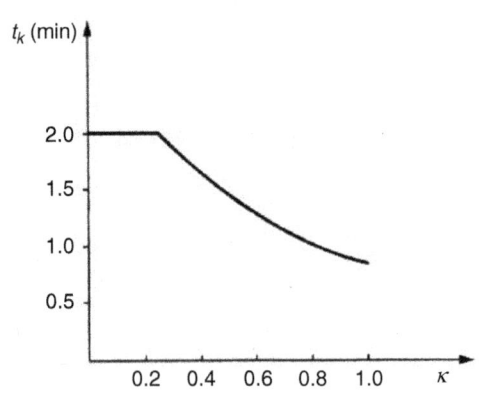

Figure 11.7: Optimal residence time for various economic values of unit activity of outlet enzyme κ.

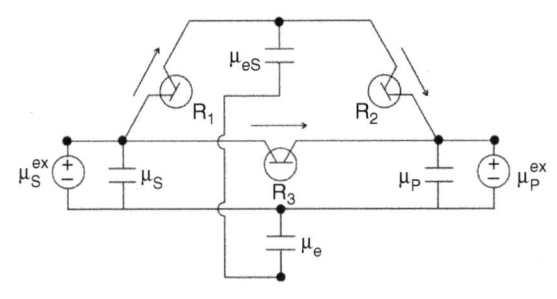

Figure 11.8: An example of a biochemical network of a reaction–diffusion system analogous to a dissipative mechanical system or an electrical network (Shiner, 1984, 1987, 1992; Shiner et al., 1996).

Chemical reactions are central to biological phenomena, including those which also involve nonchemical processes such as the action potential and muscle contraction. Thermodynamic laws act in chemical systems in particularly visible ways. For instance, Johannessen and Kjelstrup (2005) found that chemical reactors with minimum entropy production (minimum waste of energy) had constant entropy production. The reactors operated along a so-called highway. Coupling effects in chemical world lead to many unknown phenomena. With traditional formalisms phenomena with chemical–nonchemical coupling are difficult to handle; chemical reactions are treated with mass action kinetics and nonchemical processes with their appropriate formalisms. The coupling itself is then taken into account in any way possible. However, it has been shown by Shiner and Sieniutycz (1994) how the dynamics of systems of chemical reactions can be formulated in terms of formalisms common to many other sorts of dynamic processes including mechanical and electrical ones. According to Shiner and Sieniutycz (1994) the aspects of these results will be most important for biologists, particularly when systems with chemical–nonchemical coupling, are considered. The authors first develop a formulation of chemical reactions using Lagrangian dynamics extended to handle dissipative processes through a Rayleigh dissipation function. They then present a recipe for obtaining equations of motion of systems of chemical reactions based on a minimization with respect to the rates of the reactions and show that the recipe as well as the Lagrangian formulation can be justified in terms of a minimum principle. It is also shown how network representations of systems of chemical reactions which are both dynamically and thermodynamically correct and complete are equivalent to the other formulations. Other results concerning symmetry in the stationary state and the local equilibrium assumption are also reviewed. These results provide the background for a physically correct treatment of systems with chemical–nonchemical coupling.

Onsager's (1931a, b) theory gives the solution of the problem of mutual connection between the deterministic and the stochastic levels of description in the linear regime near equilibrium. Yet, for chemical reactions linear transport theory provides only a poor approximation (de Groot and Mazur, 1984). The paper by Grabert et al. (1983) proposes an approach to describe fluctuations of reversible chemical reactions in closed systems. The deterministic rate laws are cast into the form of nonlinear Onsager-type closed laws. The nonlinear chemical resistances (conductances) are found in a complete agreement with the experimentally confirmed kinetic law of mass action (Grabert et al., 1983; Shiner, 1987; Sieniutycz, 1987b).

One of the keys to a successful Lagrangian formulation of chemical reaction dynamics was the definition of these chemical resistances. Within a few years of each other several groups (Grabert et al., 1983; Shiner, 1987, 1992; Sieniutycz, 1987b) found the correct definition independently of each other:

$$R_j = \frac{RT \ln\left| \left(k_j^+ \prod_{i=1}^{N} a_i^{v_{ij}^+} \right) \Big/ \left(k_j^- \prod_{i=1}^{N} a_i^{v_{ij}^-} \right) \right|}{k_j^+ \prod_{i=1}^{N} a_i^{v_{ij}^+} - k_j^- \prod_{i=1}^{N} a_i^{v_{ij}^-}} \tag{11.1}$$

Note that the numerator is the affinity of the reaction and the denominator its rate (chapter: Power Limits in Thermochemical Units, Fuel Cells, and Separation Systems) Therefore this definition is consistent with the statement that the force driving a dissipative process = the rate of the process times its resistance. Other simple examples of this statement are electrical potential difference = current through a resistor times the resistance and (mechanical) force = velocity times the friction coefficient. Note that the chemical resistance R_j is a function of the state of the system only; it is independent of the reaction rates.

Chen and Putranto (2013) gave a comprehensive summary of the state-of-the-art and the ideas behind the reaction engineering approach (REA) to controlling food biodeterioration by researchers and industry practitioners. Starting with the formulation, modeling and applications of the lumped-REA, it goes on to detail the use of the REA to describe local evaporation and condensation, and its coupling with equations of conservation of heat and mass transfer, called the spatial-REA, to model nonequilibrium multiphase mass transfer. Some of these ideas may be applied to microbioreactors (Žnidaršič-Plazl and Plazl, 2011). Another topic where this approach might be used is the mucus distribution model in a lung with cystic fibrosis, the suitable subject matter for computational methods in medicine (Zarei et al., 2013).

In the past decade, microreactor technology (MRT) has uncovered new scientific solutions and challenges in a broad range of areas, from electronics, medical technology, fuel production, and processing, to chemical industry, environmental protection, and, last but not least, to biotechnology. According to Žnidaršič-Plazl and Plazl (2011) the objective of microreactor application is the optimization of existing (bio)chemical processes with better selectivity, enhanced heat and mass transfer efficiency, product quality and safety, as well as opening the possibilities for new and environmentally friendly process technologies and new product syntheses. Integration of the biocatalysis, biotransformation, and biodegradation processes with MRT is of a great potential in this field. Microbioreactors provide unique environments, which are excellent tool for bioprocess development and optimization, high-throughput screening and analysis, as well as for bioprocess intensification, better process control, and waste reduction. Similarly, as in conventional reactors, integration of bioprocesses with newly developed product-isolation techniques will play a crucial role in future process development within microstructured devices.

11.3 Biophysics and Bioenergetics

In the study of transport across biological membranes, extensive use is made of radioactive tracers to estimate unidirectional fluxes (Curran et al., 1967). While measurement of the flow of tracer across a membrane can usually be carried out quite accurately, the calculation of unidirectional flux of the unlabeled substance is not necessarily unequivocal. In fact, some researchers have suggested that the general approach of estimating unidirectional fluxes with tracers is incorrect and some others have called attention to possible complications in such measurements. These questions concerning certain interpretations of tracer measurements are based on the concept of interactions between flows and forces that arises within the formal framework of nonequilibrium thermodynamics. The objections raised are, in principle, correct but the possible magnitudes of the effects are not apparent from a general analysis. There are sufficient data available to obtain some insight into these phenomena in a few simple systems, and a more detailed analysis of the tracer problem is warranted. This work is concerned with such an analysis via the formalism of nonequilibrium thermodynamics and with an examination of the limited experimental data available. Supplementary, in Chapter 4 of his book on foundations of biophysics, Stanford (1975) considers various aspects of molecular transport in living systems. His text focuses on the description of random motion, diffusion equations, diffusion across membranes, membrane models, and active transport.

Mikulecky (1977) was one of the inventors of the thermodynamic network method for modeling combined coupled flows in biological systems (usually series–parallel coupled flows). In particular he developed a nonlinear network theory with applications to coupled solute and volume flow in a series membrane. Further developments of thermodynamic networks are due to Peusner (1986, 1990), Shiner (1984, 1987, 1992), and some others. Peusner (1986, 1990) presented a review of network thermodynamics in which the lumped parameter networks are specialized forms of directed graphs in which Kirchhoff's laws hold. The approach, called a topological graph approach to thermodynamics, can incorporate both the equilibrium Gibbsian theory and the theory of engines in a single representation. It yields the theory of thermodynamic transformations and the thermostatic metric, it integrates statistical and kinetic views, and it leads to nonequilibrium extensions. Examples show that a variety of reversible and irreversible problems can be presented in a unified way by using this approach.

Rotteneberg's (1979) book deals with nonequilibrium thermodynamics of energy conversion in bioenergetics. According to this author, recent studies in bioenergetics are mostly concerned with the mechanisms and energetics of membrane processes in which chemical and osmotic free energies are transformed and exchanged. While cellular chemical energy transformations not involving membranous enzymes are now relatively well understood, the study of free-energy transformation by membranous enzymes has proven harder to tackle both conceptually and experimentally. One of the goals of the book is to overcome this difficulty. In the chapter on the nonequilibrium thermodynamics of energy conversion, the focus is on the dissipation of free energy and the phenomenological rate equations, the degree of coupling, relations of stoichiometric ratios and the degree of coupling, the determination of the degree of coupling and the efficiency of the energy conversion. Further chapters deal with the linearity and reciprocity in energy-conserving membranes, respiratory control and the degree of coupling in energy-conserving membranes, and free-energy ratios at static head and stoichiometry (devoted to stoichiometry of the proton pumps and oxidative phosphorylation, and to stoichiometry and coupling in proton-linked transport systems). The next two chapters are on energy pathways in oxidative phosphorylation and on cellular energy levels, energy reserves, and their control. This book may be supplemented by Caplan's and Essig's (1983) volume on bioenergetics and linear nonequilibrium thermodynamics and Rozonoer's (1984) volume on thermodynamics and kinetics of biological processes.

One of the important issues considered by Caplan and Essig (1983) in their book is the thermodynamic definition of the active transport, usually characterized as the "demonstration of transport from a lower to a higher potential," but, as the authors state, "this criterion does not incorporate evident case of active transport in which the electrochemical potential difference may be zero or negative and does not lead naturally to a systematic and comprehensive formulation." Another valuable and constructive notion refers to "degree of coupling" which the authors define by an equation

$$q = \frac{L_{12}}{\sqrt{L_{11}L_{22}}} = -\frac{R_{12}}{\sqrt{R_{11}R_{22}}} \qquad (11.2)$$

(if q is positive, L_{12} is positive, but R_{12} is negative and vice versa).

Analyzing energy conversion in electrochemical processes and, in particular the ATP–synthesis in mitochondria, Forland (1991) pointed out that the widely used concepts, the energy of ions and electrostatic potential difference across a membrane, are nonoperational quantities; they cannot be obtained experimentally. This has created problems when dealing with the concept "liquid junction potential" and with energy conversions in biological systems. One can see how problems of electrochemistry can be dealt with using only well-defined quantities. Forland (1991) stresses that this may change radically our approach to problems in electrochemistry. To describe electrochemical processes he starts afresh from the basic concepts by considering how the electrochemistry developed over the last hundred years and analyzing how it is connected to thermodynamics. He proposes a new electrochemistry based on operational quantities and shows its application for synthesis of ATP in cells. Further reference is given to a book (Forland et al., 1988) and a review (Forland et al., 1992) where a number of corresponding examples can be found.

Westerhoff and Van Dam (1979) discuss principles, approaches, and topics for irreversible thermodynamic description of energy transduction in biomembranes. They consider a number of interesting issues, such as: dissipation function to a vesicular system, fully coupled processes, passive and active transport, oxidative phosphorylation in mitochondria, light-energized systems, transport across plasma membranes, and so forth. Further, using systematically an approach based on nonequilibrium thermodynamics Forland et al. (1992) analyze transport processes in electrolytes and membranes. Its main feature is that thermodynamic components and operationally defined quantities are used consistently. Reviewed are problems with glass electrodes and cells with liquid junctions. Analyzed is transport of two cations and water in a cation exchange membrane. The authors describe also transports combined with chemical reaction in a battery and in muscle contraction. Effects of nonisothermal transports is also considered. Common to all examples is that a better understanding can be achieved and a new information obtained through the application of this systematic approach. Muscle contraction is, in particular, analyzed. When a muscle contracts two kinds of filaments, myosin and actin move relative to each other whereby chemical energy in the form of ATP is consumed. The authors suggest a mechanism for the creation of the force between the myosin and actin filaments and, on this basis develop their own model of muscle contraction. See also Shiner and Sieniutycz (1997) and Haynie (2001).

Up to 1997 no formulation of muscle contraction has provided both the chemical kinetic equations for the reactions responsible for the contraction and the mechanical equation of motion for the muscle (Shiner and Sieniutycz, 1997). This has most likely been due to the lack of general formalisms for nonlinear systems with chemical–nonchemical coupling valid under the far from equilibrium conditions under which muscle operates physiologically. (Shiner and Sieniutycz, 1997) however, have developed such formalisms and apply them here to the formulation of muscle contraction to obtain both the chemical and the mechanical equations. The standard formulation has yielded only the dynamic equations for the chemical variables and has considered these to be functions of both time and a mechanical variable. The macroscopically observable quantities were then obtained by averaging over the mechanical variable. When attempting to derive the dynamics equations for both the chemistry and mechanics this choice of variables leads to conflicting results for the mechanical equation of motion when two different general formalisms are applied. The conflict can be resolved by choosing the variables such that both the chemical variables and the mechanical variables are considered to be functions of time alone. This adds one equation to the set of differential equations to be solved but is actually a simplification of the problem, since these equations are ordinary differential equations, not the partial differential equations of the now standard formulation, and since in this choice of variables the variables themselves are the macroscopic observables the procedure of averaging over the mechanical variable is eliminated. Furthermore, the parameters occurring in the equations at this level of description should be accessible to direct experimental determination.

Ratkje et al. (1992) also analyzed muscle contraction in the context of their lumped description of transport processes in electrolytes and membranes. The description preserves many features of the classical Onsagerian formalism, for example, local equilibrium, bilinearity of entropy production, reciprocity relations, and so forth. On the other hand, classical chemical affinity is not applied in the formalism [see also Lengyel's (1992) work giving up the affinity]. This approach synthesizes thermodynamic problems of physiology, biophysics, and physical chemistry of electrolytes. Muscle contraction is, in particular, analyzed. When a muscle contracts two kinds of filaments, myosin and actin move relative to each other whereby chemical energy in the form of ATP is consumed. The authors suggest a mechanism for the creation of the force between the myosin and actin filaments and, on this basis, develop their own model of muscle contraction. See also Shiner and Sieniutycz (1997) and Haynie (2001).

Osterle (1984) performed thermodynamic analysis of photosynthesis efficiency. Gregory (1989) described in his book the field of photosynthesis, pointing out issues such as: the capture of light energy by living organisms, biomembranes, photosynthetic bacteria, pigments, antenna complexes, reaction center complexes, and so forth. The author discusses also some topics belonging in primary photo-physics, for example: light absorption: formation of excited states of molecules, possible fates of excited states, antenna chlorophylls in chloroplasts, and so forth. A paper by Lems et al. (2010) contains the exergy analyses of the main biochemical processes of photosynthesis in green plants. In the light reactions, photosystem I, photosystem II, and ATP-synthesis are found to operate at exergy efficiencies of 50, 68, and 81%, respectively, so that 48% of the exergy of directly utilized solar photons is converted to biochemical energy carriers. Subsequently, in the dark reactions, glucose is synthesized from CO_2 at an exergy efficiency of 85%. When including the wasteful effect of photorespiration, the overall exergy efficiency of the photosynthesis of glucose is determined at 31–34%, with most of the exergy losses occurring during the initial conversion of the photon energy.

As stated by Demirel and Sandler (2002b) current bioenergetics is concerned with the energy conservation and conversion processes in a living cell, particularly in the inner membrane of the mitochondrion. This review summarizes the role of thermodynamics in understanding the coupling between the chemical reactions and the transport of substances in bioenergetics. Thermodynamics has the advantages of identifying possible pathways, providing a measure of the efficiency of energy conversion, and of the coupling between various processes without requiring a detailed knowledge of the underlying mechanisms. In the past five decades, various new approaches in thermodynamics, nonequilibrium thermodynamics and network thermodynamics have been developed to understand the transport and rate processes in physical and biological systems. For systems not far from equilibrium the theory of linear nonequilibrium thermodynamics is used, while extended nonequilibrium thermodynamics is used for systems far away from equilibrium. All these approaches are based on the irreversible character of flows and forces of an open system. Here, linear nonequilibrium thermodynamics is mostly discussed as it is the most advanced. We also review attempts to incorporate the mechanisms of a process into some formulations of nonequilibrium thermodynamics. The formulation of linear nonequilibrium thermodynamics for facilitated transport and active transport, which represent the key processes of coupled phenomena of transport and chemical reactions, is also presented. The review by Demirel and Sandler (2002b) is aimed at application of nonequilibrium thermodynamics in bioenergetics. Still the reader will have to consult extra literature references for details of the specific applications.

Bioelectrochemistry forms more a part of biophysics than of biochemistry. There is a current need for a work which will provide a background to the range of ideas which fall within its purview. Following the success of their earlier volume *Bioelectrochemistry,* Gutmann and Keyzer (1986) inform the reader of the background and present state of many areas of bioelectrochemistry, and of bridging the intellectual gaps which necessarily occur in such interdisciplinary endeavors. As the authors state in the preface, the interdisciplinary nature of bioelectrochemistry is a root of its fascination but also a cause of many actual and potential controversies. The philosophy of the cell, or organelle, being considered as a "bag of enzymes," which presumes biological processes are on the whole governed by conventional chemical rate equations—whether in the bulk or on enzymatic sites—is assaulted in part at least by the notion that many biological processes are physical rather than chemical in essence. The stunning successes of classical biochemistry make it difficult for many of its ardent practitioners to accept ideas that, for example, in biological energy transduction it is the physics of charge transfer which predominates, or that cellular communication involves radiative processes even if these are, most likely, associated with the microtrabecular lattices. Bioelectrochemistry enters the scene because of the vectorial, directed flow of electric charges and charged particles and because of the critical, nay, determining, role played by electrified surfaces in the chemistry of so many biological reactions. These are electrode processes even if there are no metallic or even classically semiconducting electrodes present. Application of electrochemical ideas to biology presupposes the existence, in the biological system under consideration, of (1) an electron donating and an electron accepting surface or site, (2) an electronically conducting "external" pathway linking the electrodes, and (3) an ionically conducting "internal" pathway supplying the electroactive species to the electrodes. However, the scope of electrochemistry includes the transfer of electrical energy by oscillatory behavior of molecules or molecular assemblies, that is, by alternating current phenomena. In the limit, direct current and alternating current effects can become indistinguishable. This book is concerned with the study of such processes. Another volume, more application oriented, refers to the electrochemistry of waste removal (Bockris et al., 1992). Still other information can be found in thermoeconomics (Corning, 2002).

11.4 Power Yield and Exergy-Valued Biofuels

Berry (1989a) presented a plausible review of methods involving the energy, resting on understanding of its role in the contemporary world.

Berthiaume et al. (2001) described exergetic evaluation of the renewability of a biofuel in the context of a method to quantify the renewability of ethanol produced from corn. The ideal CO_2–glucose–ethanol cycle is considered to show that exergy can be potentially produced through the harnessing of natural thermochemical cycles. Then exergy accounting is used to evaluate the departure from ideal behavior caused by nonrenewable resource consumption through the concept of restoration work. This procedure leads the authors to propose a renewability indicator. The different cycles and processes involved in ethanol production from corn are described. Based on the renewability indicator calculated for the overall process, for the conditions prevailing in Quebec, Canada, ethanol production is seen to be not renewable (a negative value for the renewability indicator is obtained). Nevertheless, the authors believe that the example demonstrates that the renewability indicator is a useful quantitative tool for environmental and technological decision makers, and that their approach could be applied to many other processes.

Biomass has great potential as a clean, renewable feedstock for producing modern energy carriers. Piekarczyk et al. (2013) have performed a thermodynamic evaluation of biomass-to-biofuels production systems. Biomass is a renewable feedstock for producing modern energy carriers. However, its usage is accompanied by possible drawbacks, mainly due to limitation of land and water, and competition with food production. The analysis concerns the so-called second-generation biofuels, like Fischer–Tropsch fuels or Substitute Natural Gas which are produced either from wood or from waste biomass. The most promising conversion case is the one which involves production of syngas from biomass gasification, followed by synthesis of biofuels. The thermodynamic efficiency of biofuels production is analyzed and compared using both the direct exergy analysis and the thermoecological cost.

A paper by Li et al. (2004) presents the results from biomass gasification tests in a pilot-scale (6.5-m tall × 0.1-m diameter) air-blown circulating fluidized bed gasifier, and compares them with model predictions. The operating temperature was maintained in the range of 700–850°C, while the sawdust feed rate varied from 16 to 45 kg/h. Temperature, air ratio, suspension density, fly ash reinjection and steam injection were found to influence the composition and heating value of the product gas. Tar yield from the biomass gasification decreased exponentially with increasing operating temperature for the range studied. A nonstoichiometric equilibrium model based on direct minimization of Gibbs free energy was developed to predict the performance of the gasifier. Experimental evidence indicated that the pilot gasifier deviated from chemical equilibrium due to kinetic limitations. A phenomenological model adapted from the pure equilibrium model, incorporating experimental results regarding unconverted carbon and methane to account for nonequilibrium factors, predicts product gas compositions, heating value, and cold gas efficiency in good agreement with the experimental data.

Ptasinski et al. (2007) focus on exergetic evaluation of biomass gasification, where the synthesis gas may subsequently be used for the production of electricity, fuels, and chemicals. The gasifier is one of the least-efficient unit operations in the whole biomass-to-energy chain and an efficiency analysis of the gasifier alone can substantially contribute to the efficiency improvement of this chain. The research of Ptasinski et al. (2007) is focused to compare different types of biofuels for their gasification efficiency and benchmark this against gasification of coal. Exergy-based efficiencies, defined as the ratio of chemical and physical exergy of the synthesis gas to chemical exergy of a biofuel, are proposed to quantify the value of the gasification process. Biofuels considered by Ptasinski et al. (2007) include various types of wood, vegetable oil, sludge, and manure. Exergetic efficiencies are evaluated for an idealized equilibrium gasifier with ignored heat losses. The gasification efficiencies are evaluated at the carbon-boundary point, where exactly enough air is added to avoid carbon formation and achieve complete gasification. The cold-gas efficiency of biofuels is found to be comparable to that of coal. Further, Ptasinski (2008) reviews the thermodynamic efficiency of biomass gasification and biofuels conversion. Biomass has great potential as a clean renewable feedstock for producing biofuels such as Fischer–Tropsch biodiesel, methanol, and hydrogen. The use of biomass is accompanied by possible ecological drawbacks, however, such as limitation of land or water and competition with food production. For biomass-based systems a key challenge is thus to develop efficient conversion technologies which can also compete with fossil fuels. The development of efficient technologies for biomass gasification and synthesis of biofuels requires a correct use of thermodynamics. Energy systems are traditionally analyzed by energetic analysis based on the first law of thermodynamics. However, this type of analysis shows only the mass and energy flows and does not take into account how the quality of the energy and material streams degrades through the process. In the review by Ptasinski (2008), the exergy analysis, which is based on the second law of thermodynamics, is used to analyze the biomass gasification and conversion of biomass to biofuels. The thermodynamic efficiency of biomass gasification is reviewed for air-blown as well as steam-blown gasifiers. Finally, the overall technological chains for biomass–biofuels are evaluated, including methanol, Fischer–Tropsch hydrocarbons, and hydrogen. The efficiency of biofuels production is compared with that of fossil fuels. Vitasari et al. (2011) have presented an exergy analysis of biomass-to-synthetic natural gas (SNG) production by indirect gasification of various biomass feedstock, including virgin (woody) biomass as well as waste biomass (municipal solid waste and sludge). In indirect gasification heat needed for endothermic gasification reactions is produced by burning char in a separate combustion section of the gasifier and subsequently the heat is transferred to the gasification section. The advantages of indirect gasification are no syngas dilution with nitrogen and no external heat source required. The production process involves several process units, including biomass gasification, syngas cooler, cleaning and compression, methanation reactors, and SNG conditioning. The process was simulated with a computer model using the flow-sheeting program Aspen Plus. The exergy analysis was performed for various operating conditions such as gasifier pressure, methanation pressure, and temperature. In the study by Papadikis et al. (2009) fluid–particle interaction and the impact of shrinkage on pyrolysis of biomass inside a 150 g/h fluidized bed reactor was modeled. Juraščík et al. (2010) presented the results of exergy analysis for a biomass-to-synthetic natural gas (SNG) conversion process. The main process units of biomass-to-SNG conversion technology are gasifier, gas cleaning, synthesis gas compression, CH_4 synthesis, and final SNG conditioning. The study was based on wood gasification, analyzed for different gasification conditions. The results show that the largest exergy losses take place in the biomass gasifier, CH_4 synthesis part, and CO_2 capture unit.

Szewczyk (2002, 2006) derived balances of microbial growth and the related applications in biological fuel cells. Logan (2008) described mechanism of electric current generation in microbial fuel cells. Szewczyk (2006) introduced two categories of BFCs: enzymatic

(EFCs) and microbial (MFCs). The former, due to the immobilization of biocatalysts, offer high current densities and possibilities of the device miniaturization. The latter, because of the presence of living organisms, use complex substrates and can do prolonged work. The catalytic steam reforming of bio-based alcohols, mainly bioethanol, is a new technology based on the environmental compatibility of hydrogen energy when compared with other feedstocks There are numerous BFC units working effectively at the present time (Sieniutycz and Jeżowski, 2013).

Optimization of an energy network may require different methods from those currently used in thermodynamics. Network or system thermodynamics (Peusner, 1986, 1990) can help in economic optimization of energy networks. Thermodynamics, especially as the subject of the thermal networks theory, is currently being extended to combine technical and economic approaches; the extended discipline is referred to as thermoeconomics (chapter: Thermodynamics and Optimization of Practical Processes). Ruth (1993, 1999) considered the integration of economics, ecology and thermodynamics and formulated physical principles in environmental economic analysis. A paper by Kåberger and Månsson (2001) contributes to the discussion on the relation between thermodynamics and economic theory. With respect to thermodynamic constraints on the economy, there are two diametrically opposite positions in this discussion. One claims that the constraints are insignificant ('of no immediate practical importance for modeling') and in the intermediate run, do not limit economic activity and, therefore, need not be incorporated in the economic theory. The other holds that thermodynamics tells us that there are practical limits to materials recycling, which already puts bounds on the economy and, therefore, must be included in the economic models. Using the thermodynamic concept of entropy, Kåberger and Månsson (2001) show that there are fundamental problems with both positions.

Even in the long run, entropy production associated with material dissipation need not be a limiting factor for economic development. Abundant energy resources from solar radiation may be used to recover dissipated elements (Kåberger and Månsson, 2001). With a simple, quantitative analysis it may be shown that the rate of entropy production caused by human economic activities is very small compared to the continuous natural entropy production in the atmosphere and on the Earth's surface (Aoki, 1995, 2001; Mady and Oliveira, 2013). Further, the societal entropy production is well within the range of natural variation. It is possible to replace part of the natural entropy production with societal entropy production by making use of solar energy. Society consumes resources otherwise available for coming generations. However, future generations need not have fewer resources available to them than the present generation. Human industrial activities could be transformed into a sustainable system where the more abundant elements are industrially used and recycled, using solar energy as the driving resource. An economic theory, fit to guide industrial society in that development, must not disregard thermodynamics nor must it overstate the consequences of the laws of thermodynamics (Kåberger and Månsson, 2001).

11.5 Ecology and Ecological Optimization

The thermodynamic ideas contained in the well-known book by Glansdorff and Prigogine (1971) and in the later books along this direction (Nicolis and Prigogine, 1977) diffused quickly to the ecology, and soon received its own characteristic picture. Prigogine is best known for extending the second law of thermodynamics to systems that are far from equilibrium, and implying that new forms of ordered structures could exist under such conditions. Prigogine called these forms "dissipative structures," pointing out that they cannot exist independently of their environment. Nicolis and Prigogine (1977) showed that the formation of dissipative structures allows order to be created from disorder in nonequilibrium systems. These structures have since been used to describe many phenomena of biology and ecology. Lotka (1922) proposed a contribution to the energetics of evolution. Bazykin et al. (1983) report on the evolutionary appearance of dissipative structure in economics. The reader can study the early history and the debate surrounding in numerous references (Nicolis and Portnow, 1973; Gray and Scott, 1975; Nicolis and Prigogine, 1977; Nicolis and Rouvas-Nicolis, 2007a, b). The ecological core is pointed out in many books and papers (Capra, 1996; Jorgensen, 2001; Ulanowicz, 1997; Upadhyay et al., 2000). In particular, calculations by Upadhyay et al. (2000) indicate that structural complexity is not necessary for dynamical complexity to exist. Very simple ecosystems can display dynamical behavior unpredictable in certain situations. In certain cases, when riddled basins are found, even qualitative predictability is denied.

Thermodynamics is used increasingly in ecology to understand integral properties of ecosystems because it is a basic science that describes energy transformation from a holistic viewpoint. In the past three decades, many thermodynamically oriented contributions to the ecosystem theory appeared, therefore an important current step toward integrating these contributions is to present them synthetically (Jorgensen, 2001). An ecosystem consists of interdependent living organisms that are also interdependent with their environment, all of which are involved in a continual transfer of energy and mass within a general state of equilibrium or disequilibrium. Thermodynamics can quantify in an exact way how "organized" or "disorganized" a system is, which is useful information to know when trying to understand how a dynamic ecosystem is behaving. As a part of the environmental and ecological modeling series, Jorgensen's (2001) book on thermodynamics and ecological modeling is a volume of the current thinking on how an ecosystem can be explained or predicted in terms of its thermodynamic behavior. The core part of the book explains how the thermodynamic theory can specifically be applied to the "measurement" of an ecosystem, including the evaluation of its state of entropy and enthalpy. Additionally, it shows economists how to put the theory to use when trying to quantify the movement of goods and services through a type of complex living system—a human society. This book may be supplemented by a volume on concepts and methods in ecological economics (Faber et al., 1996).

Ecological terms should be taken into account in the optimization criteria set for the improvement of the system's performance. This tendency leads to the so-called ecological optimization which involves optimization criteria with ecological components, and often requires maximizing a function representing the best compromise between the power and entropy production rate. Yan (1993) commented on ecological optimization criterion for finite-time heat engines. In particular, these criteria are used in the optimization of irreversible refrigeration cycles and irreversible power generators (engines). For example, J. Chen et al. (2002) proposes an universal irreversible model of the refrigeration cycle which includes the irreversibility of heat transfer, the heat-leak loss between the external heat reservoirs, and the internal irreversibility of the working fluid. This model is used to analyze the performance of a multistage combined refrigeration system. An ecological function is taken as an objective function for optimization; the authors derived a general expression of this objective function. The key performance parameters, such as the performance coefficient, refrigeration rate, and entropy production rate are calculated at the maximum of the ecological function. Some characteristic curves of the refrigeration system are discussed. L. Chen et al. (2005) worked out the generalized ecological optimization for generalized irreversible refrigerators. Szargut and Stanek (2007) developed a thermoecological optimization of a solar collector.

Taking an ecological optimization criterion as the objective L. Chen et al. (2004b) derived the optimal ecological performance of a Newton's-law-generalized irreversible engine with losses due to heat resistance, heat leak, and internal irreversibility. This consists of maximizing a function describing the best compromise between the power and entropy production rate of the heat engine. A numerical example is given to show the effects of heat leakage and internal irreversibility on the optimal performance of the generalized heat engine. The optimization of the ecological function reduces the entropy-generation rate of the cycle, whereas the thermal efficiency increases with a little decrease of the power output. So the ecological function represents not only a compromise between the power output and the entropy-generation rate but also a compromise between the power output and the thermal efficiency. In conclusion, by taking an ecological optimization criterion as the objective, optimal ecological performance can be derived for various irreversible engines with losses caused by heat-resistance, heat leak, and internal irreversibility. Numerical examples show the role of the effects of heat leakage and internal irreversibility on the optimal performance. Taking an ecological optimization criterion as the objective, L. Chen et al. (2010) have derived an optimal policy for the best ecological performance of a chemical engine with diffusive mass transfer. This again consists of maximizing of the best compromise between power and entropy production rate of the engine. Zhu et al. (2003, 2004) worked out the ecological optimization of generalized irreversible engines with modified and generalized heat transfer laws. Szargut and Stanek (2007) performed a thermoecological optimization of a solar collector. The depletion of nonrenewable natural exergy resources (the thermoecological cost) has been accepted as the objective function for thermoecological optimization.

Szargut (1986, 1990) was probably the first who defined the cumulative exergy consumption and cumulative exergy losses, and systematically applied the exergy function for the calculation of ecological costs. Szargut (1990) developed the theory of cumulative exergy consumption (CExC), that expresses the consumption of exergy of natural resources in all links of a technological manufacturing network leading from raw materials to the final product. When analyzing the production of materials and energy carriers, the cumulative degree of thermodynamic perfection (CDP) can be defined as a ratio of exergy to CExC. Szargut (1990) also presents a calculation method for complex processes, including the major product and byproducts. Illustrative values of CDP are cited.

The difference between CExC and exergy expresses the cumulative exergy loss. These losses have been divided into constituent exergy loss (connected with the fabrication of particular semifinished products) and partial exergy loss (connected with particular links of the technological network). Analysis of constituent and partial exergy loss provides information about a possible improvements in the technical network (Szargut, 1990). A set of equations is necessary in the analysis of cumulative exergy loses. A sequence method of analysis is introduced. Many of these issues are summarized in a book on technical and ecological applications of the exergy method (Szargut, 2005). As the supplementary sources earlier books on exergy (Szargut and Petela, 1965 and Szargut et al., 1988) may serve, which provide vast reviews of the theory and applications of the exergy function in chemical engineering, mechanical engineering, metallurgy, and in some other fields. In particular, Sec. 11.4.4 in (Szargut and Petela, 1965) deals with the exergy efficiency of plants vegetation.

An analysis illustrates the heat delivery from a heat-and-power station (Szargut, 1990). Exhaustion of unrestorable domestic natural resources can also be determined by means of CExC, but the influence of the imported raw materials and semifinished products should be taken into account. Utilization efficiency of unrestorable domestic natural resources (UDN) can be greater than 100% if the raw materials do not prevail in the export. Illustrative values of UDN are cited. Ecological cost takes into account not only CExC but also the exergy losses resulting from the deleterious impact of waste byproducts on human activity, on human health, on the productivity of the agriculture and forestry, and on the natural resources.
An ecological economy can be introduced (Szargut, 1990), based upon a postulate of minimization of the consumption of unrestorable natural resources. Szargut (2002a) describes minimization of the depletion of nonrenewable resources by means of the optimization of the design parameters, whereas Szargut (2002b) applies the exergy for the determination of the proecological tax replacing the actual personal taxes. As a supplementary source, a text on exergy analysis in thermal technology may serve (Szargut, 2001).

Exergy as an accessible work potential may be removed from a resource by loss or by transfer to other resources. In all real processes exergy transfer is always accompanied by exergy loss. Although energy transfer generally accompanies exergy transfer, the transfer of exergy may in theory occur with or without net energy transfer. Connelly and Koshland (2001a) consider what they call a central theme of industrial ecology (IE), that is, the idea of using the cyclical resource-use patterns observed in mature, biological ecosystems as a model for designing increasingly mature "industrial ecosystems" whose productivity relies less on resource extraction and waste emission. In their two-part series, they use a thermodynamic interpretation of ecosystem evolution to strengthen this biological–industrial (B–I) ecosystem analogy. They begin by describing limitations in the current analogy and discussing resulting implications for the development and implementation of IE principles. They propose that these limitations arise largely from a poor definition of resource consumption. They then show that defining resource consumption as a process of removing exergy from a resource provides a basis for interpreting ecosystem evolution as a process of allowing consumption to occur with decreasing levels of depletion, that is, a process of "delinking" consumption and depletion. They also use thermodynamic principles to deduce the limitations and interrelation of several strategies for delinking consumption and depletion. Lastly, they explore benefits and limitations of using the proposed interpretation of ecosystem evolution as an analogy for the development of industrial systems.

The thermodynamic interpretation of ecosystem evolution introduced by Connelly and Koshland (2001a) in Part 1 of their work provides a basis for quantitative analysis of strategies for reducing resource depletion. In their Part 2, Connelly and Koshland (2001b) express resource depletion rate as a product of consumption rate and the Depletion number (D_p)—a nondimensional indicator of depletion per unit consumption. They introduce two generalized models of resource use that incorporate a choice between virgin resource extraction and postconsumption resource recycling. These models are used as a basis for expressing D_p as a function of nondimensional indicators for three resource conservation strategies: exergy cycling fraction (ψ) for resource recycling, exergy efficiency (φ) for process efficiency gains, and renewed exergy fraction (Ω) for extent of renewed resource use. Connelly and Koshland (2001b) use the resulting expressions to characterize strategy interaction, strategy limitations and the roles that these strategies play in allowing resource consumption to occur with decreasing levels of depletion. They also show how to imbed this framework into an economic analysis to identify least-cost approaches to depletion avoidance in production and cycling system.

Exergetic resource depletion causes environmental change. Environmental crises such as ozone layer destruction, deforestation, air and water contamination, and global warming are the symptoms of this change (Connelly and Koshland, 2001b). Reducing resource depletion in industrial systems will reduce a major driver of ongoing environmental transformation. Resource depletion may be reduced by reducing consumption and by reducing Depletion number (D_p)—a nondimensional indicator of resource depletion per unit consumption. Biological systems have evolved to allow consumption to occur with little or no depletion. The Depletion number therefore provides one measure of system progress—or "maturity"—on an evolutionary scale (Connelly and Koshland, 2001b). The Depletion number is most useful not as measure of system sustainability, but rather as a basis for studying the evolution of industrial resource use patterns and coordinating the implementation of resource conservation strategies. The Depletion number may be expressed in terms of nondimensional indicators that measure the extent to which specific resource conservation strategies are implemented. Three indicators discussed by Connelly and Koshland (2001b) are:

1. exergy cycling fraction, a measure of recycling that accounts for both the throughput and quality change aspects of resource consumption and upgrading;
2. exergy efficiency, a universal measure of process efficiency that accounts for first and second law principles;
3. renewed exergy fraction, a measure of the extent to which resources supplied to an industrial system are derived from renewed sources.

Expressions for the depletion number serve to derive a variety of relations that provide insight into the limitations and interaction of these three strategies. Increasing exergy cycling may or may not reduce resource depletion, whereas achieving complete cycling of resources is not the only criterion for delinking consumption from depletion. In addition, as a system's renewed exergy fraction approaches one, D_p becomes decreasingly dependent on system's exergy efficiency. Practical limitations to efficiency gains may therefore be overcome by increasing renewed exergy fraction. Reductions in D_p and depletion rate may coincide with gains or reductions in a system's exergy efficiency and renewed exergy fraction. No single strategy can delink consumption from depletion (Connelly and Koshland, 2001b).

Hau and Bakshi (2004) expanded exergy analysis to account for ecosystem products and services. They apply the exergy analysis as a thermodynamic approach suitable for analyzing and improving the efficiency of chemical and thermal processes. Their extension for life cycle assessment and sustainability evaluation of industrial products and processes is particularly fruitful. Although all extensions recognize the importance of capital and labor inputs and environmental impact, most of them ignore the crucial role that ecosystems play in sustaining all industrial activity. Decisions based on approaches that take nature for granted continue to cause significant deterioration in the ability of ecosystems to provide goods and services that are essential for every human activity (Hau and Bakshi, 2004). Accounting for nature's contribution is also important for determining the impact and sustainability of industrial activity. In contrast, emergy analysis, a thermodynamic method from systems ecology, does account for ecosystems, but has encountered a lot of resistance and criticism, particularly from economists, physicists, and engineers. Hau and Bakshi (2004) expand the engineering concept of Cumulative Exergy Consumption (CEC) to include the contribution of ecosystems, which leads to the concept of Ecological Cumulative Exergy Consumption (ECEC). Practical challenges in computing ECEC for industrial processes are identified and a formal algorithm based on network algebra is proposed. ECEC is shown to be closely related to emergy (Odum, 1971, 1988), and both concepts become equivalent if the analysis boundary, allocation method, and approach for combining global energy inputs are identical. This insight permits combination of the best features of emergy and exergy analysis, and shows that most of the controversial aspects of emergy analysis need not hinder its use for including the exergetic contribution of ecosystems. Hau and Bakshi's (2004) examples illustrate the approach and highlight the potential benefits of accounting for nature's contribution to industrial activity.

High demand and consumption rates of ecological materials and services to satisfy societal needs and for the dissipation of emissions are quickly exceeding the capacity that nature can provide. To avoid a tipping point situation, where ecological services may no longer be available, society must propose a sustainable path forward. Ruiz-Mercado et al. (2012) consider sustainability indicators for chemical processes. The chemical industry's response is to incorporate a sustainability approach early into process design to reduce the quantity of goods and services needed and to prevent and minimize releases, while increasing their economic and social benefits. This approach leads to design modifications of existing and new chemical processes, which requires a complete sustainability performance assessment that can support a decision-maker to determine whether a process is becoming more or less sustainable. Hence, the development of indicators capable of assessing process sustainability becomes crucial. In fact, Ruiz-Mercado et al. (2012) present a taxonomic classification and definition of sustainability indicators according to the environmental, efficiency, energy, and economic bases proposed by the greenscope methodology for the evaluation and design of sustainable processes. In addition, Ruiz-Mercado et al. (2012) propose a general scale for measuring sustainability according to the identification and use of best possible target and worst-case scenarios as reference states, as the upper and lower bounds of a sustainability measurement scale. This taxonomy will prove valuable in evaluating chemical process sustainability in the various stages of design and optimization.

11.6 Fractals and Erythrocytes

The research carried out in the complex systems follows a fundamental perspective, stemming largely from the close interaction between statistical mechanics, thermodynamics and nonlinear dynamics. It has become one of the most dynamic and innovative orientations of the above disciplines. Among others, the research on complexity has led to the characterization of complex systems undergoing bifurcations and chaotic dynamics. Therefore, the importance of Mandelbrot's discovery that fractals occur everywhere in nature can hardly be exaggerated. According to Mandelbrot: "Clouds are not spheres, mountains are not cones, coastlines are not circles, and bark is not smooth, nor travel in a straight line" (Aharony and Feder, 1989). In their book on fractals Bak and Chen (1989) focus on physical aspects of fractals. As they conclude, fractals in nature originate from self-organized critical dynamical processes.

L. C. Cerny and E. R. Cerny (1995) have performed a fractal analysis of the erythrocyte sedimentation rate. This work extends their previous approach to the understanding of the erythrocyte sedimentation, that is, a theory of utilizing fractal analysis (Kuo et al., 1994).

In the more recent presentation, considered here, Cerny and Cerny (1995) examine the method and compare the results to several other approaches. To evaluate fractal dimensions in a variety of experimental situations, they study the sedimentation process by using whole blood, washed-resuspended red cells, and hardened erythrocytes. The analysis also treats the sedimentation process in tubes with noncircular cross sections, both convergent and divergent flow directions, the Boycott effect, as well as tubes with novel shapes. Cerny and Cerny (1995) describe a computer simulation of the sedimentation process and determine the fractal dimensions. Finally they compare the computer results with the experimental values.

In biology, forms of organs or even the shape of an organism as a whole may exhibit a self-similar structure, the principal property of an object with fractal geometry. The study of Grasman et al. (2003) deals with the multifractal nature of arterial trees in which branching of vessels repeats itself at a more and more detailed level to minimal vessel diameters of approximately 6 mm in the capillary bed. Branching is generally dichotomous. The strict hierarchical tree is lost in the capillary bed, where a mesh topology is often present. For all physical and biological systems, the repetition is finite and only in the mathematical idealization the limit to infinity can be taken. This mathematical limit procedure may lead to a simple quantification of a structure such as a power law (scaling relationship) or a fractal dimension in the case of a geometrical structure. Starting from a hemodynamic view, it has recently been shown that multifractility is applicable to vascular branching systems. In their study Grasman et al. (2003) consider the arterial tree as a multifractal structure that arises as a result of a cascade process. They analyze the fractal properties of arterial trees using the cascade model of turbulence theory. They show that the branching process leads to a nonuniform structure at the microlevel, meaning that blood supply to the tissue varies in space.

Grasman et al. (2003) conclude that, depending on the branching parameter, vessels of a specific size contribute dominantly to the blood supply of tissue. The related tissue elements form a dense set in the tissue. Furthermore, if blood flow in vessels can get obstructed with some probability, the above set of tissue elements may not be dense anymore. Then there is the risk that, spread out over the tissue, nutrient and gas exchange fall short. In computing the fractal dimension of the set of tissue elements that are supplied with blood from vessels of a specific size, Grasman et al. (2003) concentrated on the fine structure of how blood is distributed over a tissue. If there is no risk of obstruction, tissue of any shape can everywhere be supplied with blood. When there is such a risk, it may occur that no set of blood-carrying tissue elements is densely spread out over the tissue. Then a part of tissue can degenerate as the distance to such blood-carrying tissue may be too large to be bridged by diffusion.

11.7 Animal Locomotion and Pulsating Physiologies

Bejan (1997d) shows that thermodynamic optimization provides a common theoretical basis for the existence of the finely tuned frequencies of pulsating processes of animals. In respiration and circulation, the minimization of mechanical power consumption subject to finite contact area and mass transfer (metabolic) rate constraints also explains why frequencies decrease as the body size (M) increases. For example, Bejan (1997d) shows that the optimal breathing and heartbeat time intervals should increase as $M^{0.24}$, which is in excellent agreement with experimental data. He also shows that the breathing and heartbeat time intervals should be of the same order of magnitude. In ejaculation, the maximization of the mechanical power transmitted to the ejected seminal fluid explains the existence of an optimal bursting time interval. The deterministic method of thermodynamic optimization predicts temporal organization in Nature, and extends thermodynamics to the field of biology. Bejan and Marden (2006) discuss constructing animal locomotion from new thermodynamic theory. For earlier, related treatments see Ahlborn (1999) and Ahlborn and Blake (1999).

Bejan's constructal theory (Bejan, 2000) is capable of explaining flow architectures of the lungs. Reis et al. (2004) explain the reasons why we have a bronchial tree with 23 levels of bifurcation. Based on Bejan's constructal principle they found that the best oxygen access to the alveolar tissues as well as carbon dioxide removal is provided by a flow structure composed of ducts with 23 levels of bifurcation (bronchial tree) that ends with spaces (alveolar sacs) from where oxygen diffuses into the tissues. The model yields the dimensions of the alveolar sac, the total length of the airways, the total alveolar surface area, and the total resistance to oxygen transport in the respiratory tree. A constructal law also emerges: the length defined by the ratio of the square of the airway diameter to its length is constant for all individuals of the same species and is related to the characteristics of the space allocated to the respiratory process and determines univocally the branching level of the respiratory tree.

11.8 Thermostatistics of Helix-Coil Transitions

Helical conformations, such as the a-helix in polypeptides and the double helix in DNA, are common structural elements in biopolymers. As the temperature is raised or the pH is changed to extremes of acidity or alkalinity, the helix becomes disordered into a random coil state. The helix–coil transition has been extensively studied, both experimentally and theoretically, as a model for conformational transitions in biopolymers and as a way to obtain information about the intermolecular forces which stabilize biopolymer structure. Bloomfield (1999) developed three theoretical treatments that describe the helix–coil transition with increasing degrees of detail and rigor: the "all-or-none" model, the zipper model which allows initiation of the helix only once along the polymer chain, and the matrix model which places no restrictions on helix–coil junctions. The matrix model is mathematically similar to the familiar one-dimensional Ising model (Huang, 1967).

Resendis and Garcia-Colin (2001) apply the stochastic process theory to the configuration of biological systems; they extend the stochastic framework to describe the properties of the DNA chain near its melting temperature. To achieve the goal, they focus on the renaturation process; it is conceived as a one-dimensional random walk which depends on time and temperature. The fundamental units in the DNA performing during the process are the base pairs adenine–thymine (A–T) and cytosine–guanine (C–G). The main result of Resendis and Garcia-Colin (2001) is a theoretical framework that allows us to predict an important empirical relation between the melting temperature and the concentration of base pairs C–G for oligonucleotides. They conclude that the random walk model of DNA cannot only be used to analyze correlations in the DNA sequences, but it also allows one to obtain thermodynamic expressions in the renaturation process.

A chapter on statistical thermodynamics in Haynie's (2001) book provides many additional comments on helix-coil transition theory.

11.9 Biochemical Cycles in Living Cells

According to Hill (1977) free energy transduction in cyclical biochemical systems should be attributed, in general, to complete cycles and not to individual steps or transitions belonging to the cycles.

Living cells represent open, nonequilibrium, self-organizing, and dissipative systems maintained with the continuous supply of outside and inside material, energy, and information flows (Demirel, 2010). The energy in the form of adenosine triphosphate is utilized in biochemical cycles, transport processes, protein synthesis, reproduction, and performing other biological work. The processes in molecular and cellular biological systems are stochastic in nature with varying spatial and time scales, and bounded with conservation laws, kinetic laws, and thermodynamic constraints. These constraints should be taken into account by any approach to modeling of biological systems. In the component biology Demirel's (2010) review focuses on the modeling of enzyme kinetics and fluctuation of single biomolecules acting as molecular motors, while in the systems biology it focuses on modeling biochemical cycles and networks in which all the components of a biological system interact functionally over time and space.

Biochemical cycles emerge from collective and functional efforts to devise a cyclic flow of optimal energy degradation rate, which can only be described by nonequilibrium thermodynamics. Therefore Demirel (2010) emphasizes the role of nonequilibrium thermodynamics through the formulations of thermodynamically coupled biochemical cycles, entropy production, fluctuation theorems, bioenergetics, and reaction-diffusion systems. Fluctuation theorems relate the forward and backward dynamical randomness of the trajectories or paths, bridge the microscopic and macroscopic domains, and link the time-reversible and irreversible descriptions of biological systems. However, many of these approaches are in the early stages of their development and no single computational or experimental technique is able to span all the relevant and necessary spatial and temporal scales (Demirel, 2010). Novel experimental and computational techniques with high accuracy, precision, coverage, and efficiency will help for better understanding of biochemical cycles.

11.10 Elucidation of Protein Sequence–Structure Relations

Giuliani et al. (2001) consider a complexity score derived from principal components analysis of nonlinear order measures. They begin with the statement that biological systems are incredibly complex, yet our understanding of that complexity is increasing all the time. Fifty or a hundred years ago, when most biological systems were understood in far less detail, loosely defined mental models of "how things work" were enough to describe all we needed to describe in biology. Since then, "our understanding of biology has skyrocketed," and so has the need for sophisticated mathematical tools to understand and model this expanding "new" universe (Giuliani et al., 2001).

Research in biology includes a range of topics, such as the evolution of drug-resistance in HIV, the population genetics of *Escherichia coli*, blood flow through the arteries, complexity theory in biology, population dynamics, interactions of maturation delay and spatial dispersion, the spread of infectious diseases, and biological invasion (Giuliani et al., 2001). The related institutions introduce a wide range of research areas in mathematical biology. In the research by Giuliani et al. (2001) the generation of a global "complexity" score for numerical series is derived from a principal components analysis of a group of nonlinear measures of experimental as well as a simulated series. The concept of complexity is demonstrated to be independent from other descriptors of ordered series such as the amount of variance, the departure from normality and the relative nonstationarity; and to be mainly related to the number of independent elements (or operations) needed to synthesize the series. The possibility of having a univocal ranking of complexity for diverse series opens the way to a wider application of dynamical systems concepts in empirical sciences.

Giuliani et al. (2002) see the relationship between sequence-embedded information and folding behavior of proteins as the dominant concern of both theoretical and applied biochemical research. In fact, this concern redounds to areas such as: (1) sequence-based functional predictions, (2) 3D structure-based functional predictions, and (3) folding mechanism elucidation. The above issues are key points in basic research, and all of them have immediate applicable spin-offs in biotechnology, where the elucidation of new protein structures is of main interest for areas ranging from pharmaceutical industry to electronics. Moreover, the completion of the sequencing phase of the human genome project shifted the attention of researchers to so-called "structural genomics" where the structural consequences of genome data in terms of protein structure and activity are exploited (Giuliani et al., 2002). This new phase, called "postgenomic," in contrast with the previous one is of direct interest to chemists. In a fundamental 1994 paper entitled "Proteins: Where physics of simplicity and complexity meet," Frauenfelder and Wolynes focused on the peculiarity of the sequence–structure relation and on the need to have microscopic (very accurate) physical principles of "simple" systems (like atoms) cooperatively interacting to produce macroscopic principles qualitatively describing the complex systems of protein architecture. While we do have an accurate knowledge of potentials (hydrophobic interactions, hydrogen bonding, size constraints, etc.) acting at microscopic levels, the "mesoscopic" principles needed to predict the three-dimensional structure of proteins remain essentially unknown. As stated by Giuliani et al. (2002), this blend of microscopic principles and macroscopic consequences has been a typical feature of chemical sciences in the past 150 years. Proteins occupy a unique position in the hierarchy of natural systems, since they lie in a gray region between chemistry and biology. Proteins are large, complicated molecules that any polymer chemist would have difficulty in modeling. From the biological side, although any single protein would not be considered as alive, it does not take many of them (plus a bit of nucleic acid) before life-like behavior begins to emerge. For example, some of the smallest viruses, such as HIV, which might be considered on the borderline of life, are endowed with only 10 different types of proteins. From a chemist's viewpoint, proteins are linear heteropolymers that, unlike most synthetic polymers, are formed of basically nonperiodic sequences of 20 different monomers. While artificial polymers are generally very large extended molecules forming a matrix, the majority of proteins fold as self-contained structures determined by the sequence of monomers. Thus, in the opinion of Giuliani et al. (2002), one can regard the particular linear arrangement of amino acids as a sort of "recipe" for making a water-soluble polymer of a well-defined three-dimensional architecture. It is important to stress this dynamical perspective. "Well-defined three-dimensional structure" should not be intended as "fixed architecture": Many proteins appear as partially or even totally disordered when analyzed with spectroscopic methods. Yet, this apparent disorder corresponds to an efficient organization for protein physiological function. The task of being water soluble while maintaining the structural specificity necessary for a physiologically motivated activity is not easy, and only a relative minority of linear amino acid arrangements can actually accomplish this.

Thus, in opinion of Giuliani et al. (2002), the most basic problem in the sequence-structure puzzle is "What particular linear arrangement of amino acids makes a real protein?" This can be rephrased as "Is it possible to discriminate between amino acid sequences that in water acquire a well-defined three-dimensional structure, and sequences that never will?" While only specific sequences are able to generate functional proteins, there are only weak departures of real protein sequences from random strings. A "random string" should be intended in the information theoretic sense, as a series whose autocorrelation structure remains substantially invariant after random shuffling of the positions of its constituent elements. This obviously has nothing to do with the fact that particular sequence motifs carry a very peculiar "meaning" in both structural and physiological terms. Even written texts that have obvious meaningful motifs corresponding to words appear as random strings of letters. In the case of protein sequences, there is no "external reader" attaching a 3D translation to monodimensional sequences of amino acids; thus, the sequences must "embed" their own code (Giuliani et al., 2002). The possibility of sequentially denaturing/refolding a given protein indicates that the three-dimensional structure is in some way encoded in its amino acid order. Given these premises, the fact that protein sequences are only slightly different from random strings corresponds to the notion that the "code" linking a sequence to a particular structure is not emerging from simple periodicities in the amino acids' occurrence. The presence of regularities in protein's hydrophobicity profiles, taken together with the chemicophysical principles, points to hydrophobicity as the chemicophysical feature of choice for the disambiguation of the sequence-structure code. In analogy with linguistics, the rationale of this choice is the consideration that statistical correlations emerging from an otherwise quasi-random baseline point to important constraints of the studied code (Giuliani et al., 2002).

There is voluminous literature dealing with theoretical models following ab initio approaches to the sequence-structure puzzle. The great majority of the theoretical models adopt a statistical physics perspective based on proteins considered as lattices, that is, squared grids in which each residue is considered as interacting with the same number of neighbors. These interactions are modeled by carefully chosen potential functions, generally based on hydrophobicity or related properties, associating an energetic score to each amino acid pair interaction. The wide popularity of these simplified models stems from the failure of more complex (and in some sense realistic) ones, like those incorporating motion equations. Such complex models, in fact, most often failed in capturing the salient behavior of heteropolymers. For example, a so-called "glassy" behavior has been observed which typically does not occur in proteins at temperatures of interest. As a result, minimal models have been devised, and lattice models are chief among these.

In conclusion, Giuliani et al. (2002) present the general meaning and scope of the application of signal analysis methods to protein sequence-structure relationships. In their opinion, the dominant character of this field concerns the number of the contributing disciplines, both in terms of methodological (the data analysis techniques come directly from engineering, applied mathematics, computational physics) and theoretical influence (the basic assumptions and approaches coming from molecular biology, evolutionary genetics, physical chemistry, biochemistry). Now the frontiers of science are rapidly shifting from the investigation of the basic bricks of matter to the elucidation of mesoscopic principles of organization (Giuliani et al., 2002). The study of proteins is perhaps the most typical "mesoscopic" investigation field with its mixing of basic physical laws, empirical results, and qualitative descriptions (Giuliani et al., 2002). The particular approach described adopts an empirical style typical of medicinal chemistry, letting general principles emerge from the practical solution of local cases.

Signal analysis techniques offer a global and "coarse-grained" view of primary structures as opposed to the local and detailed view given by sequence alignment methods. While detailed sequence alignment strategies are expected to outperform coarse-grained approaches in cases such as homology based threading of particular structural motifs or construction of phylogenetic trees, the global view proposed by Giuliani et al. (2002) could be very important in investigating nonlocal effects of single mutations on protein structures and, in general, in all those cases in which general properties of the protein, not confined to particular motifs and/or substructures, are expected to play a dominant role. The application of a signal analysis perspective to the study of protein sequence-structure-properties relations is still in its infancy, and the obtained results are still preliminary. As stated by Giuliani et al. (2002), besides the utility of the method in solving specific, well-defined problems, what is still uncertain is the relevance of the general knowledge we can derive about protein structure and behavior from the application of time-series analysis methods. The authors stress that we cannot think of proteins as "fully optimized" systems from which it is possible to derive the "inescapable" rules of sequence-structure relations. Simply, these general rules do not exist.

Zhyuan and Thirumalai (1992) studied the folding kinetics of proteins by a model study. Guo et al. (1992) studied the kinetics of folding of a heteropolymer containing a specific sequence of hydrophobic, hydrophilic, and neutral residues. The heteropolymer, representing the alpha carbon sequence of a model protein, folds into a β–barrel structure below a characteristic folding transition temperature. Several measures are introduced to probe the dynamics of approach to a folded state with particular emphasis on the kinetics of formation of nonbonded contacts starting from a high temperature random configuration. The measures of compactness are used to argue that even below the folding temperature, nonbonded contacts are formed at a very slow rate. This result is further corroborated by analyzing the dynamics of relaxation of the nonbonded interactions which are responsible for the formation of folded structures at low temperatures. The authors also show that below the folding transition temperature and for short times, the heteropolymer undergoes large structural fluctuations. Examination of the time dependence of the radius of gyration shows that the heteropolymer is trapped in one minimum (characterized by a roughly constant gyration radius) for arbitrary times followed by sudden excursions to neighboring minima. In this regime, multistate kinetics is appropriate. However, at longer times when the heteropolymer becomes trapped in one of the minima corresponding to the folded conformation, the folding process appears to be an all-or-nothing process. The implication of these results for the kinetics of in vitro folding of proteins are discussed.

Hansmann and Okamoto (1999) developed a method of tackling the protein folding problem by a generalized ensemble approach with Tsallis statistics.

11.11 Complexity, Self-organization, Evolution, and Life

Since an early paper by Pike (1929) on the driving force in organic evolution and a theory of the origin of life, thermodynamics was used relatively seldom in next publications. Only later, Zotin and Zotina (1967) examined theoretical and experimental evidence of the validity of the Prigogine–Weame thermodynamic theory of the development and growth of living organisms. They show that for the process of development and growth a condition of linear relations is achieved between fluxes and forces. A continual reduction in the rate of specific entropy production—measured from the rate of heat production or respiration—takes place during development, growth, and aging. While this is going on, living organisms move toward the final steady state defined as the minimum rate of entropy production. The processes of injury, regeneration, and malignant growth fit the framework of the thermodynamic theory of development very well.

Proceeding from the Prigogine–Wearne theory and an equation of animal growth Zotin and Zotina (1967) derived a new relationship between the rate of specific heat production and the weight of growing animals. They indicate the main periods in the life of organisms when deviation from the steady state occurs, that is, the period of oogenesis, and the period of initial changes during regeneration and malignant growth. During these periods rejuvenation of the cells takes place (in the thermodynamic sense) and this allows them to reaccomplish the processes of development. Nicolis and Prigogine (1977) put these issues in the context of the thermodynamics of self-organization in nonequilibrium systems, and Haken (1984) did it occasionally in the context of synergetics. Ebeling and Volkenstein (1990) stressed the link of entropy and the evolution of biological information. Ebeling and Feistel (1992) developed basic principles of irreversible thermodynamics and kinetics of self-organization and evolution processes. Following Prigogine they denote by the term self-organization the spontaneous creation of order in open, entropy-exporting systems operating beyond a critical distance from equilibrium. Evolutionary processes are defined as (quasi-) infinite, branching chains of self-organization cycles. Typical processes such as self-reproduction, criticality, stochasticity, historicity, with special attention to valuation processes are discussed. Several deterministic and stochastic models for the dynamics of evolutionary processes are analyzed. Monotonic change of different types of values, such as the work value of energy, the informational value of entropy, the functional value of information, the selection value of self-reproduction rate, and other phenotypic properties, are intrinsic features of self-organization and evolution. General mathematical models were developed. Main principles and stages of early evolution and the origin of information processing were discussed.

As stated by Kay (1991), during the past 20 years our understanding of the development of complex systems has changed significantly. Two major advancements are catastrophe theory and nonequilibrium thermodynamics with its associated theory of self-organization. These theories indicate that complex system development is nonlinear, discontinuous (catastrophes), not predictable (bifurcations), and multivalued (multiple developmental pathways). Ecosystem development should be expected to exhibit these characteristics (Kay, 1991). Traditional ecological theory has attempted to describe ecosystem stress response using some simple notions such as stability and resiliency. In fact, stress-response must be characterized by a richer set of concepts. The ability of the system to maintain its current operating point in the face of the stress, must be ascertained. If the system changes operating points, there are several questions to be considered (Kay, 1991): Is the change along the original developmental pathway or a new one? Is the change organizing or disorganizing? Will the system return to its original state? Will the system flip to some new state in a catastrophic way? Is the change acceptable to humans? The integrity of an ecosystem does not reflect a single characteristic of an ecosystem. The concept of integrity must be seen as multidimensional and encompassing a rich set of ecosystem behaviors. A framework of concepts for discussing integrity is presented in Kay's (1991) article.

An inspiring article by Kay and Schneider (1992a) analyzes the link between complexity and thermodynamics and proposes an original thermodynamic paradigm for the development of ecosystems. Ecosystems are viewed as nonequilibrium structures and processes, open to material and energy flows. It is suggested that as ecosystems grow and develop, they should increase their total by developing structures and processes to assist energy degradation. Species which survive in ecosystems are those that funnel energy into their own production and reproduction and contribute to autocatalytic processes which increase the total dissipation of the ecosystem. These studies may allow for the development of measures useful to environmental management and may help the science of ecology with a much-needed theoretical framework. This article proposes a thermodynamic paradigm for the development of ecosystems. Ecosystems are viewed as nonequilibrium structures and processes, open to material and energy flows. It is suggested that as ecosystems grow and develop, they should increase their total dissipation by developing structures and processes to assist energy degradation. Species which survive in ecosystems are those that funnel energy into their own production and reproduction and contribute to autocatalytic processes which increase the total dissipation of the ecosystem. These studies may allow for the development of measures useful to environmental management and may help the science of ecology with a much needed theoretical framework. Further, Kay and Schneider (1992b) consider thermodynamics in the context of measures of ecological integrity. Kay (2000) regards cosystems as self-organizing holarchic open systems. Kay (2002) writes on complexity theory, exergy and industrial ecology, and some related implications for construction ecology.

The valoric interpretation of entropy developed already by Clausius, Helmholtz, and Ostwald is taken by Ebeling (1992) as the basis for a reinterpretation of the various entropy concepts developed by Clausius, Boltzmann, Gibbs, Shannon, and Kolmogorov. As the key points Ebeling (1992) considers the value of energy with respect to work and the value of entropy with respect to information processing. Ebeling (1993) investigates entropy and information in processes of self-organization: uncertainty and predictability. He analyzes on the macroscopic and on the microscopic level the mean uncertainties of probability distributions of discrete sets of events and those of dynamic sequences of events. First one-time distributions are studied. The key point is the specification of the state space; the coarse-grained physical phase space leads to the thermodynamic entropy and the order parameter space to the informational entropy. In a similar way the uncertainty of the state after one step ahead in time, introduced by Shannon, McMillan, and Khinchin, is related to the dynamic entropies studied by Kolmogorov and Sinai and to the gain of information. The nonequilibrium entropies and their relation to properties of attractors are discussed for several examples taken from physical and nonphysical self-organization processes.

Zainetdinov (1999) attributes the informational neg-entropy to the self-organization process in open system, and Klimontovich (1999) considers entropy, information, and relative criteria of order for states of open systems. He states that in the theory of communication two definitions of concept "information" are known. One of them coincides in form with the Boltzmann entropy. The second definition of information is defined as the difference between unconditional and conditional entropies. In the present work the latter is used for the definition of information about states of open systems with various meanings of controlling (governing) parameter. In this work two kinds of open systems are considered. Those systems belong to the first kind that with zero value of controlling parameter are in the equilibrium state. The information about the equilibrium state is equal to zero. During the self-organization in the process of escaping from the equilibrium state the information is increased. For open systems of this class the conservation law for the sum of information and entropy with all values of controlling parameter is proved.

In open systems of the second class the equilibrium is impossible. For them the concept "norm of chaoticity " is introduced. It allows one to consider two kinds of self-organization processes and to give the corresponding definitions of information. Klimontovich's (1999) treatment is carried out on a number (classical and quantum) of examples of physical systems. An example of a medical–biological system is also considered. Ebeling (2002) studies a special class of nonequilibrium systems which allows to develop an ensemble theory of some generality. For these so-called canonical-dissipative systems, the driving terms are determined only by invariants of motion. First he constructs systems which are ergodic on certain surfaces on the phase plane. These systems may be described by a nonequilibrium microcanocical ensemble, corresponding to an equal distribution on the target surface. Next he constructs and solves Fokker–Planck equations. In the last part he discusses a special realization: systems of active Brownian particles.

In conclusion, this work is devoted to the study of canonical dissipative systems which are pumped from external sources with free energy. He started his work from the Hamiltonian theory for conservative mechanical systems. In order to extend the known solutions for conservative systems to dissipative systems he used the general theory of canonical-dissipative systems. Special canonical-dissipative system were constructed which solution converges to the solution of the conservative system with given energy, or other prescribed invariants of motion. In this way he was able to generate nonequilibrium states characterized by certain prescribed invariants of mechanical motion. He postulated special distributions which are analogues of the microcanonical ensemble in equilibrium. Further he found solutions of the Fokker–Planck equation which may be considered as analogues of the canonical equilibrium ensembles. He proposed calling these distributions canonical-dissipative ensembles. By the help of the explicit nonequilibrium solutions he was able to construct for canonical-dissipative systems a complete statistical thermodynamics including an evolution criterion for the nonequilibrium. As one of the most important practical applications of a theory of canonical-dissipative type he considered problems of active Brownian motion. This class on systems is of interest for modeling biological motion and traffic. Active Brownian particles are particle-like objects with the ability to take up energy from the environment, Simple models of active Brownian particles were studied in many earlier works. These models have been used to simulate a broad variety of pattern formations in complex systems, ranging from physical to biological and social systems. At the end of the study the author states, that canonical-dissipative systems are a rather artificial class of models (Ebeling, 2002). In reality, canonical-dissipative systems, strictly according to such definition, do not exist. However there are many complex systems and among them systems of central importance which have several properties in common with canonical-dissipative systems as the mechanical character of the motion, that is, the existence of space and momentum, of quasi-Newtonian dynamics, and so forth, the support of the dynamics with free energy from internal or external reservoirs and the existence of mechanisms which convert the reservoir energy into acceleration.

Ebeling and Feistel's (2011) updated book on self-organization and evolution includes a review of more recent literature. It retains the original fascination surrounding thermodynamic systems far from equilibrium, synergetics, and the origin of life. While focusing on the physical and theoretical modeling of natural selection and evolution processes the book covers in detail experimental and theoretical fundamentals of self-organizing systems as well as such selected features as random processes, structural networks and multistable systems. The authors take examples from physics, chemistry, biology, and social systems. The result is a resource relevant for students and scientists in physics or related interdisciplinary fields, including mathematical physics, biophysics, information science and nanotechnology.

The second law of thermodynamics can be investigated with respect to its value as an indicator of the direction of the evolutionary process (Berry, 1995). Nonthermodynamic entropy concepts serve him to set the link between the entropy, irreversibility, and evolution. Possible errors in the use of thermodynamic entropy are then discussed. The importance of genuine thermodynamic potentials and their correct application for understanding processes is emphasized. There is no direct connection between evolutionary events such as speciation and thermodynamic entropy change; the irreversibility of evolution is not a consequence of thermodynamic irreversibility. In conclusion, there are two entirely different ways of bringing together thermodynamics and evolution, and these can be described as bottom–up versus top–down strategies. The top-down approach refers to evolution as a whole, which is regarded as a process that can be understood according to the laws of thermodynamics. The biological details of this global process are therefore seen as contingencies within a general thermodynamic scheme. The bottom–up approach is confined to individual organisms (and smaller subsystems like tissues or cell compartments) and tries to analyze the thermodynamic aspects of their functioning. These two strategies do not simply differ in scope or scale, they also represent different positions with respect to the relation between biological and physical theories and the question of reductionism. Thermodynamics merely serves as a discipline which, like many others as genetics or biomechanics, can make contributions for solving certain problems arising in the study of evolution. Biological evolution has characteristics quite different from those required for a rigorous treatment as a thermodynamic process.

The science of thermodynamics is concerned with understanding the properties of inanimate matter in so far as they are determined by changes in temperature. According to Demetrius (1987, 2000), an extremal principle governs the macromolecular evolution, and the Second Law asserts that in irreversible processes there is a unidirectional increase in thermodynamic entropy, a measure of the degree of uncertainty in the thermal energy state of a randomly chosen particle in the aggregate. The science of evolution is concerned with understanding the properties of populations of living matter in so far as they are regulated by changes in generation time. Directionality theory, a mathematical model of the evolutionary process, establishes that in populations subject to bounded growth constraints, there is a unidirectional increase in evolutionary entropy, a measure of the degree of uncertainty in the age of the immediate ancestor of a randomly chosen newborn. This article reviews the mathematical basis of directionality theory and analyzes the relation between directionality theory and statistical thermodynamics. Demetrius (2000) exploits an analytic relation between temperature, and generation time, to show that the directionality principle for evolutionary entropy is a nonequilibrium extension of the principle of a unidirectional increase of thermodynamic entropy. The analytic relation between these directionality principles is consistent with the hypothesis of the equivalence of fundamental laws as one moves up the hierarchy, from a molecular ensemble where the thermodynamic laws apply, to a population of replicating entities (molecules, cells, higher organisms), where evolutionary principles prevail.

Living systems imply self-reproducing constructs capable of Darwinian evolution. How such dynamics can arise from undirected interactions between simple monomeric objects remains an open question. Breivik (2001) circumvent difficulties related to the manipulation of chemical interactions, and presents a system of ferromagnetic objects that self-organize into template-replicating polymers due to environmental fluctuations in temperature. Initially random sequences of monomers direct the formation of complementary sequences, and structural information is inherited from one structure to another. Selective replication of sequences occurs in dynamic interaction with the environment, and the system demonstrates the fundamental link between thermodynamics, information theory, and life science. Corning and Kline (2000) consider thermodynamics, information, and life in the context of thermoeconomics and control information.

Pierce (2002) regards nonequilibrium thermodynamics as a tool for an alternate evolutionary hypothesis. Evolution is seen as an axiomatic consequence of organismic information obeying the second law of thermodynamics and is only secondarily connected to natural selection. As entropy increases, the information within a biological system becomes more complex or variable. This informational complexity is shaped or organized through historical, developmental, and environmental (natural selection) constraints. Biological organisms diversify or speciate at bifurcation points as the information within the system becomes too complex and disorganized. These speciation events are entirely stimulated by intrinsic informational disorganization and shaped by the extrinsic environment (in terms of natural selection). Essentially, the entropic drive to randomness underlies the phenomenon of both variation and speciation and is therefore, the ultimate cause of evolution.

Although evolution through natural selection has been vastly accepted in the scientific community as a fundamental law of biology, it has been criticized for being an incomplete interpretation of evolutionary processes. Natural selection cannot account for (1) the irreversibility of evolution, (2) the complexity-generating or anamorphic tendency of biological systems, and (3) the self-organizing behavior exhibited by all biological life forms (Wicken, 1979). Consequently, in an attempt to explain these three evolutionary trends, various scientists have strived to make a connection between physical laws and biological systems (Pike, 1929).

To elucidate the existence of complex, organized systems and the governing rules, an alternative view of biological evolution has been proposed. This unique, but controversial, evolutionary scenario suggests that the "driving force" behind biological evolution originates in the physical principle of thermodynamics and that natural selection is a microscopic process only secondarily connected to the natural selection. More specifically, proponents of a thermodynamic evolutionary theory consider the ultimate, macroscopic cause of biological evolution to be an axiomatic consequence of organismic "information" obeying the second law of thermodynamics in a nonequilibrium fashion (Wicken, 1979; Wiley and Brooks, 1988). Pierce (2002) provides an integration and summary of the literature on nonequilibrium thermodynamics and its relationship to evolutionary processes.

The second law of thermodynamics asserts that if a spontaneous process occurs, the isolated system moves toward a state of equilibrium and in the process becomes increasingly random or disordered. But, why do spontaneous processes always move the isolated system toward equilibrium? To solve the equilibrium paradigm of the second law, Boltzmann contrived a simple, yet ingenious, thought experiment.

Suppose you have a box, bisected down the center by an imaginary line, and eight distinguishable molecules; see Fig. 11.9. How many ways are there to arrange the molecules on the left and right side of the partition? Fig. 11.9 from Capra (1996) and Pierce (2002) illustrates Boltzmann's thought experiment demonstrating the probability of disorder. First, all the molecules can be arranged on one side of the box in only one way (highly organized). However, there are eight possible different arrangements if seven molecules are placed on the left side of the box and one molecule is placed on the right.

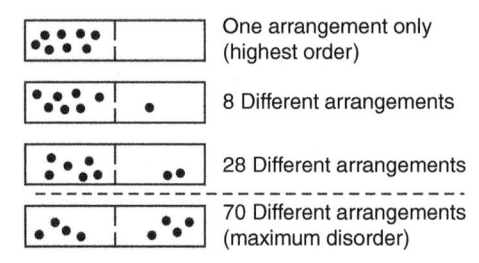

Figure 11.9: Boltzmann's thought experiment demonstrating the probability of disorder (Capra, 1996, p. 187; Pierce, 2002).

The example considered illustrates how the number of possible arrangements increases as the differences between the left and right sides of the box becomes smaller, until eventually, the left and right sides equalize (equilibrium) and the number of possible arrangements reaches seventy (highly disorganized) (Capra, 1996; Pierce, 2002). In other words, the entropy of an isolated system is always increasing because it is more probable that a system will be disordered than ordered. Boltzmann referred to all the possible arrangements of a system as "complexions" and equated them with the concept of order (Capra, 1996). Basically, the more complexions within a system, the further that system is from being ordered. But, this connection between increasing disorder and complexity contradicts the essence of biological organisms; the living world is characterized by increasing order and complexity. As a result, biological systems must be functioning at a state far from equilibrium. This observation has lead to the development of the theory of open systems and their nonequilibrium thermodynamics (Glansdorff and Prigogine, 1971; Nicolis and Prigogine, 1977; Turner, 1979; Wicken, 1979; Brooks and Wiley, 1984; Schneider and Kay, 1994a, b, and many others).

Pierce (2002) asks: How are biological organisms able to self-organize and maintain their life processes far from equilibrium? The answer to this essential question is found in the theory of "dissipative structures" (Capra, 1996). Dissipative structures are open systems, they need a continual input of free energy (or exergy in nonisothermal case) from the environment in order to maintain the capacity to do "work." It is this continual flux of energy, into and out of a dissipative structure, which leads toward self-organization and ultimately the ability to function at a state of nonequilibrium (Turner, 1979). A well-known example of a self-organizing, dissipative structure is the spontaneous organization of water due to convection. Once convection begins, the dissipative structure forms a pattern of hexagonal Bénard cells; see Fig. 5.9. As soon as the energy source (heat) is taken away, the ordered pattern disappears and the water returns to an equilibrium state. Just like convection of water, biological organisms are self-organizing dissipative structures—they take in and give off energy from the environment in order to sustain life processes and in doing so function at a state of nonequilibrium. Although biological organisms maintain a state far from equilibrium they are still controlled by the second law of thermodynamics. Like all physiochemical systems, biological systems are always generating their entropy or complexity due to the overwhelming driver toward equilibrium. But, unlike physiochemical systems, biological systems possess "information" that permits them to self-replicate and continuously amplify their complexity and organization through time (Brooks and Wiley, 1986; Wiley and Brooks, 1988; Pierce, 2002).

In conclusion, nonequilibrium thermodynamics makes sense in the evolutionary context. Proponents of nonequilibrium thermodynamics see the natural selection as a secondary cause of evolution. In their words, natural selection only tells us "how" evolution occurs, not "why" it occurs (Wicken, 1980; Pierce, 2002). To understand "why" evolution takes place, nonequilibrium theorists have tried to reduce evolution to a general physical law: the second law of thermodynamics. But, does this alternate scenario of evolution make sense? Biological systems exist as an intricate part of the universe; because of this, it seems reasonable to conclude that evolution, like all natural processes, rests ultimately on physical laws. All physical laws, with one exception, fail to differentiate between the two directions of time. The one exception is the second law of thermodynamics, which asserts that all physical processes generate entropy. Therefore, it seems logical that evolution, which more than any other natural phenomenon distinguishes between the direction of the past and the direction of the future, could ultimately derive its "arrow" from the second law.

Making this reasonable connection of asymmetry between biological processes and the second law of thermodynamics permits the nonequilibrium hypothesis to attach the concept of entropy to evolution (Pierce, 2002). More specifically, we are able to analyze the addition of information into biological systems as an entropic phenomenon. Consequently, nonequilibrium thermodynamics is able to provide us with the rationality to explain why biological systems become more complex and organized through time. Yet, two significant characteristics of evolution natural selection fails to account for. Darwin recognized that it was the generation of complexity or variation in biological systems that permitted evolution to transpire (Darwin, 1859). In other words, entropy is what produces the variability or complexity that natural selection "selects." In this sense, natural selection can be viewed, not as the fundamental law of biology, but as an auxiliary law that acts to constrain (organize) the possible variants produced by the second law of thermodynamics (Pierce, 2002).

Schneider and Kay (1994a) propose a thermodynamic paradigm for the development of ecosystems. Ecosystems are viewed as nonequilibrium structures and processes, open to material and energy flows. As ecosystems grow and develop, they increase their total dissipation by developing structures and processes to assist energy degradation. Species which survive in ecosystems are those that funnel energy into their own production and reproduction, and contribute to autocatalytic processes which increase the total dissipation of the ecosystem. Current studies in ecology may allow for the development of measures useful to environmental management and help with a much-needed theoretical framework. Schneider and Kay (1994b) treat life as manifestation of the second law of thermodynamics.

They examine the thermodynamic evolution of various evolving systems, from primitive physical systems to complex living systems, and conclude that they involve similar processes which are phenomenological manifestations of the second law of thermodynamics. They take the reformulated second law of thermodynamics of Hatsopoulos and Keenan and Kestin and extend it to nonequilibrium regions, where nonequilibrium is described in terms of gradients maintaining systems at some distance away from equilibrium.

The reformulated second law suggests that as systems are moved away from equilibrium they will take advantage of all available means to resist externally applied gradients. When highly ordered complex systems emerge, they develop and grow at the expense of increasing the disorder at higher levels in the system's hierarchy. Schneider and Kay (1994b) claim that this behavior appears universally in physical and chemical systems. They present a paradigm which provides for a thermodynamically consistent explanation of why there is life, the origin of life, biological growth, the development of ecosystems, and patterns of biological evolution observed in the fossil record.

Schneider and Kay (1994b) illustrate the use of this paradigm through a discussion of ecosystem development. The authors argue that as ecosystems grow and develop, they should increase their total dissipation, develop more complex structures with more energy flow, increase their cycling activity, develop greater diversity and generate more hierarchical levels, all to abet energy degradation. Species which survive in ecosystems are those that funnel exergy into their own production and reproduction and contribute to autocatalytic processes which increase the total dissipation of the ecosystem. In short, ecosystems develop ways to systematically increase their ability to degrade the incoming solar energy and exergy. Schneider and Kay (1994b) believe that their thermodynamic paradigm makes it possible for the study of ecosystems to be developed from a descriptive science to predictive science founded on the most basic principle of physics.

It is interesting to learn what these authors write about the origin of life. The origin of prebiotic life is the development of another route for the dissipation of induced energy gradients (Schneider and Kay, 1994b). Life with its requisite ability to reproduce, insures that these dissipative pathways continue, and it has evolved strategies to maintain these dissipative structures in the face of a fluctuating physical environment. Schneider and Kay (1994b) state that living systems are dynamic dissipative systems with encoded memories, the gene with its DNA, that allow the dissipative processes to continue without having to restart the dissipative process via stochastic events. Living systems are sophisticated mini-tornados, with a memory (its DNA), whose Aristotelian "final cause" may be the second law of thermodynamics. However, one should be clear not to overstate the role of thermodynamics in living processes. Their restated second law is a necessary but not a sufficient condition for life. This paradigm presents no contradiction with the neo-Darwinian hypothesis and the importance of the genetic process, and its role in biological evolution (Schneider and Kay, 1994b). They reject

Dawkin's selfish gene (Dawkins, 1976) as the only process in selection and would insert the gene or species as a component in competing autocatalytic ecosystems, competing to degrade available energy gradients.

The origin of life should not be seen as an isolated event. Rather it represents the emergence of yet another class of processes whose goal is the dissipation of thermodynamic gradients (Schneider and Kay, 1994b). Life should be viewed as the most sophisticated (until now) end in the continuum of development of natural dissipative structures from physical to chemical to autocatalytic to living systems. Autocatalytic chemical reactions are the backbone of chemical dissipative systems (Schneider and Kay, 1994b).

The work of Eigen (1971) and Eigen and Schuster (1979) connect autocatalytic and self-reproductive macromolecular species with a thermodynamic vision of the origin of life. According to Ishida (1981), the main conclusion of Eigen (1971) asserts that the first steps of biological self-organization occurred in a structureless soup, certainly involving functional macromolecular structures such as nucleic acids and proteins. In accordance with this assertion, he has considered a chemical system composed of a wide variety of self-reproductive macromolecules and many different energy-rich monomers required to synthesize the macromolecules. Each macromolecular and monomeric species are distributed uniformly throughout the available volume. The system is open and maintained in a disequilibrium state by continuous flows of energy-rich and energy-deficient monomers. It is thus extremely complex. In addition, metabolism, self-reproduction, and mutability are necessary for each macromolecular species to yield the behavior of Darwinian selection at the molecular level. For each macromolecular species, Eigen has proposed a simple phenomenological rate equation materializing these three necessary prerequisites. This rate equation, which we shall now call Eigen's equation, is interpreted in terms of the terminology of chemical kinetics (Ishida, 1981). Eigen (1971) himself says: "Evolution appears to be an inevitable event, given the presence of certain matter with specified autocatalytic properties and under the maintenance of the finite (free) energy flow necessary to compensate for the steady production of entropy."

From the point of view of nonequilibrium thermodynamics of chemical reactions in a homogeneous system, Eigen has shown that the Darwinian selection of such macromolecular species can be linked to Glansdorff and Prigogine's stability criterion (Schneider and Kay, 1994b). By examining the entropy changes and entropy production of such systems Ishida (1981) has formulated a nonequilibrium thermodynamics of the hypercycles. This theory is used to discuss the stability of the "quasi-species" (a solution to Eigen's equations with nonvanishing mutation terms). In principle, it is possible, using nonequilibrium thermodynamics of open systems, to put forward scenarios for the development of biochemical machinery for the duplication and translation of nucleic acids and other macromolecules essential to life (Schneider and Kay, 1994b).

Recognition that biological systems are stabilized far from equilibrium by self-organizing, informed, autocatalytic cycles and structures that dissipate unusable energy and matter has led to recent attempts to reformulate evolutionary theory. Considering an ecological approach to evolution in the thermodynamic perspective, Weber et al. (1989) hold that such insights are consistent with the broad development of the Darwinian tradition and with the concept of natural selection. Biological systems are selected that are not only more efficient than competitors but also enhance the integrity of the web of energetic relations in which they are embedded. But the expansion of the informational phase space, upon which selection acts, is also guaranteed by the properties of open informational-energetic systems. This provides a directionality and irreversibility to evolutionary processes that are not reflected in current theory (Weber et al., 1989).

For this thermodynamically based program to progress, Weber et al. (1989) believe that biological information should not be treated in isolation from energy flows, and that the ecological perspective must be given descriptive and explanatory primacy. Levels of the ecological hierarchy are relational parts of ecological systems in which there are stable, informed patterns of energy flow and entropic dissipation. Isomorphies between developmental patterns and ecological succession are revealing because they suggest that much of the encoded metabolic information in biological systems is internalized ecological information. This thermodynamic approach to evolution frees evolutionary theory from dependence on a crypto-Newtonian language more appropriate to closed equilibrium systems than to biological systems (Weber et al., 1989). The thermodynamics of complexity in biology leads to creation of order from disorder (Schrödinger, 1967; Schneider and Kay, 1995).

During the past 30 years, a broad spectrum of theories and methods have been developed in physics, chemistry and molecular biology to explain structure formation in complex systems. These methods have been applied to many different fields such as economics, sociology, and town planning. Schweitzer's (1997) book on self-organization of complex structures reflects the interdisciplinary nature of complexity and self-organization. The book presents a wide-ranging overview from fundamental aspects of the evolution of complexity, to applications in biology, ecology, sociology, economics, and urban structure formation. The main focus is on the emergence of collective phenomena from individual or microscopic interactions. In Chapter 2 of Schweitzer's (1997) book, Shiner (1997) considers the issues of self-organization, entropy and order in growing systems. The developed formalism is based on Landsberg's definition of order (Landsberg, 1984a, b) who replaced the earlier definition of Schrödinger (1967) as nonapplicable for growing systems.

To overcome the deficits of entropy as a measure for disorder when the number of states available to a system can change, Landsberg (1984a, b) defined "disorder" as the entropy normalized to the maximum entropy. In the simplest case, the maximum entropy is that of the equiprobable distribution, corresponding to a completely random system. However, depending on the question being asked and on system constraints, this absolute maximum entropy may not be the proper maximum entropy. To assess the effects of interactions on the "disorder" of a one-dimensional spin system, the correct maximum entropy is that of the paramagnet (no interactions) with the same net magnetization. For a nonequilibrium system the proper maximum entropy may be that of the corresponding equilibrium system, and for hierarchical structures, an appropriate maximum entropy for a given level of the hierarchy is that of the system which is maximally random, subject to constraints derived from the next lower level. Considerations of these examples lead Shiner and Davison (2003) to introduce the equivalent random system: "that system which is maximally random consistent with any constraints and with the question being asked. It is the entropy of the equivalent random system which should be taken as the maximum entropy in Landsberg's disorder."

Speciation is an evolutionary development of a new species, usually through the division of a single species into two separate genetically different groups. For living organisms regarded as multiorgan and multilimb systems, speciation can be treated by a complexity criterion, usually originating from the classical statistical entropy of Shannon–Boltzmann. In order to penetrate a vaster spectrum of system stability (instability) properties, Tsallis entropy S may also be used. In speciation analyses classical thermodynamic quantities do not appear, yet the model used satisfies an extremum principle that implies a maximum of entropy subject to the geometric constraint of a given thermodynamic distance. Szwast (1997) and Sieniutycz (2000g, 2004c, 2006a, 2007c) considered extremum properties of entropy to determine dynamics of growth and evolution in complex systems, and discussed the thermodynamic and variational aspects in modeling and simulation of biological evolution. Szwast et al. (2002) treated evolutionary biological processes with a variable number of states and a related extremum principle which provides a quantitative picture of evolution.

Fig. 11.10 illustrates a maximum entropy problem in the biological evolution. An optimization problem for a biological system with n states is that of maximum change of

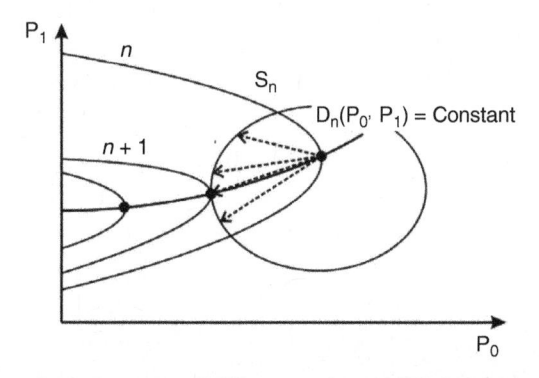

Figure 11.10: An original optimization problem for a system with n states is that of maximum change of entropy between a point and the surface of constant distance D from the point.

entropy (or a related complexity) between a point and the surface of the constant distance D from the point. A variational problem in Fig. 11.10 searches for the maximum entropy subject to geometric constraint of the constant thermodynamic distance in the non-Euclidean space of independent probabilities p_i plus possibly other constraints. Tensor form of dynamics can be obtained. Some developmental processes progress in a relatively undisturbed way, whereas others may terminate rapidly due to instabilities. This extremum principle provides a picture of evolution for processes with variable number of states. The results show that a discrete gradient dynamics governed by the entropy potential follows from the variational principles for shortest paths and suitable transversality conditions.

For an accepted complexity function, some evolutions may progress in a relatively undisturbed manner, whereas others may terminate in a rapid way due to inherent instabilities. These extremum principles may be integrated with the theory of dissipative and chaotic behavior of biochemical systems. Entropy-based models, quantifying various biological phenomena, for example, those with an increase and reduction of organs or limbs in multiorgan organisms, can involve nonclassical statistical entropies, for example, q-entropies of Tsallis or Renyi, that may modify magnitudes of unstable regions in the space of process probabilities. One can develop computer-aided simulation of evolution in systems with living organisms as an effect of extremum properties of classical statistical entropy of Gibbs–Boltzmann type or its associates, for example, Tsallis q-entropy (Tsallis, 1999a, b). Evolution of animals with multiple organs or limbs may be considered, and the substantiation for the increase of organ/limb number can be worked out (Szwast, 1997; Szwast et al., 2002). More information can be found in Szwast et al. (2002) and Sieniutycz and Jeżowski (2013).

Considering living systems (Sieniutycz, 2000i) discusses energy limits associated with the complexity and information-theoretic models, in particular those for sequentially evolving states. The information concept is not only appropriate to complex systems but is also quantitatively well defined (Shiner et al., 1999). Diverse models serve to evaluate energy limits quantitatively. In living systems nonequilibrium entropy has to be applied (Landau and Lifshitz, 1975). Living systems have developed various strategies to manipulate their self-organization in order to satisfy the principle of minimum complexity increase, ultimately, however, the physical laws set limits to their size, functioning, and rate of development. For example, the physical law of thermal conduction sets the size of warm-blooded aquatic animals that require a minimum diameter (of ca. 15 cm) in order to survive in cold oceans (Ahlborn, 1999; Ahlborn and Blake, 1999). Species that survive in ecosystems are those that funnel energy into their own production and reproduction and contribute to autocatalytic processes in the ecosystem. Also, there are data that show that poorly developed ecosystems degrade the incoming solar energy less effectively than more mature ecosystems (Schneider and Kay, 1994a, b, 1995). The cornerstone is to view living systems as stable structures increasing the degradation of the incoming solar energy, while surviving in a changing and sometimes impredictable environment. All these structures have one feature in common: they increase the system's ability to dissipate the applied gradient in accordance with the so-reformulated second law of thermodynamics (Schneider and Kay, 1994a, b, 1995). In all these situations the second law imposes constraints that are necessary but not sufficient cause for life itself. Re-examination of thermodynamics proves that the second law underlines and determines the direction of many processes of the development of living systems. As an ecosystem develops it becomes more effective in removing the exergy part in the energy it captures, and this exergy is utilized to build and support organization and structure. Time and its derivative cycling play a key role in the evolution. Optimization of pulsating physiological processes can shed light on the basic understanding of development and evolution (Bejan, 1997d). Optimal strategies of streets tree networks and urban growth can mimic the development of living systems (Bejan and Ledezma, 1998). Information theory helps to display the thermodynamic behavior of living systems during their development and evolution, and living organisms can be treated as multistage systems by a complexity criterion based on information-theoretic entropy (Szwast, 1998). Some features of living organisms can be predicted when describing their evolution in terms of extremum principles for shortest paths along with suitable transversality conditions. In their models quantities similar to entropy production are extremized, and Onsager-like symmetries are found in models of development (Szwast et al., 2002). Discrete nonlinear models are suitable to describe biological dynamics in metric spaces. Mahulikar (2005) derived modified thermodynamic principles unifying order existence and evolution.

Complex systems research is a young branch of science. It is widely recognized as a markedly interdisciplinary discipline covering a very broad spectrum of problems (Gaspard, 1998; Gaspard and Nicolis, 1990; Foster et al., 2001). The research carried out in the complex systems follows a fundamental perspective, stemming largely from the close interaction between statistical mechanics, thermodynamics and nonlinear dynamics. It has taken place in recent years and has become one of the most dynamic and innovative orientations of the above disciplines. Among others, the research on complexity has led to the characterization of complex systems undergoing bifurcations and chaotic dynamics, in terms of the spectral properties of the underlying Liouville and Frobenius–Perron operators. A classification of chaotic versus stochastic processes has also been performed on the basis of dynamical entropies. Connections with algorithmic complexity in the sense of Kolmogorov have been analyzed in a number of papers. Gaspard's (1998) book describes recent advances in the application of chaos theory to classical scattering and nonequilibrium statistical mechanics generally, and to transport by deterministic diffusion in particular. The author presents the basic tools of dynamical systems theory, such as dynamical instability, topological analysis, periodic-orbit methods, Liouvillian dynamics, dynamical randomness and large-deviation formalism. These tools are applied to chaotic scattering and to transport in systems near equilibrium and maintained out of equilibrium. The book is useful for researchers interested in chaos, dynamical systems, chaotic scattering, and statistical mechanics in theoretical, computational, and mathematical physics and also in theoretical chemistry

Current topics of interest in this area include: (1) connections between entropy, complexity, and automaticity of a sequence, with applications in the analysis of biosequences; (2) generic properties of discrete networks of interacting elements, including the stability versus complexity issue; (3) closure schemes and constitutive relations for nonlinear dynamical systems; (4) dynamical basis and characterization of coarse-graining and symbolic dynamics in nonhyperbolic systems; (5) computation of correlation times from the spectral properties of Liouville and Fokker–Planck operators or of their coarse-grained forms. Another aspect of basic complex systems research is to elucidate the mechanisms of appearance of complex behaviors. Research in the complexity has contributed to the understanding of the onset of homoclinic chaos, of its connection with multimodal oscillations and of its role in modeling phenomena arising in the context of autocatalytic chemical reactions, electrochemistry, and thermal combustion. The generic properties of coupled normal forms arising from the interference of instabilities have also been addressed. Currently, local aspects of dynamical instability in nonhyperbolic and in nonautonomous systems are being considered. This should provide access to the local stretching rates and fractal dimensions, and hence to the "most dangerous" phase space directions for instability. For dynamical systems of large spatial extension giving rise to transport phenomena, like the Lorentz gas, Gaspard and Nicolis (1990) have established a relationship between the transport coefficient and the difference between the positive Lapunov exponent and the Kolmogorov–Sinai entropy per unit time, characterizing the fractal and chaotic repeller of trapped trajectories. They have also discussed consequences of their results for nonequilibrium statistical mechanics and thermodynamics. Levy et al. (2002) explored the eigenvalue spectra of the kinetic matrix which defines the master equation for the complex kinetics of the analogous polypeptides.

As stated by Nicolis and Rouvas-Nicolis (2007b) complexity is emerging as a post-Newtonian paradigm for approaching from a unifying point of view a large body of phenomena occurring in systems constituted by several subunits, at the crossroads of physical, engineering, environmental, life and human sciences. For a long time the idea prevailed that the perception of systems of this kind as complex arises from incomplete information, in connection with the presence of a large number of variables and parameters masking the underlying regularities. Over the years experimental data and theoretical breakthroughs challenging this view have become available, showing that complexity is on the contrary rooted into the fundamental laws of physics (Nicolis and Rouvas-Nicolis, 2007a, b). This realization opens the way to a systematic study of complexity, which constitutes an interdisciplinary, fast-growing branch of science, drawing on the cross-fertilization of concepts and tools from nonlinear dynamics, statistical physics, probability and information theories, data analysis, and numerical simulation.

Traditionally, fundamental science explores the very small and the very large, both of which lie beyond man's everyday perception. The uniqueness of complex systems is that they have to do with a class of phenomena of fundamental importance in which the system and the observer may evolve on comparable time and space scales. A system perceived as complex induces a characteristic phenomenology, the principal signature of which is the multiplicity of possible outcomes, endowing it with the capacity to choose, to explore, and to adapt. This process can be manifested in different ways (Nicolis and Rouvas-Nicolis, 2007a, b). Emergent properties reflect the primordial role of interactions between parts. They are manifested by the creation of self-organized states, where order and coherence are ensured by a bottom–up mechanism rather than through a top–down design and control. Laboratory examples of this behavior are found in fluids under inhomogeneous boundary conditions (eg, Rayleigh–Benard cells in a fluid heated from below) and in open chemically reacting systems. The latter show bistability, oscillations, Turing patterns, and wave fronts in the Belousov–Zhabotinski reaction, the first chemical reaction that exhibits spatial and temporal oscillations (Zhabotinsky, 1967, 1974). Further examples, in natural systems, are provided by the communication and control networks in living matter.

The coexistence of order and disorder raises the issue of predictability of the future evolution of the system on the basis of the current record. Typical examples are provided by the atmosphere in connection with the difficulty to reliable weather forecasts beyond a horizon of a few days, as well as by extreme geological and environmental phenomena such as earthquakes and floods (Nicolis and Rouvas-Nicolis, 2007a, b). Human systems such as traders in stock markets influencing both each other and the market itself are also confronted by unexpected crises and collapses, despite the rationality that is supposed to prevail at the individual level. Fractals, deterministic chaos, and its extreme form of fully developed turbulence provide prototypes of coexistence of order and disorder in time and space. In addition to its macroscopic level of manifestations, complexity is also ubiquitous at the microscopic level. (Nicolis and Rouvas-Nicolis, 2007a, b). Systems with built-in disorder like glasses exhibit a rich variety of evolutionary processes driven by microscopic level interactions. A variety of nanosystems exhibit complex behaviors like energy transduction and anomalous transport, arising from the interplay between microscopically generated spontaneous fluctuations and systematic environmental constraints. The very origin of irreversibility is related to the intrinsic complexity of the dynamics of the atoms constituting a macroscopic system under the effect of their mutual interactions.

Self-replication is a capacity common to every living species, and simple physical intuition dictates that such a process must invariably be fueled by the production of entropy.

England (2013) undertakes to make this intuition rigorous and quantitative by deriving a lower bound for the amount of heat that is produced during a process of self-replication in a system coupled to a thermal bath. He finds that the minimum value for the physically allowed rate of heat production is determined by the growth rate, internal entropy, and durability of the replicator. He also discusses implications of this finding for bacterial cell division, as well as for the prebiotic emergence of self-replicating nucleic acids.

CHAPTER 12

Multiphase Flow Systems

12.1 Introduction

Standart (1964) has derived exact macroscopic equations for the conservation of mass, mass of constituent, momentum, and energy in heterogeneous systems. By the use of integral transformation theorems generalized for such systems, he transformed these equations to the corresponding local forms of differential equations of continuity in each phase and of local surface "phase invariants." The integral form of Bernoulli's equation and of the first law of thermodynamics have been then derived from the former local form; by use of the latter, these relations are then brought to a form independent of the absolute velocity of the phase boundary in the heterogeneous system. He also derived explicit expressions for the rate of dissipation of mechanical energy, applied to the special case of steady one-dimensional cocurrent flow of two fluid phases.

Slattery et al. (2007) have presented an extensively revised second edition of first Slattery's book on interfacial transport phenomena (Slattery, 1981), as a presentation of continuum mechanics focused on momentum, energy, and mass transfer at interfaces. The book is written for the advanced student, and designed to prepare the reader for active research. It includes exercises and answers, and can serve as a graduate-level textbook. In addition to tightening the focus of the book there are two important additions: an extended discussion of transport phenomena at common lines or three-phase lines of contact, as well as a new theory for the extension of continuum mechanics to the nanoscale region immediately adjacent to the interface. The book provides and considers in detail both applications and supporting experimental data. The emphasis is upon achieving an in-depth understanding, based upon first principles.

Bedeaux et al. (1975) applied the theory of nonequilibrium thermodynamics to a system of two immiscible fluids and their interface. A singular energy density at the interface, which is related to the phenomenon of surface tension, is taken into account. Furthermore the momentum and the heat currents are allowed to be singular at the interface. Using the conservation laws and the Gibbs's relation for the surface, an expression for the singular entropy production density at the interface is obtained. The linear phenomenological laws between fluxes and thermodynamic forces occurring in this singular entropy production density are given. Some of these linear laws are boundary conditions for the solution of the differential equations governing the evolution of the state variables in the bulk.

Podowski (2005) overviews selected physical phenomena governing two-phase and multiphase flows and heat transfer, including scale-dependent modeling issues, gas–liquid and gas–liquid–solid interfacial interactions, and theoretical and computational modeling concepts. The main emphasis is given to the assessment of the current status against ultimate expectations on the one hand, and practical needs on the other hand. Several examples are used to illustrate the state-of-the-art theoretical and numerical predictions of two-phase flow at different scale levels and geometries.

12.2 Aspects of Thermodynamics of Surfaces and Interfaces

Surface effects are essential in two-phase and multiphase flows. Bedeaux (1986, 1992) formulates rigorous nonequilibrium thermodynamics of surfaces. He considers a boundary layer between two phases, which may change its position and curvature. The analysis requires time dependent curvilinear coordinates. At the surface the field properties, densities, fluxes, and sources exhibit discontinuities which are quantitatively described by appropriate excess functions. The excesses of the densities and currents are described as singular interfacial contributions to the distributions of the densities and currents. After formulating detailed balances of mass, momentum, and energy, the interface contributions are singled out from global equations. Finally, the entropy production for interfaces, σs, is found and the phenomenological equations postulated. Onsager relations are formulated on surfaces. The general expression for σs involves a variety of fluxes and forces and the author presents reasonable simplifications. Applications to membrane transport, evaporation, surfactant physics, and chemistry are briefly discussed. The virtue of the relatively simple analysis achieved in this paper is a unified approach to bulk and interface, that is, to continuous and discontinuous properties.

In his work on nonequilibrium thermodynamics of interfacial systems, Kovac (1977) applies the theory of nonequilibrium thermodynamics to a multicomponent system containing an interface using a method developed by Bedeaux, Albano, and Mazur (Bedeaux et al., 1975). A singular mass density is allowed at the interface as well as singular densities of energy and entropy. All currents are also allowed to be singular at the interface. The conservation laws and the Gibbs relation for the entropy are used to derive the entropy production at the surface. Linear laws relating the fluxes and thermodynamic forces are presented. The theory is then applied to a two-component system where one of the phases is a liquid and the other a low-density gas and the boundary conditions at the free surface of the liquid derived. The boundary conditions include the conditions used by Levich in his theory of the damping of waves by surface-active substances, but include other effects as well.

Statistical and kinetic properties of surfaces and interfaces are essential in the theories of catalysis, grain boundaries, and adsorption. Myshlyavtsev and Yablonski (1992) offer the studies in the thermodynamics and kinetics of surface reactions. Their method is called "method of transfer matrix" and allows one to evaluate the thermostatistical and kinetic properties of surfaces. These properties are essential in the theories of catalysis, grain boundaries, and adsorption. They are also of value to a general understanding of the thermodynamics of surfaces. The transfer matrix method turns out to be the most effective for calculating with models of lattice gases. It allows one to determine the statistical thermodynamic properties from the grand partition function fitting a one-site, which is treated as a stage along a chain. Every site along the chain is found in one of m possible states where m is any nonnegative integer. It follows that the largest eigenvalue of the transfer matrix is equal to the grand partition function of a one-site along the chain. The probability that the site is in a definite state can be found. Kinetic rate constants and coefficients of surface diffusion can be determined. The reader can study sample applications of the method in a variety of involved cases: monomolecular adsorption, desorption spectra, surface diffusion coefficients, and desorption from a reconstructed surface. He may also predict applications to more complicated lattices, with several kinds of centers. This approach should be studied by everyone who wants to apply thermodynamics to practical problems of surfaces.

12.3 Heating or Cooling Policies Minimizing Entropy Production

Heat and mass exchangers may be regarded as specific two-phase systems with phases separated by an artificial membrane. In the heat exchanger optimal design, the heat transferred is usually prescribed as a "process duty," whereas the amount of entropy produced depends on the way in which the process is carried out (Andresen and Gordon, 1992b). Since the entropy produced is equivalent to availability lost, this means that an improved exchanger design which reduces the entropy generation can save the useful work consumption; the saving that could be used somewhere else in the plant. Andresen and Gordon (1992b) specify the following extensive energy-consumers equipped to exploit this available work: factories with cogeneration capabilities, distillation plants driven by heat pumps or with reuse of the process heat at lower temperatures, and the electric utility providing the energy, especially if it also delivers district heating. For these installations, minimizing entropy production can be a desirable objective.

Andresen and Gordon (1992b) consider one particular type of problem toward illustrating the optimal heating/cooling strategy that minimizes entropy production in two-phase systems. The process is constrained to proceed in a fixed, given time (as is common in industry where fixed production rates are a constraint). The problem is illustrative in that it has a simple closed-form analytic solution. It is natural to ask how close conventional designs for heating and cooling processes are to the optimal one. The researchers select the specific case of standard single-pass heat exchanger design in which three possibilities are usually considered: (1) parallel flow, (2) condensing flow (constant temperature for the effective heat reservoir), and (3) counterflow. Counterflow design is the usual choice primarily because its effectiveness (ratio of actual heat exchange rate to maximum possible heat exchange rate) is highest, and second because its lower-temperature gradients produce less entropy. The authors show that the counterflow heat exchanger can represent the optimal solution for a somewhat idealized model which nonetheless captures the essential physics of heat exchanger operation. Furthermore, common designs are often rather close to the optimal heating/cooling strategy. Empirically the counterflow design has evolved as superior. One can thus verify how beneficial it is in view of the optimal solution which minimizes the entropy production.

The objective is to find the heating/cooling strategy that minimizes entropy production when a given quantity of heat must be transferred between reservoir and system during a given time τ. The control variable is the reservoir temperature $T_0(t)$. The analysis outlined later refers to the author's heating problem, wherein $T_0(t) > T(t)$. The solution for the corresponding cooling problem involves a simple change of sign. The rate of entropy production dS_u/dt is

$$\frac{dS_u}{dt} = \kappa(T_0 - T)\left(\frac{1}{T} - \frac{1}{T_0}\right) \tag{12.1}$$

This is always nonnegative, as required by the second law of thermodynamics. The dynamic constraint for heat transfer rate $q = \kappa(T_0 - T)$ into the system is

$$\kappa(T_0 - T) = C\frac{dT}{dt} \tag{12.2}$$

Two methods can be used to determine the optimal strategy. In the first more formal and cumbersome one, one defines a modified objective or Lagrangian L with a time-dependent Lagrange multiplier $\lambda(t)$:

$$L = \kappa(T_0 - T)\left(\frac{1}{T} - \frac{1}{T_0}\right) - \lambda\left(C\frac{dT}{dt} - \kappa(T_0 - T)\right) \tag{12.3}$$

The independent variables are T, dT/dt, and T_0 (and in principle dT_0/dt which, however, does not appear in the Lagrangian L). The Euler–Lagrange equations to determine the optimal strategy then become

$$\frac{\partial L}{\partial T} - \frac{d}{dt}\left(\frac{\partial L}{\partial T/\partial t}\right) = 0 \tag{12.4}$$

and

$$\frac{\partial L}{\partial T_0} - \frac{d}{dt}\left(\frac{\partial L}{\partial T_0/\partial t}\right) = 0 \tag{12.5}$$

After some algebra (keeping in mind the time dependence of λ), one obtains the solution

$$T_0(t) = \beta T(t), \tag{12.6}$$

where β is a constant. Eq. (12.2) then yields the final solution

$$T(t) = T(0)\exp\left(\frac{k(\beta-1)t}{C}\right), \tag{12.7}$$

where the constant β is determined from the condition that $T(0)$ and $T(\tau)$ are known:

$$\beta = 1 + \frac{C}{k\tau}\ln\left(\frac{T(\tau)}{T(0)}\right). \tag{12.8}$$

The controlling reservoir temperature is then implemented according to Eqs. (12.6)–(12.8). The entropy production using this optimal strategy S^{opt} is obtained from integrating Eq. (12.1):

$$S_u^{\text{opt}} = \frac{\kappa(\beta-1)^2 \tau}{\beta}. \tag{12.9}$$

The reversible solution of $T_0(t) = T(t)$, that is, $\beta = 1$, yields vanishing entropy production but cannot satisfy the constraint of the process completion in a prescribed time τ.

The second method of solution exploits the proof of Salamon et al. (1980) that for every linear system the strategy minimizing entropy production corresponds to a constant rate of entropy production. Thus equating Eq. (12.1) to a constant yields Eq. (12.6). Using the constraint of Eq. (12.2), one again obtains the rest of the optimal solution, Eqs. (12.7) and (12.8) (Fig. 12.1).

Ideas of extended thermodynamics enter to the theory of multiphase systems.

The first example is that of Sadd and Didlake (1977) who solved the problem of melting of a semiinfinite solid subjected to a step change in temperature according to a non-Fourier

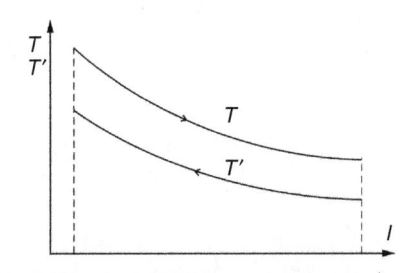

Figure 12.1: Temperature profiles of both phases in countercurrent heat exchangers are qualitatively similar to those corresponding to the optimal solution minimizing the entropy production.

heat conduction law. Unlike the classical Fourier theory which predicts an infinite speed of heat propagation, the non-Fourier theory implies that the speed of a thermal disturbance is finite. The effect of this finite thermal wave speed on the melting phenomenon is determined. The problem is solved by following a similar method as used by Carslaw and Jaeger for the corresponding Fourier problem. Non-Fourier results differ from Fourier theory only for small values of time. Comparing the temperature profiles and the solid–liquid interface location for aluminum, differences between the two theories are significant only for times on the order of 10^{-9}–10^{-11} s and in a region within approximately 10^{-4}–10^{-5} cm from the boundary surface. However, these results are based on an approximate value of the thermal relaxation time.

Other examples of application of FTT theory in multiphase systems can be found in Section 12.5.

12.4 Optimal Control in Imperfect Multiphase Systems

Poświata, (2004, 2005) and Poświata and Szwast (2003, 2006) worked out rigorous applications of optimal control in imperfect multiphase systems, especially those of fluidized drying. Their thermoeconomic optimization generalizes exergy criteria to thermoeconomic ones and enunciates the effect of bubbling in inhomogeneous fluidized beds. While the use of relatively high inlet gas velocities improves heat and mass transfer, a portion of gas flows as bubbles or canals, thus the energy carried by this gas is not exploited efficiently. This is the reason to search for best operating parameters which minimize total cost or maximize related efficiency. For inhomogeneous fluidization it is quite realistic to assume that a part of the gas is in plug flow (bubble phase) whereas its remaining part (in the "dense phase") is ideally mixed with the solid (Kunii and Levenspiel, 1991). Applications of this model to fluidized drying processes involve complex optimization computations based on the discrete maximum principle (Poświata, 2005; Poświata and Szwast, 2003, 2006). Cascade approach to the dynamical models of fluidized dryers provides a common "multistage context" capable of treating both batch fluidization and fluidization in a steady horizontal exchanger. Case B of Fig. 6.3 depicts the model proposed by Kunii and Levenspiel (1991) in which an excess gas (in reference to gas at minimum fluidization) flows in the form of bubbles. The heat and mass transfer between solid and gas and between a dense phase (with gas and solid) and bubble phase determines values of optimal gas parameters (flowrate, temperature and humidity of drying gas). The optimal parameters of both phases are complex curves whose shape depends on initial conditions of gas and solid; a vast spectrum of optimization results is described elsewhere (Poświata, 2005; Sieniutycz and Jeżowski, 2009) (Fig. 12.2).

Considering the heat transfer through an interface of a two-phase system Tan and Holland (1990) and Sieniutycz (2002e) test a Fermat-like principle of minimum time to predict the change in direction of the heat flux within each phase and through the interface between two immiscible media which differ with value of the specific thermal resistance. The basis for

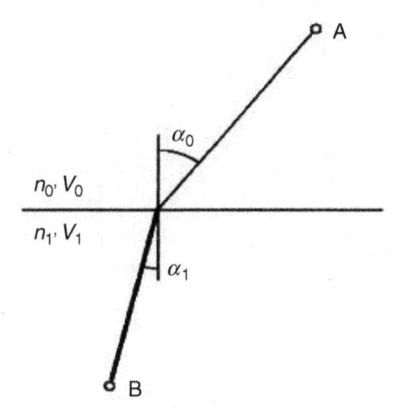

Figure 12.2: For the transfer through an interface of a two-phase system a modified Fermat principle holds as the tangent law of refraction rather than the classical sine law (Tan and Holland, 1990). Compare Fig. 7.2.

the principle are single-phase extremum properties of the entropy production and a related Onsagerian framework for heat transfer in the entropy representation. They show that a thermal ray spanned between two given points takes the shape that assures that its relatively large part resides in the region of lower thermal resistance (a "rarer" region of the medium). In other words, the energy (or mass) flux bends into the direction which ensures its shape corresponding with the longest residence time of energy in regions of larger conductivity. This property makes one possible to predict shapes of thermal rays or paths of the heat flow and to develop the corresponding Hamilton–Jacobi formalism that describes motion of thermal wavefronts.

Optimal control problems in chemically reacting systems are considered in Section 12.6.

12.5 Turbulent Mixtures, Instabilities, and Phase Transitions

Badur et al. (1997) present an analysis of algebraic and differential closure equations in the fluid subjected the first-order phase transformation. Water is a substance which occurs most commonly in nature and can exist in different states of aggregation. Phase transformations in water are a subject of our everyday experience. They happen spontaneously or are initiated by humans in various kinds of machines and devices. The most interesting phenomena, occurring in water, are condensation and flashing. Both can be initiated through a rapid pressure change, therefore there are stress-induced phase transitions brought about by the mechanism of local tension of the fluid. Condensation and flashing are often the main object of the research. In the CFD literature, there are three main types of nonequilibrium condensation models. These types of models are: the single continuum approach (also called the single-fluid approach), the Eulerian–Eulerian multiphase approach (also called the two-fluid approach) and Lagrangian–Eulerian Fluid–Solid Interaction approach. In this research, there have been attempts to compare different condensation models. Numerical calculation, done with the CFD code, were confronted with a classical experiment IMP PAN in Gdańsk in the de Laval nozzle.

Bilicki (1996) described thermodynamic disequilibrium in the two-phase system treated as a continuum with an internal structure. Downar-Zapolski and Bilicki (1998) have investigated the influence of nonequilibrium phase transition in the vapor–liquid mixture on the properties of sound waves. A dispersion relation for small disturbances has been found and analyzed. The investigation of evolution of a linear wave in the vapor–liquid mixture created the background for conclusions concerning the speed of sound in dispersive media.

Bilicki (2001) presented a method of derivation of constitutive equations based on the classical and extended irreversible thermodynamics. The method is explained for a single-phase medium and then is applied to a two-phase one-component of a liquid and its vapor. The two-phase system is described by the simplest homogeneous model (HTPM). In this model it is assumed that the velocities of both phases are equal, the two phases coexist at any point of the system, and that the transport of the phases due to diffusion is negligibly small. It is shown that the extended irreversible thermodynamics applied to the two-phase system provides an interesting equation for the evolution of the nonequilibrium dryness fraction x, comprising its second partial derivatives with respect to time and space. This equation is derived for HTPM for the first time.

Bilicki et al. (2002) explores the thermodynamic aspect of modeling two-phase systems by the methods of irreversible thermodynamics in both classical (CIT) and extended (EIT) formulation. The conservation laws for two-phase model-continuum are derived. Then, the entropy production is analyzed for two-fluid and homogeneous systems. Different equations of state are taken into consideration, namely that corresponding to the accompanying equilibrium state of physical element and a more complex one resulting from EIT. Obtained expressions for rate of entropy production per unit volume allow to identify the dissipative mechanisms in the two-phase system and suggest the forms of phenomenological relations to be adopted in the constitutive equations.

Bilicki and Badur (2003) have developed a thermodynamically consistent relaxation model to describe nonequilibrium phenomena within a turbulent binary mixture undergoing a phase transition. The model involves a one-fluid homogeneous approach. The set of governing evolution equations is given and the averaged basic balance laws are considered. The continuum under consideration consists of two constituents of which the first is undergoing a phase transition (vaporization, cavitation, flashing, condensation) during the flow with turbulent transport of mass, heat, and momentum. Starting with the Clausius–Duhem inequality for the recoverable specific entropy, several thermodynamic relationships for equilibrium and nonequilibrium fields are developed. The relaxation–retardation constitutive model for turbulent, phasic and diffusive fluxes is described. It is shown that the resulting model uncovers a few known two-phase models of turbulence and also the classical relaxation models as limiting cases. Bilicki and Badur (2003) conclude that existing models of multiphase turbulent flow cannot be called thermodynamically consistent. The multiphase flows with various predictive strategies are based on the Eulerian–Eulerian description (one-field, homogenous or two-fluid approach) or Eulerian–Lagrangian description. For the flow of a fluid that undergoes a stress-or temperature-induced phase transition, the Eulerian description is commonly adopted for the prediction of kinetics of nonequilibrium phase transition based on the experience of the metastable state description. Turbulence accompanying this phenomenon cannot be correctly described by the Reynolds averaging techniques since the turbulent fluxes arising from Reynolds averaging deal separately with the averaging procedure in one phase and a separate averaging procedure in the second phase. These studies do not seem to have advanced any theoretical reasons for the Reynolds averaging technique, but have deduced it entirely from their experiments. In reducing these results to a single model, however, many things have to be taken into account and many misleading assumptions have been made. There is one assumption which upon the face of it seems to be contrary to general experience, this is, that the concurrent phase transition does not change the level of turbulence. In this paper the authors approach the problem in another manner. Starting with the interpretation of the turbulent flux of momentum as an internal forces interchanging momentum between laminar and turbulent fluid within a continuum particle, they have endeavored to deduce from analytical considerations the model for the thermodynamically consistent laws of transport of mass, momentum, and energy between the phasic, turbulent, and diffusive microstructures. In the literature there is one main direction of reasoning—to extend the achievements of the theory of turbulence from one phase to multiphase flows. Such a line of reasoning omits a specificity of phase transition phenomena. On the basis of the authors' work on nonequilibrium phase transitions, an approach from the opposite direction has been presented—from thermodynamically substantiated models of a laminar flow undergoing a phase transition to a turbulent, phasic, and diffusive flow. But as yet, the advance in this direction has not been essential—because the discrepancy in the results of the various experiments is such that one cannot avoid the conclusion that not all important variables of the problem have yet been taken into account (Bilicki and Badur, 2003).

Imre (2000) reviews liquid–liquid equilibria in binary liquid mixtures under absolute negative pressure. Several phenomena concerning liquid–liquid equilibrium can be seen only—or can be seen better—under negative pressure; these are briefly reviewed. Negative pressure states for liquids are metastable, but existing and experimentally accessible states which can be maintained long enough to make phase equilibrium measurements. In binary liquid mixtures, several phenomena concerning liquid/liquid miscibility can be seen only (or can be seen better) under negative pressures, such as the merging of UCST (Upper Critical Solution Temperature) and LCST (Lower Critical Solution Temperature) branches or the existence of the critical temperature maximum on the UCST in weakly interacting binary mixtures, or the existence of a low pressure solubility branch (often hiding completely below $p = 0$) in some systems, like the aqueous solution of the different methylpyridines.

A unifying picture of interfacial instabilities, turbulent interfaces, and thermohydrodynamic oscillatory instabilities has been worked out for Newtonian and viscoelastic fluids (Velarde and Chu 1989, 1990, 1992). It has been shown that the competition of dissipation and damping with restoring forces leads to oscillations and limit cycles. Velarde and Chu (1992) present a clear picture of various thermohydrodynamic oscillatory instabilities in both Newtonian and viscoelastic fluids. They show that the competition of dissipation and damping with restoring forces can lead to oscillations and limit cycles. They also explain how nonlinearity helps sustain these oscillations in the form of limit cycles. First, oscillatory Rayleigh–Benard convection driven by buoyancy forces is discussed. Then, interfacial oscillators are investigated. Starting with the perturbation equations, various effects are analyzed, such as: Marangoni instabilities, the onset of oscillatory motion, transverse gravity–capillary oscillations, neutral stability conditions, longitudinal waves, interfacial electrohydrodynamics, and so forth. Solitons excited by Marangoni effect are introduced. It is shown how the Korteweg–de Vries soliton description can be used as a building block for some nonlinear traveling wave phenomena. The role played by dissipation in triggering and eventually sustaining nonlinear motions is shown. This review makes the reader familiar with the beauty of methodological unification of various nonlinear phenomena in hydrodynamics.

Javania et al. (2014) investigate heat transfer with *phase-change* materials (PCMs) in passive thermal management of electric and hybrid electric vehicles where the PCM is integrated with a Li-ion cell. When higher current is extracted from the Li-ion cells, heat is generated due to the Ohmic law. Therefore, it is vital to design a successful thermal management system (TMS) to prevent excessive temperature increase and temperature excursion in the battery pack. During the phase change process, PCMs absorb heat and create a cooling effect. In the discharging (solidification) process, stored heat is released and it creates a heating effect. The case considered includes the use of PCMs with different thicknesses around the cells.

12.6 Multiphase Chemical Reactors and Regenerators

Multiphase reactors are found in diverse applications such as in manufacture of petroleum-based fuels and products, in production of commodity and specialty chemicals, pharmaceuticals, herbicides, and pesticides, in refining of ores, in production of polymers and other materials, and in pollution abatement (Dudukovic et al., 1999). In all such applications, the knowledge of fluid dynamic and transport parameters is necessary for development of appropriate reactor models and scale-up rules. The state of the art of our understanding of the phenomena occurring in three-phase reactors such as packed beds with two-phase flow, slurry bubble columns, and ebullated beds is summarized by Dudukovic et al. (1999).

Bahroun et al. (2013) deal with the nonlinear control of a jacketed fed-batch three-phase slurry catalytic reactor to track a reference temperature trajectory. To this end, they define the so-called thermal availability function from the availability function as it has been described in the literature. This thermal availability is used as a Lapunov function in order to derive a stabilizing control law for the reactor by using the jacket temperature as the manipulated variable. This control is coupled to a high gain observer that is used to estimate the reaction rate that is considered as a time varying parameter. The o-cresol hydrogenation reaction is taken as a representative test example. A detailed model of the process as well as a simplified one that is used for the control law synthesis are described. The performance of the control scheme in presence of measurements noise and parameter uncertainty is illustrated by simulations results.

Sieniutycz (2000h, 2001c) set an optimal control framework for balance and kinetic equations of multiphase reacting systems consistently with the second law of thermodynamics. Approaches effective in the theory of nonequilibrium thermodynamics of single-phase systems were extended to include interfacial discontinuities, surface reactions and interface transports which are phenomena typical of multiphase systems. He arrived at Lagrangian and Hamiltonian structures of transport equations and laws of chemical kinetics, which constitute efficient forms serving to generalize linear models to nonlinear regimes. Local disequilibria were predicted responsible for onset of interfacial and bulk instabilities. Reti and Holderith (1980) developed models of multiphase hyperbolic chemical reactors by averaging methods.

Szwast and Sieniutycz (1998, 1999) investigated a system with moving deactivating catalyst, composed of a cocurrent tubular reactor and a catalyst regenerator with an additional flux of a fresh catalyst (Fig. 12.3). For the temperature-dependent catalyst deactivation, the optimization problem has been formulated in which a maximum of a process profit flux is achieved by a best choice of temperature profile along tubular reactor, best catalyst recycle ratio and best catalyst activity after regeneration. The set of parallel-consecutive reactions, $A + B \to R$ and $R + B \to S$, with desired product R has been taken into account. A relatively unknown, powerful discrete algorithm in which a suitably defined Hamiltonian is constant along the optimal path, has been applied for optimization. The optimal solutions have been discussed. In particular, it has been shown that an increase of the unit cost of catalyst

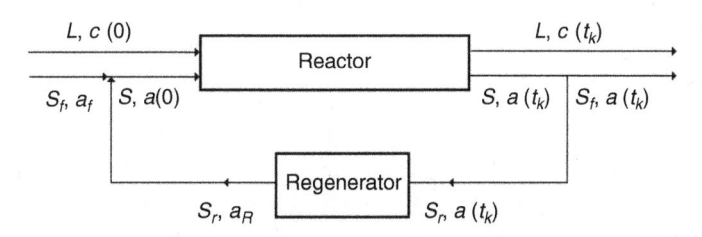

Figure 12.3: System of cocurrent tubular reactor—catalyst regenerator (Szwast and Sieniutycz, 1999).

regeneration or an increase of the catalyst recycle ratio cause such optimal temperatures in reactor which save the catalyst, as the optimal temperature profiles are then shifted toward lower temperatures. Finally these profiles reach isothermal shape at the level of minimum allowable temperature and then there is no further possibility to control the reactor process by the temperature profile. Thus the catalyst activity after regeneration, as well as an average catalyst activity in the reactor, do decrease when the unit catalyst regeneration cost increases. This is a new form of catalyst saving, as the catalyst deactivation rate becomes reduced when an average catalyst activity is allowed to decrease. It is important that this form of catalyst saving appears in the region where any saving of the catalyst by an optimal choice of temperature profile is impossible. It has been also shown that for small values of the catalyst recycle ratio, the catalyst regenerator should be removed from the system. In such a case, the renewal of catalyst takes place due to a fresh catalyst input, exclusively.

We shall discuss several specific results of this treatment.

The system considered is schematized in Fig. 12.3.

Mathematical analysis and computer calculations refer to the following set of parallel-consecutive reactions, with desired product R:

$$A + B \xrightarrow{E_1, k_{10}} R \tag{12.10}$$

$$R + B \xrightarrow{E_2, k_{20}} S \tag{12.11}$$

A temperature dependent and concentration independent catalyst deactivation was assumed. The deactivation rate can be written in the following form

$$\frac{da}{dt} = -k_{d0} \exp\left(\frac{-E_d}{RT[t]}\right) a \tag{12.12}$$

The optimal temperatures have to lie in an allowable range defined as

$$T_* \leq T[t] \leq T^* \tag{12.13}$$

where T_* and T^* describe minimum and maximum allowable temperature, respectively. It was assumed that the deactivation depends on the temperature only.

The catalytic reactions (12.10) and (12.11) are described by the following equations

$$\frac{dc_A[t]}{dt} = -k_1 \, a \, c_A \, c_B \tag{12.14}$$

$$\frac{dc_B[t]}{dt} = -k_1 \, a \, c_A \, c_B - k_2 \, a \, c_B \, c_R \tag{12.15}$$

$$\frac{dc_R[t]}{dt} = k_1 \, a \, c_A \, c_B - k_2 \, a \, c_B \, c_R \tag{12.16}$$

$$\frac{dc_S[t]}{dt} = k_2 \, a \, c_B \, c_R \tag{12.17}$$

valid for $0 \leq t \leq t_k$. The reaction rate constants k_1 and k_2 satisfy the following equation

$$k_i = k_{i0} \, \exp\left(\frac{-E_i}{RT[t]}\right) \tag{12.18}$$

valid for $i = 1,2$, where E_i = activation energy of reaction i, for $i = 1,2$.

The profit flux, P, which has to be maximized, is defined as a difference between the flux value of the desired reagent and the flux of total process cost. It is described as follows:

$$P = \frac{L}{\rho} c_R[t_k] \, M_R \, \mu_R - S_f \, \mu_f - S_r \, K_r[a[t_k], a_R] - K_c \tag{12.19}$$

The first term on the right-hand side of Eq. (12.14) describes the flux of desired reagent value, the second term describes the flux of fresh catalyst cost, the third term, flux of catalyst regeneration cost and the last one, flux of fixed cost K_c. As in virtually all contemporary works on optimization, we assume that the costs of the optimal control (here: temperature control) are the same as the control costs of process in general. Thus we assume that the heating costs can be included in the fixed costs of the process, and hence they do not affect the optimization solution. There are techniques (eg, autotransformer heating of fluids to achieve a required profile $T(t)$) which realize this concept with a considerable precision.

The following symbols are applied: ρ = mixture density, M_R = molar mass of desired reagent R, μ_R = unit price of R, μ_f = unit price of fresh catalyst, $K_r[a[t_k],a_R]$ = catalyst regeneration cost for a between $a[t_k]$ and a_R. Compare Park and Levenspiel (1976a, b), for example.

A simplest explicit form of regeneration function $K_r[a[t_k],a_R]$ is taken into account. It reads as follows

$$K_r[a[t_k], a_R] = \mu_r (a_R - a[t_k]) \tag{12.20}$$

where μ_r = unit cost of catalyst regeneration per unit catalyst activity. Applying Eq. (12.20) in Eq. (12.19) yields

$$P = \frac{L}{\rho} c_R[t_k] M_R \mu_R - S_f \mu_f - S_r \mu_r (a_R - a[t_k]) - K_c \tag{12.21}$$

which can be transformed into following dimensionless form

$$F = x_R[t_k] - S_f\lambda - S_r A(a_R - a[t_k]) - k_c \tag{12.22}$$

where F = dimensionless profit flux defined by Eq. (12.23), λ = dimensionless unit price of fresh catalyst defined by Eq. (12.24), A = dimensionless unit cost of catalyst regeneration per unit catalyst activity defined by Eq. (12.25) and k_c = dimensionless flux of constant cost defined by Eq. (12.26).

$$F = \frac{P}{(L/\rho)M_R \mu_R c_B[0]} \tag{12.23}$$

$$\lambda = \frac{\mu_f}{(L/\rho)M_R \mu_R c_B[0]} \tag{12.24}$$

$$A = \frac{\mu_r}{(L/\rho)M_R \mu_R c_B[0]} \tag{12.25}$$

$$k_c = \frac{K_c}{(L/\rho)M_R \mu_R c_B[0]} \tag{12.26}$$

Moreover, after applying the dynamical equation of the desired product R

$$\frac{dx_R[t]}{dt} = k_1 a x_A x_B - k_2 a x_B (x_B - 2x_A + 2M - 1) \tag{12.27}$$

one can write the following relationship

$$x_R[t_k] = \int_0^{t_k} \left(\frac{dx_R}{dt}\right) dt = \int_0^{t_k} (k_1 a x_A x_B - k_2 a x_B (x_B - 2x_A + 2M - 1)) dt \tag{12.28}$$

which is valid when the desired reagent R does not appear in the fresh reagents mixture.

The dimensionless profit flux, Eq. (12.22), constitutes the performance index for the optimization problem considered later. With Eq. (12.28), Eq. (12.22) can be transformed to the following final form:

$$F = \int_0^{t_k} (k_1 a x_A x_B - k_2 a x_B (x_B - 2x_A + 2M - 1)) dt - \frac{S\lambda}{R+1} - \frac{SRA(a_R - a[t_k])}{R+1} - k_c \tag{12.29}$$

The dimensionless process profit flux, Eq. (12.29), has to be maximized by an optimal choice of temperature profile along tubular reactor, catalyst recycle ratio and catalyst activity after regeneration. It has to be maximized when temperature controls are in an allowable region defined by inequality (12.13), the state (dynamic) equations (12.12), (12.14), and (12.15) as well as a catalyst mixing function are satisfied, and the residence time of reagents and catalyst in the reactor, t_k, is fixed. Since a discrete algorithm is applied to solve the optimization problem, the procedure starts with replacement of the continuous model of tubular reactor by a corresponding discrete model of the N-stage cascade of reactors with ideal mixing, $n = 1,...N$. Details of computational algorithm are available (Szwast and Sieniutycz, 1999).

The selected results are presented in Figs. 12.4–12.7.

Fig. 12.4 shows that when for A = constant catalyst recycle ratio R increases (a mean number of catalyst particles residing in the reactor increases) or for R = constant A increases (catalyst regeneration cost increases) the optimal temperatures save the catalyst as the optimal temperature profile is shifted toward lower temperatures.

Fig. 12.5 shows that, for small values of catalyst recycle ratio R, the catalyst regenerator should be removed from the system—see that the curve $a_R[R]$ starts from $R > 1$ and that a critical value of R, R_{cr}, exists which increases with A. This means that the range of values of catalyst recycle ratio R, for which the system without catalyst regenerator is the optimal one, increases with the unit cost of catalyst regeneration A. For $R > R_{cr}$ the function $a_R[R]$ is increasing one, reaching maximal catalyst activity after regenerator, the property which

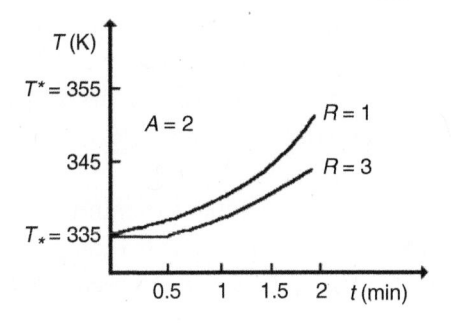

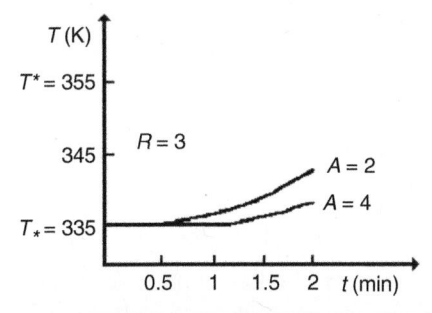

Figure 12.4: Optimal temperature profile for various values of catalyst recycle ratio R and unit regeneration cost A.

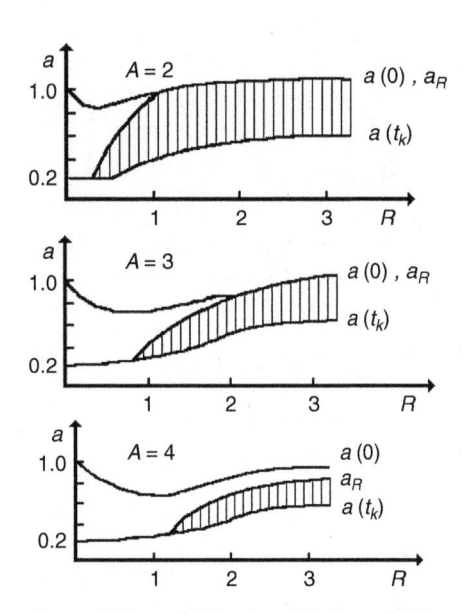

Figure 12.5: Optimal catalyst activities in inlet and outlet fluxes of the reactor and, $a[0]$ and $a[t_k]$, and the activity after regeneration a_R versus recycle ratio R.

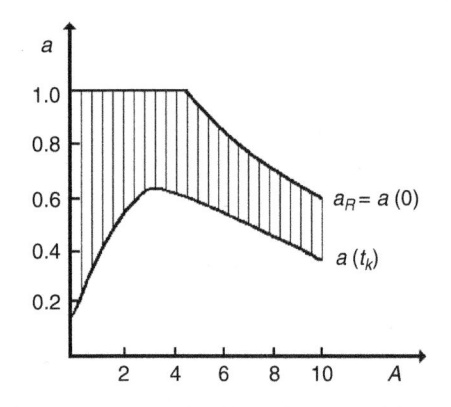

Figure 12.6: Optimal catalyst activities in inlet and outlet fluxes of the reactor versus regeneration cost A for a process without stream S_f.

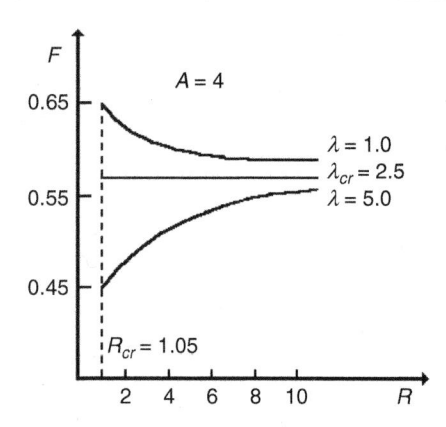

Figure 12.7: Performance index F versus catalyst ratio R for $A = 4$ and various prices of fresh catalyst λ.

can be observed for every of presented in Fig. 12.5 cases, that is, $A = 2$, $A = 3$, $A = 4$. It is important that when catalyst recycle ratio R reaches values larger than presented in Fig. 12.3, the function $a_R[R]$ becomes decreasing one. This phenomenon is explained in the text connected with Fig. 12.6.

Fig. 12.6 is valid for the case when whole catalyst flux leaving reactor is regenerated, that is, there is no fresh catalyst flux, $S_f = 0$. It shows that, for small values of unit cost of catalyst regeneration A, the catalyst is regenerated to maximum possible value, $a_R = 1$, whereas the activity of catalyst leaving reactor $a[t_k]$ increases with A. This results from the fact that, in this range of A, the optimal temperature profile is shifted in direction of lower temperatures when A increases. For larger values of A, an isothermal profile of temperature at the level of minimum allowable temperature T_* is optimal one; in such a case there is no possibility to control reactor process by temperature profile. This is a new form of catalyst saving, as the catalyst deactivation rate becomes reduced when an average catalyst activity is allowed to decrease.

Thus the catalyst activity after regeneration a_R and the activity of catalyst leaving reactor $a[t_k]$, as well as the average catalyst activity in reactor do decrease when the unit catalyst regeneration cost A increases.

It is important that this form of catalyst saving appears in the region where any saving of the catalyst by an optimal choice of temperature profile is impossible.

The following general conclusions can be formulated:

> When the catalyst recycle ratio R increases, a mean number of catalyst particles residing in the reactor increases as well. When the optimal temperatures reduce then the rate of the catalyst's decay because the optimal temperature profile is shifted toward lower temperatures. One can therefore say that optimal temperatures "save" the catalyst.

When the catalyst regeneration cost (measured by dimensionless variable A) increases, the optimal temperatures save the catalyst (regeneration is more and more expensive), again, because the optimal temperature profile is shifted toward lower temperatures.

For small values of the catalyst recycle ratio R, the catalyst regenerator should be removed from the system of reactor and regenerator (no sense to regenerate a small recycled catalyst flux S_r). The range $[0, R]$ of those values of R for which the system without catalyst regenerator is an optimal system increases with the unit cost of catalyst regeneration, A.

For the case when whole catalyst flux leaving reactor is regenerated (there is no fresh catalyst flux, $S_f = 0$) one can confirm the validity of the following statement: For small values of unit cost of catalyst regeneration A the catalyst is regenerated to maximum possible value, $a_R = 1$, whereas the activity of catalyst leaving reactor $a[t_k]$ increases with A. This statement results from the fact that, in this range of A, the optimal temperature profile is shifted in direction of lower temperatures when A increases. For larger values of A, an isothermal profile of temperature at the level of minimum allowable temperature T_* is optimal one; in such case there is no possibility to control the reactor process by a temperature profile. This is associated with decrease of value for a number of quantities (catalyst activity after regeneration, a_R, the activity of catalyst leaving the reactor, $a[t_k]$, and the average catalyst activity) when the unit regeneration cost A increases. This is a new form of catalyst saving because the catalyst deactivation rate becomes reduced when an average catalyst activity is allowed to decrease. It is important that this form of catalyst saving appears in the region where any saving of the catalyst by an optimal choice of temperature profile is impossible.

There is no optimal stationary solution for simultaneous renewal of catalyst, by input of the fresh catalyst flux and regeneration. But, in fact, the reactor–regenerator system has always to work with additional flux of fresh catalyst because the catalyst losses from the system always do exist. Thus the scheme of an industrial reactor–regenerator system is exactly like that presented in Fig. 12.1. However the catalyst recycle ratio is not a control variable; it is rather a parameter, whose value results from knowledge of the magnitude of catalyst loss from the system. Thus, the industrial optimization problem can be formulated as follows: Find the optimal temperature profile along tubular reactor and catalyst activity after regeneration for a fixed value of catalyst recycle ratio. This is exactly the same optimization problem as the one which was solved at the lower lever of the optimization described in this paper.

Closing this text we would like to stress that the Hamiltonian method of optimization applied to reacting processes with catalyst decay has proved to be an effective and powerful tool for optimization of diverse systems such as: heat exchangers, separation systems and noncatalytic chemical reactors. The implementation of this method to catalytic processes in multiphase reactors extends its realm of applications considerably, thus making possible optimization of many new catalytic systems currently investigated.

Szwast and Sieniutycz (2001) have further investigated a cocurrent tubular reactor with temperature profile control and recycle of moving deactivating catalyst. For the temperature dependent catalyst deactivation, the optimization problem has been formulated in which a maximum of a profit flux is achieved by the best choice of temperature profile and residence time of reactants for the set of catalytic reactions $A + B \rightarrow R$ and $R + B \rightarrow S$ with desired product R, the rates of reactions have been described separately for every reagent by expressions containing (temperature dependent) reaction rate constants, concentrations of reagents, catalyst activity, as well as catalyst concentration in the reacting suspension and a measure of the slip between reagents and solid catalyst particles. The algorithms of maximum principle have been used for optimization. The optimal solutions show that a shape of the optimal temperature profile depends on the mutual relations between activation energies of reactions and catalyst deactivation. It has been proved that the optimal temperature profile is a result of the compromise between the overall production rate of desired reagent R (production rate in the first reaction minus disappearance rate in the second one), necessity of saving of reagents residence time (reactor volume) and necessity of saving catalyst; the most important influence on the optimal temperature profile is associated with necessity of saving the catalyst: When catalyst recycle ratio increases (mean number of catalyst particles residing in reactor increases) optimal temperatures save the catalyst, as the optimal profile is shifted in direction of lower temperatures. The same is observed when catalyst slip increases (catalyst residence time in reactor increases). Despite of variation in catalyst concentration the optimal profile is practically the same because the decay rate is affected only by instantaneous activity of catalyst. When reactor unit volume price decreases, catalyst residence time increases, whereas optimal temperature profile is shifted to lower temperatures. When economic value of unit activity of outlet catalyst increases (catalyst with a residual activity still has an economic value), catalyst saving should be more and more intense. As far as possible, the optimal profile is shifted in direction of lower temperatures whereas the optimal residence time is still the same. Then the optimal profile is isothermal at the level of minimum allowable temperature, whereas the catalyst is saved as its residence time in reactor decreases.

Szwast and Sieniutycz (2004) have investigated a dynamic unsteady-state reactor process in a cocurrent tubular reactor with single–run reagents (continuous phase) and with multirun catalyst (dispersed phase). For the temperature-dependent catalyst deactivation the reactor process has been considered in which after optimal number B of catalyst residences in the reactor the whole amount of catalyst leaves the system; then the fresh catalyst is directed to the reactor and the next B-runs cycle of the optimal process starts. The optimization problem has been formulated in which a maximum of an average (for one cycle) process profit flux is achieved by a best choice of number B of catalyst residences in the reactor and best choice of temperature profiles along tubular reactor for each catalyst run, $b = 1,...B$, respectively.

The set of parallel–consecutive reactions, $A + B \rightarrow R$ and $R + B \rightarrow S$, with desired product R has been taken into account. The algorithms of maximum principle have been used for optimization. Optimization procedure allows for finding an optimal number of catalyst runs B for which an average process profit flux reaches a maximum for a concrete value of fresh catalyst price λ. Optimal value of B increases with λ. A shape of optimal temperature profile constitutes the effect of compromise between the overall production rate of desired reagent R (production rate in the first reaction minus disappearance rate in the second one) and the necessary savings of catalyst. The optimal solutions show that the most important influence on the optimal temperature profile is due to the need of catalyst saving; low temperatures save catalyst during its initial runs. Moreover, the optimal temperature profile is independent of fresh catalyst price λ. If optimal temperature profile for B-run reaction process is known, then, to obtain optimal profile for $(B - r)$-run process, it is enough to cut the initial part of optimal profile corresponding to the first r runs.

Currently, power production processes are often considered. Xia et al. (2009) have derived a model of a multireservoir endoreversible chemical engine which may be adopted as a multiphase power generator. They assumed an isothermal operation and used the optimal control theory to determine the optimal configuration of this engine for maximum power output. The optimal cycle consists of two constant chemical potential branches and two instantaneous constant mass-flux branches, which is independent of the number of mass reservoirs and the mass transfer law. The results show that, in order to obtain the maximum power output, some mass reservoirs should never connect to the working fluid in the mass transfer processes. A numerical example is provided for a linear mass transfer law three-mass-reservoir chemical engine. The effects of the potential changes of the intermediate mass reservoir on the optimal configuration of the chemical engine and the performance corresponding to the optimal configuration were analyzed. The obtained results were compared with those obtained for a multireservoir endoreversible heat engine. The results could provide some guidelines for optimal design and operation of real chemical engines.

12.7 Final Remarks

Flow and transport processes in porous media gained increasing attention in many natural, industrial, and even biological systems. In most cases, there is more than one fluid involved, which may lead to multiphase conditions. Then, phenomena such as mass transfer between the phases or capillary forces are obtainable (Faigle, 2009). Multiphase flow problems in porous media are manifold, and range from natural to technical problems. Taken the porous medium of the subsurface as a classical natural example, a wide range of possible substances may come into focus. Besides the exploitation of resources such as oil and gas, sustainable management of the groundwater requires profound knowledge of the underlying flow processes. Being an important resource of drinking water, the protection of groundwater from harmful pollution remains an ongoing task for applied science and research. The plentiful sources and types of contaminants range from infiltration through rainfall that is loaded with nutrients to oil spills from leaking tanks, thus remediation techniques have to be chosen in accordance to the problem, whereas they are at all possible. Consequently, multiphase modeling becomes essential to assess the long-term risks, especially in the fields of anthropogenic activities such as nuclear storage, or CO_2 injection into deep geological formations. Technical porous media problems cover large-scale transport processes such as dissolved salt intrusion through concrete which corrodes the iron reinforcement, down to small-scale gas flow composed of oil droplets and solid particles through catalytic converters, or the diffusive flow in the layers of fuel cells. All those presented problems have in common that they require an interdisciplinary approach comprising, amongst others, fluid mechanics, geology, thermodynamics, mathematics, and physics.

Van's (2008) review on Kjelstrup and Bedeaux (2008) book regards the book as a treatise about disequilibrium thermodynamics of diffuse multicomponent, chemically reacting, polarized, multiphase mixtures in local equilibrium. The authors consider several phases, and the phase boundaries (surfaces and lines) are treated as separate, autonomous thermodynamic systems. The word "heterogeneous" in the title is related to phases and phase boundaries and not to material heterogeneity. Mostly one-dimensional stationary systems are considered in the book and the transport is assumed to be perpendicular to the surfaces. Molecular dynamic simulations, kinetic theory calculations, and weakly nonlocal continuum calculations are used to provide a deeper insight and a justification of the applied basic hypotheses: local equilibrium, Onsager reciprocity, and the autonomous description of the surface in terms of the excess quantities. The application examples consider the technical details of the measurement of the necessary physical quantities (eg, transport coefficients); moreover, they give complex and real examples of important industrial applications. Elaborated exercises help the understanding of the material. The book is divided into two main parts after the introductory chapters: Part A deals with the general theory and Part B with applications. There are three introductory chapters. The first one presents the scope and the context of the book, giving a general motivation. The second introductory chapter motivates the use of nonequilibrium thermodynamics in scientific and engineering problems. The third introductory section shortly explains the basic concepts and the thermodynamic relations for discrete multiphase systems, introducing the boundaries as autonomous thermodynamic bodies (Van, 2008).

In Part A, the governing equations and the constitutive theory are introduced. The entropy production for homogeneous systems, for surfaces, and for a three-phase contact line are developed in separate sections. The flux equations and the Onsager relations are treated, and finally the characteristics of pairwise coupled systems—thermal–diffusive, thermal–conducting, diffusive–conducting—are given in separate chapters. There the emphasis is on the experimental and theoretical determination of the transport coefficients.

In Part B, the authors develop a mixture of applications: evaporation and condensation, multicomponent diffusion and heat transfer, different electrochemical applications (electrode reactions, junction potential, impedance of an electrode surface) show the power of the thermodynamic theory. The complex examples of the formation cell or the thermodynamic model of the polymer electrolyte fuel cell are based on the previous results. Some important nontrivial questions are treated in separate chapters, like the question of transported entropy or the measurement of membrane transport properties. The section about the salt power plant may be the best example of the direct engineering applications. Finally, two chapters deal with the justification of the most important hypothesis of the model. Molecular dynamic simulations demonstrate the validity range of the local equilibrium and the Onsager symmetry relations. The final chapter introduces the classical Van der Waals capillarity model—a continuous, moreover weakly nonlocal extension of the classical phase boundary model—and applies to one-dimensional phase boundary calculations, as a justification of the autonomous surface hypotheses. The authors of the book are well known in the thermodynamic community as representatives of the theory of Classical Irreversible Thermodynamics (CIT) based on local equilibrium hypothesis and linear flux–force relations.

The purpose of the book is to encourage the use of nonequilibrium thermodynamics to describe transport in complex, heterogeneous media. With large coupling effects between the transport of heat, mass, charge, and chemical reactions at surfaces, it is important to know how one should properly integrate across systems where different phases are in contact. No other book gives a prescription of how to set up flux equations for transports across heterogeneous systems. The authors apply the thermodynamic description in terms of excess densities, developed by Gibbs for equilibrium, to nonequilibrium systems. The treatment is restricted to transport into and through the surface. Using local equilibrium together with the balance equations for the surface, expressions for the excess entropy production of the surface and of the contact line are derived. Many examples are given to illustrate how the theory can be applied to coupled transport of mass, heat, charge, and chemical reactions; in phase transitions, at electrode surfaces and in fuel cells. Molecular simulations and analytical studies are used to add insight.

According to Van (2008), the considered book demonstrates the capabilities of the theory. The detailed theoretical and experimental study of the investigated systems shows very well the deep insight that one can get from clear theoretical approaches. This unity of theory and experiments is one of the most important virtues of the book. Nonequilibrium thermodynamics—as a frame theory of classical physics—is a foundation of several parts of engineering and can give a unified treatment of complex systems. The book is a great example of how new theoretical ideas—in this case, the idea of surfaces as autonomous thermodynamic systems—and clever experimental work can lead to the understanding and development of industrial technologies. The rigorous and conservative approach of CIT is an advantage of the book—a clear and solid foundation with clean-cut ranges of validity.

The broad range of different applications demonstrates the enormous unification power of nonequilibrium thermodynamics.

Others areas in the field of multiphase systems may be of interest, for example:

- Pneumatic conveying plants, transport, and pumping systems
- Mixing induced separation systems and multiphase flow generation
- Industrial multiphase, multilevel systems
- Emergency, loss prevention, and pipeline integrity management
- Combustibles, hazardous materials, and remediation of contaminated sites
- Measurement, control, and simulation of state change characteristics
- Multiphase mass transfer operations and multiphase chemical reactors

CHAPTER 13

Radiation and Solar Systems

13.1 Introduction

Radiation effects are essential during energy exchange in single-phase and multiphase systems. Radiation systems and solar radiation operations are often found in the traditional fields of mechanical engineering, chemistry, and the chemical industry as well as in other fields such as solar buildings, photochemistry, or ecosystems. Contemporary research in radiation systems accommodates scientific approaches, case histories, experience, and engineering applications. Research contributions usually take into consideration specificity of radiation thermodynamics and resulting differences in mathematical modeling, applicative properties of solar systems, and inevitability of taking quantum properties of radiation into account in all these fields. In this chapter we focus on several new results obtained in the fields of radiation and solar systems, such as: recent results of Badescu for improved expressions of power yield efficiencies, a concise treatment of the solar energy conversion into heat, and the limitations in the use of the so-called Carnot control variables in the power systems with solar radiation. Chapter: Appendix: A Causal Approach to Hydrodynamics and Heat Transfer provides some brief information regarding Lagrangians of free radiation systems.

13.2 Basic Problems in Radiation Thermodynamics

In the 1988 edition of his book Callen (1988) gives an introduction to basic properties of radiation systems including statistical fluctuations. Landsberg (1990a, b) reviews thermodynamical and statistical problems of energy conversion from radiation treated as a quantum system. He points out the need for nonequilibrium entropy, which is obtained as a statistical quantity for the ideal quantum gases. Maximization of this entropy leads to the equilibrium Bose distribution for photons. By analyzing electron transition rates he obtains the condition for population inversion and determines the nonequilibrium chemical potential of photons. Next he defines fluxes of energy and particle number using continuous spectrum and shows some applications. Various efficiency definitions for the diluted black-body radiation are applied to photothermal conversion in solar cells. Nice exposition of the thermodynamic quantum effects is achieved which can be described quantitatively by the quantum statistics.

Lavenda's (1991) book on statistical physics presents a probabilistic approach to the field. It applies Boltzmann's principle to express the entropy as the logarithm of a combinatorial factor, by showing that the entropy is the potential that determines Gauss's law of error for which the average value is the most probable value. Just as there are frequency and degree-of-belief (Bayesian) interpretations of statistical inference, the same should apply to statistical thermodynamics. The frequency interpretation applies to extensive variables, like energy and volume which can be sampled, while the degree-of-belief interpretations apply to the intensive variables, like temperature and pressure, for which sampling has no meaning. The connection between the two branches translates the Cramer–Rao inequality into thermodynamic uncertainty relations, analogous to quantum mechanical uncertainty relations, where the more knowledge we have about a thermodynamic variable the less we know about its conjugate. Since the lack of a probability distribution means the absence of its statistics, the possibility of an intermediate statistics, or what is referred to as parastatistics, between Bose–Einstein statistics and Fermi–Dirac statistics is nonexistent.

Essex (1984a) discusses those aspects of minimum entropy production at steady state which show a difference between the radiative transfer and the classical transfer of energy and mass in the matter. Essex (1984b, 1990) claims that the contemporary irreversible thermodynamics is primarily a theory of matter, not radiation. When the simple bilinear entropy production rate of irreversible thermodynamics is modified, to account for the entropy production due to radiation while preserving local equilibrium for matter, the simple bilinear form of that rate, on which the current theory depends, is lost. Essex (1990) argues that radiation is typically out of equilibrium, even if the matter which may interact with radiation is locally in equilibrium. In this context, he identifies ambiguities and paradoxes that arise in the case of radiation from the interpretation of bilinear forms expressed in terms of thermodynamic fluxes and forces. Most of these effects are caused by the anisotropy and nonlocality of nonequilibrium radiation. His analysis is directed toward showing that the entropy production is not adequately described by any of the bilinear forms thus far advanced. By showing that various physical pictures can lead to the same entropy production, he concludes that a bilinear form for entropy production does not define fluxes and forces nor does it define transfer behavior.

Holden and Essex (1997) apply to the analysis of radiative equilibrium a balanced approach for treating radiative entropy and energy, known as equal thermodynamic protocol. In this balanced context, radiative equilibrium is seen as either of two possible states: entropy or energy equilibrium. It is shown that these two states have a fundamental relationship to each other which leads to strong parallels to the relationships between energy and entropy in equilibrium thermodynamics. Their concept of parallellism in the relationships for energy and entropy seems to have a link with the idea of intrinsic photon entropy (Kirwan, 2004). In a further treatment Essex and Kennedy (1999) formulate a thermodynamic minimum principle for photon radiation. Following a parallel development of photon and neutrino statistics, this minimum principle is next extended to neutrinos in the case of zero mass and vanishing chemical potential. The principle stems more from that of Planck than that of classical Onsager–Prigogine irreversible thermodynamics. Its extension from bosons to fermions suggests that it may have a still wider validity.

Essex et al. (2003) ask How hot is radiation? and present a self-consistent approach to defining nonequilibrium radiation temperature is discussed using the distribution of the energy over microstates. They begin with ensembles of Hilbert spaces and end with practical examples based mainly on the far-from-equilibrium radiation of lasers. They show that very high, but not infinite, laser radiation temperatures depend on intensity and frequency. They show that heuristic definitions of temperature derived from a misapplication of equilibrium arguments are incorrect. More general conditions for the validity of disequilibrium temperatures are also established. Their arguments make clear that the noise temperature requires a randomized phase distribution for the noise signal, in keeping with the usual intuitive definition of noise. The results demonstrate some of the physical implications of temperature for nonequilibrium systems, in particular for radiation and laser light. More consequences can be inferred by applying the techniques and results presented there to other aspects of radiation, radiation transport, and lasers.

As stressed by Essex and Andresen (2013), functions, not dynamical equations, are the definitive mathematical objects in equilibrium thermodynamics. However, more than one function is often described as "the" equation of state for any one physical system. Usually these so named equations capture only incomplete physical content in the relationships between thermodynamic variables, while other equations, no less worthy of the name equation of state, go inconsistently by other names. While this approach to terminology can be bewildering to newcomers, it also obscures crucial properties of thermodynamic systems generally. Essex and Andresen (2013) introduce specific principal equations of state and their complements for ideal gases, photons, and neutrinos that have the complete thermodynamic content from which all other forms can be easily deduced. In addition to, as they say "effortlessly clarifying many smaller classical issues," they also make properties like the second law of thermodynamics and local thermodynamic equilibrium completely visual.

In his Letter to the Editor of *Solar Energy*, Badescu (2004a) described a number of evergreen problems in the radiation theory which he later developed in his paper on exact and approximate statistical approaches for the exergy of black-body radiation, Badescu (2008). There is a long-term debate in literature about the exergy of black-body radiation (BBR). Most authors contributing to this dispute used classical thermodynamics arguments. The objective of Badescu's (2008) paper is to propose a statistical thermodynamics approach. This gives new perspectives to previous results. Four simple statistical microscopic models are used to derive BBR exergy. They consist of combinations of quantum and classical descriptions of the state occupation number and entropy, respectively. In all four cases the

BBR exergy (or exergy flux density) is given by the internal energy (or energy flux density) times an efficiency-like factor containing the environment temperature and the black-body radiation temperature. One shows that Petela–Landsberg–Press efficiency is the "exact" result while the Jeter (Carnot) efficiency corresponds to the classical approximation. Other two (new) approximate efficiency-like factors are also reported.

Petela (1964, 1984, 2003) contributed much to derivation of the existing formulae for the thermal radiation exergy. In his papers many clarifications of terms used are given with interpretative comments. Discussed are the author's derivation procedures for three categories of radiation; enclosed system, black flux, and any arbitrary radiation flux. The formulae for radiation conversion to work or heat are established. The following peculiarities of thermal radiation are next discussed: irreversibility of emission and absorption of radiation, significance of thermal radiation exergy for a zero environment temperature, exergy of "lacking radiation," exergy at a varying environment temperature, and the analogy between exergy of substance and radiation. Utilization of solar radiation, by an absorbing surface, is evaluated and the optimum temperature of this surface is determined. The previous considerations were the basis for discussing the exergy radiation formulae by Spanner (1964) and Jeter (1981), as well as for discussing the viewpoints of Bejan and other researchers. The discrepancy between formulae by Petela and by Spanner arises because Spanner applied the absolute work instead of the useful work to express the maximum practically available work (exergy). The discrepancy between formulae by the author and by Jeter is because Petela applied the exergy analysis whereas Jeter (1981) developed the energy analysis in which the degradation of radiation to heat, at the same temperature, is not revealed. See also Szargut and Petela (1965) on the exergy of solar radiation, and, in particular, their Section 11.4.4 on exergy efficiency of plants and vegetation.

Ivanov and Volovich (2001) discuss the derivation of the bound on entropy and the so-called holographic principle. They estimate the number of quantum states inside space region on the base of uncertainty relation. The result is compared with the Bekenstein formula for entropy bound, which was initially derived from the generalized second law of thermodynamics for black holes. The holographic principle states that the entropy inside a region is bounded by the area of the boundary of that region. This principle can be called the kinematical holographic principle. The authors argue that it can be derived from the dynamical holographic principle which states that the dynamics of a system in a region should be described by a system which lives on the boundary of the region. This last principle can be valid in general relativity because the Hamiltonian reduces to the surface term.

Wiśniewski (1990) presents an Onsagerian-type and Glansdorff–Prigogine type analysis of optically active continua, capable of emitting and absorbing radiation. Radiative transfer in such continua (semitransparent media) is important in atmospheric physics, astrophysics, radiation gas dynamics and combustion. Effects such as radiation viscosity, radiative heat, and radiation pressure must be taken into account at high temperatures. Photochemical reactions can occur. All these effects require simultaneous treatment of the balance equations for matter and radiation. The author shows how the exchange of the radiative energy modifies the known classical expressions for the entropy source and the phenomenological equations. For example, the radiant energy source strength depends not only on the temperature difference but also on the velocity divergence (bulk viscosity effect) and the chemical affinity. Expressions for the excess entropy production and the local potential are obtained and the stability conditions determined.

Censor (1990) reviews the status of the theory of pulse propagation, whereas Grigoriew (1979) treats one-dimensional heat conduction problem at radiation impulses associated with wave models of heat transfer. Censor's treatment includes reversible propagation in lossless media and dissipative propagation in absorptive media. The applications of pulse propagation appear in various branches of physics involving electromagnetism, acoustics, elastodynamics, and heat transmission (eg, radar, sonar, geophysical exploration, and thermal waves, respectively). The transmission of mass, energy, and information also involves pulses. In lossless systems a physically consistent mathematical theory has been developed. It is based on the canonical Hamiltonian equations describing the rays (pulse trajectories) in physical space–time rather than in space only. The proper time is the quantity that attains an extremum on natural paths in physical space–time. This constitutes a generalization of the classical Fermat problem which was originally formulated, as a minimum time problem for stationary rays in physical space. The canonical equations result from the dispersion (or eiconal) equation describing the pulse phase in terms of a (quasilinear, homogeneous) partial differential equation of Hamilton–Jacobi type. In absorptive media, the pulse is deteriorated; it loses its identity and its information. The extension of the theory to absorptive media depends on the definition of the group velocity in the canonical equations. However, this case is important for the theory of dissipation. The author proposes a formalism based on the complex wave vector and the complex phase.

Conventional theories of irreversible thermodynamics and radiative transport predict instantaneous propagation of thermal and viscous effects. In Israel's (1976) basic paper, this defect can be traced to the omission of terms quadratic in the heat flux and viscous stresses from the expression for the entropy 4-vector in terms of the fluxes of 4-momentum and material. By a systematic restoration of the quadratic terms, this author constructs a generalized theory which is applicable to the description of transient irreversible effects and reconcilable with causality. Israel (1976) also discusses the relativistic formulation of the second law of thermodynamics for the exchange of heat and matter between bodies with arbitrary relative velocities.

Müller (1993) studies thermodynamics of light and sound and discusses several useful formulas of radiation hydrodynamics or radiation thermodynamics, in particular of the stress tensor of radiation in terms of densities of energy and momentum. He stresses the role of extended irreversible thermodynamics which permits the explicit calculation of the main part of equations of energy balance for photons and phonons.

13.3 Conversion of Solar Flux into a Heating Medium

Duffie and Beckman (1974) give in their book a comprehensive review of solar energy thermal processes. Solar energy can potentially play an important role in providing most of the heating, cooling, and electricity needs of the world. With the emergence of solar photocatalytic detoxification technology, solar energy also has the potential to solve our environmental problems. Yet, according to Goswami et al. (2004), "we do not see widespread commercial use of solar energy. Some of the emerging developments in solar may change that situation." Goswami et al. (2004) describe some of the new and emerging developments, with special emphasis on: (1) nanoscale antennas for direct conversion of sunlight to electricity with potential conversion efficiencies approaching 80–90%; (2) new thermodynamic cycles for solar thermal power that have the potential to reduce capital costs by 50%; and (3) solar photocatalytic oxidation for cleanup of industrial wastewater, drinking water, soil, and air. The discussion involves fundamentals of each of these developments, their potential, present status, and future opportunities for research.

Lund (1986, 1989) analyzes parallel-flow flat-plate solar collector absorbers and serpentine-flow flat-plate solar collector absorbers. Also, Lund (1990) treats the thermal conversion of solar power using a heat engine working between receiver and sink temperatures. FTT is applied to determine the upper limit to power output for the case of terrestrial and space-based solar-thermal engines. The differences between the optimal strategies for the receiver temperature obtained for terrestrial and space-based systems are discussed. The current design parameters are discussed for both cases considered, which leads to the conclusion that the FTT brings the theoretical limits closer to achievable practice.

Conversion of solar energy with temperature T into heat occurs in the presence of irreversible absorption of radiation. Petela (1964, 1984, 2003) describes mathematically exergy properties of heat radiation, whereas Bejan's and Badescu's attempt to unify different theories for the ideal conversion of the enclosed radiation (Bejan, 1982, 1987) and to determine upper-bound efficiencies for the endoreversible conversion of thermal radiation (Badescu, 1999, 2004). Mansoori and Patel (1979) define thermodynamic criteria for the choice of working fluids for solar absorption cooling systems. Kuran and Sieniutycz (2011) consider thermodynamics of solar energy conversion into heat, which is analyzed in some detail further on in this chapter.

The first law efficiency η can be defined as the ratio of the adsorbed heat flux to the enthalpy emission flux, whence

$$\eta \equiv \frac{\dot{q}}{h} = \varepsilon_a \left[1 - \left(\frac{T_a}{T}\right)^4\right] \tag{13.1}$$

(Kuran and Sieniutycz, 2011). This efficiency increases with the emissivity of the absorbing surface, ε_a. Its upper limit occurs for the black absorbing surface ($\varepsilon_a = 1$). The disadvantage of this definition are high efficiencies η obtained for low temperatures of absorption T_a, the property contradicting the fact that the practical value of the heat absorbed in temperature T_0 is equal to zero. This flaw is removed in the definition of the second law efficiency, η'.

The second law efficiency, η', is defined as the ratio of the useful effect expressed in terms of the exergy of heat, b_q, to the incident exergy of radiation

$$\eta' \equiv \frac{b_q}{b} \tag{13.2}$$

Applying the well-known Petela's equation (Petela, 1964, 1984)

$$\eta_P \equiv \frac{B}{E} = 1 - \frac{4}{3}\frac{T_0}{T} + \frac{1}{3}\left(\frac{T_0}{T}\right)^4 \tag{13.3}$$

in the exergy balance of absorbing surfaces we obtain

$$\eta' \equiv \frac{b_q}{b_P} 3\varepsilon_a \left(1 - \frac{T_0}{T_a}\right) \frac{T^4 - T_a^4}{3T^4 + T_0^4 - 4T_0 T^3} \tag{13.4}$$

On the other hand, if Jeter's (1981) formula (13.5) is used for the exergy flux,

$$\eta_J \equiv \frac{\dot{B}_f}{\dot{H}_f} = 1 - \frac{T_0}{T} \tag{13.5}$$

another form of the second-law efficiency equation is obtained

$$\eta' \equiv \frac{b_q}{b} = \frac{3}{4}\varepsilon_a \left(\frac{T_a - T_0}{T - T_0}\right) \frac{T^4 - T_a^4}{T_a T^3}. \tag{13.6}$$

The efficiency (13.6) achieves lower values than its Petela's counterpart (13.4), yet it shows the same dependence on the environmental temperature, T_0. The efficiency in terms of T_0 is shown graphically in Fig. 13.1, for constant values $T_a = 1000$ K and $T = 3000$ K. On the other hand Fig. 13.2 illustrates the dependence of energy efficiency η and exergy efficiency η' in terms of the radiation absorption temperature, T_a. Large energy efficiencies η (for low T_a) do not take into account low qualities of heat for temperatures T_a in the vicinity of the environment temperature, T_0, which proves that the popular first-law efficiency η is not a good performance index for the process in the regime discussed. A more correct efficiency, η', which is based on exergy and thus takes into account the quality of heat, does not exhibit the previously discussed flaw. As a result an upper limit of η' appears for some absorption temperature T_a.

Both efficiencies η and η', can be applied in analyses of solar collectors. It is the so-called stagnation temperature which plays a basic role in the theory of solar collector. The

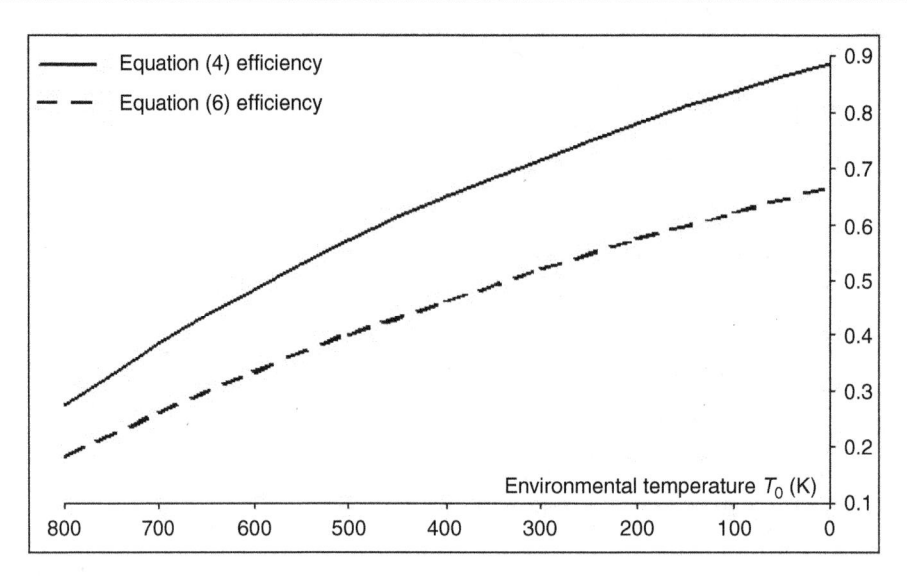

Figure 13.1: Influence of the environmental temperature on the exergy efficiency—a comparison of the implications of Petela's formula (13.4) and Jeter's formula (13.6).

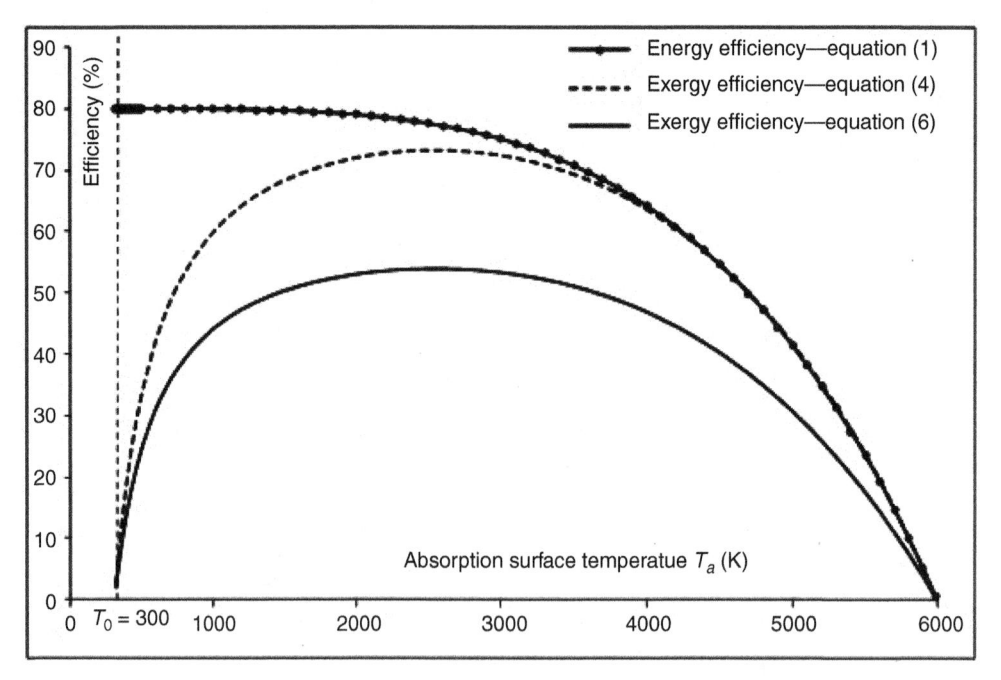

Figure 13.2: Energy efficiency η and exergy efficiency η' in terms of the absorption temperature of the radiation, T_a.

stagnation temperature is a maximum collector temperature which sets itself when the collector does not deliver any heat energy $\dot{Q}$ to the user (a receiver of the valuable energy). This is the case when $\dot{Q} = 0$ and the whole energy flux of the incoming radiation $\dot{Q}_* = \dot{q}A$ (insolation) is used to balance the heat losses to the environment, $\dot{Q}_0$. In the stagnation state these losses are maximum possible; simultaneously the stagnation temperature is the maximum possible temperature of the collector. Designating this maximum temperature as T_1 and the environment temperature as T_0 we obtain for the stagnation state of the collector

$$\dot{Q}_* = \dot{q}A = g(T_1 - T_0) \tag{13.7}$$

Here $\dot{Q}_* = \dot{q}A$ is the total insolation, whereas $g = \alpha A$ is the heat transfer conductance which determines the convective losses of heat to the environment. Eq. (13.7) thus defines the maximum collector temperature $T_1 \equiv T'_{max}$ for a prescribed insolation (the product of the radiation flux density $\dot{q}_*$ and the collector surface area A).

In an arbitrary state of the collector working, it achieves some temperature T', which is lower than the stagnation temperature T_1. The flux of solar energy delivered to the user, $\dot{Q}$, results from the energy balance

$$\dot{Q} = \dot{Q}_* - \dot{Q}_0 = g(T_1 - T_0) - g(T' - T_0) = g(T_1 - T') \tag{13.8}$$

Therefore, the energy flux delivered to the user is determined by the difference between the stagnation temperature T_1 and the current temperature of the collector, T'.

The performance criterion of the collector, which has to be maximized, is the exergy flux delivered to the user

$$\dot{B} = g(T_1 - T')\left(1 - \frac{T_2}{T'}\right) \tag{13.9}$$

The differentiation of this expression with respect to the collector temperature T' defines the derivative

$$\frac{\partial \dot{B}}{\partial T'} = g\left[(T_1 - T')\frac{T_0}{(T')^2} - \left(1 - \frac{T_0}{T'}\right)\right] = g\left(\frac{T_1 T_0}{(T')^2} - 1\right) \tag{13.10}$$

Whereas the setting of this derivative to zero leads to the optimal temperature of the collector

$$T' = \sqrt{T_1 T_0} \tag{13.11}$$

In the language of the collector's theory, this result, which is limited to the linear losses, states that the optimal collector temperature is a geometric mean of the collector's stagnation temperature T' and the environment temperature T_0.

If, however, the collector loses its energy by the radiation to the environment, then the process equations are somewhat different from those displayed earlier. The stagnation temperature satisfies then an equation

$$\dot{Q}_* = g(T_1^4 - T_0^4) \tag{13.12}$$

whereas the energy balance leads to the following energy flux delivered to the user

$$\dot{Q} = \dot{Q}_* - \dot{Q}_0 = g(T_1^4 - T_0^4) - g(T'^4 - T_0^4) = g(T_1^4 - T'^4) \tag{13.13}$$

In this case, the energy flux delivered to the user is determined by the difference between the fourth power of the stagnation temperature and the fourth power of the current collector's temperature, T'. The maximizing criterion, which is again the exergy flux, has now the form

$$\dot{B} = g(T_1^4 - T'^4)\left(1 - \frac{T_2}{T'}\right). \tag{13.14}$$

The present approach also applies to the case of more complicated forms of the function describing collector's heat losses, for example, when this function is the sum of components describing heat losses by the simultaneous convection and radiation. The stagnation temperature of the collector is then the solution of an equation

$$\dot{Q}_* = g_k(T_1 - T_2) + g_r(T_1^4 - T_2^4) \tag{13.15}$$

for a prescribed numerical value of the insolation $\dot{Q}_*$. Numerical or graphical method can be applied to solve the equation. For example, one can construct a chart $\dot{Q}_*(T_1)$, from which the solving temperature T_1 can be read off for a prescribed $\dot{Q}_*$. The heat delivered in the temperature T' to the user is

$$\dot{Q} = \dot{Q}_* - \dot{Q}_0 = g(T_1^4 - T_0^4) - g(T'^4 - T_0^4) = g(T_1^4 - T'^4). \tag{13.16}$$

The exergy flux of this heat flux, which has to be maximized with respect to the collector's temperature is described by an equation

$$\dot{B} = (g_k(T_1 - T') + g_r(T_1^4 - T'^4))\left(1 - \frac{T_2}{T'}\right) \tag{13.17}$$

From this formula an optimal collector's temperature T' can be determined graphically as the quantity maximizing the flux $\dot{B}$.

Aghbalou et al. (2006) developed exergetic optimization of a solar collector and thermal energy storage system (Garg et al., 1985, Domański, 1990). Including ecological factors, Szargut and Stanek (2007) have considered thermoecological optimization of a solar collector. The depletion of nonrenewable natural exergy resources (the thermoecological cost) was accepted as the objective function for thermoecological optimization. A detailed form of the objective function has been formulated for a solar collector producing hot water for household needs. The following design parameters have been accepted as the decision variables: the collector area per unit of the heat demand, the diameter of collector pipes, the distance of the pipe axes in the collector plate. The design parameters of the internal installation (the pipes, the hot water receiver) have not been taken into account because they are very individual. The accumulation ability of hot water comprising 1 day has been assumed. The objective function contains the following components: the thermoecological cost of copper plate, copper pipes, glass plate, steel box, thermal insulation, heat transfer liquid, electricity for driving the pump to deliver liquid fuel for the peak boiler. The duration curves of the flux of solar radiation and absorbed heat have been elaborated according to meteorological data and used in calculations. The objective function for economic optimization may have a similar form, only the cost values would be different.

We shall now pass to outline the role of Carnot control variables in solar assisted operations. Consider equations of the radiative engine, also called the Stefan–Boltzmann engine, in which both the upper and lower heat exchange undergoes in accordance with the laws of radiation (de Vos 1992; Sieniutycz, 2001e; Sieniutycz and Jeżowski, 2009)

$$q_1 = g_1(T_1^4 - T_{1'}^4) \tag{13.18}$$

$$q_2 = g_2(T_{2'}^4 - T_2^4) \tag{13.19}$$

While it is not necessary, we restrict for brevity to an endoreversible operation. Evaluating entropy production σ_s as the difference of outlet and inlet entropy fluxes yields

$$\sigma_s = \frac{q_2}{T_2} - \frac{q_1}{T_1} = \frac{(1-\eta)q_1}{T_2} - \frac{q_1}{T_1} = \frac{q_1}{T_2}\left(1 - \eta - \frac{T_2}{T_1}\right) \tag{13.20}$$

where η is the first-law efficiency. Since $\eta = 1 - T_{2'}/T_{1'}$ we obtain in terms of temperatures of circulating fluid,

$$\sigma_s = \frac{q_1}{T_2}\left(\frac{T_{2'}}{T_{1'}} - \frac{T_2}{T_1}\right) \tag{13.21}$$

Therefore after introducing an effective temperature called Carnot temperature

$$T' \equiv T_2 \frac{T_{1'}}{T_{2'}} \tag{13.22}$$

endoreversible entropy production (13.21) takes the following simple form

$$\sigma_s = q_1\left(\frac{1}{T'} - \frac{1}{T_1}\right) \tag{13.23}$$

This form is identical with the familiar expression obtained for the process of purely dissipative heat exchange between two bodies with temperatures T_1 and T'.

The endoreversible efficiency $\eta = 1 - T_2/T_{1'}$ takes in terms of T' the Carnot form

$$\eta = 1 - \frac{T_2}{T'} \tag{13.24}$$

which substantiates the name "Carnot temperature" for T'. Moreover, power produced in the endoreversible system also takes the classical form

$$p = \eta q_1 = \left(1 - \frac{T_2}{T'}\right)q_1 \tag{13.25}$$

It is essential to realize that the derivation of Eqs. (13.20)–(13.25) does not require any specific assumptions regarding the nature of heat transfer kinetics. In terms of T' description of endoreversible cycles is broken down to formally "classical" equations which contain T' in place of T_1. In fact, Carnot temperature T' efficiently represents temperature of the upper reservoir, T_1, in irreversible situations. Indeed, at the reversible Carnot point, Eq. (13.22) yields $T' = T_1$, thus returning to the classical reversible theory. Eqs. (13.23)–(13.25) then yield $\sigma_s = 0$, $\eta = \eta_C$, and $p = \eta_C q_1$, where heat flux $q_1 = 0$ for all cases in which q_1 is a function of temperature differences or the difference of some temperature functions (as, eg, in Stefan–Boltzmann equation).

Along with the efficiency $\eta = 1 - T_2/T_{1'}$ and the continuity equation for the entropy flux

$$\frac{g_1(T_1^4 - T_{1'}^4)}{T_{1'}} = \frac{g_2(T_{2'}^4 - T_2^4)}{T_{2'}} \tag{13.26}$$

the characteristics of the heat flux q_1 in terms of η may be obtained in the form (de Vos, 1992)

$$q_1(\eta, T_1, T_2) = \frac{g_1 g_2[(1-\eta)^4 T_1^4 - T_2^4]}{[g_1(1-\eta) + g_2(1-\eta)^4]}. \tag{13.27}$$

Multiplying this equation by $1 - \eta$, an expression follows for the second heat flux, q_2, so the power ouput may be calculated in terms of η as the product $q_1\eta$ or difference $q_1 - q_2$. Further, however, an alternative approach is used which applies the Carnot temperature, T'.

Since $1 - \eta = T_2/T'$, the heat flux q_1 can be expressed in terms of the temperature control T' Along with the efficiency expression we have at disposal the efficiency equality $\eta = 1 - T_2/T_{1'} = 1 - T_2/T'$, whence $T_{2'} = T_{1'}T_2/T'$, which agrees, of course, with Eq. (13.22). Substituting this expression into the continuity equation for the entropy flux (13.26) we obtain a single equation for $T_{1'}$. Then, after using this $T_{1'}$ and applying $T_{2'} = T_{1'}T_2/T'$ again, we obtain an analogous equation for $T_{2'}$. Next, with Stefan–Boltzmann equations (13.18) and (13.19) we calculate heat fluxes, q_1 and q_2. The obtained thermal characteristic that describes heat flux q_1 in terms of T' has the form

$$q_1 = g_1 g_2 \frac{T_1^4 - T'^4}{g_1 (T'/T_2)^3 + g_2} \tag{13.28}$$

Since $q_2 = (1 - \eta)q_1$ an equation is also obtained for the second heat flux, q_2. The product of Eq. (13.28) and endoreversible Carnot efficiency η yield the power expression

$$p = g_1 g_2 \frac{T_1^4 - T'^4}{g_1 (T'/T_2)^3 + g_2} \left(1 - \frac{T_2}{T'}\right) \tag{13.29}$$

Graphical search for the optimal Carnot temperature on the diagram $p(T')$ leads o the optimal power point. At the reversible Carnot point the following equalities are satisfied: $T' = T_1 = T_{1'}$, $T_2 = T_{2'}$, $\eta = 1 - T_2/T_1$, $q_1 = q_2 = p = 0$.

Closing the discussion, we recall that one of the reasons for introducing Carnot variables was their capability to yield the common equations unifying traditional and work-driven operations with heat and mass exchange (Sieniutycz, 2003a,b, 2011c; Sieniutycz and Jeżowski, 2009). This was quite successful for linear models. Here, however, the driving radiation flux, in terms of the control T' satisfies Eq. (13.28) which is similar but not identical with the equation of the corresponding process without work

$$q_1 = \frac{g_1 g_2}{g_1 + g_2}(T_1^4 - T'^4) \tag{13.30}$$

Similarly, power expression, Eq. (13.29) is not identical with an availability flux formula obtained for the processes without work. Only at the close vicinity of the short circuit point, where $T' = T_2$ and $q_1 \cong q_2$, Eq. (13.29) admits an approximation

$$p \cong g(T_1^4 - T'^4)\left(1 - \frac{T_2}{T'}\right) \tag{13.31}$$

where $g = g_1 g_2 (g_1 + g_2)^{-1}$ is the overall conductance. However, Eq. (13.29) rather than Eq. (13.31) should be used in the general case when evaluating the power output applied in solar-assisted operations. Note the exact coincidence of two considered equations in linear systems.

These results confirm the reservation which states that the formulas for overall kinetics expressed in terms of Carnot temperature T' may be identical in processes with work and without work only in linear cases. For the radiative heat transfer which is described by the nonlinear model based on the Stefan–Boltzmann formula, benefits resulting from the description in terms of the Carnot variable T' may be observed only due to the approximate similarity of the models of systems with work and without work. In terms of the Carnot temperature these models are the subject of a mutual mapping.

An additional remark is in order. In spite of the simplicity of the aforementioned model, its two "resistive parts" take rigorously into account the entropy generation caused by simultaneous emission and absorption of black-body radiation. This entropy generation may be calculated as the "classical" sum:

$$S_\sigma = q_1(T_{1'}^{-1} - T_1^{-1}) + q_2(T_2^{-1} - T_{2'}^{-1}) \tag{13.32}$$

where each q_i is given by the Stefan–Boltzmann law, Eqs. (13.18) and (13.19).

Of course, more detailed accounts of heat and power production from radiation are available (de Vos, 1992; Sieniutycz and Jeżowski, 2009). For a more comprehensive development of Carnot variables, for example, that with inclusion of mass transfer phenomena, see Sieniutycz (2011c). Because of their formal properties, approaches using Carnot controls render the mathematical modeling of energy and mass transfer in power units analogous to the modeling of energy and mass transfer in the traditional systems without explicit power fluxes, the property which facilitates the treatment of complex power systems.

Sieniutycz (2006d) analyzes contemporary trends in thermodynamics of energy generators, heat pumps, separators and other practical devices that are driven by thermal and solar energy. He focuses on applications in which the optimal control theory plays an essential role; he considers devices of engine type (generators) and of heat pump or separator type (consumers), each driven either by radiative heat or by exchanged energy or mass. He also describes difficulties in defining energy limits for transformation or consumption of solar energy and stresses links between energy limits and the classical problem of maximum work available from disequilibrium systems.

Solar drying is becoming more and more popular (Imre, 1990). Afriyie et al. (2013) have developed a simulation procedure describing the drying process within a Chimney-Dependent Solar Crop Dryer (CDSCD). The simulation follows the authors' experimental work on the effect of varying drying chamber roof inclinations on the ventilation and drying processes, and their work on the development of simulation code to help optimize ventilation in such dryers. Afriyie et al. (2013) present the modeling and subsequent validation of the drying process inside the dryer, to come out with a design tool for the CDSCD. They consider the height of the crop shelf above the drying-chamber base, crop resistance to airflow and the shading on the drying-chamber base, and the effects of these factors on the drying process.

The underload condition temperatures and velocities are predicted to within a relative difference of 1.5 and 10%, respectively of the observed values. They state that even though the heat inertia of the physical model causes deviation between the predicted drying path and the observed drying path, the two paths tend to converge at the end of each drying cycle, with a general prediction to within 10% relative difference of the observed crop moisture content. The validation results show that the simulation code can serve as an effective tool for comparing and refining the designs of the CDSCD for optimum drying performance.

A book by Mikielewicz and Cieśliński (1999) systematically treats practical systems of energy use and energy conversion, including those using unconventional energy sources in solar collectors and photovoltaics.

13.4 Maximum of Exergy or Work Fluxes in Radiation Systems

This section treats power limits in systems driven by energy of solar radiation. Power limits for exploitation and/or transformation of solar energy in industrial systems are obtained by application of the optimal control theory. One may consider devices of energy generator type (engines) and of heat pump or separator type (energy consumers), each driven either by radiative heat input or by an input of energy and/or mass fluxes from a traditional fluid. One may stress a link between a maximum power yield in an engine and the classical problem of maximum work and the exergy analysis. A similar situation is also in the realm of heat pumps which may heat the dryers, desorbers, or other separation systems. As an example, dynamic programming minimization of total work supplied to a two-stage drying system driven by heat pumps can be analyzed.

Power limits define maximum power released from energy generators or minimum work supplied to separators or heat pumps; they are important in engineering design. No design is possible that could violate these limits without changes in the system's duty. Classical thermodynamics is capable of determining work limits in terms of exergy changes. However, these limits are often too distant from reality (real energy consumption is higher than the lower bound, whereas real energy yield is lower than the upper bound). Yet, by introducing rate dependent factors, irreversible thermodynamics offers enhanced limits that are closer to reality. In examples one may focus on limits evaluated for power from the solar radiation. Limits-related analyses lead to estimates of maximum work released from a radiation engine and minimum work supplied to a heat pump. Knowing the latter one can estimate lowest supply of solar energy or microwave energy to dryers and other apparatuses.

Thermodynamic analyses are applied in modeling, simulation, and optimization of radiation engines as nonlinear energy converters. An analysis of available data for photon flux and photon density leads to an exact numerical value of photon flux constant. Basic thermodynamic principles lead to converter's efficiency and generated work in terms of driving energy flux in the system. Steady and dynamical generators of mechanical energy from radiation are investigated. In the latter, associated with an exhaust of radiation resource measured in terms of its temperature, real work is a cumulative effect obtained in a system composed of a radiation fluid, sequence of engines, and an infinite bath. This modeling leads to dynamical equations of state for radiation temperature and work output in terms of process controls. Various mathematical techniques are applied in trajectory optimization of relaxing radiation. The optimal relaxation curve is nonexponential, characteristic of radiation fluids as nonlinear systems. Algorithms for work optimization in the form of Hamilton–Jacobi–Bellman equations and dynamic programming equations lead to explicit work limits and a generalized availability. The principal performance function that expresses an optimal work depends on end thermal coordinates and a dissipation index, h, in fact a Hamiltonian of the optimization problem for extremum power or minimum entropy production. As an example of work limit from radiation a generalized available energy of radiation fluid is estimated in terms of finite rates quantified by Hamiltonian h.

In radiation fluids, which are nonlinear thermodynamic systems, fluid's properties vary along the path, and the optimal relaxation curve is nonexponential. Still the shape of the optimal curve has to be determined from the condition for the optimum power. Various differential models of controlled relaxation dynamics are studied, some of them differing with the degree of accuracy of the process description. While simpler models are easier to solve, those more complicated ones may describe the related physics in a more exact way and may be preferred. Modifications of relaxation models may also be considered, depending on the mode of energy exchange with the environment.

Gordon (1990) reviews early applications of finite time thermodynamics to problems involving applications of solar energy. By treating the Earth's atmosphere as the working fluid and investigating the solar-driven convection known as the Earth's winds, the natural convection process is modeled as an irreversible heat engine. He also analyzes solar-driven heat engines for conventional power generation. The functional form of the heat transfer is shown to have an essential effect on the performance of heat engines operating at maximum power with the most pronounced effects occurring for radiative heat transfer.

Sieniutycz and Kuran (2005) show how the nonlinear kinetics of energy transfer in propelling fluids and imperfect (non-Carnot) thermal machines can be imbedded into the contemporary theory of irreversible energy generators and heat pumps. They quantitatively describe effects of nonlinear heat transfer assuming that heat fluxes are proportional to the difference of temperature in certain power, T^a. They also show that the energy and particle transports can be treated either in a conventional way or as peculiar chemical reactions. In the latter case a recent approach distinguishes in each elementary transfer step two competitive (unidirectional) fluxes and the resulting flux follows as their difference. Nonlinear imperfect systems are investigated in the context of efficiency, heat flux, entropy production, and mechanical power, for steady and unsteady operations.

Kuran (2006) and Sieniutycz (2006d) analyze contemporary trends in thermodynamics of energy generators, heat pumps, separators and other practical devices that are driven by thermal and solar energy. They focus on applications in which the optimal control theory plays an essential role. They consider devices of engine type (generators) and of heat pump or separator type (consumers), each driven either by radiative heat or by exchanged energy or mass. They also describe difficulties in defining energy limits for transformation or consumption of solar energy, and stress links between energy limits and the classical problem of maximum work available from thermal systems. Sieniutycz and Kuran (2006) apply thermodynamic analysis in modeling thermal behavior, and work flux in finite-rate engines as nonlinear energy converters. They also perform critical analysis of available data for photon flux and photon density that leads to exact numerical value of photon flux constant. Basic thermodynamic principles lead to expressions for converter's efficiency and generated work in terms of driving energy flux in the system. Steady and dynamical processes are investigated. In the latter, associated with an exhaust of radiation resource measured in terms of its temperature decrease, real work is a cumulative effect obtained in a system composed of a radiation fluid, sequence of engines, and an infinite bath. Variational calculus is applied in trajectory optimization of relaxing radiation described by a pseudo-Newtonian model. The principal performance function that expresses optimal work depends on thermal coordinates and a dissipation index, h, in fact a Hamiltonian of the optimization problem for extremum power or minimum entropy production. As an example of work limit in the radiation system under pseudo-Newtonian approximation the generalized exergy of radiation fluid is estimated in terms of finite rates quantified by Hamiltonian h. The primary results are dynamical equations of state for radiation temperature and work output in terms of process control variables. In the second part of this paper these equations and their discrete counterparts serve to derive efficient algorithms for work optimization in the form of Hamilton–Jacobi–Bellman equations and dynamic programming equations. Significance of nonlinear analyses in dynamic optimization of radiation systems is underlined.

Sieniutycz (2007b) and Kuran and Sieniutycz (2007) treat Hamilton–Jacobi–Bellman equations and dynamic programming for power-maximizing relaxation of radiation. They develop simulation and power optimization of steady and dynamical power generators, in particular radiation engines. In dynamical cases, associated with downgrading of resources in time, real work is a cumulative effect obtained from a nonlinear fluid, set of engines, and an infinite bath. Dynamical state equations describe resources' upgrading or downgrading in terms of temperature, work output, and controls. Equations of power and converter's efficiency serve to derive Hamilton–Jacobi equations for the trajectory optimization. The relaxation curve of typical nonlinear system is nonexponential. Power extremization algorithms in the form of Hamilton–Jacobi–Bellman equations (HJB equations) lead to work limits and generalized availabilities. Optimal performance functions depend on end states and the problem Hamiltonian, h. As an example of limiting work from radiation, a generalized exergy flux of radiation fluid is estimated in terms of finite rates quantified by Hamiltonian h. In many systems governing HJB equations cannot be analytically solved. Then the use of their discrete counterparts and numerical methods is recommended. Algorithms of discrete dynamic programming (DP) are particularly effective as they lead directly to work limits and generalized availabilities. Convergence of the algorithms to solutions of HJB equations is discussed. A Lagrange multiplier helps to solve numerical algorithms of dynamic programming by eliminating the duration constraint. In discrete analyses, the Legendre transformation emerges as a significant tool.

Sieniutycz (2008a) applies thermodynamics in mathematical simulation and optimization of nonlinear energy converters, in particular radiation engines, in steady and dynamical situations. Power is a cumulative effect maximized in a system with a nonlinear fluid, an engine or a sequence of engines, and an infinite bath. Dynamical state equations are applied to describe the resource temperature and work output in terms of a process control. Recent expressions for efficiency of imperfect converters are used to derive and solve Hamilton–Jacobi equations describing resource upgrading and downgrading. Various mathematical tools are applied in trajectory optimization with special attention given to the relaxing radiation. The radiation relaxation curve is nonexponential, characteristic of a nonlinear system. Power optimization algorithms in the form of HJB equations lead to work limits and generalized availabilities. Converter's performance functions depend on end thermodynamic coordinates and a process intensity index, h, in fact, the Hamiltonian of optimization problem. As an example of limiting work from radiation, a finite rate exergy of radiation fluid is estimated in terms of finite rates quantified by the Hamiltonian h.

Many valuable research approaches to radiant heat engines in terms of the maximum power output and optimal paths come from Professor Lingen Chen's group of Chinese researchers, for example, Chen et al. (1996) and Chen et al. (2009). In particular, Li et al. (2010a) investigate maximum work output of multistage continuous Carnot heat engine system with finite reservoirs of thermal capacity and radiation between heat source and working fluid. The operation undergoes in a multistage isothermal endoreversible chemical engine system operating between a finite potential capacity high-chemical potential reservoir and an infinite potential capacity low-chemical-potential environment with the linear mass transfer law. For a fixed initial time and a fixed initial concentration of the key component in the high-chemical-potential reservoir, the continuous model was optimized by applying HJB theory and Euler–Lagrange equation, whereas the discrete model was optimized by applying dynamic programming method and discrete maximum principle. Numerical examples for the discrete model with three different boundary conditions are provided, and the results are compared with those for the continuous model. The results show that the maximum power outputs of the limiting continuous and multistage discrete isothermal endoreversible chemical engine systems are equal to the difference between their classical reversible limits and a dissipation term, in agreement with Sieniutycz, 1998b. For the maximum power outputs of both models (continuous and discrete) the relative concentration of the key component in the higher reservoir decreases linearly in time. The authors discuss the relationships among the maximum power output, the process period and the final concentration of the key component in the high-chemical-potential fluid reservoir.

Li et al. (2011) investigate the finite time exergoeconomic performance of an endoreversible Carnot heat engine propelled by a complex heat transfer. The complexity lies in generalized convective heat transfer law and generalized radiative heat transfer law, $q \propto (\Delta T^n)^m$. The finite time exergoeconomic performance optimization of the engine is tested by taking profit optimization criterion as the objective. The focus of this paper is to search the compromised optimization between economics (profit) and the utilization factor (efficiency) for endoreversible Carnot heat engine cycles. The obtained results include those obtained in many literatures and can provide some theoretical guidance for the design of practical heat engines.

Badescu (1999) derives simple upper-bound formulae for the efficiency of converting the energy of thermal radiation into mechanical work. They bounds refer to a large class of systems. They are functions of two parameters only, namely the temperatures of the energy sources. The original contributions are as follow: (1) Two sorts of "cold" thermal reservoirs are considered. (2) An endoreversible thermal engine is used to convert radiation energy into work. A number of existing theories are particular cases of the present approach Badescu (2000) derives class of accurate upper bounds for the efficiency of converting solar energy into work by taking into account (1) the irreversibilities associated with the heat transfer inside the heat engine and (2) details about the system considered, as the geometric (view) factor of the Sun, the dilution factors of solar and ambient radiation and the optical properties of the converter (concentration ratio, absorptance, reflectance, and transmittance for both transparent cover and absorber), He also gave a formula for simpler (but less accurate) upper-bound efficiency in terms of Sun and ambient temperatures. This upper bound is still more accurate than the upper-limit efficiencies usually cited in literature.

Badescu (2013) treats heat versus radiation reservoirs and concludes that relationships between lost available work and entropy generation cannot be obtained in the general case, that is, for systems in contact with arbitrary heat and/or radiation reservoirs. Reservoirs in local (full or partial) thermodynamic equilibrium are needed. The Gouy–Stodola theorem is a particular case of a more general result obtained for systems in contact with heat reservoirs in local thermal equilibrium. Radiation reservoirs cannot be assimilated to heat reservoirs. A general model of a radiation reservoir is shown. Particular cases are reservoirs containing black-body radiation, deformed black-body radiation, and nuclear radiation. Despite the case of heat reservoirs, knowledge of their temperature is not sufficient for a complete characterization of radiation reservoirs. There is a need to know other intensive thermodynamic parameters, such as the chemical potential of radiation as well as some macroscopic parameters (such as the geometric characteristics of the radiation reservoir) and microscopic parameters (such as radiation spectra or band-gaps). Some types of radiation reservoirs may behave like heat reservoirs, but only in the classical approximation.

He concludes that the popular Gouy–Stodola theorem relates the lost available work to the entropy generation in systems which interact with the environment by heat transfer. More general relationships between lost available work and entropy generation are proposed. They refer to systems in contact with heat and radiation reservoirs. No dependence between lost available work and entropy generation exists in the case of reservoirs for which intensive variables cannot be defined. Reservoirs in local thermodynamic equilibrium are needed for such purpose. Reservoirs in local thermal equilibrium are needed when systems exchanging heat are considered. The relationships between lost available work and entropy generation derived in this paper are not new and include the Gouy–Stodola theorem as a particular case. Radiation reservoirs in local thermodynamic equilibrium are considered next. The relationships between lost available work and entropy generation derived in Badescu paper are new. They are very useful when the system's environment consists of very rarefied media (for instance, in case of space solar applications) and the heat transfer between the system and the environment is negligible. Finally, systems in contact with heat and radiation reservoirs are considered. Results show that the lost available work is a relative quantity that depends on our choice of the reference heat and/or radiation reservoir. Sections 3 and 4 of the paper bring new arguments that in the general case radiation reservoirs cannot be assimilated with heat reservoirs. Examples of radiation reservoirs in thermal and chemical local equilibrium are shown in his Section 3. They include black-body radiation, deformed black-body radiation, and nuclear radiation as particular cases of a more general approach. Radiation reservoirs may be much more complex than heat reservoirs. Complete characterization of these reservoirs requires not merely knowledge of their temperature, as in the case of heat reservoirs. There is also a need to know other intensive thermodynamic parameters, such as the chemical potential of radiation as well as other macroscopic parameters (such as the geometric characteristics of the radiation reservoir) and microscopic parameters (such as radiation spectra or band-gaps). Some types of radiation reservoirs behave like heat reservoirs but only in the classical approximation. For other types of radiation reservoirs (such as those containing deformed black-body radiation), a classical approach is not possible and a quantum approach is always necessary; by their nature these reservoirs do not behave like heat reservoirs.

Work extraction from radiation reservoirs has been considered mainly in the past decades, especially in connection with the heat energy conversion. However, the radiation reservoirs are more complex than heat reservoirs. Badescu (2014) considers the simplest case of radiation reservoir (ie, black-body isotropic radiation) and the simplest case of absorber/emitter (ie, plane Lambertian black-body). Other upper-bound efficiencies are obtained for different kinds of radiation reservoirs and absorbers/emitters.

In the discussion published in Europhysics Letters Badescu (2014) asks: "Is Carnot efficiency the upper bound for work extraction from thermal reservoirs?" Since reversible upper-bound efficiencies are too high to be of practical interest, he considers irreversible engines. For radiation fluids, reservoirs are not fully characterized by their temperature. Other parameters, such as pressure and chemical potential, geometric factors, and microscopic parameters, such as the radiation spectra or band-gaps, should be known. To determine the maximum power delivery one must perform an optimization search. Badescu's (2014) treatment, which considers this issue, yields a number of new results. The most important one states that the upper bound for the reversible work extraction is not Carnot efficiency. Another result shows that all upper-bound efficiencies depend on the geometric factor of the radiation reservoir. The first result is important because it refers to reversible radiation systems; it stresses the substantial difference between the heat flux and radiation flux. The second one is easily understood for engineers in view of the role of geometric factors in problems with the radiative energy exchange.

As Badescu states, heat and radiation reservoirs are particular cases of thermal reservoirs. There are many kinds of radiation reservoirs, depending on their constitutive particles. Specific results concerning the available work transported by particle fluxes have been reported recently for black-body radiation in the classical and quantum description, deformed black-body radiation, phonons, electrons and mesons, bosons and fermions in the ultrarelativistic case, nuclear radiation particles with nonzero rest mass (ie, protons, neutrons, alpha and beta particles) in the classical approximation and in the quantum approach. These results were reviewed (Badescu, 2013). Work extraction from heat reservoirs has been considered from the early stages of thermodynamics. Work extraction from radiation reservoirs has been considered mainly in the past decades, in connection with solar energy conversion. However, there is quite a literature in this area (Sieniutycz and Jeżowski, 2013). Which is the maximum efficiency of heat conversion into work? The first law of thermodynamics may be used and the predicted upper efficiency is unity in the general case. However, when the particular case of heat reservoirs in thermal equilibrium is considered, both first and second laws of thermodynamics are used and a more accurate upper bound is found, under the additional assumption of reversible operation: it is Carnot efficiency. In Badescu research papers the maximum efficiency of work extraction from thermal reservoirs is examined in a systematic way. Both heat and radiation reservoirs are considered. The classical approach to deriving the Carnot efficiency assumes zero entropy generation rate. All upper-bound efficiencies obtained by Badescu (2014) are derived under the same assumption. Thermal reservoirs include heat and radiation reservoirs. Radiation reservoirs are more complex than heat reservoirs. Badescu considered the simplest case of radiation reservoir (ie, black-body isotropic radiation) and the simplest case of absorber/emitter (ie, plane Lambertian black-body). Other upper-bounds efficiencies would be obtained for different kinds of radiation reservoirs and absorbers/emitters. Badescu's results are summarized in Fig. 13.3.

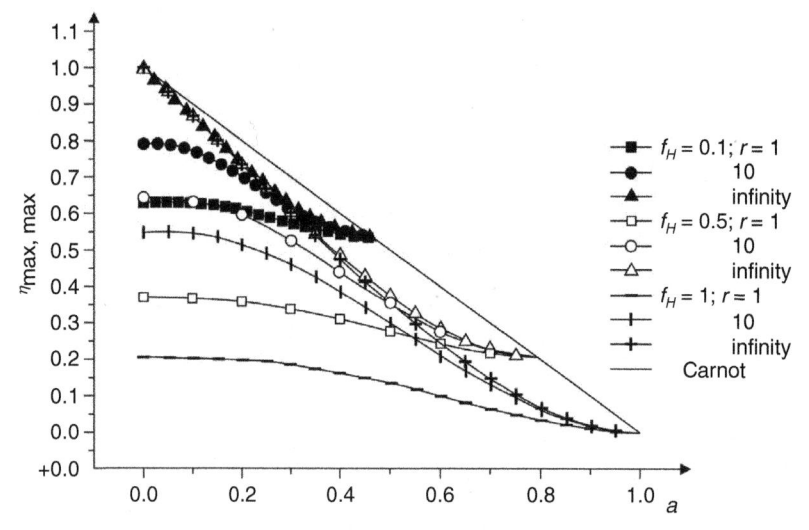

Figure 13.3: Upper-bound efficiencies obtained by Badescu (2014); $a = 1 - \eta_C$.

Here we quote in extenso the main conclusions of Badescu (2014):

1. Thermodynamic equilibrium between two bodies emitting and receiving radiation requires equality of their intensive parameters (temperature, pressure, chemical potential) as well as equality of other quantities, such as geometric factors.
2. The Carnot efficiency (Badescu' Eqs. (7) and (18)) is the upper bound for reversible work extraction from heat reservoirs; the sink may be a heat or a radiation reservoir.
3. The Carnot efficiency η_C is *not* the upper bound for work extraction from radiation reservoirs. η_C is numerically larger than the maximum efficiency $\eta_{\text{max,max}}$, Eqs. (28) and (38), obtained in Badescu (2014). His Eqs. (28) and (38) are given as Eqs. (13.33) and (13.34), respectively.

$$\eta_{\text{max,max}} = 1 - \frac{4}{3}a + \frac{1}{3}\frac{a^4}{f_H} \qquad (a \leq f_H^{1/3}) \qquad \text{Badescu' Eq.(28)} \qquad (13.33)$$

$$\eta_{\text{max,max}} = 1 - \frac{1+r}{f_H}\left(\frac{f_H + ra^3}{1+r}\right)^{4/3} + \frac{ra^4}{f_H} \qquad \text{Badescu' Eq.(38)} \qquad (13.34)$$

However, $\eta_{\text{max,max}}$ is the upper bound, since efficiency values between $\eta_{\text{max,max}}$ and η_C are not allowed. Note that η_C is positive for $a \geq f_H^{1/3}$, when in fact the work extractor does not provide a positive work rate.

4. The efficiency of Badescu's Eq. (28) is the upper bound for work extraction from a radiation reservoir, the sink being a heat reservoir. This result is new. Petela–Landsberg–Press

efficiency is a particular case of Badescu's Eq. (28): it applies only for hemispherical radiation reservoirs ($f_H = 1$).

5. The efficiency of Badescu's Eq. (38) with $r \to \infty$ is the upper bound for work extraction from a radiation reservoir, the sink being a radiation reservoir. This result is new. Petela–Landsberg–Press efficiency is a particular case of Eq. (38): it applies only for hemispherical radiation reservoirs ($f_H = 1$) and $r = (A_e/A_a) \to \infty$.

Green (2004) reviews recent developments in photovoltaics. Although photovoltaics cells have been used since the 1950s in spacecraft, the interest in their terrestrial use was heightened by the oil embargos of the early 1970s. Since then, a steadily growing terrestrial industry has developed. According to Green (2004) the photovoltaic market is booming with over 30% per annum compounded growth over the past 5 years. The government-subsidized urban–residential use of photovoltaics, particularly in Germany and Japan, is driving this sustained growth. Most of the solar cells being supplied to this market are first-generation devices based on crystalline or multicrystalline silicon wafers. Second generation thin-film solar cells based on amorphous silicon/hydrogen alloys or polycrystalline compound semiconductors are starting to appear on the market in increasing volume. Australian contributions in this area are the thin-film polycrystalline silicon-on-glass technology developed by Pacific Solar and the dye-sensitized nanocrystalline titanium cells developed by Sustainable Technologies International. In these thin-film approaches, the major material cost component is usually the glass sheet onto which the film is deposited. After reviewing the present state of development of both cell and application technologies, the likely future development of photovoltaics is outlined. Jarzebski (1990) gives in his book an advanced physical analysis of solar-energy photovoltaic conversion as the application of the quantum theory to solid states.

Sánchez-Orgaz et al. (2013) treat recuperative solar-driven multistep gas turbine power plants in terms of the recuperator influence on the effectiveness of a multistep solar-driven Brayton engine. The solar collector model includes heat losses from convection and radiation. The Brayton engine includes an arbitrary and several realistic irreversibility sources. The combination of both systems makes the evolution of the overall efficiency not trivial because of the losses arising from the coupling of the working fluid with the collector and the surroundings. The overall efficiency admits a simultaneous optimization in regards to the pressure and temperature ratios. When the design is close to the optimum values of those parameters, an increase in the effectiveness of the recuperator is configuration independent, that is, is always associated with an increase in the overall optimum efficiency.

13.5 Solar Buildings and Solar Systems

Sertorio and Tinetti (2000) investigated the strategy of the "back up" for a class of thermodynamic automata on example of solar buildings. A building that has the capacity to control the thermal flow interactions with the outside, the in- and outflows, and in addition is able to control the internal thermal flows between adjacent zones, formally acts as a thermodynamic automaton. The automaton rule minimizes the distance of the temperature of the core (living space) from a predetermined desired temperature. The automaton rule must manage the chaotic variations of the external inputs and also its own capacity to learn whether the external inputs belong to summer or winter and so forth, because according to season the automaton strategy changes (Sertorio and Tinetti, 2000). If backup injection of heating (or cooling) is necessary, how can this be operated? In fact, the building is "alive," and the injection of energy automatically perturbs the automaton logic; it may fool the automaton into making it believe that it is summer when instead it is winter (or vice versa). A problem of similar nature is known to pilots of highly automated airplanes, for instance in the process of automatic landing, where the intervention of the human action has caused disasters. It is also known in medical science where the introduction of a drug always carries basically unknown side effects (Sertorio and Tinetti, 2000).

Chwieduk (2006) evaluated solar energy availability and its natural thermal conversion in a building envelope. Solar radiation incidents on transparent and opaque elements of a building envelope and is transmitted through these elements into rooms, due to different heat and mass transfer, and optical phenomena, and can influence in a different way on energy balance of a building and in a consequence on its thermal comfort. The importance of knowledge of solar energy availability and its influence on energy balance of a building is crucial for building design and construction. In high-latitude countries, such as Poland, one focuses only on the winter heating season, which can lead to overheating of rooms in summer. The main aim of Chwieduk's (2006) study is to evaluate mathematical models of solar radiation availability and thermal conversion of solar energy in a building envelope in changing conditions in time and to analyze it in details. Nowadays, when a building envelope is designed and constructed according to energy-savings measures, with high-quality thermal insulation and building materials, the heat transfer through the opaque external walls takes the minor part in the total energy transport between the outdoor and indoor environments. An important element of the energy balance of a building is the heat needed for ventilation, even if recuperation of waste heat (from ventilation) is accomplished. However, the most important elements of the energy balance of a building are windows. Windows become the neuralgic energy element of the building envelope. The role of windows in the energy balance of a building increases with their size because of their relatively quick response to changing conditions of the outdoor environment, that is, ambient temperature and solar radiation. In middle- and high-latitude countries a lot has been already done for improving the opaque building envelope, but still not so much for windows. For the purpose of a good building design it is necessary to calculate solar radiation availability on surfaces with different inclination and orientation. The calculations of solar radiation incident on surfaces with different azimuth and inclination angles have been performed using the averaged representative hourly solar radiation data for two models of solar radiation: the isotropic diffuse sky model, Hottel–Woertz–Liu–Jordan model, and the anisotropic sky mode, for example, Hay–Davies–Klucher–Reindl (HDKR).

Results of comparative analysis have shown the distinction between two models and indicated the importance of the anisotropic model for its application in evaluation of energy balance of a building and design of a building. The results of calculations give indications for shaping of a building envelope and planning its surrounding. To describe and solve the problem of dynamics of processes in a building envelope and surroundings, the mathematical model of energy transfer phenomena in opaque and transparent elements has been developed. Focus has been put on influence of solar energy and because of that special attention has been given to energy transfer through windows. The heat transfer of the external opaque walls has been treated in less detailed manner. In order to determine the effect of window and frame construction, size, and orientation on a building's annual energy consumption for both heating and cooling, computational models have been developed. These models have taken into account the variations of ambient temperature, the direct and diffuse solar radiation, and the thermal and optical properties of construction materials, as well as actual component dimensions and orientation. In order to reduce the number of variables to a level that allows different design options to be compared in a meaningful way, it has been useful to make some simplification. The first has been to propose a representative averaged ambient temperature and solar radiation regime (taken as that of Warsaw, Poland; Chwieduk, 2006), together with a fixed room size and room temperature requirement. This means that the changes to annual energy use that are influenced by the changes in window size, orientation and inclination may be compared easily. The other simplification has been done in the modeling of energy transfer outside and inside the window and wall, as well as within the cavity formed by the glass sheets. A special quasi-three-dimensional heat-transfer model of the window edges and frame, a simplified one-dimensional model of the central part of glazing and one-dimensional model of the opaque wall was developed. They have included unsteady heat conduction in the window edges, window frame, and walls, and detailed equations to describe the radiation exchange between the ground, sky, and window, solar radiation absorption, transmission or reflection on all surfaces, and the effects of orientation and inclination on them (Chwieduk, 2006).

The developed model allows many cases to be evaluated in a given time and allows broad conclusions to be drawn. For simulation of a developed mathematical model the MATLAB program has been used (Chwieduk, 2006). The ambient temperature and solar radiation data for Warsaw have been built into it. It is possible to change the wall construction materials (their parameters) and thicknesses together with window glass and frame sizes and properties as well as the inclination and orientation (slope and azimuth angles). Because of assumed and elaborated solar radiation representative models, 12 averaged days (one for each month) are simulated (with different time steps for different elements of building envelope), each day being repeated enough times to ensure a steady solution. Conduction within solids and radiative heat transfer with surrounding are based on the fundamental equations. Convection is dealt with using the approximated correlations for the various situations: internal, external, enclosed and for different slopes. The results of considered cases for a full year are presented in a range of graphical forms showing how the energy demand changes by month, by hour in the day, and by contributory factor as required. Results of energy demand changes, for different window sizes, orientations, and inclinations show the influence of solar energy on energy balance of considered room cases. In the result of simulation studies, it has turned out that overheating in summer due to high insolation level could be a real problem for some buildings' shapes and constructions (Chwieduk, 2006). To avoid the overheating in such buildings it would be necessary to introduce air-conditioning systems. If nothing is done for improvement of windows and there role in building design, especially to windows and rooms at attics, the residential and tertiary sectors are expected to give the real fast-growing market for implementation of HVAC systems in the country. The problem is that in Poland during the designing process of residential buildings and most of the tertiary sector the analysis of the heat needs is restricted only to the winter heating season. The reduction of heat losses during the winter has become a priority. Attention to heat gains during summer is not common. It can turn out very quickly that the costs of energy delivered to the building for air conditioning, because of unacceptable summer comfort, could be higher than costs of energy supplied for space heating. The standard limits only for seasonal space heat consumption are not enough for effective annual energy conservation in buildings. The standard energy consumption indexes for space cooling (air-conditioning) need also to be introduced in the process(Chwieduk, 2006).

In the realm of photosynthesis, Pirard et al. (2014) study the kinetics of photocatalytic degradation of a pollutant. The variation of the volume of the reactor necessary to monitor the pollutant degradation might change the ratio between the amount of pollutant and of catalyst involved in the photocatalytic reaction and thus might falsify the modeling of the photocatalytic activity. However, kinetic studies published up to now in the literature do not take the volume variation into account. In this study, the influence of the experimental sampling procedure on the determination of kinetic parameters is highlighted and it is explained how to correctly determine the kinetics of a photocatalytic degradation reaction in the case of powders and films. Mass balances are developed and the particular case of a first-order reaction is treated. Furthermore, the error on the reaction rate constant induced by neglecting the volume variation is quantified. Experimental measurements with powders and films have been performed and validate theoretical statements developed in this study.

Heterogeneous semiconductor photocatalysis is an advanced oxidation process (AOP) which was proven to be a promising technology for the total mineralization of most of the organic pollutants present in water by using natural or artificial light. The most frequently used photocatalyst is titanium oxide, TiO_2, in the form of powder or thin film. Two model molecules are generally used: methylene blue under artificial UV light, or 4-nitrophenol under UV–visible light. During the photocatalytic experiments, pollutant solution aliquots are sampled at regular time intervals over a period of several hours to measure the absorbance of the solution before and after illumination using a UV–Vis spectrophotometer at a given wavelength depending on the pollutant.

In most of published studies, an apparent first-order kinetic model is adjusted on the experimental data to determine the reaction rate constant, without taking into account the variation of the volume in the reactor due to the sampling of the pollutant solution. See Pirard et al. (2014) for suitable references. The aim of their work is to highlight the influence of the operating variables as well as the experimental sampling procedure on the determination of the kinetic parameters. In order to model the system, the sampling procedure is approximated by a continuous sampling of the solution with a given flow rate.

De Vos (1995) treats thermodynamics of photochemical solar energy conversion by deriving and analyzing suitable models of endoreversible thermodynamics. These models treat stationary situations, but nevertheless are accurate enough to treat situations far from equilibrium. He investigates both a chemical engine and a photovoltaic engine. Brought together, these two engines model a photochemical engine.

Kåberger and Månsson (2001) contribute to the discussion on the relation between entropy, thermodynamics, and economic theory. With respect to thermodynamic constraints on the economy, there are two diametrically opposite positions in this discussion. One claims that the constraints are insignificant ("of no immediate practical importance for modeling") and in the intermediate run, do not limit economic activity and, therefore, need not be incorporated in the economic theory. The other holds that thermodynamics tells us that there are practical limits to materials recycling, which already puts bounds on the economy and, therefore, must be included in the economic models. Using the thermodynamic concept of entropy (Kåberger and Månsson, 2001) show that there are fundamental problems with both positions. Even in the long run, entropy production associated with material dissipation need not be a limiting factor for economic development. Abundant energy resources from solar radiation may be used to recover dissipated elements. With a simple, quantitative analysis it may be shown that the rate of entropy production caused by human economic activities is very small compared to the continuous natural entropy production in the atmosphere and on the Earth's surface. Further, the societal entropy production is well within the range of natural variation. It is possible to replace part of the natural entropy production with societal entropy production by making use of solar energy. Society consumes resources otherwise available for coming generations. However, future generations need not have fewer resources available to them than the present generation. Human industrial activities could be transformed into a sustainable system where the abundant elements are industrially used and recycled, using solar energy as the driving resource. An economic theory, to fit to guide industrial society, must not disregard theory nor must it overstate the consequences of the laws of thermodynamics (Kåberger and Månsson, 2001).

Służalec and Muskalski (1987) studied by numerical analysis heat-flow problems in biological tissues under the influence of laser radiation. They stressed, among others, the significant role of laser surgery, and, in general, great possibilities resulting from the application of laser light in various fields of technology and medicine.

Ebeling's and Feistel's (2011) book on physics of self-organization and evolution is unique in covering experimental and theoretical fundamentals of self-organizing systems as well as such selected features as random processes, structural networks, and multistable systems, while focusing on physical and theoretical modeling of natural selection and evolution processes. Role of solar fluxes in creation and sustaining of origins of life is mentioned.

13.6 Closing Remarks

Many separation systems may be supported by flows of solar energy. Drying or other separation operation may be also driven by fluxes of mechanical work (power). Suitable methodology to describe such processes is based on the definition of thermodynamic limits that is common for operations involving heat pumps, separators, energy converters, and other practical devices. We may focus on work limits for transformation or consumption of solar energy and applications in which an essential role is played by optimization and the optimal control theory. Two kinds of devices may be considered: engines (energy generators) and the units of heat pump or separator type (energy consumers). Each device may be driven either by the radiative heat flux or by the coupled fluxes of energy and mass in a traditional fluid. A link with the classical problem of maximum work and exergy analysis is worth pointing out. Among relatively new concepts, the notion of the Carnot control variables (Section 13.4) plays an important role. In particular, one can demonstrate their role in the analysis of heat pumps that drive drying systems or other separation systems. As an example, dynamic programming minimization of total work supplied to a two-stage drying system driven by heat pumps is described and analyzed (Sieniutycz, 2000c).

Work limits define maximum work released from energy generators and minimum work supplied to separators or heat pumps; they are important in engineering design. No design is possible that could violate these limits without changes in the system's duty. Classical thermodynamics is capable of determining work limits in terms of exergy changes. However, they are often too distant from reality (real energy consumption is higher than the lower bound and real energy yield is lower than the upper bound). Yet, by introducing rate dependent factors, irreversible thermodynamics offers enhanced limits that are closer to reality. In examples considered here we focus on limits evaluated for the work from the solar radiation. Limits-related analyses lead to estimates of maximum work released from a radiation engine and minimum work supplied to a heat pump. Knowing the latter, one can estimate lowest supply of solar energy or microwave energy in dryers.

Abandoning endoreversibility, Sieniutycz (2006b) and Sieniutycz and Jeżowski (2013) have shown that for a quite general class of imperfect devices and aparatuses the thermal efficiency η of an irreversible thermal machine assumes a simple, pseudo-Carnot form

$$\eta = 1 - \Phi T_2 / T' \tag{13.35}$$

which generalizes endoreversible Eq. (13.24) of Section 13.3. Certain irreversible (rate dependent) availability may be attributed to Eq. (13.33). With an appropriate definition of the Carnot temperature T' (Section 13.3), which refers, in fact, to an operator depending on the actual fluid's temperature and its rate change in time, Eq. (13.33) applies to internally irreversible systems in which Φ is the so-called internal irreversibility factor within the thermal or solar machine. In steady systems an expression for mechanical power p supplied to a heat pump heating a drying agent follows in the form

$$p = q_1 \eta = g'(\Phi, T_1, T_2)(T_1 - T')(1 - \Phi T_2 / T') \qquad (13.36)$$

Eq. (13.34) is suitable to evaluate power fluxes produced or consumed in irreversible systems. One may use this equation to optimize solar dryers and heat pumps. Krieger et al. (2016) perform parameter studies of heat storage in concentrated solar power plants required to compensate for variable availability of solar radiation. Reversible hydrogenation reactions as storage systems in solar plants are modeled. The energy density achievable with thermochemical heat storage is higher than that for molten salt which represents the state-of-the-art technology. The efficiency of different reversible hydrogenation reactions as thermochemical heat storage systems have been examined, since they can be operated at appropriate temperatures. In long-time storage, examined systems are more efficient than sensible heat storage.

CHAPTER 14

Appendix: A Causal Approach to Hydrodynamics and Heat Transfer

14.1 Introduction

The conservation equations for energy and momentum in thermohydrodynamic systems are identical for reversible and irreversible processes provided that in both cases they are considered in their general nontruncated form which does not eliminate nonequilibrium effects such as heat flux and nonequilibrium stress (Hirschfelder et al., 1954; Sieniutycz, 1994). This general form holds in the extended state space which is spanned not only on the traditional thermodynamic coordinates, such as densities of components and entropy, but also on the corresponding fluxes. The number of all coordinates of the extended state or the state dimensionality is a minimum number of variables necessary and sufficient to characterize the system. The energy-momentum tensor of a thermohydrodynamic system is thus the property of the extended state or the state described by a minimum number of thermohydrodynamic variables. Presence of both mechanical and thermal variables in the set of the state coordinates is essential for complete description of real thermohydrodynamic systems. The identity of nontruncated conservation laws in both cases (reversible and not) reflects the general property of the system in which internal contributions cancel out so that the momentum and energy balances are unaffected by irreversible sources. In particular, for reacting systems it is known from both statistical mechanics (Tolman, 1949; Hirschfelder et al., 1954; Chapman and Cowling, 1973) and thermohydrodynamics (Bird et al., 1960; Wiśniewski, 1984; Wiśniewski et al., 1973; Sieniutycz, 1994) that the presence of reaction rates does not change the general analytical structure of energy and momentum equations. The described property is, in fact, the basis in Onsager's theory of nonequilibrium thermodynamics (Onsager, 1931a, b).

In the so-called adiabatic fluid the heat flux is eliminated by the adiabaticity assumption which truncates general conservation laws or rules out heat flux as a variable; thus the state space is incomplete and the above property does not apply. The reduced number of state coordinates in the perfect fluid model causes the necessity to augment the perfect-fluid conservation laws when they are applied to real nonadiabatic systems, that is, systems with finite heat flux (reversible or not). The augmenting procedure, which is advised even in the best textbooks on theoretical physics (Landau and Lifshitz, 1974a, 1987) is purely formal. Although that procedure need not result in errors in the equations themselves, it is often understood in the form of the fundamentally incorrect conclusion that any disequilibrium flux, such as heat flux or nonequilibrium stress, is the irreversible property so that the irreversibility influences the conservation laws. This incorrectness persists in many otherwise good papers on thermohydrodynamics. In fact, however, the general (nontruncated) energy-momentum tensor in the full state space is irreversibility-independent, which is a very useful property to obtain conservation laws of real complex systems.

In the present chapter this property is exploited to obtain conservation laws in relativistic thermohydrodynamics of irreversible gravitating fluids from a reversible while extended formulation of Hamilton's principle of least action (Sieniutycz, 1992b, 1994, 2002c). With the extension, the early relativistic theory of heat transfer worked out by Schmid (1967b, 1970a, b) can be extended to include the flux of the entropy and phenomenon of thermal inertia. Sieniutycz (1998c) has constructed a generalized relativistic Lagrangian which yields the heat flux q and nonequilibrium stress τ in the energy-momentum tensor of an extended reversible fluid with thermal inertia. The actual momentum of heat (thermal momentum related to the entropy flow) follows to be many orders of magnitude larger than q/c^2 (c is the light speed) but it is consistent with Grad's (1958) kinetic theory and with experiments in heat conduction (Sieniutycz, 1994, 2002b). The resulting conservation laws exhibit the heat flux and nonequilibrium stress, thus they may be used as components of irreversible models containing explicitly irreversible equations. Such equations are, for example, Fourier's law of heat conduction or its non-causality-violating counterpart called the Cattaneo equation (Cattaneo, 1958; Vernotte, 1958; Chester, 1963). After its combining with the energy balance, the latter serves to obtain the hyperbolic (damped-wave) equation for temperature, that is, Eq. (14.3) below or its more exact extensions.

Some literature references cited in the present chapter serve to mention particular concepts and formal techniques, and do not exhaust even in a small part very large research material obtained for relativistic continua. For a coherent review of macroscopic approaches to relativistic thermomechanics of continuous media a book by Baranov and Kolpascikov (1974) is recommended. Use of composite variational principle is based on the theory of Caviglia (1988), who also showed a way of derivation of the conservation laws. Elementary calculus of variations is used here to derive the Lagrange equations of motion for the relativistic fluid (Gelfand and Fomin, 1963). With these results conservation laws and matter tensor can be found, and other problems, such as stability and causality in dissipative fluids, may be considered (Hiscock and Lindblom, 1983). Diffusion processes can be treated in curved phase space (Graham, 1977). Even dissipative phenomena can be included provided that the Riemannian geometry is replaced by the Finslerian geometry (Ingarden, 1996). Applications are broad and may lead to variational principles for stellar structures (Kennedy and Bludman, 1997); this text reformulates the four basic equations of stellar structure as two alternate pairs of variational principles. Applying the geometry of spaces with the Jacobi metric, Szydłowski et al. (1996) formulated a generalized Maupertuis principle for systems with the Lagrangian and an indefinite form of the kinetic energy. The generalization was applied to the theory of gravity and cosmology where the metric determined by the kinetic energy form has a Lorentz signature. The theorem was proved concerning the behavior of trajectories in a neighborhood of the boundary of the region admissible for motion. This region is not a smooth manifold but turns out to be a differential space of constant differential dimension. This fact allows us to use geometric methods analogous to those elaborated for smooth manifolds. It is shown that singularities of the Jacobi metric are not dangerous for the motion; its trajectories are smooth in the sense of the theory of differential spaces. Sandoval-Villalbazo and Garcia-Colin (1999) performed an analysis of the classical derivation of the continuity equation in general relativity. The consequences of the equation obtained were discussed and compared with similar results related with the conservation of particles (baryons and leptons). Emphasis was placed on the fact that different transport equations may be derived from the possible choices of the mass balance equation in nonequilibrium formalisms. Nevertheless, the continuity equation agrees with the notion of particle conservation.

Pavon et al. (1980) treat heat conduction in relativistic extended thermodynamics. Pavon et al. (1983) consider equilibrium and nonequilibrium fluctuations in relativistic fluids (see more in chapter: Thermodynamic Lapunov Functions and Stability). Pavon (1992) deals with extended relativistic thermodynamics (ERT). His paper deals with applications to astrophysics and to cosmology, rather than dealing with the general theory. The reader interested in causal relativistic theories is referred to the papers by Israel (1976) and Hiscock and Linblom (1983). The second moments of fluctuations are determined from the Einstein probability formula involving the second differential of the nonequilibrium entropy. These moments are subsequently used to determine the transport coefficients of heat conductivity, bulk viscosity, and shear viscosity of a radiative fluid. By exploiting correlation formulae, second-order coefficients of transport are determined. Nonequilibrium fluctuation theory, assuming the validity of the Einstein formula close to equilibrium, is used to determine the nonequilibrium corrections of the bulk viscous pressure. An analysis of survival of the protogalaxies in an expanding universe completes the presentation of astrophysical applications of ERT. Cosmological applications are discussed next. It is shown that nonequilibrium effects and transport properties (such as, eg, bulk viscosities) play a certain role in the cosmic evolution of the universe. In that context, the following issues are outlined: cosmological evolution (described by a Bianchi-type model with the bulk viscosity), production of entropy in the leptonic period, the inflationary universe, and FRW cosmology. Some cosmological aspects of the second law are also discussed.

In conventional radiative hydrodynamics, the dissipative fluxes are proportional to the gradients of the temperature, velocity, and so forth, with proportionality constants inferred from the linearized radiative transfer equation. The dynamical evolution is determined by the conservation of matter, energy, and momentum. As stated by Schweitzer (1988), conventional theory breaks down if the geometry under consideration is *photospheric*, that is, if emission/absorption processes are weak compared to scattering and photons can "random walk" over large distances without being absorbed. The fundamental dynamical equations of motion include the relativistic transfer equation and form the starting point for establishing extended versions of radiative hydrodynamics. In Schweitzer's (1988) treatment the transfer equation is replaced by a hierarchy of relativistic moment equations and the dissipative fluxes are treated as dynamical variables in their own right. The general moment equations involve source terms which are a priori undetermined. If the radiation is almost thermal the source terms can be expressed with Rosseland means; see, for example, Wikipedia: Opacity (optics). In this case, the moment equations reproduce the structure equations postulated by Israel (1976), for general dissipative continuous media. In the limit of slow motion, the moment equations are studied separately and interpreted as hyperbolic diffusion equations.

Sieniutycz's (1990) treatment, while nonrelativistic, deals with a macroscopic extension of the de Broglie microthermodynamics (de Broglie, 1964) so as to preserve the usual thermostatistical effects ignored in the de Broglie theory. The extension includes basic macroscopic concepts, the usual statistical temperature T and the chemical potential μ, in a generalized macroscopic theory of fluids. The derivative $d = \partial E/\partial \rho_s$ of the energy density E with respect to the entropy density ρs, is found as the sum of the intrinsic relativistic temperature of de Broglie θc^2 (a counterpart of the $c2$ term in the relativistic chemical potential $g = c_2 + \mu$), and the usual statistical T, so that $d = \theta c_2 + T$ and $\theta = m/k_B$. Nonequlibrium multicomponent fluids are investigated in the context of the Hamiltonian action and the resulting conservation laws, under a "nonrelativistic" approximation of low transport velocities and for the c_2 terms ignored. The sourceless constraints for mass and entropy (reversibility) are enough to derive the standard form of conservation laws for the energy and momentum, improved by the presence of a generalized energy of diffusion. Thermal inertia effects appear naturally due to the finite value of θ. The resulting thermohydrodynamics exhibits many new features both for the static and dynamic aspects of the theory. In particular, both the thermal equilibrium and the heat flow are affected by the gravitational field. The entropy flux is associated with a momentum (thermal momentum), and causes phenomena such as heat flow, self-diffusion, and nonequilibrium stresses, which are usually regarded as irreversible. General Gibbs' equations are obtained which take into account the role of the transport velocities or momenta, and the macroscopic averages of the quantum phases of the species. A unification is achieved for the mechanical and thermal Legendre transformations within the field hydro-thermodynamics. It is shown that for inherently irreversible cases the thermodynamic formalism must involve quantum phases explicitly and that the entropy source is incorporated in the Gibbs' equation as the derivative of the energy with respect to the thermal phase.

Further developments of this approach are formulated in the relativistic context. Sieniutycz (1992b, 1998c, 1994, 2002b) has constructed a general relativistic Lagrangian which yields the effect of heat flux q and nonequilibrium stress τ in the energy-momentum tensor of an extended reversible fluid exhibiting thermal inertia. The actual momentum of heat (thermal momentum related to the entropy flow) follows to be many orders of magnitude larger than q/c^2 (c is the light speed), but it is consistent with Grad's (1958) kinetic theory and with experiments in heat conduction. On the other hand, the net momentum of heat remains q/c^2, in agreement with the standard relativistic result, this net momentum being the result of incomplete compensation of the actual thermal momentum and the momentum associated with self-diffusion of particles. The classical densities of mass and entropy, ρ and ρ_s, cease to be natural variables of energy density E in the sense of Callen (1988) whenever inertial effects prevail. This fact necessitates the use of what may be called the thermal potential T^-, a new quantity replacing the classical temperature T. Changes in thermodynamic formalism require the replacement of T by T^-. The admission of a freely varied four-flux of entropy in an extended Hamilton principle implies all nonequilibrium corrections (q and τ) to the energy-momentum tensor, making it possible to investigate the effect of nonequilibrium phenomena on the properties of associated gravitational fields. For the instantaneous properties of this field it is inessential whether the origins of heat flux **q** and of nonequilibrium stress γ in the energy-momentum tensor **G** are reversible or not, which is the property that makes **G** directly applicable to real dissipative fluids. Selected aspects of these relativistic formulations are contained in the present chapter subsequently.

14.2 Action, Lagrangian and Thermohydrodynamic Potentials

Here we outline the role of Lagrangians (or related Hamiltonians) in development of the theory of thermohydrodynamic transformations and related Gibbs' equations which describe various thermodynamic potentials. This information is purposely condensed for the sake of brevity. For more details the reader is referred to previous publications (Sieniutycz, 1990, 1992b, 1994, 2002b). We restrict our discussion in this section to the Lorenzian frames of special relativity.

The basis of the variational theory of a reversible fluid in Eulerian representation of fluid's motion is the field action, A, based on the four-dimensional integrand over pressure P. Due to the relativistic invariance of four-volume element $dVdt$ the invariance of A is consistent with the invariance of P. For example, in the Lorenzian frames

$$A = \int P \, dV \, dt = \int P^0 \, dV^0 \, dt^0 = A^0, \qquad (14.1)$$

where the superscript 0 refers to rest-frame quantities. (For the simplicity of notation single integrals are symbolically used in place of multiple four-dimensional and three-dimensional integrals.) As four-dimensional volume element $dVdt$ is the relativistic invariant, or $dVdt = dV^0 dt^0$, Eq. (14.1) implies the invariance of P in the light of the invariance of A. Moreover, in view of the invariance of *P* and self-consistency of thermodynamic transformations, the three-dimensional integral of Mathieu thermodynamic potential (the ratio of the grand potential and temperature *T*)

$$\Omega = \int \frac{PdV}{T} = \int \frac{P^0 dV^0}{T^0} = \Omega^0 \tag{14.2}$$

is Lorentz invariant only if the temperature transforms in the same way as the volume *V*, that is, a moving body appears cooler. This means that the relativistic temperature transformation satisfies the Einstein–Planck transformation formulae rather than Ott's transformations. Indeed, a comparison of Eqs. (14.1) and (14.2) proves the frequency-like transformation for the relativistic temperature. This agrees with the interpretation in which temperature is a measure of the frequency of thermal agitation. More details on this issue can be found in Section 14.5, where transformations of a generalized temperature refer to systems with thermal inertia.

Associated with the entropy density ρ_s is the total entropy flux $\mathbf{J}_s$, whose diffusive part is defined operationally as ratio of density of heat flux $\mathbf{q}$ and *T* in the frame of fluid particles. In an arbitrary frame the ratio $\mathbf{u}_s = \mathbf{J}_s/\rho_s$ defines the velocity of entropy transfer which is composed on the sum of hydrodynamic velocity $\mathbf{u}$ and diffusive part $\mathbf{q}/T\rho_s$. The quantities ρ_s and $\mathbf{J}_s$ and constitute a relativistic four-vector. The ratio $\mathbf{J}_s/\rho_s$ describes the absolute velocity of entropy flow, $d\mathbf{x}_s/dt$, or the change of the entropy coordinate $\mathbf{x}_s$ (Lagrangian thermal coordinate) in time *t*. The idea of the Lagrangian thermal coordinate is associated with the Lagrangian formulation of the Maxwell–Cattaneo hydrodynamics (Grmela and Teichman, 1983).

The above consideration means that besides using the hydrodynamic concept of moving "fluid particles" we also employ the parallel concept of moving "thermal elements" associated with quantal "heat particles" or quasi-particles of the thermal process (Markus and Gambar, 2005; Vazquez et al., 2009), in fact relatives of phonons or quasiparticles of quantized acoustic fields. Various hypotheses for thermal mass and thermal inertia are known (Veinik, 1966; Sieniutycz and Berry, 1997). Quantum theories, such as de Broglie microthermodynamics (De Broglie, 1964, 1970), offer a reinterpretation of wave mechanics and may accommodate EIT effects (Sieniutycz, 1990, 1992b). An approach toward constructing the Hamiltonian formulation as a basis of quantized thermal process (Markus, 2005) has turned out to be quite successful. The approach is based on a generalized Hamilton–Jacobi equation for dissipative heat flow (Markus and Gambar, 2004, 2005; Vázquez et al., 2009) whose physical implications differ substantially from those ignoring dissipation (Schmid, 1967a).

Only in reversible processes is the identity $\mathbf{u}_s = \mathbf{u}$ satisfied. In the case of irreversible processes both velocities $\mathbf{u}_s$ and $\mathbf{u}$ differ. The simplest example is the heat transfer in a resting medium where only $\mathbf{u}_s$ is nonvanishing (the consequence of a finite $\mathbf{q}$), whereas the hydrodynamic velocity **u** equals zero. The velocity of the entropy transfer $\mathbf{u}_s$ is a concept more suitable and more exact than the concept of the velocity of energy transfer because the energy density and energy flux are not scalars but components of a four-tensor.

The kinetic potential density L is an important quantity in the hydrodynamic theory of heat-conducting fluid. It is essential that in the realm of thermodynamic transformations L needs to be considered as a function of densities and velocities, whereas the energy density E (which is the Legendre transform of L with respect to velocities) is a function of densities and momenta, $\mathbf{p} = \partial L / \partial \mathbf{u}$, $\mathbf{p}_s = \partial L / \partial \mathbf{u}_s$, and so forth. On the other hand, negative partial derivatives of L with respect to densities of entropy and particles are respectively the temperature T and chemical potential μ. A simpler example is the kinetic potential of black photons which represent the perfectly disordered radiation. In this case the contribution of the rest mass to L vanishes due to the diminishing mass of the photon, and only density of internal energy defined by the Stefan–Boltzmann law contributes to L. Moreover, in this case, a truncated kinetic potential holds

$$L = -\rho_e^0(\rho_s^0, \rho_n^0) = -\rho_e^0(\rho_s \sqrt{1 - u_s^2/c^2})$$

which proves that densities ρ_e^0 and L are independent of ρ_n^0. Therefore, for black photons, partial derivative $\mu = -\partial L / \partial \rho_n = 0$, which is the classical result in the thermodynamics of radiation. Moreover, partial derivative $T = -\partial L / \partial \rho_s = T^0 \sqrt{1 - u_s^2/c^2}$ defines the temperature of the black photon system that moves with velocity $\mathbf{u}_s = \mathbf{J}_s / \rho_s$. The same formulas can be derived by considering the energy density E in terms of corresponding momentum densities.

Here we give merely an introduction to the important problem of thermodynamic potentials of thermohydrodynamic systems (Sieniutycz, 1990, 1992b, 1994). Functions or functionals that can be obtained from L or E through Legendre transformations are called the thermohydrodynamic potentials (THP). Their dependence on velocities or momenta, and their general properties make them the extensions of the classical mechanical Lagrangians and usual thermostatic potentials. Their perfect differentials describe various Gibbs' equations for the thermohydrodynamic system.

One may distinguish three types of the THPs. Those obtained from the kinetic potential L by the thermodynamic Legendre transformation of thermodynamic variables (densities, intensities) are called the THPs of the Lagrangian type, or Lagrangians. These THPs always contain velocities. Those obtained from the kinetic potential L (or other Lagrangian) by mechanical Legendre transformation involving all velocities are called the THPs of the energy type, because they always contain the momenta, the natural variables of the energy. For example, the pressure function $P(\mathbf{u}, \mathbf{u}_s, \rho, \rho_s)$ is a Lagrangian type THP, whereas the potential $\Omega(\mathbf{p}, \mathbf{p}_s, \rho, \rho_s)$, which coincides with the negative of P only in the static case when the role of the kinetic terms is inessential, is a THP of the energy type. In the static situation all Lagrangians become the usual thermodynamic potentials with negative sign, and all the energy type THPs, as well as Lagrangian type THPs, become the usual THPs.

The third group is composed of mixed structures, containing both velocities and momenta (and any thermodynamic variables). They are called Routhians, *per analogiam* with such mixed structures in classical mechanics. From the viewpoint of their sign they may behave both like Lagrangians and like energy type THPs. The Routhian form is especially useful if some but not all of the extensive (coordinate-like) variables do not appear in the Lagrangian.

Further details and the corresponding examples of some THP's are given in subsequent sections where we consider systems containing massive particles and exhibiting thermal inertia.

14.3 Basic Information on Thermal Inertia

By applying equations of Cattaneo type instead of the Fourier's law, one takes into account the phenomenon of thermal inertia, which causes a gradual change of the heat flux under a rapid change of the temperature gradient (Chester, 1963). This phenomenon has attracted the researchers' attention as the way of solving the well-known paradox of infinite propagation of thermal disturbances contained in the classical Fourier's model of heat conduction. In resting systems with nonvanishing thermal inertia an extended Fourier law holds which links the heat flux operator $\mathbf{Q} \equiv \mathbf{q} + \tau \partial \mathbf{q}/\partial t$ with the negative product of the thermal conductivity and temperature gradient, $-\lambda \nabla T$ (Chester, 1963). The operator contains the thermal relaxation time τ, whose order of magnitude is that of the mean time between molecular collisions. In a stationary rigid continuum of constant specific heat capacity C, thermal conductivity λ and density ρ the combination of the extended Fourier law with the conservation law for the thermal energy leads to the hyperbolic equation for the temperature

$$\nabla^2 T - \frac{\partial^2 T}{c_0^2 \partial t^2} - \frac{\partial T}{\tau c_0^2 \partial t} = 0, \tag{14.3}$$

where $c_0 = \sqrt{\lambda/\rho C \tau}$ is the propagation speed of thermal disturbances. Thermal waves of finite speed implied by this equation, known collectively as second sound, were first detected in solid helium ^{3}He (Ackerman et al., 1966) and then in high purity dielectric crystals of sodium fluoride, NaF (Jackson et al., 1970). It was shown soon that classical irreversible thermodynamics is insufficient to describe these phenomena correctly. Only extended irreversible theories (Jou et al., 1988, 2001), which take into the contribution of the heat flux to the (disequilibrium) entropy, are apropriate. Such contributions are obtained in nonequilibrium statistical mechanics (both relativistic and nonrelativistic), in particular in Grad's moment approach to the solution of Boltzmann kinetic equation (Grad, 1958). Along with standard thermal conductivities λ, nonvanishing values of thermal relaxation times

τ and of the corresponding finite values of c_0 can be estimated from the Grad's solutions, both nonrelativistic and relativistic, as shown in many recent works on the subject (Jou et al., 2001).

It follows from the results of Grad's analysis (Sieniutycz and Berry, 1989, 1997) that, in the entropy representation, the finite speed thermal waves are associated with a local deviation internal energy from its equilibrium, $\Delta\rho_e = (1/2)(\rho_e)^{-1}c_0^{-2}\mathbf{q}^2$. In terms of the entropy flux $\mathbf{j}_s = \mathbf{q}/T$, the deviation is: $\Delta\rho_e = (1/2)(\rho_e)^{-1}Tc_0^{-2}\mathbf{j}_s^2$. (In nonequilibrium media which are by nature inhomogeneous the actual local value of the entropy density ρ_s is taken instead of ρ_s^{eq} of linerized models.) One can thus introduce and then work with the coefficient $\theta = Tc_0^{-2}$, which is constant since c^0 is of the order of the thermal speed, $(c^0)^2 \cong k_B T/(am)$ (Chester, 1963; Jou et al., 2001), where k_B is the Boltzmann's universal constant and a is close to 1. (The constant a equals 1 for an ideal gas.) The constant $\theta = am/k_B$ depends only on the mass of the particle, m. Nowadays diverse hypotheses exist which predict values of the numerical coefficient a at the quotient m/k_B in θ. These are discussed elsewhere (Sieniutycz and Berry, 1989, 1997). θ is, in fact, the thermal inertia per unit of the entropy (Sieniutycz and Berry, 1997). Occasionally the density $\rho_{s'} \equiv \rho_s\theta$, where is the entropy density in mass units, is used. The important conclusion stemming from the constancy of θ is that the statistical mechanical evaluation of the disequilibrium energy in the form $\Delta\rho_e = (1/2)(\rho_s)^{-1}\theta\mathbf{j}_s^2$, or in the equivalent form $\Delta\rho_e = (1/2)\rho_s\theta\mathbf{v}_s^2$ ($\mathbf{v}_s \equiv \mathbf{j}_s/\rho_s$) is the rest-frame-velocity of entropy transfer), is consistent with the simple two-phase model of the medium in which the massive entropy and bare matter are two independent massive entities, and the kinetic energy of each entity has the classical structure of the usual kinetic energy. The inertial entropy and bare matter may thus be regarded as two phases which flow separately of each other in a general disequilibrium situation. The classical heat conduction process in solids in which the solid skeleton does not move ($\mathbf{u}_m = 0$) and the entropy still flows due to a nonvanishing heat flux is the suitable example. In essence, there are a few basic contemporary hypotheses of thermal inertia: those of Grad–Boltzmann, de Broglie, and Veinik (Sieniutycz and Berry, 1997). While Grad's results follow from the kinetic theory, Veinik (1966, 1973) formed his ideas on the electrochemical grounds; they stem from his experimental investigations of simultaneous transfer of heat and electicity in a "thermodynamic couple." On the surface, the model based on the Grad–Boltzmann theory seems to exclude the linearity of the kinetic energy of heat with respect to θ. However, this is only an effect of the barycentric reference frame for heat flow contained in that theory: the inertial heat is defined there in the frame of the vanishing momentum, $\mathbf{\Gamma} = \mathbf{0}$, not in the frame of bare particles. While the barycentric frame causes no difficulties for the inertialess heat, in processes with pronouced thermal inertia it necessarily involves a particle drift, in the direction opposite to the heat flow. This involves extra kinetic energy of the particle motion which causes apparent values of θ, say θ', that are state dependent. Yet, the effect of inertia of pure heat flow, defined as energy flow in the frame of particles (where the particle drift vanishes), is linear with respect to θ. (When the frame is different the boundary conditions for heat flow are changed into those which may disagree with the experiment.) In fact, this consideration proves that Eckart's (1940) particle frame is the most natural frame to describe the heat transfer in the way consistent with real experiments. For an observer at rest in this frame, the flux of particles is zero, and the heat flux is the flow of energy relative to the particle stream (Landau and Lifshitz, 1987). Thus the particle frame is both suitable and essential in relativistic descriptions. When the frame is properly adjusted, it appears from the relativistic Lagrangian that the linearity and constancy of inertial coefficient θ hold also in models based on the Grad–Boltzmann theory. Therefore all analyses here use a constant value of θ, the thermal inertia per unit of entropy.

14.4 Matter Tensor of General Relativistic Theory

In this chapter we deal with general relativistic systems in motion, in which, as shown by Israel (1976), second-order terms in the entropy four-flux are sufficient to overcome the paradox of infinite propagation speeds of thermal signals. Israel's theory involves the energy-momentum tensor G^{ik} as the quantity generalizing mass in both a covariant form of the Gibbs' equation and a generalized Gibbs–Duhem formula containing the four-velocity. It follows that the corrections to the classical entropy and entropy flux must be of, at least, second order in dissipative fluxes to make the theory compatible with standard thermodynamics in the quasistatic limit.

The basic ingredients of the standard relativistic theory of a one-component fluid are the symmetric energy-momentum tensor G^{ik}, the particle four-flux J^i, and the entropy flux S^i ($i, k = 1, \ldots, 4$). Both G^{ik} and J^i are sourceless, that is,

$$G^{ik}_{;i} = 0, \quad J^i_{;i} = 0. \tag{14.4}$$

The entropy four-flux, S^i, of an irreversible process obeys the second law constraint: the positive entropy production $S^i_{;i} \geq 0$, where the equality sign refers to a limiting reversible process to which the Hamilton's principle can be applied. (This is done in Section 14.6, where Hamilton's principle of the stationary action is applied for the action functional (14.16) constructed for the standard relativistic kinetic potential (14.14).) Eq. (14.4) contains the energy and momentum balances and the balance of matter. The signature convention is $(++ + -)$. U_i is the particle frame four-velocity and $h^{ik} = g^{ik} + c^{-2}U^iU^k$ is the projection tensor. The energy-momentum tensor of the Eckart's (1940) theory has the following structure

$$G^{ik} = c^{-2}(E^0 U^i U^k + q^i U^k + q^k U^i) + \tau^{ik} + P h^{ik}, \tag{14.5}$$

where the quantities q^i and τ^{ik} satisfy $q^i U_i = 0$; $U_i \tau^{ik} = 0$.

Eqs. (14.4) and (14.5) refer to Eckart's relativistic scheme. Yet, in the framework of Eckart's (1940) theory of heat flow, thermal inertia is possible only as relativistic phenomenon. Moreover, Eq. (14.5) is not derived therein; it is rather only postulated on the basis of general covariance principles. This is not a case of the present approach. Our inclusion of thermal inertia is accompanied with the recognition of the importance of a free entropy flow (independent of the flow of the particles) in the Hamilton's principle. On this basis we shall derive Eq. (14.4) and some of its important associates from the stationary action approach. The resulting theory, which is in the spirit of an extended thermodynamics, is an extension of Ray's (1972) variational construction of the energy-momentum tensor in general relativity G^{ik}. We shall show that not only the formula (14.4) for G^{ik} can be derived in a direct way but also that definitions of heat **q** and stress **τ** can be furnished in G^{ik}.

14.5 Thermal Mass and Modified Temperatures

The hypothesis of the thermal mass incorporates the assumption that part of the observed rest mass of a macroscopic body is of purely thermal origin, meaning that it should be attributed to entropy rather than to particles. Consistently, the kinetic potential (unconstrained Lagrangian) L is used which is based on the split of the background relativistic energy $\rho^0 c^2$ into the "bare matter" part, $\rho_m^0 c^2$, and the "thermal part," $\rho_s^0 \theta c^2$, such that their sum remains equal to $\rho^0 c^2$. The zero superscript refers to the rest frame quantities. The particle of the traditional theory, where the entropy is inertialess, is "clothed in the thermal mass robe" while that of the modified theory is a bare (unclothed) particle (De Broglie, 1964, 1970; Sieniutycz and Berry, 1997). This reorientation does not change either the internal energy ρ_e of the medium or the observed density of mass ρ (and many other important thermodynamic quantities, as the pressure P), but it allows to share the inertial responsibilities between the entropy and matter in a more balanced way. Suggested by the Grad–Boltzmann theory, the mass–entropy equivalence constant θ is applied, which is the amount of the "thermal mass" per unit of the entropy. For the internal consistency of the desccription, the explicit use of thermal mass requires however some modifications in definitions of thermodynamic quantities, the most important being the redefinition of the temperature which may be simply regarded as the consequence of the passage from the conventional variables (ρ_s, ρ) to the „canonical variables (ρ_s, ρ_m) in the Gibbs' equation for the differential of the relativistic internal energy. In accord with the above considerations, the hydrodynamic velocity **u** and the total mass density ρ satisfy the relations

$$\rho u = \rho_m u_m + \theta \rho_s u_s, \quad \rho = \rho_m + \theta \rho_s. \tag{14.6}$$

Using these relationships we shall now consider the differential of the relativistic internal energy density in terms of the densities of the entropy and bare mass (ρs and ρm) or "canonical densities". The related (relativistic) intensity variables are subscripted by asterixes. Yet, since these intensities coincide with the usual temperature T and chemical potential μ only when θ equals zero and the effect of usual T is decreased, we designate them with –0 superscript. The index 0 refers to the rest fame whereas the minus sign points out their basic property according to which they transform according to Planck–Einstein when passing to moving frames (cooler intensities; Sieniutycz, 1998c). In the rest frame and in terms of canonical densities

$$dE^0 = \frac{\partial E^0}{\partial \rho_s^0} d\rho_s^0 + \frac{\partial E^0}{\partial \rho_m^0} d\rho_m^0 \equiv T_*^{-0} d\rho_s^0 + \mu_*^{-0} d\rho_m^0. \tag{14.7}$$

The energy differential can next be expressed in terms of traditional densities ρ_s^0 and ρ^0

$$dE^0 = T_*^{-0} d\rho_s^0 + \mu_*^{-0} d(\rho^0 - \theta\rho_s^0) = (T_*^{-0} - \theta\mu_*^{-0}) d\rho_s^0 + \mu_*^{-0} d\rho^0. \tag{14.8}$$

Comparison of this result with the traditional form Gibbs' equation which operates with the traditional densities ρ_s^0 and ρ^0

$$dE^0 = T_*^0 d\rho_s^0 + \mu_*^0 d\rho^0 = T^0 d\rho_s^0 + (\mu^0 + c^2) d\rho^0 \tag{14.9}$$

yields the desired connection between the relativistic intensities

$$\mu_*^{-0} = \mu_*^0 \equiv \mu^0 + c^2; \quad T_*^{-0} = T^0 + \theta\mu_*^0 \equiv T^0 + \theta(\mu^0 + c^2). \tag{14.10}$$

The asterisk-free intensities are their nonrelativistic counterparts. The canonical intensities (with minus superscript) represent then the usual relativistic chemical potential and its thermal analogue, T_*^{-0}, which we call the *thermal potential*. The ratio T_*^{-0}/θ has units of the chemical potential and, as it represents the modified temperature T_*^{-0} in the energy units, it may be regarded as the *chemical potential of the thermal matter*. Eq. (14.10) implies that the rest frame nonrelativistic canonical chemical potential coincides with the classical one, whereas the nonrelativistic thermal potential T^- differs from the classical temperature T due to the contribution of the nonrelativistic chemical potential. As shown elsewhere (Sieniutycz, 1998c) the transformations contained in Eq. (14.10) set the appropriate rest-frame inputs (nonrelativistic T^{-0} and μ^{-0}) to the relativistic transformations of thermodynamic intensities in moving systems. And, it is the *relativistic* canonical intensities, not their truncated nonrelativistic counterparts, that obey the well-known Planck–Einstein formula for the relativistic temperature transformation. At disequilibrium these relativistic intensities incorporate two different velocities of thermal and bare matter, $\mathbf{u}_s$ and $\mathbf{u}_m$. For example, in the Lorentzian frames of special relativity

$$T_*^- = T_*^{-0}\sqrt{(1-\mathbf{u}_s^2/c^2)}; \quad \mu_*^- = \mu_*^{-0}\sqrt{(1-\mathbf{u}_m^2/c^2)}. \tag{14.11}$$

When $\theta = 0$ and for small velocities, the velocity $\mathbf{u}_m$ approaches the hydrodynamic velocity $\mathbf{u}$, $T = T^0$, and only the chemical potential transforms; $\mu = \mu^0 - \mathbf{u}^2/2$. This is well known in the theory of hydrodynamic fluctuations (Keizer, 1987a). Thus, in the framework of the present theory, temperature satisfies the Einstein–Planck transformations. This result is confirmed by

De Parga et al. (2005), who derived the scheme of thermodynamic relativistic transformations inspired by the Planck–Einstein theory, but changing the relativistic transformation of energy. This change permits the form invariance of thermodynamics. The key point to obtain these results was their use of finite time thermodynamics (FTT), to demonstrate and creatively exploit the relativistic invariance of thermal efficiency.

14.6 Tensor of Matter Including Heat and Viscous Stress

We shall determine the matter tensor **G** for a general relativistic fluid with thermal inertia. The standard kinetic potential of the general theory can be expressed in terms of the canonical densities,

$$L = -\rho^0 c^2 - \rho^0 e(\rho^0, \rho_s^0) = -(\theta \rho_s^0 + \rho_m^0)[c^2 + e(\rho_m^0, \rho_s^0)]. \tag{14.14}$$

The corresponding Lagrangian density in its invariant form adjoints with the help of Lagrange multipliers relevant constraints containing the four-velocity vectors U_s^i and U_m^i (or corresponding fluxes), each obeying the same formulae as the four-vector of velocity $U^i \equiv (U^\alpha, U^4)$,

$$U^i U^i = \mathbf{U}\mathbf{U} - U^4 U^4 = -c^2, \tag{14.13}$$

where $i = 1, \ldots, 4$ and $\alpha = 1, \ldots, 3$. In this formalism any four-flux vector can be written as

$$(\mathbf{J}, J^4) = (\rho \mathbf{u}, \rho c) = \left(\rho^0 U^\alpha, \rho^0 U^4\right) = \rho^0 U^i. \tag{14.14}$$

With subscripts s and m these equations apply both to thermal and bare matter. Our approach here extends to media with heat the variational formalism first formulated by Ray for perfect or adiabatic fluids (Ray, 1972). The four-velocity constraint

$$g_{ik} U^i U^k + c^2 = 0 \tag{14.15}$$

contains in the case of gravitational fields the metric tensor g_{ik}. The signature convention is $(+++-)$. Inclusion of the both degrees of freedom (flows of the entropy and matter) and absorbing constraints by Lagrange multipliers yields the Lagrangian density:

$$\begin{aligned}
\Lambda &= \frac{c^3}{16\pi\kappa'}(-g)^{1/2} R - c^{-1}(-g)^{1/2}[(\theta\rho_s^0 + \rho_m^0)c^2 + \rho_e^0(\rho_s^0, \rho_m^0)] \\
&\quad + (-g)^{1/2}\gamma_s(g_{ik}U_s^i U_s^k + c^2) + (-g)^{1/2}\gamma_m(g_{ik}U_m^i U_m^k + c^2) \\
&\quad + (-g)^{1/2}\eta_s(\rho_s^0 U_s^i)_{;i} + (-g)^{1/2}\phi_m(\rho_m^0 U_m^i)_{;i} \\
&\quad + (-g)^{1/2}\lambda_s X_{s,i} U_s^i + (-g)^{1/2}\lambda_m X_{m,i} U_m^i
\end{aligned} \tag{14.16}$$

where κ' is the gravitational constant. R is the Riemannian curvature scalar which determines the Lagrangian of the gravitational field; its implicit form is here sufficient. The first line of the above equation represents the kinetic potential L of the gravitational field and matter. The second line contains the four-velocity constraint (14.15) applied for both thermal and bare matter. The continuity equations of the reversible evolution are adjoined to L in the third line of this equation. Its last line contains the corresponding identity constraints (Ray, 1972).

The energy-momentum tensor is obtained by varying the action with respect to g_{ik}, $\rho_s^0, \rho_m^0, U_s^i, U_m^i, \gamma_s, \gamma_m, \phi, \eta$, and λ which yield a set of equations of motion. The extremum conditions with respect to the densities are

$$\eta_i U_s^i = -c^{-1}(\theta c^2 + T^{-0}); \qquad \phi_i U_m^i = -c^{-1}(c^2 + \mu^{-0}), \qquad (14.17)$$

and those with respect to the velocity components

$$\gamma_s g_{ik} U_s^i = \rho_s^0 \eta_k - \lambda_s X_{s,k}; \qquad \gamma_m g_{ik} U_m^i = \rho_s^0 \phi_k - \lambda_m X_{m,k}. \qquad (14.18)$$

When the above conditions are combined with the velocity constraints for Us and Um, we obtain

$$\gamma_s = (2c^3)^{-1} \rho_s^0 (\theta c^2 + T^{-0}), \gamma_m = (2c^3)^{-1} \rho_m^0 (c^2 + \mu^{-0}). \qquad (14.19)$$

Thus the multipliers of four-velocities are additive components of the relativistic enthalpy density of the fluid.

Einstein's equations are contained in the extremum conditions of the action with respect to the components of the metric tensor

$$\frac{\partial \Lambda}{\partial g_{ik}} - \left(\frac{\partial \Lambda}{\partial g_{ik,r}}\right)_{,r} + \left(\frac{\partial \Lambda}{\partial g_{ik,r,l}}\right)_{,l,r} = 0. \qquad (14.20)$$

For the Lagrangian (14.18) these equations are obtained in the usual form

$$E^{ik} = (8\pi\kappa'/c^4) G^{ik}, \qquad (14.21)$$

where E^{ik} is the Einstein's tensor of the gravitational field. The energy-momentum tensor or the matter tensor, G^{ik}, is the source of this field. The matter tensor G^{ik} is affected by the heat flow, thermal inertia, and the nonequilibrium stress. Here it is obtained in the form

$$\begin{aligned} G^{ik} = & c^{-2} \rho_s^0 (\theta c^2 + T^{-0}) U_s^i U_s^k + c^{-2} \rho_m^0 (c^2 + \mu^{-0}) U_m^i U_m^k \\ & + g^{ik} [-(\theta \rho_s^0 + \rho_m^0) c^2 - \rho_e^0 + \rho_s^0 (\theta c^2 + T^{-0}) + \rho_m^0 (\theta c^2 + \mu^{-0})] \end{aligned} \qquad (14.22)$$

where the conditions (14.19) for the Lagrangian multipliers have been used. Since the expression in the last line is exactly the pressure scalar, P, we find,

$$G^{ik} = c^{-2}\rho_s^0(\theta c^2 + T^{-0})U_s^i U_s^k + c^{-2}\rho_m^0(c^2 + \mu^{-0})U_m^i U_m^k + g^{ik}P. \tag{14.23}$$

This equation takes into account effects of heat and nonequilibrium stress (the total viscous stress in the case of a purely dissipative fluid) through relative four-velocities of the entropy and bare matter with respect to the hydrodynamic four-velocity. It allows one to investigate the effect of the heat flow on the solution of the Einstein's equations. The rest mass (or relativistic energy *without* the statistical term) furnishes the definition of the hydrodynamic velocity U^i

$$\theta\rho_s^0 U_s^i + \rho_m^0 U_m^i - \rho^0 U^i = 0 \tag{14.24}$$

[equivalent to the set represented by Eq. (14.6)]. With this, Eq. (14.23) can be cast into the extended form of **G** of Eq. (14.5)

$$G^{ik} = c^{-2}(E^0 U^i U^k + q^i U^k + q^k U^i) + \tau^{ik} + Ph^{ik}, \tag{14.25}$$

which uses the projection tensor $h^{ik} = g^{ik} + c - 2U^i U^k$. An equivalent form of this result is

$$G^{ik} = c^{-2}[(\rho^0 c^2 + \rho^0 e^0)U^i U^k + q^i U^k + q^k U^i] + \tau^{ik} + h^{ik} P. \tag{14.26}$$

In these equations q^i is the four-vector of heat expressed as

$$q^i = \rho_s^0(\theta c^2 + T^{-0})U_s^i + \rho_m^0(c^2 + \mu^{-0})U_m^i - \rho^0(c^2 + h^{-0})U^i \tag{14.27}$$

and τ^{ik} is the four-tensor of nonequilibrium stresses expressed as

$$\tau^{ik} = c^{-2}[\rho_s^0(\theta c^2 + T^{-0})v_s^i v_s^k + \rho_m^0(c^2 + \mu^{-0})v_m^i v_m^k]. \tag{14.28}$$

Here $v_s^i = U_s^i - U^i$ is the relative four-velocity. The heat flux is defined as the difference between the actual energy flux and the energy flux of a corresponding perfect fluid. Eq. (14.26) is known (Ray, 1972; Sieniutycz and Berry, 1997), but the expressions (14.27) and (14.28) are relatively unknown. Note that in terms of the traditional temperature T^0 the relativistic heat flux equals $q^i = T^0 \rho_s^0 (U_s^i - U^i) = T^0 \rho_s^0 v_s^i$. Indeed, with Eq. (14.24)

$$\begin{aligned} q^i &= T^{-0}\rho_s^0 U_s^i + \mu^{-0}\rho_m^0 U_m^i - h^{-0}\rho^0 U^i = T^{-0}\rho_s^0(U_s^i - U^i) + \mu^{-0}\rho_m^0(U_m^i - U^i) \\ &= (T^{-0} - \theta\mu^{-0})\rho_s^0(U_s^i - U^i) = T^0 \rho_s^0(U_s^i - U^i) = T^0 j_s^i. \end{aligned} \tag{14.29}$$

Eqs. (14.28) and (14.29) may be regarded as macroscopic definitions of the heat flux and nonequilibrium stress tensor in barycentric frames. Unlike adiabatic fluid models, the superconductor model with a thermal inertia preserves both heat and nonequilibrium stress, which is the substantial improvement. For a special case when $U_s^i = U_m^i = U^i$, the tensor G^{ik} simplifies to the form

$$G^{ik} = c^{-2}\rho^0(c^2 + h^0)U^i U^k + g^{ik}P \qquad (14.30)$$

which describes an adiabatic, perfect relativistic fluid.

14.7 Conclusions and Final Remarks

We have derived (not just assumed or guessed) relativistic definitions of heat and nonequilibrium stress, Eqs. (14.28) and (14.29), from the extended Hamilton principle allowing thermal degrees of freedom. Flows of the matter and inertial entropy have similar effect on the matter tensor, and the split of the total mass into the thermal mass and bare mass does not change observable effects at thermal equilibrium. Nonequilibrium descriptions are benefited by the concept of the thermal mass, where both the heat flow **q** and the nonequilibrium stress **τ** emerge as effects of the entropy flow (thermal mass flow) in the fluid frame.

The substantial virtue of the approach based on the Lagrangian of a superconducting fluid is that it does not truncate terms in the tensor **G**; the obtained energy flux contains both the heat flux, **q**, and the nonequilibrium flux of momentum, **τ**. In the standard model of adiabatic (perfect) fluid, they are absent. Thus nontruncated conservation laws are found, usable even for dissipative fluids, since they contain the disequilibrium fluxes **q** and **τ**.

Clearly, the superconducting model does not admit any dissipative mechanisms for fluxes. In fact, the assumed zero entropy production admits that fluxes can only be related to purely reversible effects, such as "ballistic" nondissipative heat transfer or elastic transport of momentum. This reversibility is a typical limitation of all classical action-type approaches. Yet, for the derived structure of **G**, conservation laws and gravitational metrics g_{ik}, are immaterial whether the fluxes **q** and **τ** evolve reversibly or not. As long as models are accepted which don't eliminate the entropy flow (the case of the present model and not that of adiabatic fluid), general conservation laws are obtained with fluxes **q** and **τ** from a two-phase reversible model of a superconducting fluid. This condition is the only requirement necessary to describe gravitational metrics in general relativity, where the relativistic tensor **G** is the unique source of the gravitational field, Eq. (14.21).

Therefore, our conclusion should remove a frequent misunderstanding concerning the role of dissipative effects in the relativistic theory of gravitation. Here the effect of dissipation on gravitational fields is shown to be indirect at most: as a possible phenomenon causing definite flows, **q** and **τ**, which could otherwise be attributed to some reversible causes. This is similar to effects of electric currents which cause magnetic fields regardless of whether they are reversible (caused by the motion of the conductor) or irreversible (caused by the conductivity electrons).

In view of its link with the de Broglie thermodynamics (de Broglie, 1964), the present chapter may be seen as an approach that bridges the mainstream conservative ideas of relativistic thermodynamics (Tolman, 1949), with recent leading edge exterior works calling for exploration of alternatives to a Bigbang universe (Amoroso et al., 2002). Of interest in this realm is also Grössing's (2000) book that discusses the contemporary conflict between relativity and quantum mechanics, and regards it as the apparent conflict. The author proposes a resolution of the dilemma based on de Broglie's causal interpretation (de Broglie, 1964, 1970) as elaborated by David Bohm. He claims that a "medium" or "aether" may be introduced in a manner consistent with both relativity and quantum theory, and which allows the two theories to be unified by the identification of circularly causal processes at their core. He also describes several experiments confirming his predictions.

As an example of a healthy link between the cybernetics, control, and optimal control, series of papers by H. Rosenbrock treating, in particular, quantum mechanics via control theory, are recommended (Rosenbrock, 1986, 1995, 1997, 1999). Indeed, starting with the well-known fact that the Hamilton–Jacobi equation can be derived from Hamilton's variational principle by the methods of control theory. Rosenbrock (1986) shows that a suitable random disturbance added to the formulation of Hamilton's principle leads to Schrödinger's equation, and to some other results in quantum theory. Rosenbrock (1995) extends in several directions his earlier development of quantum mechanics from a stochastic variational principle. Extensions are given to relativistic systems, to Dirac's equation, and to elementary quantum field theory. Rosenbrock's aim is to show that results in the standard theory can be obtained from an extended form of Hamilton's principle, which has the advantage of conciseness and a relatively close relation to the classical formulations. Rosenbrock's wave function appears as a modified form of the optimal cost function, and the photon is identified with a singularity in the electromagnetic field. Interference is explained by optimization of an expected value, the ensemble over which the expectation is taken being dependent upon the information available. A note (Rosenbrock, 1997) corrects an error in a previous paper (Rosenbrock, 1995) which is set out to extend a variational principle for quantum mechanics to special relativity. Rosenbrock's next paper (Rosenbrock, 1999) gives further applications of the generalized Hamilton's principle to quantum mechanics. Namely, the wave function does not meet all the requirements in this treatment to be the state, and the way in which it can be made to do so is described and applied to the two-slit experiment. In another journal a paper may be noted on the state estimation for a class of nonlinear dynamic systems through the Hamilton–Jacobi–Bellman technique (Filippova, 2013). It would be interesting to confront the stochastic results of Rosenbrock with the theories postulating statistical origin of quantum mechanics (Kaniadakis, 2002).

Also, an approach toward constructing the Hamiltonian formulation as a basis of quantized thermal processes (Markus, 2005) has turned out to be quite successful. It involves a generalized Hamilton–Jacobi equation for dissipative heat processes (Markus and

Gambar, 2004) as the fundamental component of the variational theory developed by these authors (see also chapter: Variational Approaches to Nonequilibrium Thermodynamics). The basic results are quasiparticles in thermal processes (Markus and Gambar, 2005; Vazquez et al., 2009), siblings of phonons or quasiparticles of quantized acoustic fields.

There are also approaches in quantum cybernetics directed toward a unification of relativity and quantum theory via circularly causal modeling (Grössing, 2000). Other, independent ones are applied in the context of unified theory. The Evans wave equation of a unified theory can be derived from an appropriate Lagrangian density and action in general relativity, identifying the origin of the Planck constant h in general relativity (Evans, 2004). The action so defined is shown to be the origin both of the Planck constant h and of the Evans spin field B. The classical Fermat principle of least time, and the classical Hamilton principle of least action, are expressed in terms of a tetrad multiplied by a phase factor $\exp(iS/h)$, where S is the action in general relativity. Wave (or quantum) mechanics emerges from these classical principles of general relativity for all matter and radiation fields, giving a unified theory of quantum mechanics based on differential geometry and general relativity. The phase factor $\exp(iS/h)$ is an eigenfunction of the Evans wave equation and is the origin in general relativity and geometry of topological phase effects in physics, including the Aharonov–Bohm class of effects, the Berry phase, the Sagnac effect, related interferometric effects, and all physical optical effects through the Evans spin field B and Stokes' theorem in differential geometry. The Planck constant h is thus identified as the least amount possible of action or angular momentum or spin in the universe. This is also the origin of the fundamental Evans spin field B which is observed in any physical optical effect. It originates in torsion, spin and the second (or spin) Casimir invariant of the Einstein group. Mass originates in the first Casimir invariant of the Einstein group. These two invariants define every particle (Evans, 2004).

References

Ackerman, C.C., Bertman, B., Fainrbank, H.A., Guyer, R.A., 1966. Second sound in solid helium. Phys. Rev. Lett. 18, 789–791.

Afriyie, J.K., Rajakaruna, H., Nazha, M.A.A., Forson, F.K., 2013. Mathematical modelling and validation of the drying process in a chimney-dependent solar crop dryer. Energy Convers. Manage. 67 (3), 103–116.

Aghbalou, F., Badia, F., Illa, J., 2006. Exergetic optimization of solar collector and thermal energy storage system. Int. J. Heat Mass Transfer 49, 1255–1263.

Agorreta, E., Salvador, M., Santamaria, J., Monzon, A. 1991. Simultaneous activation and deactivation phenomena in isopropyl alcohol dehydrogenation on a Cu/SiO$_2$ catalyst. Catal. Orig. Res. 391–398.

Aharony, A., Feder, J., 1989. Fractals in Physics: Essays in honour of Benoit B. Mandelbrot. Amsterdam, North-Holland. Proceedings of the International Conference, Vance, France.

Ahlborn, B.K., 1999. Thermodynamic limits of body dimension of warm-blooded animals. J. Non-Equilib. Thermodyn. 40, 407–504.

Ahlborn, B.K., Blake, R.W., 1999. Lower size limit of aquatic mammals. Am. J. Phys. 67, 1–3.

Akpinar, K.E., 2010. Drying of mint leaves in a solar dryer and under open sun: modelling, performance analyses. Energ. Convers. Manag. 51 (12), 2407–2418.

Albers, B., 2003. Relaxation analysis and linear stability vs. adsorption in porous materials. Continuum Mech. Thermodyn. 15, 73–95.

Alemany, P.A., Zanette, D.H., 1994. Fractal random walks from a variational formalism for Tsallis entropies. Phys. Rev. E 49, R956(R).

Alhumaizi, K., 2000. Chaotic behavior of autocatalytic reaction with mutation. Chaos Soliton. Fract. 11, 1279–1286.

Alicki, R., Fannes, M., 2005. Quantum mechanics, measurement and entropy. Rep. Math. Phys. 55 (1), 47–59.

Almeida, M.P., 2005. Generalized entropy and the Hamiltonian structure of statistical mechanics. In: Sieniutycz, S., Farkas, H. (Eds.), Variational and Extremum Principles in Macroscopic Systems. Chapter 2 of Part II. Elsevier, Oxford, pp. 395–410, 2005.

Altenberger, A., Dahler, J., 1992. Statistical mechanical theory of diffusion and heat conduction in multicomponent systems. Article in: flow, diffusion and transport processes. Advances in Thermodynamics Series, 6, Taylor and Francis, New York, pp. 58–81.

Alvarez, P.I., Vega, R., Blasco, R., 2005. Cocurrent downflow fluidized bed dryer: experimental equipment and modeling. Drying Technol. 23 (7), 1435–1449.

Alvarez, T.A., Valero, A., Montes, J.M., 2006. Thermoeconomic analysis of a fuel cell hybrid power system from the fuel cell experimental data. Energy 31, 1358–1370.

Amar, F., Bernholc, J., Berry, R.S., Jellinek, J., Salamon, P., 1989. The shapes of first stage sinters. J. Appl. Phys. 65, 3219–3225.

Amelkin, S.A., Hoffmann, K.H., Sicre, B., Tsirlin, A.M., 2000. Extreme performance of heat exchangers of various hydrodynamic models of flows. Periodica Polytechnica Ser. Chem. Eng. 44 (1), 3–16.

Amelkin, S.A., Andresen, B., Burzler, J.M., Hoffmann, K.H., Tsirlin, A.M., 2005. Thermo-mechanical systems with several heat reservoirs: maximum power processes. J. Non-Equilib. Thermodyn. 30, 67–80.

Amoroso, R.L., Hunter, F.G., Katafos, F.M., Vigier, F.J.P., 2002. Gravitation and cosmology: from the Hubble radius to the Planck scale, Proceedings of a Symposium in Honour of the 80th Birthday of Jean-Pierre Vigier, Kluwer Akademic, Dordrecht, Fundamental Theories of Physics 126.

References

Andresen, B., 1983. Finite-Time Thermodynamics. University of Copenhagen, Copenhagen.

Andresen, B., 1990. Finite-time thermodynamics. Article in: finite-time thermodynamics and thermoeconomics. Advances in Thermodynamics Series, 4, Taylor and Francis, New York, pp. 66–94.

Andresen, B., 2008. The need for entropy in finite-time thermodynamics and elsewhere, Meeting the Entropy Challenge: Intern. Thermodynamics Symposium in Honor and Memory of Professor Joseph H. Keenan, AIP Conference Proceedings, vol. 1033, pp. 213–218.

Andresen, B., 2011. Current trends in finite-time thermodynamics. Angew. Chem. 50, 2690–2704.

Andresen, B., Gordon, J., 1991. Optimal paths for minimizing entropy production in a common class of finite-time heating and cooling processes. Report no. 91–10, University of Copenhagen, H.C. Ørsted Institute, Copenhagen.

Andresen, B., Gordon, J., 1992a. Optimal paths for minimizing entropy production in a common class of finite-time heating and cooling processes. Int. J. Heat Fluid Flow 13, 294–299.

Andresen, B., Gordon, J., 1992b. Optimal heating and cooling strategies for heat exchanger design. J. Appl. Phys. 71, 76–79.

Andresen, B., Gordon, J., 1994. Constant thermodynamic speed for minimizing entropy production in thermodynamic processes and simulated annealing. Phys. Rev. E 50, 4346–4351.

Andresen, B., Rubin, M.H., Berry, R.S., 1983. Availability for finite-time processes. General theory and model. J. Phys. Chem. 87, 2704.

Andresen, B., Berry, R.S., Ondrechen, M.J., Salamon, P., 1984a. Thermodynamics for processes in finite time. Acc. Chem. Res. 17, 266–271.

Andresen, B., Salamon, P., Berry, R.S., 1984b. Thermodynamics in finite time. Phys. Today September, 62–70.

Andresen, B., Zimmerman, E.C., Ross, J., 1984c. Objections to a proposal on the rate of entropy production or systems far from equilibrium. J. Chem. Phys. 81 (10), 4676–4678.

Andriuszczenko, A.I., 1965. Thermodynamic Calculations of Optimal Parameters of Electric Thermal Power Stations. WNT, Warsaw.

Angulo-Brown, F., 1991. An ecological optimization-criterion for finite-time heat engines. J. Appl. Phys. 69, 7465–7469.

Anthony, K.-H., 1981. A new approach describing irreversible processes. In: Brulin, O., Hsieh, R.K.T. (Eds.), Continuum models of Discrete Systems 4. North-Holland, Amsterdam.

Anthony, K.-H., 1986. A new approach to thermodynamics of irreversible processes by means of Lagrange-formalism. In: Kilmeister, W. (Ed.), Disequilibrium and Self-Organisation. Mathematics and Its Applications, vol. 30. D. Reidel Publishing Company, Dordrecht, pp. 75–92

Anthony, K.-H., 1988. Entropy and dynamical stability—a method due to Lagrange formalism as applied to TIP. In: Besseling, J.F., Eckhaus, F.W. (Eds.), Trends in Applications of Mathematics to Mechanics. Springer, Berlin.

Anthony, K.-H., 1989. Unification of continuum mechanics and thermodynamics by means of a Lagrange formalism. Arch. Mech. 41, 511–534.

Anthony, K.-H., 1990. Phenomenological thermodynamics of irreversible processes within Lagrange formalism. Acta Physica Hungarica 67 (3), 321–340.

Anthony, K.-H., 2000. Lagrange-Formalism and thermodynamics of irreversible processes: the 2nd law of thermodynamics and the principle of least entropy production as straightforward structures in Lagrange-Formalism. In: Sieniutycz, S., Farkas, H. (Eds.), Variational and Extremum Principles in Macroscopic Systems. Elsevier Science, Oxford, pp. 25–56, Chapter I.2.

Anthony, K.-H., 2001. Hamilton's action principle and thermodynamics of irreversible processes—a unifying procedure for reversible and irreversible processes. J. Non-Newtonian Fluid Mech. 96 (1–2), 291–339.

Antonia, R.A., Phan-Thien, N., Satyaprakash, B.R., 1981. Autocorrelation and spectrum of dissipation fluctuations in a turbulent jet. Phys. Fluids 24, 554.

Antoniouk, A., Arnaudon, M., Cruzeiroc, A.B., 2014. Generalized stochastic flows and applications to incompressible viscous fluids. Bulletin des Sciences Mathématiques 138 (4), 565–584.

Ao, Y., Chen, J., Duan, M., Shen, S., 2007. Gray modeling of heat-pump-surrounding system exergy efficiency. Proceedings of 2007 IEEE International Conference on Grey Systems and Intelligent Services, Nanjing, China, November, pp. 18–20.

Aoki, I., 1995. Entropy production in living systems: from organisms to ecosystems. Thermochim. Acta 250, 359–370.

Aoki, I., 2001. Entropy and exergy principles in living systems. In: Jorgensen, S.E. (Ed.), Thermodynamics and Ecological Modelling. Lewis (Publishers of CRC Press), Boca Raton, pp. 165–190.

Aono, O., 1975. Thermodynamic coupling of diffusion with chemical reaction. J. Statis. Phys. 13 (4), 331–335.

Aris, R., 1961. The Optimal Design of Chemical Reactors: A Study in Dynamic Programming. Academic Press, New York.

Aris, R., 1964. Discrete Dynamic Programming. Blaisdell, New York.

Aris, R., 1969. On stability criteria of chemical reaction engineering. Chem. Eng. Sci. 24 (1), 149–169.

Aris, R., 1975. The Mathematical Theory of Diffusion and Reaction in Permeable Catalysts: The Theory of the Steady State. Oxford University Press, University of Oxford.

Aris, R., Amundson, N.R., 1958. An analysis of chemical reactor stability and control—I: the possibility of local control, with perfect or imperfect control mechanisms. Chem. Eng. Sci. 7 (3), 121–131.

Arons, A.B., 1999. Development of energy concepts in introductory physics courses. Am. J. Phys. 67 (12), 1063–1067.

Arpaci, V.S., 1987. Radiative entropy production—lost heat into entropy. Int. J. Heat Mass Transfer 30, 2115.

Arpaci, V.S., 1990. Foundations of entropy production. Article in: nonequilibrium theory and extremum principles. Advances in Thermodynamics Series, vol. 3, Taylor and Francis, New York, pp. 369–392.

Arpaci, V., Selamet, A., 1988. Entropy production in flames. Combust. Flame 73, 251–259.

Artuso, R., 1997. Anomalous diffusion in classical dynamical systems. Phys. Rep. 290, 37–47.

Ascroft, N.W., 1976. Solid State Physics. Wiley, New York.

Asegehegn, T.W., Schreiber, M., Krautz, H.J., 2011. Numerical simulation of dense gas–solid multiphase flows using Eulerian–Eulerian two-fluid model. In: Zhu, J. (Ed.), Computational Simulations, Applications (ISBN: 978-953-307-430-6). InTech, Available from: http://www.intechopen.com/books/computational-simulations-and-applications/numerical-simulation-of-densegas-solid-multiphase-flows-using-eulerian-eulerian-two-fluid-model

Assaf, M., 2010. Theory of Large Fluctuations in Stochastic Populations. PhD thesis, Hebrew University of Jerusalem.

Assari, M.R., Basirat, H., Tabrizi, Najafpour, E., 2013. Energy and exergy analysis of fluidized bed dryer based on two-fluid modeling. Int. J. Thermal Sci. 64, 213–219.

Astarita, G., 1989. Thermodynamics. In: An Advanced Textbook for Chemical Engineers. Plenum Press, New York.

Atanackovic, T.M., Pilipovic, S., Zorica, D., 2007. A diffusion wave equation with two fractional derivatives of different order. J. Phys. A 40 (20), 5319.

Athans, M., Falb, P.C., 1966. Optimal Control. McGraw-Hill, New York.

Atherton, R.W., Homsy, G.M., 1975. On the existence and formulation of variational principles for nonlinear differential equations. Stud. Appl. Math 54, 31–80.

Aubert, J.H., Tirrel, M., 1980. Macromolecules in nonhomogeneous velocity gradient fields. J. Chem. Phys. 72 (4), 2694–2701.

Avnir, D., Citri, O., Farin, D., Ottolenghi, M., Samuel, J., Seri-Levy, A., 1989. Optimization of heterogeneous-catalyst structure: simulations and experiments with fractal and non-fractal systems. In: Plath, P.J. (Ed.), Optimal Structures in Heterogeneous Reaction Systems. Springer, Berlin, pp. 65–81.

Ayres, R.U., 1978. Resources Environment and Economics—Applications of the Materials/Energy Balance Principle. Wiley, New York.

Ayres, R.U., 1999. The second law, the fourth law, recycling and limits to growth. Ecol. Econ. 29, 473–483.

Ayres, R.U., Ayres, L.W., Martinas, K., 1996. Eco-thermodynamics: Exergy and life cycle analysis. Working Paper (96/./EPS). INSEAD, Fontainebleau, France.

Ayres, R.U., Ayres, L.W., Martinas, K., 1998. Exergy, waste accounting, and life-cycle analysis. Energy 23, 355–363.

Baclic, B., Seculic, D., 1978. A crossflow compact heat exchanger of minimum irreversibility. Termotechnika 4, 34.

Badescu, V., 1999. Simple upper bound efficiencies for endoreversible conversion of thermal radiation. J. Non-Equilib. Thermodyn. 24, 196–202.

References

Badescu, V., 2000. Accurate upper bound efficiency for solar thermal power generation. Int. J. Solar Energy 20 (3), 149–160.

Badescu, V., 2004a. Letter to the editor. Solar Energy 76 (4), 509–511.

Badescu, V., 2004b. Optimal paths for minimizing lost available work during usual heat transfer processes. J. Non-Equilib. Thermodyn. 29, 53–73.

Badescu, V., 2008. Exact and approximate statistical approaches for the exergy of black-body radiation. Central Eur. J. Phys. 6 (2), 344–350.

Badescu, V., 2013. Lost available work and entropy generation: heat versus radiation reservoirs. J. Non-Equilib. Thermodyn. 38, 313–333.

Badescu, V., 2014. Is Carnot efficiency the upper bound for work extraction from thermal reservoirs? Europhys. Lett. 106, 18006.

Badur, J., 2003. Modelowanie Zrównoważonego Spalania w Turbinach Gazowych. IMP PAN press, Gdańsk.

Badur, J., 2005. On the pseudo-mass, pseudo-momentum and pseudo-energy balances derived from the primary variational principles. In: Sieniutycz, S., Farkas, H. (Eds.), Variational and Extremum Principles in Macroscopic Systems. Elsevier, Oxford, pp. 267–291.

Badur, J., 2012. Energy—history, development, and current. IMP PAN press, Gdańsk.

Badur, J., Badur, J., 2005. The introduction to variational derivation of the pseudo-momentum conservation in thermohydrodynamics. In: Sieniutycz, S., Farkas, H. (Eds.), Variational and Extremum Principles in Macroscopic Systems. Elsevier, Oxford, pp. 267–291.

Badur, J., Bilicki, Z., Kwidziński, R., 1997. Operacyjna lepkość objętościowa w procesie transportu pędu ekspandującej wody i uderzeniowej kondensacji pary wodnej. Zeszyty Naukowe IMP PAN, Gdańsk.

Bag, B.C., Ray, D.S., 2000. Fluctuation–dissipation relationship in chaotic dynamics. Phys. Rev. E 62 (2), 1927–1935.

Bahroun, S., Couenne, F., Jallut, C., Valentin, C., 2013. Thermodynamics-based nonlinear control of a three-phase slurry catalytic fed-batch reactor. IEEE Transactions on Control Systems Technology 21 (2), 360–371.

Baierlein, R., 2001. The elusive chemical potential. Am. J. Phys. 69 (4), 423–434.

Bailey, C.D., 2002. The unifying laws of classical mechanics. Found. Phys. 32 (1), 159–176.

Bailie, R.C., 1978. Energy Conversion Engineering. Addison-Wesley, Reading (Mass).

Bak, P., Chen, K., 1989. The physics of fractals. Physica D 38, 5–12.

Bak, P., Chen, K., 1989. The physics of fractals. In: Aharony and Feder (Eds.), Fractals in Physics. Proceedings of the International Conference, Vance, France, Amsterdam, North-Holland, Chapter 2.

Bak, T.A., Salamon, P., Andresen, B., 2002. Optimal behavior of consecutive chemical reactions. J. Phys. Chem. (R.S. Berry's issue) 106, 10961–10964.

Bakalis, D.P., Stamatis, A.G., 2012. Full and part load exergetic analysis of a hybrid micro gas turbine fuel cell system based on existing components. Energy Convers. Manage. 64, 213–221.

Bakshi, B.R., Grubb, G.F., 2012. Implications of thermodynamics for sustainability. In: Cabezas, H., Diwekar, U. (Eds.), Sustainability: Multi-Disciplinary Perspectives. Bentham Science Publishers, The Netherlands, pp. 222–242.

Bakshi, B.R., Grubb, G.F., 2013. Implications of thermodynamics for sustainability. In: Cabezas, H., Diwekar, U. (Eds.), Sustainability: Multi-Disciplinary Perspectives. Bentham Science Publishers, Sharjah, United Arab Emirates, pp. 222–242.

Bałdyga, J., Pohorecki, R., 1995. Turbulent micromixing in chemical reactors—a review. Chem. Engin. J. 58, 183–195.

Bampi, B., Morro, A., 1982. The inverse problem of the calculus of variations applied to continuum physics. J. Math. Phys. 23, 2312–2321.

Bampi, F., Morro, A. 1984. Nonequilibrium thermodynamics: a hidden variable approach. In: Cassas Vazquez, J., Jou, D., Lebon, G. (Eds.), Lecture Notes in Physics, vol. 199. Springer, Berlin, pp. 221–232.

Banach, Z., Piekarski, S., 1992. A coordinate-free description of nonequilibrium thermodynamics. Arch. Mach. 44, 191–202.

Band, Y.B., Kafri, O., Salamon, P., 1980. Maximum work production from a heated gas in a cylinder equipped with a piston. Chem. Phys. Lett. 72, 127.

Band, Y.B., Kafri, O., Salamon, P., 1981. Optimal heating of a working fluid in a cylinder with a moving piston. J. Appl. Phys. 52, 3745.

Band, Y.B., Kafri, O., Salamon, P., 1982a. Optimal expansion of a heated working fluid. J. Appl. Phys. 53, 8.
Band, Y.B., Kafri, O., Salamon, P., 1982b. Optimization of a model external combustion engine. J. Appl. Phys. 53 (29).
Banerjec, R., Narayankhedar, K.G., Sukhatme, S.P., 1992. Exergy analysis of kinetic swing adsorption processes. Chem. Eng. Sci. 47, 1307.
Baranov, A.A., Kolpascikov, V.L., 1974. Relativistic Thermomechanics of Continuous Media. Nauka i Tekhnika, Minsk.
Baranowski, B., 1974. Nonequilibrium Thermodynamics in Physical Chemistry. PWN, Warsaw.
Baranowski, B., 1991. Non-equilibrium thermodynamics as applied to membrane transport. J. Membrane Sci. 57 (2–3), 119–159.
Baranowski, B., 1992. Diffusion in elastic media with stress fields. Flow, Diffusion and Transport Processes. Advances in Thermodynamics Series, vol. 6, pp. 168–199.
Bardi, M., Capuzzo-Dolcetta, I., 1997. Optimal Control and Viscosity Solutions of Hamilton–Jacobi–Bellman Equations (with appendices by Maurizio Falcone and Pierpaolo Soravia). Birkhauser, Boston.
Barenblatt, G.I., 2003. Scaling, Cambridge Texts in Applied Mathematics. Cambridge University Press, Cambridge, UK.
Barg, C., Secchi, A.R., Trielweiler, J.O., Ferreira, J.M.P., 2001. Simulation of an industrial PSA unit. Latin Am. Appl. Res. 31, 469–475.
Barletta, A., 2015. On the thermal instability induced by viscous dissipation. Int. J. Thermal Sci. 88, 238–247.
Barrozo, M.A.S., Murata, V.V., Costa, S.M., 1998. The drying of soybean seeds in countercurrent and concurrent moving bed dryers. Drying Technol. 16 (9–10), 2033–2047.
Barrozo, M.A.S., Murata, V.V., Assis, A.J., Freire, J.T., 2006a. Modeling of drying in moving bed. Drying Technol. 24 (3), 269–279.
Barrozo, M.A.S., Felipe, D.J.M., Sartori, D.J.M., Freire, J.T., 2006b. Quality of soybean seeds undergoing moving bed drying: Countercurrent and crosscurrent flows. Drying Technol. 24 (4), 415–422.
Barrozo, M.A.S., Mujumdar, A., Freire, J.T., 2014. Air-drying of seeds: a review. Drying Technol. 32 (10), 1127–1141.
Bartoszkiewicz, M., Miekisz, S., 1989. Some properties of generalized irreversible thermodynamics founded by Bearman–Kirkwood equations. J. Chem. Phys. 90, 1787.
Bartoszkiewicz, M., Miekisz, S., 1990. Diffusion-viscous flow coupling and global transport coefficients symmetry in matter transport through porous membranes. Ber. Bunsenges. Phys. Chem 94, 887–893.
Bass, J., 1968. Cours de Mathématiques, vol. 1. Masson, Paris.
Bateman, H., 1929. Note on a differential equation which occurs in the two-dimensional motion of a compressible fluid. Proc Roy. Soc. Lond. Ser. A 125, 598–618.
Bateman, H., 1931. On dissipative systems and related variational principles. Phys. Rev. 38, 815.
Bator, R., Sieniutycz, S., 2006. Application of artificial neural network for emission prediction of dust pollutants. Int. J. Energy Res. 30, 1023–1036.
Bauer, F., Grune, L., Semmler, W., 2006. Adaptive spline interpolation for Hamilton–Jacobi–Bellman equations. Appl. Numer. Math 56, 1196–1210.
Baumeister, K.J., Hamil, T.D., 1968. Hyperbolic heat conduction equation—a solution for the semiinfinite body problem. J. Heat Transfer Trans. ASME 91, 543–548.
Baumgärtner, S., 2004. Thermodynamic models. In: Proops, J., Safonov, P. (Eds.), Modelling in Ecological Economics. Edward Elgar, Cheltenham, UK, and Northampton, MA, USA, pp. 102–129, Chapter 6.
Bazykin, A.D., Khibnik, A.I., Aponina, E.A., 1983. A model of evolutionary appearance of dissipative structure in economics. J. Math. Biol. 18, 13–23.
Bear, J., 1972. Dynamics of Fluids in Porous Media. Elsevier, New York.
Bear, J., Verruijt, A., 1987. Modeling Groundwater Flow and Pollution. Kluwer Academics, Dordrecht.
Bear, J., Cheng, A.H.-D., Sorek, S., Ouazar, D., Herrera, I. (Eds.), 1999. Seawater Intrusion in Coastal Aquifers—Concepts, Methods and Practices. Kluwer Academics, Dordrecht, The Netherlands, p. 625.
Bearman, R.J., 1959. On the linear phenomenological equations, II: The linear statistical mechanical theory. J. Chem. Phys. 31, 751.

References

Bearman, R.J., Kirkwood, J.G., 1958. The statistical mechanical theory of transport Processes XI. Equations of transport in multicomponent systems. J. Chem. Phys. 28, 136.

Beck, C., 2002a. Generalized statistical mechanics and fully developed turbulence. Physica A 306, 5189–5198.

Beck, Ch., 2002b. Nonextensive statistical mechanics approach to fully developed turbulence. Chaos Soliton. Fract. 13, 499–506.

Bedeaux, D., 1986. Nonequilibrium thermodynamics and statistical physics. In: Prigogine, I., Rice, S.A. (Eds.), Advances in Chemical Physics, vol. 64. Wiley, New York, p. 67.

Bedeaux, D., 1992. Nonequilibrium thermodynamics of surfaces. Flow, Diffusion and Transport Processes, Advances in Chemical Physics. Advances in Thermodynamics Series, 6, Taylor and Francis, New York, pp. 430–459.

Bedeaux, D., Kjelstrup, S., 1999. Transfer coefficients for evaporation. Physica A 270, 413–426.

Bedeaux, D., Albano, A.M., Mazur, P., 1975. Boundary conditions and nonequilibrium thermodynamics. Physica A 82 (3), 438–462.

Bedeaux, D., Standaert, F., Hemmes, K., Kjelstrup, S., 1999. Optimization of processes by equipartition. J. Non-Equilib. Thermodyn. 24, 242–259.

Bedeaux, D., Kjelstrup, S., Öttinger, H.C., 2006. On a possible difference between the barycentric velocity and the velocity that gives translational momentum in fluids. Physica A 371, 177–187.

Bedeaux, D., Kjelstrup, S., Zhu, L., Koper, Ger, J.M., 2006. Nonequilibrium thermodynamics–a tool to describe heterogeneous catalysis. Phys. Chem. Chem. Phys 8, 5421–5427.

Bedeaux, D., Ortiz de Zárate, J.M., Pagonabarraga, I., Sengers, J.V., Kjelstrup, S., 2011. Concentration fluctuations in nonisothermal reaction-diffusion systems, II: the nonlinear case. J. Chem. Phys. 135, 124516.

Bednarek, G., Strumillo, Cz., Kudra, T., 1981. Mathematical modelling of dielectric drying, vol. 314. Materials of IV Symposium on Drying. IChP Press, Warsaw.

Bedringas, K.W., Ertesvag, I.S., Byggstoyl, S., Magnussen, B.F., 1997. Exergy Analysis of solid-oxide fuel cell (SOFC) systems. Energy 22, 403–412.

Beeckman, J.W., Froment, G.F., 1980. Catalyst deactivation by site coverage and pore blockage: finite rate of growth of the carbonaceous deposit. Chem. Engin. Sci. 35 (4), 805.

Bejan, A., 1982. Entropy Generation through Heat and Fluid Flow. Wiley Interscience, New York.

Bejan, A., 1987. The thermodynamic design of heat and mass transfer processes and devices: a review. Heat Fluid Flow 8, 258.

Bejan, A., 1988. Advanced Engineering Thermodynamics. Wiley Interscience, New York.

Bejan, A., 1996a. Entropy generation minimization: the new thermodynamics of finite-size devices and finite–time processes. J. Appl. Phys. 79, 1191–1218.

Bejan, A., 1996b. Entropy generation minimization. The Method of Thermodynamic Optimization of Finite-Size Systems and Finite-Time Processes. CRC Press, Boca Raton.

Bejan, A., 1996c. Notes on the history of the method of entropy generation minimization (finite time thermodynamics). J. Non-Equilib. Thermodyn. 21, 239–242.

Bejan, A., 1997a. Constructal theory of organization in nature. Advanced Engineering Thermodynamics. Wiley, New York.

Bejan, A., 1997b. Constructal theory: from thermodynamic and geometric optimization to predicting shape in nature. Energy Convers. Manage. 39, 1705–1718.

Bejan, A., 1997c. Constructal theory network of conducting paths for cooling a heat-generating volume. Int. J. Heat Mass Transfer 40, 799–816.

Bejan, A., 1997d. Theory of organization in Nature: pulsating physiological processes. Int. J. Heat Mass Transfer 40, 2097–2104.

Bejan, A., 2000. Shape and Structure from Engineering to Nature. Cambridge University Press, Cambridge, UK.

Bejan, A., 2003. Constructal comment on a Fermat-type principle for heat flow (Letter to Editor). Int. J. Heat Mass Transfer 46, 1885–1886.

Bejan, A., Ledezma, G.A., 1998. Streets tree networks and urban growth: optimal geometry for quickest access between a finite-size volume and one point. Physica A 255, 211–217.

Bejan, A., Lorente, S., 2001. Thermodynamic optimization of flow geometry in mechanical and civil engineering. J. Non-Equilib. Thermodyn. 26, 305–354.

Bejan, A., Lorente, S., 2004. The constructal law and the thermodynamics of flow systems with configuration. Int. J. Heat Mass Transfer 47, 3203–3214.
Bejan, A., Lorente, S., 2005. La Loi Constructale. L'Harmattan, Paris.
Bejan, A., Lorente, S., 2006a. Constructal theory of generation of configuration in nature and engineering. J. Appl. Phys. 100, 041301.
Bejan, A., Lorente, S., 2006b. Heterogeneous porous media as multiscale structures for maximum flow access. J. Appl. Phys. 100, 114909.
Bejan, A., Lorente, S., 2008a. Design with Constructal Theory. Wiley, Hoboken, NJ.
Bejan, A., Lorente, S., 2008b. Design with Constructal Theory. 1st International Workshop "Shape and Thermodynamics," Florence, September 25 and 26.
Bejan, A., Marden, J.H., 2006. Constructing animal locomotion from new thermodynamic theory. American Scientist, July–August issue, pp. 343–349.
Bejan, A., Schultz, W., 1983. Energy conservation in parallel thermal insulation systems. Int. J. Heat Mass Transfer 26, 335.
Bejan, A., Badescu, V., de Vos, A., 2000. Constructal theory of economics structure generation in space and time. Energy Convers. Manage. 41, 1429–1451.
Bejan, A., Dincer, I., Lorente, S., Miguel, A.F., Reis, A.H., 2005. Porous and Complex Flow Structures in Modern Technologies. Springer, New York, see also Ioan Pop's review of this book in Int. J. Heat Mass Transfer 48, 1413–1415.
Bejan, A., Mamut, E. (Eds.), 1999. Thermodynamic Optimization of Complex Energy Systems. Kluwer Academic Publishers, Dordrecht.
Bejan, A., Lorente, S., Lee, J., 2008. Unifying constructal theory of tree roots, canopies and forests. J. Theoret. Bio. 254, 529–540.
Bellman, R.E., 1957. Dynamic Programming. Princeton University Press, Princeton.
Bellman, R.E., 1961. Adaptive Control Processes: A Guided Tour. Princeton University Presss, Princeton, NJ.
Bellman, R.E., 1967. Introduction to Mathematical Theory of Control Processes. Academic Press, New York.
Bellman, R.E., Kalaba, R., 1965. Dynamic Programming and Modern Control Theory. Academic Press, New York.
Benard, H., 1900. Les tourbillons cellulaires dans une nappe liquide. Rev. Gen. Sci. Pures Appl. 11, 1261.
Benelmir, R., Feidt, M., 1997. Thermoeconomics and finite size thermodynamics for the optimization of a heat pump. Int. J. Energy Environ. Econ. 5 (1), 129–133.
Benelmir, R., Feidt, M., 1998. Energy cogeneration systems and energy management strategy. Energy Convers. Manage. 39 (16/18), 1791–1802.
Benin, D., 1970. Improved variational principles for transport coefficients. Phys. Rev. B 1, 2777.
Bennett, C.H., 1982. The thermodynamics of computation—a review. Int. J. Theor. Phys 21, 905–940.
Bensoussan, A., 1981. Lectures on stochastic control: nonlinear filtering and stochastic control. Lecture Notes in Math, vol. 972, Springer-Verlag, Berlin, pp. 1–62.
Benton, S., 1977. The Hamilton–Jacobi Equation: A Global Approach. Academic Press, New York.
Berdicevskii, V.L., 1983. Variational Principles in Mechanics of Continua. Nauka, Moscow.
Beretta, G.P., 1986. A theorem on Lyapounov stability for dynamical systems and a conjecture on a property of entropy. J. Math Phys. 27 (1), 305–398.
Berezowski, M., 2003. Fractal solutions of recirculation tubular chemical reactors. Chaos Soliton. Fract. 16, 1–12.
Berger, J.S., Perlmutter, D.D., 1965. An extended region of asymptotic reactor stability. Chem. Engin. Sci. 20 (2), 147–156.
Bergman, T.L., Faghri, A., Viskanta, R., 2008. Frontiers in transport phenomena research and education: energy systems, biological systems, security, information technology and nanotechnology. Int. J. Heat Mass Transfer 51, 4599–4613.
Beris, A.N., Edwards, B.J., 1994. Thermodynamics of Flowing Systems with Internal Microstructure. University Press, Oxford.
Beris, A.N., Öttinger, H.C., 2008. Bracket formulation of nonequilibrium thermodynamics for systems interacting with the environment. J. Non-Newtonian Fluid Mech. 152, 2–11.

References

Berkovsky, B.M., Bashtovoi, V.G., 1972. The finite velocity of heat propagation from the viewpoint of the kinetic theory. Int. J. Heat Mass Transfer 20, 621–627.

Bernstein, D.S., Haddad, W.S., Hyland, D.C., Tyan, F., 1993. Maximum-entropy-type Lyapunov functions for robust stability and performance analysis. Systems Control Lett. 21, 73–87.

Berry, R.S., 1979. Energy analysis for energy policy. J. Business Administr. 10, 83.

Berry, M.V., 1989. Falling fractal flakes. Physica D 38, 29–31.

Berry, R.S., 1989a. Understanding Energy. Chicago University Press, Chicago.

Berry, R.S., 1989b. Energy future: time horizons and instability. Environment 31, 42.

Berry, R.S., 1990. Foreword to finite-time thermodynamics and thermoeconomics. Advances in Thermodynamics Series, vol. 4, Taylor and Francis, New York, pp. vii–vii10.

Berry, S., 1995. Entropy, irreversibility and evolution. J. Theor. Biol. 175, 197–202.

Berry, R.S., Salamon, P., Heal, G., 1978. On a relation between thermodynamic and economic optima. Resource. Energy 1, 125–127.

Berry, R.S., Kazakov, V.A., Sieniutycz, S., Szwast, Z., Tsirlin, A.M., 2000. Thermodynamic Optimization of Finite Time Processes. Wiley, Chichester, UK.

Berthiaume, R., Bouchard, Ch., Rosen, M.A., 2001. Exergetic evaluation of the renewability of a biofuel. Exergy Int. J. 1 (4), 256–268.

Bertola, V., Cafaro, E., 2007. On the speed of heat. Phys. Lett. A 372, 1–4.

Beuther, H., Larson, O.A., Perrotta, A.J., 1980. The mechanism of coke formation on catalysts. Stud. Surf. Sci. Catal. 6, 271–282.

Beveridge, G.S., Schechter, S., 1970. Optimization: Theory and Practice. McGraw-Hill, New York.

Bhalekar, A.A., 1990. On the generalized phenomenological irreversible thermodynamic theory (GPITT). J. Math. Chem. 5, 187–196.

Bhalekar, A.A., 1996. Does Clausius' inequality analogue exist for open systems. J. Non-Equilib. Thermodyn. 21, 330–338.

Bhalekar, A.A., 1998. Universal inaccessibility principle. Pranama J. Phys. 50 (4), 281–294.

Bhalekar, A.A., 1999. Extended irreversible thermodynamics and the quality of temperature and pressure. Pranama J. Phys. 53 (2), 331–339.

Bhalekar, A.A., 2000a. On the time dependent entropy vis-à-vis Clausius' inequality. Asian J. Chem. 12 (2), 417–427.

Bhalekar, A.A., 2000b. On the irreversible thermodynamic framework for closed systems consisting of chemically reactive components. Asian J. Chem. 12 (2), 433–444.

Bhalekar, A.A., 2000c. On the generalized zeroth law of thermodynamics. Indian J. Phys. 74B (2), 153–157.

Bhalekar, A.A., Burande, C.S., 2005. A study of thermodynamic stability of deformation in visco-elastic fluids by Lyapunov function analysis. J. Non-Equilib. Thermodyn. 30 (1), 53–65.

Bhalekar, A.A., Garcia-Colin, L.-S., 1998. On the construction of an extended thermodynamic framework for irreversible processes. Pranama J. Phys. 50 (4), 295–305.

Bhattacharya, D.K., 1982. A variational principle for thermodynamic waves. Annal. Phys. 7 (39), 252–332.

Biegler, L.T., Grossmann, I.E., Westerberg, A.W., 1997. Systematic Methods of Chemical Process Design. Prentice-Hall, Englewood Cliffs, NJ.

Bilicki, Z., 1996. Thermodynamic nonequilibrium in the two-phase system—a continuum with an internal structure. Archiv. Thermodyn. 17 (1/2), 109–134.

Bilicki, Z., 2001. Extended irreversible thermodynamics applied to two-phase systems. Archiv. Thermodyn. 22 (1/2), 71–88.

Bilicki, Z., Badur, J., 2003. A thermodynamically consistent relaxation model for turbulent binary mixture undergoing phase transition. J. Non-Equilib. Thermodyn. 28 (4), 311–340.

Bilicki, Z., Giot, M., Kwidzinski, R., 2002. Fundamentals of twophase flow by the method of irreversible thermodynamics. Int. J. Multiphase Flow 28, 1983–2005.

Bilous, O., Amundson, N.R., 1955. Chemical reactor stability and sensitivity. AIChE J. 1 (4), 513–521.

Biot, M.A., 1970. Variational Principles in Heat Transfer. Clarendon Press, Oxford.

Bird, R.B., Öttinger, H.C., 1992. Transport properties of polymeric liquids. Ann. Rev. Phys. Chem. 43, 371–406.

Bird, R.B., Steward, W.E., Lightfoot, E.N., 1960. Transport Phenomena. Wiley, New York.

Bird, R.B., Amstrong, R.C., Hassager, D., 1977. Dynamics of Polymeric Liquids. Wiley, New York.
Bloomfield, V.A., 1999. Statistical thermodynamics of helix–coil transitions in biopolymers. Am. J. Phys. 67 (12), 1212–1215.
Bochenek, R., Sitarz, R., Antos, D., 2011. Design of continuous ion exchange process for the wastewater treatment. Chem. Eng. Sci. 66, 6209–6219.
Bockris, J. O'M., 1975. Energy: The Solar Hydrogen Alternative.Wiley, New York.
Bockris, J.O'M., Kainthla, R.C., 1986. The conversion of light and water to hydrogen and electric power. Hydrogen Energ. Process VI (2), 449.
Bockris, J.O'M., Bhardwaj, R.C., Tennakoon, C.L.K., 1992. The electrochemistry of waste removal. J. Serb. Chem. Soc 57, 799–818.
Bodvarsson, G.S., Witherspoon, P.A., 1989. Geothermal reservoir engineering, Part 1. Geotherm. Sci. Technol. 1, 1–2.
Bohm, D., 1952. A suggested interpretation of the quantum theory in terms of "hidden" variables I. Phys. Rev. 85 (2), 166–179.
Bohm, D., Vigier, P., 1954. Model of a causal interpretation of the quantum theory in terms of a fluid with irregular fluctuations. Phys. Rev. 96 (1), 208–216.
Bojic, M., 1997. Cogeneration of power and heat by using endoreversible Carnot engine. Energy Convers. Manage. 38 (18), 1877–1880.
Boltyanski, V.G., 1971. Mathematical Methods of Optimal Control. Holt, Rinehart & Winston, New York.
Boltyanski, V.G., 1973. Optimal Control of Discrete Systems. Nauka, Moscow.
Boreskov, G.K., 1986. Heterogeneous Catalysis (in Russian). Nauka, Moscow.
Bormaszenko, E., 2007. Entropy of relativistic mono-atomic gas and temperature relativistic transformation in thermodynamics. Entropy 9, 113–117.
Bosniakovic, F., 1965. Technische Thermodynamik, I und II, Dresden: Theodor Steinkopff. (See also English ed.: Technical Thermodynamics, Trans. by P.L. Blackshear, Jr.Holt, Rinehart & Winston, New York.)
Bouchaud, J.-P., Georges, A., 1990. Anomalous diffusion in disordered media: statistical mechanisms, models and physical applications. Phys. Rep. 195, 127.
Boukary, M.S., Lebon, G., 1986. A comparative analysis of binary fluid mixtures by extended thermodynamics and the kinetic theory. Physica 137A, 546–572.
Boven, R., Chen, P.J., 1973. Thermodynamic restriction on the initial slope of the stress relaxation function. Arch. Rat. Mech. Anal. 51, 278–284.
Boyle, G., 2004. Renewable Energy, second ed. The Open University Bath Press, Glasgow.
Bracco, S., Siri, S., 2010. Exergetic optimization of single level combined gas–steam power plants considering different objective functions. Energy 35 (12), 5365–5373.
Bracken, J., McCormick, G.P., 1968. Selected Applications of Nonlinear Programming. Wiley, New York.
Brauner, N., 1996. Role of interfacial shear modeling in predicting stability of stratified two-phase flow. In: Cheremisinoff, N.P., Encyclopedia of Fluid Mechanics, vol. 5. Advances in Engineering Fluid Mechanics: Boundary Conditions Required for CFD Simulation, pp. 317–378.
Breivik, J., 2001. Self-organization of template-replicating polymers and the spontaneous rise of genetic information. Entropy 3, 273–279.
Breymann, W., Tel, T., Volmer, J., 1998. Entropy balance, time reversibility and mass transport in dynamical systems. Chaos 8 (2), 396–408.
Brodowicz, K., Dyakowski, T., 1990. Heat Pumps (in Polish). PWN, Warsaw.
Brodyanskii, V.M. (Ed.), 1968. Energy and Exergy (in Russian). Mir, Moscow.
Brodyanskii, V.M., 1973. Exergy Method of Thermodynamic Analysis (in Russian). Energia, Moscow.
Brooks, D.R., Wiley, E.O., 1984. Evolution as an entropic phenomenon. In: Pollard, J.W. (Ed.), Evolutionary Theory: Paths to the Future. John Wiley and Sons, London.
Brooks, D.R., Wiley, E.O., 1986. Evolution as Entropy: Towards a Unified Theory of Biology. University of Chicago Press, Chicago.
Brownstein, K.R., 1999. The whole-partial derivative. Am. J. Phys. 67 (7), 647.
Bubnov, V.A., 1976a. On the nature of heat transfer in acoustic wave. Inzh. Fiz. Zh 31, 531–536.
Bubnov, V.A., 1976b. Wave concepts in theory of heat. Int. J. Heat Mass Transfer 19, 175–195.

References

Bubnov, V.A., 1982. Toward the theory of thermal waves. Inzh. Fiz. Zh. 43, 431–438.

Builtjes, P.J.H., 1977. Memory effects in turbulent flows. W. T. H. D. 97: 1–145.

Burande, Ch.S., Bhalekar, A.A., 2005. Thermodynamic stability of elementary chemical reactions proceeding at finite rates revisited using Lapunov function analysis. Energy 30 (6): 97–91.

Burghardt, A., 2008. Future challenges for research in chemical and process engineering. Chem. Proc. Eng. 29, 527–540.

Burghardt, A., Berezowski, M., 2003. Periodic solutions in a porous catalyst pellet—homoclinic orbits. Chem. Eng. Sci. 58, 2657–2670.

Burley, P., Foster, J., 1994. Economics and Thermodynamics—New Perspectives on Economic Analysis. Kluwer Academics, Dordrecht.

Burnett, D., 1935. The distribution of the molecular velocities and the mean motion in a nonuniform gas. Proc. Lond. Math. Soc. 40, 382–435.

Busse, F.H., 1967. The stabiliy of finite amplitude cellular convection and its relation to an extremum principle. J. Fluid. Mech. 30, 625.

Butt, J.B., Petersen, E.E., 1988. Activation, Deactivation and Poisoning of Catalysts. New York, Academic Press.

Bykov, V.I., Yablonskii, G.S., 1980. Simulation and optimization of processes with decreasing activity (processes of vinyl ethers synthesis). Stud. Surf. Sci. Catal. 6, 295–303.

Bykov, V.I., Gorban, A.N., Jablonski, G.S., 1982. Description of nonisothermal reactions in terms of Marcelin–De-Donder kinetics and its generalizations. React. Kinet. Catal. Lett. 20, 261–265.

Bykov, V.I., Gorban, A.N., Mirkes, E.M., 1990. Chemical dynamics and equilibrium. Private communication. Computational Center, Krasnoyarsk, USSR.

Cabezas, H., Diwekar, U., 2013. Sustainability: Multi-Disciplinary Perspectives. Bentham Science Publishers, The Netherlands.

Cadenillas, A., Karatzas, I., 1995. The stochastic maximum principle for linear, convex optimal control with random coefficients. SIAM J. Control Optim. 33, 590–624.

Calabro, A., Girardi, G., Fiorini, P., Sciubba, E., 2004. Exergy analysis of a "CO_2 zero emission" high-efficiency plant. Proceedings of ECOS'04, Guanajuato, Mexico, July 2004.

Calise, F., Dentice d'Accadia, M., Palombo, A., Vanoli, L., 2006. Simulation and exergy analysis of a hybrid solid oxide fuel cell (SOFC)–gas turbine system. Energy 31, 3278–3299.

Calise, F., Dentice d'Accadia, M., Palombo, A., Vanoli, L., 2007. A detailed one-dimensional finite-volume simulation model of a tubular SOFC and a pre-reformer. Int. J. Thermodyn. 10 (3), 87–96.

Callen, H., 1988. Thermodynamics and an Introduction to Thermostatistics. Wiley, New York.

Campanari, S., 2001. Thermodynamic model and parametric analysis of a tubular SOFC module. J. Power Sources 92, 26–34.

Canon, M.D., Cullun, C.D., Polak, E.R., 1972. Theory of Optimal Control and Mathematical Programming. Wiley, New York.

Caplan, S.R., Essig, A., 1983. Bioenergetics and Linear Nonequilibrium Thermodynamics. Harvard University Press, Cambridge, MA.

Capra, F., 1996. The Web of Life: A New Scientific Understanding of Living Systems. Anchor Books, New York.

Captained, G., 2003. Linear analysis of an aero thermal instability occurring in diffusion-controlled premixed catalytic combustion. Int. J. Heat Mass Transfer 46, 3927–3934.

Capuzzo Dolcetta, I.I., Falcone, M., 1989. Discrete dynamic programming and viscosity solutions of the Bellman equation. Annales de L'Instit Le Henri Poincafe. Analyse Non Lineaire 6, 303–328.

Carcaterra, A., 2014. Thermodynamic temperature in linear and nonlinear Hamiltonian Systems. Int. J. Engin. Sci. 80, 189–208.

Carrasi, M., Morro, A., 1972. A modified Navier-Stokes equation and its consequence on sound dispersion. Il Nuovo Cimento 9B, 321–343.

Carrasi, M., Morro, A., 1973. Some remarks about dispersion and adsorption of sound in monoatomic rarefied gases. Il Nuovo Cimento 13B, 249–281.

Carreau, P.J., Gmerla, M., Ait-Kadi, A., 1986. A conformational model for polymer solutions. Commun. 2nd Conference of European Rheologists, Prague, June 17–20.

Carrera-Patino, M., 1988. Adiabatic processes in monoatomic gases. J. Chem. Phys. 89, 2271–2277.
Carrera-Patino, M., Berry, R.S., 1991. A model for coupling heat and mass transfer in batch distillation. J. Phys. Chem.
Casartelli, M., Diana, E., Galgani, L., Scotti, A., 1976. Phys. Rev. A 13, 1921.
Casartelli, M., Diana, E., Galgani, L., Scotti, A., 1976. Numerical computations on a stochastic parameter related to the Kolmogorov entropy. Phys. Rev. A 13, 1921.
Casas, M., Martínez, S., Pennini, F., Plastino, A., 2002. Thermodynamics and the Tsallis variational problem. Physica A 305 (1–2), 41–47.
Casas-Vazquez, J., Jou, D., 1985. On the thermodynamic curvature of nonequilibrium gases. J. Chem. Phys. 83, 4715–4716.
Casas-Vazquez, J., Jou, D., 1992. Extended irreversible thermodynamics and fluctuation theory. Extended Thermodynamic Systems. Advances in Thermodynamics Series, vol. 7, Taylor and Francis, New York, pp. 263–288.
Casas-Vazquez, J., Jou, D., 1994. Nonequilibrium temperature versus local-equilibrium temperature. Phys. Rev. E 49 (2), 1040–1048.
Castaing, B., 1996. The temperature of turbulent flows. J. de Physique II 6 (1), 105–114.
Cattaneo, C., 1958. Sur une forme d'l'equation eliminant le paradoxe d'une propagation instantanée. C. R. Hebd. Seanc. Acad. Sci. 247, 431–433.
Caviglia, G., 1988. Composite variational principles and the determination of conservation laws. J. Math. Phys. 29, 812–816.
Celik, A.N., Muneer, T., 2013. Neural network based method for conversion of solar radiation data. Energy Convers. Manage. 67 (3), 117–124.
Censor, D., 1990. Fermat principle, Hamiltonian ray equations, group velocity and wave packets in absorptive media. Nonequilibrium Theory and Extremum Principles. Advances in Thermodynamics Series, vol. 3, Taylor and Francis, New York, pp. 448–481.
Cenusa, V.E., Badea, A., Feidt, M., Benelmir, R., 2004. Exergetic optimization of the heat recovery steam generators by imposing the total heat transfer area. Int. J. Thermodyn. 7 (3), 149–156.
Cerci, Y., 2001. The minimum work requirement for distillation processes. Exergy Int. J. 2 (1), 1–11.
Cercignani, C., 1975. Theory and Application of the Boltzmann Equation. Scottish Academic Press, Edinburgh.
Cerny, L.C., Cerny, E.R., 1995. A fractal analysis of the erythrocyte sedimentation rate. Biorheology 32 (2), 224–1224.
Chaichana, K., Patcharavorachot, Y., Chutichaia, B., Saebea, D., Assabumrungrata, S., Arpornwichanop, A., 2012. Neural network hybrid model of a direct internal reforming solid oxide fuel cell. Int. J. Hydrogen Ener. 37 (3), 2498–2508.
Chaisantikulwat, A., Diaz-Goano, C., Meadows, E.S., 2008. Dynamic modeling and control of planar anode-supported solid oxide fuel cell. Computers Chem. Engin. 32, 2365–2381.
Chandrasekhar, S., 1961. Hydrodynamic and Hydromagnetic Stability. Clarendon Press, Oxford.
Chapman, S., Cowling, T.G., 1973. The Mathematical Theory of Non-Uniform Gases, third ed. Cambridge University Press, Cambridge, UK.
Charach, Ch., Zemel, A., 1992. Irreversible thermodynamics of phase-change heat transfer: basic principles and applications to latent heat storage. Open Systems Inform. Dyn. 1 (3), 423–458.
Charbeneau, J.A., 2000. Groundwater Hydraulics and Pollutant Transport. Prentice-Hall, Upper Saddle River, NJ.
Chato, J.C., 1989. Second law limitations imposed by heat exchangers on the performance of thermodynamic cycles. Talk presented at ASME Winter Annual Meeting.
Chato, J.C., Damianides, C., 1986. Second-law based optimization of heat exchanger networks using load curves. Int. J. Heat Mass Transfer 29, 1079.
Chato, J.C., Khodadadi, J.M., 1984. Optimization of cooled shields in insulations. J. Heat Mass Transfer 106, 871.
Chen, L., 2005. Finite-Time Thermodynamic Analysis of Irreversible Processes and Cycles (in Chinese). High Education Press, Beijing.
Chen, X.D., 2007a. Simultaneous heat and mass transfer. In: Sablani, S., Rahman, S., Datta, A., Mujumdar, A.S. (Eds.), Handbook of Food and Bioprocess Modeling Techniques. CRC Press, Boca Raton, FL, pp. 179–233.

References

Chen, X.D., 2007b. Drying as a means of controlling food bio-deterioration. In: Gary, S., Tucker (Eds.), Control of Food Biodeterioration. Blackwell, London.

Chen, J., Andresen, B., 1999. Optimal performance of an endoreversible Carnot cycle used as a cooler. In: Wu, Ch., Chen, L., Chen, J. (Eds.), Advance in Recent Finite Time Thermodynamics. Nova Science, New York, pp. 37–49.

Chen, J., Chen, X., Wu, Ch., 2002. Ecological optimization of a multistage irreversible combined refrigeration system. Energy Convers. Manage. 43, 2379–2393.

Chen, X.D., Putranto, A., 2013. Modelling Drying Processes: A Reaction Engineering Approach. Cambridge University Press, Cambridge, UK.

Chen, G.-Q., Su, B., 2003. Discontinuous solutions for Hamilton–Jacobi equations: uniqueness and regularity. Discrete Continuous Dyn. Syst. 9, 167–192.

Chen, L., Sun, F., 2004. In: Lingen Chen, Fengrui Sun (Eds.), Advances in Finite Time Thermodynamics: Analysis and Optimization. Nova Science Publishers, New York.

Chen, L., Sun, F., 2004. Advances in Finite Time Thermodynamics: Analysis and Optimization. Nova Science Publishers, New York.

Chen, X.D., Xie, G.Z., 1997. Fingerprints of the drying behavior of particulate or thin layer food materials established using a reaction engineering model. Trans. IChemE C 75, 213–222.

Chen, J., Yan, Z., 1989. Unified description of endoreversible cycles. Phys. Rev. A 39, 4140.

Chen, J., Zhao, Y., 2009. Modeling and optimization of a typical fuel cell–heat engine hybrid system and its parametric design criteria. J. Power Sources 186, 96–103.

Chen, J., Wu, Ch., Kiang, R.J., 1996. Maximum specific power output of an irreversible radiant heat engine. Energy Convers. Manage. 37, 17–22.

Chen, L., Sun, F., Wu, C., 1997a. Performance characteristics of isothermal chemical engines. Energy Convers. Manage. 38, 1841–1846.

Chen, L., Sun, F., Wu, C., Gong, J., 1997b. Maximum power of a combined cycle isothermal chemical engine. Appl. Thermal Engin. 17, 629–637.

Chen, L., Sun, F., Wu, C., 1998. Performance of chemical engines with a mass leak. J. Phys. D 31, 1595–1600.

Chen, J., Bi, Y., Wu, Ch., 1999a. A generalized model of a combined heat pump cycle and its performance. In: Wu, Ch., Chen, L., Chen, J. (Eds.), Advance in Recent Finite Time Thermodynamics. Nova Science, New York, pp. 525–540.

Chen, L., Duan, H., Sun, F., Wu, C., 1999b. Performance of a combined-cycle chemical engine with mass leak. J. Non-Equibri. Thermodyn. 24, 280–290.

Chen, L., Wu, C., Sun, F., 1999c. Finite time thermodynamic optimization or entropy generation minimization of energy systems. J. Non-Equilib. Thermodyn. 24, 327–359.

Chen, J., Yan, Z., Chen, L., Andresen, B., 2000. On the performance of irreversible combined heat pump cycles. In: Hirs, G.G. (Ed.), ECOS 2000. Proceedings, University Twente, The Netherlands, pp. 269–277.

Chen, J., Yan, Z., Lin, G., Andresen, B., 2001a. On the Curzon–Ahlborn efficiency and its connection with the efficiencies of real heat engines. Energy Convers. Manage. 42 (2), 173–181.

Chen, J., Chen, X., Wu, Ch., 2001b. Optimization of the rate of exergy output of a multistage endoreversible combined refrigeration system. Energy Int. J. 0 (0), 1–7.

Chen, G., Andries, J., Spliethoff, H., Fang, M., van de Enden, P.J., 2004a. Biomass gasification integrated with pyrolysis in a circulating fluidized bed. Solar Energy 76, 345–349.

Chen, L., Zhou, J., Sun, F., Wu, Ch., 2004b. Ecological optimization for generalized irreversible Carnot engines. Appl. Energ. 77 (3), 327–338.

Chen, L., Zhou, J., Sun, F., Wu, Ch., 2005. Ecological optimization for generalized irreversible Carnot refrigerators. J. Phys. D 38 (1), 113–118.

Chen, L., Xia, D., Sun, F., 2008. Optimal performance of an endoreversible chemical engine with diffusive mass transfer law. Proc. I Mech E. J. Mech. Engin. Sci. 222, 1535–1539.

Chen, L., Xia, S., Sun, F., 2009. Optimal paths for minimizing entropy generation during heat transfer processes with a generalized heat transfer law. J. Appl. Phys. 105 (4), 044907.

Chen, L., Xia, D., Sun, F., 2010. Ecological optimization of generalized irreversible chemical engines. Int. J. Chemical Reactor Engin. 8, A121.

Chen, L., Xia, S., Sun, F., 2011. Maximum power output of multistage continuous and discrete isothermal endoreversible chemical engine system with linear mass transfer law. Int. J. Chem. Reactor Engin. 9, 1–37.

Chen, L., Xia, S., Sun, F., 2013. Maximum power output of multistage irreversible heat engines under a generalized heat transfer law by using dynamic programming. Scientia Iranica B 20 (2), 301–312.

Chester, M., 1963. Second sound in solids. Phys. Rev. 131, 2013–2115.

Chua, K.J., Chou, S.K., Hawlader, M.N.A., Ho, J.C., Mujumdar, A.S., 2002. On the study of time-varying temperature drying—effect on drying kinetics and product quality. Drying Technol. 20, 1579–1610.

Chwieduk, D., 2006. Modelowanie i analiza pozyskiwania oraz konwersji termicznej energii promieniowania słonecznego w budynku. Prace Instytutu Podstawowych Problemów Techniki PAN 11: 5–262.

Ciborowski, J., 1965. Fundamentals of Chemical Engineering. WNT, Warsaw.

Ciborowski, J., 1976. Basic Chemical Engineering (in Polish). Wydawnictwa Naukowo Techniczne, Warsaw.

Ciborowski, J., Sieniutycz, S., 1969a. Elements de cinetique applique e au sechage a contrecourant de corps finement divises. Int. J. Heat Mass Transfer 12, 479.

Ciborowski, J., Sieniutycz, S., 1969b. Untersuchungen der Kinetik der nach dem Gegenstromprinzip durch gefurten Trocknung und Adsorption bein Fallen Zerkleinerter Materiallen. Chemische Techn. 21, 750.

Ciborowski, J., Sieniutycz, S., 1969c. Fundamental problems of heat and mass exchange in fluidized beds on thermodynamic diagrams enthalpy-composition (in Polish). Chem Stos. 6 (2B), 111–125.

Ciborowski, J., Sieniutycz, S., 1969d. Investigation of kinetics of countercurrent drying and adsorption when solid particles are falling in gas (in Polish). Chem Stos. 6 (3B), 225–243.

Ciborowski, J., Sieniutycz, S., 1969e. Graphical methods for investigation of drying of falling particles on thermodynamic diagrams enthalpy-composition (in Polish). Chem Stos. 6 (3B), 245–266.

Clark, J.A., 1986. Thermodynamic optimization, interface with economic analysis. J. Non-Equilib. Thermodyn. 10, 85.

Clarke, F.H., 1997. Non-Smooth Analysis and Control Theory. Springer-Verlag, New York.

Clarke, E.C., Morgan, D., 1983. Chemical Heat Pumps for Industry, Proceedings of 16th Intersociety Energy Conversion Engineering Conference, USA.

Coleman, B.D., Gurtin, M., 1964. Thermodynamics with internal variables. J. Chem. Phys. 47, 597–613.

Compte, A., Jou, D., 1996. Nonequilibrium thermodynamics and anomalous diffusion. J. Phys. A 29, 4321–4329.

Connelly, L., Koshland, C.P., 2001a. Exergy and industrial ecology–Part 1: An exergy-based definition of consumption and a thermodynamic interpretation of ecosystem evolution. Exergy Int. J. 1 (3), 146–165.

Connelly, L., Koshland, C.P., 2001b. Exergy and industrial ecology. Part 2: A nondimensional analysis of means to reduce resource depletion. Exergy Int. J. 1 (4), 234–255.

Constantin, P., Iyer, G., 2008. A stochastic Lagrangian representation of the three-dimensional incompressible Navier–Stokes equations. Commun. Pure App. Math. LXI, 0330–0345.

Converse, A.O., Gross, G.D., 1963. Optimal distillate-rate policy in batch distillation. Ind. Eng. Chem. Fundament. 2, 217.

Corella, J., Asua, J.M., 1982. Kinetic equations of mechanistic type with nonseparable variables for catalyst deactivation by coke. Models and data analysis methods. Ind. Eng. Chem. Process Des. Dev. 21 (1), 55–61.

Corella, J., Menéndez, M., 1986. The modelling of the kinetics of deactivation of monofunctional catalysts with an acid strength distribution in their non-homogeneous surface. Application to the deactivation of commercial catalysts in the FCC process. Chem. Engin. Sci. 41 (7), 1817–1826.

Corella, J., Monzon, A., 1988. Modeling of the deactivation kinetics of solid catalysts by two or more simultaneous and different causes. Ind. Eng. Chem. Res. 27 (3), 369–374.

Corella, J., Bilbao, R., Jose, A., Molina, J.A., Artigas, A., 1985. Variation with time of the mechanism, observable order, and activation energy of catalyst deactivation by coke in the FCC process. Ind. Eng. Chem. Rroc. Des. Dev. 24 (3), 625–636.

References

Corella, J., Adanez, J., Monzon, A., 1988. Some intrinsic kinetic equations and deactivation mechanisms leading to deactivation curves with a residual activity. Ind. Eng. Chem. Res. 27 (3), 375–381.

Corning, P.A., 2002. Thermoeconomics—beyond the Second Law. J. Bioecon. 4, 57–88, Available from: www.complexsystems.org.

Corning, P.A., Kline, S.J., 2000. Thermodynamics, information and life revisited, Part II: Thermoeconomics and control information. Syst. Res. Behav. Sci. 15, 453–482.

Costa, P., Trevissoi, C., 1973. Thermodynamic stability of chemical reactors. Chem. Eng. Sci. 28, 2195.

Costea, M., Feidt, M., Petrescu 1999. Synthesis on Stirling engine optimization. A chapter in: Behan, A., and Mamut, E. (Eds.), Thermodynamic Optimization of Complex Energy Systems. Kluwer Academics, Dordrecht, Netherlands, pp. 403–410.

Coveney, P.V., 1988. The second law of thermodynamics: entropy, irreversibility and dynamics. Nature 333, 409–415.

Cownden, R., Nahon, M., Rosen, M., 2001. Exergy analysis of fuel cells for transportation applications. Exergy 1, 112–121.

Crandall, M.G., Lions, P.L., 1986. On existence and uniqueness of solutions of Hamilton–Jacobi equations. Nonlinear Anal. Theor. Methods Appl. 10, 353–370.

Crandall, M.G., Ishii, H., Lions, P.-L., 1992. User's guide to viscosity solutions of second-order partial differential equations. Bull. Am. Math. Soc. 27 (1), 1–67.

Cross, M.C., Hohenberg, P.C., 1993. Pattern formation outside of equilibrium. Rev. Mod. Phys. 65 (3), 851–1112.

Cross, A.D., Jones, P.L., Lawton, J., 1982a. Simultaneous energy and mass transfer in radiofrequency fields, Part I: Validation of the theoretical model. Trans. J. Chem. E 60, 67.

Cross, A.D., Jones, P.L., Lawton, J., 1982b. Simultaneous energy and mass transfer in radio-frequency fields, Part II: Characteristics of radiofrequency dryers. Trans. J. Chem. E 60, 75.

Cukrowski, A.S., Kolbus, A., 2005. On validity of linear phenomenological nonequilibrium thermodynamics in chemical kinetics. Acta Physica Polonica B 36, 1485–1507.

Curado, E.M., 1999. General aspects of the thermodynamic formalism. Brasilian J. Phys. 29 (1), 36–45.

Curado, E.M.F., Tsallis, C., 1991. Generalized statistical mechanics: connection with thermodynamics. J. Phys. A 24 (2), L69 [corrigenda 24, 3187 (1991); 25, 1019 (1992)].

Curran, P.F., Taylor, A.E., Solomon, K., 1967. Tracer Diffusion. Biophys. J. 7, 879–901.

Curtiss, C.F., Bird, R.B., 1996. Statistical mechanics of transport phenomena: polymeric liquid mixtures. Advances in Polymer Sciencevol. 125Springer, Berlin.

Curzon, F.L., Ahlborn, B., 1975. Efficiency of Carnot engine at maximum power output. Amer. J. Phys. 43, 22.

Cussler, E.I., 1984. Mass Transfer in Fluid Systems. Cambridge University Press, Cambridge, UK.

Cvetkovic, V., 1997. Transport of reactive solutes. In: Dagan, G., Neuman, S.P. (Eds.), Subsurface Flow and Transport: A Stochastic Approach. Cambrdige University Press, Cambridge, UK, pp. 133–145.

Dagan, G., 1989. Flow and Transport in Porous Formations. Springer, New York.

Dagan, G., Neuman, S.P. (Eds.), 1997. Subsurface Flow and Transport: A Stochastic Approach. Cambridge University Press, Cambridge, UK.

Dantzig, G.B., 1968. Linear Programming and Extensions. Princeton University Press, Princeton, NJ.

Darwin, Ch, 1859. The Origin of Species. Penguin books, London.

Datsevich, L.B., 2003. Oscillations in pores of a catalyst particle in exothermic liquid (liquid–gas) reactions: analysis of heat processes and their influence on chemical conversion, mass and heat transfer. Applied Catalysis A 250 (1), 125–141.

David, P., 1990. Thermodynamics and process design. Chem. Eng. May, 31–33.

Davies, P., 1966. Deposition from moving aerosols. In: Davies, P. (Ed.), Aerosol Science. Academic Press, London.

Davison, M., Essex, C., 2001. Fractional differential eqations and initial value problems. Math Scientist 23, 108–116.

Dawkins, R., 1976. The Selfish Gene. Oxford Press, Oxford, UK, Chapter 11.

Day, W.A., 1971. Restrictions on relaxation functions in linear viscoelasticity. Q. J. Mech. Appl. Math. 24, 487–497.

De Broglie, L.V., 1964. La thermodynamique "cache" des particules. Ann. Inst. Henri Poincare 1, 1–19.
De Broglie, L.V., 1970. The reinterpretation of wave mechanics. Found. Phys. 1 (1), 5–15.
De Donder, Th. 1920. Lecons de thermodynamique et de chimie physique. Paris: F.H. Van den Dugen and G. van Lerberghe.
De Groot, S.R., Mazur, P., 1984. Nonequilibrium Thermodynamics. Dover, New York.
De Groot, A., Woudstra, N., 1995. Exergy analysis of a FC system. J. Inst. Energ. 68, 32–39.
De la Selva, S.M.T., Garcia-Colin, L.S., 1986. Hydrodynamic fluctuations in a chemically reacting fluid. J. Chem. Phys. 85 (4), 2140–2146.
De la Selva, S.M.T., Garcia-Colin, L.S., 1986. Hydrodynamic fluctuations in a chemically reacting fluid. J. Chem. Phys. 85 (4), 2140–2146.
De Miguel, R., 2006a. On the nonequilibrium thermodynamics of large departure from Butler–Volmer behavior. J. Phys. Chem. B Lett. 110, 8176–8178.
De Miguel, R., 2006b. Thermal effects in a modified law of mass action. Chem. Phys. 321, 62–68.
De Pablo, A., Quiros, F., Rodriguez, A., Vazquez, J.L., 2011. A fractional porous medium equation. Adv. Math. 226 (2), 1378–1409.
De Parga, G.A., Lopez-Carrera, B., Angulo-Brown, F., 2005. A proposal for relativistic transformation in thermodynamics. J. Phys A 38, 2821–2834.
De Rossi, F., Mastrullo, R., Mazzei, P., Sasso, M., 1998. EASY. Exergetic Analysis of Vapour Compression Systems. A software aid in thermodynamic education. Napoli, Liguori Editore.
De Vos, A., 1985. Efficiency of some heat engines at maximum power conditions. Am. J. Phys. 53, 570–573.
De Vos, A., 1991. Endoreversible thermodynamics and chemical reactions. J. Phys. Chem. 95, 4534–4540.
De Vos, A., 1992. Endoreversible Thermodynamics of Solar Energy Conversion. Clarendon Press, Oxford.
De Vos, A., 1993a. Entropy fluxes, endoreversibility and solar energy conversion. J. Appl. Phys. 74, 3631–3637.
De Vos, A., 1993b. The endoreversible theory of solar energy conversion: a tutorial. Solar Energy Materials and Solar Cells 31, 75–93.
De Vos, A., 1995. Thermodynamics of photochemical solar energy conversion. Solar Energ. Mater. Solar Cells 38, 11–22.
De Vos, A., 1997. Endoreversible economics. Energy Convers. Manage. 38, 311–317.
Delmon, B., Froment, G.F. (Eds.), 1980. Catalyst Deactivation 1980 International Symposium Proceedings Hardcover—October 1, 1980. Elsevier, New York.
Delmon, B., Froment, G.F. (Eds.), 1987. Catalyst deactivation 1987 Proceedings of the 4th International Symposium, Antwerp, September 29–October 1, 1987. Elsevier, New York.
Delsman, E.R., Uju, C.U., de Croon, M.H.J.M., Schouten, J.C., Ptasinski, K.J., 2006. Exergy analysis of an integrated fuel processor and fuel cell (FP–FC) system. Energy 31, 3300–3309.
Demetrius, L., 1987. An extremal principle of macromolecular evolution. Phys. Scr. 36, 693–901.
Demetrius, L., 2000. Thermodynamics and evolution. J. Theor. Biol. 206, 1–16.
Demirel, Y., 2002. Nonequilibrium Thermodynamics: Transport and Rate Processes in Physical and Biological Systems. Elsevier, Amsterdam.
Demirel, Y., 2004. Exergy use in bioenergetics. Int. J. Exergy 1, 128–146.
Demirel, Y., 2005. Stability of transport and rate processes. Int. J. Thermodyn. 8 (2), 67–82.
Demirel, Y., 2006a. Non-isothermal reaction-diffusion system with thermodynamically coupled heat and mass transfer. Chem. Eng. Sci. 61 (10), 3379–3385.
Demirel, Y., 2006b. Non-isothermal reaction diffusion system with thermodynamically coupled heat and mass transfer. J. Non-Newtonian Fluid Mech. 165, 953–972.
Demirel, Y., 2008. Modeling of thermodynamically coupled system of chemical reaction and transport processes. Chem. Eng. J. 139 (1), 106–117.
Demirel, Y., 2009. Thermodynamically coupled heat and mass flows in a reaction-transport system with external resistances. Int. J. Heat Mass Transfer 52, 2018–2025.
Demirel, Y., 2010. Nonequilibrium thermodynamics modeling of coupled biochemical cycles in living cells. J. Non-Newtonian Fluid Mech. 165, 953–972.

References

Demirel, Y., 2014. Nonequilibrium Thermodynamics. Transport and Rate Processes in Physical, Chemical and Biological Systems, third ed. Elsevier, Amsterdam.

Demirel, Y., Sandler, S.I., 2001. Linear non-equilibrium thermodynamic theory for coupled heat and mass transport. Int. J. Heat Mass Transfer 44, 2439–2451.

Demirel, Y., Sandler, S.I., 2002a. Effect of concentration and temperature on the coupled heat and mass transport in liquid mixtures. Int. J. Heat Mass Transfer 45, 75–86.

Demirel, Y., Sandler, S.I., 2002b. Thermodynamics and bioenergetics. Biophysical Chemistry 97, 87–111.

Demirel, Y., Sandler, S.I., 2004. Nonequilibrium thermodynamics in engineering and science. J. Phys Chem. B 108, 31–43.

Denbigh, K.G., 1956. The second-law efficiency of chemical processes. Chem. Eng. Sci. 6, 1.

Denman, H.H., Buch, L.H., 1962. Solution of the Hamilton–Jacobi equation for a certain dissipative classical mechanical systems. J. Math. Phys. 14, 326–329.

Denn, M., 1982. Stability of Reaction and Transport Processes. Prentice-Hall, Englewood Cliffs, NJ.

Deryagin, B.V., Tzuraev, N.V., 1984. Wetted Layers. Nauka, Moscow.

Dettmann, S.P., Morris, G.P., Rondoni, G.P., 1997. Irreversibility, diffusion and multifractal measures in thermostatted systems. Chaos Soliton. Fract. 8 (5), 783–792.

Dijkema, G.P.J., Weijnen, M.P.C., 1997. On the use of fuel cells in process industry. 1st European Congress on Chemical Engineering, ECCE-1, 1. AIDIC, Florence, Italy, pp. 689–692.

Dincer, I., Cengel, Y.A., 2001. Energy, entropy and exergy concepts and their roles in thermal engineering. Entropy 3, 116–149.

Dincer, I., Rosen, M.A., 2007. Exergy: energy. Environment and Sustainable Development. Elsevier, Oxford.

D'Isep, F., Sertorio, L., 1982. Maximum mechanical work from the steady energy circulation on a finite body, Il. Nuovo Cimento 67 B, 41.

D'Isep, F., Sertorio, L., 1983. Maximum irreversible availability for continuous systems. Il. Il Nuovo Cimento 6 C, 305–319.

D'Isep, F., Sertorio, L., 1986. Irreversibility for Active Structures, Il. Nuovo Cimento 94 B, 168.

Diu, B., Guthmann, C., Lederer, D., Roulet, B., 1990. Macroscopic motion of a totally isolated system in statistical equilibrium. Am. J. Phys. 58 (10), 974–978.

Djaeni, M., Bartels, P.V., Sanders, J.P.M., van Straten, G., van Boxtel, A.J.B., 2008. Computational fluid dynamics for multistage adsorption dryer design. Drying Technol. 26 (4), 487–502.

Djaeni, M., van Straten, G., Bartels, P.W., Sanders, J.P.M., van Boxtel, A.J.B., 2009. Energy efficiency of multi-stage adsorption drying for low-temperature drying. Drying Technol. 27 (4), 555–564.

Długosz, A., 1985. Simulation and optimization of reactor-regenerator system in catalytic cracking process. PhD thesis, Warsaw Technological University, Warsaw.

Dolphin, M., Francaviglia, M., Rogolino, P., 1999. A geometric model for the thermodynamics of simple materials. Periodica Polytech. Ser. Mech. Eng 43 (1), 29–36.

Domański, R., 1978a. Analytical description of temperature field caused by heat pulses. Arch. Termod. i Spalania 9, 401–413.

Domański, R., 1978b. 6-th Int. Heat Transfer Conf. Paper CO-10: 275.

Domański, R., 1990. Storage of Thermal Energy. PWN, Warsaw.

Dong, La, Yong, Li, Yanjun, Dai, Tianshu, Ge, Ruzhu, Wang, 2013. Effect of irreversible processes on the thermodynamic performance of open-cycle desiccant cooling cycles. Energy Convers. Manage. 67, 44–56.

Donmezer, N., Graham, S., 2014. A multiscale thermal modeling approach for ballistic and diffusive heat transport in two dimensional domains. Int. J. Thermal Sci. 76, 235–244.

Douvartzides, S., Coutelieris, F., Tsiakaras, P., 2004. Exergy analysis of a solid oxide fuel cell power plant fed by either etanol or methane. J. Power Sources 131, 224–230.

Downar-Zapolski, P., Bilicki, Z., 1998. The pseudo-critical vapour-liquid flow. Archiv. Thermodyn. 17 (3/4), 13–25.

Drouot, R., Maugin, G.A., 1983. Phenomenological theory for polymer diffusion in non-homogeneous velocity gradient flows. Rheol. Acta 22, 336–347.

Drouot, R., Maugin, G.A., 1985. Continuum modeling of polyelectrolytes in solutions. Rheol. Acta 24, 474–487.

Drouot, R., Maugin, G.A., 1987. Optical and hydrodynamical regimes in dilute solutions of polyelectrolytes. Rheol. Acta 26, 350–357.

Drouot, R., Maugin, G.A., 1988. Polyelectrolytes in solutions: equilibrium conformations of the microstructure without and with external fields. Int. J. Engin. Sci. 26, 225–241.

Drouot, R., Maugin, G.A., Morro, A., 1987. Anisotropic equilibrium conformations in solutions of polymers and polyelectrolytes. J. Non-Newtonian Fluid Mech. 23, 295–304.

Dryden, I.G.C. (Ed.), 1975. The Efficient Use of Energy. IPC Science and Technology Press, Guildford.

Du Plessis, J.P., Masliyah, J.H., 1991. Flow through isotropic granular porous media. Transport in Porous Media 6, 207–221.

Dudukovic, M.P., Larachi, F., Mills, P.L., 1999. Multiphase reactors—revisited. Chemical Engineering Science 54 (13–14), 1975–1995.

Duffie, J.A., Beckman, W.A., 1974. Solar Energy Thermal Processes. Wiley, New York.

Duminil, M., 1976. Basic principles of thermodynamics as applied to heat pumps. In: Camatini, E., Kester, T. (Eds.), Heat Pumps and Their Application to Energy Conservation. Leyden, Noordhoff, p. 97.

Dunn, P.D., Reay, D.A., 1978. Heat Pipes, second ed. Pergamon, Oxford, UK.

Durbin, P.A., 1993. Application of near-wall turbulence model to boundary layers and heat transfer. Int. J. Heat Fluid Flow 14, 316–323.

Durmayaz, A., Sogut, O.S., Sahin, B., Yavuz, H., 2004. Optimization of thermal systems based on finite-time thermodynamics and thermoeconomics. Progress Energy Combustion Sci. 30 (2), 175–217.

Dykman, M.I., Smelyanskiy, V.N., Luchinsky, D.G., Mannella, R., McClintock, P.V.E., Stein, N.D., 1999. Large fluctuations and irreversibility in nonequilibrium systems. Nonlinear Phenomena Complex Syst. 2 (1), 1–7.

Ebeling, W., 1983. Discussion of the Klimontovich theory of hydrodynamic turbulence. Annalen der Physik 40 (1), 25–33.

Ebeling, W., 1985. Thermodynamics of selforganization and evolution. Biomed. Biochim. Acta 44, 831–838.

Ebeling, W., 1992. On the relation between various entropy concepts and the valoric interpretation. Physica A 182, 108–120.

Ebeling, W., 1993. Entropy and information in processes of self-organization: uncertainty and predictability. Physica A 194 (1–4), 563–575.

Ebeling, W., 2002. Canonical dissipative systems and applications to active Brownian motion. Nonlinear Phenomena Complex Syst. 5 (4), 325–331.

Ebeling, W., Engel-Herbert, H., 1986. Entropy lowering and attractors in phase space. Acta Physica Hungarica 66 (1–4), 339–348.

Ebeling, W., Feistel, R., 1992. Theory of selforganization and evolution: the role of entropy, value and information. J. Non-Equilib. Thermodyn. 17, 303–332.

Ebeling, W., Feistel, R., 2011. Physics of Self-Organization and Evolution (ISBN: 978-3-527-40963-1) Wiley-VCH, Berlin.

Ebeling, W., Volkenstein, M.V., 1990. Entropy and the evolution of biological information Physica A 163, 398–402.

Ebeling, W., Klimontovich, Y. L., 1984. Selforganization and Turbulence in Liquids. Teubner Verlagsgesellschaft, Leipzig.

Ebeling, W., Engel-Herbert, H., Herzel, H., 1986. On the entropy of dissipative and turbulent structures. Annal. Phys. (Leipzig) 498 (3–5), 187–195.

Eckart, C., 1940. The thermodynamics of irreversible processes III. Relativistic theory of the simple fluid. Phys. Rev. 58, 919–924.

Edwards, B.J., Beris, A.N., 2004. Nonequilibrium thermodynamics and complex fluids. J. Non-Newtonian Fluid Mech. 120, 1–2.

Edwards, B.J., Öttinger, H.C., Jongshaap, R.J.J., 1997. On the relationships between thermodynamic formalisms for complex fluids. J. Non-Equilib. Thermodyn. 22, 356–373.

Eigen, M., 1967. In: Claesson, S. (Ed.) Fast reactions and primary processes in chemical kinetics. Nobel Symposium No. 5. Almqvist and Wiksell, Stockholm.

Eigen, M., 1971. Self-organization of matter and the evolution of biological macro molecules. Naturwissenschaften 58 (10), 465–523.

References

Eigen, M., 1992. Steps towards Life: A Perspective on Evolution. Oxford University Press, Oxford.

Eigen, M., Schuster, P., 1979. The Hypercycle: A Principle of Natural Self-Organization. Springer-Verlag, Berlin.

Eigen, M., Winkler, R., 1993. Laws of the Game: How the Principles of Nature Govern Chance. Princeton University Press, Princeton, NJ.

Einstein, A., 1905. On the movement of small particles suspended in a stationary liquid demanded by the molecular kinetic theory of heat. Ann. Phys. 17, 549–560.

Ekman, A., Liukkonen, S., Kontturi, K., 1978. Diffusion and electric conduction in multicomponent electrolyte systems. Electrochemica Acta 23, 243–250.

El Naschie, M.S., 2000. Chaos for Engineers (Theory, Applications and Control): a review of the book by T. Kapitaniak. Chaos Soliton. Fract. 11, 641–642.

Elafif, A., Grmela, M., Lebon, G., 1999. Rheology and diffusion in simple and complex fluids. J. Non-Newtonian Fluid Mech. 86, 253–275.

El-Genk, M.S., Tournier, J.-M., 2000. Design optimization and integration of nickel/Haynes-25 AMTEC cells into radioisotope power systems. Energy Convers. Manage. 41 (16), 1703–1728.

El-Nabulsi, R.A., 2011. The fractional Boltzmann transport equation. Computers & Mathematics with Applications 62 (3), 1568–1575.

El-Sayed, Y.M., 2003. The Thermoeconomics of Energy Conversions. Pergamon, Oxford, UK, p. 4.

El-Wakil, S.A., Zahran, M.A., 2000. Fractional Fokker–Planck equation. Chaos Soliton. Fract. 12, 791–798.

El-Wakil, S.A., Elhandbay, A., Zahran, M.A., 2001. Fractional (space–time) Fokker–Planck equation. Chaos Soliton. Fract. 12, 1035–1040.

Engel-Herbert, H., Ebeling, W., 1988a. The behaviour of the entropy during transitions far from thermodynamic equilibrium: I. Sustained oscillations. Physica A 149 (1–2), 182–194.

Engel-Herbert, H., Ebeling, W., 1988b. The behaviour of entropy during transitions far from thermodynamic equilibrium II. Hydrodynamic flows. Physica 149A, 195–205.

Engel-Herbert, H., Schumann, W.M., 1987. Entropy decrease during excitation of sustained oscillations. Annal. Phys. (Leipzig) 499 (6), 393–402.

England, J.L., 2013. Statistical physics of self-replication. J. Chem. Phys. 139, 121923.

Ensinas, A.V., Nebra, S.A., Lozano, M.A., Serra, L., 2007. Design of evaporation systems in sugar can factories using a thermoeconomic optimization procedure. Int. J. Thermodyn. 10 (3), 97–105.

Enskog, D., 1917. Kinetic Theory of Processes in Moderately Dense Gases. Dissertation. Upsala University, Upsala.

Epstein, I.D., 1998. An Introduction to Nonlinear Chemical Dynamics: Oscillations, Waves, Patterns, and Chaos. Oxford University Press, Oxford, UK.

Erlach, B., Serra, L., Valero, A., 1999. Structural theory as standard for thermoeconomics. Energy Convers. Manage. 40, 1627–1649.

Ertl, G., 1991. Oscillatory kinetics and spatio-temporal self-organization in reactions at solid surfaces. Science 254 (5039), 1750–1755.

Ertl, G., 2008. Reactions at surfaces: from atoms to complexity (Nobel Lecture). Angew. Chem. Int. Ed. 47, 3524–3535.

Escher, C., Ross, J., 1983. Multiple ranges of flow rate with bistability and limit cycles for Schlögl mechanism in CSTR. J. Chem. Phys. 79, 3773–3777.

Escher, C., Ross, J., 1985. Reduction of dissipation in a thermal engine by means of periodic changes of external constraints. J. Chem. Phys. 82 (5), 2453–2456.

Escher, C., Kloczkowski, A., Ross, J., 1985. Increased power output and resonance effects in a thermal engine driven by a first or second order model reaction. J. Chem. Phys. 82, 2457–2465.

Essex, C., 1984a. Minimum entropy production at steady state and radiative transfer. Astrophys. J. 285, 279–283.

Essex, C., 1984b. Radiation and the violation of bilinearity in the thermodynamics of irreversible processes. Planet Space Sci. 32, 1035–1043.

Essex, C., 1990. Radiation and the continuing failure of the bilinear formalism. Nonequilibrium Theory and Extremum Principles. Advances in Thermodynamics Series, vol. 3, pp. 435–447.

Essex, C., Andresen, B., 2013. The principal equations of state for classical particles, photons, and neutrinos. J. Non-Equilib. Thermodyn. 38 (4), 293–312.

Essex, C., Kennedy, D.C., 1999. Minimum entropy production of neutrino radiation in the steady state. J. Statist. Phys. 94 (1/2), 253–266.

Essex, C., Kennedy, D.C., Berry, R.S., 2003. How hot is radiation. Am J. Phys. 71 (10), 969–978.

Eu, B.C., 1980. A modified moment method and irreversible thermodynamics. J. Chem. Phys. 73, 2858.

Eu, B.C., 1982. Irreversible thermodynamics of fluids. Ann. Phys. (NY) 140, 341–371.

Eu, B.C., 1986. On the modified moment method and irreversible thermodynamics. J. Phys. Chem. 85, 1592–1602.

Eu, B.C., 1987. Kinetic theory and irreversible thermodynamics of dense fluids subject to an external field. J. Phys. Chem. 87, 1220–1237.

Eu, B.C., 1988. Entropy for irreversible processes. Chem. Phys. Lett. 143, 65.

Eu, B.C., 1989. Irreversible thermodynamic theory of scalar induced melting point depression. Physica A160, 87.

Eu, B.C., 1992. Kinetic theory and irreversible thermodynamics of multicomponent fluids. Extended Thermodynamic Systems. Advances in Thermodynamics Series, vol. 7, Wiley, New York, pp. 183–222.

Evans, R., 1992. Fundamentals of Inhomogeneous Fluids. Dekker, New York, Chapter 3.

Evans, M.W., 2004. Derivation of the Evans wave equation from the Lagrangian and action: origin of the Planck constant in general relativity. Found. Phys. Lett. 17 (6), 535–558.

Evren-Selamet, E., 1995. Hysteretic foundations of entropy production. Proceedings of ECOS'95 July II-15, paper A1-182, Istanbul, pp. 81–88.

Eyink, G.L., 1990. Dissipation and large thermodynamic fluctuations. J. Stat. Phys. 61, 533–572.

Eyink, G.L., 1996. Action principle in nonequilibrium statistical dynamics. Phys. Rev. E 54, 3419.

Eyink, G.L., 1997. Fluctuations in the irreversible decay of turbulent energy. Phys. Rev. E 56 (5), 5413–5422.

Eyink, G.L., 2002. Gregory Eyink, University of Arizona. http://online.itp.ucsb.edu/online/hydrot_c00/eyink/oh/02.html.

Eyink, G.L., 2007. Turbulent diffusion of lines and circulations. Phys. Lett. A 368, 486–490.

Eyink, G.L., 2010. Stochastic least-action principle for the incompressible Navier–Stokes equation. Physica D 239 (14), 1236–1240.

Eyring, H., 1962. The transmission coefficient in reaction rate theory. Rev. Mod. Phys. 34, 616–619.

Eyring, H., Jhon, M.S., 1969. Significant Liquid Structures. Wiley, New York.

Faber, M., Jost, F., Manstetten, R., 1995. Limits and perspectives on the concept of sustainable development. Economie Appliqúee 48, 233–251.

Faber, M., Manstetten, R., Proops, J., 1996. Ecological Economics—Concepts and Methods. Edward Elgar, Cheltenham, UK.

Fabrizio, M., Morro, A., 1988. Viscoelastic relaxation functions compatible thermodynamics. J. Elasticity 19, 63–75.

Fabrizio, M., Giorgi, C., Morro, A., 1989. Minimum principles, convexity and thermodynamics in viscoelasticity. Continuum Mech. Thermodyn. 1, 197–211.

Faesa, A., Wuillemin, Z., Tanasini, P., Accardo, N., Modenac, S., Schindler, H.J., Cantonia, M., Lübbe, H., Diethel, S., Hessler-Wyser, A., Herle, J.V., 2011. Design of experiment approach applied to reducing and oxidizing tolerance of anode supported solid oxide fuel cell, Part II: Electrical, electrochemical and microstructural characterization of tape-cast cells. J. Power Sources 196 (21), 8909–8917, 2011.

Faghri, A., Guo, Z., 2005. Review: challenges and opportunities of thermal management issues related to fuel cell technology and modelling. Int. J. Heat Mass Transfer 48, 3891–3920.

Faigle, B., 2009. Two-phase flow modeling in porous media with kinetic interphase mass transfer processes in fractures. Master Thesis (Diplomarbeit). Universität Stuttgart.

Falcone, M., Makridakis, C., 2001. Numerical Methods for Viscosity Solutions and Applications. World Scientific, Singapore.

Fan, L.T., 1966. The Continuous Maximum Principle: A Study of Complex System Optimization. Wiley, New York.

Fan, L.T., Wang, C.S., 1964. The Discrete Maximum Principle: A Study of Multistage System Optimization. Wiley, New York.

References

Fang, G., 1998. Rate of liquid evaporation: statistical rate theory approach, Thesis, Graduate Department of Mechanical and Industrial Engineering, University of Toronto.

Fang, G., Ward, C.A., 1999. Temperature measured close to the interface of an evaporating liquid. Phys. Rev. E 59, 417.

Farid, M., 2003. A new approach to modelling of single droplet drying. Chem. Eng. Sci. 58, 2985–2993.

Farkas, M., 1997. Two ways of modelling cross diffusion. Nonlinear Anal. Theory Method. Appl. 30 (2), 1225–1233.

Farkas, H., Noszticzius, Z., 1992. Explosive, stable and oscillatory behavior in some chemical systems. Flow, Diffusion and Transport Processes. Advances in Thermodynamics Series, vol. 6, Taylor and Francis, New York, pp. 303–339.

Farkas, H., Sieniutycz, S., Kály-Kullai, K., 2005. The Fermat principle and chemical waves. In: Sieniutcyz, S., Farkas, H. (Eds.), Variational and Extremum Principles in Macroscopic Systems. Elsevier Science, Oxford, pp. 355–373, Chapter I.17.

Favache, A., Dochain, D., 2009. Thermodynamics and chemical systems stability: the CSTR case study revisited. J. Process Control 19, 371–379.

Fechner, W., Kotowski, W., Budner, Z., Thews, K., 1997. Energia elektryczna wprost z metanolu w ogniwie paliwowym. Karbo-Energochemia-Ekologia 12, 397–402.

Feidt, M., 1996. Optimisation d'un cycle de Brayton moteur en contact avec des capacites thermiques finies. Rev. Gen. Therm 418/419, 662–666.

Feidt, M., 1997. Sur une systématique des cycles imparfaits. Entropie, (Numero Special "De la thermotechnique à la thermodynamique"), 205, 53–61.

Feidt, M., 1999. Thermodynamics and optimization of reverse cycle machines. In: Bejan, A., Mamut, E. (Eds.), Thermodynamic Optimization of Complex Energy Systems. Kluwer Academics, Dordrecht, pp. 385–401.

Feidt, M., 2006. Energetique: concepts et applications. Dunod Editeur, Paris.

Feidt, M., Costea, M., Petre, C., Petrescu, S., 2007. Optimization of the direct Carnot cycle. Appl. Therm. Engin. 27, 829–839.

Feigenbaum, M.J., 1979. The universal metric properties of nonlinear transformations. J. Stat. Phys. 21 (6), 669–705.

Feigenbaum, M.J., 1980. The transition to aperiodic behavior in turbulent systems. Commun. Math. Phys. 77, 65–86.

Felipe, C.A.S., Barrozo, M.A.S., 2003. Drying of soybean seeds in a concurrent moving bed: heat and mass transfer and quality analysis. Drying Technol. 21 (30), 439–456.

Fernández-Pineda, C., Velasco, S., 2001. Application of thermodynamic extremum primciples. Am. J. Phys. 69 (11), 1160–1165.

Fernández-Seara, J., Piñeiro, C., Alberto Dopazo, J., Fernandes, F., Sousa, P.X.B., 2012. Experimental analysis of a direct expansion solar assisted heat pump with integral storage tank for domestic water heating under zero solar radiation conditions. Energy Convers. Manage. 59, 1–8.

Feudel, U., Feistel, R., Ebeling, W., 1984. Electrical dissipative structures in membrane-coupled compartment systems. Annal. Phys. (Leipzig) 496 (4–5), 267–279.

Field, R., Noyes, R.M., 1974. Oscillations in chemical systems, IV: Limit cycle behavior in a model of a real chemical reaction. J. Chem. Phys. 60 (5), 1877.

Filippova, T.F., 2013. State estimation for a class of nonlinear dynamic systems through HJB technique. Cy. Phys. 2 (3), 127–132.

Findeisen, W., 1974. Multilevel Control Systems (in Polish). PWN, Warsaw.

Findeisen, W., Szymanowski, J., Wierzbicki, A., 1980. Theory and Computational Methods of Optimization (in Polish). Państwowe Wydawnictwa Naukowe, Warszawa.

Finlayson, B.A., 1972. The Method of Weighted Residuals and Variational Principles. Academic Press, New York.

Finlayson, B.A., Scriven, L.E., 1965. The method of variational residuals and its relation to certain variational principles for the analysis of transport processes. Chem. Eng. Sci. 20, 395.

Finlayson, B.A., Scriven, L.E., 1967. On the search for variational principles. Int. J. Heat Mass Transfer 10, 799.

Fitts, D.D., 1962. Nonequilibrium Thermodynamics. McGraw-Hill, New York.

Fleming, W., Soner, H., 1993. Controlled Markov Processes and Viscosity Solutions. Applications of Mathematics. Springer-Verlag, Berlin.

Flumerfelt, R.W., Slattery, J.C., 1969. An experimental study of the validity of Fourier's law. AIChE J. 15 (2), 291–292.

Fogler, H.S., 2006. Elements of Chemical Reaction Engineering, forth ed. Pearson Education, New Jersey.

Forland, K.S., 1991. Energy conversion in electrochemical processes. The ATP–synthesis in mitochondria. Det Kgl. Norske Videnskabers Selskabs Forhandlinger, 161–170.

Forland, K.S., Forland, T., Ratkje, S.K., 1988. Irreversible thermodynamics. Theory and Applications. Wiley, Chichester, UK.

Forland, K.S., Forland, T., Ratkje, S.K., 1992. Transport processes in electrolytes and membranes. Flow, Diffusion and Rate Processes. Advances in Thermodynamics Series, vol. 6, pp. 340–384.

Foster, J., Kay, J., Roe, P. 2001. Teaching complexity and system thinking to engineers. 4th UICEE Annual Conference on Engineering Education. Bangkok, Thailand Feb. 7–11.

Fox, R.F., 1980. The "excess entropy" around nonequilibrium steady states, is not a Liapunov function. Proc. Natl. Acad. Sci. USA 77 (7), 3763766.

Fox, J.N., Trefny, J.U., Buchanan, J., Shen, L., Bertman, B., 1972. Onset of ballistic temperature pulses in solid ^{4}He. Phys. Rev. Lett. 28, 16.

Frangopoulos, C.A., von Spakovsky, M.R., Sciubba, E., 2002. A brief review of methods for the design and synthesis optimisation of energy systems. Proceedings of ECOS 2002, vol. I, Berlin, Germany, Jul. 3–5, 2002 306–316.

Freeze R.A., Cherry, J.A., 1979. Groundwater, p. 604. Prentice-Hall, Upper Saddle River, NJ.

Frieden, B.R., Soffer, B.H., 1995. Lagrangians of physics and the game of Fisher-information transfer. Phys. Rev. E 52, 2274.

Frieden, B.R., A. Plastino, A., Plastino, A.R., Soffer, B.H., 1990. Fisher-based thermodynamics: its Legendre transform and concavity properties. Phys Rev. 60 (1): 48–53.

Frieden, B.R., Plastino, A., Plastino, A.R., Soffer, B.H., 2002a. Non-equilibrium thermodynamics and Fisher information: an illustrative example. Phys. Lett. A 304, 73–78.

Frieden, B.R., Plastino, A., Plastino, A.R., Soffer, B.H., 2002b. Schrödinger link between nonequilibrium thermodynamics and Fisher information. Phys. Rev. E 66, 046128.

Froment, G.F., Bischoff, K.B., 1979. Chemical reactor analysis and design. Wiley, New York.

Fryda, L., Panopoulos, K.D., Karl, J., Kakaras, E., 2008. Exergetic analysis of solid oxide fuel cell and biomass gasification integration with heat pipes. Energy 33, 292–299.

Fuchs, N.A., 1955. Mechanics of aorosols (in Russian). IAN SSSR, Moscow.

Fujita, H., 1961. Diffusion in polymer-diluent eystems. Fortschritte der Hochpolymeren Forschung, Bd3. Adv. in Polymer Sci., 1–47.

Fukui, K., 1981. Towards chemodynamics. Int. J. Quant. Chem. 15, 621–632.

Fukui, K. 1981b. Variational principles in a chemical reaction. Int. J. Quant. Chem. 15, 633–642.

Fuliński, A., Grzywna, Z., Mellor, I., Siwy, Z., Usherwood, P.N.R., 1998. Non-Markovian character of ionic current fluctuations in membrane channels. Phys. Rev. E 58 (1), 919–924.

Fung, Y.C., Tong, P., 2001. Classical and Computational Solid Mechanics. World Scientific, Singapore.

Gadewar, S.B., Doherty, M.F., Malone, M.F., 2001. A systematic method for reaction invariants and mole balances for complex chemistries. Comput. Chem. Eng. 25, 1199–1217.

Gage, K.S., 1972. Viscous stability theory for thermally stratifies flow: discontinuous jets and shear layers. Phys. Fluids 15 (4), 526–533.

Gaggioli, R.A. (Ed.), 1980. Thermodynamics: Second Law Analysis. American Chemical Society, Washington, D.C.

Gambar, K., Markus, F., 1993. On the global symmetry of thermodynamics and Onsager's reciprocity relations. J. Non-Equilib. Thermodyn. 18, 51–57.

Gambar, K., Markus, F., 1994. Hamilton–Lagrange formalism of nonequilibrium thermodynamics. Phys. Rev. E 50, 1227–1231.

Gambar, K., Markus, F., 2003. Onsager's regression and the field theory of parabolic processes. Physica A 320, 193–203.

References

Gambar, K., Markus, F., Nyiri, B., 1991. A variational principle for the balance and constitutive equations in convective systems. J. Non-Equilib. Thermodyn. 16, 217–223.

Ganjeh, K.A., Mohd, J.M.N., Mat, L.T., Hassan, B., 2013. Exergoenvironmental optimization of heat-recovery steam generators in combined cycle power plant through energy and exergy analysis. Energy Convers. Manage. 67 (3), 27–33.

Garcia Colin, L.S., 1988. Comments on the kinetic mass action law revisited by thermodynamics. J. Phys. Chem. 92, 3017–3018.

Garcia Colin, L., 1992. Chemically reacting systems and extended ireversible thermodynamics. Extended Thermodynamic Systems. Advances in Thermodynamics Series, vol. 7, Taylor and Francis, New York, pp. 364–385.

Garcia-Colin, L.-S., Bhalekar, A.A., 1997. Recent trends in irreversible thermodynamics. Proc. Pakistan Acad. Sci. 50 (4), 295–305.

Garcia-Colin, L.-S., Green, M.S., 1966. Definition of temperature in the kinetic theory of dense gases. Physical Rev. 150 (1), 153–158.

Garcia Colin, L.S., de la Selva, S.M.T., Pina, E., 1986. Consistency of the kinetic mass action law with thermodynamics. J. Phys. Chem. 90, 953–956.

Garcia-Collin, L.S., Lopez de Haro, M., Rodriguez, R.F., Casas-Vazquez, J., Jou, D., 1984. On the foundations of extended irreversible thermodynamics. J. Statist. Phys. 37, 465–484.

Garfinkle, M., 1992. Mass action and the natural path. J. Non-Equilib. Thermodyn. 17 (3), 281–302.

Garg, H.P., Mullick, S.C., Bhargava, A.K., 1985. Solar Thermal Energy Storage. Reidel, Dordrecht.

Garrard, A., Fraga, E.S., 1998. Mass exchange network synthesis using genetic algorithms. Comput. Chem. Engin. 22 (12), 1837–1850.

Gaspard, P., 1998. Chaos, Scatterning and Statistical Mechanics. Cambridge University Press, Cambridge, UK.

Gaspard, P., 2004. Fluctuation theorem for nonequilibrium reactions. J. Chem. Phys. 120, 8898.

Gaspard, P., Nicolis, G., 1990. Transport properties, Lyapunov exponents, and entropy unit time. Phys. Rev. Lett. 65, 1693–1696.

Gaspard, P., Claus, I., Gilbert, T., Dorfman, J.R., 2001. The fractality of the hydrodynamic modes of diffusion. Phys. Rev. Lett. 86, 1506–1509.

Gass, S.I., 1958. Linear Programming: Methods and Applications. McGraw-Hill, New York.

Gaveau, B., Moreau, M., Toth, J., 1999a. Variational nonequilibrium thermodynamics of reaction-diffusion systems, I: The information potential. J. Chem Phys. 111 (17), 7736–7747.

Gaveau, B., Moreau, M., Toth, J., 1999b. Variational nonequilibrium thermodynamics of reaction-diffusion systems, II: Path integrals, large fluctuations, and rate constants. J. Chem Phys. 111 (17), 7748–7757.

Gaveau, B., Moreau, M., Toth, J., 2001a. Variational nonequilibrium thermodynamics of reaction-diffusion systems, III: Progress variables and dissipation of energy and information. J. Chem Phys. 115 (2), 680–690.

Gaveau, B., Martinas, K., Moreau, M., Toth, J., 2001b. Entropy, extropy and information potential in stochastic systems far from equilibrium. Physica A Statist. Mech. Appl. 305 (3–4), 445–466.

Gaveau, B., Moreau, M., Toth, J., 2005. Master equations and path-integral formulations of variational principles for reaction-diffusion problems. In: Sieniutycz, S., Farkas, H. (Eds.). Variational and Extremum Principles in Macroscopic Systems. Elsevier Science, Oxford, pp. 315–336.

Gavrilyuk, S., Gouin, H., 1999. A new form of governing equations of fluids arising from Hamilton's principle. Int. J. Engin. Sci. 37, 1495–1520, see also Gouin and Ruggeri, 2003.

Gayubo, A.G., Arandes, J.M., Aguayo, A.T., Olazar, M., Bilbao, J., 1993a. Temperature *vs.* time sequences to palliate deactivation in parallel and in series-parallel with the main reaction: parametric study. Chem. Eng. J. 51, 167–176.

Gayubo, A.G., Arandes, J.M., Aguayo, A.T., Olazar, M., Bilbao, J., 1993b. Optimization of themperature-time sequences in reaction-regeneration cycles. Application to isomerization of cis-butene. Eng. Chem. Res. 32, 2542–2547.

Gelfand, J.M., Fomin, S.W., 1963. Calculus of Variations. Prentice-Hall, Englewood Cliffs, NJ.

Gelfand, J.M., Fomin, S.W., 1970. Variational Calculus. PWN, Warsaw.

Gelhar, L.W., 1993. Stochastic Subsurface Hydrology. Prentice-Hall, Englewood Cliffs, NJ.

Geworkian, P., 2007. Sustainable Energy Systems Engineering. The Complete Green Building Design Structure. McGraw-Hill, New York.

Ghodoossi, L., 2004. Conceptual study on constructal theory. Energy Convers. Manage. 45 (9–10), 1379–1395.

Gibbs, J.W., 1948. Collected Works, vol. 1: Thermodynamics. Yale University Press, New Haven.

Giesekus, H., 1986. Constitutive models of polymer fluids: toward a unified approach. In: Kroner, E., Kirchgassner, K. (Eds.), Trends in Applications of Pure Mathematics to Mechanics. Springer, Berlin, pp. 331–348.

Gillmore, R., 1984. Length and curvature in the geometry of thermodynamics. Phys. Rev. A 30, 1994–1997.

Giona, M., 1996. Transport phenomena in fractal and heterogeneous media-input/output renormalization and exact results. Chaos Soliton. Fract. 7 (9), 1371–1396.

Giuliani, A., Colafranceschi, M., Webber, Jr., Ch.L., Zbilut, J.P., 2001. A complexity score derived from principal components analysis of nonlinear order measures. Physica A 301, 567–588.

Giuliani, A., Benigni, R., Zbilut, J.P., Webber, Jr., Ch.L., Sirabella, P., Colosimo, A., 2002. Nonlinear signal analysis methods in the elucidation of protein sequence-structure relationships. Chem. Rev. 102, 1471–1491.

Glansdorff, P., Prigogine, I., 1971. Thermodynamic Theory of Structure Stability and Fluctuations. Wiley, New York.

Glass, L., 2003. Obituary. Arthur T. Winfree (1942–2002). Nature 421, 34.

Glasstone, S., Leidler, K., Eyring, H., 1941. The Theory of Rate Processes. McGraw-Hill, New York.

Goldbeater, A., 1996. Biochemical Oscillations and Cellular Rhythms: The Molecular Bases of Periodic and Chaotic Behavior. Cambridge University Press, Cambridge.

Goldstein, S., 1953. On diffusion by discontinuous movements and on the telegraph equation. Q. Mech. Appl. Math 6, 290–312.

Goldstein, S., 2009. Bohmian Mechanics and Quantum information. Found. of Physics 40 (4), 335–355.

Gomes, D.A., 2005. A variational formulation for the Navier–Stokes equation. Commun. Math. Phys. 257, 227–234.

Gorban, A.N. 1984. Off Equilibrium: Equations of Chemical Kinetics and their Thermodynamic Analysis, Nayka, Novosibirsk.

Gorban, A.N., 2014. New universal Lyapunov functions for nonlinear kinetics. A talk delivered at Department of Mathematics University of Leicester, UK. July 21, Leicester.

Gorban, A.N., Bykov, V.I., Jablonskii, G.S. 1986. Thermodynamic function analogue for reactions proceeding without interaction of various substances. Chem. Eng. Sci. 41 (11), 2739–2745.

Gorban, A.N., Karlin, I.V., Öttinger, H.C., Tatarinova, L.L., 2001. Ehrenfest's argument extended to a formalism of nonequilibrium thermodynamics. Phys. Rev. E 63, 066124.

Gordon, J.M., 1990. Nonequilibrium thermodynamics for solar energy applications. Finite-Time Thermodynamics and Thermoeconomics. Advances in Thermodynamics Series, vol. 4, Taylor and Francis. New York, pp. 95–120.

Gordon, J.M., 1991. Generalized power versus efficiency characteristics of heat engines: the thermoelectric generator as an instructive illustration. Am. J. Phys. 59, 551–555.

Gordon, J.M., 1993. Maximum work from isothermal chemical engines. J. Appl. Phys. 73, 8–11.

Gordon, J.M., Orlov, V.N., 1993. Performance characteristics of endoreversible chemical engines. J. Appl. Phys. 74, 5303–5308.

Goswami, D.Y., Vijayaraghavan, S., Lu, S., Tamm, G., 2004. New and emerging developments in solar energy. Solar Energy 76, 33–43.

Gouin, H., Ruggeri, T., 2003. Hamiltonian principle in the binary mixtures of Euler fluids with applications to the second sound phenomena. Rend. Mat. Acc. Lincei s. 9, vol. 14, pp. 69–83, Chapter 2.

Grabert, H., Green, M.S., 1979. Fluctuations in nonlinear irreversible processes. Phys. Rev. A 19, 1747–1757.

Grabert, H., Graham, R., Green, M.S., 1980. Fluctuations in nonlinear irreversible processes II. Phys. Rev. 21, 2136–2146.

Grabert, H., Hänggi, P., Oppenheim, I., 1983. Fluctuations in reversible chemical reactions. Physica 117A, 300–316.

Grad, H., 1958. Principles of the theory of gases. In: Flugge, S. (Ed.), Handbook der Physik, vol. 12. Springer, Berlin.

References

Graham, R., 1977. Lagrangian for diffusion in curved phase space. Phys Rev. Lett. 38, 51–53.
Graham, R., 1985. Entropy-like potentials in non-equilibrium systems with coexisting attractors. In: Haken, H. (Ed.), In: Complex Systems—Operational S Approaches. Springer, Berlin.
Graham, R., Schenzle, A., 1983. Non-equilibrium potentials and stationary probability distributions of some dissipative models without manifest detailed balance. Z. Phys. B 52, 61–68.
Graham, R., Tel, T., 1984. Existence of a potential for dissipative dynamical systems. Phys. Rev. Lett. 52, 9–12.
Grandy, W.T., 1983. Principle of maximum entropy and irreversible processes. Rep. Progr. Phys. 62.
Grandy, W.T., 1988. Foundations of Statistical Mechanics, vol. I: Equilibrium Phenomena; vol. II. Nonequilibrium Phenomena. Kluwer Academics, Amsterdam.
Grasman, J., Brascamp, J.W., Van Leeuwenz, J.L., Van Putten, B., 2003. The multifractal structure of arterial trees. J. Theor. Biol. 220, 75–82.
Grassia, P., 2001. Dissipation, fluctuations, and conservation laws. Am. J. Phys. 69, 113–119.
Gray, P., Scott, O.K., 1975. Kinetics of oscillatory reactions. Reaction KineticsThe Chemical Sciety, London, Burlington House, Chapter 8.
Gray, P., Scott, O.K., 1990. Chemical Oscillations and Instabilities. Clarendon Press, Oxford.
Gray, C.G., Karl, G., Novikov, V.A., 1998. From Maupertuis to Schrödinger. Quantization of classical variational principles. Am. J. Phys. 67 (11), 959–961.
Gray, C.G., Karl, G., Novikov, V.A., 2004. Progress in classical and quantum variational principles. Reports on Progress in Physics, arXiv: physics/0312071.
Green, M.A., 2004. Recent developments in photovoltaics. Solar Energy 76, 3–8.
Gregory, R.P.F., 1989. Photosynthesis. Chapman and Hall, New York.
Greig, D.M., 1980. Optimisation. Longman, New York.
Grendar, jr, M., Grendar, M., 2001. Maximum entropy: clearing up mysteries. Entropy 3, 58–63.
Grigoriew, B.A., 1979. Simplification of one-dimensional heat conduction problem at radiation impulses. Heat Phys. High Temp. 11, 133–137.
Grmela, M., 1985. Bracket formulation for Navier–Stokes equations. Phys. Lett. 111 A, 36–40.
Grmela, M., 1993a. Weakly nonlocal hydrodynamics. Phys. Rev. E 47, 351–365.
Grmela, M., 1993b. Thermodynamics of driven systems. Phys. Rev. E 48, 919–930.
Grmela, M., 2001. Complex fluids subjected to external influences. J. Non-Newtonian Fluid Mech. 96, 221–254.
Grmela, M., Lebon, G., 1990. Hamiltonian extended thermodynamics. J. Phys. A 23, 3341–3351.
Grmela, M., Öttinger, H.C., 1997. Dynamics and thermodynamics of complex fluids, I: development of a general formalism. Phys. Rev. E 56, 6620–6632.
Grmela, M., Teichman, J., 1983. Lagrangian formulation of the Maxwell-Cattaneo hydrodynamics. Int. J. Eng. Sci. 21, 297–313.
Grössing, G., 2000. Quantum Cybernetics: Toward a Unification of Relativity and Quantum Theory via Circularly Causal Modeling (ISBN 978-0-387-98960-0). Springer Verlag, New York.
Grubecki, I., 1997. Analiza polityki temperaturowej dla reakcji enzymatycznej w bioreaktorze okresowym (PhD thesis). PhD thesis. University of Technology, Warsaw.
Grubecki, I., Wójcik, M., 2000. Comparison between isothermal and optimal temperature policy for batch reactor. Chem. Eng. Sci. 55, 5161–5163.
Guida, V., Ocone, R., Astarita, G., 1994. Micromixing and memory in chemical reactors. Chem. Eng. Sci. 49 (24b), 5215–5228.
Guldberg, C.M., Waage, P., 1867. Etudes sur les Affinités Chimiques. Christiania: Brogger et Christie.
Guo, Z.-Y., Cao, B.-Y., 2008. A general heat conduction law based on the concept of motion of thermal mass. Acta Physica Sinica 57 (7), 4273–4279.
Guo, Z.-Y., Hou, Q.-W., 2010. Thermal wave based on the thermomass model. J. Heat Transfer 132 (7), 072403.
Guo, Y., Jou, D., Wang, M., 2016. Understanding of flux-limited behaviors of heat transport in nonlinear regime. Physics Letters A 380, 452–457.
Guo, Z., Thirumalai, D., Honeycutt, J.D., 1992. Folding kinetics of proteins: a model study. J. Chem. Phys. 97 (1), 525–535, 072403.

Gustavsson, M., 2012. Residual thermodynamics: a framework for analysis of nonlinear irreversible processes. Int. J. Thermodyn. 15 (2), 69–82.

Gutman, F., Keyzer, H. (Eds.), 1986. Modern Bioelectrochemistry. Plenum Press, New York.

Guy, A., 1992. Thermodynamic analysis of thermo-electric effects in a metal–metal couple. Flow, Diffusion and Transport Processes. Advances in Thermodynamics Series, vol. 6, Taylor and Francis, New York, pp. 419–425.

Guyer, R.A., Krumhansl, J.A., 1964. Dispersion relation for second sound in solids. Phys. Rev. A 133, 1141–1415.

Gyarmati, I., 1961a. On the phenomenological basis of irreversible thermodynamics, I: Onsager theory. Period. Politechn. 5 (3), 219–243.

Gyarmati, I. 1961b. On the phenomenological basis of irreversible thermodynamics, II: on a possible nonlinear theory. Period. Politechn. 5 (4), 321–339.

Gyarmati, I., 1969. On the "Governing principle of dissipative processes" and its extension to nonlinear problems. Ann. Phys. (Leipzig) 23, 353–375.

Gyarmati, I., 1970. Nonequilibrium Thermodynamics. Springer, Berlin.

Gyarmati, I., 1977. On the wave approach of thermodynamics and some problems of nonlinear theories. J. Nonequilib. Thermodyn. 2, 233–243.

Gyftopoulos, E.P., 1998a. Maxwell and Boltzmann's triumphant contributions to and misconceived interpretation of thermodynamics. Int. J. Appl. Thermodyn. 1 (1–4), 9–19.

Gyftopoulos, E.P., 1999b. Infinite time (reversible) versus finite time (irreversible) thermodynamics: a misconceived distinction. Energy 24 (12), 1035–1039.

Gyftopoulos, E.P., 2002. On the Curzon-Ahlborn efficiency and its lack of connection to power producing processes. Energy Convers. Manage. 43, 609–615.

Gyftopoulos, E.P., 2003. A tribute to energy systems scientists and engineers. Int. J. Thermodyn. 6 (2), 49–57.

Haase, R., 1969. Thermodynamics of Irreversible Processes. Addison-Wesley, Reading, MA, Dover edition in 1990.

Haken, H., 1984. Advanced Synergetics. Springer-Verlag, Berlin.

Halkin, H., 1966. A Maximum principle of the Pontryagin type for systems described by nonlinear difference equations. SIAM J. Control ser. A (4), 528–547.

Haltchev, I.A., Denier, J.P., 2003. The stability of boundary-layer fluxes under conditions of intense interfacial mass transfer: the effect of interfacial coupling. Int. J. Heat Mass Transfer 46, 3881–3895.

Hansmann, H.E., Okamoto, Y., 1999. Tackling the protein folding problem by a generalized ensemble approach with Tsallis statistics. In: Salinas, S.R.A., Tsallis, C. (Eds.), Nonextensive Statistical Mechanics and Thermodynamics, Special issue of Brasilian J. Phys. 29 (1), 187–199.

Harada, M., Tanigaki, M., Nishimura, A., Eguchi, W., 1974. Self-diffusivity and viscosity of simple liquids. J. Chem. Eng. Japan 7 (6), 407–413.

Harland, J.R., Salamon, P., 1988. Simulated annealing: a review of the thermodynamic approach. Nuclear Physics B (proc. Suppl.) 5A, 109.

Harremoes, P., Topsoe, F., 2001. Maximum entropy fundamentals. Entropy 3, 191–226.

Hattia, M., Tioursib, M., 2009. Dynamic neural network controller model of PEM fuel cell system. Int. J. Hydrogen Energ. 34 (11), 5015–5021.

Hau, J.L., Bakshi, B.R., 2004. Expanding exergy analysis to account for ecosystem products and services. Environ. Sci. Technol. 38 (13), 3768–3777.

Hausen, H., 1976. Wärmeübertragung im Gegenstrom, second ed. Gleichstrom und Kreutzstrom. Springer, Berlin.

Haynie, D.T., 2001. Biological Thermodynamics. Cambridge University Press, Cambridge, UK.

He, J.-H., 1997. Semi-inverse method of establishing generalized variational principles for fluid mechanics with emphasis on turbomachinery aerodynamics. Int. J. Turbo Jet-Engines 14 (1), 23–28.

Herivel, J.W., 1955. The derivation of the equations of motion of an ideal fluid by Hamilton's principle. Proc. Cambridge Philos. Soc. 51, 344–349.

Hernandez, L., Kafarov, V., 2009. Use of bioethanol for sustainable electrical energy production. Int. J. Hydrogen Energ. 34 (3), 7041–7050.

References

Herzel, H., Ebeling, W., Schmitt, A.O., 1994. Entropies of biosequences: the role of repeats. Phys. Rev. E 50, 5061.

Hesse, D., 1974. Derivation of the transport equation for multicomponent diffusion through capillaries in the transition region between gas diffusion and Knudsen diffusion. Berichte dre Bunsen-Gesellschaft 78, 276–279.

Hesse, D., 1977a. Multicomponent diffusion and chemical reaction of gases in porous catalysts. I. Chem. Engin. Sci. 32, 413.

Hesse, D., 1977b. Multicomponent diffusion and chemical reaction of gases in porous catalysts II. Chem. Engin. Sci. 32, 427.

Hill, T., 1977. Biochemical cycles and free energy transduction. TIBS September, 204–207.

Hirs, G.G. (Ed.), 2000. ECOS 2000 Proceedings. Part 3 Process Integration. University Twente, The Netherlands.

Hirschfelder, J.O., Curtiss, C.F., Bird, R.B., 1954. Molecular Theory of Gases and Liquids. Wiley, New York.

Hiscock, W.A., Lindblom, L.A., 1983. Stability and causality in dissipative fluids. Ann. Phys. (NY) 151, 466–496.

Hjemfelt, A., Ross, J., 1989. Resonant response of a driven chemical system: effects on efficiency. J. Chem. Phys. 91, 2293.

Hjemfelt, A., Schreiber, I., Ross, J., 1991. Efficiency of power production in simple nonlinear electrochemical systems. J. Phys. Chem. 95, 6048–6053.

Hoang, N.H., Couenne, F., Jallut, C., Le Gorrec, Y., 2013. Thermodynamics-based stability analysis and its use for nonlinear stabilization of the CSTR. Comput.Chem. Eng. 58, 156–177.

Hoffmann, K.H., 1990. Optima and bounds for irreversible thermodynamic processes. Finite-Time Thermodynamics and Thermoeconomics. Advances in Thermodynamics Series, vol. 4, Taylor and Francis, New York, pp. 22–65.

Hoffmann, K.-H., Watowich, S.J., Berry, R.S., 1985. Optimal paths for thermodynamical systems. J. Appl. Phys. 58 (2125), 2125, (see also ibid 58, 2893).

Hoffmann, K.-H., Burzler, J.M., Schubert, M., 1997. Endoreversible thermodynamics. J. Non-Equilib. Thermodyn. 22, 311–355.

Hoffmann, K.-H., Essex, C., Schulzky, C., 1998. Fractional diffusion and entropy production. J. Non-Equilib. Thermodyn. 23, 166–175.

Hoffmann, K.H., Burzler, J., Fischer, A., Schaller, M., Schubert, S., 2003. Optimal process paths for endoreversible systems. J. Non-Equilib. Thermodyn. 28, 233–268.

Hofman, T., 2002. Termodynamika Molekularna. Oficyna Wydawnicza Politechniki Warszawskiej, Warszawa.

Holden, G., Essex, C., 1997. Radiative entropy equilibrium. J. Quant. Spectrosc. Radiat. Transfer 57 (1), 33–40.

Holderith, J., Reti, P., 1979. On the boundary conditions for hyperbolic reactors. Annal. Univers. Scient. Budapest XV, 39–43.

Holland, P.R., 1993. The de Broglie–Bohm theory of motion and quantum field theory. Phys. Reports 224 (3), 95–150.

Horsthemke, W., Moore, P.K., 2004. Turing instability in inhomogeneous arrays of diffusively coupled reactors. J. Phys. Chem. A. 108, 2225–2231.

Hougen, O.A., Watson, K.M., Ragatz, R.A., 1954. Chemical Process Principles. Wiley, New York.

Hua, J.Z., Brennecke, J.F., Stadtherr, M.A., 1996. Reliable prediction of phase stability using an interval-Newton method. Fluid Phase Equilibria 116, 52.

Huang, K., 1967. Statistical Mechanics. Wiley, New York.

Hughes, R., 1984. Deactivation of Catalysts. Academic Press, London.

Huilgol, R.R., Phan-Thien, N., 1986. Recent advances in the continuum mechanics of viscoelastic liquids. Int. J. Engin. Sci. 24, 161–261.

Hüttner, M., 1983. Investigation into the kinetics of the regeneration of catalyst in fluidized catalytic cracking. PhD Thesis. Warsaw Technological University, Warsaw.

Hwa, T., Kardar, M., 1989. Fractals and self-organized criticality in disipative dynamics. Physica D 38, 198–202.

Imre, I. (Ed.), 1990. Issue on solar drying. Drying Technol., 8.(No. 2).

Imre, A.R., 2000. Binary liquids under absolute negative pressure. Periodica Polytech. Ser. Chem. Engin. 44 (1), 39–48.

Ingarden, R.S., 1976. Quantum information theory. Rep. Math. Phys. 10 (1), 43–71.

Ingarden, R.S., 1992. Towards mesoscopic thermodynamics: small systems in higher-order states. Open Sys. Inform. Dynamics 1 (1), 75–102.

Ingarden, R.S., 1996. On physical applications of Finsler geometry. Contemp. Math. 196, 213–223.
Ingarden, R.S., Urbanik, K., 1962. Quantum informational thermodynamics. Acta Physica Polonica 21, 281–304.
Ingarden, R.S., Kawaguchi, M., Sato, Y., 1982. Information geometry of classical thermodynamical systems. Tensor, N.S., 267–278.
Ingarden, R.S., Kossakowski, A., Ohya, M., 1979. Information Dynamics and Open Systems. Kluwer Academics, Dordrecht.
Ingebritsen, S.E., Sanford, W.E., 1998. Groundwater in Geologic Processes. Cambridge University Press, Cambridge, UK.
Innocentini, M.D.M., Pardo, A.R.F., Salvini V.R., Pandolfelli, V.C., 1999. How accurate is Darcy's law for refractories? Am. Ceram. Soc. Bull. 77, 11. Data Depository File No. 325.
Ioffe, I.I., Reshetov, V.A., Dobrotvorskii, A.M., 1985. Heterogeneous catalysis (in Russian), Chimia, Leningrad: Chimia.
Ionescu, T.C., Edwards, B.J., Keffer, D.J., Mavrantzas, V.G., 2008. Energetic and entropic elasticity of nonisothermal flowing polymers: experiment, theory, and simulation. J. Rheol. 52 (1), 105–140.
Iordanidis, A.A., Annaland, M.v.S., Kronberg, A.E., Kuipers, J.A.M., 2003. A critical comparison between the wave model and the standard dispersion model. Chem. Eng. Sci. 58, 2785–2795.
Ishida, K., 1981. Nonequilibrium thermodynamics of the selection of biological macromolecules. J. Theor. Biol. 88, 257–273.
Ishida, K., Matsumoto, S., 1975. Non-equilibrium thermodynamics of temporally oscillating chemical reactions. J. Theor. Biol. 52 (2), 343–363.
Israel, W., 1976. Nonstationary irreversible thermodynamics: a causal relativistic theory. Ann. Phys. (NY) 100 (1/2), 310–331.
Ivanov, M.G., Volovich, V., 2001. Entropy bounds, holographic principle and uncertainty relation. Entropy 3, 66–75.
Izüs, G., Deza, R., Ramirez, O., Wio, H.S., Zanette, D.H., Borzi, C., 1995. Global stability of stationary patterns in bistable reaction-diffusion systems. Phys. Review E 52, 129–136, TUSK.
Jaakko, J., Saastamoinen, J.J., 2005. Comparison of moving bed dryers of solids operating in parallel and counterflow modes. Drying Technol. 23 (5), 1003–1025.
Jackson, D., 1975. Classical Electrodynamics. Wiley, New York.
Jackson, H.E., Walker, C.T., McNelly, T.F., 1970. Second sound in NaF. Phys. Rev. Lett. 25, 26–28.
Jain, R.K., Ruckenstein, E., 1976. Stability of stagnant viscous films on a solid surface. J. Colloid Interf. Sci. 54 (1), 108–116.
Jamaleddine, T.J., Ray, M.B., 2010. Application of computational fluid dynamics for simulation of drying processes: a review. Drying Technol. 28 (2), 120–154.
Jami, M., Mezrhab, A., Bouzidi, M., Lallemand, P., 2007. Lattice Boltzmann method applied to the laminar natural convection in an enclosure with a heat-generating cylinder conducting body. Int. J. Thermal Sci. 46, 38–47.
Janyszek, H., Mrugała, R., 1990. Riemannian and Finslerian geometry and fluctuations of thermodynamic systems. Nonequilibrium Theory and Extremum Principles. Advances in Thermodynamics Series, vol. 3, Taylor and Francis, New York, pp. 159–174.
Jaroniec, M., Lu, X., Madey, R., Avnir, D., 1990. Thermodynamics of gas adsorption on fractal surfaces of heterogeneous microporous solids. J. Chem. Phys. 92, 7589–7595.
Jarzebski, Z.M., 1990. Solar Energy Photovoltaic Conversion. PWN, Warsaw.
Javania, N., Dincer, I., Natererb, G.F., Yilbas, B.S., 2014. Heat transfer and thermal management with PCMs in a Li-ion battery cell for electric vehicles. Int. J. Heat Mass Transfer 72, 690–703.
Jaworski, W., 1981. Information thermodynamics with the second order temperatures for the simplest classical systems. Acta Physica Polonica. A 60, 645–659.
Jaworski, Z., Zakrzewska, B., 2008. Multiscale modelling of chemical reactors. Chem. Process Eng 29, 567–581.
Jaynes, E.T., 1957. Information theory and statistical mechanics. Phys. Rev 106, 620–630.
Jaynes, E.T., 1963. Information theory and statistical mechanics. In: Ford, W.K. (Ed.), Statistical Physics, 1962 Brandeis Lectures.

References

Jeter, J., 1981. Maximum conversion efficiency for the utilization of direct solar radiation. Solar Energy 26, 231–236.

Jezierski, J., Kijowski, J., 1990. Thermo-hydrodynamics as a field theory. Nonequilibrium Theory and Extremum Principles. Advances in Thermodynamics Series, vol. 3, Taylor and Francis, New York, pp. 282–316.

Johannessen, E., Kjelstrup, S., 2005. A highway in state space for reactors with minimum entropy production. Chem. Eng. Sci. 60, 3347–3361.

Johannessen, E., Gross, J., Bedeaux, D., 2011. Nonequilibrium thermodynamics of interfaces using classical density functional theory. J. Chem. Phys. 129 (18), 184703.

Jongschaap, R.J., 1990. Microscopic modelling of the flow properties of polymers. Rep. Prog. Phys. 53 (1), 1–55.

Jorgensen, S.E. (Ed.), 2001. Thermodynamics and Ecological Modelling. Lewis (Publishers of CRC Press), Boca Raton.

Joseph, D.D., 1976. Stability of Fluid Motions. Springer, Berlin.

Jou, D., 1988. Fluctuation theory in extended irreversible thermodynamics. Physica A, 221–234.

Jou, D., Camacho, J., 1991. On the nonequilibrium thermodynamics of non-Fickian diffusion. Macromolecules 24, 3597–3602.

Jou, D., Casas-Vazquez, J., 1992. Possible experiment to check the reality of a nonequilibrium temperature. Phys. Rev. A 45 (12), 8371–8373, see also a reply to comments; Phys. Rev. E 48 (4), 3201–3202.

Jou, D., Llebot, J.E., 1980. Electric current fluctuations in extended irreversible thermodynamics. J. Phys. A 13, 47.

Jou, D., Salhoumi, A., 2001. Legendre transforms in nonequilibrium thermodynamics: an illustration in electrical systems. Phys. Lett. A 283, 163–167.

Jou, D., Casas-Vazquez, J., Lebon, G., 1988. Extended irreversible thermodynamics. Rep. Progr. Phys. 51, 1105–1172.

Jou, D., Casas-Vazquez, J., Lebon, G. 1999. Extended irreversible thermodynamics revisited (1988–98).

Jou, D., Casas-Vazquez, J., Lebon, G., 2001. Extended irreversible thermodynamics, third ed. Springer, Berlin.

Jou, D., Sellitto, A., Alvarez, F.X., 2011. Heat waves and phonon-wall collisions in nanowires. Proc. R. Soc. A 467, 2520–2533.

Jurado, F., 2003. Power supply quality improvement with a SOFC plant by neural-network-based control. J. Power Sources 117 (1–2), 75–83.

Juraščík, M., Sues, A., Ptasinski, K.J., 2010. Exergy analysis of synthetic natural gas production method from biomass. Energy 35 (2), 880–888.

Jurkowska, J. 1982. Trajectories of the carbon monoxide oxidation in flow systems. MsD thesis (in Polish). Faculty of Chemical Engineering at the Warsaw University of Technology, Warsaw.

Kåberger, T., Månsson, B.A., 2001. Entropy and economic processes—physics perspectives. Ecol. Econ. 36, 165–179.

Kakaç, S., Bon, B., 2008. A Review of two-phase flow dynamic instabilities in tube boiling systems. Int. J. Heat Mass Transfer 51, 399–433.

Kakaç, S., Pramuanjaroenkij, A., Zhou, X.Y., 2007. A review of numerical modeling of solid oxide fuel cells. Int. J. Hydrogen Energ. 32, 761–786.

Kaleta, A., Górnicki, K., Winiczenko, R., Chojnacka, A., 2013. Evaluation of drying models of apple (var. Ligol) dried in a fluidized bed dryer. Energy Conversion and Manage 67 (3), 179–185.

Kalina, J., 2011. Modelling of fluidized bed biomass gasification in the quasiequilibrium regime for preliminary performance studies of energy conversion plants. Chem. Process Eng. 32 (2), 73–89.

Kaliski, S., 1965. Wave equation of heat conduction. Bull. Acad. Pol. Sci. Ser. Sci. Tech. 13, 211–219.

Kaniadakis, G., 2002. Statistical origin of quantum mechanics. Physica A 307 (1–2), 172–184.

Kaniadakis, G., Lissia, M., Rapisarda, A. (Eds.), 2002. Nonextensive thermodynamics and its applications. Proceedings of the International School and Worksshop, NEXT 2001, Villiasimius, Italy, May 23–30, 2001. Phys. A, vol. 305 (1 + 2).

Kantor, Y., 1989. Energetic and entropic elasticity on the Sierpiński gasket. Physica D 38, 215–220.

Kantorovich, L.V., Krylow, V.I., 1958. Approximate methods of higher analysis. Groningen: Noordhoff.

Kao, T., 1977. Non-Fourier heat conduction in thin surface layers. J. Heat Transfer Trans. ASME 99, 343–345.

Kapitaniak, T., 1998. Chaos for Engineers (Theory, Applications and Control). Springer, Berlin.

Karcz, M., Badur, J., 2005. An alternative two-equation turbulent heat diffusivity closure. Int. J. Heat Mass Transfer 48, 2013–2022.

Kargol, M., 1985. Zagadnienia osmotycznej przemiany energii swobodnej w pracę użyteczną w modelowych układach membranowych. Zagadnienia Biofizyki Współcz 10, 81–96.

Karkheck, J., 1986. Kinetic theories and ensembles of maximum entropy. Kinam 7A, 191–208.

Karkheck, J., 1989. Kinetic theories and ensembles of maximum entropy. In: Skilling, J. (Ed.), Maximum Entropy and Bayesian Methods. Kluwer Academics, pp. 491–496.

Karkheck, J., 1992. Irreversibility, kinetic equations and ensembles of maximum entropy. Flow, Diffusion and Transport Processes. Advances in Thermodynamics Series, vol. 6, Taylor and Francis, New York, pp. 23–57.

Karlin, I.V., Tatarinova, L.L., Gorban, A.N., Öttinger, H.C., 2003. Irreversibility in the short memory approximation. Physica A 327, 327–424.

Karpov, A.I., Bulgakov, V.K., Novozhilov, V.B., 2003. Quantitative estimation of relationship between the state with minimal entropy production and the actual stationary regime of flame propagation. J. Non-Equilib. Thermodyn. 28, 193–205.

Katchalsky, A., Curran, P.F., 1965. Nonequilibrium Thermodynamics in Biophysics. Harvard University Press, Cambridge, MA.

Kaushik, S.C., Kumar, S., 2001. Finite time thermodynamic evaluation of irreversible Ericsson and Stirling heat engines. Energy Convers. Manage. 42 (3), 295–312.

Kawczyński, A.L., 1990. Chemical Reactions: From Equilibrium through Dissipative Structures to Chaos (in Polish). WNT, Warsaw.

Kawczyński, A.L., Comstock, W.S., Field, R.J., 1992. The evolution of pattern in a chemically oscillating medium. Physica D 54, 220–234.

Kay, J., 1991. A nonequilibrium thermodynamic framework for discussing ecosystem integrity. Environ. Manage. 15 (4), 483–495.

Kay, J., 2000. Ecosystems as self-organizing holarchic open systems: narratives and the second law of thermodynamics. In: Joergensen, S.E., Muller, F. (Eds.), Handbook of Ecosystem Theories and Management. CRC Press–Lewis Publishers, pp. 135–160.

Kay, J.J., 2002. On complexity theory, exergy and industrial ecology: some implications for construction ecology. In: Kibert, C., Sendzimir, J., Guy, B. (Eds.), Construction Ecology: Nature as the Basis for Green Buildings. Spon Press, London.

Kay, J., Schneider, E.D., 1992a. Complexity and thermodynamics. Towards a New Ecology Futures 26 (6), 626–647.

Kay, J., Schneider, E.D., 1992b. Thermodynamics and measures of ecological integrity, Proceedings of Ecological Indicators. Elsevier, Amsterdam. pp. 159–182.

Kazim, A., 2004. Exergy analysis of a PEM fuel cell at variable operating conditions. Energy Convers. Manage. 45, 1949–1961.

Kazimi, M.S., Erdman, C.A., 1973. On the interface temperature of two suddenly contacting materials. J. Heat Transfer Trans. ASME 97, 617–619.

Keesom, W.H., 1949. Helium. Moscow, IL.

Keesom, W.H., Sarris, B.F., 1940. Further measurements of the heat conductivity of liquid helium II. Physica 7, 241–252.

Keizer, J., 1976a. On the kinetic meaning of the second law of thermodynamics. J. Chem. Phys. 64 (11), 4466–4474.

Keizer, J., 1976b. Fluctuations, stability and generalized state functions at non-equilibrium steady states. J. Chem. Phys. 65 (11), 4431–4444.

Keizer, J., 1985. Lectures on the statistical thermodynamics of nonequilibrium stady states. In: Casas-Vazquez, J., Jou, D., Rubi, J.M. (Eds.), In Recent Developments in Nonequilibrium Thermodynamics: Fluids and Related Topics, Lecture Notes in Physics 253. Springer, Berlin, p. 3.

Keizer, J., 1987a. Statistical Thermodynamics of Nonequilibrium Processes. Springer, New York.

Keizer, J., 1987b. The nonequilibrium electromotive force, II: theory for a continuously stirred tank reactor. J. Chem. Phys. 87 (7), 4074–4079.

References

Keizer, J., Chang, O.-K., 1987. The nonequilibrium electromotive force, I: measurements in a continuously stirred tank reactor. J. Chem. Phys. 87 (7), 4064–4073.

Keizer, J., Fox, R.F., 1974. Qualms regarding the range of validity of the Glansdorff–Prigogine criterion for stability of nonequilibrium states. Proc. Natl. Acad. Sci. USA 71 (1), 192–196.

Kemp, I.C., Bahu, R.E., Pasley, H.S., 1994. Model development and experimental studies of vertical pneumatic conveying dryers. Drying Technol. 12 (6), 1323–1340.

Kennedy, D.C., Bludman, S.A., 1997. Variational principles for stellar structures. Astrophysical J. 484, 329–340.

Kestin, J., 1979. In: Domingos, et al. (Eds.), Foundations of Non-Equilibrium Thermodynamics. Macmillan Press, London.

Kestin, J., 1993. Internal variables in the local equilibrium approximation. J. Non-Equilib. Thermodyn. 18, 360–379.

Kestin, J., Bataille, J., 1980. Thermodynamics of solids. Continuum Models of Discrete Systems. University of Waterloo Press, Ontario.

Kestin, J., Rice, J.M., 1970. Paradoxes in the application of thermodynamics to strained solids. In: Stuart, F.B., Gal Or, B., Brainard, A.J. (Eds.), A Critical Review of Thermodynamics. Mono Book Corp, Baltimore, pp. 275–298.

Khinast, J., Luss, D., Harold, M.P., Ostermaier, J.J., McGill, R., 1998. Continuously stirred decanting reactor—operability and stability considerations. AIChE J. 44 (2), 372–387.

Khinchin, A.I., 1957. Mathematical Foundations of Information Theory. Dover, New York.

Kiehn, R.M., 1975. Conformal invariance and Hamilton–Jacobi theory for dissipative systems. J. Math. Phys. 16, 1032–1033.

Kincaid, J.M., Lopez de Haro, M., Cohen, E.G.D., 1983. The Enskog theory for multicomponent mixtures, II: mutual diffusion. J. Chem. Phys. 79, 4509–4521.

Kincaid, J.M., Cohen, E.G.D., Lopez de Haro, M., 1987. The Enskog Theory of Multicomponent Mixtures, IV: Thermal Diffusion. J. Chem. Phys. SEE J PHYS CHEM BELOW — ARE BOTH CORRECT JOURNAL NAMES? 86, 963.

King, E.L., Altman, C., 1956. A schematic method of deriving the rate laws for enzyme catalyzed reactions. J. Phys. Chem. 60, 1375–1378.

Kirkwood, J.G., 1967. In: Zwanzig, R. (Ed.), Collected Works: Selected Topics in Statistical Mechanics. Gordon and Breach, New York.

Kirkwood, J.G., Buff, F.P., Green, M.S., 1967. The statistical mechanical theory of transport processes, III: The coefficients of shear and bulk viscosity of liquids. In: Zwanzig, R. (Ed.), J. G. Kirkwood Collected Works: Selected Topics in Statistical Mechanics. Gordon and Breach, New York.

Kirschner, I., Törös, R., 1984. Existence of stationary thermodynamic Lagrange functions with nonlinear source density. J. Non-Equilib. Thermodyn. 9, 165.

Kirwan, A.D., 2004. Intrinsic photon entropy? The dark side of light? Int. J. Eng. Sci. 42, 725–734.

Kjelstrup, S., Bedeaux, D., 1997. Jumps in electric potential and in temperature at the electrode surfaces of the solid oxide fuel cells. Physica A 244, 213–226.

Kjelstrup, S., Bedeaux, D., 2008. Non-Equilibrium Thermodynamics of Heterogeneous Systems Hardbound. World Scientific Publishing Co, Singapore.

Kjelstrup, S., Bedeaux, D., 2008. Non-Equilibrium Thermodynamics of Heterogeneous Systems. World Scientific, Singapore.

Kjelstrup, S., Sauar, E., van der Koi, H., Bedeaux, D., 1999. The driving force distribution for minimum lost work in a chemical reactor close to and far from equilibrium I Theory. Ind. Eng. Chem. Res. 38, 3046–3050.

Kleinstreuer, C.L., 2003. Two-Phase Flow: Theory and Applications. Taylor & Francis Group–CRC Press, Boca Raton.

Klimontovich, Y.L, 1982. Kinetic Theory of Nonideal Gases and Nonideal Plasmas. Pergamon, Oxford.

Klimontovich, Y.L, 1986. Statistical Physics. Harwood Academic Publishers, Chur.

Klimontovich, Y.L, 1991. Turbulent Motion and the Structure of Chaos. Kluwer Academics, Dordrecht.

Klimontovich, Y.L, 1999. Entropy, information, and criteria of order in open systems. Nonlinear Phenomena Complex Syst. 2 (4), 1–25.

Kluitenberg, G., 1966. Nonequilibrium Thermodynamics, Variational Techniques, Stability. In: Donelly, R., Herman, R., Prigogine, I. (Eds.), Application of the theory of irreversible processes to continuum mechanics. University Press, Chicago.

Kluitenberg, G.A., 1977. On dielectric and magnetic relaxation and vectorial internal degrees of freedom in thermodynamics. Physica 87a, 541–563.

Kluitenberg, G.A., 1981. On vectorial internal variables and dielectric and magnetic relaxation phenomena. Physica A 109 (1–2), 91–122.

Kobe, D.H., Reali, G., Sieniutycz, S., 1986. Lagrangians for dissipative systems. Amer. J. Phys. 54, 997–999.

Koh, S., Eringen, C., 1963. On the foundations of nonlinear thermo-viscoelasticity. Int. J. Eng. Sci. 1, 199–229.

Kondepudi, D.K., Prigogine, I., 1998. Modern thermodynamics: from heat engines to dissipative structures. Wiley, New York.

Konopka, A.K., 1984. Is the information content of DNA evolutionary significant? Letter to the Editor. J. Theor. Biol. 107, 697–704.

Konopka, A.K., 1985. Theory of degenerate coding and informational parameters of protein coding genes. Biochemie 67, 455–468.

Kopelman, R., 1988. Fractal Reaction Kinetics. Science 241 (4873), 1620–1626.

Koschmieder, E.L., 1993. Benard Cells and Taylor Vortices. Cambridge University Press, Cambridge, UK.

Kotas, T.J., 1985. Exergy Method of Thermal Plant Analysis. Butterworth, Borough Green.

Kotas, T.J., 1986. Exergy method of thermal and chemical plant analysis. Chem. Eng. Res. Des. 64, 212.

Kotowski, R., 1989. On the Lagrange functional for dissipative processes. Arch. Mech. 41, 571–581.

Kotowski, R., Radzikowska, E., 2002. Evolution criterion in the coupled fields theory. Technische Mechanik 22 (2), 141–151.

Kovac, J., 1977. Non-Equilibrium Thermodynamics of Interfacial Systems. Physica A 86 (1), 1–24.

Kowalski, S.J., Strumiłło, Cz., 1997. Moisture transport, thermodynamics, and boundary conditions in porous materials in presence of mechanical stresses. Chem. Eng. Sci. 52 (7), 1141–1150.

Kowalski, S.J., 2003. Thermomechanics of Drying Processes. Springer, Berlin.

Kowalski, S.J., Rajewska, K., Rybicki, A., 2000. Destruction of wet materials by drying. Chem. Eng. Sci. 55, 5755–5762.

Kowalski, S.J., Łechtańska, J.M., Szadzińska, J., 2011. Quality aspects of fruit and vegetables dried convectively with osmotic pretreatment. Chemical and Process Engin. 34 (1), 51–62.

Kramers, H.A., 1940. Brownian motion in a field of force and the diffusion model of chemical reactions. Physica VII (4), 284–304.

Kranys, M., 1966. Relativistic hydrodynamics without paradox of infinite velocity of heat conduction. Nuovo Cimento 42B, 1125–1194.

Kremer, G.M., 1985. Erweiterte Thermodynamik Idealer und Dichter Gase. Diss. TU Berlin.

Kremer, G.M., 1986. Extended thermodynamics of ideal gases with 14 fields. Ann. Inst. Henri Poincare 45, 419–440.

Kremer, G.M., 1986. Extended thermodynamics of ideal gases with 14 fields. Ann. Inst. Henri Poincare 45, 419–440.

Kremer, G.M., 1987. Extended thermodynamics of nonideal gases. Physica 144A, 156–178.

Kremer, G.M., 1989. Extended thermodynamic of molecular ideal gases. Continuum Mech. Thermodyn. 1, 21–45.

Kremer, G.M., 1992. On extended thermodynamics of ideal and real gases. Article in: Extended Thermodynamic Systems. Adv. Thermodyn. Ser. 7, 140–182.

Kreuzer, H.J., 1981. Nonequilibrium Thermodynamics and Its Statistical Foundations. Clarendon Press, Oxford.

Kreuzer, H.J., Chapman, R.G., March, N.H., 1988. Nonequilibrium thermodynamics of sublimation, evaporation and condensation. Phys. Rev. A 37, 582.

Krieger, Ch., Müller, K., Arlt, W., 2016. Thermodynamic analysis of reversible hydrogenation for heat storage in concentrated solar power plants. Solar Energy 123 (1), 40–50.

Krischer, O., 1956. Die Wissenschaftlichen Grundlagen der Trocknungstechnik. Springer, Berlin.

Krishna, R., Standart, G.L., 1979. Mass and energy transfer in multicomponent systems. Chem. Eng. Commun. 3, 201.

Krishna, R., Taylor, R., 1986. Multicomponent mass transfer: theory and applications. In: Cheremissinoff, N.C. (Ed.), Handbook of Heat and Mass Transfer. Gulf Publ, Houston.

References

Kuang, Y.H., Wang, R.Z., Yu, L.Q., 2003. Experimental study on solar assisted heat pump system for heat supply. Energy Convers. Manage. 44 (7), 1089–1098.

Kubo, R., 1988. Statistical Mechanics. North Holland, Amsterdam.

Kubo, R., Matsuo, K., Kitahara, K., 1973. Fluctuation and relaxation of macrovariables. J. Statis. Phys. 9 (1), 51–96.

Kuddusi, L., Egrican, N., 2008. A critical review of constructal theory. Energy Convers. Manage. 49, 1283–1294.

Kuhn, H.W., Tucker, A.W., 1951. Nonlinear programming. In: Neyman, J. (Ed.), Proceedings of the Second Berkeley Symposium on Mathematical Statistics and Probability. University of California Press, Berkeley, pp. 481–493.

Kuiken, G.D.C., 1994. Thermodynamics of Irreversible Processes. Wiley, Chichester, UK.

Kullback, S., 1959. Information Theory and Statistics. Wiley, New York.

Kundu, P., Kumar, V., Mishra, I.M., 2014. Numerical modeling of turbulent flow through isotropic porous media. Int. J. Heat Mass Transfer 75, 40–57.

Kunii, D., Levenspiel, O., 1991. Fluidization Engineering. Butterworth-Heinemann, Newton.

Kuo, C.D., Bai, J.J., Chien, S., 1994. A fractal model for erythrocyte sedimentation. Biorheology 31 (1), 77–89.

Kupershmidt, B.A., 1990a. Hamiltonian formalism for reversible nonequilibrium fluids with heat flow. J. Phys. A 23, L529–L532.

Kupershmidt, B.A., 1990b. Canonical variational theory of relativistic magneto-hydrodynamics with anisotropic pressure. Nonequilibrium Theory and Extremum Principles. Advances in Thermodynamics Series, vol. 3, Taylor and Francis, New York, pp. 318–327.

Kupershmidt, B.A., 1992. The Variational Principles of Dynamics. World Scientific, New York.

Kuran, P., 2006. Nonlinear Models of Production of Mechanical Energy in Non-Ideal Generators Driven by Thermal or Solar Energy. PhD thesis, Faculty of Chemical & Process Engineering, Warsaw University of Technology.

Kuran, P., Sieniutycz, S., 2007. Maximum power obtained from the flux of solar radiation. Chemical and Process Engineering 28, 827–839.

Kuran, P., Sieniutycz, S., 2011. Thermodynamics of solar energy conversion into heat. 5th European Solar Thermal Energy Conference ESTEC 2011, Marseille, October 20–21.

La Salle, J., Lefschetz, S., 1961. Stability by Liapunov's Direct Method. Academic, New York.

Lagnado, R.R., Phan-Thien, N., Leal, L.G., 1985. The stability of two-dimensional linear flows of a Oldroyd-type fluid. J. Non-Newtonian Fluid Mech. 18, 25–59.

Landau, L.D., Lifschitz, E.M., 1967. Electrodynamics of Moving Media. Pergamon, London.

Landau, L., Lifshitz, E., 1974a. Mechanics of Continua. Pergamon, London.

Landau, L., Lifshitz, E., 1974b. The Classical Theory of Fields. Pergamon, London.

Landau, L., Lifshitz, E., 1975. Statistical Physics. Pergamon, Oxford.

Landau, L., Lifshitz, E., 1976. Mechanics. Pergamon, Oxford.

Landau, L., Lifshitz, E., 1984. Electrodynamics of Continua. Pergamon, Oxford.

Landau, L., Lifshitz, E., 1987. Fluid Mechanics. Pergamon, Oxford.

Landauer, R., 1961. Irreversibility and heat generation in the computing process. IBM J. Res. Dev 5, 183–191.

Landsberg, P.T., 1972. The fourth law of thermodynamics. Review of the book: Thermodynamic Theory of Structure, Stability and Fluctuations by P. Glansdorff and & I. Prigogine. Nature 238, 229–231.

Landsberg, P.T., 1984a. Can entropy and order increase together? Phys. Lett 102 A, 171.

Landsberg, P.T., 1984b. Entropy and order. In: Kilmister, C.W. (Ed.), Disequilibrium and Self-Organization. Reidel, Norwell, 1984.

Landsberg, P.T., 1990a. Thermodynamics and Statistical Mechanics. Dover, New York.

Landsberg, P.T., 1990b. Statistics and thermodynamics of energy conversion from radiation. Nonequilibrium Theory and Extremum Principles. Advances in Thermodynamics Series, vol. 3, Taylor and Francis, New York, pp. 482–536.

Lapasin, R., Grassi, M., Prici, S., 1996. Fractal approach to rheological modeling of aggregates suspensions. In: Ait-Kadi, A., et al. (Eds.), Proc. XII Int. Congr. on Rheology. Canadian Rheology Group, Quebec City.

Larecki, W., Piekarski, S., 1991. Phonon gas hydrodynamics based on the maximum entropy principle and the extended field theory of a rigid conductor of heat. Arch. Mech. 43 (2–3), 163–190.

Lavenda, B.H., 1972. Letter to the Editor. Lettere Al Nuovo Cimento 3, 385–390.
Lavenda, B.H., 1978. Thermodynamics of Irreversible Processes. Macmillan, London.
Lavenda, B.H., 1985a. Nonequilibrium Statistical Thermodynamics. Wiley, Chichester, UK.
Lavenda, B.H., 1985b. Brownian motion. Scientific American 252 (2), 70–84.
Lavenda, B.H., 1990. The statistical foundations of nonequilibrium thermodynamics. Nonequilibrium Theory and Extremum Principles. Advances in Thermodynamics Series, vol. 3, Taylor and Francis, New York, pp. 175–211.
Lavenda, B.H., 1991. Statistical Physics—A Probablilistic Approach. Wiley, Chichester, UK.
Lavenda, B.H., 1995. Thermodynamics of Extremes. Horwood Publishing Limited, Chichester.
Lavenda, B.H., Santamato, 1982. Irreversible thermodynamic stability criteria. Lettere Al Nuovo Cimento 33 (17), 559–564.
Lazzaretto, A., Macor, A., 1995. Direct calculation of average and marginal costs from the productive structure of an energy system. J. Energy Resour. Technol. 117, 171–178.
Lazzaretto, A., Macor, A. Reini, M., 1993a. EXSO: Exergoeconomic symbolic optimization for energy systems, I: Description of the method. Proceedings of the Conference on Energy Systems and Ecology. Kraków, Poland, July 5–9, pp. 323–330.
Lazzaretto, A., Macor, A., Reini, M., 1993b. EXSO: Exergoeconomic symbolic optimization for energy systems, II: Application to a combined plant. Proceedings of the Conference on Energy Systems and Ecology. Kraków, Poland, July 5–9, pp. 331–338.
Le Goff, P., Rivero, R., de Oliveira, S., Schwarzer, B., 1990. Compared energetic and economic optimization of industrial systems. Finite-Time Thermodynamics and Thermoeconomics. Advances in Thermodynamics Series, vol. 4, Taylor and Francis, New York, pp. 249–277.
Lebon, G., 1976. A new variational principle for the nonlinear unsteady heat conduction problem. Q. J. Mech. Appl. Math. 29, 499–509.
Lebon, G., 1978. Derivation of generalized Fourier and Stokes–Newton equations based on thermodynamics of irreversible processes. Bull. Acad. Soc. Belg. Cl. Sci. LXIV, 456–460.
Lebon, G., 1986. Variational principles in thermomechanics. In: Lebon, G., Perzyna, P. (Eds.), Recent Developments in Thermomechanics of Solids, CISM Courses and Lectures, vol. 262. Springer, Vienna, pp. 221–415.
Lebon, G., 1992. Extended irreversible thermodynamics of rheological materials. Extended Thermodynamic Systems. Advances in Thermodynamics, vol. 7, Springer, Berline, New York, pp. 310–338.
Lebon, G., 2014. Heat conduction at micro and nanoscales: a review through the prism of extended irreversible thermodynamics. J. Non-Equilib. Thermodyn. 39 (1), 35–59.
Lebon, G., Boukary, M.S., 1982. Lyapounov functions in extended irreversible thermodynamics. Phys. Lett. 88A, 391–393.
Lebon, G., Boukary, M.S., 1988. Objectivity, kinetic theory and extended irreversible thermodynamics. Int. J. Eng. Sci. 26, 471–483.
Lebon, G., Casas-Vazquez, J., 1976. On the stability condition for heat conduction with finite wave speed. Phys. Lett. 55A, 93–94.
Lebon, G., Mathieu, P., 1983. Comparison of the diverse theories of nonequilibrium thermodynamics. Int. Chem. Engin. 23, 651–662.
Lebon, G., Rubi, J.M., 1980. A generalized theory of thermoviscous fluids. J. Non-Equilib. Thermodyn. 5 (5), 285–300.
Lebon, G., Jou, D., Casas-Vazquez, J., 1980. An extension of the local equilibrium hypothesis. J. Phys. A 13, 275–280.
Lebon, G., Perez-Garcia, C., Casas-Vazquez, J., 1986. On a new thermodynamic description of viscoelastic materials. Physica 137A, 531–545.
Lebon, G., Perez Garcia, C., Casas-Vazquez, J., 1988. On the thermodynamic foundations of viscoelasticity. J. Chem. Phys. 88, 5068–5075.
Lebon, G., Dauby, P.C., Palumbo, A., Valenti, G., 1990. Rheological properties of dilute polymer solutions: an extended thermodynamic approach. Rheol. Acta 29, 127–136.

References

Lebon, G., Casas-Vazquez, J., Jou, D., Criado-Sancho, M., 1993. Polymer solutions and chemical reactions under flow: a thermodynamic description. J. Chem. Phys. 98, 7434–7439.

Lee, E.B., Marcus, L., 1967. Foundations of Optimal Control Theory. Wiley, New York.

Leech, J.W., 1958. Classical Mechanics. Wiley, New York.

Leitman, G., 1966. An Introduction to Optimal Control. McGraw-Hill, New York.

Leitman, G., 1981. The Calculus of Variations and Optimal Control. Plenum Press, New York.

Lemaitre, J., Chaboche, J.L., 1988. Mechanics of solid materials. Cambridge University Press, Cambridge, UK.

Lemański, M., 2007. Analiza obiegów energetycznych z ogniwem paliwowym i turbiną gazowo—parową. PhD thesis. Institute of Fluid Flow Machinery, Gdańsk.

Lemański, M., Badur, J., 2005. Ogniwo paliwowe SOFC z wewnętrznym reformingiem. Inżynieria Chemiczna i Procesowa 26, 157–172.

Lemański, M., Topolski, J., Badur, J., 2004. Analysis strategies for gas turbine—solid oxide fuel cell hybrid cycles. Proceedings of Conference COMPOWER Technical Economic and Environmental Aspects of Combined Cycle Power Plants, 213–220, Gdańsk TU Press.

Lems, S., van der Kooi, H.J., de Swaan Arons, J., 2003. Thermodynamic optimization of energy transfer in (bio) chemical reaction systems. Chem. Eng. Sci. 58, 2001–2009.

Lems, S., van der Kooi, H.J., de Swaan Arons, J., 2010. Exergy analyses of the biochemical processes of photosynthesis. Int. J. Exergy 7 (3), 333–351.

Lengyel, S., 1988. Deduction of the Guldberg–Waage mass action law from Gyarmati's governing principle of dissipative processes. J. Chem. Phys. 88, 1617–1621.

Lengyel, S., 1989. Chemical kinetics and thermodynamics: a history of their relationship. Chem. Eng. Sci. 17 (1–3), 441–455.

Lengyel, S., 1992. Consistency of chemical kinetics with thermodynamics. Flow, Diffusion and Transport Processes. Advances in Thermodynamics Series, vol. 6, Taylor and Francis, New York, pp. 283–302.

Lengyel, S., Gyarmati, I., 1980. Nonlinear thermodynamic studies of homogeneous chemical kinetic systems. Periodica Polytechnica 25 (1), 63–99.

Lengyel, S., Gyarmati, I., 1986. Constitutive equations and reciprocity relations of nonideal homogeneous closed chemical systems. Acta Chem. Hung. 122, 7–17.

Lenormand, R., 1989. Applications of fractal concepts in petroleum engineering. Physica D 18, 230–234.

Leonov, A.I., 1976. Nonequilibrium thermodynamics and rheology of viscoelastic polymer media. Rheol. Acta 21, 683–691.

Levenspiel, O., 1965. Chemical Reaction Engineering. Wiley, New York.

Levine, R.D., 1986. Geometry in classical statistical thermodynamics. J. Chem. Phys. 84, 910–916.

Levine, R.D., Tribus, M. (Eds.), 1979. The Maximum Entropy Principle. MIT Press, Cambridge, MAs.

Lévy, P., 1937. Théorie de l'addition des variables aléatoires. Gauthier-Villars, Paris.

Levy, Y., Jortner, J., Berry, R.S., 2002. Eigenvalue spectrum of the master equation for hierarchical dynamics of complex systems. Phys. Chem. Chem. Phys. 4, 5052–5058.

Lewis, R.M., 1967. A unifying principle in statistical mechanics. J. Math. Phys. 8, 1448–1459.

Lhuillier, D., Grmela, M., Lebon, G., 2003. A comparative study of the coupling of flow with non-Fickean thermodiffusion, Part III: Internal variables. J. Non-Equilib. Thermodyn. 28, 51–68.

Li, J.H.C., 1958. Thermodynamics for nonisothermal systems. The classical formulation. J. Chem. Phys. 29 (4), 747–753.

Li, J.H.C., 1962a. Carathéodory's principle and the thermokinetic potential in irreversible thermodynamics. J. Phys. Chem. 66 (8), 1414–1420.

Li, J.H.C., 1962b. Stable steady state and the thermokinetic potential. J. Chem. Phys. 37 (8), 1592–1595.

Li, X., 2006. Principles of Fuel Cells. Taylor and Francis, New York.

Li, X., Essex, Ch., Davison, M., Hoffman, K.-H., Schulzky, C., 2003. Fractional diffusion, irreversibility and entropy. J. Non-Equilib. Thermodyn. 28, 279–291.

Li, X.T., Grace, J.R., Lim, C.J., Watkinson, A.P., Chen, H.P., Kim, J.R., 2004. Biomass gasification in a circulating fluidized bed. Biomass Bioenerg. 26 (2), 171–193.

Li, J., Chen, L., Sun, F., 2009a. Extremal work of an endoreversible system with two finite thermal capacity reservoirs. J. Energy Institute 82, 53–56.
Li, J., Chen, L., Sun, F., 2009b. Optimum work in real systems with a class of finite thermal capacity reservoirs. Math. Comp. Modell. 49, 542–547.
Li, C.-X., Li, C.-J., Guo, L.-J., 2009c. Effect of composition of NiO/YSZ anode on the polarization characteristics of SOFC fabricated by atmospheric plasma spraying. Int. J. Hydrogen Energ. 35 (7), 2964–2969.
Li, J., Chen, L., Sun, F., 2010a. Maximum work output of multistage continuous Carnot heat engine system with finite reservoirs of thermal capacity and radiation between heat source and working fluid. Thermal Sci. 14, 1–9.
Li, J., Chen, L., Sun, F., 2010b. Fundamental optimal relation of a generalized irreversible Carnot heat pump with complex heat transfer law. Pramana J. Phys. 74 (2), 219–230.
Li, J., Chen, L., Sun, F., 2011. Finite-time exergonomic performance of an endoreversible Carnot heat engine with complex heat transfer law. Int. J. Energy Environment 2 (1), 171–178.
Li, P.-W., Tao, G., Liu, H., 2011. Effect of the geometries of current collectors on the power density in a solid oxide fuel cell. Int. J. Energy Environmental Engin. 2 (3), 1–11.
Lifshitz, E., Pitajevsky, L., 1984. Physical Kinetics. Pergamon, London.
Lin, C.C., 1963. Liquid helium. Proc. Int. School of Physics: Course XXI. Academic Press, New York.
Lin, T., 2013. Path Probability and an Extension of Least Action Principle to Random Motion. These de Doctorat. LUNAM Universite, Universite du Maine.
Lin, T., Wang, Q.A., 2013. The extrema of an action principle for dissipative mechanical systems. Journ. Appl. Mech. 81 (3), 031002.
Lin, G., Chen, J., Bruck, E., 2004. Irreversible chemical-engines and their optimal performance analysis. Appl. Energy 78, 123–136.
Lindenberg, K., 1990. Classical and quantum mechanical fluctuation-dissipation relations. Nonequilibrium Theory and Extremum Principles. Advances in Thermodynamics Series, vol. 3, Taylor and Francis, New York, pp. 212–249.
Lindenberg, K., 1999. On the generalized Kramers problem with exponential memory friction. Physica D 133, 348–361.
Linnhoff, B., 1979. Thermodynamic analysis in the design of process networks. PhD Dissertation, Department of Chemical Engineering, University of Leeds, Leeds, UK.
Linnhoff (Ed.), 1982. A User Guide on Process Integration for the Efficient Use of Energy, IChemE, Rugby, UK.
Linnhoff, B., Ahmad, S., 1990. Cost optimum heat exchanger networks, 1: Minimum energy and capital using simple models for capital cost. Comput. Chem. Engin. 14, 729–750.
Lions, P.L., 1982. Generalized solutions of Hamilton–Jacobi equations. Research Notes in Mathematics, vol. 69, Pitman, Boston.
Lior, N., 2002. Thoughts about future power generation systems and the role of exergy analysis in their development. Energy Convers. Manage. 43, 1187–1198.
Lior, N., Dunbar, W.R., Gaggioli, R.A., 1991. Combining fuel cells with fuel-fired power plants for improved energy efficiency. Energy 16 (10), 1259–1274.
Liu, H.-M., He, J.-H., 2004. Variational approach to chemical reaction. Comput. Chem. Eng. 28, 1549.
Liu, I-Shih., 1972. Method of Lagrange multipliers for exploitation of the entropy principle. Arch. Ration. Mech. Anal. 46, 131–148, 1972.
Liu, I-S., Müller, I., 1983. Extended thermodynamics of classical and degenerate gases. Arch. Rational Mech. Anal. 83, 285–332.
Liu, J.-X., Zheng, W.-M., Hao, B.-L., 1996. From annular to interval dynamics: symbolic analysis of the periodically forced brusselator. Chaos Soliton. Fract. 7 (9), 1427–1453.
Llebot, J.E., 1992. Extended irreversible thermodynamics of electrical phenomena. Extended Thermodynamic Systems. Advances in Thermodynamics Series, vol. 7, Taylor and Francis, New York, pp. 339–363.
Llewellyn, R.W., 1963. Linear Programming. Holt, Rinehart & Winston, New York.
Logan, J.D., 1977. Invariant Variational Principles. Academic, New York.

References

Logan, B.E., 2008. Microbial fuel cells. Wiley, New York.

London, H., 1961. Separation of Isotopes. George Newnes, London.

Lopez de Haro, M., Cohen, E.G.D., 1984. The Enskog theory for multicomponent mixtures, III: transport properties of dense binary mixtures with one tracer component. J. Chem. Phys. 80, 408–415.

Lorente, S.J., 2007. Constructal view of electrokinetic transfer through porous media. Phys. D 40, 2941–2947.

Lorente, S., Bejan, A., 2006. Heterogeneous porous media as multiscale structures for maximum flow access. J. Appl. Phys. 100, 114909.

Lotka, A.J., 1920. Undamped oscillations derived from the law of mass action. J. Am. Chem. Soc. 42, 1595–1599.

Lotka, A.J., 1922. Contribution to the energetics of evolution. Proc. Natl. Acad. Sci. USA 8, 147–151.

Lozano, M.A., Valero, A., 1993b. Thermoeconomic analysis of gas turbine cogeneration systems. In: Richter, H.J. (Ed.), Thermodynamic and Design Analysis and Improvement of Energy Systems, ASME Book no. H00874. ASME, New York, pp. 311–320.

Lozano, M.A., Valero, A., Serra, L., 1993. Theory of exergetic cost and thermoeconomic optimization. In: Szargut, J., Kolenda, Z., Tsatsaronis, G., Ziebik, A. (Eds.), Proceedings of the International Conference Energy Systems and Ecology ENSEC'93, Cracow, Poland, July 5–9, pp. 339–350.

Lucia, U., 2010. Maximum entropy generation in open systems: the Fourth Law? Available from: http: arxiv.org/abs/1011.3989v1.

Luikov, A.V., 1966a. Applications of nonequilibrium thermodynamics to investigations of heat and mass transfer. Int. J. Heat Mass Transfer 9, 139–152.

Luikov, A.V., 1966b. Heat and Mass Transfer in Capillary Porous Bodies. Pergamon, Oxford.

Luikov, A.V., 1969. Analytical Heat Diffusion Theory. Academic Press, New York.

Luikov, A.V., 1975. System of differential equations of heat and mass transfer in capillary-porous bodies (review). Int. J. Heat Mass Transfer 18, 1–14.

Luikov, A.V., Berkovsky, B.M., 1974. Convection and thermal waves. Energy, Moscow.

Luikov, A.V., Mikhailov, Y.A., 1963. Theory of Energy and Mass Transfer (in Russian). Gosenergoizdat, Moscow (English transl. by L.A. Fenn, Pergamon Press, London, 1968).

Luikov, A.V., Bubnov, V.A., Soloview, I.A., 1976. On wave solutions of the heat conduction equation. Int. J. Heat Mass Transfer 19, 245–249.

Luna, F., Martinez, J., 1999. Stability analysis in multicomponent drying of homogeneous liquid mixtures. Chem. Eng. Sci. 54, 5823–5837.

Lund, K., 1986. General thermal analysis of parallel-flow flat-plate solar collector absorbers. Solar Energy 36, 443.

Lund, K., 1989. General thermal analysis of serpentine-flow flat-plate solar collector absorbers. Solar Energy 42, 133.

Lund, K., 1990. Application of finite-time thermodynamics to solar power conversion. Sieniutycz, S., Salamon, P. (Eds.), Finite-Time Thermodynamics and Thermoeconomics, Advances in Thermodynamics, vol. 4, Taylor and Francis, New York, pp. 121–138.

Luss, D., 1997. Temperature fronts and patterns in catalytic systems. Ind. Eng. Chem. Res. 36, 2931.

Luss, D., Sheintuch, M., 2005. Spatiotemporal patterns in catalytic systems. Cat. Today 105, 254.

Lutz, A.E., Larson, R.S., Keller, J.O., 2002. Thermodynamic comparison of fuel cells to the Carnot cycle. Int. J. Hydrogen Energ. 27, 1103–1111.

MacFarlane, A.G.J., 1970. Dynamical System Models. George G. Harrap, London.

Machlup, S., Onsager, L., 1953. Fluctuations and irreversible processes, II: systems with kinetic energy. Phys. Rev. 91, 1512–1515.

Machta, J., 1999. Entropy, information, and computation. Am. J. Phys. 67, 1074–1077.

MacLean, H., Lave, L.B., 2003. Evaluating automobile fuel/propulsion system technologies. Prog.Energ. Combus. Sci. 29, 1–69, lifecycle problems.

Mady, C.E.K., Oliveira, Jr., S., 2013. Human body exergy metabolism. Int. J. Thermodyn. 16 (2), 73–80.

Mahulikar, S.P., 2005. Modified thermodynamic principles unifying order existence and evolution. Proceedings of ECOS 2005. Trondheim, Norway, June 20–22.

Mahulikar, S.P., Herwig, H., 2004. Conceptual investigation of entropy principle for identification of directives for creation, existence and total destruction of order. Physica Scripta 70, 212–221.

Mainardi, F., 1996. Fractional relaxation-oscillation and fractional diffusion-wave phenomena. Chaos Soliton. Fract. 7 (9), 1461–1477.

Maini, P.K., Painter, K.J., Chau, H.N.P., 1997. Spatial pattern formation in chemical and biological systems. J. Chem. Soc.–Faraday Trans. 93, 3601–3610.

Mandarelli, Ch., Jain, R.K., Ivanov, I.B., Ruckenstein, E., 1980. Stability of symmetric and unsymmetric thin liquid films to short and long wavelength perturbations. J. Colloid Interf. Sci. 78, 118.

Mandelbrot, B.B., 1982. The Fractal Geometry of Nature. W. H. Freeman and Company, New York.

Mangasarian, O.L., 1969. Nonlinear Programming. McGraw-Hill, New York.

Mann, U., 2000. New design formulation of chemical reactors with multiple reactions, I: basic concepts. Chem. Eng. Sci. 55 (5), 991–1008.

Manning, J.C., 1997. Applied Principles of Hydrology, third ed. Prentice-Hall, Upper Saddle River, NJ, p. 276.

Mansoori, G.A., Canfield, F.B., 1969. Variational approach to the equilibrium thermodynamic properties of simple fluids. J. Chem. Phys. 51, 4958–4967.

Mansoori, G.A., Patel, V., 1979. Thermodynamic basis for the choice of working fluids for solar absorption cooling systems. Solar Energy 22, 483–491.

Månsson, B.Å.G., 1985. Contributions to Physical Resource Theory. Chalmers University of Technology, Göteborg.

Månsson, B.Å.G., 1990. Thermodynamics and economics. Finite-Time Thermodynamics and Thermoeconomics. Advances in Thermodynamics Series, vol. 4, Taylor and Francis, New York, pp. 153–174.

Månsson, B.A.G., Andresen, B., 1985. Optimal temperature profile for an ammonia reactor. In: Mansson, B.A.G. (Ed.), Contributions to Physical Resource Theory, Chalmers University of Technology, Göteborg.

Månsson, B.Å.G., Lindgren, K., 1990. Thermodynamics, information and structure. Nonequilibrium Theory and Extremum Principles. Advances in Thermodynamics Series, vol. 3, Taylor and Francis, New York, pp. 95–128.

Marangoni, C.G.M., 1865. Sull' expansione dell goccie di un liquido galleggianti sulla superficie di altro liquido, Tipografia dei fratelli Fusi, Pavia.

Markus, F., 2005. Hamiltonian formulation as a basis of quantized thermal processes. In: Sieniutycz, S., Farkas, H. (Eds.), Variational and Extremum Principles in Macroscopic Systems. Elsevier, Oxford, pp. 267–291, Chapter I.13.

Markus, F., Gambar, K., 1991. A variational principle in thermodynamics. J. Non-Equilib. Thermodyn. 16, 27–31.

Markus, F., Gambar, K., 1996. Another possible description of fluctuations. Phys. Rev. E 54, 4607.

Markus, F., Gambar, K., 2003. Fisher bound and extreme physical information for dissipative proceses. Phys. Rev. E 68, 016121.

Markus, F., Gambar, K., 2004. Generalized Hamilton–Jacobi equation for simple dissipative processes. Phys. Rev. E 70, 016123.

Markus, F., Gambar, K., 2005. Quasiparticles in a thermal process. Phys. Rev. E 71, 066117.

Martynenko, O.G., Khramtsov, P.P., 2004. Induction relaxation of thermodynamic parameters in nonstationary evaporation. Int. J. Heat Mass Transfer 47, 2449–2455.

Martynenko, O.G., Sokovishin, Yu, A., 1982. Free-Convection Heat Transfer. Nauka i Technika, Minsk.

Martyushev, L.M., Seleznev, V.D., 2006. Maximum entropy production principle in physics, chemistry and biology, a review. Phys. Reports 426, 1–45.

Maugin, G.A., 1975. On the spin relaxation in deformable ferromagnets. Physica 81a, 454–458.

Maugin, G.A., 1979. Vectorial internal variables in magnetoelasticity. J. Mecanique 18, 541–563.

Maugin, G.A., 1987. Thermodynamique a variables internes et applications. In Seminar: Thermodynamics of Irreversible Processes. Institut Francais du Petrole, Rueil-Malmaison, France.

Maugin, G.A., 1990a. Thermomechanics of plasticity and fracture. Cambridge University Press, Cambridge, UK.

Maugin, G.A., 1990b. Nonlinear dissipative effects in electrodeformable solids. In: Maugin, G.A., Collet, B., Drouout, R., Pouget, J. (Eds.), Nonlinear Mechanical Couplings. Manchester University Press, Manchester, UK, Chapter 6.

References

Maugin, G.A., 1992. Thermodynamic of hysteresis. Extended Thermodynamic Systems. Advances in Thermodynamics Series, vol. 7, Taylor and Francis, New York, pp. 25–52.

Maugin, G.A., Drouot, R., 1992. Nonequilibrium thermodynamics of solutions of macromolecules. Article in: Advances in Thermodynamics Series, vol. 7, 53–75.

Maugin, G.A., Drouot, R., 1983a. Internal variables and the thermodynamics of macromolecules solutions. Int. J. Eng. Sci. 21, 705–724.

Maugin, G.A., Drouot, R., 1983b. Thermomagnetic behavior of magnetically nonsaturated fluids. J. Magnetism and Magnetic Materials 39, 7–10.

Maugin, G.A., Drouot, R., 1992. Nonequilibrium thermodynamics of solutions of macromolecules. Article in: Advances in Thermodynamics Series, vol. 7. pp. 53–75.

Maugin, G.A., Sabir, M., 1990. Mechanical and magnetic hardening of ferromagnetic bodies: influence of residual stresses and applications to nondestructive testing. Int. J. Plasticity 6, 573–589.

Maxwell, J.C., 1867. On the dynamical theory of gases. Phil. Trans. R. Soc. Lond. 157, 49–88.

Maxwell, J.C., 2003a. In: Niven, W.D. (Ed.), The Scientific Papers of James Clerk Maxwell, vol. 1, Dover, New York.

Maxwell, J.C., 2003b. In: Niven, W.D. (Ed.), The Scientific Papers of James Clerk Maxwell, vol. 2, Dover, New York.

McGlashan, N.R., Marquis, A.J., 2007. Availability analysis of post-combustion carbon capture systems: minimum work input. Proceedings of the Institution of Mechanical Engineers, Part C: J. Mech. Eng. Sci., 221:1057–1065.

McLennan, J.A., 1974. Onsager's theorem and higher-order hydrodynamic equations. Phys. Rev. A 10, 1272–1276.

McLennan, J.A., 1989. Introduction to Nonequilibrium Statistical Mechanics. Englewood Cliffs, NJ, Prentice-Hall.

Megiris, C.E., Butt, J.B., 1990a. Effects of poisoning on the dynamics of fixed bed reactors, 1: isothermal in a cyclic policy of operation. Ind. Eng. Chem. Res. 29 (6), 1065–1072.

Megiris, C.E., Butt, J.B., 1990b. Effects of poisoning on the dynamics of fixed bed reactors, 2: constant conversion policy of operation. Ind. Eng. Chem. Res. 29 (6), 1072–1075.

Meixner, J., 1949. Thermodynamik und Relaxationserscheinungen. Z. Naturforschung 4a, 594–600.

Meixner, J., 1961. Der Drehimpulssatz in der Thermodynamik der irreversiblen Prozesse. Zeit. Phys. 16, 144–165.

Meixner, J., 1963. Thermodynamics of electrical networks and the Onsager–Casimir reciprocal relations. J. Math. Phys. 4, 154.

Meixner, J., 1966. TIP has many faces. In: Parkus, H., Sedov, L. (Eds.), UATAM Symposia Vienna, Springer.

Meixner, J., 1968. Entropie in Nichtgleichewicht. Rheol. Acta 7, 8–13.

Meixner, J., 1972. The fundamental inequality in thermodynamics. Physica 59, 305–313.

Meixner, J., 1974. Coldness and Temperature. Arch. Rat. Mech. Anal. 57, 281–286.

Mert, S.O., Dincer, I., Ozcelik, Z., 2007. Exergoeconomic analysis of a vehicular PEM fuel cell system. J. Power Sources 165, 244–252.

Meyer, G., 2001. Thermodynamics of living systems within biological systems engineering: syllabus—Spring 2001. University of Nebraska–Lincoln, Nebraska, USA.

Miguel, A.F., 2000. Contribution to flow characterization through porous media. Int. J. Heat Mass Transfer 43 (13), 2267–2272.

Mijatovich, M., Veselinovic, V., 1987. Differential geometry of equilibrium thermodynamics. Phys. Rev. A 35, 1863–1867.

Mikhailov, Yu. A., Glazunov, Yu. T., 1985. Variational Methods in Theory of Nonlinear Heat and Mass Transfer. Zinatne, Riga.

Mikielewicz, J., 1997. Podstawy klasycznej termodynamiki procesów. VII Letnia Szkoªa Termodynamiki. Stawiska 24, 7–48.

Mikielewicz, J., 2001. Termodynamika klasyczna. In: Bilicki, Z., Mikielewicz, J., Sieniutycz, S. (Eds.), Współczesne Problemy Termodynamiki (Contemporary Problems of Thermodynamics). Wydawnictwo IMP PAN, Gdańsk, Chapter 1.

Mikielewicz, J., Cieśliński, J.T., 1999. Niekonwencjonalne urządzenia i systemy konwersji energii. Ossolineum, Wrocław.

Mikulecky, D.C., 1977. A simple network thermodynamic method for modeling series-parallel coupled flows, II: the nonlinear theory, with applications to coupled solute and volume flow in a series membrane. J. Theor. Biol. 69, 511–541.

Milewski, J., Lewandowski, J., 2009. Solid oxide fuel cell fueled by biogases. Archiv. Thermodyn. 30 (4), 3–12.

Milewski, J., Lewandowski, J., Miller, A., 2009. Reducing CO_2 emissions from a coal-fired power plant by using a molten carbonate fuel cell. Chemical and Process Engineering (Inżynieria Chemiczna i Procesowa) 30, 341–350.

Milikan, C.B., 1929. On the steady motion of viscous imcompressible fluids; with particular reference to a variation principle. Phil. Mag. 7, 641–662.

Miller, D.G., 1974. The Onsager relation: experimental evidence. In: Delgado Domingos, J.J., Nina, M.N., Whitelaw, J.H. (Eds.), In Foundations of Continuum Mechanics. Macmillan, London.

Minier, J.-P., Pozorski, J., 1997. Derivation of a PDF model for turbulent flows based on principles from statistical physics. Phys. Fluids 9 (6), 1748–1753.

Mironova, V.A., Amelkin, S.A., Tsirlin, A.M., 2000. Mathematical Methods of Finite Time Thermodynamics. Chimija, Moscow.

Mitura, E., 1981. Relaxation effects of heat and mass fluxes as a basis for classification of dried solids. PhD Thesis, Lodz Technical University, Poland.

Mojtabi, A., Platten, J.K., Charrier-Mojtabi, M.-C., 2002. Onset of free convection in solutions with variable Soret coefficients. J. Non-Equilib. Thermodyn. 27, 25–44.

Molski, A., 2001. Introduction to Chemical Kinetics (in Polish). Wydawnictwa Naukowo Techniczne, Warszawa.

Monin, A.S., Yaglom, A.M., 1967. The Mathematical Problems of Turbulence. Nauka, Moscow.

Montroll, E.W., Shlesinger, M.F., 1983. Maximum entropy formalism, fractals, scaling phenomena, and 1 = f noise: a tale of tails. J. Statis. Phys. 32, 209–230.

Morbidelli, M., Gavriilidis, A., Varma, A., 2001. Catalyst design: optimal distribution of catalyst in pellets. Reactors and MembranesCambridge University Press, Cambridge, UK.

Mori, H., 1965. Transport, collective motion, and Brownian motion. Progr. Theor. Phys. 33, 423–455.

Mornev, O.A., Aliev, R.R., 1995. Local variational principle of minimum dissipation in dynamics of diffusion-reaction. Russian J. Phys. Chem. 69, 1325–1328.

Moroń, W., Czajka, K., Ferens, W., Babul, K., Szydełko, A., Rybak, W., 2013. NO_x and SO_2 emission during oxy-coal combustion. Chem. Process Eng. 34 (3), 337–346.

Morosuk, T., Morosuk, C., Feidt, M., 2004. New proposal in the thermodynamic analysis of complex regeneration system. Energy 29, 2517–2535.

Moroz, A., 2008. On a variational formulation of the maximum energy dissipation principle for nonequilibrium chemical thermodynamics. Chem. Phys. Lett. 457, 448–452.

Moroz, A., 2009. A variational framework for nonlinear chemical thermodynamics employing the maximum energy dissipation principle. J. Phys. Chem. B 113, 8086–8090.

Morris, G.P., Rondoni, L., 1999. Definition of temperature in equilibrium and nonequilibrium systems. Phys. Rev. E 59 (1), , R5-7R8.

Morro, A., 1985. Thermodynamics and Constitutive Equations. In: Grioli, G. (Ed.), Relaxation phenomena via hidden variable thermodynamics. Springer, Berlin.

Morro, A., 1992. Thermodynamics and extremum principles in viscoelasticity. Extended Thermodynamic Systems. Advances in Thermodynamics Series, vol. 7, Taylor and Francis, New York, pp. 76–106.

Morse, P.M., Feshbach, H., 1953. Methods of Theoretical Physics. McGraw-Hill, New York.

Moser, F., Schnitzer, H., 1985. Heat Pumps in Industry. Elsevier, Amsterdam.

Moszynski, J.R., 1983. Information theory revisited. Arch. Termodyn. 4, 341–352.

Mozurkewich, M., Berry, R.S., 1981. FTT: Engine performance by optimized piston motion. Proc. Natl. Acad. Sci. USA 78, 1986.

Mozurkewich, M., Berry, R.S., 1982. Optimal path for thermodynamic systems: the ideal Otto cycle. J. Appl. Phys. 53, 34, see also J. Appl. Phys. 54, 3651.

Mrugala, R., 1996. On a Riemannian metric on contact thermodynamic spaces. Rep. Math. Phys. 38, 339–348.

References

Mrugala, R., Nulton, J.D., Schon, J.C., Salamon, P., 1990. A statistical approach to the geometric structure of thermodynamics. Phys. Rev. A 41, 3156–3165.

Mujumdar, A.S., 2007. An overview of innovation in industrial drying: current status and R&D needs. Transp. Porous Media 66, 3–18.

Mujumdar, A.S., Law, Ch.L., 2010. Drying technology: trends and applications in postharvest processing. Food Bioprocess Tech. 3 (6), 843–852.

Müller, I., 1966. Zur Ausbreitungsgeschwindigkeit vor Storungen in kontinuierlichen Medien. Dissertation TH Aachen.

Müller, I., 1967. Zum Paradox der Wärmeleitungstheorie. Z. f. Physik 198.

Müller, I., 1971a. Die Kaltenfunction, eine universelle Funktion in der Thermodynamik viscoser warmeleintender Flussigkeiten. Arch. Rat. Mech. Anal. 40, 1–36.

Müller, I., 1971b. The coldness: a universal function in thermodynamic bodies. Arch. Rat. Mech. Anal. 41, 319–329.

Müller, I., 1985. Thermodynamics. Macmillan, London.

Müller, I., 1992. Extended thermodynamics-a theory with hyperbolic field equations. Extended Thermodynamic Systems. Advances in Thermodynamics Series, vol. 7, Taylor and Francis, New York, pp. 107–139.

Müller, I., 1993. Thermodynamics of light and sound. In: Ebeling, W., Muschik, W. (Eds.), Statistical physics and thermodynamics of nonlinear nonequilibrium systems. Materials of satellite meeting in Gosen near Berlin 1992. World Scientific, Singapore, p. 1993.

Müller, I., Ruggeri, 1993. Extended Thermodynamics. Springer, Berlin.

Muschik, W., 1981. Thermodynamical theories, survey and comparison. ZAMM 61, T213.

Muschik, W., 1986. Thermodynamical theories, survey, and comparison. J. Appl. Sci. 4, 189.

Muschik, W., 1990a. Aspects of non-equilibrium thermodynamics: six lectures on fundamentals and methods. World Scientific, Singapore.

Muschik, W., 1990. Internal variables in the non-equilibrium thermodynamics. J. Non-Equilib. Thermodyn. 15, 127–137.

Muschik, W., 2004. Remarks on thermodynamical terminology. J. Non-Equilib. Thermodyn. 29, 199–203.

Muschik, W., 2007. Why so many "schools" of thermodynamics? Forsch Ingenieurwes 71, 149–161.

Muschik, W., 2008. Survey of some branches of thermodynamics. J. Non-Equilib. Thermodyn. 33, 165–198.

Muschik, W., Brunk, G., 1977. A concept of non-equilibrium temperature. Int. J. Eng. Science 11, 377–389.

Muschik, W., Papenfuss, C., 1993. An evolution criterion of nonequilibrium thermodynamics and its application to liquid crystals. Physica A 201, 515–526.

Muschik, W., Restuccia, L., 2002. Changing the observer and moving materials in continuum physics: objectivity and frame-indifference. Technisch Mech 22 (2), 152–160.

Muschik, W., Trostel, R., 1983. Variational principles in thermodynamics. ZAMM 63, T190–T192.

Muschik, W., Gümbel, S., Kröger, M., Öttinger, H.-C., 2000. A simple example for comparing GENERIC with rational nonequilibrium thermodynamics. Physica A 285, 448–466.

Muschik, W., Papenfuss, C., Ehrentraut, H., 2001. A sketch of continuum thermodynamics. J. Non-Newt. Fluid Mech. 96, 255.

Muskat, M., 1937. The Flow of Homogeneous Fluids through Porous Media. McGraw-Hill, New York.

Mynarski, S., 1979. Elementy Teorii Systemów i Cybernetyki. PWN, Warszawa.

Myshlyavtsev, A.V., Yablonski, G.S., 1992. Method of transfer matrix for calculation of thermodynamics and kinetics of surface reactions. Flow, Diffusion and Transport Processes. Advances in Thermodynamics Series, vol. 6, Taylor and Francis, New York, pp. 460–481.

Nag, A., Berry, R.S., 2007. Thermodynamics and kinetics of competing aggregation processes in a simple model system. J. Chem. Phys. 127, 184503.

Naragchi, M., 2013. Introductory thermodynamics. Electronic materials of the course ENGS205 conducted at Texas A&M University, Department of Aerospace Engineering. Chapters 3–5, pp. 29–47.

Natanson, L., 1896. Uber die Gezetze nicht umkerbarer Vorgange. Z. Phys. Chem. 21, 193–200.

Naudts, J., Czachor, M., 2001. Dynamic and thermodynamic stability of nonextensive systems. In: Abe, S., Okamoto, Y. (Eds.), Nonextensive Statistical Mechanics and Its Applications. Lecture Notes in Physics, vol. 560, pp. 243–252, Chapter VI.

Neogi, P., 1996. Diffusion in Polymers. Marcel Dekker, New York.

Nettleton, R.E., 1960. Relaxation theory of heat conduction in liquids. Phys. Fluids 3, 216–226.

Nettleton, R.E., 1969. Thermodynamics of viscoelasticity in liquids. Phys. Fluids 2, 256–263.

Nettleton, R.E., 1986. Lagrangian formulation of nonlinear extended irreversible thermodynamics. J. Phys. A 19, L295–L297.

Nettleton, R.E., 1987a. Nonlinear heat conduction in dense liquids. J. Phys. A 20, 4017–4025.

Nettleton, R.E., 1987b. Application of reciprocity to nonlinear extended thermodynamics kinetic equations. Physica A 144, 219–234.

Nettleton, R.E., 1992. Reciprocity in extended thermodynamics. Extended Thermodynamic Systems. Advances in Thermodynamics Series, vol. 7, Taylor and Francis, New York, pp. 223–262.

Nettleton, R.E., Freidkin, E.S., 1989. Nonlinear reciprocity and the maximum entropy formalism. Physica A 158, 672–690.

Newman, J., 1967. Transport processes in electrolytic solutions. In: Tobias, C.W. (Ed.), Advances in Electrochemistry and Electrochemical Engineering, vol. 5, Wiley Interscience, New York., pp. 87–135, Electrochemical Engineering.

Newman, J., 1973. Electrochemical Systems. Englewood Cliffs, NJ, Prentice-Hall.

Ni, M., Leung, M.K.H., Leung, D.Y.C., 2007. Energy and exergy analysis of hydrogen production by solid oxide steam electrolyzer plant. Int. J. Hydrogen Energ. 32, 4648–4660.

Nicolin, D.J., Jorge, R.M.J., Jorge, L.M.M., 2014. Stefan problem approach applied to the diffusion process in grain hydration. Transp. Porous Media 102, 387–402.

Nicolis, G., De Wit, A., 2007. Reaction-diffusion systems. Scholarpedia 2 (9), 1475.

Nicolis, G., Portnow, J.I., 1973. Chemical oscillations. Chem. Rev. 73 (4), 365–384.

Nicolis, G., Prigogine, I., 1977. Self-Organization in Nonequilibrium Systems. Wiley, New York.

Nicolis, G., Prigogine, I., 1979. Irreversible processes at nonequilibrium steady states and Lyapounov functions. Proc. Natl. Acad. Sci. USA 76 (12), 6060–6061.

Nicolis, G., Rouvas-Nicolis, C., 2007a. Foundations of Complex Systems. World Scientific, Singapore.

Nicolis, G., Rouvas-Nicolis, C., 2007b. Complex systems. Scholarpedia 2 (11), 1473.

Nikitina, L.M., 1968. Thermodynamic Parameters and Coefficients of Mass Transfer in Moist Materials (in Russian), 52 and 54, Moscow: Energia. Page 52, Mass capacitance of gas, based on chemical potential μ; page 54, Thermal gradient coefficient, based on μ as well.

Nonenmacher, T.F., 1980. On the derivation of second-order hydrodynamic equations. J. Non-Equilib. Thermodyn. 5, 361–378.

Nonnenmacher, T.F., 1986. Functional Poisson brackets for nonlinear fluid mechanics equations. In: Casas-Vazquez, J., Jou, D., Rubi, J. M. (Eds.), Recent Developments in Nonequilibrium Thermodynamics: Fluids and Related Topics. Springer, Berlin, pp. 149–174.

Nöther, E., 1918. Invariante Variationsprobleme. Nachr. Akad. Wiss. Gottingen. Math-Phys. Kl. II: 235–257. Also, Nöther, E., 1; 1971. Invariant Variational Problems. Transport Theory of Statistis. Phys. 1. 186–207. Translation by M.A. Tavel of the original article.

Noyola, A.P.-G., 1999. Anomalous diffusion in miscible incompressible fluids. J. Non-Equilib. Thermodyn. 24, 123–146.

Nulton, J.D., Salamon, P., 1985. Geometry of ideal gas. Phys. Rev. A 31, 2520–2524.

Nulton, J., Salamon, P., 1988. Statistical mechanics of combinatorial optimization. Phys. Rev. A 37, 1351.

Nulton, J., Salamon, P., Andresen, B., Anmin, Q., 1985. Quasistatic processes as step equilibrations. J. Chem. Phys. 83, 334.

Nummedal, l., Kjelstrup, S., Costea, M., 2003. Minimizing the entropy production rate of an exothermic reactor with a constant heat-transfer coefficient: the ammonia reaction. Ind. Eng. Chem. Res. 42, 1044–1056.

Nunziato, J.W., 1971. On the heat conduction in materials with memory. Quart. Appl. Math. 29, 187–204.

References

Nyiri, B., 1991. On the construction of potentials and variational principles in thermodynamics and physics. J. Non-Equilib. Thermodyn. 16, 39–55.

Odum, H.T., 1971. Environment. Power and Society. Wiley, New York.

Odum, H.T., 1988. Self-organization, transformity, and information. Science 242, 1132–1139.

Oláh, K., 1997a. The way to thermokinetics. ACH-Models in Chemistry 134, 343–367.

Oláh, K., 1997b. Electrode Thermokinetics. Periodica Polytechnica Chem. Eng. 41, 97–114.

Oláh, K., 1998. Reciprocity relations: Maxwell, Onsager. A thermokinetic approach, Periodica Polytechnica Chem. Eng. 42, 21–32.

Ołdak, M., 2011. Thermodynamics and mathematical modeling hybrid energy systems with solid oxide fuel cells (SOFC). MsD Thesis (in Polish). Faculty of Chemical and Process Engineering, Warsaw University of Technology, Warsaw.

Ondrechen, M.J., 1990. Non-Lorentz cycles in nonequilibrium thermodynamics (in Finite-Time Thermodynamics and Thermoeconomics). Advances in Thermodynamics Series, vol. 4. Taylor and Francis, New York, pp. 139–152.

Ondrechen, M.J., Berry, R.S., Andresen, B., 1980a. Thermodynamics in finite time: a chemically driven engine. J. Chem. Phys. 72, 5118–5124.

Ondrechen, M.J., Andresen, B., Berry, R.S., 1980b. Thermodynamics in finite time: processes with temperature-dependent chemical reactions. J. Chem. Phys. 73, 5838–5843.

Onsager, L., 1931a. Reciprocal relations in irreversible processes. I. Phys. Rev. 37, 405–426.

Onsager, L., 1931b. Reciprocal relations in irreversible processes. II. Phys. Rev 38, 2265–2279.

Onsager, L., Machlup, S., 1953. Fluctuations and irreversible processes. Phys. Rev. 91, 1505–1512.

Oono, Y., 1976. Physical meaning of d^2z of Glansdorff and Prigogine. Phys. Lett. 57A, 207.

Ordonez, J.C., Bejan, A., Cherry, R.S., 2003. Designed porous media: optimally nonuniform flow structures connecting one point with more points. Int. J. Thermal Sci. 42, 857–870.

Ordonez, J.C., Chen, S., Vargas, J.V.C., Dias, F.G., Gardolinski, J.E.F.C., Vlassov, D., 2007. Constructal flow structure for a single SOFC. Int. J. Energy Res. 31, 1337–1357.

Orlov, V.N., Berry, R.S., 1990. Power output from an irreversible heat engine with a nonuniform working fluid. Phys. Rev. A 42, 7230–7235.

Orlov, V.N., Berry, R.S., 1991. Estimation of minimal heat consumption for heat-driven separation processes via methods of finite-time thermodynamics. J. Phys. Chem. 95 (14), 5624–5628.

Orlov, V.N., Berry, R.S., 1993. Power and efficiency limits for internal combustion engines via methods of finite-time thermodynamics. J. Appl. Phys. 74 (7), 4317–4322.

Ortiz de Zárate, J.M., Sengers, J.V., Bedeaux, D., Kjelstrup, S., 2007. Concentration fluctuations in nonisothermal reaction-diffusion systems. J. Chem. Phys. 127, 034501.

Osserman, R., 1986. A Survey of Minimal Surfaces. Dover, New York.

Osterle, J.F., 1984. Thermodynamic analysis of photosynthesis efficiency. J. Non-Equilib. Thermodyn. 9, 55–60.

Öttinger, H.C., 2005. Beyond Equilibrium Thermodynamics. Wiley, Hoboken, NJ.

Öttinger, H.C., Grmela, M., 1997. Dynamics and thermodynamics of complex fluids, II: Illustrations of a general formalism. Phys. Rev. E 56, 6633–6652.

Pade, L.-L., Schröder, S.T., 2013. Fuel cell based micro-combined heat and power under different policy frameworks—an economic analysis. Energy Convers. Manage. 66, 295–303.

Pakowski, Z., Mujumdar, A.S., 2006. Basic process calculations and simulations in drying. In: Mujumdar, A.S. (Ed.), Handbook of Industrial Drying, third ed. CRC Press, Boca Raton, pp. 54–79.

Palazzi, F., Autissier, N., Francois, M.A., Marechal, F.M.A., Favrat, D., 2007. A methodology for thermoeconomic modeling and optimization of solid oxide fuel cell systems. Appl. Therm. Eng. 27, 2703–2712.

Palm, E., 1975. Nonlinear thermal convection. Anual Rev. Fluid Mech. 1, 39.

Papadikis, K., Gub, S., Bridgwater, A.V., 2009. CFD modeling of the fast pyrolysis of biomass in fluidised bed reactors: modelling the impact of biomass shrinkage. Chem. Eng. J. 149, 417–427.

Park, J.Y., Levenspiel, O., 1976a. Optimum operating cycle for systems with deactivating catalysts. 1. General formulation and method of solution. Ind. Eng. Chem. Proc. Des. Dev. 15 (4), 534–538.

Park, J.Y., Levenspiel, O., 1976b. Optimum operating cycle for systems with deactivating catalysts. 2. Applications to reactors. Ind. Eng. Chem. Process Des. Dev. 15 (4), 538–544.

Pavon, D., 1992. Astrophysical and cosmological applications of extended relativistic thermodynamics. Extended Thermodynamic Systems. Advances in Thermodynamics Series, vol. 7, Taylor and Francis, New York, pp. 386–407.

Pavon, D., Jou, D., Casas-Vazquez, J., 1980. Heat conduction in relativistic extended thermodynamics. J. Phys. A 13, L67–L79.

Pavon, D., Jou, D., Casas-Vazquez, J., 1983. Equilibrium and nonequilibrium fluctuations in relativistic fluids. J. Phys. A 16, 775–782.

Peletier L.A., 1981. Applications of Nonlinear Analysis in the Physical Sciences. In: Amann, H., Bazley, N., Kirchgassner, K. (Eds). Pitman, London. p. 229.

Pellam, R., 1961. Chapter 6 in Modern Physics for Engineer. McGraw-Hill, New York.

Perez, J.P., 2000. Thermodynamic interpretation of the variational Maxwell theorem in DC circuits. Am. J. Phys. 68 (9), 860–863.

Perez-Garcia, C., 1981. Some physical mechanisms of hydrodynamical instabilities. In: Casas-Vazquez, J., Lebon, G. (Eds.), Stability of Thermodynamic Systems. Proceedings of the Meeting held at Autonomous University of Barcelona, Belleterra (Barcelona), p. 94

Perez-Garcia, C., Jou, D., 1986. Continued fraction expansion and memory functions for transport coefficients. J. Phys. A 19, 2881.

Perez-Garcia, C., Jou, D., 1992. Thermodynamic aspects of memory functions for transport coefficients. Extended Thermodynamic Systems. Advances in Thermodynamics Series, vol. 7, Taylor and Francis, New York, pp. 289–309.

Perlick, V., 1992. The Hamiltonian problem from the global viewpoint. J. Math. Phys. 33, 599–606.

Peshkov, J., 1944. Second sound in helium II. J. Phys. 8, 381 (Moscow).

Petela, R., 1964. Exergy of heat radiation. J. Heat Transfer 86, 187–192.

Petela, R., 2003. Exergy of undiluted thermal radiation. Solar Energy 74, 469–488.

Petela, R., 1984. Some exergy properties of heat radiation. Arch. Thermodyn. 5 (2), 189–193.

Petty, C.A., 1975. A statistical theory for mass transfer near interfaces. Chem. Eng. Sci. 30 (4), 413–418.

Peusner, L., 1986. Studies in Network Thermodynamics. Elsevier, Amsterdam.

Peusner, L., 1990. Lumped parameter networks: a topological graph approach to thermodynamics. Nonequilibrium Theory and Extremum Principles. Advances in Thermodynamics Series, vol. 3, Taylor and Francis, New York, pp. 39–94.

Pfeifer, P., Avnir, D., 1983a. Chemistry in noninteger dimensions between two and three, I: Fractal theory of heterogeneous surfaces. J. Chem. Phys. 79 (7), 3558–3565.

Pfeifer, P., Avnir, D., 1983b. Chemistry in noninteger dimensions between two and three, II: Fractal surfaces of adsorbents. J. Chem. Phys. 79 (7), 3566–3571.

Philipse, A.P., Schram, H.L., 1991. Non-Darcian airflow through ceramic foams. J. Am. Ceram. Soc. 74, 728–732.

Piekarczyk, W., Czarnowska, L., Ptasiński, K., Stanek, W., 2013. Thermodynamic evaluation of biomass-to-biofuels production systems. Energy 62, 95–104.

Pierce, S.E., 2002. Nonequilibrium thermodynamics: an alternate evolutionary hypothesis. Crossing Boundaries 1 (2), 49–59.

Pike, F.H., 1929. The driving force in organic evolution and a theory of the origin of life. Ecology 10 (2), 167–176.

Pirard, S., Malengreaux, Ch.M., Toye, D., 2014. How to correctly determine the kinetics of a photocatalytic degradation reaction? Chem. Eng. J. 249, 1–6.

Plastino, A., Plastino, A.R., Miller, H.G., Khanna, G.C., 1996. A lower bound for Fisher's information measure. Phys. Lett. A 221, 29–33, Ch 1.

Plastino, A., Plastino, A.R., Casas, M., 2005. Fisher variational principle and thermodynamics. In: Sieniutycz, S., Farkas, H. (Eds.), Variational and Extremum Principles in Macroscopic Systems. Elsevier, Oxford, pp. 379–394, Chapter II.1.

Plath, P.J. (Ed.), 1989. Optimal Structures in Heterogeneous Reaction Systems. Springer, Berlin.

Platten, J.K., Chavepeyer, G., 1973. Oscillatory motion in Bénard cell due to the Soret effect. J. Fluid. Mech. 60, 305.

References

Plau, F., Tarbell, J.M., 1980. Stability of the CSTR by correlation of Liapunov functions. Chem. Eng. Commun. 4, 677.

Plows, W.H., 1968. Some numerical results for two-dimensional steady laminar Benard convection. Phys. Fluids 11 (8), 1593–1599.

Podowski, M., 2005. Understanding multiphase flow and heat transfer: perception, reality. Archiv. Thermodyn. 26 (3), 3–20.

Pollak, E., Levine, R.D., 1982. Maximal entropy approach to reactivity and selectivity in elementary chemical reactions. J. Phys. Chem. 86, 4931–4937.

Pontryagin, L.S., Boltyanski, V.G., Gamkrelidze, R.V., Mishchenko, E.F., 1962. The Mathematical Theory of Optimal Processes. Interscience Publishers, New York.

Pope, S.B., 2000. Turbulent Flows. Cambridge University Press, Cambridge, UK.

Poplewski, G., Jezowski, J., 2003. Random search based approach for designing optimal water networks, 4th European Congress of Chemical Engineering, Granada, Spain P-9.2-052.

Poświata, A., 2004. Optimization of drying of solid in the second drying period in bubbling fluidized bed. Inżynieria Chemiczna i Procesowa 25, 1551–1556.

Poświata, A., 2005. Optimization of Drying Processes with Fine Solid in Bubble Fluidized Bed. PhD thesis, Faculty of Chemical & Process Engineering, Warsaw Technological University, Warsaw.

Poświata, A., Szwast, Z., 2003. Minimization of exergy consumption in fluidized drying processes. In: Proceedings of ECOS 2003, Copenhagen, Denmark, June 30–July 2, vol. 2, pp. 785–792.

Poświata, A., Szwast, Z., 2006. Optimization of fine solid drying in bubble fluidized bed. Transport Porous Med. 2, 785–792.

Potter, M.C., Graber, E., 1972. Stability of plane Poiseuille flow with heat transfer. Phys. Fluids 15 (3), 387–391.

Povstenko, Y.Z., 2005. Stresses exerted by a source of diffusion in a case of a nonparabolic diffusion equation. Int. J. Eng. Sci. 43, 977–991.

Povstenko, Y.Z., 2007. Stresses exerted by a source of diffusion in a case of a nonparabolic diffusion equation. Int J. Solids Struct. 44, 2324–2348.

Pramuanjaroenkij, A., Kakaç, S., Zhou, X.,Y., 2008. Mathematical analysis of planar solid oxide fuel cells. Int. J. Hydrogen Energ. 33, 2547–2565.

Pramuanjaroenkij, A., Kakaç, S., Zhou, X.,Y., 2008. Mathematical analysis of planar solid oxide fuel cells. Int. J. Hydrogen Energ. 32, 761–786.

Predvolitelev, A.V., 1981. New Approach to the Problems of Physical Acoustics. Nauka, Moscow (in Russian).

Prigogine, I., 1947. Etude Thermodynamique des Phenomenes Irreversibles, Liege Desoer.

Prigogine, I., 1949. On the domain of validity of the local equilibrium hypothesis. Physica 15, 1942.

Prins, M.J., Ptasinski, K.J., Janssen, F.J.J.G., 2003. Thermodynamics of gas-char reactions: first and second law analysis. Chem. Eng. Sci. 58 (3–6), 1003–1011.

Prins, M.J., Ptasinski, K.J., Janssen, F., 2007. From coal to biomass gasification: comparison of thermodynamic efficiency. Energy 32 (7), 1248–1259.

Prochovsky, E.W., Krumhansl, J.A., 1964. Second sound propagation in dielectric solids. Phys. Rev. A 133, 1411–1415.

Ptasinski, K.J., 2008. Thermodynamic efficiency of biomass gasification and biofuels conversion (Review). Biofuels. Bioprod. Biorefin. 2 (3), 239–253.

Ptasinski, K.J., Hamelinck, C., Kerkhof, P.J.A.M., 2002. Exergy analysis of methanol from the sewage sludge process. Energy Convers. Manage. 43, 1445–1457.

Ptasinski, K.J., Prins, M.J., Pierik, A., 2007. Exergetic evaluation of biomass gasification. Energy 32 (4), 568–574.

Puncochar, M., Drahos, J., 2003. Entropy of fluidized bed—a measure of particles mixing. Chem. Eng. Sci. 58, 2515–2518.

Putterman, S.J., 1974. Superfluid Hydrodynamics. North Holland, Amsterdam.

Qiao, P., Shan, L., 2005. In: Sieniutycz, S., Farkas, H. (Eds.), Variational Principles in Stability Analysis of Composite Structures. Elsevier, Oxford, Chapter 6 in part II of the book: Variational and Extremum Principles in Macroscopic Systems.

Rabaey, K., Verstraete, W., 2005. Microbial fuel cells: novel biotechnology for energy generation. Trends Biotechnol. 23 (6), 291–298.

Radulescu, M., Lottin, O., Feidt, M., Lombard, C., Le Noc, D., Ledoze, S., 2006. Experimental theoretical analysis of the operation of a natural gas cogeneration system using a polymer exchange membrane fuel cell. Chem. Eng. Sci. 61, 743–752.

Rao, A., Maclay, J., Samuelsen, S., 2004. Efficiency of electrochemical systems. J. Power Sources 134, 181–184.

Ratkje, S.K., de Swaan Arons, J., 1995. Denbigh revisited: reducing lost work in chemical processes. Chem. Eng. Sci. 50 (10), 1551–1560.

Ratkje, S.K., Forland, K.S., Forland, T., 1992. Transport processes in electrolytes and membranes. Flow, Diffusion and Transport Processes. Advances in Thermodynamics Series, vol. 6, Taylor and Francis, New York, pp. 340–384.

Ray, J.R., 1972. Lagrangian density for perfect fluids in general relativity. J. Math. Phys. 13, 1451–1453.

Ray, J.R., 1979. Lagrangians for dissipative systems. Am. J. Phys. 47, 626.

Ray, J., 1980. Perfect fluids in general relativity. Il Nuovo Cim 71B, 19–26.

Ray, J., Smalley, L.L., 1982. Spinning fluids in general relativity. Phys. Rev. D 26, 2619–2622.

Ray, J., Smalley, L.L., 1983. Spinning fluids in Einstein–Cartan theory. Phys. Rev. D 27, 1383–1385.

Rayleigh, L., 1916. On convection currents in a horizontal layer of fuid, when the higher temperature is on the under side. Phil. Mag 32, 529.

Rayleigh, J.W.S., 1945. The Theory of Sound. Dover, New York.

Razón, L.F., Schmitz, R.A., 1987. Multiplicities and instabilities in chemically reacting systems—a review. Chem. Eng. Sci. 42 (5), 1005–1047.

Reay, D.A., Mac Michael, D.B.A., 1979. Heat Pumps Design and Application. Pergamon Press, London.

Reid, R.C., Prausnitz, J.M., Poiling, B.E., 1987. The properties of gases and liquids. McGraw-Hill, New York.

Reis, A.H., 2008. Constructal view of the scaling laws of street networks—the dynamics behind geometry. Physica A 387, 617–622.

Reis, A.H., Miguel, A.F., Aydin, M., 2004. Constructal theory of flow architectures of the lungs. Med. Phys. 31, 1135–1140.

Resendis, A.,O, Garcia-Colin, L.S., 2001. Application of the theory of stochastic processes to the configuration of biological systems. Physica A 290, 203–210.

Resibois, P., de Leener, K., 1977. Classical Kinetic Theory of Fluids. Wiley, New York.

Reti, P., 1980. Stability of hyperbolic reactors. React. Kinet. Catalyst. Lett. 15, 215–220.

Reti, P., Holderith, J., 1980. The modelling of multiphase hyperbolic chemical reactors by averaging methods. Hung. J. Ind. Chem. 1, 337–382.

Richardson, L.F., 1926. Atmospheric diffusion on a distance-neighbour graph. Proc. Roy. Soc. London A 110, 709–737.

Riekert, L., 1979. Large Chemical Plants. In: Froment, G.F. (Ed.), Proceedings 4th Int. Symp., 35–64, Antwerp, October, 1979.

Riona, W.L., Van Brunta, V., 1990. Differential geometry based continuation algorithms for separation process applications. Separ. Sci. Technol. 25 (13–15), 2073–2095.

Rivero, R., Garcia, M., 2001. Exergy analysis of a reactive distillation MTBE unit Int. J. Appl. Thermodyn. 4 (2), 85–92.

Rivlin, R.S., Ericksen, J.L., 1955. Stress deformation relations for isotropic materials. J. Rat. Mech. Anal 4, 325–425.

Romero-Salazar, L., Velasco, R.M., 1997. A quantum Langevin model. Physica A 234, 792–800.

Ronis, D., Procaccia, I., 1982. Nonlinear resonant coupling between shear and heat fluctuations in the fluids far from equilibrium. Phys. Rev. A 26, 1812–1813.

Rosen, P., 1953. On the variational principles for irreversible processes. J. Chem. Phys. 21, 1220–1222.

Rosen, M.A., Dincer, I., 2003. Exergoeconomic analysis of power plants operating on various fuels. Appl. Therm. Eng. 23 (6), 643–658.

Rosenbrock, H.H., 1986. A stochastic variational principle for quantum mechanics. Phys. Lett. 110A, 343–346.

References

Rosenbrock, H.H., 1995. A variational treatment of quantum mechanics. Proc. R. Soc. A 450, 417–437.

Rosenbrock, H.H., 1997. A correction to a stochastic variational treatment of quantum mechanics. Proc. R. Soc. A 453, 983–986.

Rosenbrock, H.H., 1999. The definition of state in the stochastic variational treatment of quantum mechanics. Phys. Lett. A 254, 307–313.

Rosenbrock, H.H., Storey, C., 1966. Computational Techniques for Chemical Engineers. Pergamon, Oxford.

Ross, J., 2008. Thermodynamics and Fluctuations Far from Equilibrium. Springer, Berlin.

Ross, J., Vlad, M.O., 1999. Nonlinear kinetics and new approaches to complex reaction mechanisms. Ann. Rev. Phys. Chem. 50, 51–78.

Ross, J., Hunt, K.L.C., Hunt, P.M., 1988. Thermodynamics far from equilibrium: reactions with multiple stationary states. J. Chem. Phys. 88 (4), 2719–2729.

Ross, J., Harding, R.H., Wolff, A.N., Chu, X., 1991. The relation of fluxes and forces to work in nonequilibrium systems. J. Chem. Phys. 95, 5206–5211.

Rotteneberg, H., 1979. Nonequilibrium thermodynamics of energy conversion in bioenergetics. Biochimica et Biophysica Acta 549, 225–253.

Rouse, P.E., 1953. A theory of the viscoelastic properties of dilute solutions of coiling polymers. J. Chem. Phys. 21, 1272–1280.

Rowley, R.L., 1992. Application of nonequilibrium thermodynamics to heat and mass transport properties: measurement and prediction in nonelectrolyte liquid mixtures. Flow, Diffusion and Transport Processes. Advances in Thermodynamics Series, vol. 6, Taylor and Francis, New York, pp. 82–109.

Rowley, R.L., Gruber, V., 1978. Thermal conductivities in seven ternary mixtures at 400°C and 1 atm. J. Chem Eng. Data 33, 5–8.

Rowley, R.L., Horne, F.H., 1978. The Dufour effect: experimental confirmation of the Onsager heat-mass reciprocity relation for a binary liquid mixture. J. Chem. Phys. 68, 325–326.

Roy-Aikins, J., 2002. Thermoeconomic optimisation of a hybrid microturbine and fuel cell. ECOS 02 Conference Proceedings.

Rozonoer, L.I., 1984. Thermodynamics and Kinetics of Biological Processes. Novosibirsk, Nauka.

Rubi, J.M., Kjelstrup, S., 2003. Mesoscopic nonequilibrium thermodynamics gives the same thermodynamic basis to Butler-Volmer and Nernst Equations. J. Phys. Chem. B 107, 13471–13477.

Rubin, M., 1980. Optimal configuration of an irreversible heat engine with fixed compression ratio. Phys. Rev. A 22, 1741.

Rubin, Y., 1997. Transport of inert solutes through the groundwater: recent developments and current issues. In: Dagan, G., Neuman, S.P. (Eds.), Subsurface Flow and Transport: A Stochastic Approach. Cambridge University Press, Cambridge, UK, pp. 115–132.

Ruckenstein, E., Jain, R.K., 1974. Spontaneous rupture of thin liquid films. Faraday Trans II 70, 132–147.

Ruggeri, T., 1989. Galilean invariance and entropy principle for system of balance laws. The structure of extended thermodynamics. Cont. Mech. Thermodyn. 1, 3–20.

Ruiz-Mercado, G.J., Smith, R.L., Gonzalez, M.A., 2012. Sustainability indicators for chemical processes, I: taxonomy. Ind. Eng. Chem. Res. 51 (5), 2309–2328.

Rund, H., 1966. The Hamilton–Jacobi Theory in the Calculus of Variations. Van Nostrand, London.

Rund, H., 1970. The Hamilton–Jacobi Formalism in the Calculus of Variations. Van Nostrand, London.

Ruppeiner, G., 1979. Thermodynamics: a Riemannian geometric model. Phys. Rev. A 20, 1608.

Ruppeiner, G., 1990. Thermodynamic curvature: origin and meaning. Nonequilibrium Theory and Extremum Principles. Advances in Thermodynamics Series, vol. 3, Taylor and Francis, New York, pp. 129–159.

Ruth, M., 1993. Integrating Economics. Ecology and ThermodynamicsKluwer Academics, Dordrecht.

Ruth, M., 1999. Physical principles in environmental economic analysis. In: van den Bergh, J.C.J.M. (Ed.), Handbook of Environmental and Resource Economics. Edward Elgar, Cheltenham, pp. 855–866.

Ruth, D., Ma, H., 1992. On the derivation of the Forchheimer equation by means of the averaging theorem. Transport Porous Med. 7, 255–264.

Rymarz, C., 1999. Chaos and self-organization in living systems. J. Tech. Phys. 40, 407–504.

Sadd, M.H., Didlake, J.E., 1977. Non-Fourier melting of a semi-infinite solid. J. Heat Transfer ASME 99, 25–28.
Sahin, A.Z., 2002. Finite-time thermodynamic analysis of a solar-driven heat engine. Exergy Int. J 1, 122–126.
Sahin, B., Kodal, A., Ekmekci, I., Yilmaz, T., 1997. Exergy optimization for an endoreversible cogeneration cycle. Energy 1, 122–126.
Saied, E.A., 2000. Anomalos diffusion on fractal objects: additional analytic solutions. Chaos Soliton. Fract. 11, 1369–2376.
Salamon, P., Berry, R.S., 1983. Thermodynamic length and the dissipated availability. Phys. Rev. Lett. 51, 1127–1130.
Salamon, P., Nitzan, A., 1981. Optimizations of a finite time Carnot cycle. J. Chem. Phys. 74, 3546.
Salamon, P., Nitzan, A., Andresen, B., Berry, R.S., 1980. Thermodynamics in finite time, IV: minimum entropy production in heat engines. Phys. Rev. A 21, 2115.
Salamon, P., Band, Y.B., Kafri, O., 1982. Maximum power from a cycling working fluid. J. Appl. Phys. 53, 197.
Salamon, P., Nulton, J., Ihrig, E., 1984. On the relations between entropy and energy versions of thermodynamic length. J. Chem. Phys. 80, 436–437.
Salamon, P., Nulton, J., Berry, R.S., 1985. Length in statistical thermodynamics. J. Chem. Phys. 82, 2433–2436.
Salamon, P., Bernholc, J., Berry, R.S., Carrera-Patiño, M.E., Andresen, B., 1990. The wetted solid—a generalization of the Plateau problem and its implication for sintered materials. J. Math. Phys. 31, 610–615.
Salamon, P., Nulton, J.D., Siragusa, G., Andersen, T.R., Limon, A., 2001. Principles of control thermodynamics. Energy 26, 307–319.
Salinas, S.R.A., Tsallis, C. (Eds). 1999. Nonextensive Statistical Mechanics and Thermodynamics, Special issue of Brasilian J. Phys. 29(1): 1-214.
Salmon, R., 1987a. Hamiltonian fluid mechanics. Ann. Rev. Fluid Mech. 20, 225–256.
Salmon, R., 1987b. Hamiltonian fluid mechanics. Ann Mech. Rev. 12, 51–67.
Sánchez-Orgaz, S., Medina, A., Calvo Hernández, A., 2013. Recuperative solar-driven multi-step gas turbine power plants. Energy Convers. Manage. 67 (3), 171–178.
Sancho, P., 1999. Global and local thermodynamic stability. J. Non-Equilib. Thermodyn. 24, 372–379.
Sandler, S.I., Dahler, J.S., 1964. Nonstationary diffusion. Phys. Fluids 2, 1743–1746.
Sandoval-Villalbazo, A., Garcia-Colin, L.S., 1999. Comment on the continuity equation in general relativity. J. Non-Equib. Thermodyn. 24, 380–386.
Santilli, R.M., 1977. Necessary and sufficient conditions for the existence of a Lagrangian in field theory, I: variational appproach to self-adjointness for tensorial field equations. Ann. Phys. 103, 354–408.
Schechter, R.S., 1967. The Variational Method in Engineering. McGraw-Hill, New York.
Schechter, R.S., Prigogine, I., Hamm, J.R., 1972. Thermal diffusion and convective stability. Phys. Fluids 15 (3), 379–386.
Schindler, H., Pastushenko, V.P., Titulaer, U.M., 1998. A measure for the distance from equilibrium. Eur. Biophys. J. 27, 219–226.
Schlögl, F., 1974. On statistical thermodynamics of open systems. Proceedings of the International Research Symposium on Statistical Physics (with Special Sessions on topics related to Bose Statistics). University of Calcutta Press, Calcutta.
Schlögl, F., 1980. Stochastic measures in nonequilibrium thermodynamics. Phys. Rep. 62, 267–280 (and literature therein).
Schmandt, B., Iyer, V., Herwig, H., 2014. Determination of head change coefficients for dividing and combining junctions: a method based on the second law of thermodynamics. Chem. Eng. Sci. 111, 191–202.
Schmid, L.A., 1967a. Hamilton–Jacobi equation for relativistic charged fluid flow. Il Nuovo Cimento LIIB, 313–337.
Schmid, L.A., 1967b. Heat transfer in relativistic charged fluid flow. Il Nuovo Cimento XLVIIB, 1–27.
Schmid, L.A., 1970a. Variational formulation of relativistic fluid thermodynamics. In: Stuart, E.B., et al. (Eds.), A Critical Review of Thermodynamics. Mono Book, Baltimore, pp. 493–503.

References

Schmid, L.A., 1970b. Effects of heat exchange on relativistic fluid flow. In: Stuart, E.B., et al. (Eds.), A Critical Review of Thermodynamics. Mono Book, Baltimore, pp. 504–524.

Schmidt, G., Kohler, W., Hess, S., 1981. A treatment of a kinetic model of a dense fluid. Z. Naturforsch 36a, 545.

Schneider, E.D., Kay, J.J., 1994a. Complexity and thermodynamics: towards a new ecology. Future 26, 626–647.

Schneider, E.D., Kay, J.J., 1994b. Life as manifestation of the second law of thermodynamics. Math. Comput. Model. 19 (6–8), 25–48.

Schneider, E.D., Kay, J.J., 1995. Order from disorder: the thermodynamics of complexity in biology. In: Murphy, Michael P., O'Neill, Luke A.J. (Eds.), What Is Life: The Next Fifty Years. Reflections on the Future of Biology. Cambridge University Press, Cambridge, UK, pp. 161–172.

Schoep, D.C., 2002. A statistical development of entropy for the introductory physics course. Am. J. Phys. 70 (2), 129–135.

Schofield, P., 1966. Wavelength-dependent fluctuations in classical fluids. Proc. Phys. Soc. 88, 149–170.

Scholle, M., 1994. Hydrodynamik im Lagrangeformalismus: Untersuchungen zur Wärmeleitung in idealen Flüssigkeiten, Diplomarbeit, University of Paderborn.

Scholle, M., Anthony, K.-H. 1997. Lagrange formalism and complex fields in hydro- and thermodynamics. Workshop on Dissipation in Physical Systems, vol. 1, September 9. Kielce, Poland.

Schrodinger, E., 1967. What Is Life? Cambridge University Press, Cambridge, UK, (first ed. in 1944).

Schulzky, C., Essex, C., Davison, M., Franz, A., Hoffmann, K.H., 2000. The similarity group and anomalous diffusion equations. J. Phys. A 33, 5501–5511.

Schweitzer, A.M., 1988. Relativistic radiative hydrodynamics. Ann. Phys. 183 (1), 80–121.

Schweitzer, F. (Ed.), 1997. Self-Organization of Complex Structures: From Individual to Collective Dynamics. CRC Press, Boca Raton.

Sciubba, E., 2000. On the possibility of establishing a univocal and direct correlation between monetary price and physical value: the concept of Extended Exergy Accounting, Proceedings of the 2nd International Workshop on Advanced Energy Studies, Porto Venere, Italy.

Sciubba, E., 2001. Beyond thermoeconomics? The concept of extended exergy accounting and its application to the analysis and design of thermal systems. Exergy Int. J. 1, 68–84.

Sciubba, E., 2005. Do the Navier–Stokes equations admit a variational formulation. In: Sieniutycz, S., Farkas, H. (Eds.), Variational and Extremum Principles in Macroscopic Systems. Elsevier Science, Oxford, Chapter II.10.

Sciubba, E., Wall, G., 2007. A brief commented history of exergy from the beginnings to 2004. Int. J. Thermodyn. 10, 1–26, (see also Tsatsaronis 2007 comment).

Sęk, J., 1996. Self-diffusion in liquids predicted by a theory of flow through granular media. AIChE J. 42 (12), 3333–3339.

Sęk, J., 2003. Wykorzystanie koncepcji dwufazowej struktury płynów do wyznaczania wartości współczynników dyfuzji. Inż. i Aparat. Chemiczna 5s, 181–182.

Seliger, R.L., Whitham, G.B., 1968. Variational principles in continuum mechanics. Proc. Roy. Soc. 302A, 1–25.

Sellitto, A., Cimmelli, V.A., 2012. A continuum approach to thermomass theory. J. Heat Transfer 134 (11), 112402.

Serrin, J., 1959. Mathematical principles of classical fluid mechanics. In: Flugge, S. (Ed.), Handbook der Physik VII/1. Springer, Berlin.

Sertorio, L., Tinetti, G., 2000. Solar buildings. Thermodynamics of Energy Conversion and Transport 2000. Springer, New York, pp. 106–140, Chapter 5.

Shannon, C.E., Weaver, W., 1969. The Mathematical Theory of Communication. The University of Illinois Press, Urbana, IL.

Shen, W., Benyounes, H., Gerbaud, V., 2013a. Extension of thermodynamic insights on batch extractive distillation to continuous operation. 1. Azeotropic Mixtures with a Heavy Entrainer. Ind. Eng. Chem. Res. 52 (12), 4606–4622.

Shen, W., Gerbaud, V., 2013b. Extension of thermodynamic insights on batch extractive distillation to continuous operation. 2. Azeotropic mixtures with a light entrainer. Ind. Eng. Chem. Res. 52 (12), 4623–4637.

Shiner, J.S., 1984. A dissipative Lagrangian formulation of the network thermodynamics of (pseudo-) first-order reaction-diffusion systems. J. Chem. Phys. 81, 1455–1465.

Shiner, J.S., 1987. Algebraic symmetry in chemical reaction systems at stationary states arbitrarily far from thermodynamic equilibrium. J. Chem. Phys. 87, 1089–1094.

Shiner, J.S., 1992. A Lagrangian formulation of chemical reaction dynamics far from equilibrium. Flow, Diffusion and Transport Processes. Advances in Thermodynamics Series, vol. 6, Taylor and Francis, New York, pp. 248–282.

Shiner, J.S., 1997. Self-organization, entropy and order in growing system. In: Schweitzer, F. (Ed.), Self-Organization of Complex Structures: From Individual to Collective Dynamics. CRC Press, Boca Raton, pp. 21–26, Chapter 2.

Shiner, J., Davison, M., 2003. Many entropies, many disorders. Open Sys. Inf. Dyn. 10, 281–296.

Shiner, J.S., Sieniutycz, S., 1994. The chemical dynamics of biological systems: variational and extremal formulations. Progr. Biophys. Mol. Biol. 62, 203–221.

Shiner, J.S., Sieniutycz, S., 1996. Phenomenological macroscopic symmetry in dissipative nonlinear systems. In: Shiner, J.S. (Ed.), Entropy and Entropy Generation: Fundamentals and Applications. Kluwer Academics, Dordrecht, pp. 139–158.

Shiner, J.S., Sieniutycz, S., 1997. The mechanical and chemical equations of motion of muscle contraction. Phys. Rep. 290, 183–199.

Shiner, J.S., Fassari, S., Sieniutycz, S., 1996. Lagrangian and network formulations of nonlinear electrochemical systems. In: Shiner, J.S. (Ed.), Entropy and Entropy Generation: Fundamentals and Applications. Kluwer Academics, Dordrecht, pp. 221–239.

Shiner, J.S., Davison, M., Landsberg, P.T., 1999. Simple measure for complexity. Phys. Rev. E. 59, 1459–1464.

Shlesinger, M.F., West, B.J., Klafter, J., 1987. Lévy dynamics of enhanced diffusion: application to turbulence. Phys. Rev. Lett. 58, 1100.

Shnaid, I., 2003. Thermodynamically consistent description of heat conduction with finite speed of propagation. Int. J. Heat Mass Transfer 46, 3853–3863.

Shter, I.M., 1973. The generalized Onsager principle and its application. Inzh. Fiz. Zh. 25, 736–742.

Sieniutycz, S., 1967. Investigation of Drying Kinetics when Solid Particles are Falling in a Gas. PhD thesis, Warsaw University of Technology, Warsaw.

Sieniutycz, S., 1970. Simultaneous heat and mass transfer for drying and moistening in packed beds (in Polish). Chem. Stos. 7 (3B), 353–368.

Sieniutycz, S., 1973a. The thermodynamic approach to fluidized drying and moistening optimization. AIChE J. 19, 277–285.

Sieniutycz, S., 1973b. Calculating thermodynamic functions in gas–moisture–solid systems (in Polish). Rep. Inst. Chem. Eng. Warsaw Tech. Univ. 2, 71.

Sieniutycz, S., 1974a. The thermodynamic approach in investigation of trajectories and stability of simultaneous heat and mass exchange (in Polish). Prace Nauk. Inst. Inz. Chem. i Tech. Ciepln. Polit. Wrocł. 2A, 225.

Sieniutycz, S., 1974b. The constant Hamiltonian problem and an introduction to the mechanics of optimal discrete systems. Rep. Inst. Chem. Eng. Warsaw Tech. Univ. 3, 27–53.

Sieniutycz, S., 1974c. The minimum of total available energy (exergy) dissipation in continuous fluidized wet beds. Rep. Inst. Chem. Eng. Warsaw Tech. Univ. 3 (3), 47.

Sieniutycz, S., 1975. Thermodynamic optimization of fluidized drying and moistening in case of variable temperature and humidity of inlet gas. Inż. Chem. Proc. 5, 347.

Sieniutycz, S., 1976a. A stationary action principle for a laminar motion in a viscous fluid. Inż. Chem. Proc. 6, 673–691.

Sieniutycz, S., 1976b. Theory of thermoeconomic optimization of continuous and multistage fluidized drying. Rep. Inst. Chem. Eng. Warsaw Tech. Univ. 5 (1), 5.

Sieniutycz, S., 1977a. The variational principles of classical type for uncoupled irreversible time-dependent processes with macroscopic motion and relaxation. Int. J. Heat Mass Transfer 20, 1221–1231.

Sieniutycz, S., 1978. Optimization in Process Engineering, first ed. WNT, Warsaw.

References

Sieniutycz, S., 1979. The wave equations for simultaneous heat and mass transfer in moving media structure testing, space-time transformations and variational approach. Int. J. Heat Mass Transfer 22, 585–599.

Sieniutycz, S., 1980a. The stability of coupled heat and mass transfer in the presence of inertial (wave) effects. Phys. Lett. 78A, 433–436.

Sieniutycz, S., 1980b. The variational principle replacing the principle of minimum entropy production for coupled non-stationary heat and mass transfer processes with convective motion and relaxation. Int. J. Heat Mass Transfer 23, 1183–1193.

Sieniutycz, S., 1981a. Thermodynamics of coupled heat, mass and momentum transport with finite wave speed, I: basic ideas of theory. Int. J. Heat Mass Transfer 24, 1723–1732.

Sieniutycz, S., 1981b. Thermodynamics of coupled heat, mass and momentum transport with finite wave speed, II: examples of transformations of fluxes and forces. Int. J. Heat Mass Transfer 24, 1759–1769.

Sieniutycz, S., 1981c. The thermodynamic stability of coupled heat and mass transfer described by linear wave equations. Chem. Eng. Sci. 36, 621–624.

Sieniutycz, S., 1981d. On the applicability of classical thermodynamic stability conditions for the coupled heat and mass transfer with finite wave speed. J. Non-Equilib. Thermodyn. 6, 79–84.

Sieniutycz, S., 1981e. Action functionals for linear wave dissipative systems with coupled heat and mass transfer. Physics Lett. 84A, 98–102.

Sieniutycz, S., 1982. Entropy of flux relaxation and variational theory of simultaneous heat and mass transfer governed by non-Onsager phenomenological equations. Appl. Sci. Res. 39, 81–103.

Sieniutycz, S., 1983. The inertial relaxation terms and the variational principles of least action type for nonstationary energy and mass diffusion. Int. J. Heat Mass Transfer 26, 55–63.

Sieniutycz, S., 1984a. The variational approach to nonstationary Brownian and molecular diffusion described by wave equations. Chem. Eng. Sci. 39, 71–80.

Sieniutycz, S., 1984b. Lumped parameter modelling and an introduction of one-dimensional nonadiabatic drying systems. Int. J. Heat Mass Transfer 27, 1971–1983.

Sieniutycz, S., 1985. A synthesis of some variational theorems of extended irreversible thermodynamics of non-stationary heat and mass transfer. Appl. Sci. Res. 44, 211–228.

Sieniutycz, S., 1986. On the relativistic temperature transformations and the related energy transport problem. Int. J. Heat Mass Transfer 29, 651–654.

Sieniutycz, S., 1987a. Variational approach to the fundamental equations of heat, mass and momentum transport in strongly unsteady-state processes. I. Int. Chem. Eng. 27, 545–555.

Sieniutycz, S., 1987b. From a last action principle to mass action law and extended affinity. Chem. Eng. Sci. 42, 2697–2711.

Sieniutycz, S., 1988a. Variational approach to the fundamental equations of heat, mass and momentum transport in strongly unsteady-state processes. Int. Chem. Eng. 28, 353–361.

Sieniutycz, S., 1988b. Hamiltonian tensor of energy-momentum in extended thermodynamics of one-component fluid. Inż. Chem. Proc. 4, 839–861.

Sieniutycz, S., 1989a. Experimental relaxation times, drying–moisturizing cycles and the relaxation drying equation. Chem. Eng. Sci. 44, 727–740.

Sieniutycz, S., 1989b. Energy-momentum tensor and conservation laws for extended multicomponent fluids with the transport of heat, mass and electricity. Unpublished research report of Inst. Chem. Eng. Warsaw at WTU (in Polish).

Sieniutycz, S., 1990. Thermal momentum, heat inertia and a macroscopic extension of the de Broglie microthermodynamics. I. The multicomponent fluids with the sourceless continuity constraints. Advances in Thermodynamics Series, vol. 3, Taylor and Francis, New York, pp. 328–368.

Sieniutycz, S., 1991. Optimization in Process Engineering, second ed. Wydawnictwa Naukowo Techniczne, Warszawa.

Sieniutycz, S., 1992a. Wave equations of heat and mass transfer. Article in: Flow, Diffusion and Transport Processes. Advances in Thermodynamics Series, vol. 6, Taylor and Francis, New York, pp. 146–167.

Sieniutycz, S., 1992b. Thermal momentum, heat inertia and a macroscopic extension of de Broglie micro-thermodynamics, II: The multicomponent fluids with the sources. Advances in Thermodynamics Series, 7, Taylor and Francis, New York, pp. 408–447.

Sieniutycz, S., 1994. Conservation Laws in Variational Thermo-Hydrodynamics. Kluwer Academics, Dordrecht.

Sieniutycz, S., 1995. A variational approach to chemically reacting distributed systems, Scientific Reports of Kielce Polytechnik. Mechanika 59, 131–143.

Sieniutycz, S., 1996. Thermodynamics of heat, mass and momentum relaxation and rheological bodies. Paper presented at Symposium "Thermodynamic Approach to Rheological Modeling and Simulations," XIIth International Congress on Rheology, Quebec City, August 18–23, 1996, Congress Proceedings, pp. 333–334.

Sieniutycz, S., 1997a. Variational thermomechanical processes and chemical reactions in distributed systems. Int. J. Heat Mass Transfer 40, 3467–3485.

Sieniutycz, S., 1997b. Generalized Carnot problem of maximum work in a finite time via Hamilton–Jacobi–Bellman equation, Florence World Energy Research Symposium, FLOWERS '97, Italy, Florence, July 30–August 1, pp. 151–159.

Sieniutycz, S., 1998a. Generalized Carnot problem of maximum work in finite time via Hamilton–Jacobi–Bellman theory. Energy Convers. Manage. 39, 1735–1743.

Sieniutycz, S., 1998b. Nonlinear thermokinetics of maximum work in finite time. Int. J. Engin. Sci. 36, 577–597.

Sieniutycz, S., 1998c. Thermodynamic and relativistic aspects of thermal inertia in fluids. Phys. Review E 58, 7027–7039.

Sieniutycz, S., 1999. Endoreversible modeling and optimization of multistage thermal machines by dynamic programming. In: Wu, C., Chen, L., Chen, J. (Eds.), Recent Advances in Finite Time Thermodynamics. Nova Science Publishers, New York, pp. 189–219.

Sieniutycz, S., 2000a. Hamilton–Jacobi–Bellman framework for optimal control in multistage energy systems. Phys. Reports 326, 165–285.

Sieniutycz, S., 2000b. Dynamic programming approach to a Fermat-type principle for heat flow. Int. J. Heat Mass Transfer 43 (18), 3453–3468.

Sieniutycz, S., 2000c. Thermodynamic optimization for work-assisted heating and drying operations. Energy Convers. Manage. 41 (18), 2009–2039.

Sieniutycz, S., 2000d. Optimal control of active (work-producing) systems. Inż Chem. Proc. 21 (1), 29–55.

Sieniutycz, S., 2000e. In: Sieniutycz, S., de Vos, A. (Eds.), Preface to the book: Thermodynamics of Energy Conversion and Transport. Springer, New York, pp. 10–19.

Sieniutycz, S., 2000f. Action-type variational principles for hyperbolic and parabolic heat & mass transfer. Int. J. Appl. Thermodyn. 3 (2), 73–81.

Sieniutycz, S., 2000g. Some thermodynamic aspects of development and bistability in complex multistage systems. Open Sys. Inf. Dyn. 7, 309–326.

Sieniutycz, S., 2000h. Nonequilibrium thermodynamics for multiphase reacting systems, Proceedings of 3rd Int. Symp.: Catalysis in Multiphase Reactors, May 29–31, 2000, Naples, Italy, edited by AIDIC 2000, pp. 351–358.

Sieniutycz, S. 2000i. Thermodynamics of development of energy systems with applications to thermal machines and living organisms, Eotwos University, Workshop on Recent Developments in Thermodynamics, Budapest, 22–24 VII 01, Periodica Polytechnica, Ser. Chem. Eng. 44 (1): 49–80.

Sieniutycz, S., 2001a. Work optimization in continuous and discrete systems with complex fluids. J. Non-Newtonian Fluid Mech. 96, 341–370.

Sieniutycz, S., 2001b. Optimal work flux in sequential systems with complex heat exchange. Int. J. Heat Mass Transfer 44 (5), 897–918.

Sieniutycz, S., 2001c. Nonequilibrium thermodynamics for multiphase reacting systems: an optimal control approach. Catalysis Today 66/2–4, 453–460.

Sieniutycz, S., 2001d. Shorcut for constructing variational principles for thermal fields. Open Sys. Inf. Dyn. 8, 1–100.

Sieniutycz, S., 2001e. Thermodynamic limits for work-assisted and solar-assisted drying operations. Archiv. Thermodyn. 22 (3–4), 17–36.

References

Sieniutycz, S., 2001f. Podejście Wariacyjne w Termodynamice Nierównowagowej. In: Bilicki, Z., Mikielewicz, J., Sieniutycz, S. (Eds.), Współczesne Problemy Termodynamiki. Gdańsk: Wydawnictwo IMP PAN, pp. 119–156, Chapter 6.

Sieniutycz, S., 2002a. Optimality of nonequilibrium systems and problems of statistical thermodynamics. Int. J. Heat Mass Transfer 45 (7), 1545–1561.

Sieniutycz, S., 2002b. Relativistic thermo-hydrodynamics and conservation laws in dissipative continua with thermal inertia. Rep. Math. Phys. 49 (2/3), 361–370.

Sieniutycz, S., 2002c. Relativistic aspects of hydrodynamics and theory of heat transfer, Bulletin de la Societe des Sciences et des letters de Łódź 52, 117–131. Proceedings of International Conderence: Ideas of Albert A. Michelson in Mathematical Physics at Mathematical Conference Center in Będlewo, August 4–11, 2002, Będlewo, Poland.

Sieniutycz, S., 2002d. Steady interphase heat transfer in nonlinear media—a variational principle. Archiv. Thermodyn. 23 (3), 79–90.

Sieniutycz, S., 2002e. A Fermat-like principle for chemical reactions in heterogeneous systems. Open Sys. Inf. Dyn. 9, 257–272.

Sieniutycz, S., 2003a. Synthesis of thermodynamic models unifying traditional and work-driven operations with heat and mass exchange. Open Sys. Inf. Dyn 10, 31–49.

Sieniutycz, S., 2003b. Carnot controls to unify traditional and work-assisted operations with heat & mass transfer. Int. J. Appl. Thermodyn. 6, 59–67.

Sieniutycz, S., 2003c. Thermodynamic limits on production or consumption of mechanical energy in practical and industrial systems. Progr. Energy Comb. Sci. 29, 193–246.

Sieniutycz, S., 2003d. Quasi-canonical structure of optimal control algorithms in constrained discrete systems. Rep. Math. Phys. 51, 335–344.

Sieniutycz, S., 2004a. Nonlinear macrokinetics of heat & mass transfer and chemical or electrochemical reactions. Int. J. Heat Mass Transfer 47, 515–526.

Sieniutycz, S., 2004b. Modele nieliniowej termokinetyki w procesach wymiany masy. Inż Chem. Proc. 25, 3–14.

Sieniutycz, S., 2004c. Extremum properties of entropy to determine dynamics of growth and evolution in complex systems. Physica A 340, 356–363.

Sieniutycz, S., 2004d. Highly nonlinear transport phenomena in presence of chemical reactions. Inż Chem. Proc. 25, 2001–2009.

Sieniutycz, S., 2005a. An unifying thermodynamic framework for nonlinear macrokinetics in reaction-diffusion systems. Int. J. Thermodyn. 8, 115–125.

Sieniutycz, S., 2005b. Progress in variational formulations for macroscopic processes. In: Sieniutycz, S., Farkas, H. (Eds.), Variational and Extremum Principles in Macroscopic Systems. Elsevier Science, Oxford, pp. 3–24, Chapter I.1.

Sieniutycz, S., 2005c. Field variational principles for irreversible energy and mass transfer. In: Sieniutycz, S., Farkas, H. (Eds.), Variational and Extremum Principles in Macroscopic Systems. Elsevier Science, Oxford, pp. 497–522, Chapter II.7.

Sieniutycz, S., 2005d. Thermodynamic aspects of variational principles for fluids with heat flow. Archiv. Thermodyn. 26 (2), 1–27.

Sieniutycz, S., 2006a. Thermodynamic and variational aspects in modeling and simulation of biological evolution. Archiv. Thermodyn. 27, 83–94.

Sieniutycz, S., 2006b. Development of generalized (rate dependent) availability. Int. J. Heat Mass Transfer 49, 789–795.

Sieniutycz, S., 2006c. State transformations and Hamiltonian structures for optimal control in discrete systems. Rep. Math. Phys. 57, 289–317.

Sieniutycz, S., 2006d. Thermodynamic limits in applications of energy of solar radiation. Drying Technol. 24 (9), 1139–1146.

Sieniutycz, S., 2007a. Frictional passage of fluid through inhomogeneous porous system: a variational principle. Transp. Porous Med. 69, 239–257.

Sieniutycz, S., 2007b. Hamilton–Jacobi–Bellman equations and dynamic programming for power-maximizing relaxation of radiation. Int. J. Heat Mass Transfer 50, 2714–2732.

Sieniutycz, S., 2007c. Entropy-based modeling and simulation of evolution in biological systems. In: R. Moreno-Diaz et al. (Eds.), Lecture Notes Comput. Sci. 4739, pp. 34–41.

Sieniutycz, S., 2007d. A simple chemical engine in steady and dynamic situations. Archiv. Thermodyn. 28, 57–84.

Sieniutycz, S., 2008a. Dynamical converters with power-producing relaxation of solar radiation. Int. J. Therm. Sci. 47, 495–505.

Sieniutycz, S., 2008b. An analysis of power and entropy production in a chemical engine. Int. J. Heat Mass Transfer 51, 5859–5871.

Sieniutycz, S., 2008c. A variational approach for a slow motion of aerosol particles in the atmospheric fluid. Chem. Eng. Trans. 16, 305–312.

Sieniutycz, S., 2009a. Dynamic programming and Lagrange multipliers for active relaxation of resources in nonlinear nonequilibrium systems. Appl. Math Model 33, 1457–1478.

Sieniutycz, S., 2009b. Dynamic bounds and efficiency of nonlinear converters under nonlinear transfer laws. Energy 34, 334–340, Chapter 8.

Sieniutycz, S., 2009c. Complex chemical systems with power production driven by mass transfer. Int. J. Heat Mass Transfer 52, 2453–2465.

Sieniutycz, S., 2009d. Thermodynamics of simultaneous drying and power production. Drying Technol. 27 (3), 322–335.

Sieniutycz, S., 2010a. Upper limits for power yield in thermal, chemical and electrochemical systems. In: Korsunsky, A. (Ed.), Current Themes in Engineering Science. American Institute of Physics, Melville, NY.

Sieniutycz, S., 2010b. Thermodynamic aspects of power generation in imperfect fuel cells I. Int. J. Ambient Energy 31 (4), 195–202.

Sieniutycz, S., 2010c. Thermodynamics of power production in fuel cells. Chem. Process Eng. 31, 81–105.

Sieniutycz, S., 2010d. Finite-rate thermodynamics of power production in thermal, chemical and electrochemical systems. Int. J. Heat Mass Transfer 53, 2864–2876.

Sieniutycz, S., 2011a. Thermodynamic aspects of of power generation in imperfect fuel cells II. Int. J. Ambient Energy 32 (1), 46–56.

Sieniutycz, S., 2011b. Fuel cells as energy systems: efficiency, power limits and thermodynamic behavior. J. Energy Power Eng. 5, 17–28.

Sieniutycz, S., 2011c. Identification and selection of unconstrained controls in power systems propelled by heat and mass transfer. Int. J. Heat Mass Transfer 54, 938–948.

Sieniutycz, S., 2011d. Thermodynamic aspects of power production in energy systems. In: Chiru, A. (Ed.), The Automobile and the Environment. Mathematics in Industry 15, Springer, Berlin, Chapter 1. (Selected papers from International congress on Automotive and Transport Engineering, Brasov, Romania, October 27–29, 2010).

Sieniutycz, S., 2011e. Modeling and simulation of power yield in thermal, chemical and electrochemical systems: Fuel cell case. In: Moreno-Diaz, R., Pilcher, F., Arencibia, A. (Eds.), Computer Aided Systems Theory – EUROCAST 2011, LNCS. Springer Berlin Heidelberg, Berlin, pp. 593–600.

Sieniutycz, S., 2012a. Maximizing power yield in energy systems—a thermodynamic synthesis. Appl. Math. Model. 36, 2197–2212.

Sieniutycz, S., 2012b. Power limits and thermodynamics in fuel cells. Int. J. Modern Phys. B. 17, 1–17.

Sieniutycz, S., 2012c. Minimization of power flux driving evaporation systems propelled by thermal or solar energy. Cy. Phys. 1 (3), 213–224.

Sieniutycz, S., 2013a. An unified approach to limits on power yield and power consumption in thermo–electro–chemical systems. Entropy 15, 650–677.

Sieniutycz, S., 2013b. Power yield and power consumption in thermo-electro-chemical systems—a synthesizing approach. Energy Convers. Manage. 68, 293–304.

Sieniutycz, S., Berry, R.S., 1989. Conservation laws from Hamilton's principle for nonlocal thermodynamic equilibrium fluids with heat flow. Phys. Rev. A 40, 348–361.

References

Sieniutycz, S., Berry, R.S., 1991. Field thermodynamic potentials and geometric thermodynamics with heat transfer. Phys. Rev. A 43, 2807–2818.

Sieniutycz, S., Berry, R.S., 1992. Least-entropy generation: variational principle of Onsager's type for transient hyperbolic heat and mass transfer. Phys Rev. A 46, 6359–6370.

Sieniutycz, S., Berry, R.S., 1993. Canonical formalism, fundamental equation, and generalized thermomechanics for irreversible fluids with heat transfer. Phys. Rev. E 47, 1765–1783.

Sieniutycz, S., Berry, R.S., 1997. Thermal mass and thermal inertia: a comparison of hypotheses. Open Sys. Inf. Dyn. 4, 15–43.

Sieniutycz, S., Berry, R.S., 2000. Discrete Hamiltonian analysis of endoreversible thermal cascades. In: Sieniutycz S., de Vos, A. (Eds.), Thermodynamics of Energy Conversion and Transport. Springer, New York. pp. 143–172, Chapter 6.

Sieniutycz, S., Berry, R.S., 2002. Variational theory for thermodynamics of thermal waves, Phys. Rev. E 65, 046132: 1–11.

Sieniutycz, S., de Vos, A. (Eds.), 2000. Thermodynamics of Energy Conversion and Transport. Springer, New York.

Sieniutycz, S., Farkas, H., 1997. Chemical waves in confined regions by Hamilton–Jacobi–Bellman theory. Chem. Eng. Sci. 52, 2927–2945.

Sieniutycz, S., Farkas, H. (Eds.), 2005. Variational and Extremum Principles in Macroscopic Systems. Elsevier Science, Oxford.

Sieniutycz, S., Jeżowski, J., 2009. Energy Optimization in Process Systems. Elsevier, Oxford, UK.

Sieniutycz, S., Jeżowski, J., 2013. Energy Optimization in Process Systems and Fuel Cells. Elsevier, Oxford, Amsterdam, pp. xi–xvii and 1–800.

Sieniutycz, S., Komorowska-Kulik, J., 1974. Stability of trajectories of some flow processes with simultaneous heat and mass transfer (in Polish), I: State equations leading to definite pseudoequilibrium point. Rep. Inst. Chem. Eng., Warsaw Technical University 3: 6.

Sieniutycz, S., Komorowska-Kulik, J., 1975. Stability of trajectories of some flow processes with simultaneous heat and mass transfer (in Polish), II: Thermodynamic Liapounov functions and properties of trajectories. Rep. Inst. Chem. Eng., Warsaw Tech. Univ. 4, 5.

Sieniutycz, S., Komorowska-Kulik, J., 1978. Thermodynamic approach to qualitative properties of heat and mass transfer in gas-solid flow systems, I: nonreacting systems. Int. J. Heat Mass Transfer 21, 489–497.

Sieniutycz, S., Komorowska-Kulik, J., 1980. Thermodynamic criteria of stability of carbon monoxide oxidation in a regenerator of catalytic cracking (in Polish). Inż Chem. Proc. 1 (3), 569–590.

Sieniutycz, S., Komorowska-Kulik, J., 1982. Thermodynamic approach to qualitative properties of heat and mass transfer in gas–solid flow systems, II: reacting systems. Int. J. Heat Mass Transfer 25 (4), 575–585.

Sieniutycz, S., Kubiak, M., 2002. Dynamical energy limits in traditional and work-driven operations, II: systems with heat and mass transfer. Int. J. Heat Mass Transfer 45, 5221–5238.

Sieniutycz, S., Kuran, P., 2005. Nonlinear models for mechanical energy production in imperfect generators driven by thermal or solar energy. Int. J. Heat Mass Transfer 48, 719–730.

Sieniutycz, S., Kuran, P., 2006. Modeling thermal behavior and work flux in finite-rate systems with radiation. Int. J. Heat Mass Transfer 49, 3264–3283.

Sieniutycz, S., Poświata, A., 2011. Basic thermodynamic properties of fuel cell systems. Chem. Eng. Trans. 24, 547–552.

Sieniutycz, S., Poświata, A., 2012. Thermodynamic aspects of power production in thermal, chemical and electrochemical systems. Energy 52, 1–9.

Sieniutycz, S., Ratkje, S.K., 1996a. Variational principle for entropy in electrochemical transport phenomena. Int. J. Eng. Sci. 34, 529–560.

Sieniutycz, S., Ratkje, S.K., 1996b. Pertubational thermodynamics of coupled electrochemical heat and mass transfer. Int. J. Heat Mass Transfer 39 (15), 3293–3303.

Sieniutycz, S., Salamon, P. (Eds.), 1990. Nonequilibrium Theory and Extremum Principles, Advances in Thermodynamics Series, 3, Taylor and Francis, New York.

Sieniutycz, S., Salamon, P., 1990b. Diversity of nonequilibrium theories and extremum principles. Nonequilibrium Theory and Extremum Principles. Advances in Thermodynamics Series, vol. 3, Taylor and Francis, New York, pp. 1–38.

Sieniutycz, S., Salamon, P. (Eds.), 1990. Finite–Time Thermodynamics and Thermoeconomics, Advances in Thermodynamics Series, 4, Taylor and Francis, New York.

Sieniutycz, S., Salamon, P., 1990d. Thermodynamics and optimization. Finite-Time Thermodynamics and Thermoeconomics. Advances in Thermodynamics Series, vol. 4, Taylor and Francis, New York, pp. 1–21.

Sieniutycz, S., Salamon, P. (Eds.), 1992a. Flow, Diffusion and Transport Processes. Advances in Thermodynamics Series, 6, Taylor and Francis, New York.

Sieniutycz, S., Salamon, P., 1992b. Thermodynamics of transport and rate processes. Flow, Diffusion and Transport Processes. Advances in Thermodynamics Series, vol. 6, Taylor and Francis, New York, pp. 1–22.

Sieniutycz, S., Salamon, P. (Eds.), 1992c. Extended Thermodynamic Systems. Advances in Thermodynamics Series, vol. 7, Taylor and Francis, New York.

Sieniutycz, S., Salamon, P. (Eds.), 1992d. Thermodynamics of complex systems. Extended Thermodynamic Systems. Advances in Thermodynamics Series, vol. 7, Taylor and Francis, New York, pp. 1–24.

Sieniutycz, S., Shiner, J.S., 1992. Variational and extremum properties of homogeneous chemical kinetics, I: Lagrangian and Hamiltonian-like formulations. Open Sys. Inf. Dyn. 1 (2), 149–182.

Sieniutycz, S., Shiner, J.S., 1993. Variational and extremum properties of chemical kinetics, II: Minimum dissipation approaches. Open Sys. Inf. Dyn. 1 (3), 327–348.

Sieniutycz, S., Shiner, J.S., 1994. Thermodynamics of irreversible processes and its relation to chemical engineering: Second Law analyses and Finite Time Thermodynamics. J. Non-Equilib. Thermodyn. 19, 303–348.

Sieniutycz, S., Szwast, Z., 1982. Practice of Optimization. WNT, Warsaw.

Sieniutycz, S., Szwast, Z., 1999. Optimization of multistage thermal machines by a Pontryagin's-like discrete maximum principle. In: Wu, C. (Ed.), Advances in Recent Finite Time Thermodynamics. Nova Science, New York, pp. 221–237, Chapter 12.

Sieniutycz, S., Szwast, Z., 2003. Work limits in imperfect sequential systems with heat and fluid flow. J. Nonequilib. Thermodyn. 28, 85–114.

Sieniutycz, S., von Spakovsky, M.R., 1998. Finite time generalization of thermal exergy. Energy Convers. Manage. 39, 1423–1447.

Sieniutycz, S., Szwast, Z., Kuran, P., Poświata, A., Zalewski, M., Przekop, R., Błesznowski, M., 2011. Thermodynamics and optimization of chemical and electrochemical energy generators with applications to fuel cells. Research report NN208019434. Warsaw TU: Faculty of Chemical and Process Engineering.

Sieniutycz, S., Błesznowski, M., Zieleniak, A. Jewulski, 2012. Power generation in thermochemical and electrochemical systems—a thermodynamic theory. Int. J. Heat Mass Transfer 55, 1197–1213.

Sih, G.C., 1985. Mechanics and physics of energy density and rate of change of volume with surface. Journ. Theor. Appl. Fract. Mech. 4, 157–173.

Sih, G.C., 1992. Some problems in nonequilibrium thermomechanics. Flow, Diffusion and Transport Processes. Advances in Thermodynamics Series, vol. 6, Taylor and Francis, New York, pp. 218–247.

Simonin, O., Deutsch, E., Minier, J.P., 1993. Eulerian prediction of the fluid/particle correlated motion in turbulent two-phase flows. Appl. Sci. Res. 51, 275–283.

Simshauser, P., 2010. "Vertical integration, credit ratings and retail price settings in energy-only markets: navigating the Resource Adequacy problem,". Energy Policy 38 (11), 7427–7441.

Singh, V.P., 2013. Entropy Theory and its Application in Environmental and Water Engineering. Wiley, New York.

Slattery, J., 1981. Interfacial Transport Phenomena. Springer, Berlin.

Slattery, J., Sagis, L., Oh, E.-S., 2007. Interfacial Transport Phenomena. Springer, Berlin.

Sluckin, T.J., 1981. Influence of flow on the isotropic-nematic transition in polymer solutions: a thermodynamic approach. Macromolecules 14, 1676–1680.

References

Służalec, A., Muskalski, K., 1987. Numeryczna analiza problemów przepływu ciepła w tkankach pod wpływem działania promienia laserowego. Archiv. Thermodyn. 8, 5–13.

Smith, R.A., 1966. Matrix calculations for Liapunov quadratic forms. J. Differ. Equ. 2 (2), 208–217.

Snyder, H.A., 1968. Stability of rotating Couette flow, II: comparison with numerical results. Phys. Fluids 11 (8), 1599.

Sokolov, M., Tanner, R.I., 1972. Convective stability of a general viscoelastic fluid heated from below. Phys. Fluids 15 (4), 534–539.

Sommerfeld, J.T., Lenk, C.T., 1970. Thermodynamics helps you predict selling price. Chem. Eng. May, 136.

Soo, S.L., 1962. Analytical Thermodynamics. Prentice-Hall, Englewood Cliffs, NJ, Chapter 12.

Soo, S.L., 1969. Direct Energy Conversion. Prentice-Hall, Englewood Cliffs, NJ.

Soponronnarit, S., Prachayawarakorn, S., Sripawaatakul, O., 1996. Development of cross-flow fluidized bed paddy dryer. Drying Technol. 14 (10), 2397–2410.

Soponronnarit, S., Pongtornkulpanich, A., Prachayawarakorn, S., 1997. Drying characteristics of corn in fluidized bed dryer. Drying Technol. 15 (5), 1603–1615.

Sousa, T., 2007. Thermodynamics as a Substantive and Formal Theory for the Analysis of Economic and Biological Systems. PhD thesis carried out at the Department of Theoretical Life Sciences, Vrije Universiteit Amsterdam, The Netherlands and at the Environment and Energy Section, Instituto Superior Técnico, Lisbon, Portugal.

Spalding, D.B., 1959a. Stability of steady exotermic chemical reactions in simple nonadiabatic systems. Chem. Eng. Sci. 11 (1), 53–60.

Spalding, D.B., 1959b. Conserved property diagrams for rate-process calculations. Chem. Eng. Sci., 11, 183–193 and 225–241.

Spalding, D.B., 1961. Heat and mass transfer between the gaseous and liquid phases of a binary mixture. Int. J. Heat Mass Transfer 4, 47–62.

Spalding, D.B., 1963. Convective Mass Transfer. Edward Arnold, London.

Spalding, D.B., 1979. Combustion and Mass Transfer. Pergamon, Oxford.

Spanner, D.C., 1964. Introduction to Thermodynamics. Academic Press, London, p. 218.

Sprott, J.C., 2000. Simple chaotic systems and circuits. Am. J. Phys. 68 (8), 759–763.

Sreenivasan, K.R., 1999. Fluid turbulence. Rev. Modern Phys. 71 (2), S373–S395.

Standart, G., 1964. The mass, momentum and energy equations for heterogeneous flow systems. Chem. Eng. Sci. 19, 227–236.

Stanford, A.L., 1975. Foundations of Biophysics. Academic Press, New York.

Steeb, W.-H., Kunick, A., 1982. Lagrange functions of a class of dynamical systems with a limit cycle and chaotic behavior. Phys. Rev. A 25 (6), 2889–2892.

Stephens, J.J., 1967. Alternate forms of the Herrivel–Lin variational principle. Phys. Fluids 10, 76–77.

Stephenson, 1995. Some nonlinear diffusion equations and fractal diffusion. Physica A 222, 234–247.

Struchtrup, H., Weiss, W., 1998. Maximum of entropy production becomes minimal in stationary processes. Phys. Rev. Lett. 80 (23), 5048–5051.

Strumiłło, Cz., 1983. Fundamentals of Theory and Technology of Drying, second ed. Wydawnictwa Naukowo Techniczne, Warsaw.

Strumiłło, C., Kudra, T., 1987. Drying: Principles, Applications and Design. Gordon and Breach, New York.

Strumiłło, Cz., Lopez–Cacicedo, C., 1984. Energy aspects in drying. In: Mujumdar, A.S. (Ed.), Handbook of Industrial Drying. Marcel Dekker, New York.

Styer, D.F., 2000. Insight into entropy. Am. J. Phys. 68, 1090–1096.

Sumińska, B., 1983. Investigation of relaxation effects in drying. MD thesis. Warsaw Technical University, Poland.

Sunden, B., Faghri, M., 2005. Transport Phenomena in Fuel Cells. WTI Press, Southampton.

Sundheim, B.R., 1964. Transport properties of liquid electrolytes. In: Sundheim, B.R. (Ed.), Fused Salts. McGraw-Hill, New York, pp. 165–254.

Sureshkumar, R., 2001. Local linear stability characteristics of viscoelastic periodic channel flow. J. Non-Newtonian Fluid Mech. 97, 125–148.

Syahrul, S., Hamdullahpur, F., Dincer, I., 2002. Exergy analysis of fluidized bed drying of moist particles. Exergy Int. J., 2 (2), 97–98.

Syczew, W.W., 1973. Thermodynamics of Complex Systems. PWN, Warszawa.
Szargut, J., 1986. Application of exergy for calculation of ecological cost. Bull. Polish Acad. Sci. Tech. Sci. 34, 475.
Szargut, J., 1992. Analysis of cumulative exergy consumption and cumulative exergy losses. Finite-Time Thermodynamics and Thermoeconomics. Advances in Thermodynamics Series, vol. 4, Taylor and Francis, New York, pp. 278–302.
Szargut J., 2001. Exergy analysis in thermal technology, Ekspertyza KTiSp PAN: Współczesne Kierunki w Termodynamice. In: Bilicki, Z., Mikielewicz, J., Sieniutycz, S. (Eds.), Wydawnictwa Politechniki Gdańskiej.
Szargut, J., 2002a. Minimization of the depletion of nonrenewable resources by means of the optimization of the design parameters, Proceedings of ECOS 2002, vol. I, Berlin, Germany, July 3–5, pp. 326–333.
Szargut, J., 2002b. Application of exergy for the determination of the proecological tax replacing the actual personal taxes. Energy 27, 379–389.
Szargut, J., 2005. Exergy Method: Technical and Ecological Applications. WIT Press, Southampton.
Szargut, J., Petela, R., 1965. Exergy. WNT, Warsaw (in particular see Sec. 11.4.4 on exergy efficiency of plants vegetation).
Szargut, J., Stanek, W., 2007. Thermoecological optimization of a solar collector. Energy 32 (4), 584–590.
Szargut, J., Morris, D.R., Steward, F., 1988. Exergy Analysis of Thermal, Chemical and Metallurgical Processes. Hemisphere, New York.
Szepe, S., 1966. Catalyst deactivation and some related optimization problems. PhD thesis. Illinois Institute of Technology.
Szepe, S., Levenspiel, O., 1968. Optimal temperature policies for reactors subject to catalyst deactivation. Chem. Eng. Sci. 23, 881–894.
Szepe, S., Levenspiel, O., 1971. Catalyst deactivation. Chemical Reaction Engineering, Proceeding of the Fourth European Symp., Brussels, Pergamon Press, Oxford, p. 265.
Szewczyk, K., 2002. Balances of microbial growth (in Polish). Biotechnology 57, 15–32.
Szewczyk, K., 2006. Biological fuel cells—Biologiczne ogniwa paliwowe (in Polish). Na Pograniczu Chemii i Biologii 15, 179–208.
Szwast, Z., 1988. Enhanced version of a discrete algorithm for optimization with a constant hamiltonian. Inż. Chem. Proc. 3, 529–545.
Szwast, Z., 1990. Exergy optimization in a class of drying systems with granular solids. Finite-Time Thermodynamics and Thermoeconomics. Advances in Thermodynamics Series, vol. 4, Taylor and Francis, New York, pp. 209–248.
Szwast, Z., 1994. Optimization of the Chemical Reactions in Tubular Reactors with Moving Catalyst Deactivation. Rep. Inst. of Chem. Eng. Warsaw Tech. Univ. 21 (1–4), 5–140.
Szwast, Z., 1997. Próba opisu wybranych przypadków ewolucji organizmów żywych z wykorzystaniem entropii informacyjnej. Prace Wydziału Inżynierii Chemicznej i Procesowej Politechniki Warszawskiej XXIV (3/4), 123–143.
Szwast, Z., Sieniutycz, S., 1998. Optimization of the reactor—regenerator system with catalitic parallel—consecutive reactions. Proceedings of 2nd International Symposium Catalysis in Multiphase Reactors, Toluse, France, March 16–18, 1998, pp. 217–226.
Szwast, Z., Sieniutycz, S., 1999. Optimization of the reactor-regenerator system with catalytic parallel-consecutive reactions. Catalysis Today 48, 175–184.
Szwast, Z., Sieniutycz, S., 2001. Optimal temperature profiles for parallel-consecutive reactions with deactivating catalyst. Catalysis Today 66 (2–4), 461–466.
Szwast, Z., Sieniutycz, S., 2004. Optimal dynamical processes in tubular reactor with deactivation of multirun moving catalyst. Chem Eng J. 45, 2995–3012.
Szwast, Z., Sieniutycz, S., Shiner, J.S., 2002. Complexity principle of extremality in evolution of selected living organisms by information – theoretic entropy. Chaos Soliton. Fract. 13 (9), 1871–1888.
Szydłowski, M., Heller, M., Sasin, W., 1996. Geometry of spaces with the Jacobi metric. J. Math. Phys. 37, 346–361.
Tabiś, B., 2000. Principles of Chemical Reactor Engineering (in Polish). Wydawnictwa Naukowo Techniczne, Warszawa.

References

Tachibana, A., Fukui, K., 1979a. Intrinsic dynamism in chemically reacting systems. Theoret. Chim. Acta (Berl.) 51, 189–206.

Tachibana, A., Fukui, K., 1979b. Intrinsic field theory of chemical reactions. Theoret. Chim. Acta (Berl.) 51, 275–296.

Tan, A., Holland, L.R., 1990. Tangent law of refraction for linear conduction through an interface and related variational principle. Am. J. Phys. 58, 991–998.

Tangde, V.M., Bhalekar, A.A., Venkataramani, B., 2007. Thermodynamic stability of sulfur dioxide oxidation by Lapunov function analysis against temperature perturbation. Physica Scripta 75 (4), 460–466.

Tarbel, J.M., 1977. A thermodynamic Liapounov function for the near equilibrium CSTR. Chem. Eng. Sci. 32, 1471–1476.

Tarbell, J.M., 1982. A note on the relationship between thermodynamics and Liapunov direct method. Chem. Engin. Commun. 14, 371–378.

Tassios, D.P., 1993. Applied Chemical Engineering Thermodynamics. Springer, Berlin.

Thakur, A.K., Gupta, A.K., 2006. Stationary versus fluidized-bed drying of high-moisture paddy with rest period. Drying Technol. 24, 1443–1456.

Thomas, J.M., Thomas, W.J., 1967. Introduction to the Principles of Heterogeneous Catalysis. Academic Press, London.

Tirrell, M., Malone, F., 1977. Stress-induced diffusion of macromolecules. J. Polym. Sci. 15, 1569–1583.

Toffolo, A., Lazaretto, A., 2004. On the thermoeconomic approach to the diagnosis of energy system malfunctions indicators to diagnose malfunctions: application of a new indicator for the location of causes. Int. J. Thermodyn. 7, 41–49, (the special issue on thermoeconomic diagnosis).

Tolman, R.C., 1949. Relativity, Thermodynamics and Cosmology 72. Clarendon Press, Oxford, p. 269.

Tondeur, D., 1990. Equipartition of entropy production: a design and optimization criterion in chemical engineering. Finite-Time Thermodynamics and Thermoeconomics. Advances in Thermodynamics Series, vol. 4, Taylor and Francis, New York, pp. 175–208.

Tondeur, D., Kvaalen, E., 1987. Equipartition of entropy production: an optimality criterion for transfer and separation processes. Ind. Eng. Chem. Res. 26, 50–56.

Topsoe, F., 1993. Game theoretical equilibrium, maximum entropy and minimum information discrimination. In: Mohammad–Djafari, A., Demoments, G. (Eds.), Maximum Entropy and Bayesian Methods. Kluwer Academics, Dordrecht, pp. 15–23.

Topsoe, F., 2001. Basic concepts, identities and inequalities—the toolkit of information theory. Entropy 3, 162–190.

Tórrez, N., Gustafsson, M., Schreil, A., Martínez, J., 1998. Modeling and simulation of crossflow moving bed grain dryers. Drying Technol. 16 (9–10), 1999–2015.

Tribus, M., 1970. Thermostatics and Thermodynamics (in Russian). Energia, Moscow.

Truesdell, C., 1962. Mechanical basis of diffusion. J. Chem. Phys. 37, 2236–2344.

Truesdell, C., 1969. Rational Thermodynamics. McGraw-Hill, New York.

Trulla, L.L., Zbilut, J.P., Giuliani, A., 2003. Putting relative complexity estimates to work: a simple and general statistical methodology. Physica A 319, 591–600.

Tsallis, C., 1988. Possible generalization of Boltzmann–Gibbs statistics. J. Statis. Phys. 52, 479–487.

Tsallis, C., 1999a. Nonextensive statistical mechanics and thermodynamics: historical background and present status. In: Abe, S., Okamoto, Y. (Eds.), Nonextensive Statistical Mechanics and its Applications, Lecture Notes in Physics. Springer, Berlin.

Tsallis, C., 1999b. Nonextensive statistics: theoretical, experimental and computational evidences and connections. Brasilian J. Phys. 29 (1), 1–35.

Tsallis, C., 2011. The nonadditive entropy Sq and its applications in physics and elsewhere: some remarks. Entropy 13, 1765–1804.

Tsatsaronis, G. 1984. Combination of exergetic and economic analysis in energy conversion processes, Proceedings of the European Congress Energy Economics and Management in Industry, vol. 1. Algarve, Portugal, April 2–5. Pergamon Press, Oxford, pp. 151–157.

Tsatsaronis, G., 1987. A review of exergoeconomic methodologies. In: Moran, M., Sciubba, E. (Eds.), Second Law Analysis of Thermal System. ASME, New York, pp. 81–97.

Tsatsaronis, G., 1993. Thermodynamic analysis and optimization of energy systems. Prog. Energy Combus. Sci. 19, 227–257.

Tsatsaronis, G., 1999. Strengths and limitations of exergy analysis. In: Bejan, A., Mamut, E. (Eds.), Thermodynamic Optimization of Complex Energy Systems. Kluwer Academics, Dordrecht, pp. 93–100.

Tsatsaronis, G., 2007. Comments on the paper "A brief commented history of exergy from the beginnings to 2004" by E. Sciubba and G. Wall. Int. J. Thermodyn., 10, 187–190.

Tsatsaronis, G., Winhold, M., 1985a. Exergoeconomic analysis and evaluation of energy conversion plants, Part I: a new general methodology. Energy Int. J. 10, 69–80.

Tsatsaronis, G., Winhold, M., 1985b. Exergoeconomic analysis and evaluation of energy conversion plants, Part II: analysis of a coal-fired steam power plant. Energy Int. J. 10, 81–94.

Tsatsaronis, G., Lin, L., Pisa, J., 1993. Exergy costing in exergoeconomics. J. Energy Resourc. Technol. 115, 9–16.

Tsirlin, A.M. 1974. Averaged optimization and sliding regimes in optimal control problems. Izv. AN SSSR, ser technical cybernetics 2, 143–151.

Tsirlin, A.M., 1997. Methods of Averaged Optimization and their Applications (in Russian), Nauka–Fizmatlit.

Tsirlin, A.M., 2008. Problems and Methods of Averaged Optimization. ISSN 0081–5438. Proceedings of the Steklov Institute of Mathematics 261: 270–286. Pleiades Publishing, Ltd. Original Russian text published in Trudy Matematicheskogo Instituta imeni V.A. Steklova, 261, 276–292.

Tsirlin, A.M., 2009. Optimality conditions of sliding modes and the Maximum Principle for control problems with the scalar argument, Automation and Remote Control 70 (5), 839–854. Original Russian text published in Avtomatika i Telemekhanika 5, 106–121, 2009.

Tsirlin, A.M., Kazakov, V., 2002a. Finite-time thermodynamics: limiting possibilities of irreversible separation processes. J. Phys. Chem. 106 (R.S. Berry's issue), 10926–10936.

Tsirlin, A.M., Kazakov, V., 2002b. Realizability areas for thermodynamic systems with given productivity. J. Non-Equilib. Thermodyn. 27, 91–103.

Tsirlin, A.M., Leskov, E.E., 2007. Optimization of diffusion systems. Theor. Found. Chem. Eng. 41, 405–413.

Tsirlin, A.M., Mironova, W.A., Amelkin, S.A., Kazakov, V.A., 1998. Finite-time thermodynamics: conditions of minimal dissipation for thermodynamic processes with given rate. Phys. Rev. E 58, 215–223.

Tsirlin, A.M., Leskov, E.E., Kazakov, V., 2005. Finite time thermodynamics: limiting performance of diffusion engines and membrane systems. J. Phys. Chem. A 109, 9997–10003.

Tsotsas, E., 1994. From single particle to fluid bed drying kinetics. Drying Technol. 12 (6), 1401–1426.

Turing, A., 1952. The chemical basis of morphogenesis. Phil. Trans. Roy. Soc. B. 237, 37–72.

Turner, J.S., 1979. Nonequilibrium thermodynamics, dissipative structures and self-organization: some implications for biomedical research. In: Scott, G.P., Ames, J.M. (Eds.), Dissipative Structures and Spatiotemporal Organization Studies in Biomedical Research. Iowa State University Press.

Ugarte, S., Gao, Y., Metghalchi, H., 2005. Application of the maximum entropy principle in the analysis of a nonequilibrium chemically reacting mixture. Int. J. Thermodyn. 8 (1), 43–53.

Uhlenbeck, G.E., 1949. Transport phenomena in very dilute gases. Univ. Mich. Engng. Research Inst. Rept. CM, 579.

Ulanowicz, R.E., 1997. Ecology, the Ascendent Perspective. Columbia University Press, New York.

Upadhyay, R.K., Iyengar, S.R.K., Rai, V., 2000. Stability and complexity in ecological systems. Chaos, Solitons & Fractals 11, 533–542.

Vailati, A., Giglio, M., 1997. Giant fluctuations in a free diffusion process. Nature 390, 262–263.

Vainberg, M.M., 1964. Variational Methods for the Study of Nonlinear Operators. Holden-Day, San Francisco.

Valença, G.C., Massarani, G., 2000. Grain drying in countercurrent and concurrent gas flow—modelling, simulation and experimental tests. Drying Technol. 18 (1–2), 447–455.

Valero, A., 1995. Thermoeconomics: the meeting point of thermodynamics, economics and ecology. In: Sciubba, E., Moran, M. (Eds.), Second Law Analysis of Energy Systems: Towards the 21st Century, 293–305. Circus, Rome.

References

Valero, A., 1996. Thermoeconomics as a conceptual basis for energy-ecological analysis. Private communication.
Valero, A., Serra, L., Lozano, M.A., 1993. Structural theory of thermoeconomics. In: Richter, R. (Ed.), Proceedings of the Symposium on Thermodynamics and the Design, Analysis and Improvement of Energy Systems. November 28–December 3, New Orleans, USA. ASME Book no H00527. ASME, New York, pp. 1–10.
Valero, A., Lozano, M.A., Serra, L., Tsatsaronis, G., Pisa, J., Frangopoulos, Ch., Von Spakovsky, M.R., 1994a. CGAM problem: definition and conventional solution. Energy 19 (3), 279–286.
Valero, A., Lozano, M.A., Torres, L.C., 1994b. Application of the exergetic cost theory to the CGAM Problem. Energy 19 (3), 365–381.
Valero, A., Serra, L., Uche, J., 2006. Fundamentals of exergy cost accounting and thermoeconomics. Part 1: theory. J. Energy Resour. Technol. 128, 1–8.
Van, P., 2008. A review on the book. In: Kjelstrup S., Bedeaux D. (Eds.), Non-Equilibrium Thermodynamics of Heterogeneous Systems. World Scientific Publishing Co., Singapore.
Van, P., Nyiri, B., 1999. Hamilton formalism and variational principle construction. Ann. Phys. (Leipzig) 8, 331–354.
Van Baak, D.A., 1999. Variational alternatives to Kirchhoff's loop theorem in DC circuits. Am. J. Phys. 67, 36–44.
Van Dantzig, D., 1940. On the thermo-hydrodynamics of the perfectly perfect fluids. Proc. Kon. Ned. Acad. Wet. 43, 387–402 and 609–618.
Van Kampen, N.G., 2001. Stochastic Processes in Physics and Chemistry. North-Holland, Amsterdam.
Van Saarloos, W.V., 1982. On Non-Linear Hydrodynamic Fluctuations (PhD Thesis). NKB Offset, Leiden.
Van Zeggeren, F., Storey, S., 1970. The Computation of Chemical Equilibria. Cambridge University Press, Cambridge, UK.
Varaiya, P.P., 1972. Optimization. Van Nostrand Rheinhold, New York.
Vargas, J.V.C., Bejan, A., 2004. Thermodynamic optimization of internal structure in a fuel cell. Int. J. Energy Res. 28, 319–339.
Vargas, J.V.C., Ordonez, J.C., Bejan, A., 2005. Constructal PEM fuel cell stack design. Int. J. Heat Mass Transfer 48, 4410–4427.
Vázquez, F., delRío, J.A., Gambár, K., Márkus, F., 1996. Comments on the existence of Hamiltonian principles for non-self-adjoint operators. J. Non-Equilib. Thermodyn. 21, 357–360.
Vázquez, F., Márkus, F., Gambár, K., 2009. Quantized heat transport in small systems: a phenomenological approach. Phys. Rev. E 79, 031113.
Veinik, A., 1966. Thermodynamics of Irreversible Processes. Nauka i Tekhnika, Minsk.
Veinik, A.I., 1973. Thermodynamic Couple (in Russian). Nauka i Technika, Minsk.
Velarde, M.G., Chu, X.-L., 1989. Waves and turbulence at interfaces. Physica Scripta T25, 231–237.
Velarde, M.G., Chu, X.-L., 1990. Interfacial Instabilities. World Scientific, London.
Velarde, M.G., Chu, X.-L., 1992. Dissipative thermohydrodynamic oscillators. Flow, Diffusion and Transport Processes. Advances in Thermodynamics Series, vol. 6, Taylor and Francis, New York, pp. 110–145.
Velasco, S., Roco, J.M.M., Medina, A., Hernandez, A.C., 2000. Optimization of heat engines including the saving of natural resources and the reduction of thermal pollution. J. Phys. D 33, 355–359.
Verhas, J., 1983. An extension of the "Governing principle of dissipative processes" to nonlinear constitutive equations. Ann. Phys. (Leipzig) 40, 189–195.
Verhas, J., 1990. The governing principle of dissipative processes: theory, applications and generalizations. Nonequilibrium Theory and Extremum Principles. Advances in Thermodynamics Series, vol. 3, Taylor and Francis, New York, pp. 250–281.
Vernotte, P., 1958. Les paradoxes de la theorie continue de l'equation de la chaleur. C. R. Hebd. Seanc. Acad. Sci. 246, 3154–3155.
Vidales, A.M., Miranda, E.N., 1996. Fractal porous media: relation between macroscopic properties. Chaos Soliton. Fract. 7 (9), 1365–1369.
Vilar, J.M.G., Rubi, J.M., 2001. Thermodynamics beyond local equilibrium. Proceedings of the National Academy of Sciences of the USA (PNAS) 98, 11081–11084.

References

Villers, D., Platten, J.K., 1984. Heating curves in the two-component Benard problem. J. Non-Equilib. Thermodyn. 27, 91–103.
Vitasari, C.R., Jurascik, M., Ptasinski, K.J., 2011. Exergy analysis of biomass-to-synthetic natural gas (SNG) process via indirect gasification of various biomass feedstock. Energy 36 (6), 3825–3837.
Vlad, M.O., Arkin, A., Ross, J., 2004. Response experiments for nonlinear systems with application to reaction kinetics and genetics. Proc. Natl. Acad. Sci. 101, 7223–7228.
Vojta, G., 1967. Hamiltonian formalism in the thermodynamic theory of irreversible processes in continuous systems. Acta. Chimi. Acad. Sci. Hung. 54, 55–64.
Volford, A., Simon, P.L., Farkas, H., Noszticzius, Z., 1999. Rotating chemical waves: theory and experiments. Physica A 274, 30–49.
Volpert, V., Petrovskii, S., 2009. Reaction–diffusion waves in biology. Phys. Life Rev. 6 (4), 267–310.
Von Stockar, U., 2013. Thermodynamics in Biochemical Engineering. CRC Press, Boca Raton.
Von Stockar, U., van der Wielen, L.A.M., 1997. Minireview on thermodynamics in biochemical engineering. J. Biotechnol. 59, 25–37.
Vrentas, J.S., Duda, J.L., 1979. Molecular diffusion in polymer solutions. AIChE J. 25 (1), 1–24, (see also Fujita, 1961).
Vujanovic, B., 1971. An approach to linear and nonlinear heat transfer problem. A.I.A.A. J. 9, 131–134.
Vujanovic, B., 1992. A variational approach to transient heat transfer theory. Flow, Diffusion and Transport Processes. Advances in Thermodynamics Series, vol. 6, Taylor and Francis, New York, pp. 200–217.
Vujanovic, B., Djukic, D.J., 1972. On the variational principle of Hamilton's type for nonlinear heat transfer problem. Int. J. Heat Mass Transfer 15, 1111–11115.
Vujanovic, B., Jones, S.E., 1988. Variational Methods in Nonconservative Phenomena. Academic Press, New York.
Vujanovic, B.D., Jones, S.D., 1989. Variational Methods in Nonconservative Phenomena. Academic Press, Boston.
Vulpiani, A., 1989. Lagrangian chaos and small scale structure of passive scalars. Physica D 38, 372–376.
Wagialla, K.M., Fakeeha, A.H., Elnashaie, S.S.E.H., Almaktary, A.Y., 1991. Modelling and simulation of energy storage in fluidized beds using the two-phase model. Energy Sources 13, 189.
Wagner, H.-J., 2005. Variational principles for the linearly damped flow of barotropic and Madelung-Type fluids. In: Sieniutycz, S., Farkas, H. (Eds.), Variational and Extremum Principles in Macroscopic Systems. Elsevier, Oxford, pp. 227–244.
Waldmann, L. 1958. Transporterscheinungen in Gase von mittlerem Druck. In Hanbuch der Physik, vol 12, Thermodynamics of Gases. Ed. S. Flugge. Springer, Berlin.
Wang, L., 1998. Minimum heat to environment and entropy. Int. J. Heat Mass Transfer 51, 1869–1871.
Wang, Q.A., 2003. Extensive Generalization of statistical mechanics based on incomplete information theory. Entropy 5 (2), 220–232.
Wang, Q.A., 2004. Maximum path information and the principle of least action for chaotic system. Chaos Soliton. Fract. 23, 1253.
Wang, Q.A., 2006. Probability distribution and entropy as a measure of uncertainty, arXiv:cond-mat/0612076.
Wang, Q.A., Wang, R., 2012. Is it possible to formulate least action principle for dissipative system? Technical report, arXiv:1201.6309, 2012.
Wang-Chang, C.S., Uhlenbeck, G.E., 1951. Transport phenomena in polyatomic gases. Univ. Mich. Engin. Research Inst. Rept. CM, 681.
Warmuzinski, K., Buzek, J., Podkański, J., 1995. Marangoni instability during absorption accompanied by chemical reaction. Chem. Eng. J. 58, 151–160.
Warner, G.H., 1966. Statistical Physics. Wiley, New York.
Weber, B.H., Depew, D.J., Dyke, C., Salthe, S.N., Schneider, E.D., Ulanowicz, R.E., Wicken, J.S., 1989. Evolution in thermodynamic perspective: an ecological approach. Biol. Phil. 4 (4), 373–405.
Weinhold, F., 1975. The metric geometry of equilibrium thermodynamics. J. Chem. Phys. 63, 2479, 2484, 2488, 2496.

References

Weinhold, F., 1976. The metric geometry of equilibrium thermodynamics. J. Chem. Phys. 65, 559.
Weitz, D.A., 1997. Diffusion in a different direction. Nature 390, 234, (see also Vailati and Giglio, 1997).
Werle, J., 1967. Phenomenological Thermodynamics. PWN, Warsaw.
West, B.J., Deering, W., 1994. Fractal physiology for physicists: Lévy statistics. Phys. Rep. 246 (1–2), 1–100.
Westerhoff, H.V., Van Dam, K., 1979. Irreversible thermodynamic description of energy transduction in biomembranes. Curr. Topics Bioenerg. 9, 1–62.
White, W.B., Johnson, S.H., Dantzig, G., 1958. Chemical equilibrium in complex mixtures. J. Chem. Phys. 28, 751–755.
Whitney, R.S., 2014. Most efficient quantum thermoelectric at finite power output. Phys. Rev. Lett. 112, 130601.
Wicken, J.S., 1979. The generation of complexity in evolution: a thermodynamic and information-theoretical discussion. J. Theor. Biol. 77, 349–365.
Wicken, J.S., 1980. A thermodynamic theory of evolution. J. Theor. Biol. 87, 9–23.
Wierzbicki, M., 2009. Optimization of SOFC based energy system using Aspen Plus™. MsD Thesis, Faculty of Chemical and Process Engineering, Warsaw University of Technology.
Wiggert, D.C., 1977. Analysis of early time transient heat conduction by method of characteristics. J. Heat Transfer Trans. ASME 99, 35–40.
Wiley, E.O., Brooks, D.R., 1988. Victims of history—a nonequilibrium approach to evolution. Syst. Zool. 31 (1), 1–24.
Wilk, G., Włodarczyk, Z., 2000. Interpretation of nonextensivity parameter q in some applications of Tsallis statistics and Levy distributions. Phys. Rev. Lett. 84 (13), 2770.
Williams, P.E., 2001. Mechanical entropy and its implications. Entropy 3, 76–115.
Williams, D., Durie, B., McMullan, P., Paulson, C., Smith, A. (Eds.), 2001. Greenhouse Gas Control Technologies. Proceedings of the Fifth International Conference on Greenhouse Gas Control Technologies. CSIRO Publishing, Collingwood (Australia).
Winfree, A.T., 1972. Spiral waves of chemical activity. Science 175, 634–676.
Winfree, A.T., 1973. Scroll-shaped waves of chemical activity in three dimensions. Science 181, 937.
Winfree, A.T., 1974. Rotating chemical reactions. Sci. Am. 230, 82–95.
Winkiel, E., 1976. Analiza Termodynamiczna i Liniowe Zadania Stabilności dla Suszarek i Reaktrów (Thermodynamic Analysis and Linear Stability Problems for Dryers and Reactors). MsD Thesis. Wydawnictwa Politechniki Warszawskiej, Warszawa.
Wisniak, J., 1983. Liquid-liquid phase splitting, I: analytical models for critical mixing and azeotropy. Chem. Eng. Sci. 38 (6), 969–978.
Wiśniewski, S., 1984. Balance equations of extensive quantities for a multicomponent fluid taking into account diffusion stresses. Archiwum Termodynamiki 5, 201–213.
Wiśniewski, S., 1990. Nonequilibrium thermodynamics of emitting and absorbing media. Nonequilibrium Theory and Extremum Principles. Advances in Thermodynamics Series, vol. 3, Taylor and Francis, New York, pp. 393–434.
Wiśniewski, S., 1992. Thermodynamics of ionized systems. Flow, Diffusion and Transport Processes. Advances in Thermodynamics Series, vol. 6, Taylor and Francis, New York, pp. 385–422.
Wiśniewski, S., 1995. Termodynamika Techniczna. Wydawnictwa Naukowo Techniczne, Warszawa.
Wiśniewski, S., Staniszewski, B., Szymanik, R., 1973. Thermodynamics of Nonequilibrium Processes. PWN, Warsaw.
Wu, C., Chen, L., Chen, J. (Eds.), 2000. Recent Advances in Finite Time Thermodynamics. Nova Science Publishers, New York.
Wu, X.-J., Zhu, X.-J., Cao, G.-Y., Tu, H.-Y., 2008. Predictive control of SOFC based on a GA-RBF neural network model. J. Power Sources 179, 232–239.
Xia, S., Chen, L., Sun, F., 2009. Maximum power configuration for multireservoir chemical engines. J. Appl. Phys. 105, 124905.
Xia, D., Chen, L., Sun, F., 2010a. Optimal performance of a generalized irreversible chemical engine with diffusive mass transfer law. Math. Comp. Modell 51, 127–136.

Xia, S., Chen, L., Sun, F., 2010b. Hamilton–Jacobi–Bellman equations and dynamic programming for power-optimization of multistage heat engine system with generalized convective heat transfer law. Chin. Sci. Bull. 55 (29), 2874–2884.

Xia, S., Chen, L., Sun, F., 2010c. Finite-time exergy with a finite heat reservoir and generalized radiative heat transfer law. Revista Mexicana de Fisica 56, 287–296.

Xia, S., Chen, L., Sun, F., 2011a. Endoreversible modeling and optimization of multistage heat engine system with a generalized heat transfer law via HJB equations of dynamic programming. Acta Physica Polonica A 6, 747–760.

Xia, S., Chen, L., Sun, F., 2011b. Power optimization of non-ideal energy converters under general convective heat transfer law via Hamilton–Jacobi–Bellman theory. Energy 36, 633–646.

Xu, W., Scott, K., Basu, S., 2011. Performance of a high temperature polymer electrolyte membrane water electrolyser. J. Power Sources 196 (21), 8918–8924.

Yablonskii, G.S., Bykov, V.I., Gorban, A.N., 1983. Kinetic Models of Catalytic Reactions (in Russian). Nauka, Novosibirsk.

Yajima, T., Nagahama, H., 2015. Finsler geometry for nonlinear path of fluids flow through inhomogeneous media. Nonlinear Anal. Real World Appl. 25, 1–8.

Yamano, T., 2001. A possible extension of Shannon's information theory. Entropy 3, 280–292.

Yan, Z., 1993. Comment on "Ecological optimization criterion for finite-time heat engines". J. Appl. Phys. 73, 3583.

Yang, C.H., 1974. On the explosion, glow and oscillation phenomena in the oxidation of cabon monoxide. Comb. Flame 23, 97–108.

Yang, L., Epstein, I.R., 2003. Oscillatory Turing patterns in reaction-diffusion systems with two-coupled layers. Phys. Rev. Lett. 90, 178303-1-178303-4.

Yates, F.E., 1987. Self-Organizing Systems: The Emergence of Order, Broken Symmetry, Emergent Properties, Dissipative Structures, Life: Are They Related? Plenum Press, New York, pp. 445–457.

Yourgrau, Y., Mandelstam, S., 1968. Variational Principles in Dynamics and Quantum Mechanics, third ed. W.B. Saunders, Philadelphia.

Zahed, A.H., Zhu, J.-X., Grace, J.R., 1995. Modeling and simulation of batch and continuous fluidized bed dryers. Drying Technol. 13 (1-2), 1–28.

Zainetdinov, R.I., 1999. Dynamics of informational negentropy associated with self-organization process in open system. Chaos Soliton. Fract. 10, 1425–1435, Ch 11.

Zakari, M., Jou, D., 1998. A generalized Einstein equation for flux-limited diffusion. Physica A 253, 205–210.

Zalewski, M., 2005. Chaos i Oscylacje w Reaktorach Chemicznych, Praca doktorska, Politechnika Warszawska, 2005.

Zalewski, M., Szwast, Z., 2006. Obszary chaosu dla heterogenicznej katalitycznej reakcji chemicznej. Inżynieria i Aparatura Chemiczna 6s, 251–252.

Zalewski, M., Szwast, Z., 2007. Chaos and oscillations in chosen biological arrangements of the prey–predator type. Chem. Process Eng. 28, 929–939.

Zalewski, M, Szwast Z., 2008. Chaotic behaviour of chemical reaction in a deactivating catalyst particles, Proceedings of the 18th International Congress of Chemical and Process Engineering CHISA August 24–28, 2008, Praha, Abstr. P722, pp 1–10, on CD. http://www.chisa.cz/2008/.

Zamora, I., San Martín, J.I., San Martín, J.J., Aperribay,V., Eguía, P., 2009. Neural Network Based Model for a PEM Fuel Cell System. International Conference on Renewable Energies and Power Quality (ICREPQ'09). Valencia (Spain), April 15th to 17th , 2009.

Zanette, D.H., 1999. Statistical-thermodynamical foundations of anomalous diffusion. Brasilian J. Phys. 29 (1), 108–124.

Zangwill, W.I., 1969. Nonlinear Programming: A Unified Approach. Prentice-Hall, New York.

Zarei, S., Mirtar, A., Rohwer,F., Conrad, D.J., Theilmann, R.T., Salamon, P., 2012. Mucus distribution model in a lung with cystic fibrosis. Comput. Math. Method. Med. 2012, Article ID 970809.

Zarei, S., Mirtar, A., Andresen, B., Salamon, P., 2013. Modeling the airflow in a lung with cystic fibrosis. J. Non-Equilib. Thermodyn. 38 (2), 119–140.

Zhabotinsky, A.M., 1967. Oscillatory Processes in Biological and Chemical Systems. Science Publishing, Moscow.

References

Zhabotinsky, A.M., 1974. Concentrational Oscillations. Nauka, Moscow.

Zhang, N., Chao, D.F., Yang, W.J., 2013. Convective instability in transient evaporating thin liquid layers. J. Non-Equilib. Thermodyn. 27, 71–89.

Zhao, Z., Ou, C., Chen, J., 2008. A new analytical approach to model and evaluate the performance of a class of irreversible fuel cells. Int. J. Hydrogen Energ. 33, 4161–4170.

Zhu, F., 2014. Energy and Process Optimization for the Process Industries. Wiley, New York.

Zhu, R., Li, Q.S., 2002. Linear stability analysis of a reaction-diffusion model of solid-phase combustion. Theoretical Chem. Accounts 107, 357–361.

Zhu, X., Chen, L., Sun, F., Wu, C., 2001. Optimal performance of a generalized irreversible Carnot heat pump with a generalized heat transfer law. Physica Scripta 64 (6), 584–587.

Zhu, X., Chen, L., Sun, F., Wu, Chih., 2003. The ecological optimization of a generalized irreversible Carnot engine with a generalized heat transfer law. Int. J. Ambient Energy 24 (4), 189–194.

Zhu, X., Chen, L., Sun, F., Wu, C., 2004. The ecological optimization of a generalized irreversible Carnot engine in the case of another linear heat transfer law. Advances in Finite Time Thermodynamics: Analysis and Optimization, Editor: Lingen Chen and Fengrui Sun, Nova Science Publishers, New York, pp. 29–40.

Zhuravlev, V.A., 1979. Thermodynamics of Irreversible Processes: Problems and Solutions. Nauka, Moscow, 1979 (in Russian).

Zhyuan, G., Thirumalai, D., 1992. Folding kinetics of proteins: a model study. J. Chem. Phys. 97 (1), 525–535.

Ziębik, A., 1995. Applications of second law analysis in industrial energy and technological systems, Proceedings of the Conference on Second Law Analysis of Energy Systems: Towards the 21st Century, Rome, pp. 349–359.

Ziegler, H., 1963. Some extremum principles in irreversible thermodynamics with application to continuum mechanics. In: Sneddon, I.N., Hill, R. (Eds.), Progress in Solid Mechanics, vol. IV, North-Holland, Amsterdam.

Ziegler, Th., Richter, L.G., 2000. Analysing deep-bed drying based on enthalpy-water diagrams for air and grain. Comput. Electr. Agric. 26, 105–122.

Ziman, J.M., 1960. Electrons and Phonons, The Theory of Transport Phenomena in Solids. Clarendon, Oxford.

Zimm, B.H., 1956. Dynamics of polymer molecules in dilute solution: viscoelasticity, flow birefringence and dielectric loss. J. Chem. Phys. 24, 269–278.

Žnidaršič-Plazl, P., Plazl, I., 2011. Microbioreactors. In: Comprehensive Biotechnology (Second Edition), vol. 2: Engineering Fundamentals of Biotechnology, pp. 289–301.

Zotin, A.I., Zotina, R.S., 1967. Thermodynamic aspects of developmental biology. J. Theor. Biol. 17 (1), 57–75.

Zumofen, G., Blumen, A., Klafter, J., 1989. Reaction kinetics in disordered systems: hierarchical models. In: Plath, P.J. (Ed.), Optimal Structures in Heterogeneous Reaction Systems. Springer, Berlin, pp. 65–81.

Żyła, R., Strumiłło, Cz., 1981. Application of heat pumps for the purpose of decreasing the energy use in drying. In: Stopinski, T. (Ed.), Materials of IV Symposium on Drying. Instutut Chemii Przemysłowej, Warsaw.

Glossary

Principal Symbols

A Vector of chemical affinities (J/mol)
A Matrix in linear set of perturbed OD equations (of an indefinite sign)
A Variable area perpendicular to fluid flow (m²)
A_i ith Substance as a chemical reagent (mol)
A_0 Constant transfer area projected on axis y (m²)
A, A^{class} Physical action, generalized and classical exergy (J, J/m, kJ/kg)
A_j Chemical or electrochemical affinity of jth reaction (J/mol)
A_j^s Affinity of jth reaction in the entropy representation (J/mol per K)
A^k Total exchange area at stage k (m²)
$\mathcal{A}^n(\mathbf{x}^n, \mathbf{u}^n, t^n, \theta^n)$ Criterion of stage optimality (various units possible)
a Thermostatic matrix of capacities (units of entries depend on variables used)
a Temperature power exponent (–)
a Universal constant (–)
a_i Activity of ith species (mol/m)
a_v Exchange area per unit volume (m⁻¹), solid surface area per unit volume (m⁻¹)
$a_0 = 4\sigma/c$ Radiation constant related to the Stefan–Boltzmann constant (J/m per K⁴)
B Dissipation matrix of a definite (eg, negative) sign
B Spalding's driving force (–)
B Exergy content J (kJ)
b_g, b_s Specific exergy of gas and gas in equilibrium with solid (kJ/kg)
b_v Exergy per unit volume (kJ/m³)
C Molar concentrations vector (mol/m), matrix of second derivatives of entropy
C, c Unit prices of controlled and controlling phase, respectively ($/kg)
C_g Specific heat of moist gas per unit mass of dry gas (J/g per K)
$C_{n+1},...C_r$ Invariants determining the equilibrium (pseudo-equilibrium) surface
C_v Heat capacity at a constant volume (kJm⁻³K⁻¹)
c Specific heat (J/g per K, J/m³ per K, J/mol per K), bending constant for a frictional ray
c light speed in vacuum (ms⁻¹)
c_i Molar concentration of ith component (mol/m)
c_s Specific heat of absolute dry solid (J/g per K)
c_m Mass capacity of active component of fuel (J⁻¹)
c_w Specific heat of moisture in liquid state (J/g per K)
c_0 Propagation speed of thermal waves (m/s)
D Unknown symmetric matrix
D Diffusion coefficient (m²/s)
D Diameter (m⁻¹), diffusion coefficient (m²/s), profit variable ($)
E Electric field vector (kg m A/s³)
E Total energy or energy type function (J)
$E(\mathbf{X}, \mathbf{D})$ Energy in terms of classical variables $\mathbf{X}$ and nonequilibrium variables $\mathbf{D}$

Glossary

E^0 Relativistic rest energy in proper frame (J)
E^0, E_0 Nernst ideal voltage and idle run voltage, respectively (V)
e Unit energy, specific internal energy (J/kg), economic value of unit exergy ($/J)
e_v Energy of unit volume (J/m)
F Force (N), specific external force (N/kg), vector of transfer potentials
$F, \mathcal{F}$ Free energy and extended free energy, respectively (J)
F_v, f Density of total free energy and specific free energy, respectively (J/m, J/kg)
$\mathcal{F}$ Area of surface perpendicular to flux direction (m^2)
F_k Potentials or components of vector $\mathbf{F} = (T^{-1}, \mu_1 T^{-1}, \ldots -\mu_n T^{-1})$
$F^n(x^n, \lambda)$ Optimal drying function at stage n (kJ/kg)
$f^n(x^n)$ Optimal performance function of a dynamic programming algorithm
f Rate vector with components $f_1, \ldots f_k \ldots f_s$
f_0, f_i Intensity of generalized profit and process rates
G, G^{ik} Energy–momentum tensor
G Dry gas mass flux (kg/s), molar flux (mol/s), shear modulus (Pa)
G Gibbs energy flux driving chemical engine (J/s)
$\dot{G}$ Mass flux density (kg/m per s), resource flux (chapter: Power Limits in Thermochemical Units, Fuel Cells, and Separation Systems) (g/s, mols^{-1})
g Gradient vector, constraining vector function
g Spalding's coefficient (Reynolds flux) (kg/m per s), gravity acceleration (m/s^2)
g_1, g Partial and overall conductances (J/s per K^a)
$g/\dot{G}$ Nondimensional Stanton number as ratio of Reynolds flux g and mass flux $\dot{G}$
$g_1, g_2, \ldots g_n$ Process rates corresponding with irreducible coordinates $x_1, x_2, \ldots x_n$
g_k, g_{jk} kth Constraint and metric tensor, respectively
H Magnetic field, flux density of radiative entropy (J/m^2 per K per s)
$\mathbf{H}_i$ Biot vector for flux $\mathbf{J}_i$
$\mathbf{H} = (h_{jk})$ Hessian matrix of second derivatives
H Thermodynamic enthalpy of the system (J, kJ)
$H(\mathbf{x}, \mathbf{u}, \mathbf{p}, t)$ Hamiltonian function of a continuous process
H_σ "Thermodynamic" Hamiltonian based on entropy dissipation functions (J/K per s)
$H^{n-1}(\mathbf{x}^n, \mathbf{u}^n, \mathbf{p}^{n-1}, t^n)$ Hamiltonian function of a discrete process at stage n
H_{TU} Height of transfer unit (m)
h Height of fluidizing layer, distance from inlet (m), hydraulic head $P(\rho_m g)^{-1} + z$
h Specific enthalpy (kJ/kg), Planck constant, numerical Hamiltonian (J/m^3 per K)
h_v Enthalpy of unit volume (J/m^3)
h_σ Hamiltonian density in entropy units (J/m^3 per K)
h^{ik} Projection tensor
I Inertial matrix
I Flux of inert component (mol/s), fluid flux passing through the system (kg/s)
$\mathbf{I}_k$ Unidirectional flux of kth component (kg/s)
I Solid enthalpy per unit mass of absolute dry solid (kJ/kg)
I_0 Enthalpy of inlet solid, dry solid base (kJ/kg)
I^n Solid enthalpy at nth stage of a cascade (kJ/kg)
I_s Specific enthalpy of solid phase (kJ/kg)
I_0 Enthalpy of inlet solid (kJ/kg)
i Molar flux density of electric current (mol/m^2 per s)
i Gas enthalpy per unit mass of absolute dry gas (kJ/kg)
i_g Enthalpy of controlling gas (kJ/kg)
i_0 Inlet gas enthalpy (dry gas base) (kJ/kg)
i_s Enthalpy of gas in equilibrium with solid (kJ/kg)
i Electric current density (Am^{-2})

Glossary

i Molar density of electric current (chapter: Variational Approaches to Nonequilibrium Thermodynamics) (mol/s)
i_B Enthalpy of pole B, enthalpy of two-phase mixture (kJ/kg)
i_c Partial enthalpy of moisture in liquid state (kJ/kg)
i_p Partial enthalpy of moisture in vapor state (kJ/kg)
i_{ps} Partial enthalpy of vapor in equilibrium with solid (kJ/kg)
i_{ts} Enthalpy of gas with parameters t and X_s (kJ/kg)
i'_s Effective enthalpy of gas in equilibrium with solid (kJ/kg)
i_0 Inlet gas enthalpy (kJ/kg)
i_s Enthalpy of gas at equilibrium with solid (kJ/kg)
i_g, i_s Specific enthalpy of gas and gas at equilibrium with solid (kJ/kg)
J Column matrix of independent fluxes
J Mass flux density, conserved mass current (kg/m² per s)
$\mathbf{J}_k$ Mass flux density of kth component (kg/m² per s)
$\mathbf{J}_i$ Molar flux density of ith component (mol/m² per s)
$\mathbf{J}_i$ Relativistic particle four-flux density (mol/m² per s)
$\mathbf{J}_e, \mathbf{J}_q$ Energy flux density and heat flux density, respectively (J/m² per s)
$\mathbf{J}_s, \mathbf{j}_s$ Densities of total and diffusive entropy flux, respectively (J/K per m² per s)
j Conductance ratio, g_1/g_2 (–)
K Wave four-vector ($\mathbf{k}, k_t$)
K'_g Modified overall mass transfer coefficient (kg/m² per s)
K Chemical equilibrium constant, kinetic energy, cost, drying coefficient (h^{-1})
k Hydraulic conductivity related to gradient of hydraulic head
k Thermal conductivity (W/K per m), chemical rate constant
k_B Boltzmann constant ($1.380,658 \times 10^{-23}$ Nm/K)
$\mathbf{k}^i$ Direction vector in ith iteration
$\mathbf{L} = L_{ik}$ Onsager's matrix of phenomenological coefficients
L Length (m), Lagrangian (J), latent heat (kJ/kg), vertical distance in Fig. 7.2
$Lq = \lambda T^2$ Onsager coefficient for heat transfer
L_σ Dissipative kinetic potential
$Le = K'_g C_g \alpha^{-1}$ Lewis factor (–)
l Transfer area coordinate (m), length coordinate (m), Lagrangian cost
M_i Molar mass of ith species (kg/kmol)
m Mass, mass of particle (kg)
$\dot{m}$ Mass flux (kg/s)
$\dot{m}'' = gB$ Spalding's mass flux or the product of g and B
N Number of moles, number of particles (–)
N Total number of stages, number of transfer units (–)
N^k Cumulative flux of mole number for stages 1,2 ...k (mol/s)
n Current stage number (–), number density (m^{-3}), molar fuel flux (mol/s) (chapter: Thermodynamic Controls in Chemical Reactors)
n Flux of mole numbers (mol/s), number of components (–), refraction coefficient
P Pressure tensor (Pa)
$P(\mathbf{X}, t)$ Probability distribution function (probability density p)
P Thermodynamic pressure (Pa), momentum variable, gauged optimal profit (chapter: Variational Approaches to Nonequilibrium Thermodynamics)
P, p Cumulative and local power output, respectively (J/s), momentum type integral $\partial R/\partial y$
P^n, p^n Cumulative and local power output for nth stage subprocess (J/s)
P^n Optimal gauged profit at nth stage in terms of the state $\mathbf{x}^n$ and time t^n (chapter: Variational Approaches to Nonequilibrium Thermodynamics)
p Vector of costate variables in Maximum Principle
$p_s = \partial L / \partial v_s$ Diffusive thermal momentum (Js/m)

Glossary

P Power output (J/s), photon flux constant (cm^{-2} K^{-3} s^{-1})
p_{i_0} Adjoint (costate) variables, phase space variables, probabilities
P_m^0 Molar constant of photons density (molm^{-2}K^{-3}s^{-1})
Q Electric charge (C), quenching flux (kg/s), mass of transferred fluid (kg)
Q Generalized (total) heat flux including effect of mass transfer (J/s)
Q Total heat flux per unit surface area of apparatus (kJ/m^2 per s)
Q Heat power (Js^{-1})
Q_H, Q_L Heat flux density at higher and lower temperature, respectively (kJ/m^2 per s)
Q^f Optimal performance function in terms of the complete final state ($x_0^f, \mathbf{x}^f, t^f$)
$\mathbf{q}$ Heat flux density vector (J/s per m^2)
$q = Q(S+G)^{-1}$ Heat flux per unit mass of gas–solid mixture (kJ/m^2 per s)
q Sensible heat flux between a stream and power generator (J/s)
q_1 Driving heat in the engine mode of the stage (J/s)
R Rayleigh number
$\mathbf{R}$ Matrices of chemical and electrochemical resistances
R Universal gas constant (J/K per mol)
$R^{1,2}$ Total cumulative resistance for frictional path between points 1 and 2
$R(x, y)$ Minimum resistance potential in terms of end coordinates
$R^n(\mathbf{x}^n, t^n)$ Optimal function of cost type in terms of state, time, and number of stages
$R(\mathbf{X}, t)$ Optimal work function of cost type in a continuous process (J/kg)
$\mathbf{r}$ Radius vector (m), vector of reaction rates (mol/m^3 per s), chemical flux vector (g/m^3 per s)
$r_j, \tilde{r}_j$ Chemical and electrochemical rates of jth reaction, respectively (mol/m^3 per s)
S Solid mass flux density, dry solid base (kg/m^2 per s)
S_o Dry solid mass in the system (kg, kg/m^2)
S Entropy (J/K), specific entropy (J/kg per K), performance criterion
S^T Legendre transform of entropy (J/K)
S^i Entropy four-flux (J/K per s)
S Solid entropy, entropy of controlled phase (J/K)
S_σ Specific entropy production per unit flow (J/K per kg) (J/K per mol)
S_σ Specific entropy production rate (J/K per kg per s)
s, s_v Specific and volumetric entropy, respectively, (J/K per g, J/K per m^3)
$\mathbf{T}(T, \tau)$ Vector composed of the temperature and the number of heat transfer units
$\mathbf{T}^n(\mathbf{x}^n, \mathbf{u}^n)$ Vector state transformation at nth stage
T Temperature of controlled phase (eg, solid, resource or fuel stream) (K)
T absolute temperature (K)
T_1, T_2 Bulk temperatures of reservoirs 1 and 2 (K)
T_H, T_L Higher and lower temperature, respectively (K)
T_*^{-0} Relativistic thermal potential (K)
T^- Nonrelativistic thermal potential (K)
T_s Temperature of solid phase (K)
T^e Temperature of environment (K)
T^n Temperature of phase leaving stage n (K)
T^n Temperature of phase leaving stage n (K)
T_1, T_2 Bulk temperatures of reservoirs 1 and 2 (K)
$T_{1'}, T_{2'}$ Temperatures of fluid circulating in engine, active temperatures (K)
T^e Constant equilibrium temperature of environment (K)
T' Carnot temperature, effective temperature of controlling phase (K)
$\dot{T} = u$ Rate of temperature change in nondimensional time (K)
t Temperature of gaseous phase (°C, K)
t Physical time, contact time, holdup time (s)
t_s Temperature of solid phase (°C)

Glossary

t_w Wet bulb temperature (°C, K)
u   Vector of hydrodynamic (barycentric) velocity (m/s)
u $= (u_1, u_2, ..., u_r)^T$ Control vector, vector of independent transfer potentials
U Velocity of basic flow (m/s)
U^i Relativistic four-velocity in particle frame
u Controlled direction dy/dx, internal energy (J), variable of fuel consumption
$u^n = \Delta T^n/\Delta \tau^n$ Rate of temperature change in nondimensional time (K)
$\mathbf{u}_s \equiv \mathbf{J}_s/\rho_s$ Absolute velocity of entropy transfer (m/s)
V Lapunov function, overall or specific (J/K, J/K per kg)
V Volume (m³), operational voltage (V), potential, maximum work function (J/mol)
$V^n(\mathbf{x}^n, t^n) = \max S$ Optimal function of profit type in terms of state, time, and number of stages
v   Diffusion velocity, relative velocity, velocity of resource stream (m/s)
$\mathbf{v}_s = \mathbf{j}_s/\rho_s$ Velocity of entropy transfer in fluid frame (m/s)
$v_s^i = U_s^i - U^i$ relative four-velocity of entropy (m/s)
v Value function (chapter: Variational Approaches to Nonequilibrium Thermodynamics), scalar velocity (m/s)
$v = \rho^{-1}$ Specific volume (m³/kg)
$W, \mathcal{W}$ Work and cumulative work, respectively (J), negative entropy excess
W Absolute moisture content of solid, dry solid base (g/kg) (kg/kg)
$W = P/G$ Total specific work or total power per unit mass flux (J/kg)
W Molar work at flow, total power per unit mass flux of inert (J/mol)
W_e Moisture content of solid in equilibrium with gas (g/kg) (kg/kg)
W_0 Moisture content of inlet solid (gk/g) (kg/kg)
W_p Initial moisture content of inlet solid (g/kg) (kg/kg)
w   Vector of Lagrangian multipliers of balance laws
w Specific or normalized work (J/kg)
X   Vector of initial Lagrangian coordinates in thermomechanical problem
X   Vector of independent thermodynamic forces, set of variables
X Concentration of active component in fuel, moles per mole of inert (mol/mol)
X Absolute gas humidity (g/kg) (kg/kg)
X_B Humidity of pole (g/kg) (kg/kg)
X_g Humidity of controlling gas (kJ/kg)
X_0 Inlet gas humidity (g/kg) (kg/kg)
X_s Humidity of gas in equilibrium with solid (g/kg) (kg/kg)
X_s Moisture content in solid (kg/kg)
X_α Thermodynamic force
$\dot{X} = dX/d\tau_1$ Rate change of fuel concentration in time τ_1
x   Radius vector and enlarged radius vector, respectively, vector of spatial coordinates (m)
$\mathbf{x} = (x_1, x_2, \ldots x_i \ldots x_s)$ State vector of a dynamical process
$\tilde{x} = (x_1, x_2, \ldots x_i \ldots x_s, t)$ Enlarged state vector of a control process
x Coordinate perpendicular to the resistance gradient
x Mass fraction (kg/kg), reaction extent, transfer coordinate
x Molar fraction of active component in the fuel (mol/mol)
$\mathbf{x}_i$ Concentration of ith component in moles per unit mass (mol/kg)
x_i Concentration of ith species (mol/kg)
$x_{1'}$ Molar fraction of reactant in chemically active part of the system (mol/mol)
$x_{2'}$ Molar fraction of product in chemically active part of the system (mol/mol)
x_0 Performance coordinate in dynamic optimization problem
Y_b Absolute humidity in bubble phase in Fig. 6.3 (kg/kg)
Y_g Absolute humidity of controlling phase (gas) in Fig. 6.3 (kg/kg)
Y_s Humidity of gas at equilibrium with solid in Fig. 6.3 (kg/kg)

Glossary

- y Coordinate tangent to the resistance gradient (m)
- $y_i = g_i/g_0$ Stream flux fraction of ith stream (chapter: Thermodynamics and Optimization of Practical Processes) (mol/mol)
- z Adjoint variable, vertical coordinate (m)

Greek Symbols

- α Disturbance wave number (m^{-1}), angle
- α Overall heat transfer coefficient (J/m^2 per s per K)
- α_1, α' Partial and overall heat coefficients referred to cross-sections (J/m^2 per s per K)
- α' Modified overall heat transfer coefficient (J/m^2 per s per K)
- β Resistance ratio, ρ_2/ρ_1, frequency constant (s^{-1}), radiation coefficient $\beta = \sigma a_v c_h^{-1}(p_m^0)^{-1}$ (s^{-1})
- β Relative humidity of gas (s^{-1})
- β_s Relative humidity of gas in equilibrium with solid (s^{-1})
- β' Overall mass transfer coefficient (mol/m^2 per s)
- β_i Multipliers, parameters
- ϕ Electric potential (V); molar action (–)
- Γ Complexity (–)
- γ Cumulative conductance of the system (J/s per K^a), coefficient of exponential change (–)
- Δ Disequilibrium correction, increment, deviation angle (–)
- δ Thickness (m), effective diameter (m), perturbation, variation, spatial operator
- ε Total energy flux (J/s), mechanical displacement (m), coefficient of phase changes (–)
- ε Emissivity (–)
- ε_i Displacements, parameters
- Φ Internal irreversibility (–), dissipation function (J/K per s), dimensionless phase $\eta k_B/h$
- φ Relative concentration (–)
- χ Chemical displacement, time constant $\chi = \rho c_v(\alpha' a_v)^{-1}$ (s)
- η First-law thermal efficiency p/q_1 (–), thermal eiconal (Ks)
- θ Free interval of an independent variable (s, –)
- θ^n Time interval at stage n (s, –)
- κ_j Modified progress variable of jth reaction
- Λ_σ Dissipative Lagrangian (J/K per s)
- λ Heat conductivity, time adjoint, characteristic length
- μ_k Chemical potential of kth component (J/kg)
- μ_1 Molar chemical potential of active component (J/mol)
- μ' Carnot chemical potential for active component of fuel (J/mol)
- μ' Coefficient of outlet gas utilization (–)
- ν Stoichiometric coefficient (–), frequency constant, viscosity (Pa s)
- $\nu_{ij} = \nu_{ij}^b - \nu_{ij}^f$ Resulting stoichiometric coefficient for ith species in jth reaction (–)
- Ω Mathieu thermodynamic potential PV/RT (–)
- ω Frequency constant (s^{-1})
- Π, Π_j Reaction potential vector and potential of jth reaction (J/mol, –)
- Π_j Unidirectional part of chemical affinity for jth reaction (J/mol, –)
- π_σ Dissipative momentum
- ρ Mass density (kg/m^3), specific frictional resistance
- ρ_e Energy density (J/m^3)
- ρ_s Entropy density (J/K per m^3)
- ρ_e^0 Rest frame energy density (J/m^3)
- ρ_s^0 Rest frame entropy density (J/K per m^3)
- ρ_n^0 Rest frame particle density (m^{-3})
- ρ_m Bare mass density (g/m^3)

Glossary

σ Stefan–Boltzmann radiation constant (J/m² per s per K⁴), production density (g/m³per s)
σ_{ik} Total stress tensor
σ Entropy production (J/K per s), surface tension (Pa)
σ_s Density of entropy production (J/K per m³ per s)
τ Characteristic time, relaxation time, average time between collisions(s)
τ Nondimensional time, number of transfer units (x/H_{TU}) (–)
τ^{ik} Viscous stress (Pa), disequilibrium relativistic stress (Pa)
Ξ Imperfection fraction (–)
ξ Process intensity factor (–)
$\zeta = \mu_{1'} - \mu_{2'}$ Efficiency of chemical engine as active part of chemical affinity (J/mol)
ζ_{mp} Efficiency of chemical engine at maximum power point (J/mol)
ξ_i Adjoint variable for mass flux $\mathbf{J}_i$
ξ_j Reaction progress variable (mol), internal variable,
Ω Order variable (–)
∇ Nabla differential operator
$\delta^2 S$ Second differential of entropy

Subscripts

b   Bubbles, backward
C   Carnot state (open circuit)
e   Energy, equilibrium
F   Feed
g   Gas bulk
H   High
i ith state variable
j   Reaction number
k kth component
L   Low, L-surface
m   Molar quantity
P   Product
p   interphase
S S-surface
s   Solid bulk, entropy
W   Water
v   Per unit volume
σ Dissipative quantity
0   Inlet state, idle run, reference state, profit variable
1, 2   First and second, initial and final
*   Modified cost or profit

Superscripts

a   Power exponent
e   Environment, equilibrium
f   Final state and time
i   Initial state and time
int   Internal production
k or n   Number of kth or nth stage
m   Power exponent
s   Entropy representation

Glossary

T   Transpose matrix, transform
'   Carnot variable, Carnot state, modified quantity
0   Ideal (equilibrium) voltage

Abbreviations and Acronyms

CIT   Classical irreversible thermodynamics
CNCA   Chambadal–Novikov–Curzon–Ahlborn engine
DMFC   Direct methanol fuel cells
DP   Dynamic programming
EGM   Entropy generation minimization
EIT   Extended irreversible thermodynamics
FC   Fuel cells
FTT   Finite time thermodynamics
GA   Genetic algorithms
GEM   Gibbs equilibrium manifold
GES   Gibbs equilibrium submanifold
HEN   Heat exchanger network
HJ   Hamilton–Jacobi equation
HJB   Hamilton–Jacobi–Bellman equation
LP   Linear problem/programming
MEN   Mass exchanger network
NLP   Nonlinear problem/programming
OCT   Optimal control theory
PEMFC   Polymer electrolyte membrane (PEM) fuel cells
THP   Thermohydrodynamic potential

Subject Index

A

Absolute stoichiometry, 425
Absurd physical property, 163
Abundant energy resources, 609
Acoustic dispersion, 83, 168, 185
Action approach(es), 123, 131, 156, 622
  complementary, 187
Activated state, 438
Activation energy, 572
Activation–passivation reaction(s), 486
Active transport, 88, 178, 196, 249, 446, 524, 526
Adiabatic fluid, 613
Adiabatic process(es)
  in monoatomic gases, 10
Adiabatic sorption, in countercurrent multistage fluidized bed, 303
Advanced composites, 267
Advanced oxidation process (AOP), 608
Affinities, 12, 142, 266, 426, 435, 441, 487, 586
Aggregation process(es), 412
Air-conditioning systems, 607
Algorithm(s) for work optimization, 597
Alkali metal thermal-to-electric conversion (AMTEC) cell, 493
Analytical solution, 47, 66, 293, 318, 334, 371, 399
Animal locomotion and pulsating physiologies, 537
Anisotropy, 584
Anodic and cathodic currents, 433, 487

Anomalous diffusion, 218
  coefficient of, 212
Anthony's variational method, 134
AOP. *See* Advanced oxidation process (AOP)
Approximation method(s), 100, 129
Arrhenius–Kramers description, 431
Arrhenius-law reaction, 452
Artificial neural network (ANN), 413
Asterisk-free intensities, 624
ATP synthesis, 524, 526
Augmenting procedure, 613
Autocatalytic reaction(s), 86, 249, 255, 424, 442, 450, 451
AVCO approach, 414
Averaged optimization, 364

B

Balance equation(s), 299, 358
Ballistic regime, 171
Bare matter, 200, 621, 623–625, 627
Batch distillation, 348
  coupling between heat and mass transfer, 348
Batch fluidization dryer, 302
Bateman's exponential relaxation, 187
Bateman's structure(s), 118
Bead–spring–type macromolecular models, 178
Bearman-Kirkwood equations, 179, 196
  generalized transport, 179
  ionic partial viscosities, 179
  statistical mechanical theory of transport equations, 196

Bear's refraction law, 332
Bejan's constructal theory, 321, 343, 388–393
  of organization in Nature, 388
Bejan's optimizing, 354
Bekenstein formula, for entropy, 586
Bellman principle
  dynamic programming (DP), 367
Bellman's dynamic programming (DP), 323, 364, 399, 475, 479
Bellman's recurrence equation, 333
Belousov-Zhabotinsky (BZ) reaction, 255, 269, 559
Bénard cells, 278, 281
Benard instability, 278
Benard problem, 281
Bernoulli's equation, 561
Beta-alumina solid electrolyte (BASE) tubes, 493
B-form DNA, 515
Bianchi-type model, 110, 615
Biocalorimetry, 515
Biochemical cycles in living cells, 539
Biochemical engineering, 515
Biochemical network, 520
Bioelectrochemistry, 526
Bioenergetics, 522–526
  and linear nonequilibrium thermodynamics, 523
Biofuel(s), 360, 411, 528
Biomass, 410
  for energy generation, 404
  gasification, 405
Biomass-to-synthetic natural gas (SNG) conversion process, 411, 528
  exergy analysis for, 411

Subject Index

Biophysics, 522–526
Biopolymer(s), 515
Biotechnology, 515
Biothermodynamics, 515, 516
  of live cells, 515
Biot's heat vector, 133
Biot's number(s), 183, 297
Biot's potentials, 154
Biot's thermal displacement vector, 139
Birth-death process, 228
Black-body radiation, 585, 602
Black-radiation fluid, 72
Bogolyubov inequality, 225
Bohmian mechanics, 30
Boltzmann constant, 45, 205, 222
Boltzmann definition, for entropy, 209
Boltzmann distribution, 104, 222
Boltzmann entropy, 259
Boltzmann equation, 34, 76, 78, 100, 127, 186
Boltzmann equation, for phonon gas, 165
Boltzmann–Gibbs exponential formula, 207
Boltzmann–Gibbs formulation, 225
  of thermodynamics, 216
Boltzmann–Gibbs–Shannon (BGS), 246
  entropy, 84
Boltzmann–Gibbs statistical mechanics, 224
Boltzmann–Gibbs statistics, 224, 227
Boltzmann kinetic equation, 166, 620
  by Grad's or Enskog's methods, 61
Boltzmann principle, kinetic analogues of, 43
Boltzmann's constant, 206
Boltzmann's H-theorem, 76
Boltzmann's thought experiment demonstrating the probability of disorder, 550
Boltzmann's universal constant, 621
Boltzmann transport equation (BTE), 189
Bose–Einstein statistics, 584

Boundary condition(s), 31
Boundary layer, 39, 86, 184, 264, 287, 308, 345, 401, 457, 562
Boycott effect, 536
Brayton engine, 605
Broglie microthermodynamics, 117, 616
Brownian motion, 194
  of nonlinear oscillators, 260
Brownian particle(s), 546
Brownian ratchets, 262
Bulk viscosity, 27, 28, 95, 111, 615
Butler–Vollmer equation, 433, 488
Butler–Vollmer kinetics, 421, 486
BZ reaction, 86

C

Caldirola–Kanai equation, 118
Callaway's approximation, to collision, 165
Callen's postulates, 31, 36, 61
Callen's postulational thermodynamics to nonequilibrium, 149
Callen's principles, 41
Canonical densities, 623
Canonical Hamiltonian equations, 53
CAPD. See Computer-aided progress design (CAPD)
Carathéodory's principle, 244
Carbon monoxide oxidation, 446
Carnot control variable(s), 610
Carnot cycle, 10, 354, 355, 397
Carnot efficiency, 69, 347, 355, 414, 595, 603
Carnot engine(s), 9, 348
Carnot heat engine, 399
  cycles, 601
  system, 397
Carnot heat pump(s), 354
Carnot temperature, 67, 593–595, 610
Carnot-type heat engine, 399
Carnot variable(s), 417, 484, 595
Catalyst activity after regeneration, 576
Catalyst deactivation rate, 571
Catalyst recycle ratio, 578
Catalysts deactivation, 354
Catalytic gas–solid systems, 246

Catalytic reaction(s), 572
Cattaneo equation, 151, 154
  for diffusive flux density, 268
Cattaneo form, 58
CExC. See Cumulative exergy consumption (CExC)
CFD. See Computational fluid dynamics (CFD)
CGAM problem, 362
Chambadal–Novikov–Curzon–Ahlborn engine, 476
Chaos theory, 14, 274
Chaotic, 14, 90, 108, 117, 193, 216, 218, 227, 231, 256, 268, 274, 275, 450, 451, 536, 556–558
Chapman-Enskog procedure, 79, 186
Chemical and process engineering (ChPE), 413
Chemical coupling(s), 516
Chemical efficiency, 378, 452, 459, 460, 461, 463, 467, 469, 470, 471, 496
Chemical equilibria, 423
  calculation of, 423
Chemical flux vs. Carnot reaction potential, 496
Chemical lens, 273
Chemical–nonchemical coupling, 521, 525
Chemical Ohm's law, 435
Chemical potential(s), 18, 25, 51, 83, 88, 179, 196, 237, 335, 384, 429, 432, 460, 517, 602, 619, 624
Chemical process(es)
  optimization problems, 347
Chemical reactors, 521
Chemical resistance(s) (CR), 435, 438
Chemical storage, of hydrogen in liquid fuels, 489
Chemical wave(s), 271
Chemicophysical feature, 541
Chimney-Dependent Solar Crop Dryer (CDSCD), 596
Chlorite–iodide–malonic acid reaction, 247
ChPE. See Chemical and process engineering (ChPE)
CH_4 synthesis, 528

Subject Index

Circulating fluid, 66, 317, 404, 405, 506, 593
CIT. *See* Classical irreversible thermodynamics (CIT)
Classical equilibrium thermodynamics, 347
Classical irreversible thermodynamics (CIT), 581
Classical mechanics, 25, 129, 130, 142, 261, 369, 620
Classical Onsagerian formalism, 488, 525
Clausius–Duhem inequality, 98, 197
  for the recoverable specific entropy, 568
Clausius entropy, 259
Clausius inequalities, 28, 253
  for entropy, 253
Clebsch variables, 156
Coal-fired steam power plant, 361
Cocurrent dryer(s), 299
Coefficient of performance (COP), 5
Cogeneration, 353, 362, 397, 402, 407, 490–492, 563
Cogeneration-based distric energy system, 407
CO_2–glucose–ethanol cycle, 527
Coldness, 22
Collector's theory, 591
Collector temperature, 591
Combustion, 586
Combustion engine(s), 291, 318, 350, 378, 501, 513
Complex system(s), 17, 51, 91, 193, 274, 321, 379, 536, 546, 554, 557, 558, 581
Composite variational principles, 134
Comprehensive thermodynamic theory of stability of irreversible processes (CTTSIP), 285
Computational fluid dynamics (CFD)
  two-dimensional, 316
Computer-aided progress design (CAPD), 315
Conservation law(s), 25, 29, 41, 76, 107, 137, 154, 174, 199, 561, 613, 614, 616, 620, 628

Fourier law, 620
  gravitational metrics, 628
  relativistic Lagrangian, 614
Constructal theory, 390, 393
Consumption of resources, 347
Contemporary thermodynamics, for engineering systems, 1
  classical/quasi-classical complex systems, 91–94
  complex systems, extended thermodynamics of, 95–112
  complex thermodynamic systems, 91
  energy conversion and efficiency, 2
  heat engines, 4
  irreversible phenomena, 7
  property, 3
  state of system, 4
  thermodynamic path, 4
  thermodynamic process, 4
  thermodynamic system, 3
  energy transformation laws, 2
  extended irreversible thermodynamics (EIT), 17
  finite time thermodynamics (FTT), 9, 10
  Gibbsian theory, 11
  heat quasiparticles, classical motion of, 53–59
  kinetic equation(s), 16
    significance of, 75–79
  macroscopic principle, 13
  macroscopic theory, 1
  nonequilibrium thermodynamics, 15
    theory, 17–30
  property relationship laws, 2
  steam power-plant, as heat engine, 5
  thermodynamic geometries, 59
    equilibrium metrics, 59
    finite rate exergy, 65–74
    nonequilibrium metrics, 61–64
  thermodynamic systems,
    extremum properties of, 30
    conserved irreversible hamiltonian, 50–51
    dissipative dynamics, 43
    equilibrium problems, 31
    irreversible motions, 48–50

Jaynes' MEP approach, 31–38
  kinetic potentials/optimization criteria/maximum principle, 46
  nonstationary nonequilibrium problems, 42–43
  reversible motions, 48
  steady-state problems, 38–41
  variational model, 51
  transport/rate phenomena, 79–90
  transport/rate processes, thermodynamics of, 75
  vapor-compression refrigeration cycle, 5
Continuation algorithm, 416
Continuous HJB equations, 399
Continuous stirred tank reactor (CSTR), 247, 256
Continuum thermodynamics, 7, 13, 190
Contravariant thermodynamic tensors, 64
Controlling the magnitude of limit cycles, 447
Control of biological reactions, and decaying enzymes, 517–522
Cooling effect, 570
Cooling policies, 563
Cooling problem, 564
Cooling rate, 354
CO oxidation, 251
  on phase diagram, 272
COP. *See* Coefficient of performance (COP)
Copper pipes, 592
Copper plate, 592
Corn, drying characteristics of, 305
Cost-effective energy system, 362
Cost of catalyst regeneration, 573, 575–577
Cost-type optimization, 357
Countercurrent operation cocurrent operation, 250
Counterflow, 563
Coupling effects, 521
Covariant thermodynamic tensors, 64
CO_2 zero-emission, 353
Cramer–Rao inequality, 584
o-Cresol hydrogenation reaction, 571

705

Subject Index

Critical value, 86, 106, 255, 276, 282, 575
Crude oil distillation systems, 407
Cryogenic engines, of rockets, 291
Cryogenics, 358
CSTR. *See* Continuous stirred tank reactor (CSTR)
CTTSIP. *See* Comprehensive thermodynamic theory of stability of irreversible processes (CTTSIP)
Cumulative degree of thermodynamic perfection (CDP), 533
Cumulative exergy consumption (CExC), 353, 357, 533, 535
  and thermoeconomics, 415
Current design parameters, 588
Curse of dimensionality, 334
Curvature, 15, 31, 33, 59–61, 86, 185, 261, 416, 562
Curzon–Ahlborn efficiency, 9, 349, 473
Curzon–Ahlborn–Novikov (CAN) processes, 401
Cyclic model, of chemical engines, 454
Cylindrical pellets, 246

D

Dantzig's methods, for linear programming, 364
Darcy model, 491
Darcy's equation, 343
Darcy's flow(s), 328
Darcy's flux, 326
Darcy's law, 322, 341
Darcy's path problem, 334
Darcy's rays, 328, 330, 331, 338, 340
  law of bending, 333
Darcy–Weisbach equation, 341
Darwinian evolution, 548
Deactivation rate, 572
de Broglie microthermodynamics, 190
de Broglie theory, 616
Deep-bed drying, in enthalpy–water diagrams, 306

de Groot and Mazur, 7, 8, 18, 22, 41, 77, 83, 94, 131, 135, 137, 148, 208, 210, 242, 259, 425, 427, 521
Dense gas–solid multiphase flows, 233
Dense-phase fluidization(s), 308
Density functional theory (DFT), 237
Depletion number, 534
Detailed balance, 38, 77, 86, 274, 562
Deterministic equations, for nonlinear irreversible processes, 258
DFE. *See* Distance from equilibrium (DFE)
DFT. *See* Density functional theory (DFT)
Diffusion, classical/anomalous, 193
  anomalous diffusion, 202
  Boltzmann's constant, 205, 206
  classical picture of, 194–202
  Compte's/Jou's treatment
    conclusions of, 215
    fractal entropy/generalized diffusion equation, 208–213
    special cases, 214
    Tsallis theory, application of, 206–208
  fractal structure, 231
  introducing anomalous diffusion, 202–205
  multifractal systems, 202
  nonextensive thermodynamic system(s), 206
  statistical mechanics (SM), 227–228
  statistical physics, 229–232
  stochastic systems, 227–228
  Zanette's treatment, 216
    anomalous diffusion
      generalized formalism, 224–227
      maximum-entropy formalism, 222–223
    in nature, 216
    random-walk models, 219–222
    traditional formalism, 223–224

conclusions of, 227
Lévy flights, anomalous diffusion, 216
Diffusion dynamic, 221
Diffusion equations, 88, 196
Diffusion, of gas particle(s), 194
Diffusion process(es), 11, 136, 201, 203, 262, 429, 431, 471, 614
Diffusive mass transport
  non-Fickian theories of, 83
Direct methanol fuel cells (DMFC), 492
Direct numerical simulation (DNS), 345
Disequilibrium entropy, 163
Disequilibrium fields, stability analyses of, 236
Disequilibrium system(s), 1, 172, 195, 234, 236, 262, 263, 284
Disequilibrium thermodynamics, 247, 292
Dissipative fluid(s), 616
  stability and causality, 614
Dissipative flux(es), 615
Dissipative variational formulations, 163
Distance from equilibrium (DFE), 33
Distance-from-randomness, 110
Distillation column, 347, 348
  scheme of, 348
Diverse pseudo-Carnot structures, 417
DMFC. *See* Direct methanol fuel cells (DMFC)
DNA sequencing, 232, 262
DNS. *See* Direct numerical simulation (DNS)
DP. *See* Dynamic programming (DP)
Driving forces of transport, and rate processes, 426–430
Drying modeling, 295
Drying operation(s), in thermodynamic diagrams, 291, 294, 297
  associated drying problems, 315–319
  energy-consuming process, 291
  evaporation and condensation, 291

Subject Index

experimental data, graphical classification of, 311–314
extraction of moisture, 291
gas–moisture–solid system, 293
grain drying, in countercurrent, 308
graphical approach, 297
   fluidized drying of solid, 299–307
   single grain, 297–299
kinetic properties, 292
moisture extraction modeling, 294–297
outlet stream
   fluidizing systems with thermodynamic equilibrium, 307
Drying paths, qualitative picture of, 307
Drying phenomena, intrinsic "fingerprint" of, 294
Drying process(es)
   gas and granular solid, 234
   spatial-REA, 295
Drying thermomechanics, 295
Dupuit–Forchheimer equation, 341
Dynamical approaches formulate stability problems, 235
Dynamical energy, 478
Dynamical radiation engine(s), 478
Dynamical state equations, 600
Dynamical thermal systems selected results for, 477
Dynamic programming (DP), 367, 419, 599
   equation, 334
      numerical aspects of algorithms, 334
Dynamic programming method, 375
Dynamics of oscillatory trajectories terminating at a stable focus point or falling into a stable limit cycle in process of CO oxidation, 452

E

Earth's atmosphere, 598
Earth's crust, 322
Earth's winds, 598
Eckart's principles, 120
Eckart's theory
   energy-momentum tensor, 622
Ecological applications of exergy, 353
Ecology and ecological optimization, 531–536
Economic theory, 609
Ecosystem(s), 360, 531, 535, 544, 545, 552, 556, 583
EEA. See Extended exergy accounting (EEA)
EEA, to cogenerating plant, 415
EER. See Energy, efficiency rating (EER)
Efficiency equation(s), 419, 589, 590
EGM. See Entropy, generation minimization (EGM)
Ehrenfest's idea, of coarse-graining, 78
Einstein formula, 615
Einstein–Planck transformations, 624
   formulae, 618
Einstein probability, 615
Einstein's equations, 626, 627
Einstein's mass–energy relation, 188
Einstein's relation, 205
Einstein's relativistic transformation, 25
Einstein transport equation, 203
EIT. See Extended irreversible thermodynamics (EIT)
EIT-based approach, 96
EIT problems, 101
EIT theory, 107
Electrochemical cell, 485
Electrochemical engine, 350
Electrochemistry, 432
Electrolytes, 524
Electromagnetism, 587
Electromotive force (EMF), 488
End of mission (EOM) electric power, 493
Endoreversible cycle(s), 594
Endoreversible efficiency, 594
Endoreversible heat-to-power conversion systems, 109
Endoreversible thermal engine, 349
Endoreversible thermal machine(s), 417
Endoreversible thermodynamics, 609
Energy and electron concentration, 489
Energy balance, 591
Energy cogeneration system(s), 353
Energy conservation, 516
   equations for, 613
Energy-conserving membranes, 523
Energy consumers, 474
Energy-consuming process, 291
Energy conversion plants, 361
Energy-effcient production, 400
Energy efficiency, 590
   rating (EER), 6
Energy flow, 491
Energy limit, 599
Energy management, 353, 357
Energy-momentum tensor, 26, 626
   of thermohydrodynamic system, 613
Energy production, 349
Energy system(s), 8, 358, 362, 407
Energy transduction in biomembranes, 524
Energy transformation laws, 2
ENET. See Extended nonequilibrium thermodynamics (ENET)
Engineering processes, 347
Engines, 353
Enskog method, 77
Enthalpy, 362, 485
   emission flux, 588
Enthalpy–composition chart
   fluidized drying, steady nonadiabatic process of, 302
   for gas-water-solid system, 301
   two-stage, steady fluidized drying, 303
Enthalpy–concentration diagram, 296
Enthalpy diagram(s)
   gas and solid paths, 311
   steady nonadiabatic process of fluidized drying in, 302
Entropy, 2, 349, 358, 363, 417, 428, 562, 583, 585
   approaches, 11

Subject Index

Entropy (cont.)
  based models, 556
  of the diffusive fluxes, 117
  function, 115, 146
  generation, 344, 596
    in steady systems, 460
  generation minimization (EGM), 352
  related phase, 55
  source analysis, 163
Entropy production, 437, 561, 563, 565, 568, 580, 593, 609
  associated with material dissipation, 530
  and power yield in an electrochemical engine, 496
  in quantum dynamical systems, 108
  in thermal engine, 349
Entropy yields, maximization of, 223
Environmental temperature, influence on exergy efficiency, 590
Equilibrium conditions, 31
Equilibrium Gibbsian theory, 523
Equilibrium Gibbs surface, 187
Equilibrium problems, 31
  of thermodynamic stability, 235
Equilibrium statics, 17
Equipartition of entropy production, 38
Erdmann–Weierstrass type, 371
Ergun's equation, 341
Ericsson heat engines, 399
ERT. See Extended relativistic thermodynamics (ERT)
Escherichia coli, 540
Estimate of minimum power, supplied to power consumers, 508
Euclidean diffusion equation, for diffusion, 203
Euclidean geometry, 330
Euler equations, 56, 183, 264
  of variational calculus, 371
  of variational problem, 46
Eulerian description, of heat conduction, 123
Eulerian–Eulerian multiphase approach, 567
Eulerian representation, of fluid's motion, 617

Euler-Lagrange equation, 116, 120, 140, 141, 145, 150–152, 336, 338, 397, 564, 600
  of entropy functional, 138
  of variational problem, 124
Euler solutions, 126
Evaporation, 292
Evaporation–condensation process, 319
Evaporator, distillation column, 348
Evolution, 543
Exact polarization curve, 404
Exergetic cost theory (ECT), 362
Exergetic performance, 409
Exergoeconomics. See Thermoeconomics
Exergoeconomic symbolic optimization (EXSO)
  for energy system, 362
Exergoenvironmental optimization, 353
Exergy analysis, 396, 586
Exergy and second law analyses of FC systems, 500–504
Exergy-based efficiencies, 528
Exergy consumption, 359
Exergy cost theory (ECT), 414
Exergy flux, 592
Exergy optimization, 357
Exothermic reaction, 452
EXSO. See Exergoeconomic symbolic optimization (EXSO)
Extended chemical systems, 100
Extended exergy accounting (EEA), 415
  theory, 415
Extended irreversible thermodynamics (EIT), 17, 99, 112, 163, 285
  Gibbs equation of, 35
Extended nonequilibrium thermodynamics (ENET), 196
Extended relativistic thermodynamics (ERT), 615
Eyring's absolute rate theory, 435

F

Falling particle(s), 299
Falling solid particle(s)
  countercurrent drying of, 309
Farid's model, 297, 313

FC application(s), 407
FC–heat engine hybrid systems, 494
FC system, 407
Fermat-like principle, 322
Fermat principle, 55, 56, 270, 342
  for propagation of light, 331
Fermat problem, 587
Fermat-type principle, 325, 389
  illustration of, 324
Fermi–Dirac statistics, 584
Fiber-reinforced polymer (FRP) structural shapes, 124
Fick diffusion, 324
Fick's analogy, 341
Fick's law, 22, 198
Fick theories, 169
FIM. See Fisher's information measure (FIM)
Finite rate exergy (FTE), 65
Finite-size flow system, 389, 390
Finite-time exergy, of engine, 66
Finite time thermodynamics (FTT), 8, 9, 16, 109, 350, 352, 355, 362, 393, 624
Finite volume discrete ordinates model (FVDOM) solution, 189
Finslerian geometry, 330, 614
First law of thermodynamics (FLT), 2, 361
First (stable) steady state in CSTR in temperature-concentration chart, 444
Fischer–Tropsch biodiesel, 528
Fischer–Tropsch fuel, 528
Fischer–Tropsch hydrocarbons, 528
Fisher–Schrödinger technique, 103
Fisher's information measure (FIM), 103, 104, 126
Fisher's variational principle, 104
Fixed-end boundary conditions, 73
Flashing, 159
Fluctuation–dissipation postulates, 257
Fluctuation theory, 102, 363
Fluidization modeling, 299, 317
Fluidized bed(s)
  drying processes, 299
  steady and dynamical, 293
Fluidized catalytic cracking (FCC) unit, 394

Subject Index

Fluid's frictional flow, 328
Flux relaxation effects, 94
Fokker–Planck approximation, 127
Fokker–Planck coefficients, 259, 443
Fokker–Planck description, 275
Fokker–Planck equation, 106, 190, 200, 229, 258, 262, 443, 546
 for stochastic process, 259
Fokker–Planck operators, 558
Fokker–Planck regime, 228
Forchheimer-extended Darcy law, 345
Forward equation, 373
Fossil-fuel power plant(s), 358
Fourier heat conduction, 29
Fourier heat transfer, 324
Fourier–Kirchhoff equation, 149
Fourier–Kirchhoff type matrix equation, 138
Fourier–Onsager transports, 158
Fourier representation, 220
Fourier's law, 164, 620
 of heat conduction, 614
Fourier's laws, 396
Fourier's model, of heat conduction, 620
Fourier transform, 220
Fox analysis, 249
Fractal model, of catalysts/porous systems, 378
Fractals, and erythrocytes, 536
Fractal structure(s), 391
Fractional Boltzmann transport equation, 83
Fractional Fokker–Planck equation, 450
Frechet derivative, 116
Free-energy transformation by membranous enzymes, 523
Frictional flows, in isotropic media, 322
 steady trajectory of nonlinear, 322
Frobenius–Perron operators, 557
FRP (fiber-reinforced polymer) structural shapes, 267
FTT. *See* Finite time thermodynamics (FTT)
Fuel cell hybrid power system(s), 404
Fuel cells, 522
 biological, 529
 systems, 407
Fuel processor (FP), 489
FVDOM solution. *See* Finite volume discrete ordinates model (FVDOM) solution

G

Gadolinia-doped ceria (CGO), 493
Galerkin method, 128
Gas
 countercurrent contacting of, 240
 enthalpy, 298
 equilibrium enthalpy of, 299
Gas/granular solid
 countercurrent flow of, 310
 drying/sorption, experimental trajectories of, 310
 geometric similarities of, 311
Gasification efficiencies, 528
Gasification process exergy-based efficiencies, 410
Gas–liquid–solid interfacial interactions, 562
Gas–moisture–solid enthalpy diagram for, 306
Gas–moisture–solid system, 293
Gas parameters, 299
Gas-solid equilibria, 293, 298
Gass plate, 592
Gas-to-particle mass transfer, 306
Gas turbine cogeneration systems, 362
Gas–water–solid chart, 314
 balance and kinetic constructions, 313
Gaussian approximation, 261
Gaussian central limit theorem, 220
Gaussian distribution, 254
Gaussian jump distribution, 223
Gaussian stochastic motion, 116
 of dissipative systems, 126
Gauss's law, 584
 of errors, 24
GEM. *See* Gibbs equilibrium manifold (GEM)
General chemical reactions stoichiometry of, 424
General exchange equation, 433
Generalized heat flux, 419
Generalized hydrodynamics (GH), 105, 173
Generalized power-efficiency diagrams, 353
Generalized reciprocity relations (GRR), 440
General purpose heat source (GPHS), 493
GENERIC formalism, 13
GENERIC modeling, 198
Genomic DNA sequences, 218
GES. *See* Gibbs equilibrium submanifold (GES); Gibbs equilibrium surface (GES)
GH. *See* Generalized hydrodynamics (GH)
Gibbs differential, for energy density, 148
Gibbs–Duhem equation, 28, 384
Gibbs–Duhem formula, 622
Gibbs–Duhem inequality, 22
Gibbs equation, 7, 10, 15, 17, 19, 21, 29, 42, 62, 64, 81, 97, 115, 173, 174, 616, 623, 624
 for entropy differential, 94
Gibbs equilibrium manifold (GEM), 36
Gibbs equilibrium submanifold (GES), 36, 37
Gibbs equilibrium surface (GES), 59, 64
Gibbs free energy, 85, 419, 430, 441, 516, 517
Gibbs fundamental equation, 184
Gibbsian intensities, 138
Gibbsian theory, 8, 11
Gibbs manifold, 138
Glansdorff–Prigogine criterion, 249, 442
Glansdorff–Prigogine type analyses, 26
Glansdorff–Prigogine universal criterion, 248
Gouya-Stodola law, 71, 147, 419, 474
Gouy–Stodola theorem, 601, 602
Governing Principle of Dissipative Processes (GPDP), 21
Governing principle of thermodynamics, 13

Subject Index

GPHS. *See* General purpose heat source (GPHS)
Grad–Boltzmann theory, 621, 623
Gradient representations, of thermal field(s), 139
Grad's analysis, 621
Grad's formalism, 77
Grad's moment method, 27, 75, 620
Grad's 13-moment method, 99
Grad's solution, 166
Graphical method, 291
GRR. *See* Generalized reciprocity relations (GRR)
Guldberg–Waage kinetics, of mass action, 132
Guldberg–Waage law, 131
Guldberg–Waage mass action law, 13, 441
  from Gyarmati's governing principle, 85
Gyarmati-Li reciprocal relations, 12
Gyarmati–Li theory, 441
Gyarmati's principle, 12
Gyftopoulos, 111

H

Hamiltonian
  action, 616
  approach, 25, 119, 123, 227
  based approaches, 480
  and canonical equations, 480–483
  defined, 369
  dynamical systems, 218
  energy–momentum tensor, 117
  equations, 587
  fluid mechanics, 119
  formalisms, 154
  formula, 373
  formulation, 55, 618
  function, 47, 52
  like formulations, 124
  optimal Hamiltonian function, 370
  principle, 616
  representation, of nonequilibrium isokinetic steady, 81
  system, 129, 263
Hamiltonian/Lagrangian mechanics, 116, 129
Hamilton–Jacobi approximation, 127, 261
Hamilton-Jacobi-Bellman (HJB) equation, 323, 372, 374, 397, 399, 597, 599
Hamilton-Jacobi-Bellman equation (HJB equation), 332, 333, 419
Hamilton–Jacobi–Bellman (HJB) theory, 397, 454
Hamilton-Jacobi (HJ) equation, 29, 105, 117, 127, 129, 130, 157, 190, 200, 372
  for dissipative heat flow, 618
Hamilton–Jacobi expressions, 157
Hamilton–Jacobi form, 54
Hamilton–Jacobi method, 257
Hamilton–Jacobi structure, 156
Hamilton-Jacobi theory, 323, 336, 340, 371
  for dissipative systems, 128
  for wavefronts, 323
Hamilton's action principle, 121
Hamilton's kinetic potential, 46
Hamilton's principle, 122, 614, 622
  yields, 120
Hamilton's stationary action, 29
Hamilton's variational principle, 120
Hay–Davies–Klucher–Reindl (HDKR) model, 606
H/C ratio, 410
Heat, 349
  consumption, 347
Heat-driven binary separation process
  schematic diagram of, 294
Heat engine(s), 4, 378, 399, 494, 532, 588, 600, 601
  steam power-plant, 5
Heat flux, 594, 613
  density, 165
  distribution, 164
Heating, 563
Heating/cooling policies
  minimizing entropy production, 563–566
Heating/cooling strategy, 563
Heat pump(s) (HP), 6, 354, 597, 599
  finite-time thermodynamics, 354
Heat-pump-surrounding (HPS) system, 413
Heat-recovery steam, 353
Heat transfer, 358, 613
  conductance, 591
  rate, 564
Height of transfer units (HTU), 305
Helix-coil transition theory, 516
Hessian matrix, 59
Heterogeneities, 435
Heterogeneous-catalyst structure, 449
Heterogeneous fluidized bed model, 299
Heterogeneous kinetics, 436
Heterogeneous semiconductor photocatalysis, 608
Hexagonal Bénard cells, 273
Hidden variable approache(s), 23
High-efficiency plants, 353
HJB equations, 109, 397, 417, 419
HJB theory, 401
Homogeneous model, 568
Homogeneous reaction rate, 425
Hopf bifurcation(s), 236, 247, 255, 276
Hottel–Woertz–Liu–Jordan model, 606
HP. *See* Heat pumps (HP)
HPS system. *See* Heat-pump-surrounding (HPS) system
H-theorem, 112, 265
HTU. *See* Height of transfer units (HTU)
HVAC systems, 607
Hybrid micro gas turbine fuel cell systems, 409
Hybrid SOFC-gas turbine (GT) systems, 407
Hydraulic head, 322
Hydrodynamics, 613
  adiabatic fluid, 613
  composite variational principle, 614
  conservation equations, for energy/momentum, 613
  conventional radiative, 615
  de Broglie microthermodynamics, macroscopic extension of, 616
  heat/viscous stress, tensor of matter, 625–627
  kinetic potential density, 619
  Lorentz signature, 614

matter tensor, of general
relativistic theory, 622
nonequilibrium stress, 616
relativistic extended
thermodynamics, 615
relativistic thermohydrodynamics
irreversible gravitating fluids,
conservation laws, 614
role of Lagrangians, 617–620
thermal inertia, basic
information, 620–621
thermal mass/modified
temperatures, 623–624
thermohydrodynamic potentials
(THP), 617–620
Hydrodynamic stability theory, 287
Hydrodynamic theory, of diffusion, 199
Hydrophobicity, 541
Hyperbolic diffusion equation(s), 615
Hyperbolic equation(s), 58, 93, 117, 163, 166, 241, 620
heat problem, 164
of mass transfer, 166

I

Ideal gas, 63
IM. *See* Information measures (IM)
Industrial ecology (IE), 533
Industrial energy, 353
Industrial reactor–regenerator system, 577
Information measures (IM), 103
Information-theoretic model, 36
Inhomogeneous media
flows of fluids, 343
Inhomogeneous porous bed,
frictional fluid flow, 321
bending of fluid paths, 328–330
Darcy's Law, 322
discrete model of mass transport, 324
discrete model to bending law, 325–328
example of flow of fluid, 338–340
extensions to other systems, 335–338
Hamilton–Jacobi–Bellman equation (HJB equation), 332–334

Hamilton-Jacobi theory, 323
moving porous systems, 321
nonequilibrium thermodynamics and modern integral transformations, 321
nonlinear Darcy's flow, variational approach, 331–332
Inlet air temperature, 318
Inlet gas, 302
parameters, 298
Innovation in drying designs, 354
Instabilities, 446, 567
Interfacial instabilities, 82
Interfacial transfer coefficients, 292, 293
Internal dissipation in steady thermal systems, 476
Internal irreversibilities, 65
Intrinsic reaction coordinate (IRC), 435
Invert pseudo-Carnot formula, 67
Ion exchange process(es), 404
Irreversibility analysis, 358
Irreversible motions, 48, 50
Irreversible thermodynamics, 347
Isoenergy density theory, 80
Isothermal chemical engine, 350
scheme of, 350
Israel's theory, 622

J

Jacobian matrix, 63
Jaynes' MEP approach, 31
Jump probability distribution, 226

K

Kantorovich method, 183
Karkheck's research, 80
Karush–Kuhn–Tucker conditions, 366
Keizer's discussion, 253
Kinetic analysis, 248
Kinetic law of mass action (KMAL), 259
Kinetic rate constant, 562
Kinetic relationships, 358
Kinetic theory, 95
King–Altman method, 517
Kirchhoff's laws, 8, 11
Klimontovich's hydrodynamic theory, 260, 265

KMAL. *See* Kinetic law of mass action (KMAL)
Knetic law of mass action (KMAL), 429
Kolmogorov entropy, 259
Kolmogorov–Sinai entropy, 274
Korteweg-de Vries soliton, 82
Korteweg de Vries stress tensor equations, 198
Kostin's nonlinear Schrödinger–Langevin equation, 118
Kramers–Moyal expansion, 259, 443
Kuhn–Tucker method, 367
Kunii–Levenspiel type, 300

L

Lagrange equations, 123
Lagrange formalism, 121, 240
of TIP, 121
Lagrange kinetic potential, 62
Lagrange multipliers, 37, 124, 147–149, 152, 159, 225, 226, 364, 369, 419, 625
application of, 419
of balance equation, 135
in constructing variational adjoints, 151
for energy, 143
yields, 625
Lagrangian approach, 29, 97
Lagrangian chaos, 274
Lagrangian corresponding, 336
Lagrangian definition of velocity, 117
Lagrangian density, 625
Lagrangian derivative(s), 39
Lagrangian description, 257
Lagrangian dynamics, 84
Lagrangian formulation, 12, 438, 521
of chemical dynamics, 84
Lagrangian function, 126, 334
Lagrangian relaxation path, 484
Lagrangian thermal, 618
coordinate, 618
Landsberg's citation, 248
Langevin features, 24
Langmuir–Hinshelwood–Hougen–Watson kinetics (L–H–H–W kinetics), 381, 382

Subject Index

Langmuir–Hinshelwood kinetics, 437
Langmuir–Hinshelwood mechanism, 436
Lapunov function, 233, 239, 241, 245, 248, 254, 256, 258, 263, 284, 285
  thermodynamic, 247
Lapunov's direct method, 235
Lapunov's method, 242
Lapunov stability theory, 283
Lapunov's theorem, 236
Large-eddy simulations (LES), 266
Last-in-first-out (LIFO) approach, 414
Lattice Boltzmann method, 280
Legendre transformation, 38, 156, 224, 619
  of energy density, 122
  of reversible autonomous, 45
Lenard-Jones fluids, 81
LES. *See* Large-eddy simulations (LES)
LES-based simulation models, 345
Lévy distributions, 220, 222
Lévy flights, 221, 224
Lévy-like superdiffusion, 202
Lévy stable distributions, 232
Lévy walks, 217
Lewis factor, 299
Liapounov functions, 64, 179
Liapunov method, 180
LIFO approach, 414. *See also* Last-in-first-out (LIFO) approach
Limit cycles, 446, 570
  controlling the magnitude of, 447
  quenching, 447
Linear nonequilibrium thermodynamics (LNET), 196, 526
Linear programming problem, 365
Linear stability analysis, 253
Linear thermo–electro–chemical systems, 497
Liouville equation, 35
Liouville operator, 182
Liouvillian dynamics, 557
Liquid helium, 170
Liquid junction potential, 524
Liquid/liquid miscibility, 569

LNET. *See* Linear nonequilibrium thermodynamics (LNET)
Lorentz invariant, 618
Lorentz signature, 614
Lorentz transformation, 176
Lorenzian frames, 617
Lorenz model, 256
Loschmidt's paradox, 229
Lotka–Volterra models, 86, 87, 269
Low-loss designs, 396
Lyapunov function, 571

M

Macroeconomic growth model, 359
Macroscopic equations, for energy density, 165
Mandelbrot's intermittency, 217
Marangoni effect, 286, 570
Marcelin-de-Donder kinetics, 246, 428
Marcelin–Kohnstamm–de Donder form (MKD form), 431
Markovian character, 487
Markovian dissipation, 439
Markovian process, 257
Markov processes, 274
Mass exchanger networks, 353
Mass transfer, 532, 596
  components, 358
Mass transfer law, 579
Mathematical analysis, 572
Mathematical methods of optimization, 364–365
  dynamic methods, 367–375
  static methods, 365–367
  viscosity solutions, 376
Mathematical model(s), 606
Mathematical theory of optimal control, 354
Mathieu thermodynamic potential, 617
MATLAB program, 607
Maupertuis's principle, 128
MaxEnt, 34
Maximization of profit, 347
Maximum energy dissipation (MED) principle nonequilibrium thermodynamics, 125
Maximum entropy principle (MEP), 33

Maximum entropy production principle (MEPP), 34
Maximum-entropy-type formalism, 248
Maximum of exergy or work fluxes in radiation systems, 597–605
Maximum power data, 419
Maximum useful time, 423
Maxwell–Cattaneo equation, 168, 175
Maxwell–Cattaneo hydrodynamics, 119, 618
  Lagrangian formulation, 25
Maxwell–Cattaneo's type, 198
Maxwell distributions, 260
Maxwell equation, of viscoelastic body, 177
Maxwell equilibrium distribution, 77
Maxwell fluid, 281
Maxwellian distribution, 77
Maxwell–Stefan equations, 168
  of diffusion, 171
  in Lagrangian representation of fluid's motion, 117
Mayer formulation, 368
MCFC. *See* Molten carbonate fuel cell (MCFC)
Mechanical energy, 597
Mechanical engineering, 354
Mechanical power, 610
Medical–biological system, 546
MED principle. *See* Maximum energy dissipation (MED) principle
Membrane model(s), 523
Memory function(s), 173
Memory functions, for transport coefficients, 80
MEP. *See* Maximum entropy principle (MEP)
MEPP. *See* Maximum entropy production principle (MEPP)
Mesoscopic nonequilibrium thermodynamics, 431
Methanol, 528
Method of transfer matrix, 87
Methods solving kinetic equation, 293
Methyl ter-butyl ether (MTBE) production, 394

Subject Index

Microreactor technology (MRT), 522
Microwave energy, 597
Minimal catalyst deactivation sorption models for, 376–388
Minimization here referred to (MFI), 103
Minimization of costs, 347
Minimization of entropy production, 357
Models of chemical and mechanical engineering processes, 351
Model system(s), 16
Moderately asymmetric reactor, 270
Modified Butler–Volmer equation, 431
Modified Fermat principle, 567
Moist solid
  dried solid in the enthalpy-concentration chart for, 312
  enthalpy-concentration chart, 312
Molecular diffusion, 26
Molecular dynamic simulations, 580, 581
Molecular simulations, 581
Molten carbonate FC technology, 402
Molten carbonate fuel cell (MCFC), 378, 491
  reduce CO_2 emissions, 491
Momentum, 562
Monomolecular adsorption, 585
Monte Carlo simulation, 341
Mori operator, 79
MRT. See Microreactor technology (MRT)
MTBE plant, 394
Müller approaches, 22
Müller's later approaches, 95
Multicomponent system, 562
Multiphase chemical reactors, 570
Multiphase hyperbolic chemical reactors, 571
Multiphase systems
  pseudocontinuous description, 144
Multireaction system(s)
  described by the traditional stoichiometry, 426
Multireservoir endoreversible heat engine, 579
Muschik's classification, 113

Muscle contraction, 524
Mutual diffusion coefficients, 81
Myosin, 525

N

Natural gas (NG), 491
  cogeneration system, 491
Natural resource(s), 357
Natural selection, 551
Navier–Stokes equations, 126, 127, 159, 168, 345
Navier-Stokes model, 92, 168
Navier–Stokes solutions, 127, 264
neo-Darwinian hypothesis, 552
Nernst–Einstein equation, 199
Nernst equation, 493
Neural network (NN) based method, 473
Neutrinos, 585
Newtonian fluid, 166
Newtonian heat transfer, 399
Newtonian path, 116
Newtonian process, 73
Newton's cooling law, 399
Ni catalyst, 377
NiO/YSZ anode, 491
Noether's theorem, 121
Noise induced transitions, 274
Nonadiabatic fluidized drying
  steady-state operation of, 300
Nonequilibrium behavior, 80
Nonequilibrium fluctuation theory, 110, 615
Nonequilibrium heat energy
  volumetric charge of, 181
Nonequilibrium stress, 613, 616
Nonequilibrium stresses
  four-tensor of, 627
Nonequilibrium system, 36
Nonequilibrium thermodynamics, 15, 321, 355, 363, 561, 562, 571, 581
  finite-time thermodynamics (FTT), 355
Nonequilibrium thermodynamic theory, 166
Nonequilibrium multicomponent fluids, 616
Nonextensive system(s)
  thermodynamic stability conditions of, 288

Non-Fickian diffusion, 82, 207, 215
Non-Fourier phenomenological equation, of heat, 53
Non-Gaussian fluctuations, 195
Non-Gulberg and Waage kinetics, 184
Nonisothermal drying systems, 321
Nonisothermal reaction-diffusion (RD) systems, 197, 442
Nonlinear flux-force relations, 400, 401
Nonlinear hydrodynamic fluctuations, 257
Nonlinear macrokinetics of chemical processes, 430–435
Nonlinear Onsager-type closed laws, 443
Nonlinear programming problems, 366
Nonlinear resistance, 429
Nonlocal diffusion effects, 196
Non-Markovian character, 487
Nonphysical situations, domain of, 213
Nonrenewable resources, 353
Nöther's fundamental theorems, 115
NO_x emissions, 402
NTU approach
  thermal efficiency, 409
Nuclear power plants, 358
Number of mass transfer units, 301
Numerical algorithm(s), 367, 419, 599
Numerical calculation(s), 222, 421, 567
Numerical method(s), 12, 274, 376, 472, 479, 599
Numerical predictions, 562
Numerical solutions of an optimal control problem, 424
Nusselt number, 39
  for steady two-dimensional laminar convection, 278

O

O/C ratio, 410
Ohm's electric conductivity, 158
Ohm's law, 145, 296, 428
Ohm's relation, 131
Oláh's thermodynamic theory, 90

Subject Index

Oldroyd-type fluid, 282
One-way affinity, 495
Onsager–Casimir reciprocal relations, for dielectric tensor, 39
Onsager coefficient, 41
Onsagerian formalism, 88
Onsagerian functional, 138
Onsagerian thermodynamics, 18, 108
Onsager-like symmetry, 430
Onsager–Machlup analysis, 24
Onsager–Rayleigh quadratic dissipation, 100
Onsager reciprocal relations (ORR), 440
Onsager reciprocity, 439, 580
  relations, 75
  theorem, 225
Onsager relation(s), 20, 86, 400, 562, 580
Onsager's driving force, 427
Onsager's extremum principles, 133
Onsager's linear theory, 121
Onsager's local principle, of minimum dissipation, 132
Onsager's matrix, 177
Onsager's principle, 133, 264
Onsager's regression, 124
Onsager's relationship, 175
Onsager's symmetry, 77
Onsager's theory, 426, 428
Onsager's thermodynamic Hamiltonian, 155
Onsager symmetry relations, 581
Optimal catalyst activities, 575, 576
Optimal control in imperfect multiphase systems, 566
Optimal control theory, 350
Optimal distillate-rate policy, 348
Optimal temperature profile(s), 399, 518, 575, 578
  values of enzyme, 519, 520
Optimization methods, for energy system, 362
Optimization, of heat pumps, 353
Optimization problems for chemical processes, 347
Optimization results, 351
Optimization techniques, 347
Optimization theory in thermodynamic, 354
Optimizing in thermodynamic systems
  backgrounds for, 356
Optimum configuration, 359
Original optimization problem for a system with n states, 555
ORR. See Onsager reciprocal relations (ORR)
Orr–Sommerfeld equations, 279, 281
Oscillation(s), 570
Oscillatory trajectories terminate at a stable focus point in the process
  of CO oxidation, 448
Outer and internal oscillatory trajectories fall into a stable limit cycle
  in the process of CO oxidation, 448

P

Packed bed
  exit conversion, 387
  motion of activity, 386
Packed bed reactor(s), 186
Parabolic equation
  heat problem, 164
Parallel flow, 563
Partial derivative vector, 497
Path-analyzing approach, 239
PCMs. See Phase-change materials (PCMs)
PDF. See Probability density-function (PDF)
PDF model, for turbulent flows, 345
Peclet number, 39
PEMFC. See Proton exchange membrane fuel cells (PEMFC)
PEMFC modeling, 491
PEM fuel cells, 407, 408
Perfect fluid model, 613
Performance coefficient, 532
Performance criteria, 358
Petela–Landsberg–Press efficiency, 585, 604
Petela's equation, 589
Phase-change heat transfer, 353
Phase-change materials (PCMs), 570

Phase changes, 130
Phase transformations, 567
Phase transitions, 567
Photocatalytic degradation reaction, 608
Photocatalytic detoxification technology, 588
Photochemical reactions, 26
Photosynthesis, 515
  efficiency, 526
Photosystem I, 526
Photosystem II, 526
Photovoltaic engine, 609
Planck chemical potentials, 135
Planck–Einstein formula, 624
Planck–Einstein theory, 624
Plasma formation, 89
Plastic deformations, 88
Poisson-bracket structure, 124, 156, 158
  of thermohydrodynamics, 119
Polymer electrolyte membrane fuel cell (PEMFC), 392
Polytropic gas, motion of, 210
POMCE. See Principle of minimum cross entropy (POMCE)
POME. See Principle of maximum entropy (POME)
Ponchon–Savarite graphical method, 308
Pontryagin's maximum principle, 47, 56, 332, 368, 370
Pontryagin type approaches, 480
Porous media, 321
  equation, 211
Porous medium, 325
Potentials, of thermal field, 136
Power-current density, 498
Power density vs. current density, 498
Power maximization approaches, 492
Power maximizing curve, 478
Power-maximizing efficiency, 477
Power models, of diffusion, 177
Power output, 349
Power production, in simple nonlinear electrochemical system, 486
Power yield, 474
  in chemical engines, 452–454

and exergy-valued biofuels, 527–530
and power limits in FCs, 487
Pressure drop, 424
Pressure swing adsorption (PSA), 412
  units, 412
Prigogine's principle, 35
Prigogine's theorem, on entropy production, 196
Prigogine–Weame thermodynamic theory, 543, 544
Principle of maximum entropy (POME), 363
Principle of minimum cross entropy (POMCE), 363
Principles of modeling of power yield in chemical systems, 455–457
Probability density-function (PDF), 264
Production size, 347
Propagation speed, of thermal wave, 169
Property, definition, 3
Protein sequence–structure relations
  elucidation of, 540–543
Proton exchange membrane fuel cells (PEMFC), 492
PSA. See Pressure swing adsorption (PSA)
Pseudo-Carnot formula, 66
Pseudo-homogeneous model, 299
Pseudo-Newtonian formalism, 67
Pseudo-Newtonian model, 66, 68–70, 599
Pseudo-Newtonian relaxation, 478
Pseudo-phase model, 305
Pyrolysis, 404

Q

Quantization, of classical variational principles, 128
Quantum theories, 29, 618
Quasi-steady-state drying—moisturizing cycles, 184
Quasi-three-dimensional heat-transfer model, 606
Quenching limit cycles, 447

R

Radial base function (RBF), 413
Radiation at flow, 72
Radiation engines, 417
  by the Stefan–Boltzmann equations, 478–480
Radiation flux, 586, 595
  density, 591
Radiation reservoirs, 601
Radiation systems, 583
Radiation thermodynamics
  basic problems in, 583–587
Radiative energy, 586
Radiative equilibrium, 584
RANS. See Reynolds-averaged Navier-Stokes (RANS)
Rate-controlled constrained-equilibrium (RCCE) method, 423
Rate-to-capital cost ratios, 406
Rayleigh-Benard convection, 82, 280, 570
Rayleigh-Bénard instability, 236, 248, 280
Rayleigh dissipation function, 133
Rayleigh number, 277
Rayleigh–Onsager dissipation function, 253
Rayleigh–Ritz method, 264, 265
RCCE calculations, 423
REA. See Reaction engineering approach (REA)
Reaction-diffusion systems, 200, 255, 429, 488
Reaction engineering approach (REA), 294, 366, 522
Reactivation, of catalyst, 383
Reactive distillation system, 348
Reduced chemical potential, 137
Refrigerant, 5
Refrigerated space (cooling effect), 5
Refrigeration rate, 532
Refrigerators, 4
Regeneration function, 573
Regenerators, 570
Re k–ε model, 345
Relativistic extended thermodynamics (RET), 110
Relativistic thermal inertia systems, 107
Relativistic transformations, of thermodynamic intensities, 624
Relaxation–diffusive version, 266
Relaxation–retardation constitutive model, 568
Relaxation times, 170
Relaxation times, in rarefied gases, 170
Renewed exergy fraction, 534
RET. See Relativistic extended thermodynamics (RET)
Reversible Carnot-like chemical efficiency, 484
Reversible motions, 48
Reversible thermodynamics, 347
Reversible thermohydrodynamics, 42
Reynolds-averaged Navier–Stokes (RANS), 345
Reynolds averaging technique, 568
Reynolds flux, 245, 296
Reynolds numbers, 106, 261, 279, 287, 292
Reynolds shear stress, 260
Rheology of viscoelastic media, 92
Richardson's equation, 217
Riemannian geometry, 330, 614
Riemannian metric tensor, 33
Riemannian scalar, 60
Rigorous definitions, 92
Ritz-Kantorovich direct method, 133
Rouse–Zimm relaxation models, 96, 166

S

Saturation effects, in mobility, 203
Schrödinger wave equation (SWE), 103, 104, 126
Schrödinger withdrew, 128
Second-generation biofuels, 411
Second-law-based optimization, 353
  of heat, 353
Second law of thermodynamics (SLT), 2, 34, 121, 348, 361
Second Lyapunov theorem, 446
Second (unstable) steady state in CSTR in temperature-concentration chart, 444
Self-diffusivity, 197

Subject Index

Self-organization, 543, 550
Self-replication, 559
Semiinverse method, 128
Separation systems, 347
Sequence-structure code, 541
Sequential models, for computation of dynamic generators, 466–469
Shannon's information theory, 228
Short-time effects, 163
Simple chemical, and electrochemical systems, 483–486
Simplified flow sheet of typical system in chemical industry, 422
SLT. *See* Second law of thermodynamics (SLT)
SM. *See* Statistical mechanics (SM)
Smoluchowski–Chapman–Kolmogorov functional equation, 487
Snell's law, 325
   of refraction, 340
SNG conditioning, 411
SNG production, 411
SOFC. *See* Solid oxide fuel cell (SOFC)
SOFC energy generators, 419
SOFC reactor model, 409
Solar-assisted HP system, 354
Solar buildings, and solar systems, 605–610
Solar cells, 583
Solar-driven convection, 598
Solar energy, 556, 588, 597, 599
   photovoltaic conversion, 588, 605
Solar flux
   conversion into a heating medium, 588–597
Solar power plants, 358
Solar radiation, 606
Solid catalyst particle(s), 578
Solid enthalpy diagram, 305
Solid oxide fuel cell (SOFC), 378, 407, 489
   design, 490
   principle of, 475
   tube, 409
   voltage-current density and power-current density characteristics of, 498

Solid oxide fuel cell-gas turbine (SOFC-GT) power system, 409
Solid oxide steam electrolyzer (SOSE) plant, 493
   hydrogen production, 493
Soret coefficient, 280
Soret effect, 278
SOSE plant. *See* Solid oxide steam electrolyzer (SOSE) plant
Spalding's model, of mass transfer, 296
Stability
   distributions, 220
   and fluctuations in chemical reactions, 442
   problems, 233
   of protein solutions, 515
Stanton number, 296
State of system, 4
State postulate, 2
States of gaseous and adsorbed reactants, 438
Statistical fluctuations, 583
Statistical mechanical theory, 179
Statistical mechanics (SM), 227
Statistical thermodynamics, 99
Steady fluidized drying
   with external heating, 303
Steady isothermal engines
   characteristics of, 461–466
Steady-state fluidized drying, 299
Steel box, 592
Stefan–Boltzmann engine, 593
Stefan–Boltzmann equations, 66, 477, 478, 595
Stefan–Boltzmann formula, 596
Stefan–Boltzmann law, 619
Stefan–Boltzmann radiation engine, 477
Stefan problem, 318
Stirling heat engines, 399
Stochastic diffusion theory, 43
Stoichiometric coefficients, 427
Stoichiometry scheme, of steady, multireaction system, 425
Stress, 353
Structural theory, 414
Supercomputer simulations of reactions, 449
SWE. *See* Schrödinger wave equation (SWE)

Synthesizing thermodynamic approach, 419
System of cocurrent tubular reactor—catalyst regenerator, 571

T

Taylor–Couette flow, 237
Taylor's correlated-walk approach, 217
TEA. *See* Thermoeconomic functional analysis (TFA)
Temperature-dependent catalyst deactivation, 571, 578
Temperature dependent enzyme deactivation, 518
TEO. *See* Thermoeconomic optimization (TEO)
TFA. *See* Thermoeconomic functional analysis (TFA)
Theoretical modeling, 610
Theory of exergoeconomic, 361
Theory of FC electrolyzers, 475
Thermal analysis of complex regeneration system, 354
Thermal-drying technologies, 317
Thermal efficiency, 349, 610
Thermal engine driving, heat, power output, and entropy production, 349
Thermal engineering, 361
Thermal equilibria, 358
Thermal inertia effects, 616
Thermal machine, 355
Thermal management system (TMS), 570
Thermal potential, 624
Thermal radiation, 601
Thermal technology, 353
Thermal transport, 189
Thermal-viscous coefficients, 172
Thermochemical systems
   limits on power consumption in, 504–508
Thermoconvective instabilities, in liquid layers, 277
Thermodynamic analyses in engineering, 347
Thermodynamic approach, 384

Subject Index

Thermodynamic equilibrium, 274, 443, 604
  based models, of gasification process, 405
Thermodynamic functionals, 154
Thermodynamic Hamiltonian, 155
Thermodynamic irreversibility, 358
Thermodynamic Lagrangian, 154
Thermodynamic lapunov functions
  Belousov–Zhabotynski (BZ) reaction, 255
  chaotic solutions, 268–277
  chemically reacting systems, fluctuations, and turbulent flows, 244–267
  catalytic gas–solid systems, 246
  chemical reactor stability, 245
  complex chemistries, diversity of, 245
  disequilibrium thermodynamics, 247
  Gibbs formulation of heterogeneous equilibrium, 244
  Hopf instability, 247
  Lapunov functions, 247
  maximum-entropy-type formalism, 248
  nonlinear reaction mechanisms, 246
  reaction–diffusion systems, 247
  Reynold's flux, 245
  Turing instability, 247
Clausius inequalities, 253
CSTRs (continuously stirred tank reactors), 256
density functional theory (DFT), 237
deterministic equations, for nonlinear irreversible processes, 258
diffusion and chemical instabilities, 234
disequilibrium fields, stability analyses, 236
disequilibrium systems, 236
dynamical equations, linearization of, 254
equilibrium, in liquid phases, 235
equilibrium/pseudoequilibrium points, qualitative properties of, 238
fluctuation theory, 257
Fokker–Planck equation, 257
Gaussian distribution, 254
Glansdorff–Prigogine criterion, 249
Glansdorff–Prigogine universal criterion of evolution, 248
Hopf bifurcation, 255
lapunov functions, 282–285
Lapunov's theorem, 236
Lotka–Volterra models, 255
nonequilibrium unstable transitions
  Kinetic analysis and thermodynamic interpretation of, 248
oscillatory kinetics and spatio-temporal selforganization, 252
oscillatory systems, 268–277
periodic states, 268–277
Rayleigh–Onsager dissipation function, 253
reaction–diffusion systems, 255
research and development, 285–288
shadowgraph images, 262
solid-phase combustion, reaction-diffusion model for, 253
stability of thermal fields, 277–282
stability problems, 257
and stability problems, 233
steady states stability
  close to equilibrium, 239–244
thermodynamic equilibrium point, 236
thermodynamic stability of chemical reactors, 249
turbulence problems, 264
Yang's model, 249, 251
Thermodynamic limits
  of industrial processes, 355
Thermodynamic networks (TNs), 8
  theory, 8
Thermodynamic optimization, 358
  approaches, 378
Thermodynamics
  of living systems, 516, 517
  in multiphase catalysis, 515
  in optimization of controlled systems, 357
  path, 4
  phenomenological theory of, 166
  of power production in chemical engines, 457
  process, 4
  of surfaces and interfaces, 562
  system, 3
  used in ecology, 531
Thermoecological optimization, 592
Thermoeconomic functional analysis (TFA), 414
Thermoeconomic optimization (TEO), 355, 361
  generalizes exergy criteria, 566
  problem, 357
Thermoeconomics, 362
  economic analyses, 351
  of energy conversions, 353, 354
Thermohydrodynamic potentials (THP)
  Gibbs' equations, 619
Thermohydrodynamic transformations, 617
Thermostatics, 517
Thermostatistics of helix-coil transitions, 538
Thin-layer drying rate equation, 317
Third law of thermodynamics, 2
Third (stable) steady state in CSTR in temperature-concentration chart, 445
THP. See Thermohydrodynamic potentials (THP)
Titanium oxide, 608
TMS. See Thermal management system (TMS)
TNs. See Thermodynamic networks (TNs)
Transfer coefficients, 292
Transfer matrix method, 562
Transferring graphical methods, 314
Transfer through an interface of a two-phase system, 567
Transmission, 354

Subject Index

Triangular isomerization reaction, 380
Tsallis formulation, 216
Tsallis's generalized thermostatistics, 227
Tsallis statistics, 202, 225
Tubular reactor, 574
Turbulence problems, 264
Turbulent flow through porous media, 345
Turbulent interfaces, 570
Turbulent mixtures, 567
Turing's morphological models, 248
Two-phase model
dynamical state equations of, 305
Two-phase system, 568

U

UDF. See User defined function (UDF)
Unidirectional chemical flux, 430
Unidirectional flux, 522
Universe Hamiltonian, 24, 439
Unrestorable domestic natural resources (UDN), 533
Usable energy, 354
User defined function (UDF), 405
UV–visible light, 608
UV–Vis spectrophotometer, 608

V

Vacuum-assisted resin transfer molding (VARTM), 267
Valuable variational principles, 107
Van der Waals capillarity model, 581
Vapor-compression refrigeration cycle, 5
Vapor-generation systems, 318
Vapor–liquid mixture, 568
Variational approaches, to nonequilibrium thermodynamics, 115
action functional
for hyperbolic transport, 151
adjoined constraints to potential representations, 149–151
Anthony variational approach, 122, 134
Bateman's structure, 118
conservation laws, in variational thermohydrodynamics, 120
differential equations, 134
entropy, of diffusive fluxes, 117
Euler–Lagrange equations, 116
evolution by single poissonian brackets, 154–158
Fisher's information measure (FIM), 126
Fokker–Planck approximation, 127
Fourier's transient heat conduction, 133
gradient representations, of thermal field, 139–142
Guldberg–Waage kinetics of mass action, 132
Guldberg–Waage law, 131
Hamilton–Jacobi equations, 117, 130
Hamilton–Lagrange formalism within nonequilibrium thermodynamics, 124
heat/mass transfer without chemical reaction, 137
hydrothermodynamics of, 119
inclusion, of chemical process, 142–146
irrotational motion, 117
Lagrange-formalism, 121
Lagrangian formulations of second order systems, 119
Lagrangian function, 129
lumped systems and continuous systems, 135
Maupertuis's principle, 128
method of weighted residuals, 128
multiphase systems
extension of theory, 158–159
Navier–Stokes equation, 126
Newtonian path, 116
Noether's theorem, 121
nonequilibrium thermodynamics, 130
maximum energy dissipation (MED) principle in, 125
nonlinear chemical resistance, 131
novel variational principles of, 136
Onsager's extremum principles, 133
reversible nonequilibrium fluids, 123
search for variational principle, 115
semiinverse method, 128
state of art, 115
stochastic least-action principle, 127
thermodynamic potential, change of, 146–149
variational principles, 124
variational problem, 120
Wentzel–Kramers–Brillouin method, 127
Variational principles, 115, 322, 324, 328
VARTM. See Vacuum-assisted resin transfer molding (VARTM)
Vernotte model, 188
Viscoelastic bodies, 99
Viscoelastic fluids, 570
Viscoplastic Bingham fluid, 92
Viscosity solutions, 376
Voltage-current density, 498
von-Kármán–Pohlhausen method, 128
von Neumann equation, 225
von Neumann's work, 108
Vujanovic's papers, 93

W

Wave-drying equation, 269
Wave equations, of heat/mass transfer, 163
applications, of heat wave equations, 185
for coupled heat/mass transfer, 176–179
coupled heat/mass transfer, extended thermodynamics of, 171–175
Fourier's, Fick's, or Newton's constitutive equations, 163
nonequilibrium thermodynamic theory, 166
relaxation theory, of heat flux, 166–171
research and development, 186–191

stability of dissipative wave systems, 180–181
steep temperature gradient, on heat conductivity, 164
telegrapher's equation, 164
thermal fluctuations, critical frequency, 165
thermodynamic systems, high-frequency behavior of, 185–186
variational principles, 181–184

Wave PD diffusion equation, 268
Weissenberg numbers, 282
Wentzel–Kramers–Brillouin method, 200
Weyl's quantum principle, 111
Wheat drying, 306
Winfree demonstrated theoretical links, 270
Work-driven operations, 595
Work-energy theorem, 6

Y

Yang's equations, 251
Yttria-stabilized zirconium (YSZ), 493

Z

Zeroth law of thermodynamics, 2
Zwietering reactor, 451